# Universitext

D0881615

# Universitext

*Universitext* is a series of textbooks that presents material from a wide variety of mathematical disciplines at master's level and beyond. The books, often well class-tested by their author, may have an informal, personal even experimental approach to their subject matter. Some of the most successful and established books in the series have evolved through several editions, always following the evolution of teaching curricula, to very polished texts.

Thus as research topics trickle down into graduate-level teaching, first textbooks written for new, cutting-edge courses may make their way into *Universitext*.

More information about this series at http://www.springer.com/series/223

George M. Bergman

# An Invitation to General Algebra and Universal Constructions

Second Edition

 Springer

George M. Bergman
Department of Mathematics
University of California, Berkeley
Berkeley, CA, USA

An invitation to general algebra and universal constructions. Henry Helson, Berkeley, CA, 1998. ii+398 pp. ISBN: 0-9655211-4-1

ISSN 0172-5939          ISSN 2191-6675 (electronic)
ISBN 978-3-319-11477-4          ISBN 978-3-319-11478-1 (eBook)
DOI 10.1007/978-3-319-11478-1
Springer Cham Heidelberg New York Dordrecht London

Library of Congress Control Number: 2014954583

Printed on acid-free paper

Springer is part of Springer Science+Business Media (www.springer.com)

*To Mary, Jeff, Michael, Rebecca and Clifford*

# Contents

# Chapter 1
# About the Course, and These Notes

## 1.1. Aims and prerequisites

This course will develop some concepts and results which occur repeatedly throughout the various areas of algebra, and sometimes in other fields of mathematics, and which can provide valuable tools and perspectives to those working in these fields. There will be a strong emphasis on motivation through examples, and on instructive exercises.

I will assume only an elementary background in algebra, corresponding to an honors undergraduate algebra course or one semester of graduate algebra, plus a moderate level of mathematical sophistication. A student who has seen the concept of free group introduced, but isn't sure he or she thoroughly understood it would be in a fine position to begin. On the other hand, anyone conversant with fewer than three ways of proving the existence of free groups has something to learn from Chapters 2 and 3.

As a general rule, we will pay attention to petty details when they first come up, but take them for granted later on. So students who find the beginning sections devoted too much to "trivia" should be patient!

In preparing this published version of my course notes, I have not removed remarks about homework, course procedures etc., addressed to students who take the course from me at Berkeley, which take up the next three and a half pages, since there are some nonstandard aspects to the way I run the course, which I thought would be of interest to others. Anyone else teaching from this text should, of course, let his or her students know which, if any, of these instructions apply to them. In any case, I hope readers elsewhere find these pages more amusing than annoying.

© Springer International Publishing Switzerland 2015
G.M. Bergman, *An Invitation to General Algebra and Universal Constructions*, Universitext, DOI 10.1007/978-3-319-11478-1_1

## 1.2. Approach

Since I took my first graduate course, it has seemed to me that there is something wrong with our method of teaching. Why, for an hour at a time, should an instructor write notes on a blackboard and students copy them into their notebooks—often too busy with the copying to pay attention to the content—when this work could be done just as well by a photocopying machine? If this is all that happens in the classroom, why not assign a text or distribute duplicated notes, and run most courses as reading courses?

One answer is that this is not all that happens in a classroom. Students ask questions about confusing points and the instructor answers them. Solutions to exercises are discussed. Sometimes a result is developed by the Socratic method through discussion with the class. Often an instructor gives motivation, or explains an idea in an intuitive fashion he or she would not put into a written text.

As for this last point, I think one should not be embarrassed to put motivation and intuitive discussion into a text, and I have included a great deal of both in these notes. In particular, I often first approach general results through very particular cases. The other items—answering questions, discussing solutions to exercises, etc.—which seem to me to contain the essential human value of class contact, are what I would like classroom time to be spent on in this course, while these notes will replace the mechanical copying of notes from the board.

Such a system is not assured of success. Some students may be in the habit of learning material through the process of copying it, and may not get the same benefit by reading it. I advise such students to read these notes with a pad of paper in their hands, and try to anticipate details, work out examples, summarize essential points, etc., as they go. My approach also means that students need to read each day's material *before* the class when it will be covered, which many students are not accustomed to doing.

## 1.3. A question a day

To help the system described above work effectively, I require every student taking this course to hand in, on each day of class, one question concerning the reading for that day. I strongly encourage you to get your question to me by e-mail by at least an hour before class. If you do, I will try to work the answer into what I say in class that day. If not, then hand it in at the start of class, and I will generally answer it by e-mail if I feel I did not cover the point in class.

The e-mail or sheet of paper with your question should begin with your name, the point in these notes that your question refers to, and the classifying word "urgent", "important", "unimportant" or "pro forma". The first three choices of classifying word should be used to indicate how important it is to you to have the question answered; use the last one if there was nothing in the reading that you really felt needed clarification. In that case, your "pro forma" question should be one that some reader might be puzzled by; perhaps something that puzzled you at first, but that you then resolved. If you give a "pro forma" question, you must give the answer along with it!

You may ask more than one question; you may ask, in addition to your question on the current reading, questions relating to earlier readings, and you are encouraged to ask questions in class as well. But you must always submit in writing at least one question related to the reading assignment for the day.

## 1.4. Homework

These notes contain a large number of exercises. I would like you to hand in solutions to an average of one or two problems of medium difficulty per week, or a correspondingly smaller number of harder problems, or a larger number of easier problems. Choose problems that interest you. But please, look at *all* the exercises, and at least think about how you would approach them. They are interspersed through the text; you may prefer to think about some as you come to them, and to come back to others after you finish the section. We will discuss many of them in class. I recommend spending at least one to five minutes thinking about each exercise, unless you see a solution immediately.

Grades will be based largely on homework. The amount of homework suggested above, reasonably well done, will give an A. I will give partial credit for partial results, as long as you show you realize that they are partial. I would also welcome your bringing to the attention of the class interesting related problems that you think of, or find in other sources.

It should hardly need saying that a solution to a homework exercise in general requires a proof. If a problem asks you to find an object with a certain property, it is not sufficient to give a description and say, "This is the desired object"; you must prove that it has the property, unless this is completely obvious. If a problem asks whether a calculation can be done without a certain axiom, it is not enough to say, "No, the axiom is used in the calculation"; you must prove that no calculation not using that axiom can lead to the result in question. If a problem asks whether something is true in all cases, and the answer is no, then to establish this you must, in general, give a counterexample.

I am worried that the amount of "handwaving" (informal discussion) in these notes may lead some students to think handwaving is an acceptable substitute for proof. If you read these notes attentively, you will see that handwaving does not *replace* proofs. I use it to guide us to proofs, to communicate my understanding of what is behind some proofs, and at times to abbreviate a proof which is similar to one we have already seen; but in cases of the last sort there is a tacit challenge to you, to think through whether you can indeed fill in the steps. Homework is meant to develop and demonstrate *your* mastery of the material and methods, so it is not a place for you to follow this model by challenging the instructor to fill in steps!

Of course, there is a limit to the amount of detail you can and should show. Most nontrivial mathematical proofs would be unreadable if we tried to give every substep of every step. So truly obvious steps can be skipped, and familiar methods can be abbreviated. But more students err in the direction of incomplete proofs than of excessive detail. If you have doubts whether to abbreviate a step, think out (perhaps with the help of a scratch-pad) what would be involved in a more complete argument. If you find that a step is more complicated than you had thought, then it should *not* be omitted! But bear in mind that "to show or not to show" a messy step may not be the only alternatives—be on the lookout for a simpler argument, that will avoid the messiness.

I will try to be informative in my comments on your homework. If you are still in doubt as to how much detail to supply, come to my office and discuss it. If possible, come with a specific proof in mind for which you have thought out the details, but want to know how much should be written down.

There are occasional exceptions to the principle that every exercise requires a proof. Sometimes I give problems containing instructions of a different sort, such as "Write down precisely the definition of ...", or "State the analogous result in the case ...", or "How would one motivate ...?" Sometimes, once an object with a given property has been found, the verification of that property is truly obvious. However, if direct verification of the property would involve 32 cases each comprising a 12-step calculation, you should, if at all possible, find some argument that simplifies or unifies these calculations.

Exercises frequently consist of several successive parts, and you may hand in some parts without doing others (though when one part is used in another, you should if possible do the former if you are going to do the latter). The parts of an exercise may or may not be of similar difficulty—one part may be an easy verification, leading up to a more difficult part; or an exercise of moderate difficulty may introduce an open question. (Open questions, when included, are always noted as such.)

Homework should be legible and well-organized. If a solution you have figured out is complicated, or your conception of it is still fuzzy, outline it first on scratch paper, and revise the outline until it is clean and elegant before writing up the version to hand in. And in homework as in other mathematical writing, when an argument is not completely straightforward, you should help the reader see where you are going, with comments like, "We shall now prove ...", "Let us first consider the case where ...", etc.

If you hand in a proof that is incorrect, I will point this out, and it is up to you whether to try to find and hand in a better proof. If, instead, I find the proof poorly presented, I may require that you redo it.

If you want to see the solution to an exercise that we haven't gone over, ask in class. I may postpone answering, or just give a hint, if other people still want to work on it. In the case of an exercise that asks you to supply details for the proof of a result in the text, if you cannot see how to do it you should certainly ask to see it done.

You may also ask for a hint on a problem. If possible, do so in class rather than in my office, so that everyone has the benefit of the hint.

If two or more of you solve a problem together and feel you have contributed approximately equal amounts to the solution, you may hand it in as joint work.

If you turn in a homework solution which is inspired by material you have seen in another text or course, indicate this, so that credit can be adjusted to match your contribution.

## 1.5. The name of the game

The general theory of algebraic structures has long been called Universal Algebra, but in recent decades, many workers in the field have come to dislike this term, feeling that "it promises too much", and/or that it suggests an emphasis on universal constructions. Universal constructions *are* a major theme of *this* course, but they are not all that the field is about.

The most popular replacement term is General Algebra, and I have used it in the title of these notes; but it has the disadvantage that in some contexts, it may not be understood as referring to a specific area. Below, I mostly say "General Algebra", but occasionally refer to the older term.

## 1.6. Other reading

Aside from these notes, there is no recommended reading for the course, but I will mention here some items in the list of references that you might like to look at. The books [1, 6, 8, 13, 21, 23] are other general texts in General (a.k.a. Universal) Algebra. Of these, [13] is the most technical and encyclopedic. Both [21] and [23] are, like these notes, aimed at students not necessarily having advanced prior mathematical background; however [23] differs from this course in emphasizing *partial* algebras. Like this course, [8] takes the view that this subject is an important tool for algebraists of all sorts, and it gives some interesting applications to groups, division rings, etc.

The texts [32] and [34] have both been used many times to teach Berkeley's basic graduate algebra course. (Some subset of Chapters 2–7 of the present notes can, incidentally, be useful supplementary reading for students taking such a course.) Though we will not assume the full material of such a course (let alone the full contents of those books), you may find them useful references. Of these, [34] is more complete and rigorous; [32] is sometimes better on motivation. The three volumes [26–28] include similar material. A presentation of the core material of such a course at approximately an honors undergraduate level, with detailed explanations and examples, is [29].

Each of [1, 6, 8, 13, 21, 23, 27] gives a little of the theory of *lattices*, introduced in Chapter 6 of these notes. Extensive treatments of this subject can be found in [4, 14].

Chapter 7 of these notes introduces *category theory*. [9] is the paper that created that discipline, and still very stimulating reading; [20] is a general text on the subject. An important area of category theory that our course leaves out is treated in [11]. For the thought-provoking paper from which the ideas we develop in Chapter 10 come, see [12].

An amusing parody of some of the material we shall be seeing in Chapters 5–10 is [19].

## 1.7. Numeration; history; advice; web access; request for corrections

These notes are divided into chapters, and each chapter into sections. In each section, I use two numbering systems: one that embraces lemmas, theorems, definitions, numbered displays, etc.; the other for exercises. The number of each item begins with the chapter-and-section numbers. This is followed by a ".", and the number of the result, display, etc., or a ":" and the number of the exercise. For instance, in section $m.n$, i.e., section $n$ of Chapter $m$, we might have display $(m.n.1)$, followed by Definition $m.n.2$, followed by Theorem $m.n.3$, and interspersed among these, Exercises $m.n$:1, $m.n$:2, $m.n$:3, etc. The reason for using a common numbering system for results, definitions, and displays is that it is easier to find Proposition 3.2.**9** if it is between Lemma 3.2.**8** and display (3.2.**10**) than it would be if it were Proposition 3.2.**3**, located between Lemma 3.2.**5** and display (3.2.**1**). The exercises form a separate system. There is a List of Exercises toward the back of these notes, with telegraphic descriptions of their subjects.

These notes began around 1971, as mimeographed outlines of my lectures, which I handed out to the class, and gradually improved in successive teachings of the course. With the advent of computer word-processing, such improvements became much easier to make, and the outline evolved into a readable development of the material. In Spring and Summer 1995 they were published by the short-lived Berkeley Lecture Notes series, and for several years after that, by my late colleague Henry Helson. I have continued to revise them each time I taught the course; I never stop finding points that call for improvement. But hopefully, they are now in a state that justifies publication in book form.

In recent decades, I have kept the notes available online, as an alternative to buying a paper copy. Though this will not be feasible for the final published version, I intend to keep the version that I submit to the Springer editorial staff available via my website `http://math.berkeley.edu/~gbergman`.

To other instructors who may teach from these notes (and myself, in case I forget), I recommend moving quite fast through the easy early material, and much more slowly toward the end, where there are more concepts new to the students, and more nontrivial proofs. Roughly speaking, the hard material begins with Chapter 8. A finer description of the hard parts would be: §§7.9–7.11, 8.3, 8.9–8.12, 9.9–9.10, and Chapter 10. However, this judgement is based on teaching the course to students most of whom have relatively advanced backgrounds. For students who have not seen ordinals or categories before (the kind I had in mind in writing these notes), the latter halves of Chapters 5 and 7 would also be places to move slowly.

The last two sections of each of Chapters 7, 8 and 9 are sketchy (to varying degrees), so students should be expected either to read them mainly for the ideas, or to put in extra effort to work out details.

After many years of editing, reworking, and extending these notes, I know one reason why the copy-from-the-blackboard system has not been generally replaced by the distribution of material in written form: A good set of notes takes an *enormous* amount of time to develop. But I think that it is worth the effort.

Comments and suggestions on any aspect of these notes—organizational, mathematical or other, including indications of typographical errors—are welcome at `gbergman@math.berkeley.edu`.

## 1.8. Acknowledgements

Though I had picked up some category theory here and there, the first extensive development of it that I read was Mac Lane [20], and much of the material on categories in these notes ultimately derives from that book. I can no longer reconstruct which category-theoretic topics I knew before reading [20], but my debt to that work is considerable. Cohn's book [8] was similarly my first exposure to a systematic development of General Algebra; and Freyd's fascinating paper [12] is the source of the beautiful result of §10.4, which I consider the climax of the course. I am also indebted to more people than I can name for help with specific questions in areas where my background had gaps.

For the development and maintenance of the locally enhanced version of the text-formatting program `troff`, in which I prepared earlier versions of these notes, I am indebted to Ed Moy, to Fran Rizzardi, and to D. Mark

Abrahams, and for help with the conversion to LaTeX, to George Grätzer, Paul Vojta, and, especially, Arturo Magidin.

Finally, I am grateful to the many students who have pointed out corrections to these notes over the years—in particular, to Arturo Magidin, David Wasserman, Mark Davis, Joseph Flenner, Boris Bukh, Chris Culter, and Lynn Scow.

# Part I. Motivation and Examples

In the next three chapters, we shall look at particular cases of algebraic structures and universal constructions involving them, so as to get some sense of the general results we will want to prove in the chapters that follow.

The construction of free groups will be our first example. We prepare for it in Chapter 2 by making precise some concepts such as that of a group-theoretic expression in a set of symbols; then, in Chapter 3, we construct free groups by several mutually complementary approaches. In Chapter 4, we look at a large number of other constructions—from group theory, semigroup theory, ring theory, etc.—which have, to greater or lesser degrees, the same spirit as the free group construction, and also, for variety, at two such constructions from topology.

# Chapter 2
# Making Some Things Precise

## 2.1. Generalities

Most notation will be explained as it is introduced. I will assume familiarity with basic set-theoretic and logical notation: $\forall$ for "for all" (universal quantification), $\exists$ for "there exists" (existential quantification), $\wedge$ for "and", and $\vee$ for "or". Functions will be indicated by arrows $\to$, while their behavior on elements will be shown by flat-tailed arrows, $\mapsto$. That is, if a function $X \to Y$ carries an element $x$ to an element $y$, this may be symbolized $x \mapsto y$ ("$x$ goes to $y$"). If $S$ is a set and $\sim$ an equivalence relation on $S$, the set of equivalence classes under this relation will be denoted $S/\sim$.

We will (with rare exceptions, which will be noted) write functions on the left of their arguments, i.e., $f(x)$ rather than $xf$, and understand composite functions $fg$ to be defined so that $(fg)(x) = f(g(x))$.

## 2.2. What is a group?

Loosely speaking, a group is a set $G$ given with a *composition* (or *multiplication*, or *group operation*) $\mu : G \times G \to G$, an *inverse* operation $\iota : G \to G$, and a *neutral* element $e \in G$, satisfying certain well-known laws. (We will say "neutral element" rather than "identity element" to avoid confusion with the other important meaning of the word "identity", namely an equation that holds identically.)

The most convenient way to make precise this idea of a set "given with" three operations is to define the group to be, not the set $G$, but the 4-tuple $(G, \mu, \iota, e)$. In fact, from now on, a letter such as $G$ representing a group

© Springer International Publishing Switzerland 2015
G.M. Bergman, *An Invitation to General Algebra and Universal Constructions*, Universitext, DOI 10.1007/978-3-319-11478-1_2

will stand for such a 4-tuple, and the first component, called the *underlying set* of the group, will be written $|G|$. Thus

$$G = (|G|, \mu, \iota, e).$$

For simplicity, many mathematicians ignore this formal distinction, and use a letter such as $G$ to represent both a group and its underlying set, writing $x \in G$, for instance, where they mean $x \in |G|$. This is okay, as long as one always understands what precise statement such a shorthand statement stands for. Note that to be entirely precise, if $G$ and $H$ are two groups, we should use different symbols, say $\mu_G$ and $\mu_H$, $\iota_G$ and $\iota_H$, $e_G$ and $e_H$, for the operations of $G$ and $H$. How precise and formal one needs to be depends on the situation. Since the aim of this course is to abstract the concept of algebraic structure and study what makes these things tick, we shall be somewhat more precise here than in an ordinary algebra course.

(Many workers in General Algebra use a special type-font, e.g., boldface, to represent algebraic objects, and regular type for their underlying sets. Thus, where we will write $G = (|G|, \mu, \iota, e)$, they might write $\mathbf{G} = (G, \mu, \iota, e)$.)

Perhaps the easiest exercise in the course is:

**Exercise 2.2:1.** Give a precise definition of a homomorphism from a group $G$ to a group $H$, distinguishing between the operations of $G$ and the operations of $H$.

We will often refer to a homomorphism $f : G \to H$ as a "map" from $G$ to $H$. That is, unless the contrary is mentioned, "maps" between mathematical objects mean maps between their underlying sets which respect their structure. Note that if we wish to refer to a set map not assumed to respect the group operations, we can call this "a map from $|G|$ to $|H|$".

The use of letters ($\mu$ and $\iota$) for the operations of a group, and the functional notation $\mu(x, y)$, $\iota(z)$ which this entails, are desirable for stating results in a form which will *generalize* to a wide class of other sorts of structures. But when actually working with elements of a group, we will generally use conventional notation, writing $x \cdot y$ (or $xy$, or sometimes, in abelian groups, $x + y$) for $\mu(x, y)$, and $z^{-1}$ (or $-z$) for $\iota(z)$. When we do this, we may either continue to write $G = (|G|, \mu, \iota, e)$, or write $G = (|G|, \cdot, {}^{-1}, e)$.

Let us now recall the conditions which must be satisfied by a 4-tuple $G = (|G|, \cdot, {}^{-1}, e)$, where $|G|$ is a set, "$\cdot$" is a map $|G| \times |G| \to |G|$, "$^{-1}$" is a map $|G| \to |G|$, and $e$ is an element of $|G|$, for $G$ to be called a group:

(2.2.1)
$$(\forall\, x, y, z \in |G|)\ (x \cdot y) \cdot z = x \cdot (y \cdot z),$$
$$(\forall\, x \in |G|)\ e \cdot x = x = x \cdot e,$$
$$(\forall\, x \in |G|)\ x^{-1} \cdot x = e = x \cdot x^{-1}.$$

There is another definition of group that you have probably also seen: In effect, a group is defined to be a pair $(|G|, \cdot)$, such that $|G|$ is a set, and $\cdot$ is a map $|G| \times |G| \to |G|$ satisfying

$$(\forall\, x, y, z \in |G|)\ (x \cdot y) \cdot z = x \cdot (y \cdot z),$$

(2.2.2) $\quad (\exists\, e \in |G|)\, ((\forall\, x \in |G|)\ e \cdot x = x = x \cdot e)\ \wedge$
$$((\forall\, x \in |G|)\,(\exists\, y \in |G|)\ y \cdot x = e = x \cdot y).$$

It is easy to show that given $(|G|, \cdot)$ satisfying (2.2.2), there exist a unique operation $^{-1}$ and a unique element $e$ such that $(|G|, \cdot, {}^{-1}, e)$ satisfies (2.2.1)—remember the standard results saying that neutral elements and two-sided inverses are unique when they exist. Thus, the versions (2.2.1) and (2.2.2) of the concept of a group provide equivalent information. Our description of groups as 4-tuples may therefore seem "uneconomical" compared with one using pairs, but we will stick with it. We shall eventually see that, more important than the number of terms in the tuple, is the fact that condition (2.2.1) consists of identities, i.e., universally quantified equations, while (2.2.2) does not. But we will at times acknowledge the idea of the second definition; for instance, when we ask (imprecisely) whether some semigroup "is a group".

**Exercise 2.2:2.**
(i)   If $G$ is a group, let us define an operation $\delta_G$ on $|G|$ by $\delta_G(x, y) = x \cdot y^{-1}$. Does the pair $G' = (|G|, \delta_G)$ determine the group $(|G|, \cdot, {}^{-1}, e)$? (I.e., if $G_1$ and $G_2$ yield the same pair, $G_1' = G_2'$, must $G_1 = G_2$? Some students have asked whether by "$=$" I here mean "$\cong$". No, I mean "$=$".)
(ii)   Suppose $|X|$ is any set and $\delta\colon |X| \times |X| \to |X|$ any map. Can you write down a set of axioms for the pair $X = (|X|, \delta)$, which will be necessary and sufficient for it to arise from a group $G$ in the manner described above? (That is, assuming $|X|$ and $\delta$ given, try to find convenient necessary and sufficient conditions for there to exist a group $G$ such that $G'$, defined as in (i), is precisely $(|X|, \delta)$.)
   If you get such a set of axioms, then try to see how brief and simple you can make it.

I don't know the full answers to the following variant question:

**Exercise 2.2:3.** Again let $G$ be a group, and now let us define $\sigma_G(x, y) = x \cdot y^{-1} \cdot x$. Consider the same questions for $(|G|, \sigma_G)$ that were raised for $(|G|, \delta_G)$ in the preceding exercise.

My point in discussing the distinction between a group and its underlying set, and between groups described using (2.2.1) and using (2.2.2), was not to be petty, but to make us conscious of the various ways we use mathematical language—so that we can use it without its leading us astray. At times we will bow to convenience rather than trying to be consistent. For instance, since we distinguish between a group and its underlying set, we should logically distinguish between the *set* of integers, the *additive group* of integers, the *multiplicative semigroup* of integers, the *ring* of integers, etc.; but we shall in fact write all of these $\mathbb{Z}$ unless there is a real danger of ambiguity, or a need to emphasize a distinction. When there is such a need, we can write $(\mathbb{Z}, +, -, 0) = \mathbb{Z}_{\mathrm{add}}$, $(\mathbb{Z}, \cdot, 1) = \mathbb{Z}_{\mathrm{mult}}$, $(\mathbb{Z}, +, \cdot, -, 0, 1) = \mathbb{Z}_{\mathrm{ring}}$, etc. We may likewise use "ready made" symbols for some other objects, such as $\{e\}$ for the trivial subgroup of a group $G$, rather than interrupting a discussion to set up a notation that distinguishes this subgroup from its underlying set.

The approach of regarding sets with operations as tuples, whose first member is the set and whose other members are the operations, applies, as we have just noted, to other algebraic structures than groups—to semigroups, rings, lattices, and the more exotic beasties we will meet on our travels. To be able to discuss the general case, we must make sure we are clear about what we mean by such concepts as "$n$-tuple of elements" and "$n$-ary operation". We shall review these in the next two sections.

## 2.3. Indexed sets

If $I$ and $X$ are sets, an $I$-*tuple* of elements of $X$, or a *family of elements of $X$ indexed by $I$* will be defined formally as a function from $I$ to $X$, but we shall write it $(x_i)_{i \in I}$ rather than $f : I \to X$. The difference is one of viewpoint. We think of such families as arrays of elements of $X$, which we keep track of with the help of an index set $I$, while when we write $f : A \to B$, we are most often interested in some properties relating an element of $A$ and its image in $B$. But the distinction is not sharp. Sometimes there *is* an interesting functional relation between the indices $i$ and the values $x_i$; sometimes typographical or other reasons dictate the use of $x(i)$ rather than $x_i$.

There will be a minor formal exception to the above definition when we speak of an $n$-*tuple* of elements of $X$ $(n \geq 0)$. In these beginning chapters, I will take this to mean a function from $\{1, \ldots, n\}$ to $X$, written $(x_1, \ldots, x_n)$ or $(x_i)_{i=1,\ldots,n}$, despite the fact that set theorists define the natural number $n$ inductively to be the set $\{0, \ldots, n-1\}$. Most set theorists, for consistency with that definition, write their $n$-tuples $(x_0, \ldots, x_{n-1})$; and we shall switch to that notation after reviewing the set theorist's approach to the natural numbers in Chapter 5.

If $I$ and $X$ are sets, then the set of *all* functions from $I$ to $X$, equivalently, of all $I$-tuples of members of $X$, is written $X^I$. Likewise, $X^n$ will denote the set of $n$-tuples of elements of $X$, defined as above for the time being.

## 2.4. Arity

An *n-ary* operation on a set $S$ means a map $f \colon S^n \to S$. For $n = 1, 2, 3$ the words are *unary*, *binary*, and *ternary* respectively. If $f$ is an $n$-ary operation, we call $n$ the *arity* of $f$. More generally, given any set $I$, an $I$-ary operation on $S$ is defined as a map $S^I \to S$.

Thus, the definition of a group involves one binary operation, one unary operation, and one distinguished element, or "constant", $e$. Likewise, a ring can be described as a 6-tuple $R = (|R|, +, \cdot, -, 0, 1)$, where $+$ and $\cdot$ are binary operations on $|R|$, " $-$ " is a unary operation, and $0, 1$ are distinguished elements, satisfying certain identities.

One may make these descriptions more homogeneous in form by treating "distinguished elements" as 0-*ary operations* of our algebraic structures. Indeed, since an $n$-ary operation on $S$ is something that turns out a value in $S$ when we feed in $n$ arguments in $S$, it makes sense that a 0-ary operation should be something that gives a value in $S$ without our feeding it anything. Or, looking at it formally, $S^0$ is the set of all maps from the empty set to $S$, of which there is exactly one; so $S^0$ is a one-element set, so a map $S^0 \to S$ determines, and is determined by, a single element of $S$.

We note also that distinguished elements show the right *numerical* behavior to be called "zeroary operations". Indeed, if $f$ and $g$ are an $m$-ary and an $n$-ary operation on $S$, and $i$ a positive integer $\leq m$, then on inserting $g$ in the $i$-th place of $f$, we get an operation $f(-, \ldots, -, g(-, \ldots, -), -, \ldots, -)$ of arity $m + n - 1$. Now if, instead, $g$ is an element of $S$, then when we put it into the $i$-th place of $f$ we get $f(-, \ldots, -, g, -, \ldots, -)$, an $(m - 1)$-ary operation, as we should if $g$ is thought of as an operation of arity $n = 0$.

Strictly speaking, elements and zeroary operations are in one-to-one correspondence rather than being the same thing: one must distinguish between a map $S^0 \to S$, and its (unique) value in $S$. But since they give equivalent information, we can choose between them in setting up our definitions.

So we shall henceforth treat "distinguished elements" in the definition of groups, rings, etc., as zeroary operations, and we will find that they can be handled essentially like the other operations. I say "essentially" because there are some minor ways in which zeroary operations differ from operations of positive arity. Most notably, on the empty set $X = \emptyset$, there is a *unique* $n$-ary operation for each *positive* $n$, but *no zeroary* operation. Sometimes this trivial fact will make a difference in an argument.

## 2.5. Group-theoretic terms

One is often interested in talking about what *relations* hold among the members of one or another tuple of elements of a group or other algebraic structure. For example, *every* pair of elements $(\xi, \eta)$ of a group satisfies the relation $(\xi \cdot \eta)^{-1} = \eta^{-1} \cdot \xi^{-1}$. Some *particular* pair $(\xi, \eta)$ of elements of some group may satisfy the relation $\xi \cdot \eta = \eta \cdot \xi^2$.

In general, a "group-theoretic relation" in a family of elements $(\xi_i)_I$ of a group $G$ means an equation $p(\xi_i) = q(\xi_i)$ holding in $G$, where $p$ and $q$ are *expressions* formed from an $I$-tuple of symbols using formal group operations $\cdot$, $^{-1}$ and $e$. So to study relations in groups, we need to define the set of all "formal expressions" in the elements of a set $X$ under symbolic operations of multiplication, inverse and neutral element.

The technical word for such a formal expression is a "term". Intuitively, a *group-theoretic term* is a set of instructions on how to apply the group operations to a family of elements. E.g., starting with a set of three symbols, $X = \{x, y, z\}$, an example of a group-theoretic term in $X$ is the symbol $(y \cdot x) \cdot (y^{-1})$; or we might write it $\mu(\mu(y, x), \iota(y))$. Whichever way we write it, the idea is: "apply the operation $\mu$ to the pair $(y, x)$, apply the operation $\iota$ to the element $y$, and then apply the operation $\mu$ to the pair of elements so obtained, taken in that order". The idea can be "realized" when we are given a map $f$ of the set $X$ into the underlying set $|G|$ of a group $G = (|G|, \mu_G, \iota_G, e_G)$, say $x \mapsto \xi$, $y \mapsto \eta$, $z \mapsto \zeta$ $(\xi, \eta, \zeta \in |G|)$. We can then define the result of "evaluating the term $\mu(\mu(y, x), \iota(y))$ using the map $f$" as the *element* $\mu_G(\mu_G(\eta, \xi), \iota_G(\eta)) \in |G|$, that is, $(\eta \cdot \xi) \cdot (\eta^{-1})$.

Let us try to make the concept of group-theoretic term precise. "The set of all terms in the elements of $X$, under formal operations $\cdot$, $^{-1}$ and $e$" should be a set $T = T_{X, \cdot, ^{-1}, e}$ with the following properties:

(a$_X$)     For every $x \in X$, $T$ contains a symbol representing $x$.

(a.)        For every $s, t \in T$, $T$ contains a "symbolic combination of $s$ and $t$ under $\cdot$".

(a$_{-1}$)   For every $s \in T$, $T$ contains an element gotten by "symbolic application of $^{-1}$ to $s$".

(a$_e$)      $T$ contains an element symbolizing $e$.

(b)         Each element of $T$ can be written in *one and only one way* as one and only one of the following:

    (b$_X$)     The symbol representing an element of $X$.

    (b.)        The symbolic combination of two members of $T$ under $\cdot$.

    (b$_{-1}$)   The symbol representing the result of applying $^{-1}$ to an element of $T$.

    (b$_e$)      The symbol representing $e$.

(c)         Every element of $T$ can be obtained from the elements of $X$ via the given symbolic operations. That is, $T$ has no proper subset satisfying (a$_X$)–(a$_e$).

In functional language, (a$_X$) says that we are to be given a function $X \to T$ (the "symbol for $x$" function); (a.) says we have another function, which we call "formal product", from $T \times T$ to $T$; (a$_{-1}$) gives a function $T \to T$, the "formal inverse", and $(a_e)$ a distinguished element of $T$. Translating our definition into this language, we get

**Definition 2.5.1.** *By "the set of all terms in the elements of $X$ under the formal group operations $\mu$, $\iota$, $e$" we shall mean a set $T$ which is:*

(a) *given with functions*

$$\text{symb}_T : X \to T, \quad \mu_T : T^2 \to T, \quad \iota_T : T \to T, \quad and \quad e_T : T^0 \to T,$$

*such that*

(b) *each of these maps is one-to-one, their images are disjoint, and $T$ is the union of those images, and*

(c) *$T$ is generated by $\text{symb}_T(X)$ under the operations $\mu_T$, $\iota_T$, and $e_T$; i.e., it has no proper subset which contains $\text{symb}_T(X)$ and is closed under those operations.*

The next exercise justifies the use of the word "the" in the above definition.

**Exercise 2.5:1.** Assuming $T$ and $T'$ are two sets given with functions that satisfy Definition 2.5.1, establish a natural one-to-one correspondence between the elements of $T$ and $T'$. (You must, of course, show that the correspondence you set up is well-defined, and is a bijection. Suggestion: Let $Y = \{ (\text{symb}_T(x), \text{symb}_{T'}(x)) \mid x \in X\} \subseteq T \times T'$, and let $F$ be the closure of $Y$ under componentwise application of $\mu$, $\iota$ and $e$. Show that $F$ is the graph of a bijection. What properties will characterize this bijection?)

**Exercise 2.5:2.** Is condition (c) of Definition 2.5.1 a consequence of (a) and (b)?

How can we obtain a set $T$ with the properties of the above definition? One approach is to construct elements of $T$ as finite *strings of symbols* from some alphabet which contains symbols representing the elements of $X$, additional symbols $\mu$ (or $\cdot$), $\iota$ (or $^{-1}$), and $e$, and perhaps some symbols of punctuation. But we need to be careful. For instance, if we defined $\mu_T$ to take a string of symbols $s$ and a string of symbols $t$ to the string of symbols $s \cdot t$, and $\iota_T$ to take a string of symbols $s$ to the string of symbols $s^{-1}$, then condition (b) would not be satisfied! For a string of symbols of the form $x \cdot y \cdot z$ (where $x, y, z \in X$) could be obtained by formal multiplication either of $x$ and $y \cdot z$, or of $x \cdot y$ and $z$. In other words, $\mu_T$ takes the pairs $(x, y \cdot z)$ and $(x \cdot y, z)$ to the same string of symbols, so it is not one-to-one. Likewise, the expression $x \cdot y^{-1}$ could be obtained either as $\mu_T(x, y^{-1})$ or as $\iota_T(x \cdot y)$, so the images of $\mu_T$ and $\iota_T$ are not disjoint. (It happens that in the first case, the two interpretations of $x \cdot y \cdot z$ come to the same thing in any group, because of the associative law, while in the second, the two interpretations do not: $\xi \cdot (\eta^{-1})$ and $(\xi \cdot \eta)^{-1}$ are generally distinct for elements $\xi, \eta$ of a group $G$. But the point is that in both cases condition (b) fails, making these expressions ambiguous as *instructions* for applying group operations. Note that a notational system in which "$x \cdot y \cdot z$" was ambiguous in the above way could never be used in *writing down* the associative law; and writing down identities is one of the uses we will want to make of these expressions.)

On the other hand, it is not hard to show that by introducing parentheses among our symbols, and letting $\mu_T(s, t)$ be the string of symbols $(s \cdot t)$, and $\iota_T(s)$ the string of symbols $(s^{-1})$, we can get a set of expressions satisfying the conditions of our definition.

**Exercise 2.5:3.** Verify the above assertion. (How, precisely, will you define $T$? What assumptions must you make on the set of symbols representing elements of $X$? Do you allow some elements $\text{symb}_T(x)$ to be strings of other symbols?)

Another symbolism that will work is to define the value of $\mu_T$ at $s$ and $t$ to be the string of symbols $\mu(s, t)$, and the value of $\iota_T$ at $s$ to be the string of symbols $\iota(s)$.

**Exercise 2.5:4.** Assuming the elements $\text{symb}_T(x)$ are distinct single characters, and that $\mu$, $\iota$ and $e$ are distinct characters distinct from the characters $\text{symb}_T(x)$, let us define the value of $\mu_T$ on elements $s$ and $t$ to be the symbol $\mu st$, and the value of $\iota_T$ on $s$ to be the symbol $\iota s$. Will the resulting set of strings of symbols satisfy Definition 2.5.1?

Though the strings-of-symbols approach can be extended to other kinds of algebras with finitary operations, such as rings, lattices, etc., a disadvantage of that method is that one cannot, in any obvious way, use it for algebras with operations of *infinite* arities. Even if one allows infinite strings of symbols, indexed by the natural numbers or the integers, one cannot string two or more such infinite strings together to get another string of the same sort. One can, however, for an infinite set $I$, create $I$-tuples which have $I$-tuples among their members, and this leads to the more versatile *set-theoretic* approach. Let us show it for the case of group-theoretic terms.

Choose any set of four elements, which will be denoted $*$, $\cdot$, $^{-1}$ and $e$. For each $x \in X$, define $\text{symb}_T(x)$ to be the ordered pair $(*, x)$; for $s, t \in T$, define $\mu_T(s, t)$ to be the ordered 3-tuple $(\cdot, s, t)$; for $s \in T$ define $\iota_T(s)$ to be the ordered pair $(^{-1}, s)$, and finally, define $e_T$ to be the 1-tuple $(e)$. Let $T$ be the smallest set closed under the above operations.

Now it is a basic lemma of set theory that no element can be written as an $n$-tuple in more than one way; i.e., if $(x_1, \ldots, x_n) = (x'_1, \ldots, x'_{n'})$, then $n' = n$ and $x_i = x'_i$ $(i = 1, \ldots, n)$. It is easy to deduce from this that the above construction will satisfy the conditions of Definition 2.5.1.

**Exercise 2.5:5.** Would there have been anything wrong with defining $\text{symb}_T(x) = x$ instead of $(*, x)$? If so, can you find a way to modify the definitions of $\mu_T$ etc., so that the definition $\text{symb}_T(x) = x$ can always be used?

I leave it to you to decide (or not to decide) which construction for group-theoretic terms you prefer to assume during these introductory chapters. We shall only need the *properties* given in Definition 2.5.1. From now on, we shall often use conventional notation for such terms, e.g., $(x \cdot y) \cdot (x^{-1})$. In particular, we shall often identify $X$ with its image $\text{symb}_T(X) \subseteq T_{X, \cdot, {}^{-1}, e}$. We will use the more formal notation of Definition 2.5.1 mainly when we want to emphasize particular distinctions, such as that between the formal operations $\mu_T$ etc., and the operations $\mu_G$ etc. of a particular group.

## 2.6. Evaluation

Now suppose $G$ is a group, and $f : X \to |G|$ a set map, in other words, an $X$-tuple of elements of $G$. Given a term in an $X$-tuple of symbols,

$$s \in T = T_{X, \cdot, {}^{-1}, e}$$

we wish to say how to *evaluate* $s$ at this family $f$ of elements, so as to get a value $s_f \in |G|$. We shall do this inductively (or more precisely, "recursively"; we will learn the distinction in § 5.3).

If $s = \text{symb}_T(x)$ for some $x \in X$ we define $s_f = f(x)$. If $s = \mu_T(t, u)$, then assuming inductively that we have already defined $t_f$, $u_f \in |G|$, we define $s_f = \mu_G(t_f, u_f)$. Likewise, if $s = \iota_T(t)$, we assume inductively that $t_f$ is defined, and define $s_f = \iota_G(t_f)$. Finally, for $s = e_T$ we define $s_f = e_G$. Since every element $s \in T$ is obtained from $\text{symb}_T(X)$ by the operations $\mu_T$, $\iota_T$, $e_T$, and in a unique manner, this construction gives one and only one value $s_f$ for each $s$.

We have not discussed the general principles that allow one to make recursive definitions like the above. We shall develop these in Chapter 5, in preparation for Chapter 9 where we will do rigorously and in full generality what we are sketching here. Some students might want to look into this question for themselves at this point, so I will make this:

**Exercise 2.6:1.** Show rigorously that the procedure loosely described above yields a unique well-defined map $T \to |G|$. (Suggestion: Adapt the method suggested for Exercise 2.5:1.)

In the above discussion of evaluation, we fixed $f \in |G|^X$, and got a function $T \to |G|$, taking each $s \in T$ to $s_f \in |G|$. If we vary $f$ as well as $T$, we get a two-variable evaluation map,

$$(T_{X, \cdot, {}^{-1}, e}) \times |G|^X \longrightarrow |G|,$$

taking each pair $(s, f)$ to $s_f$. Finally, we might fix an $s \in T$, and define a map $s_G \colon |G|^X \to |G|$ by $s_G(f) = s_f$ $(f \in |G|^X)$; this represents "substitution into $s$." For example, suppose $X = \{x, y, z\}$, let us identify $|G|^X$ with $|G|^3$, and let $s$ be the term $(y \cdot x) \cdot (y^{-1}) \in T$. Then for each group $G$, $s_G$ is the operation taking each 3-tuple $(\xi, \eta, \zeta)$ of elements of $G$ to the element $(\eta \xi) \eta^{-1} \in G$. Such operations will be of importance to us, so we give them a name.

**Definition 2.6.1.** *Let $G$ be a group and $n$ a nonnegative integer. Let $T = T_{n, \, -1, \, \cdot, \, e}$ denote the set of group-theoretic terms in $n$ symbols. Then for each $s \in T$, we will let $s_G \colon |G|^n \to |G|$ denote the map taking each $n$-tuple $f \in |G|^n$ to the element $s_f \in |G|$. The $n$-ary operations $s_G$ obtained in this way from terms $s \in T$ will be called the* derived $n$-ary operations *of $G$. (Some authors call these* term operations.*)*

Note that *distinct terms* can induce the same *derived operation.* E.g., the associative law for groups says that for any group $G$, the derived ternary operations induced by the terms $(x \cdot y) \cdot z$ and $x \cdot (y \cdot z)$ are the same. As another example, in the particular group $S_3$ (the symmetric group on three elements), the derived binary operations induced by the terms $(x \cdot x) \cdot (y \cdot y)$ and $(y \cdot y) \cdot (x \cdot x)$ are the same, though this is not true in all groups. (It is true in all dihedral groups.)

Some other examples of derived operations on groups are the binary operation of *conjugation*, commonly written $\xi^\eta = \eta^{-1} \xi \eta$ (induced by the term $y^{-1} \cdot (x \cdot y)$), the binary *commutator* operation, $[\xi, \eta] = \xi^{-1} \eta^{-1} \xi \eta$, the unary operation of *squaring*, $\xi^2 = \xi \cdot \xi$, and the two binary operations $\delta$ and $\sigma$ of Exercises 2.2:2 and 2.2:3. Some trivial examples are also important: the *primitive* group operations—group multiplication, inverse, and neutral element—are by definition also *derived* operations; and finally, one has very trivial derived operations such as the ternary "second component" function, $p_{3,\,2}(\xi, \eta, \zeta) = \eta$, induced by $y \in T_{\{x,\,y,\,z\},\,-1,\,\cdot,\,e}$. (Here $p_{3,\,2}$ stands for "projection of 3-tuples to their second component".)

## 2.7. Terms in other families of operations

The above approach can be applied to more general sorts of algebraic structures. Let $\Omega$ be an ordered pair $(|\Omega|, \text{ari})$, where $|\Omega|$ is a set of symbols (thought of as representing operations), and ari is a function associating to each $\alpha \in |\Omega|$ a nonnegative integer $\text{ari}(\alpha)$, the intended *arity* of $\alpha$ (§ 2.4). (For instance, in the group case which we have been considering, we have effectively taken $|\Omega| = \{\mu, \iota, e\}$, $\text{ari}(\mu) = 2$, $\text{ari}(\iota) = 1$, $\text{ari}(e) = 0$. Incidentally, the commonest symbol, among specialists, for the arity of an operation $\alpha$ is $n(\alpha)$, but I will use $\text{ari}(\alpha)$ to avoid confusion with other uses of the letter $n$.) Then an $\Omega$-*algebra* will mean a system $A = (|A|, (\alpha_A)_{\alpha \in |\Omega|})$, where $|A|$ is a set, and for each $\alpha \in |\Omega|$, $\alpha_A$ is some $\mathrm{ari}(\alpha)$-ary operation on $|A|$ :

$$\alpha_A : |A|^{\mathrm{ari}(\alpha)} \longrightarrow |A|.$$

For any set $X$, we can now mimic the preceding development to get a set $T = T_{X, \Omega}$, the set of "terms in elements of $X$ under the operations of $\Omega$"; and given any $\Omega$-algebra $A$, we can get substitution and evaluation maps as before, and so define *derived operations* of $A$.

The long-range goal of this course is to study algebras $A$ in this general sense. In order to discover what kinds of results we want to prove about them, we shall devote Chapters 3 and 4 to looking at specific situations involving familiar sorts of algebras. But let me give here a few exercises concerning these general concepts.

**Exercise 2.7:1.** On the set $\{0, 1\}$, let $M_3$ denote the ternary "majority vote" operation; i.e., for $a, b, c \in \{0, 1\}$, let $M_3(a, b, c)$ be $0$ if two or more of $a$, $b$ and $c$ are $0$, or $1$ if two or more of them are $1$. One can form various terms in a symbolic operation $M_3$ (e.g., $p(w, x, y, z) = M_3(x, M_3(z, w, y), z)$) and then evaluate these in the algebra $(\{0, 1\}, M_3)$ to get operations on $\{0, 1\}$ derived from $M_3$.

*General problem*: Determine which operations (of arbitrary arity) on $\{0, 1\}$ can be expressed as derived operations of this algebra.

As steps toward answering this question, you might try to determine whether each of the following can or cannot be so expressed:

(a) The 5-ary majority vote function $M_5 : \{0, 1\}^5 \to \{0, 1\}$, defined in the obvious manner.

(b) The binary operation sup. (i.e., $\sup(a, b) = 0$ if $a = b = 0$; otherwise $\sup(a, b) = 1$.)

(c) The unary "reversal" operation $r$, defined by $r(0) = 1$, $r(1) = 0$.

(d) The 4-ary operation $N_4$, described as "the majority vote function, where the first voter has extra tie-breaking power"; i.e., $N_4(a, b, c, d) =$ the majority value among $a, b, c, d$ if there is one, while if $a + b + c + d = 2$ we set $N_4(a, b, c, d) = a$.

Advice: (i) If you succeed in proving that some operation $s$ is *not* derivable from $M_3$, try to abstract your argument by establishing a general property that all operations derived from $M_3$ must have, but which $s$ clearly does not have. (ii) A mistake some students make is to think that a formula such as $s(\xi, \eta) = M_3(0, \xi, \eta)$ defines a derived operation. But since our system $(\{0, 1\}, M_3)$ does not include the *zeroary operation* $0$ (nor $1$), "$M_3(0, x, y)$" is not a term.

**Exercise 2.7:2** (Question raised by Jan Mycielski, letter of Jan 17, 1983). Let $\mathbb{C}$ denote the set of complex numbers, and exp the exponential function $\exp(x) = e^x$, a unary operation on $\mathbb{C}$.

(i)   Does the algebra $(\mathbb{C}, +, \cdot, \exp)$ have any automorphisms other than the identity and complex conjugation? (An *automorphism* means a bijection of the underlying set with itself, which respects the operations.) I don't know the answer to this question.

It is not hard to prove using the theory of transcendence bases of fields ([32, § VI.1], [34, § VIII.1]) that the automorphism group of $(\mathbb{C}, +, \cdot)$ is infinite (cf. [32, Exercise VI.6(b)], [34, Exercise VIII.1]). A couple of easy results in the opposite direction, which you may prove and hand in, are

(ii)   The algebra $(\mathbb{C}, +, \cdot)$ has no *continuous* automorphisms other than the two mentioned.

(iii)  If we write "cj" for the unary operation of complex conjugation, then the algebra $(\mathbb{C}, +, \cdot, \mathrm{cj})$ has no automorphisms other than id and cj.

(iv)  A map $\mathbb{C} \to \mathbb{C}$ is an automorphism of $(\mathbb{C}, +, \cdot, \exp)$ if and only if it is an automorphism of $(\mathbb{C}, +, \exp)$.

**Exercise 2.7:3.** Given operations $\alpha_1, \ldots, \alpha_r$ (of various finite arities) on a finite set $S$, and another operation $\beta$ on $S$, describe a test that will determine in a finite number of steps whether $\beta$ is a derived operation of $\alpha_1, \ldots, \alpha_r$.

The arities considered so far have been finite; the next exercise will deal with terms in operations of possibly *infinite* arities. To make this reasonable, let us note some naturally arising examples of operations of countably infinite arity on familiar sets:

*On the real unit interval* $[0, 1]$ :

(a)  the operation $\limsup$ ("limit superior"), defined by

$$\limsup_i x_i \;=\; \lim_{i \to \infty} \sup_{j \geq i} x_j,$$

(b)  the operation defined by $s(a_1, a_2, \ldots) = \sum 2^{-i} a_i$.

*On the set of real numbers* $\geq 1$ :

(c)  the continued fraction operation, $c(a_1, a_2, \ldots) = a_1 + 1/(a_2 + 1/(\ldots))$.

*On the class of subsets of the set of integers*:

(d)  the operation $\bigcup a_i$,

(e)  the operation $\bigcap a_i$.

**Exercise 2.7:4.** Suppose $\Omega$ is a pair $(|\Omega|, \mathrm{ari})$, where $|\Omega|$ is again a set of operation symbols, but where the arities $\mathrm{ari}(\alpha)$ may now be finite or infinite cardinals; and let $X$ be a set of variable-symbols. Suppose we can form a set $T$ of terms satisfying the analogs of conditions (a)–(c) of Definition 2.5.1. For $s, t \in T$, let us write $s > t$ if $t$ is "immediately involved" in $s$, that is, if $s$ has the form $\alpha(u_1, u_2, \ldots)$ where $\alpha \in |\Omega|$, and $u_i = t$ for *some* $i$.

(i)    Show that if all the arities ari($\alpha$) are *finite*, then for each term $s$ we can find a finite bound $B(s)$ on the lengths $n$ of sequences $s_1, \ldots, s_n \in T$ such that $s = s_1 > \ldots > s_n$.

(ii)   If not all ari($\alpha$) are finite, and $X$ is nonempty, show that there exist terms $s$ for which no such finite bound exists.

(iii) In the situation of (ii), is it possible to have a right-infinite chain $s = s_1 > \ldots > s_n > \ldots$ in $T$?

(iv)  Show that one cannot have a "cycle" $s_1 > \ldots > s_n > s_1$ in $T$.

Until we come to Chapter 9, we shall rarely use the word "algebra" in the general sense of this section. But the reader consulting the index should keep this sense in mind, since it is used there with reference to general concepts of which we will be considering specific cases in the intervening chapters.

# Chapter 3
# Free Groups

In this chapter, we introduce the idea of universal constructions through the particular case of free groups. We shall first motivate the free group concept, then develop three ways of constructing such groups.

## 3.1. Motivation

Suppose $G$ is a group and we take (say) three elements $a, b, c \in |G|$, and consider what group-theoretic relations these satisfy. That is, letting $T$ be the set of all group-theoretic terms in three symbols $x$, $y$ and $z$, we look at pairs of elements $p(x, y, z)$, $q(x, y, z) \in T$, and if $p_G(a, b, c) = q_G(a, b, c)$ in $|G|$, we say that $(a, b, c)$ satisfies the relation $p = q$. We note:

**Lemma 3.1.1.** *Suppose $F$ and $G$ are groups, such that $F$ is generated by three elements $a, b, c \in |F|$, while $\alpha, \beta, \gamma$ are any three elements of $G$. Then the following conditions are equivalent:*

(a) *Every group-theoretic relation $p = q$ satisfied by $(a, b, c)$ in $F$ is also satisfied by $(\alpha, \beta, \gamma)$ in $G$.*

(b) *There exists a group homomorphism $h \colon F \to G$ under which $a \mapsto \alpha$, $b \mapsto \beta$, $c \mapsto \gamma$.*

*Further, when these conditions hold, the homomorphism $h$ of (b) is unique.*

*If the assumption that $a$, $b$ and $c$ generate $F$ is dropped, one still has (b) $\Longrightarrow$ (a).*

*Proof.* Not yet assuming that $a$, $b$ and $c$ generate $F$, suppose $h$ is a homomorphism as in (b). Then I claim that for all $p \in T$,

$$h(p_F(a, b, c)) = p_G(\alpha, \beta, \gamma).$$

© Springer International Publishing Switzerland 2015
G.M. Bergman, *An Invitation to General Algebra and Universal Constructions*, Universitext, DOI 10.1007/978-3-319-11478-1_3

Indeed, the set of $p \in T$ for which the above equation holds is easily seen to contain $x$, $y$ and $z$, and to be closed under the operations of $T$, hence it is all of $T$. Statement (a) follows, giving the final assertion of the lemma. If, further, $a$, $b$ and $c$ generate $F$, then every element of $|F|$ can be written $p_F(a, b, c)$ for some $p$, so the above formula shows that given such $a$, $b$ and $c$, the homomorphism $h$ is determined by $\alpha$, $\beta$ and $\gamma$, yielding the next-to-last assertion.

Finally, suppose $a$, $b$ and $c$ generate $F$ and (a) holds. For each $g = p_F(a, b, c) \in |F|$, define $h(g) = p_G(\alpha, \beta, \gamma)$. To show that this gives a well-defined map from $|F|$ to $|G|$, note that if we have two ways of writing an element $g \in |F|$, say $p_F(a, b, c) = g = q_F(a, b, c)$, then the relation $p = q$ is satisfied by $(a, b, c)$ in $F$, hence by (a), it is satisfied by $(\alpha, \beta, \gamma)$ in $G$, hence the two values our definition prescribes for $h(g)$, namely $p_G(\alpha, \beta, \gamma)$ and $q_G(\alpha, \beta, \gamma)$, are the same.

That this set map is a homomorphism follows from the way evaluation of group-theoretic terms is defined. For instance, given $g \in |F|$, suppose we want to show that $h(g^{-1}) = h(g)^{-1}$. We write $g = p_F(a, b, c)$. Then $(\iota_T(p))_F(a, b, c) = g^{-1}$, so our definition of $h$ gives $h(g^{-1}) = (\iota_T(p))_G(\alpha, \beta, \gamma) = p_G(\alpha, \beta, \gamma)^{-1} = h(g)^{-1}$. The same reasoning applies to products and to the neutral element.                                                               □

**Exercise 3.1:1.** Show by example that if $\{a, b, c\}$ does not generate $F$, then condition (a) of the above lemma can hold and (b) fail, and also that (b) can hold but $h$ not be unique. (You may replace $(a, b, c)$ with a smaller family, $(a, b)$ or $(a)$, if you like.)

Lemma 3.1.1 leads one to wonder: Among all groups $F$ given with generating 3-tuples of elements $(a, b, c)$, is there one in which these three elements satisfy the *smallest* possible set of relations? We note what the above lemma would imply for such a group:

**Corollary 3.1.2.** *Let $F$ be a group, and $a, b, c \in |F|$. Then the following conditions are equivalent:*

(a) *$a, b, c$ generate $F$, and the only relations satisfied by $a, b, c$ in $F$ are those relations satisfied by every 3-tuple $(\alpha, \beta, \gamma)$ of elements in every group $G$.*

(b) *For every group $G$, and every 3-tuple of elements $(\alpha, \beta, \gamma)$ in $G$, there exists a unique homomorphism $h \colon F \to G$ such that $h(a) = \alpha$, $h(b) = \beta$, $h(c) = \gamma$.*                                                               □

Only one point in the deduction of this corollary from Lemma 3.1.1 is not completely obvious; I will make it an exercise:

**Exercise 3.1:2.** In the situation of the above corollary, show that (b) implies that $a$, $b$ and $c$ generate $F$. (Hint: Let $G$ be the subgroup of $F$ generated by those three elements.)

I've been speaking of 3-tuples of elements for concreteness; the same observations are valid for $n$-tuples for any $n$, and generally, for $X$-tuples for any
set $X$. An $X$-tuple of elements of $F$ means a set map $X \to |F|$, so in this
general context, condition (b) above takes the form given by the next definition. (But making this definition does not answer the question of whether
such objects exist!)

**Definition 3.1.3.** *Let $X$ be a set. By a* free group *$F$ on the set $X$, we
shall mean a pair $(F, u)$, where $F$ is a group, and $u$ a set map $X \to |F|$,
having the following* universal *property:*

*For every group $G$, and every set map $v \colon X \to |G|$, there exists a unique
homomorphism $h \colon F \to G$ such that $v = hu$; i.e., making the diagram below
commute.*

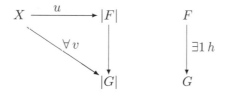

(In the above diagram, the first vertical arrow also represents the homomorphism $h$, regarded as a map on the underlying sets of the groups.)

Corollary 3.1.2 (as generalized to $X$-tuples) says that $(F, u)$ is a free group
on $X$ if and only if the elements $u(x)$ $(x \in X)$ generate $F$, and satisfy no
relations *except* those that hold for every $X$-tuple of elements in every group.
In this situation, one says that these elements "freely" generate $F$, hence
the term *free group*. Note that if such an $F$ exists, then by definition, *any*
$X$-tuple of members of any group $G$ can be obtained, in a unique way, as
the image, under a group homomorphism $F \to G$, of the *particular* $X$-tuple
$u$. Hence that $X$-tuple can be thought of as a "universal $X$-tuple of group
elements", so the property characterizing it is called a universal property.

We note a few elementary facts and conventions about such objects. If
$(F, u)$ is a free group on $X$, then the map $u \colon X \to |F|$ is one-to-one.
(This is easy to prove from the universal property, plus the well-known fact
that there exist groups with more than one element. The student who has
not seen free groups developed before should think this argument through.)
Hence given a free group, it is easy to get from it one such that the map
$u$ is actually an inclusion $X \subseteq |F|$. Hence for notational convenience, one
frequently assumes that this is so; or, what is approximately the same thing,
one often uses the same symbol for an element of $X$ and its image in $|F|$.

If $(F, u)$ and $(F', u')$ are both free groups on the same set $X$, there is a
unique isomorphism between them *as* free groups, i.e., respecting the maps
$u$ and $u'$. (Cf. diagram below.)

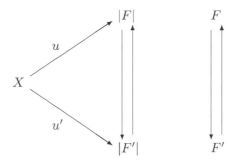

(If you haven't seen this result before, again see whether you can work out the details. For the technique you might look ahead to the proof of Proposition 4.3.3.) As any two free groups on $X$ are thus "essentially" the same, one sometimes speaks of *the* free group on $X$.

One also often says that a group $F$ "is free" to mean "there exists some set $X$ and some map $u: X \to |F|$ such that $(F, u)$ is a free group on $X$." When this holds, $X$ can always be taken to be a subset of $|F|$, and $u$ the inclusion map.

But it is time we proved that free groups exist. We will show three different ways of constructing them in the next three sections.

**Exercise 3.1:3.** Suppose one replaces the word "group" by "finite group" throughout Definition 3.1.3. Show that for any nonempty set $X$, *no* finite group exists having the stated universal property.

## 3.2. The logician's approach: construction from group-theoretic terms

We know from Corollary 3.1.2 that if a free group $F$ on three generators $a$, $b$, $c$ exists, then each of its elements can be written $p_F(a, b, c)$ for some group-theoretic term $p$, and that *two* such elements, $p_F(a, b, c)$ and $q_F(a, b, c)$, are equal if and only if the equation "$p = q$" is satisfied by every three elements of every group, i.e., follows from the group axioms. This suggests that we may be able to construct such a group by taking the set of all group-theoretic terms in three variables, constructing an equivalence relation "$p \sim q$" on this set which means "the equality of $p$ and $q$ is a consequence of the group axioms", taking for $|F|$ the quotient of our set of terms by this relation, and defining operations $\cdot$, $^{-1}$ and $e$ on $|F|$ in some natural manner. This we shall now do!

Let $X$ be any set, and $T = T_{X, \cdot, \, ^{-1}, \, e}$ the set of all group-theoretic terms in the elements of $X$. What conditions must a relation "$\sim$" satisfy for $p \sim q$ to be the condition "$p_v = q_v$" for *some* map $v$ of $X$ into *some* group $G$? Well, the group axioms tell us that it must satisfy

(3.2.1)     $(\forall\, p,\, q,\, r \in T)\ \ (p \cdot q) \cdot r \sim p \cdot (q \cdot r),$

(3.2.2)     $(\forall\, p \in T)\ \ (p \cdot e \sim p) \wedge (e \cdot p \sim p),$

(3.2.3)     $(\forall\, p \in T)\ \ (p \cdot p^{-1} \sim e) \wedge (p^{-1} \cdot p \sim e).$

Also, just the well-definedness of the operations of $G$ tells us that

(3.2.4)     $(\forall\, p,\, p',\, q \in T)\ \ (p \sim p') \implies ((p \cdot q \sim p' \cdot q) \wedge (q \cdot p \sim q \cdot p')),$

(3.2.5)     $(\forall\, p,\, p' \in T)\ \ (p \sim p') \implies (p^{-1} \sim p'^{-1}).$

Finally, of course, $\sim$ must be an equivalence relation:

(3.2.6)     $(\forall\, p \in T)\ \ p \sim p,$

(3.2.7)     $(\forall\, p,\, q \in T)\ \ (p \sim q) \implies (q \sim p),$

(3.2.8)     $(\forall\, p,\, q,\, r \in T)\ \ ((p \sim q) \wedge (q \sim r)) \implies (p \sim r).$

So let us take for "$\sim$" the *least* binary relation on $T$ satisfying conditions (3.2.1)–(3.2.8).

Let us note what this means, and why it exists: Recall that a binary relation on a set $T$ is formally a subset $R \subseteq T \times T$; when we write $p \sim q$, this is understood to be an abbreviation for $(p, q) \in R$. "Least" means smallest with respect to set-theoretic inclusion. Our conditions (3.2.1)–(3.2.8) are in the nature of closure conditions, and, as with all sets defined by closure conditions, the existence of a least set satisfying them can be established in two ways:

We may capture this set "from above" by forming the *intersection* of all binary relations on $T$ satisfying (3.2.1)–(3.2.8)—the set-theoretic intersection of these relations as subsets of $T \times T$. (Note, incidentally, that if we think of such relations as predicates rather than as sets, this intersection $\bigcap$ becomes a (generally infinite) conjunction $\bigwedge$.) The key point to observe is that each of these conditions is such that an intersection of relations satisfying it again satisfies it. Hence the intersection of *all* relations satisfying (3.2.1)–(3.2.8) will be the least such relation.

Or we can "build it up from below". Let $R_0$ denote the empty relation $\emptyset \subseteq T \times T$, and recursively construct the $i+1$-st relation $R_{i+1}$ from the $i$-th, by adding to $R_i$ those elements that conditions (3.2.1)–(3.2.8) say must *also* be in $R$, given that the elements of $R_i$ are there. Precisely, we let

$$
\begin{aligned}
R_{i+1} \;=\; & R_i && \text{(el'ts already constructed)}\\
& \cup\ \{((p \cdot q) \cdot r,\ p \cdot (q \cdot r)) \mid p,\, q,\, r \in T\} && \text{(el'ts arising by (3.2.1))}\\
& \cup\ \ldots && \ldots\\
& \cup\ \{(p, r) \mid (\exists\, q)\ (p, q) \in R_i \wedge (q, r) \in R_i\}. && \text{(el'ts arising by (3.2.8))}
\end{aligned}
$$

We now define $R = \bigcup_i R_i$. It is straightforward to show that $R$ satisfies (3.2.1)–(3.2.8), and that any subset of $T \times T$ satisfying (3.2.1)–(3.2.8) must contain $R$; so $R$, looked at as a binary relation $\sim$ on $T$, is the desired least relation.

By (3.2.6)–(3.2.8), $\sim$ is an equivalence relation; so let $|F| = T/\sim$, the set of equivalence classes of this relation; i.e., writing $[p]$ for the equivalence class of $p \in T$, $|F| = \{[p] \mid p \in T\}$. We map $X$ into $|F|$ by the function

$$u(x) \;=\; [x]$$

(or, if we do not identify $\mathrm{symb}_T(x)$ with $x$ in our construction of $T$, by $u(x) = [\mathrm{symb}_T(x)]$). We now define operations $\cdot$, $^{-1}$ and $e$ on $|F|$ by

(3.2.9)                                    $[p] \cdot [q] \;=\; [p \cdot q],$

(3.2.10)                                   $[p]^{-1} \;=\; [p^{-1}],$

(3.2.11)                                        $e \;=\; [e].$

That the first two of these are *well-defined* follows respectively from properties (3.2.4) and (3.2.5) of $\sim$! (With the third there is no problem.) From properties (3.2.1)–(3.2.3) of $\sim$, it follows that $(|F|, \cdot, {}^{-1}, e)$ satisfies the group axioms. E.g., given $[p], [q], [r] \in |F|$, if we evaluate $([p] \cdot [q]) \cdot [r]$ and $[p] \cdot ([q] \cdot [r])$ in $|F|$, we get $[(p \cdot q) \cdot r]$ and $[p \cdot (q \cdot r)]$ respectively, which are equal by (3.2.1). Writing $F$ for the group $(|F|, \cdot, {}^{-1}, e)$, it is clear from our construction of $\sim$ that every relation satisfied by the images in $F$ of the elements of $X$ is a consequence of the group axioms; so by Corollary 3.1.2 (or rather, the generalization of that corollary with $X$-tuples in place of 3-tuples), $F$ has the desired universal property.

To see this universal property more directly, suppose $v$ is any map $X \to |G|$, where $G$ is a group. Write $p \sim_v q$ to mean $p_v = q_v$ in $G$. Clearly the relation $\sim_v$ satisfies conditions (3.2.1)–(3.2.8), hence it contains the *least* such relation, our $\sim$ . So a well-defined map $h\colon |F| \to |G|$ is given by $h([p]) = p_v \in |G|$, and it follows from the way the operations of $F$, and the evaluation of terms in $G$ at the $X$-tuple $v$, are defined, that $h$ is a homomorphism, and is the unique homomorphism such that $hu = v$. Thus we have

**Proposition 3.2.12.** *$(F, u)$, constructed as above, is a free group on the given set $X$.*                                                                                          □

So a free group on every set $X$ does indeed exist!

*Remark 3.2.13.* There is a viewpoint that goes along with this construction, which will be helpful in thinking about universal constructions in general. Suppose that we are given a set $X$, and that we know that $G$ is a group, with a map $v\colon X \to |G|$. How much can we "say about" $G$ from this fact alone? We can *name* certain elements of $G$, namely the $v(x)$ $(x \in X)$, and all the elements that can be obtained from these by the group operations of $G$ (e.g., $(v(x) \cdot v(y))^{-1} \cdot ((v(y)^{-1} \cdot e)^{-1} \cdot v(z))$ if $x$, $y$, $z \in X$). A particular $G$ may contain more elements than those obtained in such ways, but we have no way of getting our hands on them from the given information. We can also derive from the identities for groups certain relations that these elements satisfy, (e.g., $(v(x) \cdot v(y))^{-1} = v(y)^{-1} \cdot v(x)^{-1}$). The elements $v(x)$ may, in particular cases, satisfy more relations than these, but again we have no way of deducing these additional relations. If we now gather together this limited "data" that we have about such a group $G$—the quotient of a set of labels for certain elements by a set of identifications among these—we find that this collection of "data" itself forms a group with a map of $X$ into it; and is, in fact, a universal such group!

*Remark 3.2.14.* At the beginning of this section, I motivated our construction by saying that "$\sim$" should mean "equality that follows from the group axioms". I then wrote down a series of eight rules, (3.2.1)–(3.2.8), all of which are clearly valid procedures for deducing equations which hold in all groups. What was not obvious was whether they would be sufficient to yield *all* such equations. But they were—the proof of the pudding being that $(T/\sim, \cdot, ^{-1}, e)$ was shown to *be* a group.

This is an example of a very general type of situation in mathematics: Some class, in this case, a class of pairs of group-theoretic terms, is described "from above", i.e., is defined as the class of all elements satisfying certain restrictions (in this case, those pairs $(p, q) \in T \times T$ such that the relation $p = q$ holds on all $X$-tuples of elements of all groups). We seek a way of describing it "from below", i.e., of *constructing* or *generating* all members of the class. Some procedure which produces members of the set is found, and one seeks to show that this procedure yields the whole set—or, if it does not, one seeks to extend it to a procedure that does.

The inverse situation is equally important, where we are given a construction which "builds up" a set, and we seek a convenient way of characterizing the elements that result. Exercise 2.7:1 was of that form. You will see more examples of both situations throughout this course, and, in fact, in most every mathematics course you take.

**Exercise 3.2:1.** Prove directly from (3.2.1)–(3.2.8) that for any $x, y \in X$, $(x \cdot y)^{-1} \sim y^{-1} \cdot x^{-1}$. (Your solution should show explicitly each application you make of each of those conditions.)

**Exercise 3.2:2.** Does the relation of the preceding exercise follow from (3.2.1)–(3.2.3) and (3.2.6)–(3.2.8) alone?

Note that in our recursive construction of the set $R$ (that is, the relation $\sim$), repeated application of (3.2.1)–(3.2.3) was really unnecessary; these conditions give the same elements of $R$ each time they are applied, so we might as well just have applied them the first time, and only applied (3.2.4)–(3.2.8) after that. Less obvious is the answer to:

**Exercise 3.2:3** (A. Tourubaroff). Can the construction of $R$ be done in three stages: First take the set $P$ of elements given by (3.2.1)–(3.2.3), then form the closure $Q$ of this set under applications of (3.2.4)–(3.2.5) (as before, by recursion or as an intersection), and finally, obtain $R$ as the closure of $Q$ under applications of (3.2.6)–(3.2.8) (another recursion or intersection)? This procedure will yield some subset of $T \times T$; the question is whether it is the $R$ we want.

   What if we do things in a different order—first (3.2.1)–(3.2.3), then (3.2.6)–(3.2.8), then (3.2.4)–(3.2.5)?

## 3.3. Free groups as subgroups of big enough direct products

Another way of getting a group in which some $X$-tuple of elements satisfies the smallest possible set of relations is suggested by the following observation. Let $G_1$ and $G_2$ be two groups, and suppose we are given elements

$$\alpha_1, \beta_1, \gamma_1 \in |G_1|, \quad \alpha_2, \beta_2, \gamma_2 \in |G_2|.$$

Then in the direct product group $G = G_1 \times G_2$ we have the elements

$$a = (\alpha_1, \alpha_2), \quad b = (\beta_1, \beta_2), \quad c = (\gamma_1, \gamma_2),$$

and we find that the set of relations satisfied by $a$, $b$, $c$ in $G$ is precisely the *intersection* of the set of relations satisfied by $\alpha_1, \beta_1, \gamma_1$ in $G_1$ and the set of relations satisfied by $\alpha_2, \beta_2, \gamma_2$ in $G_2$. This may be seen from the fact that for any $s \in T$,

$$s_G(a, b, c) = (s_{G_1}(\alpha_1, \beta_1, \gamma_1), \, s_{G_2}(\alpha_2, \beta_2, \gamma_2)),$$

as is easily verified by induction.

   More generally, if we take an arbitrary family of groups $(G_i)_{i \in I}$, and in each $G_i$ three elements $\alpha_i$, $\beta_i$, $\gamma_i$, then in the product group $G = \prod G_i$, we can define the elements

$$a = (\alpha_i)_{i \in I}, \quad b = (\beta_i)_{i \in I}, \quad c = (\gamma_i)_{i \in I},$$

and the relations that these satisfy will be just those relations satisfied simultaneously by our 3-tuples in all of these groups.

   This suggests that by using a *large enough* such family, we could arrive at a group with three elements $a$, $b$, $c$ which satisfy a *smallest possible* set of relations.

How large a family $(G_i, \alpha_i, \beta_i, \gamma_i)$ should we use?

Well, we could be sure of getting the least set of relations if we could use the class of *all* groups and *all* 3-tuples of elements of these. But taking the direct product of such a family would give us set-theoretic indigestion.

We can cut down this surfeit of groups a bit by noting that for any group $G_i$ and three elements $\alpha_i, \beta_i, \gamma_i$, if we let $H_i$ denote the subgroup of $G_i$ generated by these three elements, it will suffice for our product to involve the group $H_i$, rather than the whole group $G_i$, since the relations satisfied by $\alpha_i, \beta_i$ and $\gamma_i$ in the whole group $G_i$ and in the subgroup $H_i$ are the same. Now a finitely generated group is countable (meaning finite *or* countably infinite), so we see that it would be enough to let $(G_i, \alpha_i, \beta_i, \gamma_i)$ range over all *countable* groups, and all 3-tuples of elements thereof.

However, the class of all countable groups is still not a set. Indeed, even the class of one-element groups is not a set, because we get a different (in the strict set-theoretic sense) group for each choice of that one element. (For those not familiar with such considerations: In set theory, every element of a set is a set. If we had a set of *all* one-element groups, then we could form from this the set of all *members* of their underlying sets, which would be the set of *all* sets; and one knows that this does not exist.) But this is clearly just a quibble—obviously, if we choose any one-element set $\{x\}$, and take the unique group with this underlying set, it will serve as well as any other one-element group so far as honest group-theoretic purposes are concerned. In the same way, I claim we can find a genuine *set* of countable groups that *up to isomorphism* contains *all* the countable groups. Namely, let $S$ be a fixed countably infinite set. Then we can form the set of all groups $G$ whose underlying sets $|G|$ are subsets of $S$. Or, to hit more precisely what we want, let

(3.3.1) $$\{(G_i, \alpha_i, \beta_i, \gamma_i) \mid i \in I\}$$

be the set of all 4-tuples such that $G_i$ is a group with $|G_i| \subseteq S$, and $\alpha_i$, $\beta_i$ and $\gamma_i$ are members of $|G_i|$. Now for any countable group $H$ and three elements $\alpha, \beta, \gamma \in |H|$, we can clearly find an isomorphism $\theta$ from one of these groups, say $G_j$ $(j \in I)$, to $H$, such that $\theta(\alpha_j) = \alpha$, $\theta(\beta_j) = \beta$, $\theta(\gamma_j) = \gamma$; so (3.3.1) is "big enough" for our purpose.

So taking (3.3.1) as above, let $P$ be the direct product group $\prod_I G_i$, let $a, b, c$ be the $I$-tuples $(\alpha_i), (\beta_i), (\gamma_i) \in |P|$, and let $F$ be the subgroup of $P$ generated by $a, b$ and $c$. I claim that $F$ is a free group on $a, b$ and $c$.

We could prove this by considering the set of relations satisfied by $a, b,$ $c$ in $F$ as suggested above, but let us instead verify directly that $F$ satisfies the universal property characterizing free groups (Definition 3.1.3). Let $G$ be any group, and $\alpha, \beta, \gamma$ three elements of $G$. We want to prove that there exists a unique homomorphism $h \colon F \to G$ carrying $a, b, c \in |F|$ to $\alpha, \beta, \gamma \in |G|$ respectively. Uniqueness will be no problem—by construction $F$ is generated by $a, b$ and $c$, so if such a homomorphism exists it is unique.

To show the existence of $h$, note that the subgroup $H$ of $G$ generated by $\alpha$, $\beta$, $\gamma$ is countable, hence as we have noted, there exists for some $j \in I$ an isomorphism $\theta : G_j \cong H$ carrying $\alpha_j$, $\beta_j$, $\gamma_j \in |G_j|$ to $\alpha$, $\beta$, $\gamma \in |H|$. Now the projection map $p_j$ of the product group $P = \prod G_i$ onto its $j$-th coordinate takes $a$, $b$ and $c$ to $\alpha_j$, $\beta_j$, $\gamma_j$, hence composing this projection with $\theta$, we get a homomorphism $h : F \to G$ having the desired effect on $a$, $b$, $c$, as shown in the diagram below.

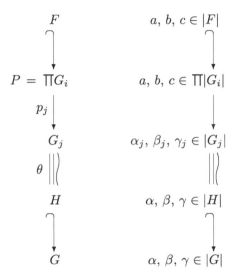

For a useful way to picture this construction, think of $P$ as the group of all functions on the base-space $I$, taking at each point $i \in I$ a value in $G_i$ :

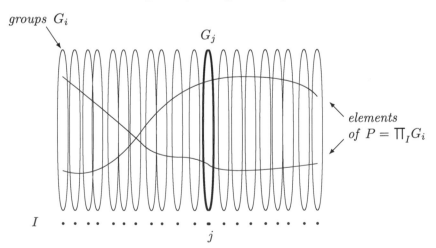

Then $F$ is the subgroup of functions generated by $a$, $b$ and $c$. Now given $\alpha$, $\beta$, $\gamma$ in any group $G$, identify the subgroup of $G$ that they generate with an appropriate $G_j$ $(j \in I)$. Then the homomorphism $h$ that we constructed above may be thought of as taking each element of $F$ to its value at the point $j$. We have chosen our space $I$ and values for $a$, $b$ and $c$ sufficiently eclectically so that it is possible to choose points at which $a$, $b$ and $c$ take on (up to isomorphism) *any* 3-tuple of values in *any* group. Thus, the functions $a$, $b$ and $c$ are a "universal" 3-tuple of group-elements.

The same argument works if we replace "3-tuple" by "$X$-tuple", where $X$ is any *countable* set. Here we use the observation that a group generated by a countable family of elements is countable. For $X$ of arbitrary cardinality, one can easily show that a group $H$ generated by an $X$-tuple of elements has cardinality $\leq \max(\mathrm{card}(X), \aleph_0)$. Hence we get:

**Proposition 3.3.2.** *Let $X$ be any set. Take a set $S$ of cardinality $\max(\mathrm{card}(X), \aleph_0)$, and let $\{(G_i, u_i) \mid i \in I\}$ be the set of all pairs such that $G_i$ is a group with $|G_i| \subseteq S$, and $u_i$ is a map $X \to |G_i|$ (i.e., an $X$-tuple of elements of $G_i$). Let $P = \prod_I G_i$, and map $X$ into $P$ by defining $u(x)$ $(x \in X)$ to be the element with component $u_i(x)$ at each $i$. Let $F$ be the subgroup of $P$ generated by $\{u(x) \mid x \in X\}$.*
*Then the pair $(F, u)$ is a free group on the set $X$.* ☐

Digression: Let $S_3$ be the symmetric group on three letters. Suppose we had begun the above investigation with a less ambitious goal: merely to find a group $J$ with three elements $a, b, c$ such that

(3.3.3) For every choice of three elements $\alpha, \beta$, $\gamma \in |S_3|$, there exists a unique homomorphism $h: J \to S_3$ taking $a, b, c$ to $\alpha$, $\beta$, $\gamma$ respectively.

$$
\begin{array}{cc}
J & a, b, c \in |J| \\
\downarrow{\scriptstyle h} & \downarrow \\
S_3 & \alpha, \beta, \gamma \in |S_3|
\end{array}
$$

Then we could have performed the above construction just using 4-tuples $(S_3, \alpha, \beta, \gamma)$ $(\alpha, \beta, \gamma \in |S_3|)$ as our $(G_i, \alpha_i, \beta_i, \gamma_i)$. There are $6^3 = 216$ such 4-tuples, so $P$ would be the direct product of 216 copies of $S_3$, and $a, b, c$ would be elements of this product which, as one runs over the 216 coordinates, take on all possible combinations of values in $S_3$. The subgroup $J$ they generate would indeed satisfy (3.3.3). This leads to:

**Exercise 3.3:1.** Does condition (3.3.3) characterize $(J, a, b, c)$ up to isomorphism? If not, is there some additional condition that $(J, a, b, c)$ satisfies which together with (3.3.3) determines it up to isomorphism?

**Exercise 3.3:2.** Investigate the structure of the group $J$, and more generally, of the analogous groups constructed from $S_3$ using different numbers of generators. To make the problem concrete, try to determine, or estimate as well as possible, the *orders* of these groups, for 1, 2, 3 and generally, for $n$ generators.

The methods by which we have constructed free groups in this and the preceding section go over essentially word-for-word with "group" replaced by "ring", "lattice", or a great many other types of mathematical objects. The determination of just what classes of algebraic structures admit this and related sorts of universal constructions is one of the themes of this course. The next exercise concerns a negative example.

**Exercise 3.3:3.** State what would be meant by a "free field on a set $X$", and show that no such object exists for any set $X$. If one attempts to apply the two methods of this and the preceding section to prove the existence of free fields, where does each of them fail?

**Exercise 3.3:4.** Let $\mathbb{Z}[x_1, \ldots, x_n]$ be the polynomial ring in $n$ indeterminates over the integers ( = the free commutative ring on $n$ generators— cf. § 4.12 below). Its field of fractions $\mathbb{Q}(x_1, \ldots, x_n)$, the field of "rational functions in $n$ indeterminates over the rationals", looks in some ways like a "free field on $n$ generators". E.g., one often speaks of evaluating a rational function at some set of values of the variables. Can some concept of "free field" be set up, perhaps based on a *modified* universal property, or on some concept of comparing relations in the *field* operations satisfied by $n$-tuples of elements in two fields, in terms of which $\mathbb{Q}(x_1, \ldots, x_n)$ is indeed the free field on $n$ generators?

**Exercise 3.3:5.** A *division ring* (or *skew field* or *sfield*) is a ring (associative but not necessarily commutative) in which every nonzero element is invertible. If you find a satisfactory answer to the preceding exercise, you might consider the question of whether there exists in the same sense a *free division ring* on $n$ generators. (This was a longstanding open question, which was finally answered in 1966, and then again, by a very different approach, about five years later.)

There are many hybrids and variants of the two constructions we have given for free groups. For instance, we might start with the set $T$ of terms in $X$, and define $p \sim q$ (for $p, q \in T$) to mean that for every map $v$ of $X$ into a group $G$, one has $p_v = q_v$ in $G$. Now for each pair $(p, q) \in T \times T$ such that $p \sim q$ *fails* to hold, we can choose a map $u_{p,q}$ of $X$ into a group $G_{p,q}$ such that $p_{u_{p,q}} \neq q_{u_{p,q}}$. We can then form the direct product group $P = \prod G_{p,q}$, take the induced map $u \colon X \to |P|$, and check that the subgroup $F$ generated by the image of this map will satisfy condition (a) of Corollary 3.1.2. Interestingly, for $X$ countable, this construction uses a product of fewer groups $G_{p,q}$ than we used in the version given above.

Finally, consider the following construction, which suffers from severe set-theoretic difficulties, but is still interesting. (I won't try to resolve these difficulties here, but will talk sloppily, as though they did not occur.)

Define a "generalized group-theoretic operation in three variables" as any function $p$ which associates to every group $G$ and three elements $\alpha, \beta, \gamma \in$

$|G|$ an element $p(G, \alpha, \beta, \gamma) \in |G|$. We can "multiply" two such operations $p$ and $q$ by defining

$$(p \cdot q)(G, \alpha, \beta, \gamma) \ = \ p(G, \alpha, \beta, \gamma) \cdot q(G, \alpha, \beta, \gamma) \in |G|.$$

for all groups $G$ and elements $\alpha, \beta, \gamma \in |G|$. We can similarly define the multiplicative inverse of such an operation $p$, and the constant operation $e$. We see that the class of generalized group-theoretic operations will satisfy the group axioms under the above three operations. Now consider the three generalized group-theoretic operations $a$, $b$ and $c$ defined by

$$a(G, \alpha, \beta, \gamma) = \alpha, \quad b(G, \alpha, \beta, \gamma) = \beta, \quad c(G, \alpha, \beta, \gamma) = \gamma.$$

Let us define a "derived generalized group-theoretic operation" as one obtainable from $a$, $b$ and $c$ by the operations of product, inverse, and neutral element defined above. Then the set of derived generalized group-theoretic operations will form a free group on the generators $a$, $b$ and $c$. (This is really just a disguised form of our naive "direct product of *all* groups" idea.)

**Exercise 3.3:6.** Call a generalized group-theoretic operation $p$ *functorial* if for every homomorphism of groups $f : G \to H$, one has $f(p(G, \alpha, \beta, \gamma)) = p(H, f(\alpha), f(\beta), f(\gamma))$ $(\alpha, \beta, \gamma \in |G|)$. (We will see the reason for this name in Chapter 7.) Show that all derived group-theoretic operations are functorial. Is the converse true?

**Exercise 3.3:7.** Same question for functorial generalized operations on the class of all *finite groups*.

# 3.4. The classical construction: free groups as groups of words

The constructions discussed above have the disadvantage of not giving very explicit descriptions of free groups. We know that every element of a free group $F$ on the set $X$ arises from a term in the elements of $X$ and the group operations, but we don't know how to tell whether two such terms—say $(b(a^{-1}b)^{-1})(a^{-1}b)$ and $e$—yield the same element; in other words, whether $(\beta(\alpha^{-1}\beta)^{-1})(\alpha^{-1}\beta) = e$ is true for all elements $\alpha, \beta$ of all groups. If it is, then by the results of § 3.2 one can obtain this fact *somehow* by the procedures corresponding to conditions (3.2.1)–(3.2.8); if it is not, then the ideas of § 3.3 suggest that we should try to prove this by looking for some particular elements for which it fails, in some particular group in which we know how to calculate. But these approaches are hit-and-miss.

In this section, we shall construct the free group on $X$ in a much more explicit way. We will then be able to answer such questions by calculating in the free group.

We first recall an important consequence of the associative identity: that "products can be written without parentheses". For example, given elements $a$, $b$, $c$ of a group, the elements $a(c(ab))$, $a((ca)b)$, $(ac)(ab)$, $(a(ca))b$ and $((ac)a)b$ are all equal. It is conventional, and usually convenient, to say, "Let us therefore write their common value as $a\,c\,a\,b$." However, we will soon want to relate these expressions to group-theoretic *terms*; so instead of dropping parentheses, let us agree to take $a(c(ab))$ as the common form to which we shall reduce the above five expressions, and generally, let us note that any product of elements can be reduced by the associative law to one with parentheses clustered to the right: $x_n\,(x_{n-1}(\ldots(x_2\,x_1)\ldots))$.

In particular, given two elements written in this form, we can write down their product and reduce it to this form by repeatedly applying the associative law:

$$(3.4.1) \qquad \begin{aligned} &(x_n\,(\ldots(x_2\,x_1)\ldots)) \cdot (y_m\,(\ldots(y_2\,y_1)\ldots)) \\ &= x_n\,(\ldots(x_2\,(x_1\,(y_m\,(\ldots(y_2\,y_1)\ldots))))\ldots). \end{aligned}$$

If we want to find the inverse of an element written in this form, we may use the formula $(x\,y)^{-1} = y^{-1}\,x^{-1}$, another consequence of the group laws. By induction this gives $(x_n(\ldots(x_2 x_1)\ldots))^{-1} = (\ldots(x_1^{-1} x_2^{-1})\ldots)x_n^{-1}$, which we may reduce, again by associativity, to $x_1^{-1}(\ldots(x_{n-1}^{-1} x_n^{-1})\ldots)$.

More generally, if we started with an expression of the form

$$x_n^{\pm 1}(\ldots(x_2^{\pm 1} x_1^{\pm 1})\ldots),$$

where each factor is either $x_i$ or $x_i^{-1}$, and the exponents are independent, then the above method together with the fact $(x^{-1})^{-1} = x$ (another consequence of the group axioms) allows us to write its inverse as $x_1^{\mp 1}(\ldots(x_{n-1}^{\mp 1} x_n^{\mp 1})\ldots)$, which is of the same form as the expression we started with; just as (3.4.1) shows that the *product* of two expressions of the above form reduces to an expression of the same form.

Note further that if two successive factors $x_i^{\pm 1}$ and $x_{i+1}^{\pm 1}$ are respectively $x$ and $x^{-1}$ for some element $x$, or are respectively $x^{-1}$ and $x$ for some $x$, then by the group axioms on inverses and the neutral element (and again, associativity), we can drop this pair of factors—unless they are the only factors in the product, in which case we can rewrite the product as $e$.

Finally, easy consequences of the group axioms tell us what the inverse of $e$ is (namely $e$), and how to multiply anything by $e$. Putting these observations together, we conclude that *given any set $X$ of elements of a group $G$, the set of elements of $G$ that can be written in one of the forms*

$$e \quad \text{or} \quad x_n^{\pm 1}(\ldots(x_2^{\pm 1}x_1^{\pm 1})\ldots),$$

(3.4.2)      where $n \geq 1$, each $x_i \in X$, and no two successive factors are an element of $X$ and the inverse of the same element, in either order,

is closed under products and inverses. So this set must be the whole subgroup of $G$ generated by $X$. In other words, any member of the subgroup generated by $X$ can be *reduced* by the group operations to an expression (3.4.2).

In the preceding paragraph, $X$ was a subset of a group. Now let $X$ be an arbitrary set, and as in §3.2, let $T$ be the set of all group-theoretic terms in elements of $X$ (Definition 2.5.1). For convenience, let us assume $T$ chosen so as to contain $X$, with $\text{symb}_T$ being the inclusion map. (If you prefer not to make this assumption, then in the argument to follow, you should insert "$\text{symb}_T$" at appropriate points.) Let $T_{\text{red}} \subseteq T$ ("red" standing for "reduced") denote the set of terms of the form (3.4.2). If $s, t \in T_{\text{red}}$, we can form their product $s \cdot t$ in $T$, and then, as we have just seen, rearrange parentheses to get an element of $T_{\text{red}}$ which is equivalent to $s \cdot t$ so far as evaluation at $X$-tuples of elements of groups is concerned. Let us call this element $s \odot t$. Thus, $s \odot t$ has the properties that it belongs to $T_{\text{red}}$, and that for any map $v \colon X \to |G|$ ($G$ a group) one has $(s \cdot t)_v = (s \odot t)_v$. In the same way, given $s \in T_{\text{red}}$, we can obtain from $s^{-1} \in T$ an element we shall call $s^{(-)} \in T_{\text{red}}$, such that for any map $v \colon X \to |G|$, one has $(s^{-1})_v = (s^{(-)})_v$.

Are any further reductions possible? For a particular $X$-tuple of elements of a particular group there may be equalities among the values of different expressions of the form (3.4.2); but we are only interested in reductions that can be done in all groups. No more are obvious; but can we be sure that some sneaky application of the group axioms wouldn't allow us to prove some two distinct terms (3.4.2) to have the same evaluations at all $X$-tuples of elements of all groups? (In such a case, we should not lose hope, but should introduce further reductions that would always replace one of these expressions by the other.)

Let us formalize the consequences of the preceding observations, and indicate the significance of the question we have asked.

**Lemma 3.4.3.** *For each $s \in T$, there exists an $s' \in T_{\text{red}}$ (i.e., an element of $T$ of one of the forms shown in (3.4.2)) such that*

(3.4.4)      *for every map $v$ of $X$ into any group $G$, $s_v = s'_v$ in $|G|$.*

*Moreover, if one of the following statements is true, all are:*

(a) *For each $s \in T$, there exists a* unique *$s' \in T_{\text{red}}$ satisfying (3.4.4).*

(b) *If $s$, $t$ are distinct elements of $T_{\text{red}}$, then "$s = t$" is not an identity for groups; that is, for some $G$ and some $v \colon X \to |G|$, $s_v \neq t_v$.*

(c) *The 4-tuple $F = (T_{\text{red}}, \odot, ^{(-)}, e_T)$ is a group.*

(d) *The 4-tuple $F = (T_{\text{red}}, \odot, ^{(-)}, e_T)$ is a free group on $X$.*

*Proof.* We get the first sentence of the lemma by an induction, which I will sketch briefly. The assertion holds for elements $x \in X$: we simply take $x' = x$. Now suppose it true for two terms $s, t \in T$. To establish it for $s \cdot t \in T$, define $(s \cdot t)' = s' \odot t'$. One likewise gets it for $s^{-1}$ using $s'^{(-)}$, and it is clear for $e$. It follows from condition (c) of the definition of "group-theoretic term" (Definition 2.5.1) that it is true for all elements of $T$.

The equivalence of (a) and (b) is straightforward. Assuming these conditions, let us verify that the 4-tuple $F$ defined in (c) is a group. Take $p, q, r \in T_{\text{red}}$. Then $p \odot (q \odot r)$ and $(p \odot q) \odot r$ are two elements of $T_{\text{red}}$, call them $s$ and $t$. For any $v \colon X \to |G|$, $s_v = t_v$ by the associative law for $G$. Hence by (b), $s = t$, proving that $\odot$ is associative. The other group laws for $F$ are deduced in the same way.

Conversely, assuming (c), we claim that for distinct elements $s, t \in T_{\text{red}}$, we can prove, as required for (b), that the equation "$s = t$" is not an identity by getting a counterexample to that equation in this very group $F$. Indeed, if we let $v$ be the inclusion $X \to T_{\text{red}} = |F|$, we can check by induction on $n$ in (3.4.2) that for all $s \in T_{\text{red}}$, $s_v = s$. Hence $s \neq t$ implies $s_v \neq t_v$, as desired.

Since (d) certainly entails (c), our proof will be complete if we can show, assuming (c), that $F$ has the universal property of a free group. Given any group $G$ and map $v \colon X \to |G|$, we map $|F| = T_{\text{red}}$ to $|G|$ by $s \mapsto s_v$. From the properties of $\odot$ and $^{(-)}$, we know that this is a homomorphism $h$ such that $h|X$ (the restriction of $h$ to $X$) is $v$; and since $X$ generates $F$, $h$ is the unique homomorphism with this property, as desired.                    □

Well—are statements (a)–(d) true, or not??

The usual way to answer this question is to test condition (c) by writing down precisely how the operations $\odot$ and $^{(-)}$ are performed, and checking the group axioms on them. Since a term of the form (3.4.2) is uniquely determined by the integer $n$ (which we take to be 0 for the term $e$) and the $n$-tuple of elements of $X$ and their inverses, $(x_n^{\pm 1}, \ldots, x_1^{\pm 1})$, one describes $\odot$ and $^{(-)}$ as operations on such $n$-tuples. E.g., one multiplies two tuples $(w, \ldots, x)$ and $(y, \ldots, z)$ (where each of $w, \ldots, z$ is an element of $X$ or a symbolic inverse of such an element) by uniting them as $(w, \ldots, x, y, \ldots, z)$, then dropping pairs of factors that may now cancel (e.g., $x$ and $y$ above if $y$ is $x^{-1}$); and repeating this last step until no such cancelling pairs remain.

But checking the associative law for this recursively defined operation turns out to be very tedious, involving a number of different cases. (E.g., you might try checking associativity for $(v, w, x) \cdot (x^{-1}, w^{-1}, y^{-1}) \cdot (y, w, z)$, and for $(v, w, x) \cdot (x^{-1}, z^{-1}, y^{-1}) \cdot (y, w, z)$, where $w, x, y$ and $z$ are four distinct elements of $X$. Both cases work, but they are different computations.)

But there is an elegant trick, not as well known as it ought to be, which rescues us from the toils of this calculation. We construct a certain $G$ which we *know* to be a group, using which we can verify condition (b)—rather than condition (c)—of the above lemma.

To see how to construct this $G$, let us go back to basics and recall where the group identities, which we need to verify, "come from". They are identities which are satisfied by *permutations* of any set $A$, under the operations of composing permutations, inverting permutations, and taking the identity permutation. So let us try to describe a set $A$ on which the group we want to construct should act by permutations in as "free" a way as possible, specifying the permutation of $A$ that should represent the image of each $x \in X$.

To start our construction, let $a$ be any symbol not in $\{x^{\pm 1} \mid x \in X\}$. Now define $A$ to be the set of all strings of symbols of the form:

$$x_n^{\pm 1} x_{n-1}^{\pm 1} \ldots x_1^{\pm 1} a$$

(3.4.5)      where $n \geq 0$, each $x_i \in X$, and no two successive factors $x_i^{\pm 1}$ and $x_{i+1}^{\pm 1}$ are an element of $X$ and the inverse of that same element, in either order.

In particular, taking $n = 0$, we see that $a \in A$.

Let $G$ be the group of *all* permutations of $A$. Define for each $x \in X$ an element $v(x) \in |G|$ as follows. Given $b \in A$,

> if $b$ does *not* begin with the symbol $x^{-1}$, let $v(x)$ take $b$ to the symbol $x\, b$, formed by placing an $x$ at the beginning of the symbol $b$;

> if $b$ does begin with $x^{-1}$, say $b = x^{-1}c$, let $v(x)$ take $b$ to the symbol $c$, formed by removing $x^{-1}$ from the beginning of $b$.

It is immediate from the definition of $A$ that $v(x)(b)$ belongs to $A$ in each case. To check that $v(x)$ is invertible, consider the map which sends a symbol $x\, b$ to $b$, and a symbol $c$ not beginning with $x$ to the symbol $x^{-1}c$; we find that this is a two-sided inverse to $v(x)$.

So we now have a map $v\colon X \to |G|$. As usual, this induces an evaluation map $s \mapsto s_v$ taking the set $T$ of terms in $X$ into $|G|$. Now consider any $s = x_n^{\pm 1}(\ldots (x_2^{\pm 1}x_1^{\pm 1})\ldots) \in T_{\mathrm{red}}$. It is easy to verify by induction on $n$ that the permutation $s_v \in |G|$ takes our "base" symbol $a \in A$ to the symbol $x_n^{\pm 1} \ldots x_1^{\pm 1} a$ (or if $s = e$, to $a$ itself). It follows that if $s$ and $t$ are distinct elements of $T_{\mathrm{red}}$, then $s_v(a)$ and $t_v(a)$ are distinct elements of $A$, so $s_v \neq t_v$ in $G$, establishing (b) of Lemma 3.4.3. By that lemma we now have

**Proposition 3.4.6.** $F = (T_{\mathrm{red}}, \odot, {}^{(-)}, e)$ *is a group; in fact, letting $u$ denote the inclusion $X \to T_{\mathrm{red}}$, the pair $(F, u)$ is a free group on $X$.*

*Using parenthesis-free notation for products, and identifying each element of $X$ with its image in $F$, this says that every element of the free group on $X$ can be written uniquely as*

$$e, \quad or \quad x_n^{\pm 1} \ldots x_2^{\pm 1} x_1^{\pm 1},$$

*where in the latter case, $n \geq 1$, each $x_i \in X$, and no two successive factors $x_i^{\pm 1}$ and $x_{i+1}^{\pm 1}$ are an element of $X$ and the inverse of that same element, in either order.*                                                                          □

What we have just obtained is called a *normal form* for elements of a free group on $X$ —a set of expressions which contains a unique expression for each member of the group, such that we can algorithmically reduce any expression to one in this set. This indeed allows us to calculate explicitly in the free group. For example, you should find it straightforward to do

**Exercise 3.4:1.** Determine whether each of the following equations holds for all elements $x$, $y$, $z$ of all groups:

(i)    $(x^{-1}y\,x)^{-1}(x^{-1}z\,x)(x^{-1}y\,x) = (yx)^{-1}z\,(yx)$.

(ii)   $(x^{-1}y^{-1}xy)^2 = x^{-2}y^{-1}x^2y$.

In the next exercise, we will use the group theorists' abbreviations

$$x^y = y^{-1}x\,y \quad \text{and} \quad [x,\,y] = x^{-1}y^{-1}x\,y$$

for the *conjugate* of an element $x$ by an element $y$ in a group $G$, respectively the *commutator* of two elements $x$ and $y$. If $H_1$, $H_2$ are subgroups of $G$, then $[H_1, H_2]$ denotes the *subgroup* of $G$ *generated* by all commutators $[h_1,\,h_2]$ $(h_1 \in H_1, h_2 \in H_2)$.

**Exercise 3.4:2.**

(i)    Show that the commutator operation is not associative; i.e., that it is not true that for all elements $a$, $b$, $c$ of every group $G$ one has $[a, [b, c]] = [[a, b], c]$.

(ii)   Prove a group identity of the form

$$[[x^{\pm 1},\, y^{\pm 1}],\, z^{\pm 1}]^{y^{\pm 1}} \,[[y^{\pm 1},\, z^{\pm 1}],\, x^{\pm 1}]^{z^{\pm 1}} \,[[z^{\pm 1},\, x^{\pm 1}],\, y^{\pm 1}]^{x^{\pm 1}} = e,$$

for some choice of the exponents $\pm 1$. (There is a certain amount of leeway in these exponents; you might try to adjust your choices so as to get maximum symmetry. The result is known as the *Hall-Witt identity*; however its form may vary with the text; in particular, we get different identities depending on whether the above definition of $[x, y]$, preferred by most contemporary group theorists, is used, or the less common definition $x\,y\,x^{-1}\,y^{-1}$.)

(iii) Deduce that if $A$, $B$ and $C$ are subgroups of a group $G$ such that two of $[[A, B], C]$, $[[B, C], A]$, $[[C, A], B]$ are trivial, then so is the third. (The "three subgroups theorem".)

(As noted above, $[A, B]$ means the subgroup of $G$ *generated* by elements $[a, b]$ with $a \in A$, $b \in B$. Thus, $[[A, B], C]$ means the subgroup generated by elements $[g, c]$ with $c \in C$, and $g$ in the subgroup generated as above. You will need to think about the relation between the condition $[[A, B], C] = \{e\}$ and the condition that $[[a, b], c] = e$ for all $a \in A$, $b \in B$, $c \in C$.)

(iv) Deduce that if $A$ and $B$ are two subgroups of $G$, and $[A, [A, B]]$ is trivial, then so is $[[A, A], B]$. Is the converse true?

Incidentally, group theorists often abbreviate $[[x, y], z]$ to $[x, y, z]$. If I worked with commutators every day, I might do the same, but as an occasional visitor to the subject, I prefer to stick with more transparent notation.

The idea of finding normal forms, or other explicit descriptions, of objects defined by universal properties is a recurring one in algebra. The form we have found is specific to free groups. It might appear at first glance that corresponding forms could be obtained mechanically from any finite system of operations and identities; e.g., those defining rings, lattices, etc.; and thus that the results of this section should generalize painlessly (as those of the two preceding sections indeed do!) to very general classes of structures. But this is not so. An example we shall soon see (§ 4.5) is that of the Burnside problem, where a sweet and reasonable set of axioms obstinately refuses to yield a normal form. Other nontrivial cases are free Lie algebras [85] (cf. § 9.7 below) and free lattices [4, § VI.8], for which normal forms are known, but complicated; free modular lattices, for which it has been proved that the word problem is undecidable (no recursive normal form can exist); and groups defined by particular families of generators and relations (§ 4.3 below), for which the word problem has been proved undecidable in general, though nice normal forms exist in many cases. In general, normal form questions must be tackled case by case, but for certain large families of cases there *are* interesting general methods [46].

The trick that we used to show that the set of terms $T_{\text{red}}$ constitutes a normal form for the elements of the free group is due to van der Waerden, who introduced it in [142] to handle the more difficult case of coproducts of groups (§ 4.6 below). Though the result we proved is, as we have said, specific to groups, the idea behind the proof is a versatile one: If you can reduce all expressions for elements of some universal structure $F$ to members of a set $T_{\text{red}}$, and wish to show that this gives a normal form, then look for a "representation" of $F$ (in whatever sense is appropriate to the structure in question—in the group-theoretic context this was "an action of the group $F$ on a set $A$") which distinguishes the elements of $T_{\text{red}}$. A nice twist which often occurs, as in the above case, is that the object on which we "represent" $F$ may be the set $T_{\text{red}}$ itself, or some closely related object.

My development of Proposition 3.4.6 was full of motivations, remarks, etc. You might find it instructive to write out for yourself a concise, direct, self-contained proof that the set of terms indicated in Proposition 3.4.6, under the operations described, forms a group, and that this has the universal property of the free group on $X$.

**Exercise 3.4:3.** If $X$ is a set, and $s \neq t$ are two reduced group-theoretic terms in the elements of $X$ (as in Lemma 3.4.3(b)), will there in general exist a *finite* group $G$, and a map $v : X \to |G|$, such that $s_v \neq t_v$? (In other words, are the only identities satisfied by all finite groups those holding in all groups?)

If you succeed in answering the above question, you might try the more difficult ones in the next exercise.

**Exercise 3.4:4.**

(i)  If $X$ is a set, $F$ the free group on $X$, $H$ a subgroup of $F$, and $s$ an element of $F$ such that $s \notin |H|$, will there in general exist a finite group $G$ and a homomorphism $f : F \to G$ such that $f(s) \notin f(|H|)$?

(ii)  Same question, under the assumption that the subgroup $H$ is finitely generated.

Free groups can also be represented by matrices:

**Exercise 3.4:5.** Let $\mathrm{SL}(2, \mathbb{Z})$ denote the group of all $2 \times 2$ matrices of integers with determinant 1, and let $H$ be the subgroup thereof generated by the two matrices $x = \begin{pmatrix} 1 & 3 \\ 0 & 1 \end{pmatrix}$ and $y = \begin{pmatrix} 1 & 0 \\ 3 & 1 \end{pmatrix}$. Show that $H$ is free on $\{x, y\}$. (Hint: Let $c$ be the column vector $\begin{pmatrix} 1 \\ 1 \end{pmatrix}$. Examine the form of the column vector obtained by applying an arbitrary reduced group-theoretic word in $x$ and $y$ to $c$.)

If you do the above, you might like to think further about what pairs of (possibly distinct) integers, or for that matter, what pairs of real or complex numbers can replace the two "3"s in the above matrices. For integers the answer is known; for rational, real and complex numbers, there are many partial results (see [76]), but nothing close to a complete answer at present.

# Chapter 4
# A Cook's Tour of Other Universal Constructions

We shall now examine a number of other constructions having many similarities to that of free groups. In each case, the construction can be motivated by a question of the form, "Suppose we have a structure about which we know only that it satisfies such and such conditions. How much can we say about it based on this information alone?" In favorable cases, we shall find that if we collect the "data" we can deduce about such an object, this data itself can be made into an object $F$, which satisfies the given conditions, and satisfies no relations not implied by them (cf. Remark 3.2.13). This $F$ is then a "universal" example of these conditions, and that fact can be translated into a "universal mapping property" for $F$.

Although the original question, "What can we say about such an object?", and the "least set of relations" property, are valuable as motivation and intuition, the universal mapping property gives the characterization of these constructions that is most useful in applications. So though I will sometimes, but not always, refer to those motivating ideas, I will always characterize our constructions by universal properties.

The existence of these universal objects can in most cases be proved from scratch by either of the methods of §§3.2 and 3.3: construction from below, as sets of terms modulo necessary identifications, or construction from above, as subobjects of big direct products. But often, as a third alternative, we will be able to combine previously described universal constructions to get our new one.

Where possible, we will get explicit information on the structure of the new object—a normal form or other such description. It is a mark of the skilled algebraist, when working with objects defined by universal properties, to know when to use the universal property, and when to turn to an explicit description.

As we move through this chapter, I shall more and more often leave standard details for the reader to fill in: the precise meaning of an object "universal for" a certain property, the verification that such an object exists, etc.

© Springer International Publishing Switzerland 2015
G.M. Bergman, *An Invitation to General Algebra and Universal Constructions*, Universitext, DOI 10.1007/978-3-319-11478-1_4

In the later sections, commutative diagrams illustrating universal properties will often be inserted without explanation. These diagrams are not substitutes for assertions, but aids to the reader in visualizing the situation of the assertion he or she needs to formulate.

Constructions of *groups* will receive more than their rightful share of attention here because they offer a wide range of interesting examples, and are more familiar to many students than lattices, noncommutative rings (my own love), Lie algebras, etc.

Let us begin by noting how some familiar elementary group-theoretic constructions can be characterized by universal properties.

## 4.1. The subgroup and normal subgroup of $G$ generated by $S \subseteq |G|$

Suppose we are explicitly given a group $G$, and a subset $S$ of $|G|$.

Consider a subgroup $A$ of $G$ about which we are told only that it contains all elements of the set $S$. How much can we say about $A$?

Clearly $A$ contains all elements of $G$ that can be obtained from the elements of $S$ by repeated formation of products and inverses, and also contains the neutral element. This is all we can deduce, for it is easy to see that the set of elements which can be so obtained will form the underlying set of a subgroup of $G$, namely the subgroup $\langle S \rangle$ *generated by* the set $S$. This description builds $\langle S \rangle$ up "from below". We can also obtain it "from above", as the intersection of all subgroups of $G$ containing $S$. Whichever way we obtain it, the defining universal property of $\langle S \rangle$ is that it is a subgroup which contains $S$, and is contained in every subgroup $A$ of $G$ that contains $S$:

$$
\begin{array}{ccc}
S & \subseteq & |\langle S \rangle| \qquad \langle S \rangle \\
  & \diagdown & \cap \qquad\quad \cap \\
  &  & |A| \qquad\quad A
\end{array}
$$

(In the second part of the above display, we symbolize the group homomorphism given by an inclusion map of underlying sets by an inclusion sign between the symbols for the groups; a slight abuse of notation.)

We know a somewhat better description of the elements of $\langle S \rangle$ than the one I just gave: Each such element is either $e$ or the product of a sequence of elements of $S$ and their inverses. A related observation is that $\langle S \rangle$ is the image of the map into $G$ of the free group $F$ on $S$ induced by the inclusion-map $S \to |G|$. In particular cases one may get still better descriptions. For instance, if $S = \{a, b, c\}$ and $a$, $b$ and $c$ commute, then $\langle S \rangle$ consists of all elements $a^m b^n c^p$; if $G$ is the additive group of integers, then the subgroup generated by $\{1492, 1974\}$ is the subgroup of all even integers; if $G$ is a symmetric group $S_n$ $(n \geq 2)$, and $S$ consists of the two permutations $(12)$ and $(12\ldots n)$, then $\langle S \rangle$ is all of $G$.

There is likewise a least *normal* subgroup of $G$ containing $S$. This is called "the normal subgroup of $G$ generated by $S$", and has the corresponding universal property, with the word "normal" everywhere inserted before "subgroup".

**Exercise 4.1:1.** Show that the normal subgroup $N \subseteq G$ generated by $S$ is the subgroup of $G$ generated by $\{g\,s\,g^{-1} \mid g \in |G|,\ s \in S\}$.

Can $|N|$ also be described as $\{g\,h\,g^{-1} \mid g \in |G|,\ h \in |\langle S\rangle|\}$?

**Exercise 4.1:2.** Let $G$ be the free group on two generators $x$ and $y$, and $n$ a positive integer. Show that the normal subgroup of $G$ generated by $x^n$ and $y$ is generated as a subgroup by $x^n$ and $\{x^i\,y\,x^{-i} \mid 0 \le i < n\}$, and is in fact a *free* group on this $(n+1)$-element set. Also describe the normal subgroup we get if we let $n = 0$.

## 4.2. Imposing relations on a group. Quotient groups

Suppose next that we are given a group $G$, and are interested in homomorphisms of $G$ into other groups, $f\colon G \to H$, which make certain specified pairs of elements fall together. That is, let us be given a family of pairs of elements $\{(x_i,\ y_i) \mid i \in I\} \subseteq |G| \times |G|$ (perhaps only one pair, $(x, y)$) and consider homomorphisms $f$ from $G$ into other groups, which satisfy

$$(4.2.1) \qquad\qquad (\forall i \in I)\ \ f(x_i) = f(y_i).$$

Note that given one homomorphism $f\colon G \to H$ with this property, we can get more such homomorphisms $G \to K$ by forming composites $g\,f$ of $f$ with arbitrary homomorphisms $g\colon H \to K$. It would be nice to know whether there exists one pair $(H, f)$ which satisfies (4.2.1) and is *universal* for this condition, in the sense that given any other pair $(K, h)$ satisfying it, there is a unique homomorphism $g\colon H \to K$ making the diagram below commute. (In that diagram, "$\forall h \ldots$" is short for, "For all homomorphisms $h\colon G \to K$ such that $(\forall i \in I)\ h(x_i) = h(y_i)$", while "$\exists 1$" is a common abbreviation for "there exists a unique".)

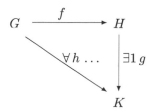

It is not hard to prove the existence of such a universal pair directly, either
by a "group-theoretic terms modulo an equivalence relation" construction, as
in §3.2, or by an "image in a big direct product" construction, analogous to
that of §3.3. But let us look at the problem another way. Condition (4.2.1)
is clearly equivalent to

(4.2.2)                           $(\forall i \in I) \ f(x_i \, y_i^{-1}) = e.$

So we are looking for a universal homomorphism which annihilates (sends
to $e$) a certain family of elements of $|G|$. We know that the set of elements
annihilated by a group homomorphism is always a normal subgroup, so this
is equivalent to saying that $f$ should annihilate the normal subgroup of $G$
generated by $\{x_i \, y_i^{-1} \mid i \in I\}$, referred to at the end of the preceding section.
And in fact, the pair $(G/N, q)$, where $N$ is this normal subgroup, $G/N$ is
the quotient group, and $q : G \to G/N$ is the quotient map, has precisely the
universal property we want:

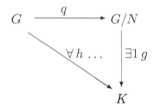

So this quotient group is the solution to our problem.

If we had never seen the construction of the quotient of a group by a
normal subgroup, an approach like the above would lead to a motivation
of that construction. We would ask, "What do we know about a group $H$,
given that it has a homomorphism of $G$ into it satisfying (4.2.1)?" We would
observe that $H$ contains an image $f(a)$ of each $a \in G$, and that two such
images are equal *if* they belong to the same coset of the normal subgroup
generated by the $x_i \, y_i^{-1}$'s. We would discover how the group operations must
act on these images-of-cosets, and conclude that this set of cosets, under these
operations, was a universal example of this situation.

Let us assume even a little more naiveté in

**Exercise 4.2:1.** Suppose in the above situation that we had not been so
astute, and had only noted that $f(a) = f(b)$ must hold in $H$ whenever
$a \, b^{-1}$ lies in the *subgroup* generated by $\{x_i y_i^{-1}\}$. Attempt to describe the
group operations on the set of equivalence classes under this relation, show
where this description fails to be well-defined, and show how this "failure"
could lead us to discover the *normality* condition needed.

The construction we have described is called *imposing the relations* $x_i = y_i$
$(i \in I)$ on $G$. We can abbreviate the resulting group by $G/(x_i = y_i \mid i \in I)$.

For the next exercise, recall that if $G$ is a group, then a $G$-set means a pair $S = (|S|, m)$, where $|S|$ is a set and $m: |G| \times |S| \to |S|$ is a map, which we shall abbreviate by writing $m(g, s) = g\,s$, satisfying

$$(4.2.3) \qquad \begin{aligned} (\forall\, s \in |S|,\, g,\, g' \in |G|) \ \ g\,(g's) &= (g\,g')\,s, \\ (\forall\, s \in |S|) \ \ e\,s &= s; \end{aligned}$$

in other words, a set on which $G$ acts by permutations [34, §I.5], [32, §II.4] [29, §1.7]. (We remark that this structure on the set $|S|$ can be described in two other ways: as a homomorphism from $G$ to the group of permutations of $|S|$, and alternatively, as a system of unary operations $t_g$ on $|S|$, one for each $g \in |G|$, satisfying *identities* corresponding to all the *relations* holding in $G$.)

A *homomorphism* $S \to S'$ of $G$-sets (for a *fixed* group $G$) means a map $a: |S| \to |S'|$ satisfying

$$(4.2.4) \qquad\qquad (\forall\, s \in |S|,\, g \in |G|) \ \ a\,(g\,s) = g\,a\,(s).$$

If $H$ is a subgroup of the group $G$, let $|G/H|$ denote the set of left cosets of $H$ in $G$. We shall write a typical left coset as $[g] = gH$. Then $|G/H|$ can be made the underlying set of a left $G$-set $G/H$, by defining $g\,[g'] = [g\,g']$.

**Exercise 4.2:2.** Let $H$ be any subgroup of $G$. Find a universal property characterizing the pair $(G/H, [e])$. In particular, what form does this universal mapping property take in the case where $H = \langle x_i^{-1} y_i \mid i \in I \rangle$ for some set $\{(x_i, y_i) \mid i \in I\} \subseteq |G| \times |G|$?

With the concept of imposing relations on a group under our belts, we are ready to consider

## 4.3. Groups presented by generators and relations

To start with a concrete example, suppose we are curious about groups $G$ containing two elements $a$ and $b$ satisfying the relation

$$(4.3.1) \qquad\qquad a\,b = b^2 a.$$

One may investigate the consequences of this equation with the help of the group laws. What we would be investigating is, I claim, the structure of the group with a *universal* pair of elements satisfying (4.3.1).

More generally, let $X$ be a set of symbols (in the above example, $X = \{a, b\}$), and let $T$ be the set of all group-theoretic terms in the elements of $X$. Then formal group-theoretic *relations* in the elements of $X$ mean formulae "$s = t$", where $s, t \in T$. Thus, given any set $R \subseteq T \times T$ of pairs $(s, t)$ of terms, we may consider groups $H$ with $X$-tuples of elements $v: X \to |H|$ satisfying the corresponding set of relations

(4.3.2)                               $(\forall\,(s,\,t)\in R)\ \ s_v = t_v.$

(So (4.3.1) is the case of (4.3.2) where $X = \{a,\,b\}$ and $R$ is the singleton $\{(a\,b,\,b^2a)\}$.) In this situation, we have

**Proposition 4.3.3.** *Let $X$ be a set, $T$ the set of all group-theoretic terms in $X$, and $R$ a subset of $T \times T$. Then there exists a universal example of a group with an $X$-tuple of elements satisfying the relations "$s = t$" $((s,\,t)\in R)$. I.e., there exists a pair $(G,\,u)$, where $G$ is a group, and $u$ a map $X \to |G|$ such that*

$$(\forall\,(s,\,t)\in R)\ \ s_u = t_u,$$

*and such that for any group $H$, and any $X$-tuple $v$ of elements of $H$ satisfying (4.3.2), there exists a unique homomorphism $f\colon G \to H$ satisfying $v = f\,u$ (in other words, having the property that the $X$-tuple $v$ of elements of $H$ is the image under $f$ of the $X$-tuple $u$ of elements of $G$).*

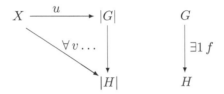

*Further, the pair $(G,\,u)$ is determined up to canonical isomorphism by these properties, and the group $G$ is generated by $u(X)$.*

*Three Methods of Proof.* Clearly, two methods that we may use are the constructions of §§3.2 and 3.3, applied essentially word-for-word, with the further condition (4.3.2) added to the group axioms throughout. (Note that unlike (3.2.1)–(3.2.8), the set of equations (4.3.2) involves no universal quantification over $T$; we only require the relations in $R$ to hold for the particular $X$-tuple $v$ of elements of each group $H$.)

However, we can now, alternatively, get the pair $(G,\,u)$ with less work. Let $(F,\,u_F)$ be the *free* group on $X$, let $N$ be the normal subgroup of $F$ generated by $\{s_{u_F}t_{u_F}^{-1}\mid (s,\,t)\in R\}$, i.e., by the set of elements of $F$ that we want to annihilate. Let $G = F/N$, let $q\colon F \to F/N$ be the canonical map, and let $u = q\,u_F$. That $(G,\,u)$ has the desired universal property follows immediately from the universal properties of free groups and quotient groups.

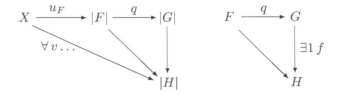

Having constructed $(G, u)$ by any of these three methods, let us now prove the final sentence of the proposition. If $(G', u')$ is another pair with the same universal property, then by the universal property of $G$ there exists a homomorphism $i: G \to G'$ such that $iu = u'$, and by the universal property of $G'$, an $i': G' \to G$ such that $i'u' = u$. These are inverses of one another; indeed, note that $i'i\,u = i'u' = u$, hence by the uniqueness condition in the universal property of $G$, $i'i$ equals the identity map of $G$; by a like argument, $ii'$ is the identity of $G'$, so $i$ is invertible, yielding the asserted isomorphism.

That $G$ is generated by $u(X)$ can be seen directly from each of our constructions, but let us also show from the universal property that this *must* be so. Consider the subgroup $\langle u(X) \rangle$ of $G$ generated by $u(X)$. The universal property of $G$ gives a homomorphism $j: G \to \langle u(X) \rangle$ which is the identity on elements of $u(X)$. Following it by the inclusion of $\langle u(X) \rangle$ in $G$ yields an endomorphism of $G$ which agrees with the identity map on $u(X)$, and so, by the uniqueness assertion in the universal property, *is* the identity. So the inclusion of $\langle u(X) \rangle$ in $G$ is surjective, as desired. □

Though we implied above that the advantage of getting our construction by combining two known constructions was that this was less work than constructing it from scratch, another general advantage of that approach, which we shall see in later sections, is that we can apply results about the known constructions to the new one.

The group $G$ of the preceding proposition is called the group *presented* by the *generators* $X$ and *relations* $R$. A common notation for this is

$$(4.3.4) \qquad\qquad G = \langle X \mid R \rangle.$$

For example, the universal group with a pair of elements satisfying (4.3.1) would be written

$$\langle a, b \mid ab = b^2 a \rangle.$$

In a group presented by generators and relations (4.3.4), one often uses the same symbols for the elements of $X$ and their images in $|G|$, even if the map $u$ is not one-to-one. For instance, from the well-known lemma saying that if an element $\eta$ of a group (or monoid) has both a left inverse $\xi$ and a right inverse $\zeta$, then $\xi = \zeta$, it follows that in the group $\langle x, y, z \mid xy = e = yz \rangle$, one has $u(x) = u(z)$. Unless there is a special need to be more precise, we may express this by saying "in $\langle x, y, z \mid xy = e = yz \rangle$, one has $x = z$."

Given a group $G$, one can find a family of generators for $G$, and then take a family of relations which imply all the group-theoretic relations satisfied in $G$ by those generators; in other words, an expression for $G$ in the form (4.3.4). This is called a *presentation* of the group $G$.

Recall that the concept of group-theoretic term was introduced both for the consideration of what relations hold among all families of elements of all groups, and to write specific relations that hold among particular families of

elements in particular groups. For the purpose of discussing identities holding in all groups, it was necessary to distinguish between expressions such as
$(x\,y)^{-1}$ and $y^{-1}x^{-1}$, between $(x\,y)\,z$ and $x\,(y\,z)$, etc. But in considering
relations in particular groups we can generally take for granted the group
identities, i.e., not distinguish pairs of expressions that have the same evaluations in all groups. For example, in (4.3.1), the right hand side could be
replaced by $b\,(b\,a)$ without changing the effect of the condition. Hence in
considering groups presented by generators and relations, one often considers the relations to be given by pairs, not of *terms*, but of their equivalence
classes under the relation of having equal values in all groups—in other words
pairs $(s,\,t) \in |F| \times |F|$, where $F$ is the free group on $X$. For such $(s,\,t)$,
an $X$-tuple $v$ of elements of a group $G$ is considered to "satisfy $s = t$"
if $h(s) = h(t)$, for $h$ the homomorphism $F \to G$ induced by $v$ as in
Definition 3.1.3.

Whether $s$ and $t$ are group-theoretic terms as in Proposition 4.3.3, or
elements of a free group as in the above paragraph, we should note that
there is a certain abuse of language in saying that a family $v$ of elements of
a group $G$ "satisfies the relation $s = t$", and in writing equations "$s = t$"
in presentations of groups. What we mean in such cases is that a certain
equation *obtained from* the pair $(s,\,t)$ and the $X$-tuple $v$ holds in $G$; but
the equality $s = t$ between terms or free group elements is itself generally
false! As with other convenient but imprecise usages, once we are conscious of
its impreciseness, we may use it, but should be ready to frame more precise
statements when imprecision could lead to confusion (for instance, if we also
want to discuss which of certain terms, or elements of a free group, are really
equal).

We have noted that a relation $(s,\,t)$ is satisfied by an $X$-tuple $v$ of
elements of a group $G$ if and only if $(s\,t^{-1})_v = e$ in $G$; in other words,
if and only if the relation $(s\,t^{-1},\,e)$ is satisfied by $v$. Thus, every presentation of a group can be reduced to one in which the relations occurring all
have the form $(r,\,e)$ for terms (or free-group elements) $r$. The elements $r$
are then called the *relators* in the presentation, and in expressing the group,
one may list relators rather than relations. E.g., the group we wrote earlier
as $\langle a,\,b \mid ab = b^2a \rangle$ would, in this notation, be written $\langle a,\,b \mid aba^{-1}b^{-2} \rangle$.
However, I will stick to equational notation in these notes.

**Exercise 4.3:1.** Show that the three groups described below are isomorphic
(as groups, ignoring the maps "$X \to |G|$" etc. coming from the presenta-
tions of the first two).

(a) $G = \langle a,\,b \mid a^2 = e,\,ab = b^{-1}a \rangle$.

(b) $H = \langle s,\,t \mid s^2 = t^2 = e \rangle$.

(c) $D = \{\text{distance-preserving permutations of the set } \mathbb{Z}\}$, i.e., the group
consisting of all translation-maps $n \mapsto n + c \ (c \in \mathbb{Z})$ and all reflection-
maps $n \mapsto -n + d \ (d \in \mathbb{Z})$.

The universal property of a group presented by generators and relations is extremely useful in considerations such as that of

**Exercise 4.3:2.** Find all endomorphisms of the group of the preceding exercise. Describe the structure of the monoid of these endomorphisms.

Returning to the example with which we started this section,

**Exercise 4.3:3.** Find a normal form or other convenient description for the group presented by two generators $a$, $b$ and the one relation (4.3.1): $ab = b^2a$.

The following question, suggested by a member of the class some years ago, is harder, but has a nice solution:

**Exercise 4.3:4.** (D. Hickerson.) Do the same for $\langle a, b \mid ab = b^2a^2 \rangle$.

Any group $G$ can be presented by some system of generators and relations. E.g., take $|G|$ itself for generating set, and the multiplication table of $G$ as a set of relations. But it is often of interest to find concise presentations for given groups. Note that the *free* group on a set $X$ may be presented by the generating set $X$ and the empty set of relations!

**Exercise 4.3:5.** Suppose $f(x, y)$ and $g(y)$ are group-theoretic terms in two and one variables respectively. What can you prove about the group with presentation

$$\langle w, x, y \mid w = f(x, y), \ x = g(y) \rangle ?$$

Generalize if you can.

**Exercise 4.3:6.** Consider the set $\mathbb{Z} \times \mathbb{Z}$ of "lattice points" in the plane. Let $G$ be the group of "symmetries" of this set, i.e., maps $\mathbb{Z} \times \mathbb{Z} \to \mathbb{Z} \times \mathbb{Z}$ which preserve distances between points.

(i)   Find a simple *description* of $G$. (Cf. the description of the group of symmetries of the set $\mathbb{Z}$ in terms of translations and reflections in Exercise 4.3:1(c).)

(ii)   Find a simple *presentation* for $G$, and a *normal form* for elements of $G$ in terms of the generators in your presentation.

**Exercise 4.3:7.** Suppose $G$ is a group of $n$ elements. Then the observation made above, on how to present any group by generators and relations, yields upper bounds on the minimum numbers of generators and relations needed to present $G$. Write down these bounds; then see to what extent you can improve on them.

The above exercise shows that every *finite* group is *finitely presented*, i.e., has a presentation in terms of finitely many generators and finitely many relations. Of course, there are also finitely presented groups which are infinite.

The next two exercises concern the property of finite presentability. (The first is not difficult; the second requires some ingenuity, or experience with infinite groups.)

**Exercise 4.3:8.** Show that if $G$ is a group which has some finite presentation, and if $\langle x_1, \ldots, x_n \mid R \rangle$ is any presentation of $G$ using finitely many generators, then there is a finite subset $R_0 \subseteq R$ such that $\langle x_1, \ldots, x_n \mid R_0 \rangle$ is also a presentation of $G$.

**Exercise 4.3:9.** Find a finitely generated group that is not finitely presented.

Another kind of question one can ask is typified by

**Exercise 4.3:10.** Is the group

$$\langle x, y \mid x\,y\,x^{-1} = y^2, \ y\,x\,y^{-1} = x^2 \rangle$$

trivial $(= \{e\})$? What about

$$\langle x, y \mid x\,y\,x^{-1} = y^2, \ y\,x\,y^{-1} = x^3 \rangle ?$$

(If you prove either or both of these groups trivial, you should present your calculation in a way that makes it clear, at each stage, which defining relation you are applying, and to what part of what expression.)

For the group-theory buff, here are two harder, but still tractable examples. In part (ii) below, note that $Z$ is a common notation in group theory for the infinite cyclic group. (The similarity to $\mathbb{Z}$ as a symbol for the integers is a coincidence: The latter is based on German *Zahl*, meaning "number" while the group-theoretic symbol is based on *zyklisch*, meaning "cyclic". Although the additive group of integers *is* an infinite cyclic group, a group denoted $Z$ can either be written additively, or multiplicatively, e.g. as $\{x^i \mid i \in \mathbb{Z}\}$.) The finite cyclic group of order $n$ is likewise denoted $Z_n$.

**Exercise 4.3:11.** (J. Simon [116].)
   (i)  Is either of the groups $\langle a, b \mid (b^{-1}a)^4\,a^{-3} = e = b^{10}\,(b^{-1}a)^{-3} \rangle$, or $\langle a, b \mid (b\,a^{-1})^{-3}a^{-2} = e = b^9\,(b\,a^{-1})^4 \rangle$ trivial?
   (ii) In the group $\langle a, b \mid b\,a^{-4}\,b\,a\,b^{-1}a = e \rangle$, is the subgroup generated by $b\,a\,(b^{-1}a)^2$ and $a^3\,b^{-1}$ isomorphic to $Z \times Z$?

Suppose $G$ is a group with presentation $\langle X \mid R \rangle$. An interesting consequence of the universal property characterizing $G$ in Proposition 4.3.3 is that for any group $H$, the set of homomorphisms $\mathrm{Hom}(G, H)$ is in natural one-to-one correspondence with the set of $X$-tuples of elements of $H$ satisfying the relations $R$.

For instance, if $n$ is a positive integer, and we write $Z_n = \langle x \mid x^n = e \rangle$ for the cyclic group of order $n$, then we see that for any group $H$, we have a natural bijection between $\mathrm{Hom}(Z_n, H)$ and $\{a \in |H| \mid a^n = e\}$, each $a$ in the latter set corresponding to the unique homomorphism $Z_n \to H$ carrying $x$ to $a$. (Terminological note: a group element $a \in |H|$ which satisfies $a^n = e$ is said to have *exponent* $n$. This is equivalent to its having order *dividing* $n$.)

Similarly, one finds that $\langle x, y \mid x\,y = y\,x \rangle$ is isomorphic to $Z \times Z$, hence $\mathrm{Hom}(Z \times Z, H)$ corresponds to the set of all ordered pairs of commuting elements of $H$.

Thus, presentations of groups by generators and relations provide a bridge between the internal structure of groups, and their "external" behavior under homomorphisms. This will be particularly valuable when we turn to category theory, which treats mathematical objects in terms of the homomorphisms among them.

The last exercise of this section describes a very interesting group, though most of its striking properties cannot be given here.

**Exercise 4.3:12.** Let $G = \langle x, y \mid y^{-1}x^2y = x^{-2},\ x^{-1}y^2x = y^{-2} \rangle$.

(i)   Find a normal form or other convenient description for elements of $G$. Verify from this description that $G$ has no nonidentity elements of finite order.

(ii)   Calling the group characterized in several ways in Exercise 4.3:1 "$D$", show that $G$ has exactly three normal subgroups $N$ such that $G/N \cong D$, and that the intersection of these three subgroups is $\{e\}$.

(iii) It follows from (ii) above that $G$ can be identified with a subgroup of $D \times D \times D$. Give a criterion for an element of $D \times D \times D$ to lie in this subgroup, and prove directly from this criterion that no element of this subgroup has finite order.

The study of groups presented by generators and relations is called Combinatorial Group Theory, and there are several books with that title. An interesting text which assumes only an undergraduate background, but goes deep into the techniques of the subject, is [33]. There is also a web-page on the area, [149], including a list of open questions.

Though group presentations *often* yield groups for which a normal form can be found, it has been proved by Novikov, Boone and Britton that there exist finitely presented groups $G$ such that no algorithm can decide whether an arbitrary pair of terms of $G$ represent the same element. A proof of this result is given in the last chapter of [35].

## 4.4. Abelian groups, free abelian groups, and abelianizations

An *abelian group* is a group $A$ satisfying the further identity

$$(\forall\, x, y \in |A|)\ x\,y = y\,x.$$

The discussion of §3.1 carries over without essential change and gives us the concept of a *free abelian group* $(F,\, u)$ on a set $X$; the method of §3.2 establishes the existence of such groups by constructing them as quotients of sets $T$ of terms by appropriate equivalence relations, and the method of §3.3 yields an alternative construction as subgroups of direct products of large enough families of abelian groups. We may clearly also obtain the free abelian group on a set $X$ as the group presented by the generating set $X$ and the relations $s\,t = t\,s$, where $s$ and $t$ range over all elements of $T$. This big set of relations is easily shown to be equivalent, for any $X$-tuple of elements of any group, to the smaller family $x\,y = y\,x$ $(x,\, y \in X)$; so the free abelian group on $X$ may be presented as

$$\langle X \mid x\,y = y\,x \ (x,\, y \in X)\rangle.$$

To investigate the *structure* of free abelian groups, let us consider, say, three elements $a$, $b$, $c$ of an arbitrary abelian group $A$, and look at elements $g \in A$ that can be obtained from these by group-theoretic operations. We know from §3.4 that any such $g$ may be written either as $e$, or as a product of the elements $a$, $a^{-1}$, $b$, $b^{-1}$, $c$, $c^{-1}$. We can now use the commutativity of $A$ to rearrange this product so that it begins with all factors $a$ (if any), followed by all factors $a^{-1}$ (if any), then all factors $b$ (if any), etc. Now performing cancellations if both $a$ and $a^{-1}$ occur, or both $b$ and $b^{-1}$ occur, or both $c$ and $c^{-1}$ occur, we can reduce $g$ to an expression $a^i\, b^j\, c^k$, where $i$, $j$ and $k$ are integers (positive, negative, or 0; exponentiation by negative integers and by 0 being defined by the usual conventions). Let us call the set of such expressions $T_{\text{ab-red}}$, and define composition, inverse, and an identity element on this set by

(4.4.1)              $(a^i\, b^j\, c^k) \odot (a^{i'}\, b^{j'}\, c^{k'}) = a^{i+i'}\, b^{j+j'}\, c^{k+k'},$

(4.4.2)                      $(a^i\, b^j\, c^k)^{(-)} = a^{-i}\, b^{-j}\, c^{-k},$

(4.4.3)                                      $e = a^0\, b^0\, c^0.$

Note that $\odot$ and $^{(-)}$ are here different operations from those represented by the same symbols in §3.4, but that the idea is as in that section; in particular, it is clear that for any map $v$ of $\{a, b, c\}$ into an *abelian* group, one has $(s \cdot t)_v = (s \odot t)_v$ and $(s^{-1})_v = (s^{(-)})_v$. It is now easy to verify that under these operations, $T_{\text{ab-red}}$ itself forms an abelian group $F$. This verification does not require any analog of "van der Waerden's trick" (§3.4); rather, the result follows from the known fact that the integers (which appear as exponents) do form an abelian group under $+$, $-$, and $0$.

It follows, as in §3.4, that this $F$ is the *free* abelian group on $\{a, b, c\}$, and thus that the set $T_{\text{ab-red}}$ of terms $a^i\,b^j\,c^k$ is a normal form for elements of the free abelian group on three generators.

The above normal form is certainly simpler than that of the free group on $\{a, b, c\}$. Yet there is a curious way in which it is more complicated: It is based on our choice to use alphabetic order on the generating set $\{a, b, c\}$. Using different orderings, we get different normal forms, e.g., $b^j\,c^k\,a^i$, etc. If we want to generalize our normal form to the free abelian group on a finite set $X$ without any particular structure, we must begin by ordering $X$, say writing $X = \{x_1, x_2, \ldots, x_n\}$. Only then can we speak of "the set of all expressions $x_1^{i_1} \ldots x_n^{i_n}$ ". If we want a normal form in the free abelian group on an *infinite* set $X$, we must again choose a total ordering of $X$, and then either talk about "formally infinite products with all but finitely many factors equal to $e$", or modify the normal form, say to "$e$ or $x^{i(x)}\,y^{i(y)} \ldots z^{i(z)}$ where $x < y < \cdots < z \in X$, and all exponents shown are nonzero" (the last two conditions to ensure uniqueness!).

We may be satisfied with one of these approaches, or we may prefer to go to a slightly different kind of representation for $F$, which we discover as follows: Note that if $g$ is a member of the free abelian group $F$ on $X$, then for each $x \in X$, the exponent $i(x)$ to which $x$ appears in our normal forms for $g$ is the same for these various forms; only the position in which $x^{i(x)}$ is written (and if $i(x) = 0$, whether it is written) changes from one normal form to another. Clearly, any of our normal forms for $g$, and hence the element $g$ itself, is determined by the $X$-tuple of exponents $(i(x))_{x \in X}$. So let us "represent" $g$ by this $X$-tuple; that is, identify $F$ with a certain set of integer-valued functions on $X$. It is easy to see that the group operations of $F$ correspond to componentwise addition of such $X$-tuples, componentwise additive inverse, and the constant $X$-tuple $0$; and that the $X$-tuple corresponding to each generator $x \in X$ is the function $\delta_x$ having value $1$ at $x$ and $0$ at all other elements $y \in X$. The $X$-tuples that correspond to members of $F$ are those which are nonzero at only finitely many components. Thus we get the familiar description of the free abelian group on $X$ as the subgroup of $\mathbb{Z}^X$ consisting of all functions having finite support in $X$. (The *support* of a function $f$ means $\{x \mid f(x) \neq 0\}$.)

**Exercise 4.4:1.** If $X$ is infinite, it is clear that the whole group $\mathbb{Z}^X$ is *not* a free abelian group on $X$ under the map $x \mapsto \delta_x$, since it is not generated by the $\delta_x$. Show that $\mathbb{Z}^X$ is in fact not a free abelian group on *any* set of generators. You may assume $X$ countable if you wish.

(For further results on $\mathbb{Z}^X$ and its subgroups when $X$ is countably infinite, see Specker [133]. Among other things, it is shown there that the uncountable group $\mathbb{Z}^X$ has only countably many homomorphisms into $\mathbb{Z}$, though its countable subgroup $F$ clearly has uncountably many! It is also shown that the subgroup of *bounded* functions on $X$ is free abelian, on uncountably many generators. This fact was generalized to not necessarily countable $X$ by Nöbeling [120]. For a simpler proof of this result, using ring theory, see [45, §1].)

The concept of the *abelian group* presented by a system of generators and relations may be formulated exactly like that of a group presented by generators and relations. It may also be constructed analogously: as the quotient of the free abelian group on the given generators by the subgroup generated by the relators $s\,t^{-1}$ (we don't have to say "normal subgroup" because normality is automatic for subgroups of abelian groups); or alternatively, as the *group* presented by the given generators and relations, together with the additional relations saying that all the generators commute with one another.

Suppose now that we start with an arbitrary group $G$, and impose relations saying that for all $x, y \in |G|$, $x$ and $y$ commute: $x\,y = y\,x$. That is, we form the quotient of $G$ by the normal subgroup generated by the elements $(y\,x)^{-1}(x\,y) = x^{-1}y^{-1}x\,y$. As noted in the paragraph introducing Exercise 3.4:2, these elements are called *commutators*, and often written

$$x^{-1}y^{-1}x\,y = [x,\,y].$$

(Another common notation is $(x,\,y)$, but we will not use this, to avoid confusion with ordered pairs.) The normal subgroup that these generate is called the *commutator subgroup*, or *derived subgroup* of $G$, written $[G,\,G]$, and often abbreviated by group theorists to $G'$. The quotient group, $G^{\mathrm{ab}} = G/[G,\,G]$, is an abelian group with a homomorphism $q$ of the given group $G$ into it which is *universal* among homomorphisms of $G$ into abelian groups $A$, the diagram for the universal property being

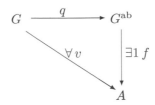

This group $G^{ab}$ (or more precisely, the pair $(G^{ab}, q)$, or any isomorphic pair) is called the *abelianization* or *commutator factor group* of $G$.

Suppose now that we write down any system of generators and relations for a group, and compare the *group* $G$ and the *abelian group* $H$ that these same generators and relations define. By the universal property of $G$, there will exist a unique homomorphism $r: G \to H$ taking the generators of $G$ to the corresponding generators of $H$. It is easy to check that $(H, r)$ has the universal property characterizing the abelianization of $G$. So this gives another way of describing abelianization. Note, as a consequence, that given an arbitrary system of generators and group-theoretic relations, the *group* these present will determine, up to natural isomorphism, the abelian group that they present (but not vice versa).

**Exercise 4.4:2.** Find the structures of the abelianizations of the groups presented in Exercises 4.3:1, 4.3:3, 4.3:4, 4.3:10, and 4.3:11(i). (This is easier than determining the structures of the groups themselves, hence I am giving as one exercise the abelianized versions of those many earlier exercises.)

**Exercise 4.4:3.** Show that any group homomorphism $f: G \to H$ induces a homomorphism of abelian groups $f^{ab}: G^{ab} \to H^{ab}$. State precisely the condition relating $f$ and $f^{ab}$. Show that for a composite of group homomorphisms, one has $(f\,g)^{ab} = f^{ab}g^{ab}$. Deduce that for any group $G$, there is a natural homomorphism of monoids, $\mathrm{End}(G) \to \mathrm{End}(G^{ab})$, and a natural homomorphism of automorphism groups, $\mathrm{Aut}(G) \to \mathrm{Aut}(G^{ab})$.

(Here $\mathrm{End}(G)$ denotes the set of endomorphisms of $G$, regarded as a monoid under composition, while $\mathrm{Aut}(G)$ denotes the group of automorphisms of $G$, i.e., the group of invertible elements of $\mathrm{End}(G)$.)

**Exercise 4.4:4.** For $G$ as in Exercises 4.3:1 and 4.3:2, is the natural homomorphism $\mathrm{Aut}(G) \to \mathrm{Aut}(G^{ab})$ of the above exercise one-to-one?

**Exercise 4.4:5.** If $H$ is a subgroup of $G$, what can be said about the relation between $H^{ab}$ and $G^{ab}$? Same question if $H$ is a homomorphic image of $G$.

**Exercise 4.4:6.** Let $K$ be a field, $n$ a positive integer, and $\mathrm{GL}(n, K)$ the group of invertible $n \times n$ matrices over $K$. Determine as much as you can about the structure of $\mathrm{GL}(n, K)^{ab}$.

**Exercise 4.4:7.** If $G$ is a group, will there exist a universal homomorphism of $G$ into a *solvable* group, $G \to G^{\mathrm{solv}}$? What if $G$ is assumed finite?

Does there exist a "free solvable group" on a set $X$, or some similar construction?

**Exercise 4.4:8.** Show that the free abelian group on $n$ generators cannot be presented *as a group* by fewer than $n$ generators and $n(n-1)/2$ relations.


## 4.5. The Burnside problem

In 1902, Burnside [65] asked whether a finitely generated group, all of whose elements have finite order, must be finite. This problem was hard to approach because, with nothing assumed about the *values* of the finite orders of the elements, one had no place to begin a calculation. So Burnside also posed this question under the stronger hypothesis that there be a *common* finite bound on the orders of all elements of $G$.

The original question with no bound on the orders was suddenly answered negatively in 1964, with a counterexample arising from the Golod-Shafarevich construction [83]; there is a short and fairly self-contained presentation of this material in the last chapter of [31]. In the opposite direction, Burnside himself proved that if $G$ is a finitely generated group of *matrices* over a field, and all elements of $G$ have finite order, then $G$ is finite [66]. On the other hand, his question is still open for *finitely presented* groups [150].

Turning to the question for a general group $G$ with a common bound on the orders of its elements, note that if $m$ is such a bound, then $m!$ is a common *exponent* for these elements; while if $n$ is a common exponent, it is also a bound on their orders. So "there is a common bound on the orders of all elements" is equivalent to "all elements have a common exponent". The latter condition is more convenient to study, since the statement that $x$ has exponent $n$ has the form of an equation. So for any positive integer $n$, one defines the *Burnside problem for exponent $n$* to be the question of whether every finitely generated group satisfying the identity

$$(4.5.1) \qquad\qquad (\forall x) \; x^n = e$$

is finite.

For $n = 1$, the answer is trivially yes, for $n = 2$ the same result is an easy exercise, for $n = 3$ it is not very hard to show, and it has also been proved for $n = 4$ and $n = 6$. On the other hand, it has been shown in recent decades that the answer is negative for all odd $n \geq 665$ [37], and for all $n > 8000$ [111] (cf. [94]). This leaves a large but finite set of cases still open: all odd values from 5 to 663, and all even values from 8 to 8,000. We won't go into these hard group-theoretic problems here. But the concept of universal constructions does allow us to understand the nature of the question better. Call a group $G$ an $n$-Burnside group if it satisfies (4.5.1). One may define the *free $n$-Burnside group* on any set $X$ by the obvious universal property, and it will exist for the usual reasons. In particular, it can be presented, as a group, by the generating set $X$, and the infinite family of relations equating the $n$-th powers of all terms in the generators to $e$. I leave it to you to think through the following relationships:

**Exercise 4.5:1.** Let $n$ and $r$ be positive integers.

(i)   What implications can you prove among the following statements?

(a) Every $n$-Burnside group which can be generated by $r$ elements is finite.

(b) The free $n$-Burnside group on $r$ generators is finite.

(c) The group $\langle x_1, \ldots, x_r \mid x_1^n = \cdots = x_r^n = e \rangle$ is finite.

(d) There exists a finite $r$-generator group having a finite presentation (a presentation by those $r$ generators and finitely many relations) in which all relators are $n$-th powers, $\langle x_1, \ldots, x_r \mid w_1^n = \cdots = w_s^n = e \rangle$ (where each $w_i$ is a term in $x_1, \ldots, x_r$. Cf. Exercises 4.3:7 and 4.3:8.)

(e) There exists an integer $N$ such that all $n$-Burnside groups generated by $r$ elements have order $\leq N$.

(f) There exists an integer $N$ such that all *finite* $n$-Burnside groups generated by $r$ elements have order $\leq N$ ("the restricted Burnside problem").

(ii)   What implications can you prove among cases of statement (a) above involving the same value of $n$ but different values of $r$? involving the same value of $r$ but different values of $n$?

(iii) I described [111] above as proving a negative answer to the Burnside problem for *all* $n > 8000$. Actually, the result proved there only applies to those $n > 8000$ which are multiples of 16. Show, however, that this result, and the corresponding result for odd $n \geq 665$ proved in [37], together imply the asserted result for all $n > 8000$.

Note that if for a given $n$ and $r$ we could find a *normal form* for the free $n$-Burnside group on $r$ generators, we would know whether (b) was true! But except when $n$ or $r$ is very small, such normal forms are not known. For further discussion of these questions, see [30, 38, Chapter 18]. Recent results, including a solution to the restricted Burnside problem ((f) above), some negative results on the *word problem* for free Burnside groups, and the result that for $p > 10^{75}$ there exist infinite groups of exponent $p$ all of whose proper subgroups are cyclic ("Tarski Monsters"), can be found in [74, 101, 117, 121, 140, 141], and references given in those works.

A group $G$ is called *residually finite* if for any two elements $x \neq y \in |G|$, there exists a homomorphism $f$ of $G$ into a *finite* group such that $f(x) \neq f(y)$.

**Exercise 4.5:2.** Investigate implications involving conditions (a)–(f) of the preceding exercise, together with

(g) The free $n$-Burnside group on $r$ generators is residually finite.

**Exercise 4.5:3.**

(i)    Restate Exercise 3.4:3 as a question about residual finiteness (showing, of course, that your restatement is equivalent to the original question).

(ii)   If $G$ is a group, does there exist a universal homomorphism $G \to G^{\mathrm{rf}}$, of $G$ into a residually finite group?

## 4.6. Products and coproducts of groups

Let $G$ and $H$ be groups. Consider the following two situations:

(a) a group $P$ given with a homomorphism $p_G : P \to G$ and a homomorphism $p_H : P \to H$, and

(b) a group $Q$ given with a homomorphism $q_G : G \to Q$, and a homomorphism $q_H : H \to Q$.

(Diagrams below.)

Note that if in situation (a) we choose a homomorphism $a$ of any other group $P'$ into $P$, then $P'$ also acquires homomorphisms into $G$ and $H$, namely $p_G\,a$ and $p_H\,a$. Similarly, if in situation (b) we choose any homomorphism $b$ of $Q$ into a group $Q'$, then $Q'$ acquires homomorphisms $b\,q_G$ and $b\,q_H$ of $G$ and $H$ into it:

So we may ask whether there exists a *universal* example of a $P$ with maps into $G$ and $H$, that is, a 3-tuple $(P, p_G, p_H)$ such that for any group $P'$, every pair of maps $p'_G : P' \to G$ and $p'_H : P' \to H$ arises by composition of $p_G$ and $p_H$ with a unique homomorphism $a : P' \to P$; and, dually, whether there exists a universal example of a group $Q$ with maps of $G$ and $H$ into it.

In both cases, the answer is yes. The universal $P$ is simply the direct product group $G \times H$, with its projection maps $p_G$ and $p_H$ onto the two factors; the universal property is easy to verify. The universal $Q$, on the other hand, can be constructed by generators and relations. It has to have for each $g \in |G|$ an element $q_G(g)$—let us abbreviate this to $\bar{g}$—and for each $h \in |H|$ an element $q_H(h)$—call this $\tilde{h}$. So let us take for generators a set of symbols

(4.6.1)                         $\{\bar{g}, \tilde{h} \mid g \in |G|,\ h \in |H|\}.$

The relations these must satisfy are those saying that $q_G$ and $q_H$ are homomorphisms:

(4.6.2)     $\overline{g}\,\overline{g'} = \overline{gg'}$  $(g,\, g' \in |G|)$,   $\widetilde{h}\widetilde{h'} = \widetilde{hh'}$  $(h,\, h' \in |H|)$.

It is immediate that the group presented by generators (4.6.1) and relations (4.6.2) has the desired universal mapping property. (We might have supplemented (4.6.2) with the further relations $\overline{e_G} = e$, $\widetilde{e_H} = e$, $\overline{g^{-1}} = \overline{g}^{-1}$, $\widetilde{h^{-1}} = \widetilde{h}^{-1}$. But these are implied by the relations listed, since, as is well known, any set map between groups which preserves products also preserves neutral elements and inverses.) More generally, if $G$ is a group which can be presented as $\langle X \mid R \rangle$, and if, similarly, $H = \langle Y \mid S \rangle$, then we may take for generators of $Q$ a disjoint union $X \sqcup Y$, and for relations the union of $R$ and $S$. For instance, if

$$G = Z_3 = \langle x \mid x^3 = e \rangle \quad \text{and} \quad H = Z_2 = \langle x \mid x^2 = e \rangle,$$

then $Q$ may be presented as

$$\langle x,\, x' \mid x^3 = e,\, x'^{\,2} = e \rangle,$$

with $q_G$ and $q_H$ determined by the conditions $x \mapsto x$ and $x \mapsto x'$, respectively. You should be able to verify the universal property of $Q$ from this presentation.

(If you are not familiar with the concept of a "disjoint union" $X \sqcup Y$ of two sets $X$ and $Y$, I hope that the above context suggests the meaning. Explicitly, it means the union of a bijective copy of $X$ and a bijective copy of $Y$, chosen to be disjoint. So, if $X = \{a,\, b,\, c\}$, $Y = \{b,\, c,\, d,\, e\}$ where $a,\, b,\, c,\, d,\, e$ are all distinct, then their ordinary set-theoretic union is the five-element set $X \cup Y = \{a,\, b,\, c,\, d,\, e\}$, but an example of a "disjoint union" would be any set of the form $X \sqcup Y = \{a,\, b,\, c,\, b',\, c',\, d,\, e\}$ where $a,\, b,\, c,\, b',\, c',\, d,\, e$ are distinct, given with the obvious maps taking $X$ to the three-element subset $\{a,\, b,\, c\}$ of this set and $Y$ to the disjoint four-element subset $\{b',\, c',\, d,\, e\}$. Though there is not a unique way of choosing a disjoint union of two sets, the construction is unique in the ways we care about. E.g., note that in the above example, any disjoint union of $X$ and $Y$ will have $|X| + |Y| = 7$ elements. Hence one often speaks of "the" disjoint union. We will see, a few sections from now, that disjoint union of sets is itself a universal construction of set theory.)

To see for general $G$ and $H$ what the group determined by the above universal property "looks like", let us again think about an arbitrary group $Q$ with homomorphisms of $G$ and $H$ into it, abbreviated $g \mapsto \overline{g}$ and $h \mapsto \widetilde{h}$. The elements of $Q$ which we can name in this situation are, of course, the products

(4.6.3)     $x_n^{\pm 1} x_{n-1}^{\pm 1} \cdots x_1^{\pm 1}$  with  $x_i \in \{\overline{g},\, \widetilde{h} \mid g \in |G|,\, h \in |H|\}$,  and $n \geq 0$.

(Notational remark: In §3.4, I generally kept $n \geq 1$, and introduced "$e$" as a separate kind of expression. Here I shall adopt the convenient convention that the product of the empty (length 0) sequence of factors is $e$, so that the case "$e$" may be absorbed in the general case.)

Now for any $g \in |G|$ or $h \in |H|$ we have noted that $\overline{g}^{-1} = \overline{g^{-1}}$ and $\widetilde{h}^{-1} = \widetilde{h^{-1}}$ in $Q$; hence the inverse of any member of the generating set $\{\overline{g}, \widetilde{h} \mid g \in |G|, h \in |H|\}$ is another member of that set. So we may simplify any product (4.6.3) to one in which all exponents are $+1$, and so write it without showing these exponents. We also know that $\overline{e} = \widetilde{e} = e$, so wherever instances of $\overline{e}$ or $\widetilde{e}$ occur in such a product, we may drop them. Finally, if two factors belonging to $\{\overline{g} \mid g \in |G|\}$ occur in immediate succession, the relations (4.6.2) allow us to replace these by a single such factor, and, likewise, we may do the same if there are two adjacent factors from $\{\widetilde{h} \mid h \in |H|\}$. So the elements of $Q$ that we can construct can all be reduced to the form

$$x_1 \ldots x_n$$

(4.6.4)    where $n \geq 0$, $x_i \in \{\overline{g} \mid g \in |G| - \{e\}\} \cup \{\widetilde{h} \mid h \in |H| - \{e\}\}$, and no two successive $x$'s come from the same set, $\{\overline{g} \mid g \in |G| - \{e\}\}$ or $\{\widetilde{h} \mid h \in |H| - \{e\}\}$.

We can express the *product* of two elements (4.6.4) as another such element, by putting the sequences of factors together, and reducing the resulting expression to the above form as described above; likewise it is clear how to find expressions of that form for inverses of elements (4.6.4), and for the element $e$. In any particular group $Q$ with homomorphisms of $G$ and $H$ into it, there may be other elements than those expressed by (4.6.4), and there may be some equalities among such products. But as far as we can see, there don't seem to be any cases left of two expressions (4.6.4) that must represent the same element in *every* such group $Q$. *If* in fact there are none, then, as in §3.4, the expressions (4.6.4) will correspond to the distinct elements of the *universal* $Q$ we are trying to describe, and thus will give a normal form for the elements of this group.

To show that there are no undiscovered necessary equalities, we can use the same stratagem as in §3.4 – it was for this situation that van der Waerden devised it!

**Proposition 4.6.5** (van der Waerden [142]). *Let $G$, $H$ be groups, and $Q$ the group with a universal pair of homomorphisms $G \to Q$, $H \to Q$, written $g \mapsto \overline{g}$, $h \mapsto \widetilde{h}$. Then every element of $Q$ can be written uniquely in the form* (4.6.4).

*Proof.* Let us, as before, introduce an additional symbol $a$, and now denote by $A$ the set of all symbols

(4.6.6)     $x_n \ldots x_1\, a$, where $x_1, \ldots, x_n$ are as in (4.6.4) (the $n = 0$ case being interpreted as the bare symbol $a$).

We would like to describe actions of $G$ and $H$ on this set. It is clear what these actions *should* be, but an explicit description is a bit messy, because of the need to state separately the cases where the element of $A$ on which we are acting does or does not begin with an element of the group we are acting by, and if it does, the cases where this beginning element is or is not the inverse of the element by which we are acting. This messiness in the definition makes still more messy the verification that the "actions" give homomorphisms of $G$ and of $H$ into the group of permutations of $A$.

We shall get around these annoyances (which are in any case minor compared with the difficulties of doing things *without* van der Waerden's method) by another trick. Let us describe a set $A_G$ which is in bijective correspondence with $A$: For those elements $b \in A$ which already begin with a symbol $\bar{g}$ $(g \in |G| - \{e\})$, we let $A_G$ contain the same element $b$. For elements $b$ which do not, let the corresponding element of $A_G$ be the expression $\bar{e}\, b$. Thus *every* element of $A_G$ begins with a symbol $\bar{g}$ $(g \in |G|)$, and we can now describe the action of $g' \in |G|$ on $A_G$ as simply taking an element $\bar{g}\, c$ to $\overline{g'g}\, c$. It is trivial to verify that *this* is a homomorphism of $G$ into the group of permutations of $A_G$. This action on $A_G$ now *induces*, in an obvious way, an action on the bijectively related set $A$.

Likewise, an action of $H$ on $A$ can be defined, via an action on the analogously constructed set $A_H$.

Thus we have homomorphisms of both $G$ and $H$ into the permutation group of $A$; this is equivalent to giving a homomorphism of the group $Q$ we are interested in into this group of permutations. Further, given any element (4.6.4) of $Q$, it is easy to see by induction on $n$ that its image in the group of permutations of $A$ sends the "starting point" element $a$ to precisely $x_n \ldots x_1\, a$. Hence two distinct expressions (4.6.4) correspond to elements of $Q$ having distinct actions on $a$, hence these elements of $Q$ are themselves distinct. So not only can every element of $Q$ be written in the form (4.6.4), but distinct expressions (4.6.4) correspond to distinct elements of $Q$, proving the proposition.     □

For a concrete example, again let $G = Z_3 = \langle x \mid x^3 = e \rangle$ and let $H = Z_2 = \langle y \mid y^2 = e \rangle$. Then $A$ will consist of strings such as $a$, $y\, a$, $x\, y\, x^2 a$, etc. (We can drop "$-$" and "$\sim$" here because $|G| - \{e\}$ and $|H| - \{e\}$ use no symbols in common.) The element $x$ of $G = Z_3$ will act on this set

by three-cycles, $b \begin{smallmatrix} \nearrow x\, b \\ \downarrow \\ \searrow x^2 b \end{smallmatrix}$ , one for each string $b$ not beginning with $x$, while

the element $y$ of $H = Z_2$ acts by transposing pairs of symbols $b \rightleftarrows y\, b$, where $b$ does not begin with $y$. If we want to see that say, $y\, x\, y\, x^2$ and $x^2\, y\, x\, y$ have distinct actions on $A$, we simply note that the first sends the

symbol $a$ to the symbol $y\,x\,y\,x^2 a$, while the second takes it to $x^2\,y\,x\,y\,a$.
A picture of the $Q$-set $A$, for this $G$ and $H$, looks like some kind of seaweed:

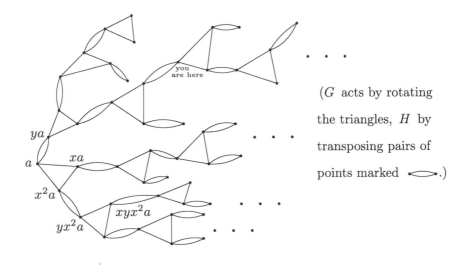

($G$ acts by rotating
the triangles, $H$ by
transposing pairs of
points marked ⬬.)

We recall that the universal group "$P$" considered at the beginning of
this section turned out to be the direct product of $G$ and $H$. Since $Q$ is
characterized by a dual universal property, we shall call it the *coproduct* of
$G$ and $H$.

Because of the similarity of the normal form of this construction to that
of *free groups*, group-theorists have long called it the *free product* of the
given groups. However, the constructions for *sets, commutative rings, abelian
groups, topological spaces*, etc. characterized by this same universal property
show a great diversity of forms, and have been known under different names
in the respective disciplines. The use of the term "coproduct", introduced
by category theory (Chapter 7 below), unifies the terminology, and we shall
follow it. On the other hand, the "$P$" constructions look very similar in all
these cases, and have generally all had the name "direct product", which is
retained (shortened to "product") by category theory.

In both our product and coproduct constructions, the pair of groups $G$ and
$H$ may be replaced by an arbitrary family $(G_i)_{i \in I}$. The universal example of
a group $P$ given with an $I$-tuple of maps $p_i : P \to G_i$ is again the classical
direct product $\prod_I G_i$ with its $I$-tuple of projection maps. The coproduct
$Q = \coprod_I G_i$, generated by the images of a universal family of maps $G_i \to Q$
$(i \in I)$ can be constructed, as above, using strings of nonidentity elements
from a disjoint union of the underlying sets of these groups, such that two
factors from the same group $G_i$ never occur consecutively. The coproduct
symbol $\coprod$ is, of course, the direct product symbol $\prod$ turned upside-down.

**Exercise 4.6:1.** If $X$ is a set, then a coproduct of copies of the infinite
  cyclic group $Z$, indexed by $X$, $\coprod_X Z$, will be a free group on $X$. Show

this by universal properties, and describe the correspondence of normal forms. Can you find any other families of groups whose coproduct is a free group?

**Exercise 4.6:2.** Let us here (following group-theorists' notation) write coproducts of finite families of groups as $Q = G * H$, $Q = F * G * H$, etc. Prove that for any three groups $F$, $G$ and $H$, one has $(F * G) * H \cong F * G * H \cong F * (G * H)$, using (a) universal properties, and (b) normal forms.

**Exercise 4.6:3.** For any two groups $G$ and $H$, show how to define natural isomorphisms $i_{G,H} : G \times H \cong H \times G$, and $j_{G,H} : G * H \cong H * G$. What form do these isomorphisms take when $G = H$? (Describe them on elements.)

    It is sometimes said that "We may identify $G \times H$ with $H \times G$, and $G * H$ with $H * G$, by treating the isomorphisms $i_{G,H}$ and $j_{G,H}$ as the identity, and identifying the corresponding group elements." Is this reasonable when $G = H$?

**Exercise 4.6:4.** Show that in a coproduct group $G * H$, the only elements of finite order are the conjugates of the images of elements of finite order of $G$ and $H$. (First step: Find how to determine, from the normal form of an element of $G * H$, whether it is a conjugate of an element of $G$ or $H$.)

    Can you similarly describe all finite *subgroups* of $G * H$?

There is a fact about the direct product group which one would not at first expect from its universal property: It also has two natural maps *into* it: $f_G : G \to G \times H$ and $f_H : H \to G \times H$, given by $g \mapsto (g, e)$ and $h \mapsto (e, h)$. (Note that there are no analogous maps into a direct product of *sets.*) To examine this phenomenon, we recall that the universal property of $G \times H$ says that to map a group $A$ into $G \times H$ is equivalent to giving a map $A \to G$ and a map $A \to H$. Looking at $f_G$, we see that the two maps it corresponds to are the identity map $\mathrm{id}_G : G \to G$, defined by $\mathrm{id}_G(g) = g$, and the trivial map $e : G \to H$, defined by $e(g) = e$. The map $f_H$ is characterized similarly, with the roles of $G$ and $H$ reversed.

    The group $G \times H$ has, in fact, a second universal property, in terms of this pair of maps. The 3-tuple $(G \times H, f_G, f_H)$ is universal among 3-tuples $(K, a, b)$ such that $K$ is a group, $a : G \to K$ and $b : H \to K$ are homomorphisms, and the images in $K$ of these homomorphisms *centralize* one another:

$$(\forall g \in |G|, \, h \in |H|) \quad a(g)\, b(h) \; = \; b(h)\, a(g),$$

equivalently:

$$[a(G), b(H)] = \{e\}.$$

(The notation $[-, -]$ for commutators of elements and subgroups of a group was defined in the paragraph preceding Exercise 3.4:2.)

If $P = \prod_I G_i$ is a direct product of arbitrarily many groups, one similarly has natural maps $f_i \colon G_i \to P$, but if the index set $I$ is infinite, the images of the $f_i$ will not in general generate $P$, and it follows that $P$ cannot have the same universal property. But one finds that the subgroup $P_0$ of $P$ generated by the images $f_i(G_i)$ (which consists of those elements of $P$ having only finitely many coordinates $\neq e$) is again a universal group with maps of the $G_i$ into it having images that centralize one another.

**Exercise 4.6:5.**

    (i)   Prove the above new universal property of $G \times H$.

    (ii)  Describe the map

$$m \colon G * H \to G \times H$$

which the universal property of $G * H$ associates to the above pair of maps $f_G$, $f_H$, and deduce that this map $m$ is surjective, and that its kernel is the normal subgroup of $G * H$ generated by the commutators $[\bar{g}, \tilde{h}]$ $(g \in |G|, h \in |H|)$.

(iii) Give versions of the above results for products and coproducts of possibly infinite families $(G_i)_{i \in I}$.

One may wonder why commutativity suddenly came up like this, since the original universal property by which we characterized $G \times H$ had nothing to do with it. The following observation throws a little light on this. The set of relations that will be satisfied in $G \times H$ by the images of elements of $G$ and $H$ under the two maps $f_G$ and $f_H$ defined above will be the *intersection* of the sets of relations satisfied by their images in $K$ under $a \colon G \to K$, $b \colon H \to K$, in the two cases

(4.6.7)      $K = G;$   $a = \mathrm{id}_G,$   $b = e,$

(4.6.8)      $K = H;$   $a = e,$   $b = \mathrm{id}_H.$

(Why?)

And what are such relations? Clearly $a(g)b(h) = b(h)a(g)$ holds in each case. The above second universal property of $G \times H$ is equivalent to saying that no relations hold in both cases *except* these relations and their consequences.

A coproduct group $G * H$ similarly has natural maps $u_G \colon G * H \to G$ and $u_H \colon G * H \to H$, constructed from the identity maps of $G$ and $H$ and the trivial maps between them; but $u_G$ and $u_H$ have no unexpected properties that I know of.

**Exercise 4.6:6.** For every group $G$, construct a map $G \to G \times G$ and a map $G * G \to G$ using universal properties, and the identity map of $G$, but *not* using the trivial map of $G$. Describe how these maps behave on elements.

**Exercise 4.6:7.** Suppose $(G_i)_{i \in I}$ is a family of groups, and we wish to consider groups $G$ given with homomorphisms $G_i \to G$ such that the images of *certain* pairs $G_i$, $G_{i'}$ commute, while no condition is imposed on the remaining pairs. To formalize this, let $J \subseteq I \times I$ be a symmetric antireflexive relation on our index set $I$ (antireflexive means $(\forall i \in I) \, (i, \, i) \notin J$); and let $H$ be the universal group with homomorphisms $r_i \colon G_i \to H$ $(i \in I)$ such that for $(i, \, i') \in J$, $[r_i(G_i), \, r_{i'}(G_{i'})] = \{e\}$.

Study the structure of this $H$, and obtain a normal form if possible. You may assume the index set $I$ finite if this helps.

## 4.7. Products and coproducts of abelian groups

Let $A$ and $B$ be abelian groups. Following the model of the preceding section, we may look for abelian groups $P$ and $Q$ having universal pairs of maps:

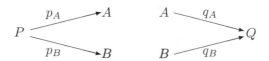

Again abelian groups with both these properties exist—but this time, they turn out to be the same group, namely $A \times B$! (The reader should verify both universal properties.) To look at this another way, if we construct abelian groups $P$ and $Q$ with the universal properties of the direct product and coproduct of $A$ and $B$ respectively, and then form the homomorphism $m \colon P \to Q$ analogous to that of Exercise 4.6:5, this turns out to be an isomorphism.

Note that though $A \times B$ is the universal *abelian* group with homomorphisms of $A$ and $B$ into it, this is not the same as the universal *group* with homomorphisms of $A$ and $B$ into it—that group, $A * B$, constructed in the preceding section, will generally not be abelian when $A$ and $B$ are. Thus, the coproduct of two abelian groups $A$ and $B$ as *abelian groups* is generally not the same as their coproduct as *groups*. Rather, we can see by comparing universal properties that the coproduct as abelian groups is the abelianization of the coproduct as groups: $A \times B = (A * B)^{\mathrm{ab}}$.

Hence, in using the coproduct symbol "$\amalg$", we have to specify what kind of coproduct we are talking about, $\amalg_{\mathrm{gp}} A_i$ or $\amalg_{\mathrm{ab\,gp}} A_i$, unless this is clear from context. On the other hand, a *direct product* of abelian groups as abelian groups is the same as their direct product as groups.

For a not necessarily finite family $(A_i)_{i \in I}$ of abelian groups, the coproduct still *embeds in* the direct product under the map $m$. It can in fact be described as the subgroup of that direct product group consisting of those elements almost all of whose coordinates are $e$. When abelian groups are written additively, this coproduct is generally called the "direct sum" of the groups, and denoted $\bigoplus_I A_i$. In the case of two groups, this is written $A \oplus B$, which thus has the same meaning as $A \times B$.

*Notes on confused terminology:* Some people extend the term "direct sum" to mean "coproduct" in all contexts: groups, rings, etc. Other writers, because of the form that "direct sum" has for finite families of abelian groups, use the phrase "direct sum" as a synonym of "direct product", even in the case of infinite families of groups. The coproduct of an infinite family of abelian groups is sometimes called their "restricted direct product" or "restricted direct sum", the direct product then being called the "complete direct product" or "complete direct sum". In these notes, we shall stick with the terms "product" and "coproduct", as defined above (except that we shall often expand "product" to "direct product", to avoid possible confusion with meanings such as a product of elements under a multiplication).

What is special about abelian groups, that makes finite products and coproducts come out the same; and why only *finite* products and coproducts? One may consider the key property to be the fact that homomorphisms of abelian groups can be *added*; i.e., that given two homomorphisms $f, g \colon A \to B$, the map $f + g \colon A \to B$ defined by $(f + g)(a) = f(a) + g(a)$ is again a homomorphism. (The corresponding statement is not true for nonabelian groups.) Temporarily writing $*_{\mathrm{ab\,gp}}$ for the coproduct of two abelian groups, one finds, in fact, that the map $m \colon G *_{\mathrm{ab\,gp}} H \to G \times H$ referred to in the second paragraph of this section has an inverse, given by the sum

$$q_G\, p_G + q_H\, p_H \colon\ G \times H\ \to\ G *_{\mathrm{ab\,gp}} H,$$

hence it is an isomorphism, allowing us to identify the above two groups. For coproducts of noncommutative groups, the corresponding map is not a group homomorphism, while for coproducts of infinite families of abelian groups, no analog of the above map can be constructed because one cannot in general make sense of an infinite sum of homomorphisms. So it is only when the coproduct is taken in the class of abelian groups, and the given family is finite, that we get this identification.

Part (ii) of the next exercise concerns a subtle but interesting distinction.

**Exercise 4.7:1.**

(i)   Show that for any groups $G$ and $H$ one has $(G * H)^{\mathrm{ab}} \cong (G \times H)^{\mathrm{ab}} \cong G^{\mathrm{ab}} \times H^{\mathrm{ab}}$.

(ii)   Given an infinite family of groups $(G_i)$, is it similarly true that $(\coprod_{\mathrm{gp}} G_i)^{\mathrm{ab}} \cong \coprod_{\mathrm{ab\,gp}} (G_i^{\mathrm{ab}})$ (i.e., $\bigoplus G_i^{\mathrm{ab}}$), and that $(\prod G_i)^{\mathrm{ab}} \cong \prod (G_i^{\mathrm{ab}})$? If one of these isomorphisms is not always true, can you establish any general results on when it holds, and when it fails?

## 4.8.  Right and left universal properties

The universal property of direct products differs in a basic way from the other universal properties we have looked at so far. In all other cases, we constructed an object (e.g., a group) $F$ with specified "additional structure" or conditions (c.g., a map of a given set $X$ into it) such that any instance of a structure of that sort on any object $A$ could be obtained by a unique homomorphism *from* the universal object $F$ *to* the object $A$. A direct product $P = G \times H$ is an object with the opposite type of universal property: all groups with the specified additional structure (a map into $G$, and a map into $H$) are obtained by mapping arbitrary groups $A$ *into* the universal example $P$. Thus, while the free group on a set $X$, the abelianization of a group $G$, the coproduct of two groups $G$ and $H$, etc., can be thought of as "first" or diagrammatically "leftmost" groups with given kinds of structure, the direct product $G \times H$ is the "last" or "rightmost" group with maps into $G$ and $H$. We shall refer to these two types of conditions as "left" and "right" universal properties respectively. (This terminology is based on thinking of arrows as going from left to right, though it happens that in most of the diagrams in sections before §4.6, the arrow from the left universal object to the general object was drawn downward.)

The philosophy of how to construct objects with properties of either kind is in broad outline the same: Figure out as much information as possible about an *arbitrary* object (not assumed universal) with the given sort of "additional structure", and see whether that information can itself be considered as a description of an object. If it can, this object will in general turn out to be universal for the given structure! In the case of "left universal" constructions (free groups, coproducts, etc.), this "information" means answers to the question, "What elements do we know exist, and what equalities must hold among them?" (Cf. Remark 3.2.13). In the right universal case, on the other hand, the corresponding question is, "*Given* an element of our object, what data can one describe *about it* in terms of the additional structure?"

Let us illustrate this with the case of the direct product of groups. Given groups $G$ and $H$, consider any group $P$ with specified homomorphisms $p_G$, $p_H$ into $G$ and $H$ respectively. What data can we find about an element $x$ of $P$ using these maps? Obviously, we can get from $x$ a pair of elements $(g, h) \in |G| \times |H|$, namely

$$g = p_G(x) \in |G|, \quad h = p_H(x) \in |H|.$$

Can we get any more data? We can also obtain elements $p_G(x^2)$, $p_H(x^{-1})$, etc.; but these can be found by group operations from the elements $g = p_G(x)$ and $h = p_H(x)$, so they give no new information about $x$. All right then, let us agree to classify elements of $P$ according to the pairs $(g, h) \in |G| \times |H|$ which they determine.

Now suppose $x \in |P|$ gives the pair $(g, h)$, and $y$ gives the pair $(g', h')$. Can we find from these the pair given by $x\,y \in |P|$? the pair given by $x^{-1}$? Clearly so: these will be $(g\,g', h\,h')$, and $(g^{-1}, h^{-1})$ respectively. And we can likewise write down the pair that $e \in |P|$ yields: $(e_G, e_H)$.

Very well, let us take the "data" by which we have classified elements of our arbitrary $P$, namely the set of pairs $(g, h)$ $(g \in |G|, h \in |H|)$— together with the law of composition we have found for these pairs, namely $(g, h) \cdot (g', h') = (g\,g', h\,h')$, the inverse operation $(g, h) \mapsto (g^{-1}, h^{-1})$, and the neutral element pair $(e_G, e_H)$—and ask whether this data forms a group. It does! And, because of the way this group was constructed, it will have homomorphisms into $G$ and $H$, and we find it is universal for this property. It is, of course, the product group $G \times H$.

Here is a pair of examples we have not yet discussed. Suppose we are given a homomorphism of groups

$$f: G \to H.$$

Now consider

(a) homomorphisms $a: A \to G$, from arbitrary groups $A$ into $G$, whose composites with $f$ are the trivial homomorphism, i.e., which satisfy $f\,a = e$; and

(b) homomorphisms $b: H \to B$, from $H$ into arbitrary groups $B$, whose composites with $f$ are the trivial homomorphism, i.e., which satisfy $bf = e$.

Given a homomorphism of the first sort, one can get further homomorphisms with the same property by composing with homomorphisms $A' \to A$, for arbitrary groups $A'$; so one may look for a pair $(A, a)$ with the *right* universal property that every such pair $(A', a')$ arises from $(A, a)$ via a unique homomorphism $A' \to A$. For (b), one would want a corresponding *left* universal $B$.

To try to find the right-universal $A$, we ask: Given an arbitrary homomorphism $A \to G$ with $f\,a = e$ as in (a), what data can we attach to any element $x \in |A|$? Its image $g = a(x)$, certainly. This must be an element which $f$ carries to the neutral element, since $f\,a = e$; thus the set of possibilities is $\{g \in |G| \mid f(g) = e\}$. We find that this set forms a group (with a map into $G$, namely the inclusion) having the desired universal property. This is the *kernel* of $f$.

We get the left universal example of (b) by familiar methods: Given arbitrary $b\colon H \to B$ with $bf = e$ as in (b), $B$ must contain an image $\bar{h} = b(h)$ of each element $h \in |H|$. The fact that $bf = e$ tells us that the images in $B$ of all elements of $f(G)$ must be the neutral element, and we quickly discover that the universal example is the quotient group $B = H/N$, where $N$ is the normal subgroup of $H$ generated by $f(G)$. This group $H/N$ is called the *cokernel* of the map $f$.

Right universal constructions are not as conspicuous in algebra as left universal constructions. When they occur, they are often fairly elementary and familiar constructions (e.g., the direct product of two groups; the kernel of a homomorphism). However, we shall see less trivial cases in later chapters; some of the exercises below also give interesting examples.

**Exercise 4.8:1.** Let $G$ be a group, and $X$ a set. Show that there exist

(i)  a $G$-set $S$ with a universal map $f\colon |S| \to X$, and

(ii)  a $G$-set $T$ with a universal map $g\colon X \to |T|$,

and describe these $G$-sets. Begin by stating the universal properties explicitly.

(Hint to (i): Given any $G$-set $S$ with a map $f\colon |S| \to X$, an element $s \in |S|$ will determine not only an element $x = f(s) \in X$, but for every $g \in |G|$ an element $x_g = f(g\,s) \in X$. From the *family* of elements, $(x_g)_{g \in |G|}$ determined by an $s \in S$, can one describe the family determined by $hs$ for any $h \in |G|$?)

One can carry the idea of the above exercise further in several directions:

(a) Given a group homomorphism $\varphi\colon G_1 \to G_2$, note that from any $G_2$-set $S$ one can get a $G_1$-set $S_\varphi$, by taking the same underlying set, and defining for $g \in |G_1|$, $s \in |S|$

$$g\,s = \varphi(g)\,s.$$

Now given a $G_1$-set $X$, one can look for a $G_2$-set $S$ with a universal homomorphism of $G_1$-sets $S_\varphi \to X$, or for a $G_2$-set $T$ with a universal homomorphism of $G_1$-sets $X \to T_\varphi$. The above exercise corresponds to the cases where $G_1 = \{e\}$, since an $\{e\}$-set is essentially a set with no additional structure. You should verify that for $G_1 = \{e\}$, the universal questions just mentioned reduce to those of that exercise.

(b) Instead of looking at *sets* $S$ on which a group $G$ acts by *permutations*, one can consider abelian groups or vector spaces on which $G$ acts by *automorphisms*. Such structures are called *linear representations* of $G$. In this case, the universal constructions analogous to those of (a) above are still possible, and they give two concepts of "induced representations" of a group, important in modern group theory.

(c) The preceding point introduced extra structure on the *sets* on which our groups act. One can also consider the situation where one's groups $G$ have additional structure, say topological or measure-theoretic, and restrict attention to continuous, measurable, etc., $G$-actions on appropriately structured spaces $S$. The versions of "induced representation" that one then obtains are at the heart of the modern representation theory of topological groups.

**Exercise 4.8:2.** Let $G$ be a group. As discussed in the last two sentences of point (a) above, the ideas described there, applied to the unique homomorphism $\{e\} \to G$, lead to the two universal constructions of Exercise 4.8:1. Apply the same ideas to the unique homomorphism $G \to \{e\}$ (again combining them with the observation that an $\{e\}$-set is essentially the same as a set) and describe the resulting constructions explicitly.

**Exercise 4.8:3.** Formulate right universal properties analogous to the left universal property defining free groups and the abelianization of a group, and show that no constructions exist having these properties. What goes wrong when we attempt to apply the general approach of this section?

**Exercise 4.8:4.** If $X$ is a set and $S$ a subset of $X$, then given any set map $f : Y \to X$, one gets a subset of $Y$, $T = f^{-1}(S)$. Does there exist a *universal* pair $(X, S)$, such that for any set $Y$, every subset $T \subseteq Y$ is induced in this way via a unique set map $f : Y \to X$?

**Exercise 4.8:5.** Let $A$, $B$ be fixed sets. Suppose $X$ is another set, and $f : A \times X \to B$ is a set map. Then for any set $Y$, and map $m : Y \to X$, a set map $A \times Y \to B$ is induced. (How?) Does there exist, for each $A$ and $B$, a universal set $X$ and map $f$ as above, i.e., an $X$ and an $f$ such that for any $Y$, all maps $A \times Y \to B$ are induced by unique maps $Y \to X$?

**Exercise 4.8:6.** Let $R$ be a ring with 1. (Commutative if you like. If you consider general $R$, then for "module" understand "left module" below.) Before attempting each of the following questions, formulate precisely the universal property desired.

(i)   Given a set $X$, does there exist an $R$-module $M$ with a universal set map $|M| \to X$?

(ii)   If $M$ is an $R$-module, let $M_{\mathrm{add}}$ denote the underlying additive group of $M$. Given an abelian group $A$, does there exist an $R$-module $M$ with a universal homomorphism of abelian groups $M_{\mathrm{add}} \to A$?

(iii) and (iv) What about the left universal analogs of the above right universal questions?

## 4.9. Tensor products

Let $A$, $B$ and $C$ be abelian groups, which we shall write additively. Then by a *bilinear map* $\beta \colon (A,\, B) \to C$ we shall mean a set map $\beta \colon |A| \times |B| \to |C|$ such that

(i)    for each $a \in |A|$, the map $\beta(a, -) \colon |B| \to |C|$ (that is, the map taking each element $b \in |B|$ to $\beta(a,\, b) \in |C|$) is a *linear* map (homomorphism of abelian groups) from $B$ to $C$, and

(ii)    for each $b \in |B|$, the map $\beta(-,\, b) \colon |A| \to |C|$ is a linear map from $A$ to $C$.

This is usually called a bilinear map "from $A \times B$ to $C$." (I usually call it that myself.) However, that terminology misleads some students into thinking that it has something to do with the *group* $A \times B$. In fact, although the definition of bilinear map involves the group structures of $A$ and $B$, and involves the set $|A| \times |B|$, it has nothing to do with the structure of direct product group that one can put on this set. This is illustrated by:

**Exercise 4.9:1.** Show that for any abelian groups $A$, $B$, $C$, the only map $|A| \times |B| \to |C|$ which is both a linear map $A \times B \to C$, and a bilinear map $(A,\, B) \to C$ is the zero map.

As examples to keep in mind, take any ring $R = (|R|,\, +,\, \cdot,\, -,\, 0,\, 1)$, and let $R_{\mathrm{add}}$ denote the additive group $(|R|,\, +,\, -,\, 0)$. Then the maps $(x,\, y) \mapsto x + y$ and $(x,\, y) \mapsto x - y$ are *group homomorphisms* $R_{\mathrm{add}} \times R_{\mathrm{add}} \to R_{\mathrm{add}}$, but not bilinear maps; while the multiplication map $(x,\, y) \mapsto x \cdot y$ is a *bilinear map* $(R_{\mathrm{add}},\, R_{\mathrm{add}}) \to R_{\mathrm{add}}$, but not a group homomorphism $R_{\mathrm{add}} \times R_{\mathrm{add}} \to R_{\mathrm{add}}$.

I am speaking about abelian groups to keep the widest possible audience. However, abelian groups can be regarded as $\mathbb{Z}$-modules, and everything I have said and will say about bilinear maps of abelian groups applies, more generally, to bilinear maps of modules over an arbitrary commutative ring, and in particular, of vector spaces over a field, with the adjustment that "linear map" in (i) and (ii) above should be understood to mean module homomorphism. (There are also extensions of these concepts to left modules, right modules, and bimodules over noncommutative rings, which we will look at with the help of a more sophisticated perspective in §10.8; but we won't worry about these till then.)

Given two abelian groups $A$ and $B$, let us construct an abelian group $A \otimes B$ (called the *tensor product* of $A$ and $B$) as follows: We present it using a set of generators which we write $a \otimes b$, one for each $a \in |A|$, $b \in |B|$, and defining relations which are precisely the conditions required to make the map $(a,\, b) \mapsto a \otimes b$ bilinear; namely

$$\begin{aligned}
(a + a') \otimes b &= a \otimes b + a' \otimes b, \\
a \otimes (b + b') &= a \otimes b + a \otimes b'.
\end{aligned} \qquad (a,\, a' \in |A|,\ b,\, b' \in |B|).$$

(If we are working with $R$-modules, we also need the $R$-module relations

$$(r\,a)\otimes b \;=\; r(a\otimes b) \;=\; a\otimes(r\,b) \qquad\qquad (a\in |A|,\, b\in |B|,\, r\in |R|).$$

To indicate that one is referring to the tensor product as $R$-modules rather than the tensor product as abelian groups, one often writes this $A\otimes_R B$.)

By construction, $A\otimes B$ will be an abelian group with a bilinear map $\otimes\colon (A,\,B)\to A\otimes B$; and the universal property arising from its presentation translates to say that the map $\otimes$ will be universal among bilinear maps on $(A,\,B)$.

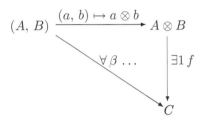

We can get a simpler presentation of this group if we are given presentations of $A$ and $B$. To describe this, let us write our presentation of $A$ as a representation $A = F(X)/\langle S\rangle$, where $F(X)$ is the free abelian group on the set $X$ of generators, and $\langle S\rangle$ is the subgroup of $F(X)$ generated by the family $S$ of relators (elements that are required to go to 0). If $A$ is so presented, and likewise $B$ is written as $F(Y)/\langle T\rangle$, then it is not hard to show (and you may do so as the next exercise) that

$$(4.9.1) \qquad\qquad A\otimes B \;\cong\; F(X)\otimes F(Y)/\langle S\otimes Y \cup X\otimes T\rangle,$$

where $S\otimes Y$ means $\{s\otimes y \mid s\in S,\, y\in Y\}\subseteq |F(X)\otimes F(Y)|$, and $X\otimes T$ is defined analogously. One finds that $F(X)\otimes F(Y)$ is a free abelian group on its subset $X\otimes Y$ (more precisely: it is a free abelian group on $X\times Y$ via the mapping $(x,\,y)\mapsto u(x)\otimes v(y)$, where $u\colon X\to |F(X)|$ and $v\colon Y\to |F(Y)|$ are the universal maps associated with the free groups $F(X)$ and $F(Y)$). Hence (4.9.1) is equivalent to a presentation of $A\otimes B$ by the generating set $X\times Y$ and a certain set of relations.

In the following exercises, unless the contrary is stated, you can, if you wish, substitute "$R$-module" for "abelian group", and prove the results for this more general case.

**Exercise 4.9:2.** Prove (4.9.1), and the assertion that $F(X)\otimes F(Y)$ is free abelian on $X\times Y$. Can the "denominator" of (4.9.1) be replaced simply by $\langle S\otimes T\rangle$?

**Exercise 4.9:3.**

(i)   Given abelian groups $A$ and $C$, is there a universal pair $(B, \beta)$, of an abelian group $B$ and a bilinear map $\beta \colon (A, B) \to C$?

(ii)   Given an abelian group $C$, is there a universal 3-tuple $(A, B, \beta)$, such that $A$ and $B$ are abelian groups and $\beta$ a bilinear map $(A, B) \to C$?

Before answering each part, say what the universal property would be and whether it would be a right or left universal property. Try the approach suggested in the preceding section for finding such objects.

Why have we defined bilinear maps only for *abelian* groups? This is answered by

**Exercise 4.9:4.** Let $F$, $G$ and $H$ be not necessarily abelian groups (so this exercise has no generalization to $R$-modules), and suppose $\beta \colon |F| \times |G| \to |H|$ is a map such that

(4.9.2)
$$(\forall\, f \in |F|)\ \text{the map } g \mapsto \beta(f, g) \text{ is a group homomorphism } G \to H,$$
$$(\forall\, g \in |G|)\ \text{the map } f \mapsto \beta(f, g) \text{ is a group homomorphism } F \to H.$$

(i)   Show that the subgroup $H_0$ of $H$ generated by the image of $\beta$ is abelian.

(ii)   Deduce that the map $\beta$ has a natural factorization

$$|F| \times |G| \;\longrightarrow\; |F^{\mathrm{ab}}| \times |G^{\mathrm{ab}}| \xrightarrow{\ \beta'\ } |H_0| \;\longhookrightarrow\; |H|,$$

where $\beta'$ is bilinear. Thus, the study of maps satisfying (4.9.2) is reduced to the study of bilinear maps of *abelian* groups. This makes it easy to do

(iii)   For general groups $F$ and $G$, deduce a description of the group $H$ with a universal map $\beta$ satisfying (4.9.2), in terms of tensor products of abelian groups.

(iv)   Deduce from (iii) or show directly that if from the definition of a ring one drops the assumption the addition is commutative, this will in fact follow from the other assumptions. (Recall that we assume throughout that rings have 1.)

(Nonetheless, there are sometimes ways of generalizing a concept other than the obvious ones, and some group-theorists have introduced a version of the concept of bilinear map which does not collapse in the manner described above in the noncommutative case; see [64] and papers referred to there.)

Although the image of every group homomorphism is a subgroup of the codomain group, this is not true of images of bilinear maps:

**Exercise 4.9:5.**

(i)   Let $U$, $V$, $W$ be finite-dimensional vector spaces over a field, and consider composition of linear maps as a set map $\mathrm{Hom}(U,\,V) \times \mathrm{Hom}(V,\,W) \to \mathrm{Hom}(U,\,W)$. Note that if we regard these hom-sets as additive groups, this map is bilinear. Suppose $V$ is one-dimensional; describe the range of this composition map. Is it a subgroup of $\mathrm{Hom}(U,\,W)$?

(ii)   If $A$ and $B$ are abelian groups, does every element of $A \otimes B$ have the form $a \otimes b$ for some $a \in |A|$, $b \in |B|$? (Prove your answer, of course.)

Another important property of tensor products is noted in

**Exercise 4.9:6.** If $A$, $B$ and $C$ are abelian groups, show that there is a natural isomorphism $\mathrm{Hom}(A \otimes B,\,C) \cong \mathrm{Hom}(A,\,\mathrm{Hom}(B,\,C))$.
State an analogous result holding for sets $A$, $B$, $C$ and set maps.

A class of tensor products that is easy to describe is noted in

**Exercise 4.9:7.** Show that for any abelian group $A$ and any nonnegative integer $n$, one has $A \otimes Z_n \cong A/nA$, where $Z_n$ denotes the cyclic group of order $n$.

An interesting problem is

**Exercise 4.9:8.** Investigate conditions on abelian groups (or $R$-modules) $A$ and $B$ under which $A \otimes B = \{0\}$.

Although I pointed out earlier that the condition that a set map $\beta\colon |A| \times |B| \to |C|$ be a bilinear map $(A,\,B) \to C$ is *not* defined in terms of the group structure of the direct product group, $A \times B$, there are certain relations between these concepts:

**Exercise 4.9:9.**

(i)   Show that if $A$ and $B$ are abelian groups, and $\beta\colon (A,\,B) \to C$ a bilinear map, then $\beta$, regarded as a map on underlying sets of groups, $|A \times B| \to |C|$, satisfies nontrivial identities. That is, show that for some $m$ and $n$ one can find a derived $n$-ary operation $s$, and $n$ derived $m$-ary operations $t_1, \ldots, t_n$, for abelian groups, such that

$$s(\beta(t_1(x_1, \ldots, x_m)), \ldots, \beta(t_n(x_1, \ldots, x_m))) = 0$$

holds for all $x_1, \ldots, x_n \in |A \times B|$, with the $t$'s evaluated using the group structure of $A \times B$; but such that the corresponding equation does not hold for arbitrary maps $\beta\colon |D| \to |C|$ of underlying sets of abelian groups.

(ii)   On the other hand, show that the bilinearity of $\beta$ cannot be characterized in terms of such identities; in other words, that there exist maps $\beta\colon |A| \times |B| \to |C|$ which are not bilinear maps $(A,\,B) \to C$, but which satisfy all identities that are satisfied by bilinear maps.

(iii) Can you find a list of identities which imply all identities satisfied by bilinear maps $\beta$, in the sense described in (i)?

In subsequent sections, we shall occasionally refer again to bilinear maps. In those situations, we may use either the notation "$(A,\ B) \to C$" introduced here, or the more standard notation "$A \times B \to C$". (Of course, if all we have to say is something like "this map $|A| \times |B| \to |C|$ is bilinear", we will not need either notation.)

## 4.10. Monoids

So far, we have been moving within the realm of groups. It is time to broaden our horizons. We begin with semigroups and monoids, objects which are very much like groups in some ways, and quite different in others.

We recall that a *semigroup* means an ordered pair $S = (|S|,\ \cdot)$ such that $|S|$ is a set and $\cdot$ a map $|S| \times |S| \to |S|$ satisfying the associative identity, while a *monoid* is a 3-tuple $S = (|S|,\ \cdot,\ e)$ where $|S|$ and $\cdot$ are as above, and the third component, $e$, is a *neutral element* for the operation $\cdot$. As with groups, the multiplication of semigroups and monoids is most often written without the "$\cdot$" when there is no need to be explicit. A *homomorphism* of semigroups $f \colon S \to T$ means a set map $f \colon |S| \to |T|$ which respects "$\cdot$"; a *homomorphism of monoids* is required to respect neutral elements as well: $f(e_S) = e_T$.

(I have long considered the use of two unrelated terms, "semigroup" and "monoid", for these very closely related types of objects to be an unnecessary proliferation of terminology. In most areas of mathematics, distinctions between related concepts are made by modifying phrases, e.g., "abelian group" versus "not necessarily abelian group", "ring with 1" versus "ring without 1", "manifold with boundary" versus "manifold without boundary". The author of a paper considering one of these concepts will generally begin by setting conventions, such as "In this note, unless the contrary is stated, rings will have unit element, and ring homomorphisms will be understood to respect this element". In papers of mine where monoids came up, I followed the same principle for a long time, calling them "semigroups with neutral element" or, after saying what this would mean, simply "semigroups". I did the same in these notes through 1995. However, it seems the term "monoid" is here to stay, and I now follow standard usage, given above.)

The concept of monoid seems somewhat more basic than that of semigroup. If $X$ is any set, then the set of all maps $X \to X$ has a natural monoid structure, with functional composition as the multiplication and the identity map as the neutral element, and more generally, this is true of the set of endomorphisms of any mathematical object. Sets whose natural structure is one of semigroup and not of monoid tend to arise as subsidiary constructions, when one considers those elements of a naturally occurring monoid that satisfy some restriction which excludes the neutral element; e.g., the set of maps $X \to X$ having finite range, or the set of even integers under multiplication. However, "semigroup" is the older of the two terms, so the study of semigroups and monoids is called "semigroup theory".

If $(|S|, \cdot, e)$ is a monoid, one can, of course, look at the semigroup $(|S|, \cdot)$, while if $(|S|, \cdot)$ is a semigroup, one can "adjoin a neutral element" and get a monoid $(|S| \sqcup \{e\}, \cdot, e)$. Thus, most results on monoids yield results on semigroups, and vice versa. To avoid repetitiveness, I will focus here on monoids, and mention semigroups only when there is a contrast to be made. Most of the observations we will make about monoids have obvious analogs for semigroups, the exceptions being those relating to invertible elements.

The concept of a free monoid $(F, u)$ on a set $X$ is defined using the expected universal property (diagram below).

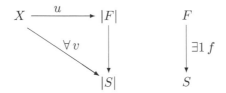

Free monoids on all sets exist, by the general arguments of §3.2 and §3.3. One also has a normal form in the free monoid on $X$, analogous to that of §3.4 but without any negative exponents. That is, every element can be written uniquely as a product,

$$x_n \ \ldots \ x_1,$$

where $x_1, \ldots, x_n \in |X|$, and $n \geq 0$ (the product of $0$ factors being understood to mean the neutral element). Multiplication is performed by juxtaposing such products. "Van der Waerden's trick" is not needed to establish this normal form, since there is no cancellation to complicate a direct verification of associativity. From this normal form and that of free groups, we see that the free monoid on $X$ is in fact isomorphic to the submonoid generated by $X$ within the free group on $X$.

If $X$ is a set, and $R$ a set of pairs of *monoid terms* in the elements of $X$, there will likewise exist a monoid determined by "generators $X$ and relations $R$", i.e., a monoid $S$ with a map $u \colon X \to |S|$ such that for each of the pairs $(s, t) \in R$, one has $s_u = t_u$ in $S$, and which is universal for this property. As in the group case, this $S$ can be obtained by a direct construction, using terms modulo identifications deducible from the monoid laws and the set of relations $R$, or as a submonoid of a large direct product, or by taking the free monoid $F$ on the set $X$, and imposing the given relations.

But how does one "impose relations" on a monoid? In a group, we noted that any relation $x = y$ was equivalent to $x\,y^{-1} = e$, so to study relations satisfied in a homomorphic image of a given group $G$, it sufficed to study the set of elements of $G$ that went to $e$; hence, the construction of imposing relations reduced to that of dividing out by an appropriate normal subgroup. But for monoids, the question of which elements fall together does not come down to that of which elements go to $e$. For instance, let $S$ be the free monoid on $\{x, y\}$, and map $S$ homomorphically to the free monoid on $\{x\}$ by sending both $x$ and $y$ to $x$. Note that any product of $m$ $x$'s and $n$ $y$'s goes to $x^{m+n}$ under this map. So though the only element going to $e$ is $e$ itself, the homomorphism is far from one-to-one.

So to study relations satisfied in the image of a monoid homomorphism $f\colon S \to T$, one should look at the whole set

$$K_f \;=\; \{(s, t) \mid f(s) = f(t)\} \;\subseteq\; |S| \times |S|.$$

We note the following properties of $K_f$ :

(4.10.1)    $(\forall\, s \in S)\ (s, s) \in K_f.$

(4.10.2)    $(\forall\, s, t \in S)\ (s, t) \in K_f \implies (t, s) \in K_f.$

(4.10.3)    $(\forall\, s, t, u \in S)\ (s, t) \in K_f,\ (t, u) \in K_f \implies (s, u) \in K_f.$

(4.10.4)    $(\forall\, s, t, s', t' \in S)\ (s, t) \in K_f,\ (s', t') \in K_f \implies (ss', tt') \in K_f.$

Here (4.10.1)–(4.10.3) say that $K_f$ is an equivalence relation, and (4.10.4) says that it "respects" the monoid operation.

I claim, conversely, that if $S$ is a monoid, and $K \subseteq |S| \times |S|$ is any subset satisfying (4.10.1)–(4.10.4), then there exists a homomorphism $f$ of $S$ into a monoid $T$ such that $K_f = K$. Indeed, since $K$ is an equivalence relation on $|S|$, we may define $|T| = |S|/K$ and let $f\colon |S| \to |T|$ be the map taking each $x \in |S|$ to its equivalence class $[x] \in |T|$. It is easy to see from (4.10.4) that the formula $[s] \cdot [t] = [s\,t]$ defines an operation on $|T|$, and to verify that this makes $T = (|T|, \cdot, [e])$ a monoid such that $f$ is a homomorphism, and $K_f = K$.

### Exercise 4.10:1.

(i)    Compare this construction with that of §3.2. Why did we need the conditions (3.2.1)–(3.2.3) in that construction, but not the corresponding conditions here?

(ii)    Given two monoid homomorphisms $f\colon S \to T$ and $f'\colon S \to T'$, show that there exists an isomorphism between their images making the diagram below commute if and only if $K_f = K_{f'}$.

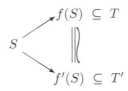

**Definition 4.10.5.** *For any monoid $S$, a binary relation $K$ on $|S|$ satisfying* (4.10.1)–(4.10.4) *above is called a* congruence *on $S$. The equivalence class of an element is called its* congruence class *under $K$; the monoid $T$ constructed above is called the* quotient *or* factor monoid *of $S$ by $K$, written $S/K$.*

Given a set $R$ of pairs of elements of a monoid $S$, it is clear that one can construct the *least* congruence $K$ containing $R$ by closing $R$ under four operations corresponding to conditions (4.10.1)–(4.10.4). The quotient $S/K$ has the correct universal property to be called the monoid obtained by imposing the relations $R$ on the monoid $S$. We shall sometimes denote this $S/R$, or $S/(s = t \mid (s, t) \in R)$, or, if the elements of $R$ are listed as $(s_i, t_i)$ $(i \in I)$, as $S/(s_i = t_i \mid i \in I)$.

Returning to the point that led us to this discussion of congruences, by imposing relations on a *free* monoid, we can get a monoid presented by any family of generators $X$ and family of relations $R$. Like the corresponding construction for groups, this is written $\langle X \mid R \rangle$. When there is danger of ambiguity, the group- and monoid-constructions can be distinguished as $\langle X \mid R \rangle_{\mathrm{gp}}$ and $\langle X \mid R \rangle_{\mathrm{md}}$.

**Exercise 4.10:2.** Given congruences $K$ and $K'$ on a monoid $S$, will there exist a least congruence containing both $K$ and $K'$? A greatest congruence contained in both? Will set-theoretic union and intersection give such congruences? If not, what useful descriptions can you find for them? Is there a least congruence on $S$? A greatest?

If $K$ is a congruence on $S$, characterize congruences on $T = S/K$ in terms of congruences on $S$.

**Exercise 4.10:3.** If $S$ is a monoid and $X$ a subset of $|S| \times |S|$, will there be a largest congruence contained in $X$? If not, will this become true under additional assumptions, such as that $X$ is an equivalence relation on $|S|$, or is the underlying set of a submonoid of $S \times S$?

Some general observations: One can speak similarly of congruences on *groups, rings, lattices,* etc. They are defined in each case by conditions (4.10.1)–(4.10.3), plus a family of conditions analogous to (4.10.4), one for each operation of positive arity on our algebras. The special fact that allowed us to give a simpler treatment in the case of groups can now be reformulated, "A congruence $K$ on a group $G$ is uniquely determined by the congruence class of the neutral element $e \in |G|$, which can be any *normal subgroup* $N$ of $G$. The congruence classes of $K$ are then the cosets of $N$ in $G$." Hence in group theory, rather than considering congruences, one almost always talks about normal subgroups.

Since a ring $R$ has an additive group structure, a congruence on a ring will in particular be a congruence on its additive group, and hence will be determined by the congruence class $J$ of the additive neutral element 0. The possibilities for $J$ turn out to be precisely the *ideals* of $R$, so in ring theory, one works with ideals rather than congruences. (However, historically, the congruence relation "$a \equiv b \pmod{n}$" on the ring $\mathbb{Z}$ of integers was talked about before one had the concept of the ideal $n\mathbb{Z}$. Ring theorists still sometimes write $a \equiv b \pmod{J}$ rather than $a - b \in J$.)

On the other hand, on objects such as monoids and lattices, congruences cannot be reduced to anything simpler, and are studied as such.

As usual, questions of the *structure* of monoids presented by generators and relations must be tackled case by case. For example:

**Exercise 4.10:4.** Find a normal form or other description for the monoid presented by two generators $a$ and $b$ and the one relation $ab = e$.

(Note that in the above and the next few exercises, letters $a$ through $d$ denote general monoid elements, but $e$ is always the neutral element. If you prefer to write 1 instead of $e$ in your solutions, feel free to do so, but point out that you are using that notation.)

**Exercise 4.10:5.**
(i) Same problem for generators $a, b, c, d$ and relations

$$a b = a c = d c = e.$$

(ii) Same problem for generators $a, b, c, d$ and relations

$$a b = a c = c d = e.$$

**Exercise 4.10:6.** Same problem for generators $a, b, c$ and relations

$$a b = a c, \quad b a = b c, \quad c a = c b.$$

**Exercise 4.10:7.** Same problem for generators $a, b$ and the relation $a b = b^2 a$.

**Exercise 4.10:8.**
(i) Find a normal form for the monoid presented by two generators $a, b$, and the one relation $abba = baab$. (This is hard, but can be done.)

(ii) (Victor Maltcev) Does there exist a normal form or other useful description for the monoid presented by generators $a, b$ and the relation $abbab = baabb$? (I do not know the answer.)

One may define the *direct product* and the *coproduct* of two (or an arbitrary family of) monoids, by the same universal properties as for groups,

These turn out to have the same descriptions as for groups: The direct product of an $I$-tuple of monoids consists of all $I$-tuples such that for each $i \in I$, the $i$-th position is occupied by a member of the $i$-th monoid, with operations defined componentwise; the coproduct consists of formal products of strings of elements, other than the neutral element, taken from the given monoids, such that no two successive factors come from the same monoid. Van der Waerden's method *is* used in establishing this normal form, since multiplication of two such products can involve "cancellation" if any of the given monoids have elements satisfying $ab = e$.

On monoids, as on groups, one has the construction of *abelianization*, gotten by imposing the relations $ab = ba$ for all $a, b \in |S|$.

One may also define the *kernel* and *cokernel* of a monoid homomorphism $f \colon S \to S'$ as for groups:

$$(4.10.6) \qquad \begin{aligned} \mathrm{Ker}\, f \;&=\; \text{submonoid of } S \\ &\quad \text{with underlying set } \{s \in |S| \mid f(s) = e\}, \end{aligned}$$

$$(4.10.7) \qquad \mathrm{Cok}\, f \;=\; S'/(f(s) = e \mid s \in |S|).$$

But we have seen that the structure of the image of a monoid homomorphism $f$ is not determined by the kernel of $f$, and it follows that not every homomorphic image $T$ of a monoid $S'$ can be written as the cokernel (4.10.7) of a homomorphism of another monoid $S$ into $S'$ (e.g., the image of $S'$ under a non-one-to-one homomorphism with trivial kernel cannot). Hence these concepts of kernel and cokernel are not as important in the theory of monoids as in group theory.

We have noted that for $f$ a homomorphism of monoids, a better analog of the group-theoretic concept of kernel is the congruence

$$(4.10.8) \qquad K_f \;=\; \{(s, t) \mid f(s) = f(t)\} \subseteq |S| \times |S|.$$

Note that $K_f$ is the underlying set of a submonoid of $S \times S$, which we may call $\mathrm{Cong}\, f$. Likewise, since to impose relations on a monoid we specify, not that some elements should go to $e$, but that some pairs of elements should fall together, it seems reasonable that a good generalization of the cokernel concept should be, not an image $q(S)$ universal for the condition $q f = e$, where $f$ is a given monoid homomorphism into $S$, but an image $q(S)$ universal for the condition $q f = q g$, for some *pair* of homomorphisms

(4.10.9)                              $f, g : T \to S.$

Given $f$ and $g$ as above, $q(S)$ may be constructed as the quotient of the monoid $S$ by the congruence generated by all pairs $(f(t), g(t))$ $(t \in |T|)$. Postponing till the end of this paragraph the question of what $q(S)$ will be called, let us note that there is a dual construction: Given $f, g$ as in (4.10.9), one can get a universal map $p$ into $T$ such that $f p = g p$. This will be given by in inclusion in $T$ of the submonoid whose underlying set is $\{t \mid f(t) = g(t)\}$, called the *equalizer* of $f$ and $g$. Dually, the $q(S)$ constructed above is called the *coequalizer* of $f$ and $g$.

**Exercise 4.10:9.** Let $f : S \to T$ be a monoid homomorphism.

(i)   Note that there is a natural pair of monoid homomorphisms from Cong $f$ to $S$. Characterize the 3-tuple formed by Cong $f$ and these two maps by a universal property.

(ii)   What can be said of the equalizer and coequalizer of the above pair of maps?

(iii)   Can you construct from $f$ a monoid CoCong $f$ with a pair of maps into it, having a dual universal property? If so, again, look at the equalizer and coequalizer of this pair.

**Exercise 4.10:10.** The definition of equalizer can be applied to groups as well as monoids. If $G$ is a group, investigate which subgroups of $G$ can occur as equalizers of pairs of homomorphisms on $G$.

## 4.11. Groups to monoids and back again

If $S$ is a monoid, we can get a group $S^{\mathrm{gp}}$ from $S$ by "adjoining inverses" to all its elements in a universal manner. Thus, $S^{\mathrm{gp}}$ is a *group* $G$ having a map $q : |S| \to |G|$ which respects products and neutral elements, and is universal among all such maps from $S$ to groups.

But what kind of a map, exactly, is $q$? Since $S = (|S|, \cdot, e)$ is a monoid while $S^{\mathrm{gp}} = G = (|G|, \cdot, {}^{-1}, e)$ is a group, we cannot call it a group homomorphism or a monoid homomorphism from $S$ to $G$. But it is more than just a set map, since it respects $\cdot$ and $e$. The answer is that $q$ is a monoid homomorphism from $S$ to the *monoid* $(|G|, \cdot, e)$ (i.e., $(|G|, \mu_G, e_G)$). So for an arbitrary group $H = (|H|, \mu_H, \iota_H, e_H)$, let us write $H_{\mathrm{md}}$ for $(|H|, \mu_H, e_H)$, that is, "$H$ considered as a monoid". We can now state the universal property of $S^{\text{gp}}$ and $q$ neatly: $S^{\text{gp}}$ is a group $G$, and $q$ is a monoid homomorphism from $S$ to $G_{\text{md}}$, such that for any group $H$ and any monoid homomorphism $a\colon S \to H_{\text{md}}$, there exists a unique group homomorphism $f\colon G \to H$ such that $a = fq\colon S \to H_{\text{md}}$.

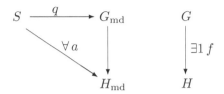

We shall call $S^{\text{gp}}$ the *universal enveloping group* of the monoid $S$. It may be presented as a *group* by taking a generator for each element of $|S|$, and taking for defining relations the full multiplication table of $S$. More generally, if we are given some presentation of $S$ by generators and relations as a monoid, $G$ will be the group presented by the same generators and relations.

**Exercise 4.11:1.** Show that a monoid $S$ is "embeddable in a group" (meaning embeddable in the monoid $H_{\text{md}}$ for some group $H$) if and only if the universal map $q\colon |S| \to |S^{\text{gp}}|$ is one-to-one.

**Exercise 4.11:2.** Describe the universal enveloping groups of the monoids of Exercises 4.10:4–4.10:7, and also of the monoid presented by generators $a$, $b$, $c$ and the one relation $ab = ac$.

The last part of the above exercise reveals one necessary condition for the one-one-ness of the exercise preceding it to hold: The monoid $S$ must have the *cancellation* property $xy = xy' \implies y = y'$. An interesting way of obtaining a full set of necessary and sufficient conditions for the universal map of a given monoid into a group to be one-to-one was found by A. I. Mal'cev ([113, 114]; also described in [8, §VII.3]).

**Exercise 4.11:3.** Let $G$ be a group and $S$ a *submonoid* of $G_{\text{md}}$, which generates $G$ as a group. Observe that the inclusion of $S$ in $G_{\text{md}}$ induces a homomorphism $S^{\text{gp}} \to G$. Will this in general be one-to-one? Onto?

If you have done Exercise 4.3:3, consider the case where $G$ is the group of that exercise, and $S$ the submonoid generated by $a$ and $ba$. Describe the structures of $S$ and of $S^{\text{gp}}$.

Suppose $S$ is an *abelian* monoid. In this situation, important applications of the universal enveloping group construction have been made by A. Grothendieck; the group $S^{\text{gp}}$ for $S$ an abelian monoid is therefore often called "the Grothendieck group $K(S)$". This group is also abelian, and has a simple description: Using additive notation, and writing $\bar{a}$ for $q(a)$, one finds that every element of $K(S)$ can be written $\bar{a} - \bar{b}$ $(a, b \in |S|)$, and that one has equality $\bar{a} - \bar{b} = \bar{a'} - \bar{b'}$ between two such elements if and only if there exists $c \in |S|$ such that $a + b' + c = a' + b + c$ [34, p. 40]. (If you have seen the construction of the *localization* $R\,S^{-1}$ of a commutative ring at a multiplicative subset $S$, you will see that these constructions are closely related. In particular, the multiplicative group of nonzero elements of the field of fractions $F$ of a commutative integral domain $R$ is the Grothendieck group of the multiplicative monoid of nonzero elements of $R$.) The application of this construction to the abelian monoid of isomorphism classes of finite-dimensional vector bundles on a topological space $X$, made a monoid under the operation corresponding to the construction "$\oplus$" on vector bundles, is the starting point of $K$-*theory*. But perhaps this idea has been pushed too much—it is annoyingly predictable that when I mention to a fellow algebraist a monoid of isomorphism classes of modules under "$\oplus$", he or she will say, "Oh, and then you take its Grothendieck group," when in fact I wanted to talk about the monoid itself.

Given a monoid $S$, there is also a *right-universal* way of obtaining a group: The set of *invertible elements* ("units") of $S$ can be made a group $U(S)$ in an obvious way, and the inclusion $i\colon U(S) \to S$ is universal among "homomorphisms of groups into $S$", in the sense indicated in the diagram below.

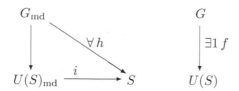

**Exercise 4.11:4.** Let $S$ be the monoid defined by generators $x$, $y$, $z$ and relations $x\,y\,z = e$, $z\,x\,y = e$. Investigate the structures of $S$ and its abelianization $S^{\mathrm{ab}}$. Describe the groups $U(S)$, $U(S)^{\mathrm{ab}}$, and $U(S^{\mathrm{ab}})$.

The two constructions that relate *semigroups* to monoids mentioned near the beginning of the preceding section are related to each other in a way paralleling the relation between $(\ )^{\mathrm{gp}}$ and $(\ )_{\mathrm{md}}$:

**Exercise 4.11:5.**

(i)   If $S = (|S|, \cdot)$ is a semigroup, describe how to extend the multiplication "$\cdot$" to $|S| \sqcup \{e\}$ so that $(|S| \sqcup \{e\}, \cdot, e)$ becomes a monoid.

Let us call the monoid resulting from the above construction $S^{\mathrm{md}}$, while if $S' = (|S'|, \cdot, e)$ is a monoid, let us write $S'_{\mathrm{sg}}$ for the semigroup $(|S'|, \cdot)$.

(ii)   Show that given a semigroup $S$, the monoid $S^{\mathrm{md}}$ is universal among monoids $T$ given with semigroup homomorphisms $S \to T_{\mathrm{sg}}$.

(iii)   Given a monoid $S = (|S|, \cdot, e)$, what is the relation between the monoids $S$ and $(S_{\mathrm{sg}})^{\mathrm{md}}$? Is there a natural homomorphism in either direction between them?

## 4.12. Associative and commutative rings

An *associative ring* $R$ means a 6-tuple

$$R = (|R|, +, \cdot, -, 0, 1)$$

such that $(|R|, +, -, 0)$ is an abelian group, $(|R|, \cdot, 1)$ is a monoid, and the monoid operation $\cdot : |R| \times |R| \to |R|$ is *bilinear* with respect to the additive group structure. (Dropping the "1" from this definition, one gets a concept of "ring without 1", but we shall not consider these except in one exercise near the end of this section.) A *ring homomorphism* is a map of underlying sets respecting all the operations, including 1. (Some writers, although requiring their rings to have 1, perversely allow "homomorphisms" that may not preserve 1; but we shall stick to the above more sensible definition.) An associative ring is called *commutative* if the multiplication $\cdot$ is so.

   "Commutative associative ring" is usually abbreviated to "commutative ring". Depending on the focus of a given work, *either* the term "associative ring" *or* the term "commutative ring" is usually shortened further to "ring"; an author should always make clear what his or her usage will be. Here, I shall generally shorten "associative ring" to "ring"; though I will sometimes retain the word "associative" when I want to emphasize that commutativity is not being assumed.

   (When one deals with *nonassociative* rings—which we shall not do in this chapter—it is the associativity condition on the *multiplication* that is removed. Frequently one then considers in its place other identities, which may involve both addition and the multiplication; for instance, the identities of Lie rings, which we will introduce in §9.7, or of Jordan rings, mentioned at the end of that section. In the definition of a given kind of nonassociative ring, it may or may not be natural to have a 1 or other distinguished element. The assumption that $(|R|, +, -, 0)$ is an abelian group, and that multiplication is bilinear with respect to this group structure, is made in all versions of ring theory: commutative, associative and nonassociative. If weaker assumptions are made, in particular, if this abelian group structure is replaced by a monoid or semigroup structure, and/or if multiplication is only assumed linear in one of its two arguments, the resulting structures are given names such as "semiring", "half-ring" or "near-ring".)

If $k$ is a fixed commutative ring, then $k$-modules form a natural generalization of abelian groups, on which a concept of bilinear map is also defined, as
noted parenthetically in §4.9 above. Hence one can generalize the definition of
associative ring by replacing the abelian group structure by a $k$-module structure, and the bilinear map of abelian groups by a bilinear map of $k$-modules.
The result is the definition of an *associative algebra* over $k$. The reader familiar with these concepts may note that everything I shall say below for rings
remains valid, mutatis mutandis, for $k$-algebras. (An associative $k$-algebra is
sometimes defined differently, as ring $R$ given with a homomorphism of $k$
into its center; but the two formulations are equivalent: Given a $k$-algebra
$R$ in the present sense of a ring with appropriate $k$-module structure, the
map $c \mapsto c 1_R$ $(c \in k)$ is easily shown to be a homomorphism of $k$ into
the center of $R$, while given a homomorphism $g$ of $k$ into the center of a
ring $R$, the definition $c \cdot r = g(c)\, r$ gives an appropriate module structure,
and these constructions are inverse to one another. For algebras without 1,
and for nonassociative algebras, this equivalence does not hold, and the "ring
with $k$-module structure" definition is then the useful one.)

The subject of universal constructions in ring theory is a vast one. In
this section and the next, we will mainly look at the analogs of some of the
constructions we have considered for groups and monoids.

First, free rings. Let us begin with the commutative case, since that is the
more familiar one. Suppose $R$ is a commutative ring, and $x$, $y$, $z$ are three
elements of $R$. Given any ring-theoretic combination of $x$, $y$ and $z$, we can
use the distributive law of multiplication (i.e., bilinearity of $\cdot$) to expand this
as a sum of products of $x$, $y$ and $z$ (monomials) and additive inverses of
such products. Using the commutativity and associativity of multiplication,
we can write each monomial so that all factors $x$ come first, followed by
all $y$'s, followed by all $z$'s. We can then use commutativity of addition
to bring together all occurrences of each monomial (arranging the distinct
monomials in some specified order), and finally combine occurrences of the
same monomial using integer coefficients. If we now consider all ring-theoretic
terms in *symbols* $x$, $y$ and $z$, of the forms to which we have just shown we
can reduce any combination of *elements* $x$, $y$ and $z$ in any ring, we see, by
the same argument as in §3.4, that the set of these "reduced terms" should
give a normal form for the free commutative ring on three generators $x$,
$y$ and $z$—*if* they form a commutative ring under the obvious operations.
It is, of course, well known that the set of such expressions *does* form a
commutative ring, called the *polynomial ring* in three indeterminates, and
written $\mathbb{Z}[x,\, y,\, z]$.

So polynomial rings over $\mathbb{Z}$ are free commutative rings. (More generally,
the free commutative $k$-algebra on a set $X$ is the polynomial algebra $k[X]$.)
The universal mapping property corresponds to the familiar operation of
*substituting values* for the indeterminates in a polynomial.

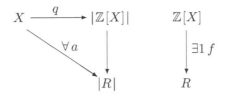

When we drop the commutativity assumption and look at general associative rings, the situation is similar, except that we cannot rewrite each monomial so that "all $x$'s come first" etc. Thus we end up with linear combinations (with coefficients in $\mathbb{Z}$) of arbitrary products of our generators. We claim that formal linear combinations of such products give a normal form for elements of the free associative ring on the set $X$. This ring is written $\mathbb{Z}\langle X \rangle$, and sometimes called the ring of *noncommuting polynomials* in $X$.

We were sketchy in talking about $\mathbb{Z}[X]$ because it is a well-known construction, but let us stop and sort out just what we mean by the above description of $\mathbb{Z}\langle X \rangle$, before looking for a way to prove it.

We could choose a particular way of arranging the parentheses in every monomial term (say, nested to the right), a particular way of arranging the different monomials, and of arranging the parentheses in every sum or difference, and so obtain a set of *ring-theoretic terms* to which every term could be reduced, which we would prove constituted a normal form for the free ring. But observe that the question of putting parentheses into monomial terms is really just one of how to write elements in a *free monoid*, while the question of expressing sums and differences is that of describing an element of the *free abelian group* on a set of generators. Let us therefore assume that we have chosen one or another way of calculating in free abelian groups—whether using a normal form, or a representation by integer-valued functions with only finitely many nonzero values, or whatever—and likewise that we have chosen a way of calculating in free monoids. Then we can calculate in free rings! A precise statement is

**Lemma 4.12.1.** *Let $\mathbb{Z}\langle X \rangle$ denote the free ring on the set $X$. Then the additive group of $\mathbb{Z}\langle X \rangle$ is a free abelian group on the set of products in $\mathbb{Z}\langle X \rangle$ of elements of $X$ (including the empty product, 1), and this set of products forms, under the multiplication of $\mathbb{Z}\langle X \rangle$, a free monoid on $X$.*

*Proof.* Let $S$ denote the free monoid on $X$, and $F(|S|)$ the free abelian group on the underlying set of this monoid. We shall begin by describing a map $|F(|S|)| \to |\mathbb{Z}\langle X \rangle|$.

If we write $u$ for the universal map $X \to |\mathbb{Z}\langle X \rangle|$, then by the universal property of free monoids, $u$ induces a homomorphism $u'$ from the free monoid $S$ into the multiplicative monoid of $\mathbb{Z}\langle X \rangle$. Hence by the universal property of free abelian groups, there exists a unique abelian group homomorphism $u''$ from the free abelian group $F(|S|)$ into the additive group of $\mathbb{Z}\langle X\rangle$ whose restriction to $|S|$ is given by $u'$. Clearly the image of the monoid $S$ in $\mathbb{Z}\langle X\rangle$ is closed under multiplication and contains the multiplicative neutral element; it is easy to deduce from this and the distributive law that the image of the abelian group $F(|S|)$ is closed under all the ring operations. (Note that our considerations so far are valid with $\mathbb{Z}\langle X\rangle$ and $u$ replaced by any ring $R$ and set map $X \to |R|$.) Since this image contains $X$, and $\mathbb{Z}\langle X\rangle$ is generated as a ring by $X$, the image is all of $|\mathbb{Z}\langle X\rangle|$, i.e., $u''$ is surjective. (The above argument formalizes our observation that every element of the subring generated by an $X$-tuple of elements of an arbitrary ring $R$ can be expressed as a linear combination of products of elements of the given $X$-tuple.)

We now wish to show that $u''$ is one-to-one. To do this it will suffice to show that there is *some* ring $R$ with an $X$-tuple $v$ of elements, such that under the induced homomorphism $\mathbb{Z}\langle X\rangle \to R$, any two elements of $\mathbb{Z}\langle X\rangle$ which are images of distinct elements of $F(|S|)$ are mapped to distinct elements of $R$.

How do we find such an $R$? Van der Waerden's trick for groups suggests that we should obtain it from some natural *representation* of the desired free ring. We noted in §3.4 that the *group* operations and identities arise as the operations and identities of the permutations of a set, so for "representations" of groups, we used actions on sets. The operations and identities for associative rings arise as the natural structure on the set of all endomorphisms of an abelian group $A$—one can compose such endomorphisms, and add and subtract them, and under these operations they form a ring $\mathrm{End}(A)$. So we should look for an appropriate family of endomorphisms of some abelian group to represent $\mathbb{Z}\langle X\rangle$.

Let us, as in (3.4.5), introduce a symbol $a$; let $Sa$ denote the set of symbols $x_n \ldots x_1 a$ $(x_i \in X, n \geq 0)$; and this time let us further write $F(Sa)$ for the free abelian group on this set $Sa$. For every $x \in X$, let $\bar{x}: Sa \to Sa$ denote the map carrying each symbol $b \in Sa$ to the symbol $x\,b$. This extends uniquely (by the universal property of free abelian groups) to an additive group homomorphism $\bar{\bar{x}}: F(Sa) \to F(Sa)$. Thus $(\bar{\bar{x}})_{x\in X}$ is an $X$-tuple of elements of the associative ring $\mathrm{End}(F(Sa))$.

Taking $R = \mathrm{End}(F(Sa))$, the above $X$-tuple induces a homomorphism

$$f: \mathbb{Z}\langle X\rangle \to R.$$

Now given any element of $F(|S|)$, which we may write

$$(4.12.2) \qquad r = \sum_{s\in|S|} n_s\, s \quad (n_s \in \mathbb{Z}, \text{ almost all } n_s = 0),$$

we verify easily that the element $f(u''(r)) \in \mathrm{End}(F(Sa))$ carries $a$ to $\sum n_s\, s\, a$. Hence distinct elements (4.12.2) must give distinct elements $u''(r) \in \mathbb{Z}\langle X\rangle$, which proves the one-one-ness of $u''$ and establishes the lemma. $\qquad\square$

For many fascinating results and open problems on free algebras, see [72, 73]. For a smaller dose, you could go to my paper [44], which answers the question, "When do two elements of a free algebra commute?" That problem is not of great importance itself, but it leads to the development of a number of beautiful and useful ring-theoretic tools.

**Exercise 4.12:1.** Let $\alpha$ denote the automorphism of the polynomial ring $\mathbb{Z}[x, y]$ which interchanges $x$ and $y$. It is a standard result that the fixed ring of $\alpha$, i.e., $\{p \in \mathbb{Z}[x, y] \mid \alpha(p) = p\}$, can be described as the polynomial ring in the two elements $x + y$ and $x\,y$.

(i)   Consider analogously the automorphism $\beta$ of the free associative ring $\mathbb{Z}\langle x, y\rangle$ interchanging $x$ and $y$. Show that the fixed ring of $\beta$ is generated by the elements $x+y$, $x^2+y^2$, $x^3+y^3$, $\ldots$, and is a free ring on this infinite set.

(ii)   Observe that the homomorphism $\mathbb{Z}\langle x, y\rangle \to \mathbb{Z}[x, y]$ taking $x$ to $x$ and $y$ to $y$ must take the fixed ring of $\beta$ into the fixed ring of $\alpha$. Will it take it *onto* the fixed ring of $\alpha$?

(iii) If $G$ is the free group on generators $x$ and $y$, and if $\gamma$ is the automorphism interchanging $x$ and $y$ in this group, describe the fixed subgroup of $\gamma$. Do the same for the free *abelian* group on $x$ and $y$. (The analog of (ii) for groups is trivial to answer when this has been done.)

The preceding description of the free ring on a set $X$ involved the free monoid on $X$, and we can see that our earlier description of the free commutative ring (the polynomial ring) bears an analogous relationship to the free commutative monoid. These connections between rings and monoids can be explained in terms of another universal construction:

If $R = (|R|, +, \cdot, -, 0, 1)$ is an associative ring, let $R_{\mathrm{mult}}$ denote its multiplicative monoid, $(|R|, \cdot, 1)$. Then for any monoid $S$, there will exist, by the usual arguments, a ring $R$ with a *universal* monoid homomorphism $u \colon S \to R_{\mathrm{mult}}$.

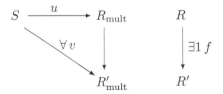

To study this object, let us fix $S$, and consider any ring $R'$ with a homomorphism $S \to R'_{\mathrm{mult}}$. The elements of $R'$ that we can capture using this map are the linear combinations of images of elements of $S$, with integer coefficients. (Why is there no need to mention products of such linear combinations?) One finds that the *universal* such ring $R$ will have as additive structure the free abelian group on $|S|$, with multiplicative structure determined by the condition that the given map $|S| \to |R|$ respect multiplication, together with the bilinearity of multiplication. The result is called the *monoid ring* on $S$, denoted $\mathbb{Z} S$.

Given a presentation of $S$ by generators and relations (written multiplicatively), a presentation of $\mathbb{Z} S$ as a ring will be given by the same generators and relations. In particular, if we take for $S$ the *free* monoid on a set $X$, presented by the generating set $X$ and no relations, then $\mathbb{Z} S$ will be presented as a *ring* by generators $X$ and no relations, and so will be the free ring on $X$, which is just what we saw in Lemma 4.12.1. If we take for $S$ a free *abelian* monoid, then $S$ may be presented as a monoid by generators $X$ and relations $x \, y = y \, x$ $(x, \, y \in X)$, hence this is also a presentation of $\mathbb{Z} S$ as a ring. Since commutativity of a set of generators of a ring is equivalent to commutativity of the whole ring, the above presentation makes $\mathbb{Z} S$ the free commutative ring on $X$.

If $S$ is a monoid, then a "linear action" or "representation" of $S$ on an abelian group $A$ means a homomorphism of $S$ into the multiplicative monoid of the endomorphism ring $\mathrm{End}(A)$ of $A$. By the universal property of $\mathbb{Z} S$, this is equivalent to a ring homomorphism of $\mathbb{Z} S$ into $\mathrm{End}(A)$, which is in turn equivalent to a structure of left $\mathbb{Z} S$-module on the abelian group $A$. In particular, to give an action of a *group* $G$ by automorphisms on an abelian group $A$ corresponds to making $A$ a left module over the *group ring* $\mathbb{Z} G$. Much of modern group theory revolves around linear actions, and hence is closely connected with the properties of $\mathbb{Z} G$ (and more generally, with group *algebras* $k \, G$ where $k$ is a commutative ring, so that left $k \, G$-modules correspond to actions of $G$ on $k$-modules). For some of the elementary theory, see [34, Chapter XVIII]. A major work on group algebras is [122].

In the above discussion, we "factored" the construction of the free associative or commutative ring on a set $X$ into two constructions: the free (respectively, free abelian) monoid construction, which universally closes $X$ under a multiplication with a neutral element, and the monoid-ring construction, which brings in an additive structure in a universal way. These constructions can also be factored the other way around! Given a set $X$, we can first map it into an abelian group in a universal way, getting the free abelian group $A$ on $X$, then form a ring (respectively a commutative ring) $R$ with a universal additive group homomorphism $A \to R_{\mathrm{add}}$. For any abelian group $A$, the associative ring with such a universal homomorphism is called the *tensor ring* on $A$, because its additive group structure turns out to have the form

$$\mathbb{Z} \oplus A \oplus (A \otimes A) \oplus (A \otimes A \otimes A) \oplus \dots,$$

though we shall not show this here. The corresponding universal *commutative* ring is called the *symmetric ring* on $A$; its structure for general $A$ is more difficult to describe. For more details see [34, §§XVI.7, 8] or [58]. Thus, a free associative ring can be described as the tensor ring on a free abelian group, and a polynomial ring as the symmetric ring on a free abelian group.

On to other constructions. Suppose $R$ is a commutative ring, and $(f_i, g_i)$ $(i \in I)$ a family of pairs of elements of $R$. To impose the relations $f_i = g_i$ on $R$, one forms the factor-ring $R/J$, where $J$ is the ideal generated by the elements $f_i - g_i$. This ideal is often written $(f_i - g_i)_{i \in I}$. Another common notation, preferable because it is more explicit, is $\sum_{i \in I} R\,(f_i - g_i)$, or, if we set $Y = \{f_i - g_i \mid i \in I\}$, simply $RY$. It consists of all sums

$$(4.12.3) \quad \sum r_i\,(f_i - g_i) \quad (r_i \in |R|, \text{ nonzero for only finitely many } i \in I).$$

The construction of imposing relations on a *noncommutative* ring $R$ is of the same form, but with "ideal" taken to mean a *two-sided ideal*—an additive subgroup of $R$ closed under both left and right multiplication by members of $R$. The two-sided ideal generated by $\{f_i - g_i \mid i \in I\}$ is also often written $(f_i - g_i)_{i \in I}$, and again there is a more expressive notation, $\sum_{i \in I} R\,(f_i - g_i)\,R$, or $RYR$. This ideal consists of all sums of products of the form $r\,(f_i - g_i)\,r'$ $(i \in I,\ r,\ r' \in R)$. Note, however, that in the noncommutative case, it is not in general enough to have, as in (4.12.3), *one* such summand for each $i \in I$. For instance, in $\mathbb{Z}\langle x, y\rangle$, the ideal generated by the one element $x$ contains the element $x\,y + y\,x$, which cannot be simplified to a single product $r\,x\,r'$.

**Exercise 4.12:2.** Let $R$ be a commutative ring. Will there, in general, exist a universal homomorphism of $R$ into an *integral domain* $R'$? If not, can you find conditions on $R$ for such a homomorphism to exist? Suggestion: Consider the cases $R = \mathbb{Z},\ \mathbb{Z}_6,\ \mathbb{Z}_4$.

**Exercise 4.12:3.**

(i)   Obtain a normal form for elements of the associative ring $A$ presented by two generators $x$, $y$, and one relation $y\,x - x\,y = 1$.

(ii)   Let $\mathbb{Z}[x]_{\mathrm{add}}$ be the *additive group* of polynomials in one indeterminate $x$. Show that there exists a homomorphism $f$ of the ring $A$ of part (i) into the endomorphism ring of this abelian group, such that $f(x)$ is the operation of multiplying polynomials by $x$ in $\mathbb{Z}[x]$, and $f(y)$ the operation of differentiating with respect to $x$. Is this homomorphism one-to-one?

The ring of the above example, or rather the corresponding algebra over a field $k$, is called the *Weyl algebra*. It is of importance in quantum mechanics, where multiplication of the wave-function of a particle by the coordinate function $x$ corresponds to measuring the particle's $x$-coordinate, while differentiating with respect to $x$ corresponds to measuring its momentum in the $x$-direction. The fact that these operators do not commute leads, via the mysterious reasoning of quantum mechanics, to the impossibility of measuring those two quantities simultaneously, the simplest case of the "Heisenberg uncertainty principle".

*Direct products* $\prod_I R_i$ of associative rings or commutative rings turn out, as expected, to be gotten by taking direct products of underlying sets, with componentwise operations.

**Exercise 4.12:4.** (Andreas Dress.)

(i)   Find all subrings of $\mathbb{Z} \times \mathbb{Z}$. (Remember: a subring must have the same multiplicative neutral element 1. Try to formulate your description of each such subring $R$ as a necessary and sufficient condition for an arbitrary $(a, b) \in \mathbb{Z} \times \mathbb{Z}$ to lie in $|R|$.)

A much harder problem is:

(ii)   Is there a similar characterization of all subrings of $\mathbb{Z} \times \mathbb{Z} \times \mathbb{Z}$?

**Exercise 4.12:5.** Show that the commutative ring presented by one generator $x$, and one relation $x^2 = x$, is isomorphic (as a ring) to the direct product ring $\mathbb{Z} \times \mathbb{Z}$.

**Exercise 4.12:6.** Given generators and relations for two rings, $R$ and $S$, show how to obtain generators and relations for $R \times S$.

**Exercise 4.12:7.** Describe

(i)   the commutative ring $A$ presented by one generator $x$, and one relation $2x = 1$, and

(ii)   the commutative ring $B$ presented by one generator $x$ and two relations $4x = 2$, $2x^2 = x$. (Note that both of these relations are implied by the relation of (i).)

Your descriptions in parts (i) and (ii) should make it clear whether these rings are isomorphic.

(iii) Show that each of these rings has the property that for any ring $R$ (commutative if you wish) there is *at most* one homomorphism of the indicated ring ($A$, respectively $B$) into $R$.

**Exercise 4.12:8.** Suppose $R$ is a ring whose underlying abelian group is finitely generated. Show that as a ring, $R$ is finitely presented. (You may use the fact that every finitely generated abelian group is finitely presented.)

If you are comfortable with algebras over a commutative ring $k$, try to generalize this result to algebras over some or all such $k$.

In discussing universal properties, I have neglected to mention some trivial cases. Let me give these in the next two exercises. Even if you do not write them up, think through the "ring" cases of parts (i) and (ii) of the next exercise, since some later exercises use them.

**Exercise 4.12:9.**

(i)   Consider the free group, the free monoid, the free associative ring, and the free commutative ring on the *empty* set of generators. Reformulate the universal properties of these objects in as simple a form as possible. Display the group, monoid, ring, and commutative ring characterized by these properties, if they exist.

(ii)   State, similarly, the universal properties that would characterize the product and coproduct of an empty family of groups, monoids, rings, or commutative rings, and determine these objects, if any.

(iii)  Give as simple as possible a system of defining generators and relations for the rings $\mathbb{Z}$ and $\mathbb{Z}/n\mathbb{Z}$.

The next exercise concerns semigroups, and rings without neutral elements. Note that when we say "without 1" etc., this does not *forbid* the existence of an element 1 satisfying $(\forall x)\ 1x = x = x1$. It just means that we don't *require* the existence of such elements, and that when they exist, we don't give them a special place in the definition, or require homomorphisms to respect them.

**Exercise 4.12:10.** Same as parts (i) and (ii) of the preceding exercise, but for semigroups, and for rings without 1. Same for sets. Same for $G$-sets for a fixed group $G$.

**Exercise 4.12:11.** Suppose that, as some (misguided) authors do, we required rings to have 1 but did not require ring homomorphisms to preserve 1. Show that under this definition, there would exist no free commutative ring on one generator. (In fact, there wouldn't be free rings on any positive number of generators; but this case is enough to prove.)

Now back to rings with 1, and homomorphisms preserving 1.

## 4.13. Coproducts and tensor products of rings

We have noted that the descriptions of *coproducts* vary from one sort of algebraic object to another, so it will not be surprising to find that they have different forms for commutative and noncommutative rings. Let us again start with the commutative case.

Suppose $S$ and $T$ are fixed commutative rings, and we are given homomorphisms $s \mapsto \bar{s}$ and $t \mapsto \tilde{t}$ of these into a third commutative ring $R$. What elements of $R$ can we capture? Obviously, elements $\bar{s}$ $(s \in |S|)$ and $\tilde{t}$ $(t \in |T|)$. From these we can form products $\bar{s}\tilde{t}$, and we can then form sums of elements of all these sorts:

$$(4.13.1) \qquad \bar{s} + \tilde{t} + \bar{s}_1\tilde{t}_1 + \cdots + \bar{s}_n\tilde{t}_n.$$

We don't get more elements by multiplying such sums together, because a product $(\bar{s}\,\tilde{t}\,)(\bar{s}'\,\tilde{t}')$ reduces to $\overline{ss'}\,\widetilde{tt'}$. Let us also note that the lone summands $\bar{s}$ and $\tilde{t}$ in (4.13.1) can actually be written in the same form as the other summands, because $\overline{1_S} = \widetilde{1_T} = 1_R$, hence $\bar{s} = \bar{s}\,\widetilde{1_T}$ and $\tilde{t} = \overline{1_S}\,\tilde{t}$. So the subring of $R$ that we get is generated as an additive group by the image of the map

$$(4.13.2) \qquad\qquad (s,\, t) \;\mapsto\; \bar{s}\,\tilde{t}$$

of $|S| \times |T|$ into $|R|$. If we look for equalities among sums of elements of this form, we find

$$\overline{(s + s')}\,\tilde{t} \;=\; \bar{s}\,\tilde{t} + \bar{s}'\,\tilde{t}, \quad \text{and} \quad \bar{s}\,\widetilde{(t + t')} \;=\; \bar{s}\,\tilde{t} + \bar{s}\,\tilde{t}',$$

in other words, relations saying that (4.13.2) is bilinear. These relations and their consequences turn out to be *all* we can find, and one can show that the *universal* $R$ with ring homomorphisms of $S$ and $T$ into it, that is, the coproduct of $S$ and $T$ as commutative rings, has the additive structure of the *tensor product* of the additive groups of $S$ and $T$. The elements that we have written $\bar{s}$ and $\tilde{t}$ are, as the above discussion implies, $s \otimes 1_T$ and $1_S \otimes t$ respectively; the multiplication is determined by the formula

$$(4.13.3) \qquad\qquad (s \otimes t)(s' \otimes t') \;=\; s\,s' \otimes t\,t'$$

which specifies how to multiply the additive generators of the tensor product group. For a proof that this extends to a bilinear operation on all of $S \otimes T$, and that this operation makes the additive group $S \otimes T$ into a ring, see Lang [34, §XVI.6]. (Note: Lang works in the context of algebras over a ring $k$, and he defines such an algebra as a homomorphism $f$ of $k$ into the center of a ring $R$—what I prefer to call, for intuitive comprehensibility, a ring $R$ given with a homomorphism of $k$ into its center; cf. parenthetical remark near the beginning of §4.12 above. Thus, when he defines the coproduct of two commutative $k$-algebras to be a certain *map*, look at the *codomain* of the map to see the ring that he means. Or, instead of looking in Lang for this construction, you might do Exercise 4.13:5 below, which gives a generalization of this result.)

This coproduct construction is called the "tensor product of commutative rings".

**Exercise 4.13:1.** If $m$ and $n$ are integers, find the structure of the tensor product ring $(\mathbb{Z}/m\mathbb{Z}) \otimes (\mathbb{Z}/n\mathbb{Z})$ by two methods:

(i)  By constructing the tensor product of the abelian groups $Z_m$ and $Z_n$, and describing the multiplication characterized above.

(ii)  By using the fact that a presentation of a coproduct can be obtained by "putting together" presentations for the two given objects. (Cf. Exercise 4.12:9.)

**Exercise 4.13:2.** Let $\mathbb{Z}[i]$ denote the ring of *Gaussian integers* (complex numbers $a + bi$ such that $a$ and $b$ are integers). This may be presented as a commutative ring by one generator $i$, and one relation $i^2 = -1$. Examine the structures of the rings $\mathbb{Z}[i] \otimes (\mathbb{Z}/p\mathbb{Z})$ ($p$ a prime). E.g., will they be integral domains for all $p$? For some $p$?

The next two exercises concern tensor products of algebras over a field $k$, for students familiar with this concept. Tensor products of this sort are actually simpler to work with than the tensor products of rings described above, because every algebra over a field $k$ is free as a $k$-module (since every $k$-vector-space has a basis), and tensor products of free modules are easily described (cf. paragraph containing (4.9.1) above).

**Exercise 4.13:3.** Let $K$ and $L$ be extensions of a field $k$. A *compositum* of $K$ and $L$ means a 3-tuple $(E, f, g)$ where $E$ is a field extension of $k$, and $f : K \to E$, $g : L \to E$ are $k$-algebra homomorphisms such that $E$ is generated by $f(|K|) \cup g(|L|)$ as a field (i.e., under the ring operations, and the *partial operation* of multiplicative inverse).

(i)   Suppose $K$ and $L$ are *finite-dimensional* over $k$, and we form their tensor product algebra $K \otimes_k L$, which is a commutative $k$-algebra, but not necessarily a field. Show that up to isomorphism, all the composita of $K$ and $L$ over $k$ are given by the factor rings $(K \otimes_k L)/P$, for prime ideals $P \subseteq K \otimes_k L$. (First write down what should be meant by an isomorphism between composita of $K$ and $L$.)

(ii)   What if $K$ and $L$ are not assumed finite-dimensional?

**Exercise 4.13:4.**

(i)   Determine the structure of the tensor product $\mathbb{C} \otimes_{\mathbb{R}} \mathbb{C}$, where $\mathbb{C}$ is the field of complex numbers and $\mathbb{R}$ the field of real numbers. In particular, can it be written as a nontrivial direct product of $\mathbb{R}$-algebras?

(ii)   Do the same for $\mathbb{Q}(2^{1/3}) \otimes_{\mathbb{Q}} \mathbb{Q}(2^{1/3})$.

(iii)   Relate the above results to the preceding exercise.

You can carry this exercise much farther if you like—find a general description of a tensor product of a finite Galois extension with itself; then of two arbitrary finite separable field extensions (by taking them to lie in a common Galois extension, and considering the subgroups of the Galois group they correspond to); then try some examples with inseparable extensions.... In fact, one modern approach to the whole subject of Galois theory is via properties of such tensor products. (E.g., see [109], starting around §11 (p. 105).)

If $S$ and $T$ are arbitrary (not necessarily commutative) associative rings, one can still make the tensor product of the additive groups of $S$ and $T$ into a ring with a multiplication satisfying (4.13.3). It is not hard to verify that this will be universal among rings $R$ given with homomorphisms $f : S \to R$, $g : T \to R$ such that all elements of $f(S)$ *commute with* all elements of $g(T)$.

(Cf. the "second universal property" of the direct product of two groups, end of §4.6 above. In fact, some early ring-theorists wrote $S \times T$ for what we now denote $S \otimes T$, considering this construction as the ring-theoretic analog of the direct product construction on groups.) This verification is the exercise mentioned earlier as an alternative to looking in Lang for the universal property of the tensor product of commutative rings:

**Exercise 4.13:5.** Verify the above assertion that given rings $S$ and $T$, the universal ring with mutually commuting homomorphic images of $S$ and $T$ has the additive structure of $S \otimes T$, and multiplication given by (4.13.3). (Suggestion: map the additive group $S \otimes T$ onto that universal ring $R$, then use "van der Waerden's trick" to show the map is an isomorphism.) Obtain as a corollary the characterization of the coproduct of commutative rings referred to earlier.

**Exercise 4.13:6.** Show that if $S$ and $T$ are monoids, then the monoid ring construction (§4.12) satisfies $\mathbb{Z}S \otimes \mathbb{Z}T \cong \mathbb{Z}(S \times T)$.

**Exercise 4.13:7.** Suppose $S$ and $T$ are associative rings, and we form the additive group $R_{\mathrm{add}} = S_{\mathrm{add}} \otimes T_{\mathrm{add}}$. Is the multiplication described above in general the unique multiplication on $R_{\mathrm{add}}$ which makes it into a ring $R$ such that the maps $s \mapsto s \otimes 1_T$ and $t \mapsto 1_S \otimes s$ are ring homomorphisms? You might look, in particular, at the case $S = \mathbb{Z}[x]$, $T = \mathbb{Z}[y]$.

Let us now look at coproducts of not necessarily commutative rings, writing these $S * T$ as for groups and monoids. They exist by the usual general nonsense, and again, a presentation of $S*T$ can be gotten by putting together presentations of $S$ and $T$. But the explicit description of these coproducts is more complicated than for the constructions we have considered so far. For $S$ and $T$ arbitrary associative rings, there is no neat explicit description of $S * T$. Suppose, however, that the additive group of $S$ is free as an abelian group on a basis containing the unit element, $\{1_S\} \cup B_S$, and that of $T$ is free as an abelian group on a basis of the same sort, $\{1_T\} \cup B_T$. (For example, the rings $S = \mathbb{Z}[x]$ and $T = \mathbb{Z}[i]$ have such bases, with $B_S = \{x, x^2, \ldots\}$ and $B_T = \{i\}$.) Then we see that given a ring $R$ and homomorphisms $S \to R$, $T \to R$, written $s \mapsto \bar{s}$ and $t \mapsto \tilde{t}$, the elements of $R$ that we get by ring operations from the images of $S$ and $T$ can be written as linear combinations, with integer coefficients, of products $x_n \ldots x_1$ where $x_i \in \overline{B_S} \cup \widetilde{B_T}$ (i.e., $\{\bar{b} \mid b \in B_S\} \cup \{\tilde{b} \mid b \in B_T\}$), and no two factors from the same basis-set occur successively. (In thinking this through, note that a product of two elements from $\overline{B_S}$ can be rewritten as a linear combination of single elements from $\overline{B_S} \cup \{\overline{1_S}\}$, and that occurrences of $\overline{1_S}$ can be eliminated because in $R$, $\overline{1_S} = 1_R$; and the same considerations apply to elements from $\widetilde{B_T}$. In this description we are again considering $1_R$ as the "empty" or "length 0" product.) And in fact, the coproduct of $S$ and $T$ as associative rings turns out to have precisely the set of such products as an additive basis.

**Exercise 4.13:8.** Verify the above assertion, using an appropriate modification of van der Waerden's trick.

**Exercise 4.13:9.**

(i)   Study the structure of the coproduct ring $\mathbb{Z}[i] * \mathbb{Z}[i]$, where $\mathbb{Z}[i]$ denotes the Gaussian integers as in Exercise 4.13:2. In particular, try to determine its center, and whether it has any zero-divisors.

(ii)   In general, if $S$ and $T$ are rings free as abelian groups on *two-element* bases of the forms $\{1, s\}$ and $\{1, t\}$, what can be said about the structure and center of $S * T$?

The next part shows that the above situation is exceptional.

(iii)  Suppose as in (ii) that $S$ and $T$ each have additive bases containing 1, and neither of these bases consists of 1 alone, but now suppose that at least one of them has more than two elements. Show that in this situation, the center of $S * T$ is just $\mathbb{Z}$.

When the rings in question are not free $\mathbb{Z}$-modules, the above description does not work, but in some cases the result is nonetheless easy to characterize.

**Exercise 4.13:10.**

(i)   Describe the rings $\mathbb{Q} * \mathbb{Q}$ and $\mathbb{Q} \otimes \mathbb{Q}$.

(ii)   Describe the rings $(\mathbb{Z}/n\mathbb{Z}) * \mathbb{Q}$ and $(\mathbb{Z}/n\mathbb{Z}) \otimes \mathbb{Q}$, where $n$ is a positive integer.

Some surprising results on the module theory of ring coproducts are obtained in [47]. (That paper presumes familiarity with basic properties of semisimple artin rings and their modules. The reader who is familiar with such rings and modules, but not with homological algebra, should not be deterred by the discussion of homological properties of coproducts in the first section; that section gives homological applications of the main result of the paper, but the later sections, where the main result is proved, do not require homological methods.)

## 4.14. Boolean algebras and Boolean rings

Let $S$ be a set, and let $\mathbf{P}(S)$ denote the power set of $S$, that is, $\{T \mid T \subseteq S\}$. There are various natural operations on $\mathbf{P}(S)$: union, intersection, complement (i.e., $^cT = \{s \in S \mid s \notin T\}$), and the two zeroary operations, $\emptyset \in \mathbf{P}(S)$ and $S = {}^c\emptyset \in \mathbf{P}(S)$. Thus we can regard $\mathbf{P}(S)$ as the underlying set of an algebraic structure

(4.14.1)                     $(\mathbf{P}(S), \cup, \cap, {}^c, \emptyset, S)$.

This structure, or more generally, any 6-tuple consisting of a set and five operations on that set, of arities 2, 2, 1, 0, 0, satisfying all the identities satisfied by structures of the form (4.14.1) for sets $S$, is called a *Boolean algebra*.

Such 6-tuples do not quite fit any of the pigeonholes we have considered so far. For instance, neither of the operations $\cup$, $\cap$ is the composition operation of an abelian group, hence a "Boolean algebra" is not a ring.

However, there is a way of looking at $\mathbf{P}(S)$ which reduces us to ring theory. There is a standard one-to-one correspondence between the power set $\mathbf{P}(S)$ of a set $S$ and the set of functions $2^S$, where $2$ means the two-element set $\{0, 1\}$; namely, the correspondence associating to each $T \in \mathbf{P}(S)$ its characteristic function (the function whose value is $1$ on elements of $T$ and $0$ on elements of $^cT$). If we try to do arithmetic with these functions, we run into the difficulty that the sum of two $\{0, 1\}$-valued functions is not generally $\{0, 1\}$-valued. But if we identify $\{0, 1\}$ with the underlying set of the ring $\mathbb{Z}/2\mathbb{Z}$ rather than treating it as a subset of $\mathbb{Z}$, this problem is circumvented: $2^S$ becomes the ring $(\mathbb{Z}/2\mathbb{Z})^S$ —the direct product of an $S$-tuple of copies of $\mathbb{Z}/2\mathbb{Z}$. Moreover, it is possible to describe union, intersection, etc., of subsets of $S$ in terms of the ring operations of $(\mathbb{Z}/2\mathbb{Z})^S$. Namely, writing $\bar{a}$ for the characteristic function of $a \subseteq S$, we have

$$(4.14.2) \qquad \overline{a \cap b} = \bar{a}\,\bar{b}, \quad \overline{a \cup b} = \bar{a} + \bar{b} + \bar{a}\,\bar{b}, \quad \overline{^c a} = 1 + \bar{a}, \quad \bar{\emptyset} = 0, \quad \bar{S} = 1.$$

Conversely, each ring operation of $(\mathbb{Z}/2\mathbb{Z})^S$, translated into an operation on subsets of $S$, can be expressed in terms of our set-theoretic Boolean algebra operations. The expressions for multiplication, for $0$, and for $1$ are clear from (4.14.2); additive inverse is the identity operation, and $+$ is described by

$$(4.14.3) \qquad\qquad \bar{a} + \bar{b} = \overline{(a \cap {}^cb) \cup ({}^ca \cap b)}.$$

(The set $(a \cap {}^cb) \cup ({}^ca \cap b)$ is called the "symmetric difference" of the sets $a$ and $b$.)

Note that the ring $B = (\mathbb{Z}/2\mathbb{Z})^S = (2^S, +, \cdot, -, 0, 1)$, like $\mathbb{Z}/2\mathbb{Z}$, satisfies

$$(4.14.4) \qquad (\forall x \in |B|) \ x^2 = x,$$

from which one easily deduces the further identities,

$$(4.14.5) \qquad \begin{aligned} &(\forall x, y \in |B|) \ xy = yx, \\ &(\forall x \in |B|) \ x + x = 0 \quad \text{(equivalently: } 1 + 1 = 0 \text{ in } B\text{).} \end{aligned}$$

An associative ring satisfying (4.14.4) (and so also (4.14.5)) is called a *Boolean ring*. We shall see below (Exercise 4.14:2) that the identities defining a Boolean ring, i.e., the identities of associative rings together with (4.14.4), imply *all* identities satisfied by rings $(\mathbb{Z}/2\mathbb{Z})^S$. Hence we shall see that Boolean rings and Boolean algebras are essentially equivalent—we can turn each into the other using (4.14.2) and (4.14.3).

**Exercise 4.14:1.** The *free Boolean ring* $F(X)$ on any set $X$ exists by the usual general arguments. Find a normal form for the elements of $F(X)$ when $X$ is finite. To prove that distinct expressions in normal form represent distinct elements, you will need some kind of representation of $F(X)$; use a representation by subsets of a set $S$.

**Exercise 4.14:2.** Assume here the result which follows from the indicated approach to the preceding exercise, that the free Boolean ring on any finite set $X$ can be embedded in the Boolean ring of subsets of some set $S$.

(i)   Deduce that all identities satisfied by the rings $(\mathbb{Z}/2\mathbb{Z})^S$ ($S$ a set) follow from the identities by which we defined Boolean rings.

(ii)   Conclude that the free Boolean ring on an *arbitrary* set $X$ can be embedded in the Boolean ring of $\{0, 1\}$-valued functions on some set (if you did not already prove this as part of your proof of (i)).

(iii) Deduce that there exists a finite list of identities for Boolean *algebras* which implies all identities holding for such structures (i.e., all identities holding in sets $\mathbf{P}(S)$ under $\cup$, $\cap$, $^c$, 0 and 1).

**Exercise 4.14:3.** An element $a$ of a ring (or semigroup or monoid) is called *idempotent* if $a^2 = a$. If $R$ is a *commutative* ring, let us define

$$\mathrm{Idpt}(R) \;=\; (\{a \in R \mid a^2 = a\}, \; \dotplus, \; \cdot, \; \dotminus, \; 0, \; 1),$$

where $a \dotplus b = a + b - 2\,a\,b$ and $\dotminus a = a$.

(i)   Verify that each of the above operations carries the set $|\mathrm{Idpt}(R)|$ into itself.

(ii)   Show that if $a \in |\mathrm{Idpt}(R)|$, then $R$ can (up to isomorphism) be written $R_1 \times R_2$ for some rings $R_1$, $R_2$, in such a way that the element $a$ has the form $(0, 1)$ in this direct product. Deduce that if $a_1, \ldots, a_i \in |\mathrm{Idpt}(R)|$, then $R$ can be written as a finite direct product in such a way that each $a_i$ has each coordinate 0 or 1. This result can be used to get a proof of the next part that is conceptual rather than purely computational:

(iii) Show that for any commutative ring $R$, $\mathrm{Idpt}(R)$ is a Boolean ring.

(iv) Given any Boolean ring $B$, show that there is a universal pair $(R, f)$ where $R$ is a commutative ring, and $f : B \to \mathrm{Idpt}(R)$ a homomorphism.

(v)   Investigate the structure of the $R$ of the above construction in some simple cases, e.g., $B = \mathbb{Z}/2\mathbb{Z}$, $B = (\mathbb{Z}/2\mathbb{Z})^2$, $B = (\mathbb{Z}/2\mathbb{Z})^S$.

(Students familiar with algebraic geometry will recognize that the idempotent elements of a commutative ring $R$ correspond to the continuous $\{0, 1\}$-valued functions on $\mathrm{Spec}(R)$. Thus the Boolean rings $\mathrm{Idpt}(R)$ of the above exercise have natural representations as Boolean rings of $\{0, 1\}$-valued functions on sets.)

**Exercise 4.14:4.**

(i)  If $f\colon U \to V$ is a set map, describe the homomorphism it induces between the Boolean rings $(\mathbb{Z}/2\mathbb{Z})^U$ and $(\mathbb{Z}/2\mathbb{Z})^V$. (You first have to decide which way the homomorphism will go.)

(ii)  Let $B$ be a Boolean ring. Formulate universal properties for a "universal representation of $B$ by subsets of a set", in each of the following senses:

    (a) A universal pair $(S, f)$, where $S$ is a set, and $f$ a Boolean ring homomorphism $B \to (\mathbb{Z}/2\mathbb{Z})^S$.

    (b) A universal pair $(T, g)$, where $T$ is a set, and $g$ a Boolean ring homomorphism $(\mathbb{Z}/2\mathbb{Z})^T \to B$.

(iii)  Investigate whether such universal representations exist. If such representations are obtained, investigate whether the maps $f$, $g$ will in general be one-to-one and/or onto.

**Exercise 4.14:5.**

(i)  Show that every finite Boolean ring is isomorphic to one of the form $2^S$ for some finite set $S$.

(ii)  For what finite sets $S$ is the Boolean ring $2^S$ free? How is the number of free generators determined by the set $S$?

**Exercise 4.14:6.** A subset $T$ of a set $S$ is said to be *cofinite* in $S$ if $^c T$ (taken relative to $S$, i.e., $S - T$) is finite. Show that $\{T \subseteq \mathbb{Z} \mid T$ is finite or cofinite $\}$ is the underlying set of a Boolean subring of $2^{\mathbb{Z}}$, which is neither free, nor isomorphic to a Boolean ring $2^U$ for any set $U$.

**Exercise 4.14:7.** It is not hard to see (as for groups, monoids, rings, and commutative rings) that any two Boolean rings $B_1$ and $B_2$ will have a *coproduct* as Boolean rings.

(i)  Will this coproduct in general coincide with the coproduct of $B_1$ and $B_2$ as rings? As commutative rings?

(ii)  Suppose $B_1$ and $B_2$ are finite, so that we can take $B_1 = 2^S$, $B_2 = 2^T$ for finite sets $S$ and $T$. Can you describe the coproduct of these two Boolean rings, and the canonical maps from those rings into their coproduct, in terms of $S$ and $T$?

    For additional credit you might see whether the result you get in (ii) extends, in one way or another, to Boolean rings $2^S$ and $2^T$ for infinite sets $S$ and $T$, or to other sorts of infinite Boolean rings, such as those described in the preceding exercise.

    Above, I have for purposes of exposition distinguished between the power set $\mathbf{P}(S)$ of a set $S$ and the function-set $2^S$. But these notations are often used interchangeably, and I may use them that way myself elsewhere in these notes.

## 4.15. Sets

The objects we have been studying have been sets with additional operations. Let us briefly note the forms that some of the sorts of constructions we have discovered take for plain sets.

Given a family of sets $(S_i)_{i \in I}$, the object with the universal property characterizing products is the usual direct product, $\prod_I S_i$, which may be described as the set of functions on $I$ whose value at each element $i$ belongs to the set $S_i$. The projection map $p_i$ in the statement of the universal property takes each such function to its value at $i$. Note that the product of the vacuous family of sets (indexed by the empty set!) is a one-element set.

The coproduct of a family of sets $(S_i)_{i \in I}$ is their *disjoint union* $\bigsqcup_I S_i$, to which we referred in passing in §4.6. If the $S_i$ are themselves disjoint, one can take for this set their ordinary union; the inclusions of the $S_i$ in this union give the universal family of maps $q_j: S_j \to \bigsqcup_I S_i$ ($j \in I$). A construction that will work without any disjointness assumption is to take

$$(4.15.1) \qquad \bigsqcup_I S_i = \{(i, s) \mid i \in I, \ s \in S_i\}$$

with universal maps given by

$$(4.15.2) \qquad q_i(s) = (i, s) \quad (i \in I, s \in S_i).$$

A frequent practice in mathematical writing is to assume ("without loss of generality") that a family of sets is disjoint, if this would be notationally convenient, and if there is nothing logically forcing them to have elements in common. When this disjointness condition holds one can, as noted, take the universal maps involved in the definition of a coproduct of sets to be inclusions. But in other cases—for instance if we want to consider a coproduct of a set with itself, or of a set and a subset—a construction like (4.15.1) is needed. Note that when a construction is described "in general" under such a disjointness assumption, and is later applied in a situation where one cannot make that assumption, one must be careful to insert $q_i$'s where appropriate.

**Exercise 4.15:1.** Investigate laws such as associativity, distributivity, etc. which are satisfied up to natural isomorphism by the constructions of pairwise product and coproduct of sets.

Examine which of these laws are also satisfied by products and coproducts of groups, and which are not.

Sets can also be constructed by "generators and relations". If $X$ is a set, then relations are specified by a set $R$ of ordered pairs of elements of $X$, which we want to make fall together. The universal image of $X$ under a map making the components of each of these pairs fall together is easily seen to be the quotient of $X$ by the least equivalence relation containing $R$.

The constructions named in this section—direct product of sets, disjoint union, and quotient by the equivalence relation generated by a given binary relation—were taken for granted in earlier sections. So the point of this section was not to introduce the reader to those constructions, but to note their relation to the general patterns we have been seeing.

## 4.16. Some algebraic structures we have not looked at

...lattices ([4], [14], and Chapter 6 below), modular lattices, distributive lattices; partially ordered sets (Chapter 5 below); cylindric algebras [87]; heaps (Exercise 9.6:10 below); loops [8, p. 52]; Lie algebras ([95], §9.7 below), Jordan algebras [96], general nonassociative algebras; rings with polynomial identity [129], rings with involution, fields, division rings, Hopf algebras [136]; modules, bimodules (§§10.8–10.9 below); filtered groups, filtered rings, filtered modules; graded rings, graded modules; ordered groups, right-ordered groups, lattice-ordered groups [82], ....

As noted, we will consider some of these in later chapters.

On the objects we *have* considered here, we have only looked at basic and familiar universal constructions. Once we develop a general theory of universal constructions, we shall see that they come in many more varied forms.

For diversity, I will end this chapter with two examples for those who know some general topology.

## 4.17. The Stone-Čech compactification of a topological space

We know that the real line $\mathbb{R}$, as a topological space, is not compact. But when studying the limit-behavior of $\mathbb{R}$-valued functions or sequences, it is frequently convenient to adjoin to $\mathbb{R}$ an additional point, "$\infty$", obtaining the compact space $\mathbb{R} \cup \{\infty\}$ shown below.

$\mathbb{R} \cup \{\infty\}$ :

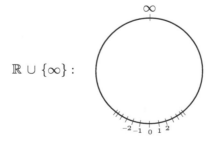

At other times, one adjoins to $\mathbb{R}$ two points, $+\infty$ and $-\infty$, getting a compact space

$$\mathbb{R} \cup \{+\infty, -\infty\} :$$

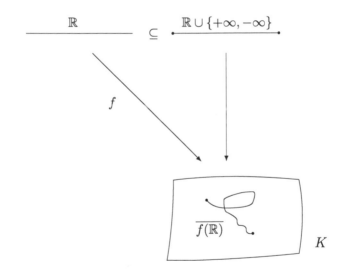

Note that $\mathbb{R} \cup \{\infty\}$ may be obtained from $\mathbb{R} \cup \{+\infty, -\infty\}$ by an *identification*. Hence $\mathbb{R} \cup \{+\infty, -\infty\}$ can be thought of as making finer distinctions in limiting behavior than $\mathbb{R} \cup \{\infty\}$.

One might imagine that $\mathbb{R} \cup \{+\infty, -\infty\}$ makes "the finest possible distinctions". A precise formulation of this would be a conjecture that for any continuous map $f$ of $\mathbb{R}$ into a compact Hausdorff space $K$, the closure of the image of $\mathbb{R}$ should be an image of $\mathbb{R} \cup \{+\infty, -\infty\}$; i.e., that the map $f$ should factor through the inclusion $\mathbb{R} \subseteq \mathbb{R} \cup \{+\infty, -\infty\}$. Here is a picture of an example for which this is true:

But by thinking about either of the following pictures, you can see that the above conjecture is not true in general:

(4.17.1)

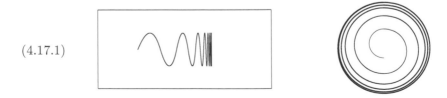

However, we can still ask whether there is *some* compactification of $\mathbb{R}$ which makes "the most possible distinctions". Let us raise this question with $\mathbb{R}$ replaced by a general topological space $X$, and give the desired object a name.

**Definition 4.17.2.** *Let $X$ be a topological space. A Stone-Čech compactification of $X$ will mean a pair $(C, u)$, where $C$ is a compact Hausdorff space and $u$ a continuous map $X \to C$, universal among all continuous maps of $X$ into compact Hausdorff spaces $K$ (diagram at right).*

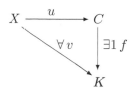

**Exercise 4.17:1.** Show that if a pair $(C, u)$ as in the above definition exists, then $u(X)$ is dense in $C$. In fact, show that if $(C, u)$ has the indicated universal property but without the condition of uniqueness of factoring maps $g$ (see above diagram), then

(i)    uniqueness of such maps holds if and only if $u(X)$ is dense in $C$; and

(ii)   if $C'$ is the closure of $u(X)$ in $C$, the pair $(C', u)$ has the full universal property.

We now want to determine whether such compactifications always exist.

The analog of our construction of free groups from terms as in §3.2 would be to adjoin to $X$ some kinds of "formal limit points". But limit points of what? Not every sequence in a compact Hausdorff space $K$ converges, nor need every point of the closure of a subset $J \subseteq K$ be the limit of a sequence of points of $J$ (unless $K$ is first countable); so adjoining limits of *sequences* would not do. The approach of adjoining limit points *can* in fact be made to work, but it requires considerable study of how such points may be described; the end result is a construction of the Stone-Čech compactification of $X$ in terms of *ultrafilters*. We shall not pursue that approach here; it is used in [138, Theorem 17.17 et seq.]. (NB: The compactification constructed there may not be Hausdorff when $X$ is "bad", so in such cases it will not satisfy our definition.)

The "big direct product" approach is more easily adapted. If $v_1 : X \to K_1$ and $v_2 : X \to K_2$ are two continuous maps of $X$ into compact Hausdorff spaces, then the induced map $(v_1, v_2) : X \to K_1 \times K_2$ will "make all the distinctions among limit points made by either $v_1$ or $v_2$", since the maps $v_1$ and $v_2$ can each be factored through it; further, if we let $K'$ denote the *closure* of the image of $X$ in $K_1 \times K_2$ under that map, and $v' : X \to K'$ the induced map, then all these distinctions are still made in $K'$, and the image of $X$ is dense in this space. We can do the same with an arbitrary family of maps $v_i : X \to K_i$ $(i \in I)$, since Tychonoff's Theorem tells us that the product space $\prod_I K_i$ is again compact.

As in the construction of free groups, to obtain our Stone-Čech compactification by this approach we have to find some *set* of pairs $(K_i, v_i)$ which are "as good as" the class of *all* maps $v$ of $X$ into all compact Hausdorff spaces

$K$. For this purpose, we want a bound on the cardinalities of the *closures* of all images of $X$ under maps into compact Hausdorff spaces $K$. To get this, we would like to say that every point of the closure of the image of $X$ somehow "depends" on the images of elements of $X$, in such a fashion that different points "depend" on these differently; and then bound the number of kinds of "dependence" there can be, in terms of the cardinality of $X$. The next lemma establishes the "different points depend on $X$ in different ways" idea, and the corollary that follows gives the desired bound.

**Lemma 4.17.3.** *Let $K$ be a Hausdorff topological space, and for any $k \in |K|$, let $N(k)$ denote the set of all open neighborhoods of $k$ (open sets in $K$ containing $k$). Then for any map $v$ from a set $X$ into $K$, and any two points $k_1 \neq k_2$ of the closure of $v(X)$ in $K$, one has $v^{-1}(N(k_1)) \neq v^{-1}(N(k_2))$ (where by $v^{-1}(N(k))$ we mean $\{v^{-1}(U) \mid U \in N(k)\}$, a subset of $\mathbf{P}(X)$).*

*Proof.* Since $k_i$ $(i = 1, 2)$ is in the closure of $v(X)$, every neighborhood of $k_i$ in $K$ has nonempty intersection with $v(X)$, i.e., every member of $v^{-1}(N(k_i))$ is nonempty. Since $N(k_i)$ is closed under pairwise *intersections*, so is $v^{-1}(N(k_i))$. But since $K$ is Hausdorff and $k_1 \neq k_2$, these two points possess disjoint neighborhoods, whose inverse images in $X$ will have empty intersection. If the sets $v^{-1}(N(k_1))$ and $v^{-1}(N(k_2))$ were the same, this would give a contradiction to the above nonemptiness observation.    □

Thus, we can associate to distinct points of the closure of $v(X)$ distinct sets of subsets of $X$. Hence,

**Corollary 4.17.4.** *In the situation of the above lemma, the cardinality of the closure of $v(X)$ in $K$ is $\leq 2^{2^{\operatorname{card} X}}$.*    □

So now, given any topological space $X$, let us choose a set $S$ of cardinality $2^{2^{\operatorname{card}|X|}}$, and let $A$ denote the set of all pairs $a = (K_a, u_a)$ such that $K_a$ is a compact Hausdorff space with underlying set $|K_a| \subseteq S$, and $u_a$ is a continuous map $X \to K_a$. (We no longer need to keep track of cardinalities, but if we want to, $\operatorname{card} A \leq 2^{2^{2^{2^{\operatorname{card}|X|}}}}$, assuming $X$ infinite. The two additional exponentials come in when we estimate the number of topologies on a set of $\leq 2^{2^{\operatorname{card}|X|}}$ elements.) Thus, if $v$ is any continuous map of $X$ into a compact Hausdorff space $K$, and we write $K'$ for the closure of $v(X)$ in $K$, then the pair $(K', v)$ will be "isomorphic" to some pair $(K_a, u_a) \in A$, in the sense that there exists a homeomorphism between $K'$ and $K_a$ making the diagram below commute.

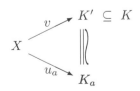

We now form the compact Hausdorff space $P = \prod_{a \in A} K_a$, and the map $u \colon X \to P$ induced by the $u_a$'s, and let $C \subseteq P$ be the closure of $u(X)$. It is easy to show, as we did for groups in §3.3, that the pair $(C, u)$ satisfies the universal property of Definition 4.17.2. Thus:

**Theorem 4.17.5.** *Every topological space $X$ has a Stone-Čech compactification $(C, u)$ in the sense of Definition 4.17.2.*               □

**Exercise 4.17:2.** Show that in the above construction, $u(X)$ will be homeomorphic to $X$ under $u$ if and only if $X$ can be *embedded* in a compact Hausdorff space $K$ (where an "embedding" means a continuous map $f \colon X \to K$ inducing a homeomorphism between $X$ and $f(X)$, the latter set being given the topology induced by that of $K$). Examine conditions on $X$ under which these equivalent statements will hold. Show that for any topological space $X$, there exists a universal map into a space $Y$ embeddable in a compact Hausdorff space, and that this map is always onto, but that it may not be one-to-one. Can it be one-to-one and onto but not a homeomorphism?

Note: Most authors use the term "compactification" to mean a dense *embedding* in a compact space. Hence, they only consider a space $X$ to have a Stone-Čech compactification if the map $u$ that we have constructed *is* an embedding.

**Exercise 4.17:3.** Suppose we leave off the condition "Hausdorff"—does a space $X$ always have a universal map into a *compact* space $C$? A compact $T_1$ space $C$? ...

**Exercise 4.17:4.** Let $C$ be the Stone-Čech compactification of the real line $\mathbb{R}$, and regard $\mathbb{R}$ as a subspace of $C$.

(i)  Show that $C - \mathbb{R}$ has exactly two connected components.

(The above shows that there was a grain of truth in the naive idea that $\mathbb{R} \cup \{+\infty, -\infty\}$ was the universal compactification of $\mathbb{R}$. Exercise 4.17:5 will also be relevant to that idea.)

(ii)  What can you say about path-connected components of $C - \mathbb{R}$?

(iii)  Show that no *sequence* in $\mathbb{R}$ converges to a point of $C - \mathbb{R}$.

A continuous map of $\mathbb{R}$ into a topological space $K$ may be thought of as an open *curve* in $K$. If $K$ is a *metric space* one can define the *length* (possibly infinite) of this curve.

**Exercise 4.17:5.** Show that if $v \colon \mathbb{R} \to K$ is a curve of *finite length* in a compact (or more generally, a complete) metric space $K$, then $v$ factors through the inclusion of $\mathbb{R}$ in $\mathbb{R} \cup \{+\infty, -\infty\}$.

Is the converse true? That is, must every map $\mathbb{R} \to K$ which factors through the inclusion of $\mathbb{R}$ in $\mathbb{R} \cup \{+\infty, -\infty\}$ have finite length?

**Exercise 4.17:6.** (Exploring possible variants of Exercise 4.17:4 and 4.17:5.) It would be nice to get a result like the first assertion of the preceding exercise, but with a purely topological hypothesis on the map $v$, rather than a condition involving a metric on $K$. Consider, for instance, the following condition on a map $v$ of the real line into a compact Hausdorff space $K$:

> (4.17.6)    For every closed set $V \subseteq K$, and open set $U \supseteq V$, the set $v^{-1}(U) \subseteq \mathbb{R}$ has only finitely many connected components that contain points of $v^{-1}(V)$.

(You should convince yourself that this fails for the two cases shown in (4.17.1).)

(i)    Can we replace the assumptions in Exercise 4.17:5 that $K$ is a metric space and $v$ has finite length by (4.17.6) or some similar condition?

(ii)    In the plane $\mathbb{R}^2$, let $X$ be the open unit disc, $C$ the closed unit disc, and $u \colon X \to C$ the inclusion map. Does the pair $(C, u)$ have any universal property with respect to $X$, like that indicated for $\mathbb{R} \cup \{+\infty, -\infty\}$ with respect to $\mathbb{R}$ in the preceding exercise?

(iii)    Does the open disc have a universal path-connected compactification?

(iv)    In general, if $C$ is the Stone-Čech compactification of a "nice" space $X$, what can be said about connected components, path components, homotopy, cohomotopy, etc. of $C - X$?

In §3.4 we saw that we could improve on the construction of the free group on $X$ from "terms" by noting that a certain subset of the terms would make do for all of them. For the Stone-Čech compactification, the "big direct product" construction is subject to a similar simplification. In that construction, we made use of all maps (up to homeomorphism) of $X$ into compact Hausdorff spaces of reasonable size. I claim that we can in fact make all the "distinctions" we need using maps into the closed unit interval, $[0, 1]$! The key fact is that any two points of a compact Hausdorff space $K$ can be separated by a continuous map into $[0, 1]$ (Urysohn's Lemma). I will sketch how this is used.

Let $X$ be any topological space, let $W$ denote the set of all continuous maps $w \colon X \to [0, 1]$, let $u \colon X \to [0, 1]^W$ be the map induced by $(w)_{w \in W}$, and let $C \subseteq [0, 1]^W$ be the closure of $u(X)$. It is immediate that $C$ has the property

> (4.17.7)    Every continuous function of $X$ into $[0, 1]$ is the composite of $u$ with a unique continuous function $C \to [0, 1]$ (namely, the restriction to $C$ of one of the projections $[0, 1]^W \to [0, 1]$).

To show that $C$ has the universal property of the Stone-Čech compactification of $X$, let $K$ be a compact Hausdorff space. We can separate points of $K$ by some set $S$ of continuous maps $s\colon K \to [0, 1]$, hence we can embed $K$ in a "cube" $[0, 1]^S$. (The map $K \to [0, 1]^S$ given by our separating family of functions is one-to-one; hence, as $K$ is compact Hausdorff, it will be a topological embedding [99, Theorem 5.8, p. 141].) Let us therefore assume, without loss of generality, that $K$ is a subspace of $[0, 1]^S$. Now given any map $v\colon X \to K$, we regard it as a map into the overspace $[0, 1]^S$, and get a factorization $v = g\,u$ for a unique map $g\colon C \to [0, 1]^S$ by applying (4.17.7) to each coordinate. Because $K$ is compact, it is closed in $[0, 1]^S$, so $g$ will take $C$, the closure of $u(X)$, into $K$, establishing the universal property of $C$. Cf. [99, pp. 152–153].

Another twist: Following the idea of Exercise 3.3:6, we may regard a point $c$ of the Stone-Čech compactification $C$ of a space $X$ as determining a function $\widetilde{c}$ which associates to every continuous map $v$ of $X$ into a compact Hausdorff space $K$ a point $\widetilde{c}(v) \in K$ —namely, the image of $c$ under the unique extension of $v$ to $C$. This map $\widetilde{c}$ will be "functorial", i.e., will respect continuous maps $f\colon K_1 \to K_2$, in the sense indicated in the diagram below.

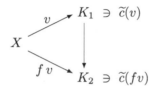

From Urysohn's Lemma one can deduce that $\widetilde{c}$ is determined by its behavior on maps $w\colon X \to [0, 1]$, hence, more generally, by its behavior on maps $w$ of $X$ into closed intervals $[a, b] \subseteq \mathbb{R}$. We carry this observation further in

**Exercise 4.17:7.** A *bounded* real-valued continuous function on $X$ can be regarded as a continuous map from $X$ into a compact subset of $\mathbb{R}$, and our $\widetilde{c}$ can be applied to this map.

(i)  Show that in this way one may obtain from $\widetilde{c}$ a function from the set $B(X)$ of all bounded real-valued continuous functions on $X$ to the real numbers $\mathbb{R}$. (To prove this function well-defined, i.e., that the result of applying $\widetilde{c}$ to a bounded function is independent of our choice of compact subset of $\mathbb{R}$ containing the range of this function, use the functoriality property of $\widetilde{c}$.)

(ii)  Show that this map is a ring homomorphism $B(X) \to \mathbb{R}$ (with respect to the obvious ring structure on $B(X)$).

One can show, further, that every ring homomorphism $B(X) \to \mathbb{R}$ is continuous, and deduce that each such homomorphism is induced by a point of $C$. So one gets another description of the Stone-Čech compactification $C$ of $X$, as the space of homomorphisms into $\mathbb{R}$ of the ring $B(X)$ of bounded continuous real-valued functions on $X$. The topology of $C$ is the function topology on maps of $B(X)$ into $\mathbb{R}$.

Perhaps I have made this approach sound too esoteric. A simpler way of putting it is to note that every bounded continuous real function on $X$ (i.e., every continuous function which has range in a compact subset of $\mathbb{R}$) extends uniquely to a bounded continuous real function on its Stone-Čech compactification $C$, so $B(X) \cong B(C)$; and then to recall that for any compact Hausdorff space $C$, the homomorphisms from the function-ring $B(C)$ into $\mathbb{R}$ are just the evaluation functions at points of $C$.

One can use this approach to get another proof of the existence of the Stone-Čech compactification of a topological space [80, Chapter 6]. This homomorphism space can also be identified with the space of all *maximal ideals* of $B(X)$, equivalently, of all *prime ideals* that are closed in the topology given by the sup norm.

**Exercise 4.17:8.** Suppose $B'$ is any $\mathbb{R}$-*subalgebra* of $B(X)$. Let $C'$ denote the set of all maximal ideals of $B'$. Show that there is a natural map $m \colon C \to C'$. Show by examples that this map can fail to be one-to-one (even if $B'$ separates points of $X$), or to be onto. Try to find conditions for it to be one or the other.

The Stone-Čech compactification of the topological space $\mathbb{R}$ is enormous, since examples like (4.17.1) show that it has to be compatible with distinctions among points made by a vast class of closures of images of $\mathbb{R}$. One may wonder whether any noncompact Hausdorff space has a Stone-Čech compactification that is more modest. Can it add only one point to the space, for instance? The next exercise finds conditions for this to happen. We shall see in Exercise 5.5:17 how to get a space $X$ that satisfies these conditions.

**Exercise 4.17:9.** Let $X$ be a noncompact topological space which can be embedded in a compact Hausdorff space. Show that the following conditions are equivalent.

(a) The Stone-Čech compactification of $X$ has the form $u(X) \cup \{y\}$, where $u$ is the universal map of $X$ into that compactification, and $y$ is a single point not in $u(X)$.

(b) Of any two disjoint closed subsets $F, G \subseteq X$, at least one is compact.

(c) Every continuous function $X \to [0, 1]$ is constant on the complement of some compact subset of $X$.

One can also consider universal constructions which mix topological and algebraic structure:

**Exercise 4.17:10.** Let $G$ be any *topological group* (a group given with a Hausdorff topology on its underlying set, such that the group operations are continuous). Show that there exists a universal pair $(C, h)$, where $C$ is a compact topological group, and $h: G \to C$ a continuous group homomorphism.

This is called the *Bohr compactification* of $G$.

Show that $h(G)$ is dense in $C$. Is $h$ generally one-to-one? A topological embedding? What will be the relation between $C$ and the Stone-Čech compactification of the underlying topological space of $G$?

If it helps, you might consider some of these questions in the particular case where $G$ is the additive group of the real line.

In [110, §41], the Bohr compactification of a topological group $G$ is obtained as the maximal ideal space of a subring of $B(G)$, the subring of "almost periodic" functions.

Most often, complex- rather than real-valued functions are used in the ring-of-bounded-functions constructions we have discussed.

## 4.18. Universal covering spaces

Let $X$ be a pathwise connected topological space with a basepoint (distinguished point) $x_0$. (Formally, this would be defined as a 3-tuple $(|X|, T, x_0)$, where $|X|$ is a set, $T$ is a pathwise connected topology on $|X|$, and $x_0$ is an element of $|X|$.)

A *covering space* of $X$ means a pair $(Y, c)$, where $Y$ is a pathwise connected space with a basepoint $y_0$, and $c$ is a continuous basepoint-preserving map $Y \to X$, such that every $x \in X$ has a neighborhood $V$ such that $c^{-1}(V)$ is homeomorphic, as a space mapped to $V$, to a direct product of $V$ with a discrete space. (Draw a picture!) Such a $c$ will have the *unique path-lifting property*. Given any continuous map $p: [0, 1] \to X$ taking $0$ to $x_0$, there will exist a unique continuous map $\widetilde{p}: [0, 1] \to Y$ taking $0$ to $y_0$ such that $p = c\widetilde{p}$. Further, $\widetilde{p}$ will depend continuously on $p$ in the appropriate function-space topology.

Given $X$, consider any covering space $(Y, c)$ of $X$, and let us ask what points of $Y$ we can "describe" in a well-defined manner.

Of course, we have the basepoint, $y_0$. Further, for every path $p$ in $X$ starting at the basepoint $x_0$, we know there will be a unique lifting of $p$ to a path $\widetilde{p}$ in $Y$ starting from $y_0$; so $Y$ also has all points of this lifted path. It is enough, however, to note that we have the endpoint $\widetilde{p}(1)$ of each such lifted path, since all the other points of $\widetilde{p}$ can be described as endpoints of liftings of "subpaths" of $p$. In fact, every $y \in Y$ will be the endpoint $\widetilde{p}(1)$ of a lifted path in $X$. For $Y$ was assumed pathwise connected, hence for any $y \in Y$ we can find a path $q$ in $Y$ with $q(0) = y_0$, $q(1) = y$. Letting $p = cq$, a path in $X$, we see that $q = \widetilde{p}$, so $y = \widetilde{p}(1)$.

Suppose $p$ and $p'$ are two paths in $X$; when will $\widetilde{p}(1)$ and $\widetilde{p}'(1)$ be the same point of $Y$? Clearly, a necessary condition is that these two points have the same image $x$ in $X : p(1) = p'(1) = x$. Assuming this condition, note that if $p$ and $p'$ are homotopic in the class of paths in $X$ from $x_0$ to $x$, then as one continuously deforms $p$ to $p'$ in this class, the lifted path in $Y$ will vary continuously, hence its endpoint in $c^{-1}(x)$ will vary continuously. But $c^{-1}(x)$ is discrete, so the endpoint must remain constant. Thus, $p$'s being homotopic to $p'$ in the class of paths with these specified endpoints implies $\widetilde{p}(1) = \widetilde{p}'(1)$.

So in general, we get a point of $Y$ for every homotopy class $[p]$ of paths in $X$ with initial point $x_0$ and common final point. In a particular covering space $Y$, there may or may not be further equalities among these points of $Y$; but we can ask whether, if we write $U$ for the set of such homotopy classes of paths, and $u$ for the map from $U$ to $X$ defined by $u([p]) = p(1)$, we can make $U$ a topological space in such a way that the pair $(U, u)$ is a covering space for $X$. Under appropriate assumptions on the topology of $X$ (the hypotheses used in [90] are that $X$ is connected, locally pathwise connected, and semi-locally simply connected), this can indeed be done. The resulting covering space $U$ has a unique continuous map onto each covering space $Y$ of $X$, which respects basepoints and respects the maps into $X$. Hence $(U, u)$ is called the *universal covering space* of $X$.

The universal covering space is a versatile animal—like the direct product of groups, it has, in addition to the above left universal property, a right universal one:

It is not hard to show that $U$ is simply connected. Consider, now, pairs $(S, c)$, where $S$ is a simply connected pathwise connected topological space with basepoint $s_0$, and $c\colon S \to X$ a basepoint-respecting continuous map. Let us ask, for such a space $S$, the question that we noted in §4.8 as leading to *right universal* constructions: If $s$ is an arbitrary point of $S$, what data will it determine that can be formulated in terms of the given space $X$? Well, obviously $s$ determines the point $c(s) \in X$. To get more information, note that since $S$ is pathwise connected, there will be some path $q$ in $S$ connecting $s_0$ to $s$; and since $S$ is *simply* connected, all such paths $q$ are homotopic. Applying $c$ to these paths, we see that $s$ determines a *homotopy class* of paths in $X$ from $x_0$ to $c(s)$. But as we have just noted, the set of homotopy classes of paths from $x_0$ to points of $X$ can (under appropriate conditions) itself be made into a simply connected space, the universal covering space of $X$. One deduces that this space $U$ is right universal among simply connected spaces with basepoint, given with maps into $X$ (diagram below).

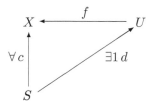

We could also look for a *right* universal covering space for $X$, or a simply connected space with basepoint having a *left* universal map into $X$. But these turn out to be uninteresting: They are $X$ itself, and the one-point space.

There are many other occurrences of universal constructions in topology. Some, like the two considered in this and the preceding section, can be approached in the same way as universal constructions in algebra. Others, used in algebraic topology, are different in that one is interested, not in maps being equal, unique, etc., but *homotopic*, unique *up to homotopy*, etc. These conditions can be brought into the same framework as our other universal properties via the formalism of category theory (Chapters 7 and 8 below), but the tasks of constructing and studying the objects these conditions characterize require different approaches, which we will not treat in this course.

# Part II. Basic Tools and Concepts

In the next five chapters we shall assemble the concepts and tools needed for the development of a general theory of algebras and of universal constructions among them.

In Chapters 5 and 6, we discuss ordered sets, lattices, closure operators, and related concepts, since these will be used repeatedly. Because of the relation between well-ordering and the Axiom of Choice, after discussing well-ordered sets, I take the occasion to review briefly the Zermelo-Fraenkel axioms for set theory, and several statements equivalent to the Axiom of Choice.

Clearly, the general context for studying universal constructions should be some model of "a system of mathematical objects and the maps among them". This is provided by the concept of a *category*. We develop the basic concepts of category theory in Chapter 7, and in Chapter 8 we formalize universal properties in category-theoretic terms.

Finally, in Chapter 9 we introduce the categories that will be of special interest to us: the *varieties of algebras*.

# Chapter 5
# Ordered Sets, Induction, and the Axiom of Choice

## 5.1. Partially ordered sets

We began Chapter 2 by making precise the concept of a group. Let us now do the same for that of a partially ordered set.

A partial ordering on a set is an instance of a "relation". This is a different sense of the word from that of the last two chapters. These notes will deal extensively with both kinds of "relations"; which sense is meant will generally be clear from context. When there is danger of ambiguity, I will make the distinction explicit, as I do, for instance, in the index.

Intuitively, a relation on a family of sets $X_1, \ldots, X_n$ means a *condition* on $n$-tuples $(x_1, \ldots, x_n)$ $(x_1 \in X_1, \ldots, x_n \in X_n)$. Since the information contained in the relation is determined by the set of $n$-tuples that satisfy it, the relation is taken to *be* this set in the formal definition, given below. That the relation is viewed as a "condition" comes out in the notation and language used.

**Definition 5.1.1.** *If* $X_1, \ldots, X_n$ *are sets, a* relation *on* $X_1, \ldots, X_n$ *means a subset* $R \subseteq X_1 \times \cdots \times X_n$. *Relations are often written as predicates; i.e., the condition* $(x_1, \ldots, x_n) \in R$ *may be written* $R(x_1, \ldots, x_n)$, *or* $R\,x_1 \ldots x_n$, *or, if* $n = 2$, *as* $x_1 R x_2$.

*A relation on* $X, \ldots, X$, *i.e., a subset* $R \subseteq X^n$, *is called an* $n$-ary *relation on* $X$.

*If* $R$ *is an* $n$-ary *relation on* $X$, *and* $Y$ *is a subset of* $X$, *then the* restriction *of* $R$ *to* $Y$ *means* $R \cap Y^n$, *regarded as an* $n$-ary *relation on* $Y$.

We now recall

**Definition 5.1.2.** *A* partial ordering *on a set* $X$ *means a binary relation* "$\leq$" *on* $X$ *satisfying the conditions*

© Springer International Publishing Switzerland 2015
G.M. Bergman, *An Invitation to General Algebra and Universal Constructions*, Universitext, DOI 10.1007/978-3-319-11478-1_5

$$(\forall\, x \in X) \quad x \leq x \qquad\qquad\qquad\qquad \text{(reflexivity)},$$
$$(\forall\, x, y \in X) \quad x \leq y \,\wedge\, y \leq x \implies x = y \qquad \text{(antisymmetry)},$$
$$(\forall\, x, y, z \in X) \quad x \leq y, \ y \leq z \implies x \leq z \qquad \text{(transitivity)}.$$

A total ordering *on* $X$ *means a partial ordering which also satisfies*

$$(\forall\, x, y \in X) \quad x \leq y \ \text{or} \ y \leq x.$$

A partially (*respectively* totally) ordered set *means a set* $X$ *given with a partial (total) ordering* $\leq$. (*"Partially ordered set" is often shortened to "poset", though we will not do so here.*)

If $X$ *is partially ordered by* $\leq$, *and* $Y$ *is a subset of* $X$, *then* $Y$ *will be understood to be partially ordered by the restriction of* $\leq$, *which will be denoted by the same symbol unless there is danger of ambiguity. This is called the* induced ordering *on* $Y$.

A more formal definition would make a partially ordered set a pair $P = (|P|, \leq)$ where $\leq$ is a partial ordering on $|P|$. But for us, partially ordered sets will in general be tools rather than the objects of our study, and it would slow us down to always maintain the distinction between $P$ and $|P|$; so we shall usually take the informal approach of understanding a partially ordered set to mean a set $P$ for which we "have in mind" a partial ordering relation $\leq$. Occasionally, however, we shall be more precise and refer to the pair $(|P|, \leq)$.

Standard examples of partially ordered sets are the set of real numbers with the usual relation $\leq$, the set $\mathbf{P}(X)$ of subsets of any set $X$ under the inclusion relation $\subseteq$, and the set of positive integers under the relation "$\,|\,$", where $m \mid n$ means $m$ divides $n$.

A *total ordering* is also called a *linear ordering*. The term "ordered" without any qualifier is used by some authors as shorthand for "partially ordered", and by others for the stronger condition "totally ordered"; we will here generally specify "partially" or "totally".

The versions of the concepts of homomorphism and isomorphism appropriate to partially ordered sets are given by

**Definition 5.1.3.** *If* $X$ *and* $Y$ *are partially ordered sets, an* isotone *map from* $X$ *to* $Y$ *means a function* $f : X \to Y$ *such that* $x_1 \leq x_2 \implies f(x_1) \leq f(x_2)$.

*An invertible isotone map whose inverse is also isotone is called an* order isomorphism.

**Exercise 5.1:1.** Give an example of an isotone map of partially ordered sets which is invertible as a set map, but which is not an order isomorphism.

Some well-known notation: When $\leq$ is a partial ordering on a set $X$, one commonly writes $\geq$ for the opposite relation; i.e., $x \geq y$ means $y \leq x$. Clearly the relation $\geq$ satisfies the same conditions of reflexivity, antisymmetry and transitivity as $\leq$.

This leads to a semantic problem: As long as $\geq$ is just an auxiliary notation used in connection with the given ordering $\leq$, one thinks of an element $x$ as being "smaller" (or "lower") than an element $y \neq x$ if $x \leq y$. But the fact that $\geq$ is also reflexive, antisymmetric and transitive means that one can take it as a new partial ordering on $X$, i.e., consider the partially ordered set $(X, \geq)$, and one should consider $x$ as "smaller" than $y$ in this partially ordered set if the pair $(x, y)$ belongs to this *new* ordering. Such properties as which maps $X \to Y$ are isotone (with respect to a fixed partial ordering on $Y$) clearly change when one goes from considering $X$ under $\leq$ to considering it under $\geq$.

The set $X$ under the opposite of the given partial ordering is called the *opposite* of the original partially ordered set. When one uses the formal notation $P = (|P|, \leq)$ for a partially ordered set, one can write $P^{\mathrm{op}} = (|P|, \geq)$. One may also replace the symbol $\geq$ by $\leq^{\mathrm{op}}$, writing $P^{\mathrm{op}} = (|P|, \leq^{\mathrm{op}})$. Thus, if $x$ is smaller than $y$ in $P$, i.e., $x \leq y$, then $y$ is smaller than $x$ in $P^{\mathrm{op}}$, i.e., $y \leq^{\mathrm{op}} x$. ("Dual ordering" is another term often used, and $*$ is sometimes used instead of $^{\mathrm{op}}$.)

In these notes we shall rarely make explicit use of the opposite partially ordered set construction. But once one gets past the notational confusion, the symmetry in the theory of partially ordered sets created by that construction is a useful tool: After proving any result, one can say "By duality ...", and immediately deduce the corresponding statement with all ordering relations reversed.

One also commonly uses $x < y$ as an abbreviation for $(x \leq y) \wedge (x \neq y)$, and of course $x > y$ for $(x \geq y) \wedge (x \neq y)$. These relations do *not* satisfy the same conditions as $\leq$. The conditions they satisfy are noted in

**Exercise 5.1:2.** Show that if $\leq$ is a partial ordering on a set $X$, then the relation $<$ is transitive and is *antireflexive*, i.e., satisfies $(\forall\, x \in X)\; x \not< x$. Conversely, show that any transitive antireflexive binary relation $<$ on a set $X$ is induced in the above way by a unique partial ordering $\leq$.

A relation $<$ with these properties (transitivity and antireflexivity) might be called a "partial strict ordering". One can thus refer to "the partial strict ordering $<$ corresponding to the partial ordering $\leq$," and "the partial ordering $\leq$ corresponding to the partial strict ordering $<$". Of course, for a partial ordering denoted by a symbol such as "$|$" ("divides"), or $R$ (a partial ordering written as a binary relation), there is no straightforward symbol for the corresponding partial strict ordering.

**Exercise 5.1:3.** For partially ordered sets $X$ and $Y$, suppose we call a function $f\colon X \to Y$ a *strict isotone map* if $x < y \implies f(x) < f(y)$. Show that

$$\text{one-to-one and isotone} \implies \text{strict isotone} \implies \text{isotone,}$$

but that neither implication is reversible.

In contexts where "$\leq$" already has a meaning, if another partial ordering has to be considered, it is often denoted by a variant symbol such as $\preccurlyeq$. One then uses corresponding symbols $\succcurlyeq$, $\prec$, $\succ$ for the opposite order, the strict order relation, etc. (However, order-theorists dealing with a partial ordering $\leq$ sometimes write $y \succ x$ to mean "$y$ covers $x$", that is "$y > x$ and there is no $z$ between $y$ and $x$". When the symbol is used this way, it cannot be used for the strict relation associated with a second ordering. We shall not use the concept of covering in these notes.)

A somewhat confused situation is that of symbols for the *subset* relation. Most often, the notation one would expect from the above discussion is followed: $\subseteq$ is used for "is a subset of", $\supseteq$ for the opposite relation, and $\subset$, $\supset$ for strict inclusions. We will follow these conventions here. However, many authors, especially in Eastern Europe, write $\subset$ for "is a subset of", a usage based on the view that since this is a more fundamental concept than that of a proper subset, it should be denoted by a primitive symbol, not by one obtained by adding an extra mark to the symbol for "proper subset". Such authors use $\subsetneqq$ (or typographical variants) for "proper subset" (and the reversed symbols for the reversed relations). There was even at one time a movement to make "$<$" mean "less than or equal to", with $\lneqq$ for strict inequality. Together with the above set-theoretic usage, this would have formed a consistent system, but the idea never got off the ground. Finally, many authors, for safety, use a mixed system: $\subseteq$ for "subset" and $\subsetneqq$ for "proper subset". (That was the notation used in the first graduate course I took, and I sometimes follow it in my papers. However, I rarely need a symbol for strict inclusion, so the question of how to write it seldom comes up.)

Although partially ordered sets are not algebras in the sense in which we shall use the term, many of the kinds of universal constructions we have considered for algebras can be carried out for them. In particular

**Definition 5.1.4.** *Let $(X_i)_{i \in I}$ be a family of partially ordered sets. Then their* direct product *will mean the partially ordered set having for underlying set the direct product of the underlying sets of the $X_i$, ordered so that $(x_i)_{i \in I} \leq (y_i)_{i \in I}$ if and only if $x_i \leq y_i$ for all $i \in I$.*

**Exercise 5.1:4.**

(i)  Verify that the above relation is indeed a partial ordering on the product set, and that the resulting partially ordered set has the appropriate universal property to be called the direct product of the partially ordered sets $X_i$.

(ii)  Let $X$ be a set and $R$ a binary relation on $X$. Show that there exists a universal example of a partially ordered set $(Y, \leq)$ with a map $u : X \to Y$ such that for all $x_1, x_2 \in X$, one has $(x_1, x_2) \in R \implies u(x_1) \leq u(x_2)$ in $Y$. This may be called the partially ordered set *presented* by the generators $X$ and the relation-set $R$ (analogous to the presentations of groups, monoids and rings we saw in §§ 4.3, 4.10 and 4.12). Will the map $u$ in general be one-to-one? Onto?

(iii) Determine whether there exist constructions with the universal properties of the *coproduct* of two partially ordered sets, and of the *free* partially ordered set on a set $X$. Describe these if they exist.

(iv) Discuss the problem of *imposing* a set $R$ of further relations on a given partially ordered set $(X, \leq)$; i.e., of constructing a universal isotone map of $X$ into a partially ordered set $Y$ such that the images of the elements of $X$ also satisfy the relations comprising $R$. If this can be done, examine the properties of the construction.

We have noted that for any set $X$, the set $\mathbf{P}(X)$ of subsets of $X$ is partially ordered by $\subseteq$. Given a partially ordered set $S$, we may look for universal ways of representing $S$ by subsets of a set $X$. Note that if $f : X \to Y$ is a map between sets, then $f$ induces, in natural ways, both an isotone map $\mathbf{P}(X) \to \mathbf{P}(Y)$ and an isotone map $\mathbf{P}(Y) \to \mathbf{P}(X)$, the first taking subsets of $X$ to their images under $f$, the second taking subsets of $Y$ to their inverse images. Let us call these the "direction-preserving construction" and the "direction-reversing construction" respectively. Thus, given a partially ordered set $S$, there are four universal sets we might look for: a set $X$ having an isotone map $S \to \mathbf{P}(X)$ universal in terms of the direction-preserving construction of maps among power sets, a set $X$ with such a map universal in terms of the direction-reversing construction, and sets $X$ with isotone maps in the reverse direction, $\mathbf{P}(X) \to S$, universal for the same two constructions of maps among power sets.

**Exercise 5.1:5.**

(i)   Write out the universal properties of the four possible constructions indicated.

(ii)  Investigate which of the four universal sets exist, and describe these as far as possible.

**Definition 5.1.5.** *Let $X$ be a partially ordered set, $S$ a subset of $X$, and $s$ an element of $S$. Then $s$ is said to be* minimal *in $S$ if there is no $t \in S$ with $t < s$, while $s$ is said to be the* least *element of $S$ if for all $t \in S$, $s \leq t$. The terms* maximal *and* greatest *are used for the dual concepts.*

(There was really no need to refer to $X$ in the above definition, since the properties in question just depend on the set $S$ and the induced order relation on it; but these concepts are often applied to subsets of larger partially ordered sets, so I included this context in the definition.)

**Exercise 5.1:6.** Let $X$ be a partially ordered set.

(i)   Show that if $X$ has a least element $x$, then $x$ is the unique minimal element of $X$.

(ii)  If $X$ is finite, show conversely that a unique minimal element, if it exists, is a least element.

(iii)  Give an example showing that if $X$ is not assumed finite, this converse is false.

(I have included this exercise as a warning. I have many times found myself unwittingly writing or saying "unique minimal element" when I meant "least element". It somehow sounds more precise; but it doesn't mean the same thing.)

**Exercise 5.1:7.** Let  $(X, \leq)$  be a partially ordered set. Then the pair  $(X, \leq)$  constitutes a presentation of itself as a partially ordered set in the sense of Exercise 5.1:4(ii); but of course, there may be proper subsets  $R$  of the relation  $\leq$  such that  $(X, R)$  is a presentation of the same partially ordered set. (I.e., such that  $R$  "generates"  $\leq$  in an appropriate sense.)

(i)   If  $X$  is finite, show that there exists a *least* subset of  $R$  which generates  $\leq$ .

(ii)  Show that this is not in general true for infinite  $X$ .

Point (i) of the above exercise is the basis for the familiar way of diagraming finite partially ordered sets. One draws a picture with vertices representing the elements of the set, and edges corresponding to the members of the least relation generating the partial ordering; i.e., the smallest set of order-relations from which all the others can be deduced. The higher point on each edge represents the larger element under the partial ordering. This picture is called the *Hasse diagram* of the given partially ordered set.

For example, the picture below represents the set of all nonempty subsets of  $\{0, 1, 2\}$ , partially ordered by inclusion. The relation  $\{1\} \leq \{0, 1, 2\}$  is not shown explicitly, because it is a consequence of the relations  $\{1\} \leq \{0, 1\} \leq \{0, 1, 2\}$  (and also of  $\{1\} \leq \{1, 2\} \leq \{0, 1, 2\}$ ).

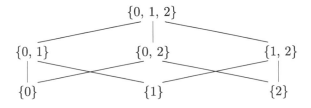

Here are a few more pieces of commonly used terminology.

**Definition 5.1.6.** *Let*  $\leq$  *be a partial ordering on a set*  $X$ .

*If*  $x$ ,  $y$  *are elements of*  $X$  *with*  $x \leq y$ , *then the* interval  $[x, y]$  *means the subset*  $\{z \in X \mid x \leq z \leq y\}$ , *with the induced partial ordering*  $\leq$ .

A *subset C of a partially ordered set X which is totally ordered under the induced ordering is called a* chain *in X.*

*Elements x and y of X are called* incomparable *if neither $x \leq y$ nor $y \leq x$ holds. A subset $Y \subseteq X$ is called an* antichain *if every pair of distinct elements of Y is incomparable.*

*An element $x \in X$ is said to* majorize *a subset $Y \subseteq X$ if for all $y \in Y$, $y \leq x$. One similarly says x* majorizes *an element y if $y \leq x$.*

*A subset Y of X is said to be* cofinal *in X if every element of X is majorized by some element of Y.*

Note that in addition to the above order-theoretic meaning of "chain", there is a nonspecialized use of the word; for instance, one speaks of a "chain of equalities $x_1 = x_2 = \cdots = x_n$." We shall at times use the term in this nontechnical way, relying on context to avoid ambiguity.

Unfortunately, there are no standard terms for the duals of the concepts "majorize" and "cofinal". One occasionally sees "minorize" and "coinitial", but these seem awkward. I often write "downward cofinal" for the dual of "cofinal". The best circumlocution I know for the dual of "majorizes" is, "is majorized by every element of".

The concept of cofinality, noted in the last sentence of Definition 5.1.6, probably originated in topology: If $s$ is a point of a topological space $S$, and $N(s)$ the set of all neighborhoods of $s$, then a *neighborhood basis* of $s$ means a subset $B \subseteq N(s)$ cofinal in that set, under the ordering by reverse inclusion. The virtue of this concept is that one can verify that a function on $S$ approaches some limit at $s$ by checking its behavior on members of such a $B$. E.g., one generally checks continuity of a function at a point $s$ of the real line using the cofinal system of neighborhoods $\{(s - \varepsilon, s + \varepsilon) \mid \varepsilon > 0\}$.

**Exercise 5.1:8.**

(i)  Show that if $X$ is a *finite* partially ordered set, then a subset $Y$ is cofinal in $X$ if and only if it contains all maximal elements of $X$.

(ii)  Show by example that this is not true for infinite partially ordered sets. Is one direction true?

**Exercise 5.1:9.** Let $X$ be a finite partially ordered set. One defines the *height* of $X$ as the maximum of the cardinalities of all chains in $X$, and the *width* of $X$ as the maximum of the cardinalities of all antichains in $X$.

(i)  Show that $\operatorname{card}(X) \leq \operatorname{height}(X) \cdot \operatorname{width}(X)$.

(The above result fails for infinite partially ordered sets, as will be shown in Exercise 5.6:12(ii).)

(ii)  Must every (or some) chain in $X$ of maximal cardinality have nonempty intersection with every (or some) antichain of maximal cardinality?

**Definition 5.1.7.** *Let $\leq$ and $\preccurlyeq$ be partial orderings on a set X. Then one says $\leq$ is an* extension *or* strengthening *(or sometimes, a* refinement*) of $\preccurlyeq$ if it contains the latter, as subsets of $X \times X$; that is, if $x \preccurlyeq y \implies x \leq y$.*

The relation of "extension" is a partial ordering on the set of partial orderings on $X$. This fact can be looked at as follows. If we regard each partial ordering on $X$ as a subset $R \subseteq X \times X$, and partially order the class of all subsets of $X \times X$ by inclusion (the relation $\subseteq$), then the relation of extension is the *restriction* of this partial ordering to the subclass of those $R \subseteq X \times X$ which are partial orders. This observation saves us the work of verifying that the concept of extension satisfies the conditions for being a partial order, since we know that the restriction of a partial order on a set to any subset is again a partial order. Many of the partial orderings that arise naturally in mathematics are, similarly, restrictions of the inclusion relation or of some other natural partial ordering on a larger set.

**Exercise 5.1:10.** Consider the set of all partial orderings on a set to be partially ordered as above.

(i)   Show that the *maximal* elements in the set of all partial orderings on a set $X$ are precisely the *total* orderings.

(ii)   How many maximal elements does the set of partial orderings of a set of $n$ elements have?

(iii)   How many minimal elements does the set of partial orderings of a set of $n$ elements have?

(iv)   Show that every partial ordering on a finite set $X$ is the set-theoretic intersection of a set of total orderings.

If $\preccurlyeq$ is a partial ordering on a finite set, the smallest number of total orderings that can be intersected to get $\preccurlyeq$ is called the "order dimension" of the partially ordered set. The next question is open-ended.

(v)   What can you say about the order dimension function? (You might look for general bounds on the order dimension of a partially ordered set of $n$ elements, try to evaluate the order dimensions of particular partially ordered sets, look at the behavior of order dimension under various constructions, etc.)

Here is an outstanding open problem.

**Exercise 5.1:11.** Let $(X, \preccurlyeq)$ be a finite partially ordered set. Let $N$ denote the number of total orderings "$\leq$" on $X$ extending $\preccurlyeq$ ("linearizations of $\preccurlyeq$") and for $x, y \in X$, let $N_{x,y}$ denote the number of these extensions "$\leq$" which satisfy $x \leq y$.

(i)   Prove or disprove, if you can,

*Fredman's conjecture*: For any $(X, \preccurlyeq)$ such that $\preccurlyeq$ is not a total order, there exist elements $x, y \in X$ such that

(5.1.8) $$1/3 \leq N_{x,y}/N \leq 2/3.$$

If you cannot settle this open question, here are some special cases to look at:

(ii) Let $r$ be a positive integer, and let $X$ be the partially ordered set consisting of a chain of $r$ elements, $p_1 \prec \cdots \prec p_r$, and an element $q$ incomparable with all the $p_i$. What are $N$ and the $N_{p_i, q}$ in this case? Verify Fredman's conjecture for this partially ordered set.

(iii) Is the above example consistent with the stronger assertion that if $X$ has no greatest element, then an $x$ and a $y$ satisfying (5.1.8) can be chosen from among the *maximal* elements of $X$? With the assertion that for every two maximal chains in $X$, one can choose an $x$ in one of these chains and a $y$ in the other satisfying (5.1.8)? If one or the other of these possible generalizations of Fredman's Conjecture is not excluded by the above example, can you find an example that does exclude it?

(iv) Let $r$ again be a positive integer, and let $X$ be the set $\{1, \ldots, r\}$ partially ordered by the relation $\prec$ under which $i \prec j$ if and only if $j - i \geq 2$ (where in this definition $\geq$ has the usual meaning for integers). Verify the conjecture in this case as well. For how many two-element subsets $\{i, j\}$ are $i$ and $j$ incomparable under $\prec$, and of these, how many satisfy (5.1.8)?

(v) If $X$ is any partially ordered set such that the function $N_{x, y}/N$ never takes on the value $1/2$, define a relation $\leq_!$ on $X$ by writing $x \leq_! y$ if either $x = y$, or $N_{x, y}/N > 1/2$. Determine whether this is *always*, *sometimes* or *never* a (total) ordering on $X$. Show that for any $X$ which is a counterexample to Fredman's Conjecture, $\leq_!$ must be a total ordering on $X$.

Fredman's conjecture arose as follows. Suppose that $(X, \leq)$ is a finite *totally* ordered set, but that one has only partial information on its ordering; namely, one knows for certain pairs of elements $x, y$ which element is greater, but not for all pairs. This partial information is equivalent to a partial ordering $\prec$ on $X$ weaker than $\leq$. Suppose one is capable of "testing" pairs of elements to determine their relation under $\leq$, and one wants to fully determine $\leq$ using a small number of such tests. One would like to choose each test so that it approximately halves the number of candidate orderings. Examples show that one cannot do as well as that; but Fredman's Conjecture would imply that one can always reduce this number by at least a third at each step. For some literature on the subject, see [63] and papers referred to there, and more recently, [123].

Conceivably, one might be able to prove Fredman's conjecture by assuming one had a counterexample, and considering the peculiar place the relation $\leq_!$ of part (v) of the above exercise would have to have among the total orderings on $X$ extending $\prec$. One can see something of the structure of the set of all total orderings on a set from the next exercise.

**Exercise 5.1:12.** Define the *distance* between two total orderings $\leq_i$, $\leq_j$ on a finite set $X$ as

$$d(\leq_i, \leq_j) = \text{number of pairs of elements } (x, y)$$
$$\text{such that } x <_i y \text{ but } x >_j y.$$

Show that $d$ is a metric on the set of all total orderings, and that for any partial ordering $\preccurlyeq$ on $X$, any two total orderings extending $\preccurlyeq$ can be connected by a chain (*not* meant in the order-theoretic sense!) $\leq_1, \ldots, \leq_n$ where each $\leq_i$ is a total ordering extending $\preccurlyeq$, and $d(\leq_i, \leq_{i+1}) = 1$ for $i = 1, \ldots, n - 1$.

Here is another open question.

**Exercise 5.1:13** (*Reconstruction problem for finite partially ordered sets.*) Let $P$ and $Q$ be finite partially ordered sets with the same number $n > 3$ of elements, and suppose they can be indexed $P = \{p_1, \ldots, p_n\}$, $Q = \{q_1, \ldots, q_n\}$ in such a way that for each $i$, $P - \{p_i\}$ and $Q - \{q_i\}$ are isomorphic as partially ordered sets. Must $P$ be isomorphic to $Q$?

(Here nothing is assumed about what bijections give the isomorphisms $P - \{p_i\} \cong Q - \{q_i\}$. We are definitely not assuming that they are the correspondences $p_j \longleftrightarrow q_j$ $(j \neq i)$; if we did, the question would have an immediate positive answer. A way to state the hypothesis without referring to such a correspondence is to say that the families of isomorphism classes of partially ordered $(n-1)$-element subsets of $P$ and of $Q$, counting multiplicities, are the same.)

If the above question has an affirmative answer, then "one can reconstruct $P$ from its $(n-1)$-element partially ordered subsets", hence the name of the problem. What is known on the subject is surveyed in [127].

(The analogous reconstruction problem for *graphs* is also open, and better known [86].)

## 5.2. Digression: preorders

One sometimes encounters binary relations which, like partial orderings, are reflexive and transitive, but which do not satisfy the antisymmetry condition. For instance, although the relation "divides" on the positive integers is a partial ordering, the relation "divides" on the set of all integers is not antisymmetric, since every $n$ divides $-n$ and vice versa. More generally, on the elements of any commutative integral domain, "divides" is a reflexive transitive relation, but for every element $x$ and invertible element $u$, $x$ and $ux$ each divide the other. Similarly, on a set of *propositions* (sentences in some formal language) about a mathematical situation, the relation $P \implies Q$ is reflexive and transitive, but not generally antisymmetric: Distinct sentences can each imply the other, i.e., represent equivalent conditions.

To cover such situations, one makes

**Definition 5.2.1.** *A reflexive transitive* (not *necessarily antisymmetric*) *binary relation on a set* $X$ *is called a* preorder *on* $X$.

The concept of a preordered set can be reduced in a natural way to a combination of two sorts of structure that we already know:

**Proposition 5.2.2.** *Let* $X$ *be a set. Then the following data are equivalent.*

(i)    *A preorder* $\preccurlyeq$ *on* $X$.

(ii)   *An equivalence relation* $\approx$ *on* $X$, *and a partial ordering* $\leq$ *on the set* $X/\approx$ *of equivalence classes.*

   *Namely, to go from* (i) *to* (ii), *given the preorder* $\preccurlyeq$ *define* $x \approx y$ *to mean* $(x \preccurlyeq y) \wedge (y \preccurlyeq x)$, *and for any two elements* $[x], [y] \in X/\approx$, *define* $[x] \leq [y]$ *in* $X/\approx$ *to hold if and only if* $x \preccurlyeq y$ *holds in* $X$.

   *Inversely, given, as in* (ii), *an equivalence relation* $\approx$ *and a partial ordering* $\leq$ *on* $X/\approx$, *one gets a preorder* $\preccurlyeq$ *by defining* $x \preccurlyeq y$ *to hold in* $X$ *if and only if* $[x] \leq [y]$ *in* $X/\approx$.                                           □

**Exercise 5.2:1.** Prove the above proposition. (This requires one verification of well-definedness, and some observations showing why the two constructions, performed successively in either order, return the original data.)

This is neat: A reflexive transitive relation (a preorder) decomposes into a reflexive transitive *symmetric* relation (an equivalence relation) and a reflexive transitive *antisymmetric* relation (a partial ordering).

As an example, if we take the set of elements of a commutative integral domain $R$, preordered by divisibility, and divide out by the equivalence relation of mutual divisibility, we get a partially ordered set, which can be identified with the set of principal ideals of $R$ partially ordered by reverse inclusion.

In view of Proposition 5.2.2, there is no need for a *theory* of preorders—that is essentially subsumed in the theory of partial orderings. But it is valuable to have the term "preorder" available when such relations arise.

The remainder of this section consists of some exercises on preorders which will not be used in subsequent sections. Exercises 5.2:2–5.2:9 concern a class of preorders having applications to ring theory, group theory, and semigroup theory. (Dependencies within that group of exercises: All later exercises depend on 5.2:2–5.2:3, and 5.2:5 is also assumed in 5.2:6–5.2:9. If you wish to hand in one of these exercises without writing out the details of others on which it depends, you should begin with a summary of the results from the latter that you will be assuming. You might check that summary with your instructor first.) The last exercise of this section, in contrast, will relate preorders and topologies.

**Exercise 5.2:2.** If $f$ and $g$ are nondecreasing functions from the positive integers to the nonnegative integers, let us write $f \preccurlyeq g$ if there exists a positive integer $N$ such that for all $i$, $f(i) \leq g(Ni)$.

(i)   Show that $\preccurlyeq$ is a preorder, but not a partial order, on the set of nondecreasing functions.

(ii)   On the subset of functions consisting of all polynomials with nonnegative integer coefficients, get a description of $\preccurlyeq$ in terms of the expressions for these polynomials, and determine its "decomposition" as in Proposition 5.2.2.

(iii) Do the same for the union of the set of polynomials of (ii), and the set of exponential functions $i \mapsto n^i$ for all integers $n > 1$.

(iv) Show that the partial ordering $\leq$ on equivalence classes induced by the above preordering $\preccurlyeq$ on nondecreasing functions from positive integers to nonnegative integers is not a total ordering.

(v)   Regarding the nondecreasing functions from positive integers to nonnegative integers as a monoid under addition, show that the equivalence relation $\approx$ induced by $\preccurlyeq$ is a congruence on this monoid, so that the factor set again becomes an additive monoid.

**Exercise 5.2:3.** Let $S$ be any monoid, and $x_1, \ldots, x_n$ elements of $S$, and for each positive integer $i$, let $g_{x_1, \ldots, x_n}(i)$ denote the number of distinct elements of $S$ which can be written as words of length $\leq i$ in $x_1, \ldots, x_n$ (where factors may occur more than once). This is a nondecreasing function from the positive integers to the nonnegative integers, the *growth function* associated with $x_1, \ldots, x_n$.

Show that if $S$ is generated by $x_1, \ldots, x_n$, and if $y_1, \ldots, y_m$ is any finite family of elements of $S$, then in the notation of the preceding exercise, $g_{y_1, \ldots, y_m} \preccurlyeq g_{x_1, \ldots, x_n}$. Deduce that if $x_1, \ldots, x_n$ and $y_1, \ldots, y_m$ are two generating sets for the same monoid, then $g_{x_1, \ldots, x_n} \approx g_{y_1, \ldots, y_m}$, where $\approx$ is the equivalence relation determined as in Proposition 5.2.2 by the preorder $\preccurlyeq$.

Thus, if $S$ is finitely generated, the equivalence class $[g_{x_1, \ldots, x_n}]$ is the same for all finite generating sets $x_1, \ldots, x_n$ of $S$. This equivalence class is therefore an invariant of the finitely generated monoid $S$, called its *growth rate*.

We see also that if a finitely generated monoid $S$ is embeddable in another finitely generated monoid $T$, then the growth rate of $S$ must be $\leq$ that of $T$.

**Exercise 5.2:4.**

(i)   Determine the structure of the partially ordered set consisting of the growth rates of the *free abelian* monoids on finite numbers of generators together with those of the *free* monoids on finite numbers of generators.

(ii)   With the help of the result of (i), show that the free abelian monoid on $m$ generators is embeddable in the free abelian monoid on $n$ generators if and only if $m \leq n$.

(iii)   Verify that for any positive integer $n$, the map from the *free monoid* on $n$ generators $x_1, \ldots, x_n$ to the free monoid on two generators $x, y$ taking $x_i$ to $x\,y^i$ $(i = 1, \ldots, n)$ is an embedding. Is this consistent with the results of (i)?

This concept of growth rate is more often studied for groups and rings than for monoids. Note that elements $x_1, \ldots, x_n$ of a group $G$ generate $G$ as a group if and only if $x_1, x_1^{-1}, \ldots, x_n, x_n^{-1}$ generate $G$ as a monoid, so the group-theoretic growth function of $G$ with respect to $\{x_1, \ldots, x_n\}$ may be defined to be the growth function of $G$ as a monoid with respect to the generating set $\{x_1, x_1^{-1}, \ldots, x_n, x_n^{-1}\}$. The equivalence class of the growth functions determined in this way by generating sets for $G$ is called the *growth rate* of the group $G$, which is thus the same as the growth rate of $G$ as a monoid. This concept of growth rate has been applied, in particular, to fundamental groups of manifolds [147].

If $R$ is an algebra over a field $k$, then in defining its growth rate as an algebra, one lets $g_{x_1, \ldots, x_n}(i)$ denote, not the *number* of distinct elements of $R$ that can be written as products of $\leq i$ factors taken from $\{x_1, \ldots, x_n\}$, but the *dimension* of the $k$-vector space spanned in $R$ by such products. The remainder of the definition is as for monoids. Though it is a bit of a digression from the subject of preorders, I will sketch in the next few exercises an important invariant obtained from these growth rates, and some of its properties.

**Exercise 5.2:5.** If $S$ is a monoid with finite generating set $x_1, \ldots, x_n$, the *Gel'fand-Kirillov dimension* of $S$ is defined as

$$(5.2.3) \qquad \mathrm{GK}(S) = \limsup_i \frac{\ln(g_{x_1, \ldots, x_n}(i))}{\ln(i)}.$$

(Here "ln" denotes the natural logarithm, and $\limsup_i a(i)$ means

$$\lim_{j \to \infty} \sup_{i \geq j} a(i).$$

Thus, if $a$ is a nonnegative function, $\limsup_i a(i)$ is a nonnegative real number or $+\infty$.)

(i)   Show that the right hand side of (5.2.3) is a function only of the *growth rate* $[g_{x_1, \ldots, x_n}]$, hence does not depend on the choice of generators $x_1, \ldots, x_n$; hence that the Gel'fand-Kirillov dimension of a monoid is well defined.

(ii)   Determine the Gel'fand-Kirillov dimensions of the free abelian monoid and the free monoid on $n$ generators.

**Exercise 5.2:6.** In the early literature on Gel'fand-Kirillov dimension, it was often stated (in effect) that for monoids $S_1$, $S_2$, one had $\mathrm{GK}(S_1 \times S_2) = \mathrm{GK}(S_1) + \mathrm{GK}(S_2)$. Sketch an argument that seems to give this result, then point out the fallacy, and if you can, find a counterexample. (Actually, the statement was made for tensor products of algebras, rather than direct products of monoids, but either case would imply the other.)

**Exercise 5.2:7.**

(i)   Show that if $S$ is a finitely generated monoid and $\mathrm{GK}(S) < 2$, then $\mathrm{GK}(S) = 0$ or $1$.

(ii)   Show, on the other hand, that there exist finitely generated monoids having for Gel'fand-Kirillov dimensions all real numbers $\geq 2$, and $+\infty$. (Suggestion: begin by showing that for any finite or infinite set $Y$ of elements of a free monoid $F$, one can construct a homomorphic image $S$ of $F$ in which all elements not having members of $Y$ as subwords have distinct images, while all elements that do have subwords in $S$ have a common image, "$0$".)

(iii)   Show that there exist finitely generated monoids with distinct growth rates, but the same finite Gel'fand-Kirillov dimension.

We haven't seen any results on growth rates of $k$-algebras yet. If one is only concerned with what growth rates occur, there is essentially no difference between the cases of $k$-algebras and of monoids, as shown in

**Exercise 5.2:8.** Let $k$ be any field.
     Show that for every monoid $S$ with generating set $s_1, \ldots, s_n$, there exists a $k$-algebra $R$ with a generating set $r_1, \ldots, r_n$ such that for all $i$, $g_{r_1, \ldots, r_n}(i) = g_{s_1, \ldots, s_n}(i)$. Similarly, show that for every $k$-algebra $R$ with generating set $r_1, \ldots, r_n$, there exists a monoid $S$ with a generating set $s_1, \ldots, s_{n+1}$ such that for all $i$, $g_{s_1, \ldots, s_{n+1}}(i) = g_{r_1, \ldots, r_n}(i) + 1$. Deduce that the same sets of growth rates occur for monoids and for nonzero $k$-algebras.

However, if one is interested in the growth of algebras with particular ring-theoretic properties, these do not in general reduce to questions about monoids. For instance, students familiar with transcendence degrees of field extensions might do.

**Exercise 5.2:9.** Show that if $k$ is a field and $R$ a finitely generated commutative $k$-algebra without zero-divisors, then the Gel'fand-Kirillov dimension of $R$ as a $k$-algebra (defined, as for monoids, by (5.2.3)) equals the transcendence degree over $k$ of the field of fractions of $R$ (hence is an integer).

For more on Gel'fand-Kirillov dimension in ring theory, see [104].

For students familiar with the definitions of general topology, another instance of the concept of preorder is noted in:

**Exercise 5.2:10.**

(i)   Show that if $X$ is a topological space, and if for $x, y \in X$, we define $y \leq x$ to mean "the closure of $\{x\}$ contains $y$", then $\leq$ is a preorder on $X$.

(ii)   Show that if $X$ is *finite*, the above construction gives a bijection between topologies and preorders on $X$.

(iii)   Under the above bijection, what classes of preorders correspond to $T_0$, respectively $T_1$, respectively $T_2$ topologies?

(iv)   If $X$ is *infinite*, is the above map from topologies to preorders one-to-one? Onto? Can one associate to every preorder on $X$ a strongest and/or a weakest topology yielding the given preorder under this construction?

## 5.3. Induction, recursion, and chain conditions

The familiar principle of induction on the natural numbers (nonnegative integers) that one learns as an undergraduate is based on the order properties of that set. In this and the next two sections, we shall examine more general kinds of ordered sets over which one can perform inductive proofs. We shall also see that analogous to *inductive proofs* there is a concept of *recursive constructions*, which can be performed under similar hypotheses.

(Any students to whom the distinction between "minimal" and "least" elements in a partially ordered set was new should review Definition 5.1.5 before going on.)

**Lemma 5.3.1.** *Let $(X, \leq)$ be a partially ordered set. Then the following conditions are equivalent:*

(i)   *Every nonempty subset of $X$ has a minimal element.*

(ii)   *For every descending chain $x_0 \geq x_1 \geq \cdots \geq x_i \geq \ldots$ in $X$ indexed by the natural numbers, there is some $n$ such that $x_n = x_{n+1} = \ldots$.*

(ii$'$)   *Every strictly descending chain $x_0 > x_1 > \ldots$ indexed by an initial subset of the natural numbers (that is, either by $\{0, 1, \ldots, n\}$ for some $n$, or by the set of all nonnegative integers) is finite (that is, is in fact indexed by $\{0, 1, \ldots, n\}$ for some $n$).*

(ii$''$)   *$X$ has no strictly descending chains $x_0 > x_1 > \ldots$ indexed by the full set of natural numbers.*

*Proof.* (i) $\Longrightarrow$ (ii$''$) $\Longleftrightarrow$ (ii$'$) $\Longleftrightarrow$ (ii) is straightforward. Now assume (ii$''$), and suppose that (i) failed, i.e., that we had a nonempty subset $Y \subseteq X$ with no minimal element. Take any $x_0 \in Y$. Since this is not minimal, we can find

$x_1 < x_0$. Since this in turn is not minimal, we can find $x_2 < x_1$. Continuing this process, we get a contradiction to (ii″).                                    □

**Definition 5.3.2.** *A partially ordered set* $X$ *is said to have* descending chain condition (*abbreviated "DCC"; called "minimum condition" by some authors*) *if it satisfies the equivalent conditions of the above lemma.*

*Likewise, a partially ordered set* $X$ *with the dual condition* (*every non-empty subset has a maximal element, equivalently,* $X$ *has no infinite ascending chains*) *is said to have* ascending chain condition, *or* ACC, *or* maximum condition.

*A* well-ordered *set* *means a totally ordered set with descending chain condition.*

*Remark:* A *chain* in $X$, as defined following Definition 5.1.2, is a totally ordered subset, and it is meaningless to call such a subset "increasing" or "decreasing". In the above lemma and definition, the phrases "descending chain" and "ascending chain" are used as shorthand for a chain which can be indexed in a descending, respectively in an ascending manner by the natural numbers. (One may consider this a mixture of the order-theoretic meaning of "chain" and the informal meaning, referring to a sequence of elements indexed by consecutive integers with a specified relationship between successive terms.) But though this shorthand is used in the fixed phrases "ascending chain condition" and "descending chain condition", we have made the meanings of these phrases explicit, via the above lemma and definition.

That the natural numbers are well-ordered has been known in one form or another for millennia, but the importance of ACC and DCC for more general partially ordered sets was probably first noted in ring theory, in the early decades of the twentieth century. Rings with these conditions on their sets of *ideals*, partially ordered by inclusion, are called "Noetherian" and "Artinian" respectively, after Emmy Noether and Emil Artin who studied them.

One does not need to formally state a "principle of induction over partially ordered sets with ACC (or DCC)". Rather, when one wishes to prove a result for all elements of a partially ordered set $X$ with, say, DCC, one can simply begin, "Suppose there are elements of $X$ for which the statement is false. Let $x$ be minimal for this property", since if the set of such elements is nonempty, it must have a minimal member. Then one knows the statement is *true* for all $y < x$; and if one can show from this that it is true for $x$ as well, one gets a contradiction, proving the desired result. Since this is a familiar form of argument, one often abbreviates it and says, "Assume inductively that the statement is true for all $y < x$", proves from this that it is true for $x$ as well, and concludes that it is true for all elements of $X$.

In the most familiar sort of induction on the natural numbers, one starts by proving the desired result for 0 (or 1). Why was there no corresponding step in the schema described above? The analog of the statement that our desired result holds for 0 would be the statement that it holds for all minimal elements of $X$. But if one proves that a statement is true for an element $x$ whenever it is true for all smaller elements, then in particular, one has it in the case where the set of smaller elements is empty. Depending on the situation, the proof that a result is true for $x$ if it is true for all smaller elements may or may not involve a different argument when $x$ is minimal.

**Exercise 5.3:1.** A noninvertible element of a commutative integral domain $C$ is called *irreducible* if it cannot be written as a product of two noninvertible elements. Give a concise proof that if $C$ is a commutative integral domain with ascending chain condition on ideals (or even just on principal ideals), then every nonzero noninvertible element of $C$ can be written as a product of irreducible elements.

In addition to *proofs* by induction, one often performs *constructions* in which each step requires that a set of preceding steps already have been done. The definition of the Fibonacci numbers $f_i$ $(i = 0, 1, 2, \ldots)$ by the conditions

$$(5.3.3) \qquad f_0 = 0, \quad f_1 = 1, \quad f_{n+2} = f_n + f_{n+1} \text{ for } n \geq 0$$

is of this sort. These are called *recursive* definitions or constructions, and we shall now see that, like inductive proofs, they can be carried out over general partially ordered sets with chain conditions.

Let us analyze what such a construction involves in general, and then show how to justify it. Suppose $X$ is a partially ordered set with DCC, and suppose that we wish to construct recursively a certain function $f$ from $X$ to a set $T$. To say that for some $x \in X$ the value of $f$ has been determined for all $y < x$ is to say that we have a function $f_{<x} : \{y \mid y < x\} \to T$. So "a rule defining $f$ at each $x$ if it has been defined at all $y < x$" can be formalized as a $T$-valued function $r$ on the set of all pairs $(x, f_{<x})$ where $x \in X$ and $f_{<x}$ is a function $\{y \mid y < x\} \to T$. In most applications, our rule defining $f$ at $x$ in terms of the values for $y < x$ actually requires that these values satisfy some additional conditions, and we verify these conditions inductively, as the construction is described recursively. But to avoid complicating our abstract formalization, let us assume $r$ defined for *all* pairs $(x, f_{<x})$ where $x \in X$ and $f_{<x}$ is a function $\{y \mid y < x\} \to T$. For if we have a definition of $r$ in "good" cases, we can extend it to other cases in an arbitrary way (e.g., assume $0 \in T$ and send $(x, f_{<x})$ to 0 if $f_{<x}$ is not "good"). Then the inductive proof that $f$ is "good" can be formally considered to come *after* the recursive construction of $f$.

We see that the property characterizing the function $f$ constructed recursively as above is that for each $x \in X$, $f(x)$ is a certain function of the *restriction* of $f$ to $\{y \mid y < x\}$. For any function $f : X \to Y$, and any subset $Z$ of $X$, let us denote the restriction of $f$ to $Z$, a function $Z \to Y$, by $f \mid Z$. (A variant symbol which is sometimes used, but which we will not use, is $f \upharpoonright Z$.)

We now justify recursive constructions by proving

**Lemma 5.3.4.** *Let $X$ be a partially ordered set with descending chain condition, $T$ any set, and $r$ a function associating to every pair $(x, f_{<x})$ such that $x \in X$, and $f_{<x}$ is a function $\{y \in X \mid y < x\} \to T$, an element $r(x, f_{<x}) \in T$. Then there exists a unique function $f : X \to T$ such that for all $x \in X$,*

$$f(x) = r(x, f \mid \{y \mid y < x\}).$$

*Proof.* Let $X' \subseteq X$ denote the set of all $x \in X$ for which there exists a unique function $f_{\leq x} : \{y \mid y \leq x\} \to T$ with the property that

$$(5.3.5) \qquad (\forall y \leq x) \ f_{\leq x}(y) = r(y, f_{\leq x} \mid \{z \mid z < y\}).$$

We claim, first, that for any two elements $x_0, x_1 \in X'$, the functions $f_{\leq x_0}$, $f_{\leq x_1}$ agree on $\{y \mid y \leq x_0 \wedge y \leq x_1\}$. For if not, choose a minimal $y$ in this set at which they disagree. Then by (5.3.5), $f_{\leq x_0}(y) = r(y, f_{\leq x_0} \mid \{z \mid z < y\})$, and $f_{\leq x_1}(y) = r(y, f_{\leq x_1} \mid \{z \mid z < y\})$. But by choice of $y$, the restrictions of $f_{\leq x_0}$ and $f_{\leq x_1}$ to $\{z \mid z < y\}$ are equal, hence by the above equations, $f_{\leq x_0}(y) = f_{\leq x_1}(y)$, contradicting our choice of $y$.

Next, suppose that $X'$ were not all of $X$. Let $x$ be a minimal element of $X - X'$. Since, as we have just seen, the functions $f_{\leq y}$ for $y < x$ agree on the pairwise intersections of their domains, they piece together into one function $f_{<x}$ on the union of their domains. (Formally, this "piecing together" means taking the union of these functions, as subsets of $X \times T$.) If we now define $f_{\leq x}$ to agree with this function $f_{<x}$ on $\{y \mid y < x\}$, and to have the value $r(x, f_{<x})$ at $x$, we see that this function satisfies (5.3.5), and is the unique function on $\{y \mid y \leq x\}$ which can possibly satisfy that condition. This means $x \in X'$, contradicting our choice of $x$.

Hence $X' = X$. Now piecing together these functions $f_{\leq x}$ defined on the sets $\{y \mid y \leq x\}$, we get the desired function $f$ defined on all of $X$. $\square$

Note that most of the above proof consisted of an *inductive* verification that for every $x \in X$, there exists a unique function $f_{\leq x}$ satisfying (5.3.5). So recursion is justified by induction.

Example: The Fibonacci numbers are defined recursively by using for $X$ the ordered set of nonnegative integers, and defining $r(n, (f_0, \ldots, f_{n-1}))$ to be 0 if $n = 0$, to be 1 if $n = 1$, and to be $f_{n-2} + f_{n-1}$ if $n \geq 2$.

The next exercise shows that recursive constructions are not in general possible if the given partially ordered set does not satisfy descending chain condition.

**Exercise 5.3:2.** Show that there does not exist a function $f$ from the interval $[0, 1]$ of the real line to the set $\{0, 1\}$ determined by the following rules:

(a) $f(0) = 0$.

(b) For $x > 0$, $f(x) = 1$ if for all $y \in [0, x)$, $f(y) = 0$; otherwise, $f(x) = 0$.

If you prefer, you may replace the interval $[0, 1]$ in this example with the countable set $\{0\} \cup \{1/n \mid n = 1, 2, 3, \ldots\}$.

**Exercise 5.3:3.** Generalizing the above exercise, show that if $(X, \leq)$ is any partially ordered set *not* satisfying descending chain condition, then

(i)  There exists a function $r$ as in the statement of Lemma 5.3.4, with $T = \{0, 1\}$, such that no function $f$ satisfies the conditions of the conclusion of that lemma.

(ii)  There exists a function $r$ as in the statement of Lemma 5.3.4, with $T = \{0, 1\}$, such that more than one function $f$ satisfies the conditions of the conclusion of that lemma.

In a way, solving a differential equation with given initial conditions is like a "recursive construction over an interval of the real numbers". But precisely because the real numbers do not have descending chain condition, the conditions for existence and uniqueness of a solution, and the arguments needed to prove these, are subtle. (A different sort of property of the real line, which plays a role in such arguments, is connectedness.)

There is a situation at the very foundation of mathematics which can be interpreted in terms of a partially ordered system with descending chain condition. The Axiom of Regularity of set theory (which will be stated formally in the next section) says that there is no "infinite regress" in the construction of sets; that is, that there are no left-infinite chains of sets under the membership relation:

$$\cdots \in S_n \in \cdots \in S_2 \in S_1 \in S_0.$$

This is not a difficult axiom to swallow, since if we had a set theory for which it was not true, we could pass to the "smaller" set theory consisting of those sets which admit no such chain to the left of them. The class of such sets would be closed under all the constructions required by the remaining axioms of set theory, and the "new" set theory would satisfy the Axiom of Regularity.

To interpret Regularity in the terms we have just been discussing, let us write $A \prec B$, for sets $A$ and $B$, if there is a chain of membership-relations, $A = S_0 \in S_1 \in \cdots \in S_n = B$ $(n > 0)$. This relation is clearly transitive. The Regularity Axiom implies that $\prec$ is antireflexive (if we had $A \prec A$, then a chain of membership relations connecting $A$ with itself could be iterated to give an infinite chain going to the left), hence $\prec$ is the partial strict ordering corresponding to a partial ordering $\preceq$; and Regularity applied again says that this partial ordering has descending chain condition. (Well, almost. We have only defined the concepts of partial ordering and chain condition for *sets*, and the class of all sets is not a set. To get around this problem we can translate these observations more precisely as saying that for each set $A$, the collection $\{B \mid B \preccurlyeq A\}$ is itself a set, and has descending chain condition under $\preccurlyeq$.) This allows one to prove set-theoretic results by induction with respect to this ordering, and likewise use it in recursive set-theoretic definitions.

We had another such situation in Chapter 2, when we talked about the set $T = T_{X, \mu, \iota, e}$ of group-theoretic terms in a set of symbols $X$. These also satisfy a principle of regularity, in terms of the relation "$s$ occurs in $t$", which we denoted $t > s$ in Exercise 2.7:4. To show this, let $T'$ denote the set of elements of $T$ admitting no infinite descending $>$-chains to the right of them. One verifies that $T'$ is closed under the operations of conditions (a) and (b) of the definition of $T$ (Definition 2.5.1), and concludes that if $T'$ were properly smaller than $T$, one would have a contradiction to condition (c) of that definition. We only sketched the construction of $T$ in Chapter 2, but in §9.3 below we will introduce the concept of "term" for general classes of algebras, and the above argument will then enable us to perform recursion and induction on such terms.

One can, of course, do inductive proofs and recursive constructions over partially ordered sets with *ascending* as well as descending chain condition. These come up often in ring theory, where Noetherian rings, i.e., rings whose partially ordered set of ideals has ACC, are important. In proving that a property holds for every ideal $I$ of such a ring, one may consider an ideal $I$, and assume inductively that the desired property holds for all strictly *larger* ideals.

To get the result allowing us to perform recursive constructions in such situations, i.e., the analog of Lemma 5.3.4 with $>$ replacing $<$, it is not necessary to repeat the proof of that lemma; we can use duality of partially ordered sets. I will give the statement and sketch the argument this once, to show how an argument by duality works. After this, if I want to invoke the dual of an order-theoretic result previously given, I shall consider it sufficient to say "by duality", or "by the dual of Proposition #.#.#", or the like.

**Corollary 5.3.6.** *Let $X$ be a partially ordered set with ascending chain condition, $T$ any set, and $r$ a function associating to every pair $(x, f_{>x})$ consisting of an element $x \in X$ and a function $f_{>x} : \{y \mid y > x\} \to T$ an element $r(x, f_{>x}) \in T$. Then there exists a unique function $f : X \to T$ such that for all $x \in X$, $f(x) = r(x, f \mid \{y \mid y > x\})$.*

Sketch of proof. The *opposite* of the partially ordered set $X$ (the structure with the same underlying set but the opposite ordering) is a partially ordered set $X^{\mathrm{op}}$ with *descending chain condition*, and $r$ can be considered to be a function $r'$ with exactly the properties required to apply Lemma 5.3.4 to that partially ordered set. That lemma gives us a unique function $f'$ from $X^{\mathrm{op}}$ to $T$ satisfying the conclusions of that lemma relative to $r'$, and this is equivalent to a function $f$ from $X$ to $T$ satisfying the desired condition relative to $r$.                                                                                   □

Example: The Fibonacci numbers $f_n$ were defined above for $n \geq 0$. With the help of downward recursion on the set of negative integers, one can now easily verify that there is also a unique way of defining $f_n$ for negative $n$, such that combining the values for negative and nonnegative $n$, we get a sequence $(f_i)_{i \in \mathbb{Z}}$ which satisfies $f_{n+2} = f_n + f_{n+1}$ for all $n$.

Often the key to making an inductive argument or a recursive construction work is a careful choice of a parameter over which to carry out the induction or recursion, and an appropriate ordering on the set of values of that parameter. The next definition describes a way of constructing partial orderings that is frequently useful for such purposes. The well-ordered index set $I$ in that definition can be as simple as $\{0, 1\}$.

**Definition 5.3.7.** *Let $(X_i)_{i \in I}$ be a family of partially ordered sets, indexed by a* well-ordered *set $I$. Then* lexicographic order *on $\prod_I X_i$ is defined by declaring $(x_i) \leq (y_i)$ to hold if and only if either $(x_i) = (y_i)$, or for the least $j \in I$ such that $x_j \neq y_j$, one has $x_j < y_j$ in $X_j$.*

Note that if $I = \{1, \ldots, n\}$ with its natural order, then this construction orders $n$-tuples $(x_1, \ldots, x_n) \in \prod_I X_i$ by the same "left-to-right" principle that is used to arrange words in the dictionary; hence the name of the construction. The usefulness of this construction in obtaining orderings with descending chain condition is indicated in part (iii) of

**Exercise 5.3:4.** Let $(X_i)_{i \in I}$ be as in Definition 5.3.7.

(i)  Verify that the relation on $\prod_I X_i$ given by that definition is indeed a partial order.

(ii)  Show that if each $X_i$ is *totally* ordered, then so is their direct product under that ordering. Show, in contrast, that the corresponding statement is not in general true for the *product ordering*, described in Definition 5.1.4.

(iii)  Show that if $I$ is *finite*, and each of the partially ordered sets $X_i$ has descending chain condition, then so does their product under lexicographic ordering.

(iv)  Comparing lexicographic ordering with the product ordering, deduce that given a finite family of partially ordered sets, each of which has descending chain condition, their direct product, under the product ordering, also has descending chain condition.

(v)  Show that the product of a family of copies of the two-element totally ordered set $\{0, 1\}$, indexed by the natural numbers, does not have descending chain condition under the product ordering. Deduce that lexicographic ordering on products of infinite families of partially ordered sets with descending chain condition also fails, in general, to have descending chain condition.

(By duality, analogs of (iii)–(v) also hold for ascending chain condition.)

In the next exercise, a lexicographic ordering is used to give a concise proof of a standard result on symmetric polynomials.

**Exercise 5.3:5.** Let $R$ be a commutative ring, and $R[x_1, \ldots, x_n]$ the polynomial ring in $n$ indeterminates over $R$. Given any nonzero polynomial $f = \sum c_{i(1), \ldots, i(n)} x_1^{i(1)} \ldots x_n^{i(n)}$ (almost all $c_{i(1), \ldots, i(n)}$ zero), let us define the *leading term* of $f$ to be the nonzero summand in this expression with the largest exponent-string $(i(1), \ldots, i(n))$ under lexicographic ordering on the set of all such strings. (Since the set of nonzero summands is finite, no chain condition is needed to make this definition.)

(i)   Let $f$ and $g$ be nonzero elements of $R[x_1, \ldots, x_n]$, and suppose that the coefficient occurring in the leading term of $f$ is not a zero-divisor in $R$. (E.g., this is automatic if $R$ is an integral domain.) Show that the leading term of $fg$ is the product of the leading terms of $f$ and of $g$.

An element of $R[x_1, \ldots, x_n]$ is called *symmetric* if it is invariant under the natural action of the group of all permutations of the index set $\{1, \ldots, n\}$ on the indeterminates $x_1, \ldots, x_n$. For $1 \leq d \leq n$, the $d$-th *elementary* symmetric function $s_d$ is defined to be the sum of all products of exactly $d$ distinct members of $\{x_1, \ldots, x_n\}$. Our goal will be to show that the ring of symmetric polynomials in $n$ indeterminates over $R$ is generated over $R$ by the elementary symmetric polynomials.

(ii)   For nonnegative integers $j(1) \ldots j(n)$, find the leading term of the product $s_1^{j(1)} \ldots s_n^{j(n)}$.

(iii)   Show that the following sets are the same:

   (a) The set of all $n$-tuples $(i(1), \ldots, i(n))$ of nonnegative integers such that $i(1) \geq \cdots \geq i(n)$.

   (b) The set of all exponent-strings $(i(1), \ldots, i(n))$ of leading terms $c_{i(1), \ldots, i(n)} x_1^{i(1)} \ldots x_n^{i(n)}$ of symmetric polynomials.

   (c) The set of all exponent-strings of leading terms of products $s_1^{j(1)} \ldots s_n^{j(n)}$, as in (ii) above.

(iv)   Deduce that any nonzero symmetric polynomial can be changed to a symmetric polynomial with lower exponent-string-of-the-leading-term, or to the zero polynomial, by subtracting a scalar multiple of a product of elementary symmetric polynomials. Conclude, by induction on this exponent-string, that the ring of symmetric polynomials in $n$ indeterminates over $R$ is generated over $R$ by the elementary symmetric polynomials.

(For standard proofs of the above result, see [32, pp. 252–255], or [34, Theorem IV.6.1, p. 191]. For some related results on noncommutative rings, see [61].)

**Exercise 5.3:6.** For nonnegative integers $i$ and $j$, let $n_{i,j}$ be defined recursively as the least nonnegative integer not equal to $n_{i, j'}$ for any $j' < j$, nor to $n_{i',j}$ for any $i' < i$. (What ordering of the set of pairs $(i, j)$ of nonnegative integers can one use to justify this recursion?)

Find and prove a concise description of $n_{i,j}$. (Suggestion: Calculate some values and note patterns. To find the "pattern in the patterns", write numbers to base 2.)

We end with two miscellaneous exercises.

**Exercise 5.3:7.** For $X$ a topological space, show that the following conditions are equivalent. (We do not understand "compact" to entail Hausdorff or nonempty.)

(a) Every subset of $X$ is compact in the induced topology.

(b) Every open subset of $X$ is compact in the induced topology.

(c) The partially ordered set of open subsets of $X$ has ascending chain condition.

**Exercise 5.3:8.** One may ask whether Exercise 5.3:1 has a converse: that if $C$ is a commutative integral domain in which every nonzero noninvertible element can be written as a product of irreducibles, then $C$ has ascending chain condition on principal ideals. Show by example that this is *not* true.

## 5.4. The axioms of set theory

We are soon going to look at some order-theoretic principles equivalent to the powerful Axiom of Choice. Hence it is desirable to review the statement of that axiom, and its status in relation to the other axioms of set theory. For completeness, I will record in this section the whole set of axioms most commonly used by set theorists.

Let us begin with some background discussion. In setting up a rigorous foundation for mathematics, one might expect the theory to require several sorts of entities: "primitive" elements such as numbers, additional *sets* formed out of these, *ordered pairs* of elements, *functions* from one set to another, etc. But as the theory was developed, it turned out that one could get everything one wanted from a single basic concept, that of set, and a single relation among sets, that of membership. The result is a set theory in which the only members of sets are themselves sets.

As an important example of how other "primitives" are reduced to the set concept, we recall the case of the *natural numbers* (nonnegative integers). The first thing we learn in our childhood about these numbers is that they are used to count things; to say how many objects there are in a collection. The early set theorists observed that one can formalize the concept of two sets having the "same number" of elements set-theoretically, as meaning that there exists a bijection between them. This is clearly an equivalence relation on sets. Hence the natural numbers ought be some entities which one could associate to finite sets, so that two sets would get the same entity associated to them if and only if they were in the same equivalence class. The original plan was to use, as those entities, the equivalence classes themselves, i.e., to *define* the natural numbers 0, 1, 2, etc., to be the corresponding equivalence classes. Thus, the statement that a finite set had $n$ elements would mean that it was a *member* of the number $n$. (Cardinalities of infinite sets were to be treated similarly.) This is good in principle—don't create new entities to index the equivalence classes if the equivalence classes themselves will do. But in this case, the equivalence classes turned out not to be a good choice: they are too big to be sets. So the next idea was to choose one easily described member from each such class, call these chosen elements the natural numbers 0, 1, 2, ..., and define a set to have $n$ elements if it could be put in bijective correspondence with the "sample" set $n$.

Where would one get these "sample" finite sets from, using pure set theory? There is no problem getting a sample zero-element set—there is a unique set with zero elements, the empty set $\emptyset$. Having taken this step, we have *one* set in hand: $\emptyset$. This means that we are in a position to create a sample one-element set, the set with that element as its one member, i.e., $\{\emptyset\}$. Having found these two elements, $\emptyset$ and $\{\emptyset\}$, we can define a two-element set $\{\emptyset, \{\emptyset\}\}$ to use as our next sample—and so on. After the first couple of steps, we are not so limited in our options. (For example, one might, instead of following the pattern illustrated, take for 4 the set of all subsets of 2.) However, the above approach, of always taking for the next number the set of numbers found so far, due to John von Neumann, is an elegant way of manufacturing one set of each natural-number cardinality, and it is taken as the definition of these numbers by modern set theorists:

$$0 = \emptyset, \quad 1 = \{\emptyset\}, \quad 2 = \{\emptyset, \{\emptyset\}\},$$
$$3 = \{\emptyset, \{\emptyset\}, \{\emptyset, \{\emptyset\}\}\},$$

(5.4.1) $\qquad\qquad\qquad . \quad . \quad .$

$$i + 1 = i \cup \{i\} = \{0, 1, 2, \ldots, i\},$$
$$. \quad . \quad .$$

Another basic concept which was reduced to the concepts of set and membership is that of *ordered pair*. If $X$ and $Y$ are sets, then one can deduce from the axioms (shortly to be listed) that $X$ and $Y$ can each be determined uniquely from the set $\{\{X\}, \{X, Y\}\}$. Since all one needs about ordered pairs is that they are objects which specify their first and second components unambiguously, one *defines* the ordered pair $(X, Y)$ to mean the set $\{\{X\}, \{X, Y\}\}$.

One then goes on to define the *direct product* of two sets in terms of ordered pairs, binary relations in terms of direct products, functions in terms of binary relations, etc. From natural numbers, ordered pairs, and functions, one constructs the integers, the rational numbers, the real numbers, the complex numbers, etc., by well-known techniques, which I won't review here.

(One also wants to define ordered $n$-tuples. The trick by which ordered pairs were defined turns out not to generalize in an easy fashion; the most convenient approach is to define an ordered $n$-tuple to mean a function whose domain is the set $n$. However, this conflicts with the definition of ordered pair! To handle this, a careful development of set theory must use different symbols, say $\langle X, Y \rangle$ for the concept of "ordered pair" first described, and $(X_0, X_1, \ldots, X_{n-1})$ for the ordered $n$-tuples subsequently defined.)

The above examples should give some motivation for the "sort" of set theory described by the axioms which we shall now list. Of course, a text on the foundations of mathematics will first develop language allowing one to state these axioms precisely, and, since a statement in such language is not always easy to understand, it will precede or follow many of the precise statements by intuitive developments. I have tried below to give formulations that make it as clear as possible what the axioms assert, and have added some further remarks after the list. But for a thorough presentation, and for more discussion of the axioms, the student should see a text on the subject. Two recommended undergraduate texts are [15] and [24]. Written for a somewhat more advanced audience is [22].

Here, now, are the axioms of *Zermelo–Fraenkel Set Theory with the Axiom of Choice*, commonly abbreviated ZFC.

**Axiom of Extensionality.** *Sets are equal if and only if they have the same members. That is,* $X = Y$ *if and only if for every set* $A$, $A \in X \iff A \in Y$.

**Axiom of Regularity** (or **Well-foundedness,** or **Foundation**). *For every nonempty set* $X$, *there is a member of* $X$ *which is disjoint from* $X$.

**Axiom of the Empty Set.** *There exists a set with no members.* (Common notation: $\emptyset$.)

**Axiom of Separation.** *If* $X$ *is a set and* $P$ *is a condition on sets, there exists a set* $Y$ *whose members are precisely the members of* $X$ *satisfying* $P$. (Common notation: $Y = \{A \in X \mid P(A)\}$.)

**Axiom of Doubletons** (or **Pairs**). *If* $X$ *and* $Y$ *are sets, there is a set* $Z$ *whose only members are* $X$ *and* $Y$. (Common notation: $Z = \{X, Y\}$.)

**Axiom of Unions.** *If* $X$ *is a set, there is a set* $Y$ *whose members are precisely all members of members of* $X$. (Common notations: $Y = \bigcup X$ or $\bigcup_{A \in X} A$.)

**Axiom of Replacement.** *If* $f$ *is an operation on sets (formally characterized by a set-theoretic proposition* $P(A, B)$ *such that for every set* $A$ *there is a unique set* $f(A)$ *such that* $P(A, f(A))$ *holds), and* $X$ *is a set, then there exists a set* $Y$ *whose members are precisely the sets* $f(A)$ *for* $A \in X$. (Common notation: $Y = \{f(A) \mid A \in X\}$. When there is no danger of confusion, this is sometimes abbreviated to $Y = f(X)$.)

**Axiom of the Power Set.** *If* $X$ *is a set, there exists a set* $Y$ *whose members are precisely all subsets of* $X$. *(Common notations:* $Y = \mathbf{P}(X)$ *or* $2^X$.)

**Axiom of Infinity.** *There exists a set having* $\emptyset$ *as a member, closed under the construction* $i \mapsto i \cup \{i\}$ *(cf. (5.4.1)), and minimal for these properties.* (Common name: The set of natural numbers.)

**Axiom of Choice.** *If* $X$ *is a set, and* $f$ *is a function associating to every* $x \in X$ *a nonempty set* $f(x)$, *then there exists a function* $g$ *associating to every* $x \in X$ *an element* $g(x) \in f(x)$.

Explanations of some of the names: *Extensionality* means that a set is determined by its *extent*, not its *intent*. *Separation* says that one can form new sets by using any well-defined criterion to "separate out" certain elements of an existing set. The Axiom of *Infinity* is so called because if we did not assume it, the collection of all sets which can be built up from the empty set in finitely many steps would satisfy our axioms, giving an example of a set theory in which all sets are finite. One can show that, given the preceding axioms, that axiom is equivalent to the existence of an infinite set.

We described *Regularity* earlier as saying that there was no infinite regress under "$\in$". That formulation requires one to have the set of natural numbers to index such a regress, so we chose a formulation that can be expressed independently of the Axiom of Infinity. In the presence of the other axioms one can prove the two formulations equivalent. (Roughly, if one had an infinite chain $\cdots \in S_2 \in S_1 \in S_0$, then $\{S_i\}$ would be a counterexample to Regularity, while if a set $X$ were a counterexample to Regularity, one could select such a chain from its elements.)

Actually, the Axiom of Regularity makes little substantive difference for areas of mathematics other than set theory itself (e.g., see [24, p. 92 et seq.]). Without it, one can have sets with exotic properties such as being members of themselves, but the properties of set-theoretic concepts used by most of mathematics—bijections, direct products, cardinality arguments, etc.— are little affected. Its absence would simply make it a bit trickier, say, to construct, given sets $X$ and $Y$, a copy of $Y$ *disjoint* from $X$. The Regularity Axiom seems to have crept into the Zermelo–Fraenkel axioms by the back door: It was not in the earlier formulations of those axioms, and still does not appear in some listings, such as that in [15]. But it is generally accepted, and we will count it among the axioms here, and rely on the convenience it provides. It gives one a comforting assurance that sets are built up from earlier sets with no "vicious circles" in the process; hence the name "Well-Foundedness". (By extension, many set-theorists call the condition of descending chain condition on any partially ordered set "well-foundedness".)

Observe that the Axioms of Extensionality and Regularity essentially clarify what we intend to *mean* by a "set". The next seven axioms each say that certain sets exist. In each case these are sets which are *uniquely determined* by the conditions assumed. Those seven axioms can all be considered cases of a single axiom proposed by Frege in 1893, the *Axiom of Abstraction*, saying that "Given any property, there exists a set whose members are just those entities possessing that property". That axiom nicely embodies the idea of a set, but it turned out to be too strong to be consistent: it allowed one to define things like "the set $S$ of all sets which are not members of themselves," which led to contradictions. (Russell's Paradox: "Is that $S$ a member of itself?" Either a positive or a negative answer implies its own negation.) The difficulty was, somehow, that the Axiom of Abstraction assumed "all sets" to be known. In particular, the set $S$ we were constructing was already "there", to be chosen or rejected in choosing the members of that same set $S$, allowing one to create a self-contradictory criterion for that choice. Subsequent experience suggested that such contradictions could be avoided by requiring every set to be constructed from sets "constructed before it"; and the seven axioms in question represent sub-cases of the rejected "Axiom of Abstraction" which meet this condition.

The Axiom of Regularity was probably another reaction against those paradoxes. Though adding an axiom can't remove a contradiction, the encounter with Russell's Paradox very likely led mathematicians to feel that sets that could not be "built up from scratch" were unhealthy, and should be excluded.

The last axiom of ZFC, that of Choice, is of a different sort from those that precede it. It asserts the existence of something not uniquely defined by the given data: a function that chooses, *in an unspecified way*, one element from each of a family of sets. This was very controversial in the early decades of the twentieth century, both because it led to consequences which seemed surprising then (such as the existence of nonmeasurable sets of real numbers), and because of a feeling by some that it represented an unjustifiable assumption that something one could do in the finite case could be done in the infinite case as well. It is a standard assumption in modern mathematics; such basic results as that every vector space has a basis, that a direct product of compact topological spaces is compact, and that a countable union of countable sets is countable, cannot be proved without it. But there have been, and still are, mathematicians who reject it: the *intuitionists* of the early 1900s, and the *constructivists* today.

Even accepting the Axiom of Choice, as we shall, it is at times instructive to note whether a result or an argument depends on it, or can be obtained from the other axioms. (This is like the viewpoint that, even if one does not accept the constructivists' extreme claim that proofs of existence that do not give explicit constructions are *worthless*, one may consider constructive proofs to be desirable when they can be found.)

In the next two sections we shall develop several set-theoretic results whose proofs require the Axiom of Choice, and we will show that each of these statements is, in fact, equivalent to that axiom, in the presence of the other axioms. Hence, in those sections, we shall not assume the Axiom of Choice except when we state this assumption explicitly, and the arguments we give to show these equivalences will all be justifiable in terms of the theory given by the other axioms, called "Zermelo–Fraenkel Set Theory", abbreviated ZF. (However, we shall not in general attempt to show explicitly how the familiar mathematical techniques that we use are justified by those axioms—for that, again see a text in set theory.) In all later chapters, on the other hand, we shall freely use the Axiom of Choice, i.e., we will assume ZFC.

In the handful of results proved so far in this chapter, we have implicitly used the Axiom of Choice just once: in Lemma 5.3.1, in showing (ii″) $\implies$ (i). Hence for the remainder of this chapter, we shall forgo assuming that implication, and will understand the descending chain condition to refer to condition (i) of that lemma (which still *implies* (ii)–(ii″)).

Let us note explicitly one detail of set-theoretic language we have already used: Since all sets satisfying a given property may not together form a set, one needs a word to refer to "collections" of sets that are not necessarily themselves sets. These are called *classes*. An example is the *class of all sets*. One can think of classes which are not sets, not as actually being *mathematical objects*, but as providing a convenient *language* to use in making statements about all sets having one or another property.

Since classes are more general than sets, one may refer to any set as a "class", and this is sometimes done for reasons not involving the logical distinction, but just to vary the wording. E.g., rather than saying "the set of those subsets of $X$ such that ...", one sometimes says "the class of those subsets of $X$ such that ...". And, for some reason, one always says "equivalence class", "conjugacy class", etc., though they are sets.

**Exercise 5.4:1.** Show that for every set $X$ there exists a set $Y \supseteq X$ such that $a \in b \in Y \implies a \in Y$; in fact, that there exists a *least* such set. (Thus, $Y$ is the "closure" of $X$ under passing to members of sets.)

## 5.5. Well-ordered sets and ordinals

Recall (Definition 5.3.2) that a partially ordered set $(X, \leq)$ is called *well-ordered* if it is totally ordered and has descending chain condition. In a totally ordered set, a minimal element is the same as a least element, so the condition of well-ordering says that every nonempty subset of $X$ has a *least* member.

This condition goes a long way toward completely determining the structure of $X$. Applied first to $X$ as a subset of itself, it tells us that if $X$ is nonempty, it has a least element, $x_0$. If $X$ does not consist of $x_0$ alone, then $X - \{x_0\}$ is nonempty, hence this set has a least element, which we may call $x_1$. We can go on in this fashion, and unless $X$ is finite, we will get a uniquely determined sequence of elements $x_0 < x_1 < x_2 < x_3 < \ldots$ at the "bottom" of $X$. This list may exhaust $X$, but if it does not, there will necessarily be a least element in the complement of the subset so far described, which we may call $x_{1,0}$, and if this still does not exhaust $X$, there will be a least element greater than it, $x_{1,1}$, etc. We can construct in this way successive hierarchies, and hierarchies of hierarchies—I will not go into details—on the single refrain, "If this is not all, there is a least element of the complement".

A couple of concrete examples are noted in

**Exercise 5.5:1.** If $f$ and $g$ are real-valued functions on the real line $\mathbb{R}$, let us in this exercise write $f \leq g$ to mean that there exists some real number $N$ such that $f(t) \leq g(t)$ for all $t \geq N$.

(i) Show that this relation $\leq$ is a preordering, that its restriction to the set of polynomial functions is a total ordering, and that on polynomials with *nonnegative integer* coefficients, it in fact gives a well-ordering. Determine, if they exist, the elements $x_0, x_1, \ldots, x_n, \ldots, x_{1,0}, x_{1,1}$ of this set, in the notation of the preceding paragraphs.

(ii) Show that the set consisting of all polynomials with nonnegative integer coefficients, and, in addition, the function $e^t$, is still well-ordered under the above relation.

(iii) Find a subset of the rational numbers which is order-isomorphic (under the standard ordering) to the set described in (ii).

To make precise the idea that the order structure of a well-ordered set is "unique, as far as it goes", let us define an "initial segment" of any totally ordered set $X$ to mean a subset $I \subseteq X$ such that $x \leq y \in I \implies x \in I$. Then we have

**Lemma 5.5.1.** *Let $X$ and $Y$ be well-ordered sets. Then exactly one of the following conditions holds:*

(a) *$X$ and $Y$ are order-isomorphic.*

(b) *$X$ is order-isomorphic to a proper initial segment of $Y$.*

(c) *$Y$ is order-isomorphic to a proper initial segment of $X$.*

*Further, in (b) and (c) the initial segments in question are unique, and in all three cases the order-isomorphism is unique.*

*Proof.* We shall construct an order isomorphism of one of these three types by a recursive construction on the well-ordered set $X$. Let me first describe the idea intuitively: We start by pairing the least element of $X$ with the least element of $Y$; and we go on, at every stage pairing the least not-yet-paired-off element of $X$ with the least not-yet-paired-off element of $Y$, until we run out of elements of either $X$ or $Y$ or both.

Now in our formulation of recursive constructions in Lemma 5.3.4, we said nothing about "running out of elements". But we can use a trick to reduce the approach just sketched to a recursion of the sort characterized by that lemma.

Form a set consisting of the elements of $Y$ and one additional element which we shall denote DONE. Given any $x \in X$, and any function $f_{<x} : \{x' \in X \mid x' < x\} \to Y \cup \{\text{DONE}\}$, we define $r(x, f_{<x}) \in Y \cup \{\text{DONE}\}$ as follows:

> If the image of $f_{<x}$ is a proper initial segment of $Y$, let $r(x, f_{<x})$ be the least element of $Y$ not in that segment. Otherwise, let $r(x, f_{<x}) = \text{DONE}$.

By Lemma 5.3.4, this determines a function $f : X \to Y \cup \{\text{DONE}\}$. It is straightforward to verify inductively that for those $x$ such that $f(x) \neq$ DONE, the restriction $f_{\leq x}$ of $f$ to $\{x' \in X \mid x' \leq x\}$ will be the only order isomorphism between that initial segment of $X$ and any initial segment of $Y$. From this we easily deduce that if the range of $f$ does not contain the value DONE, exactly one of conclusions (a) or (b) holds, but not (c); while if the range of $f$ contains DONE, (c) holds but not (a) or (b). In each case, $f$ determines the unique order isomorphism with the indicated properties. $\quad\square$

**Exercise 5.5:2.** Give the details of the last paragraph of the above proof.

Since the well-ordered sets fall into such a neat array of isomorphism classes, it is natural to look for a way of choosing one "standard member" for each of these classes, just as the natural numbers are used as "standard members" for the different sizes of finite sets. Recall that in the von Neumann construction of the natural numbers, (5.4.1), each number arises as the set of all those that precede it, so that we have $i < j$ if and only if $i \in j$, and $i \leq j$ if and only if $i \subseteq j$. Clearly, each natural number, being finite and totally ordered, is a well-ordered set under this ordering. Let us take these von Neumann natural numbers as our standard examples of finite well-ordered sets, and see whether we can extend this family in a natural way to get models of infinite well-ordered sets.

Continuing to use the principle that each new object should be the set of all that precede, we take *the set of natural numbers* as the standard example chosen from among the well-ordered sets which when listed in the manner discussed at the beginning of this section have the form $X = \{x_0, x_1, \dots\}$, with subscripts running over the natural numbers but nothing beyond those. Set theorists write this object

$$\omega = \{0, 1, 2, \dots, i, \dots\}.$$

The obvious representative for those sets having an initial segment isomorphic to $\omega$, and just one element beyond that segment, is written

$$\omega + 1 = \omega \cup \{\omega\} = \{0, 1, 2, \dots, i, \dots; \omega\}.$$

We likewise go on to get $\omega + 2$, $\omega + 3$ etc. The element coming after all the $\omega + i$'s $(i \in \omega)$ is denoted $\omega + \omega$ or $\omega 2$. (We will see later why it is not written "more naturally" as $2\omega$.) After the elements $\omega 2 + i$ $(i \in \omega)$ comes $\omega 3$; ... after all the elements of the form $\omega i$ $(i \in \omega)$ one has $\omega \omega = \omega^2$. In fact, one can form arbitrary "polynomials" in $\omega$ with natural number coefficients, and the set of these has just the order structure that was given to the polynomials with natural number coefficients in Exercise 5.5:1 (though in our "polynomials" in $\omega$, the coefficients are, as noted, written on the right). Then the set of *all* these polynomials in $\omega$ is taken as the next standard sample well-ordered set ....

So far, we have been sketching an idea; let us make it precise. First, a small terminological point: If $X$ is any well-ordered set, and $\alpha$ the "standard" well-ordered set (to be constructed below) that is order-isomorphic to it, then Lemma 5.5.1 shows us how to index the elements of $X$ by the members of $\alpha$—i.e., by the "standard" well-ordered sets smaller than $\alpha$. Thus, the well-ordered sets less than $\alpha$ serve as translations, and (starting with $\omega$) generalizations, of the sequence of words "first, second, third, ..." which are used in ordinary language to index the elements of finite totally ordered sets. Hence, the term *ordinal*, used by grammarians for those words, is used by mathematicians for the "standard samples" of isomorphism types of well-ordered sets. Let us now give the formal definition of these objects, and investigate their properties.

**Definition 5.5.2.** *An* ordinal *(or* von Neumann ordinal) *is a set* $\alpha$ *such that* $\gamma \in \beta \in \alpha \implies \gamma \in \alpha$, *and such that if* $\beta \in \alpha$ *and* $\gamma \in \alpha$, *then either* $\beta = \gamma$, *or* $\beta \in \gamma$, *or* $\gamma \in \beta$.

**Proposition 5.5.3.**

(i)    *Every member* $\beta$ *of an ordinal* $\alpha$ *is an ordinal.*

(ii)   *If* $\alpha$ *is an ordinal and* $S$ *is a subset of* $\alpha$, *then the following conditions are equivalent:*

   (a)  $S$ *is an ordinal.*

   (b)  *Every member of a member of* $S$ *is a member of* $S$,

   (c)  $S \in \alpha$ *or* $S = \alpha$.

(iii)  *For any two ordinals* $\alpha$ *and* $\beta$, *either* $\alpha$ *is a subset of* $\beta$, *or* $\beta$ *is a subset of* $\alpha$.

*Proof.* In proving (i), the one nonobvious condition needed is that $\delta \in \gamma \in \beta \in \alpha$ implies $\delta \in \beta$. Now in that situation, the first part of the definition of $\alpha$ being an ordinal shows that both $\beta$ and $\delta$ are members of $\alpha$, so by the last part of that definition, either $\delta \in \beta$ or $\beta = \delta$ or $\beta \in \delta$. Either of the last two alternatives, combined with the relations $\delta \in \gamma \in \beta$, would give a contradiction to the Regularity Axiom. So $\delta \in \beta$, as required.

In (ii), the implication (c) $\implies$ (a) follows from (i), while (a) $\implies$ (b) follows from the definition of ordinal. To get (b) $\implies$ (c), assume (b), and let us again call on the Axiom of Regularity, this time to give us an element $\alpha' \in \alpha \cup \{\alpha\}$ which contains $S$ as a subset, but no member of which does so. (Since $S \subseteq \alpha$, the class of members of $\alpha \cup \{\alpha\}$ which contain $S$ as a subset is indeed nonempty.) Now if $\alpha' \neq S$, then there is some $\beta \in \alpha' - S$. Since both $\beta$ and all the elements of $S$ are elements of $\alpha$, the last condition in the definition of $\alpha$ being an ordinal tells us that each $\gamma \in S$ is either equal to $\beta$, or has $\beta$ as a member, or is a member of $\beta$. If either of the first two possibilities occurred for some $\gamma \in S$, we would have $\beta \in S$, contradicting our choice of $\beta$. Hence every member of $S$ must be a member of $\beta$, i.e., $S \subseteq \beta$. But this contradicts our choice of $\alpha'$ as having no member which contained $S$ as a subset. So in fact we must have $S = \alpha' \in \alpha \cup \{\alpha\}$, proving (c).

Finally, let us get (iii) by showing that the set $S = \alpha \cap \beta$ must equal either $\alpha$ or $\beta$. Clearly, $S$ is a subset of both $\alpha$ and $\beta$ satisfying (b), hence since (b) $\implies$ (c), $S$ either equals or is a member of each of $\alpha$ and $\beta$. If it were a *member* of both, then we would have $S \in \alpha \cap \beta = S$, again contradicting Regularity; so it must equal one of them.                                    □

We would like to summarize statement (iii) above as saying that the ordinals form a totally ordered set under inclusion—and in fact, since by (ii), inclusion among ordinals is equivalent to "$\in$ or $=$", and by Regularity, the latter relation satisfies descending chain condition, we would like to say that the ordinals form a well-ordered set. The only trouble is that they do not form a set!

Here are some things we *can* say:

## Proposition 5.5.4.

(i)    *Every nonempty class of ordinals has a "$\subseteq$-least" member. (In other words, the* class *of ordinals satisfies under $\subseteq$ the analog of the set-theoretic property of well-orderedness.) In particular, every ordinal, and more generally, every set of ordinals, is well-ordered under $\subseteq$.*

(ii)    *The union of any set of ordinals is an ordinal.*

(iii)    *The class of all ordinals is not a set.*

(iv)    *Every well-ordered set has a unique order isomorphism with an ordinal.*

*Proof.* To get (i), let $C$ be a nonempty class of ordinals, take any $\beta \in C$, and note that $C' = \{\alpha \subseteq C \mid \alpha \subseteq \beta\}$ is a nonempty *set* of ordinals. The axiom of Regularity says that $C'$ has a member $\gamma$ which is disjoint from $C'$. One can see that $\gamma$ will be a least member of $C'$ under $\subseteq$, and since members of $C'$ (which are contained in $\beta$) are less than other members of $C$ (since these are not), $\gamma$ must in fact be a least member of $C$. The final sentence of (i) clearly follows.

With the help of part (iii) of the preceding proposition, it is easy to check that the union of any set $S$ of ordinals satisfies the definition of an ordinal, giving (ii). Moreover, if we call this union $\alpha$, then $\alpha$ will majorize all members of $S$, hence its successor, $\alpha \cup \{\alpha\}$, cannot be a member of $S$. So for any set $S$ of ordinals we have an ordinal not in $S$, proving (iii).

To show (iv), let $S$ be a well-ordered set. For convenience, let us form a new ordered set $T$ consisting of the elements of $S$, ordered as in $S$, and one additional element $z$, greater than them all. It is immediate that $T$ will again be well-ordered. I claim that for every $t \in T$, there is a unique order-isomorphism between $\{s \in T \mid s < t\}$ and some (unique) ordinal. Indeed, if not, there would be a least $t$ for which this failed, so each $s < t$ would be isomorphic to a unique ordinal $\beta_s$, by a unique order-isomorphism. From this it is easy to deduce that the set $\alpha = \{\beta_s \mid s < t\}$ would be order-isomorphic to $\{s \in T \mid s < t\}$, that it would be the unique ordinal with this property, and that the isomorphism would be unique, contradicting our choice of $t$, and establishing our claim. So in particular, there is a unique order-isomorphism between $S = \{s \in T \mid s < z\}$ and an ordinal, as required.                     $\square$

The proofs of the above two propositions make strong use of the Axiom of Regularity. How do set-theorists who do not assume that axiom define ordinals so that the same results will hold? The easy way is to add to the definition of ordinal the case of Regularity that we need; namely, that every nonempty $\beta \in \alpha \cup \{\alpha\}$ have a member which is disjoint from $\beta$. A different approach is taken in [24, § 7.2]. Rather than starting by *defining* "ordinal", that developments starts with a construction that, from any well-ordered set $S$, builds by recursion on the elements of $S$ a certain set $\alpha$, order-isomorphic to $S$. It is the sets constructed in this way that are then named "ordinals". Finally, a result is proved characterizing those sets as the sets satisfying Definition 5.5.2, modified as just noted to compensate for the lack of the Axiom of Regularity.

**Exercise 5.5:3.** Let $\alpha$ and $\beta$ be ordinals. Show that there exists a one-to-one isotone map $f : \alpha \to \beta$ if and only if $\alpha \leq \beta$.

**Exercise 5.5:4.** If $P$ is a partially ordered set with DCC, let the *height* $\mathrm{ht}(p)$ of an element $p \in P$ be defined, recursively, as the least ordinal greater than the heights of all elements $q < p$. Define $\mathrm{ht}(P)$ to be the least ordinal greater than the heights of all elements of $P$.

(i)    Show that the height function on elements is the least strictly isotone ordinal-valued function on $P$, and that it has range precisely $\mathrm{ht}(P)$.

(ii)    Show that for every ordinal $\alpha$ there exists a partially ordered set containing no infinite chains, and having height $\alpha$.

(iii) Suppose we define the *chain height* of $P$, $\mathrm{chht}(P)$, to be the least ordinal which cannot be embedded in $P$ by an isotone map, and for $p \in P$ define $\mathrm{chht}(p) = \mathrm{chht}(\{q \in P \mid q < p\})$. What relationships can you establish between the functions $\mathrm{ht}$ and $\mathrm{chht}$?

**Exercise 5.5:5.** Show that the two conditions in Definition 5.5.2 (the definition of an ordinal) are independent, by giving examples of sets satisfying each but not the other.

For additional credit, you can show that each of your examples has smallest cardinality among sets with the desired property.

Since one considers ordinals to be ordered under the relation $\subseteq$, equivalently, "$\in$ or $=$", one has the choice, in speaking about them, between writing $\leq$ and $\subseteq$, and likewise between $<$ and $\in$. Both the order-theoretic and the set-theoretic notation are used, sometimes mixed together.

For every ordinal $\alpha$, there is a least ordinal greater than $\alpha$, namely $\alpha \cup \{\alpha\}$. This is called the *successor* of $\alpha$, and written $\alpha + 1$. "Most" ordinals are successor ordinals. Those, such as $0$, $\omega$, $\omega 2$, etc., which are not, are called *limit ordinals*. (Although, as I just stated, $0$ is logically a limit ordinal, and I will consider it such here, it is sometimes treated as a special case, neither a successor nor a limit ordinal.)

**Exercise 5.5:6.** Show that an ordinal is a limit ordinal if and only if it is the least upper bound of all strictly smaller ordinals; equivalently, if and only if, as a set, it is the union of all its members.

Now that we understand why ordinals (and so in particular, natural numbers) are defined so that each is equal to the set of those that precede it, let us rescind the convention we set up in §2.3, where, for the sake of familiarity, I said that an $n$-tuple of elements of a set $S$ would mean a function $\{1, \ldots, n\} \to S$:

**Definition 5.5.5.** *Throughout the remainder of these notes, $n$-tuples will be defined in the same way as $I$-tuples for other sets $I$. That is, for $n$ a natural number, an $n$-tuple of elements of a set $S$ will mean a function $n \to S$, i.e., a family $(s_0, s_1, \ldots, s_{n-1})$ $(s_i \in S)$. The set of all such functions will be denoted $S^n$.*

We have referred to ordinals denoted by symbols such as $\omega 2$, $\omega^2 + 1$, etc. As this suggests, there is an arithmetic of ordinals. If $\alpha$ and $\beta$ are ordinals, $\alpha + \beta$ represents the ordinal which has an initial segment $\alpha$, and the remaining elements of which form a subset order-isomorphic to $\beta$. This exists, since by putting an order-isomorphic copy of $\beta$ "above" the ordinal $\alpha$, one gets a well-ordered set, and we know that there is a unique ordinal order-isomorphic to it. Similarly, $\alpha\beta$ represents an ordinal which is composed of a family of disjoint well-ordered sets, each order-isomorphic to $\alpha$, one above the other, with the order structure of the set of copies being that of $\beta$. These operations are (of course) formally defined by recursion, as we will describe below.

Unfortunately, the formalization of recursion that we proved in Lemma 5.3.4 is not quite strong enough for the present purposes, because in constructing larger ordinals from smaller ones, we will not easily be able to give *in advance* a codomain set corresponding to the $T$ of that lemma, and as a result, we will not be able to precisely specify the function $r$ required by that lemma either. However, there is a version of recursion based on the Replacement Axiom (Fraenkel's contribution to Zermelo–Fraenkel set theory) which gets around this problem. Like that axiom, it assumes we are given a construction which is not necessarily a function, because its range and domain are not assumed to be sets, but which nonetheless uniquely determines one element given another. I will not discuss this concept, but will state the result below. The proof is exactly like that of Lemma 5.3.4, except that the Axiom of Replacement is used to carry out the "piecing together" of partial functions.

**Lemma 5.5.6** (Cf. [24, Theorem 7.1.5, p. 74]). *Let $X$ be a partially ordered set with descending chain condition, and $r$ a construction associating to every pair $(x, f_{<x})$, where $x \in X$ and $f_{<x}$ is a function with domain $\{y \in X \mid y < x\}$, a uniquely defined set $r(x, f_{<x})$. Then there exists a unique function $f$ with domain $X$ such that for all $x \in X$, $f(x) = r(x, f \mid \{y \mid y < x\})$.*

$\square$

We can now define the operations of ordinal arithmetic. For completeness we start with the (nonrecursive) definition of the successor operation. Note that in each of the remaining (recursive) definitions, the ordinal $\alpha$ is held "constant", and the ordinal over which we are doing the recursion is written $\beta$ or $\beta + 1$.

*Definition of the successor of an ordinal:*

$$(5.5.7) \qquad \beta + 1 = \beta \cup \{\beta\}.$$

*Definition of addition of ordinals:*

$$(5.5.8) \qquad \begin{aligned} \alpha + 0 &= \alpha, & \alpha + (\beta + 1) &= (\alpha + \beta) + 1, \\ \alpha + \beta &= \textstyle\bigcup_{\gamma < \beta} \alpha + \gamma & \text{for } \beta \text{ a limit ordinal} &> 0. \end{aligned}$$

*Definition of multiplication of ordinals:*

$$(5.5.9) \qquad \begin{aligned} \alpha\, 0 &= 0, & \alpha(\beta + 1) &= (\alpha\beta) + \alpha, \\ \alpha\beta &= \textstyle\bigcup_{\gamma < \beta} \alpha\gamma & \text{for } \beta \text{ a limit ordinal} &> 0. \end{aligned}$$

*Definition of exponentiation of ordinals:*

$$(5.5.10) \qquad \begin{aligned} \alpha^0 &= 1, & \alpha^{(\beta+1)} &= (\alpha^\beta)\alpha, \\ \alpha^\beta &= \textstyle\bigcup_{\gamma < \beta} \alpha^\gamma & \text{for } \beta \text{ a limit ordinal} &> 0. \end{aligned}$$

**Exercise 5.5:7.** Definitions (5.5.8) and (5.5.9) do not look like the descriptions of ordinal addition and multiplication sketched in words above. Show that they do in fact have the properties indicated there.

Although the operations defined above agree with the familiar ones on the finite ordinals (natural numbers), they have unexpected properties on infinite ordinals. Neither addition nor multiplication is commutative:

$$1 + \omega = \omega, \quad \text{but} \quad \omega + 1 > \omega,$$

$$2\omega = \omega, \quad \text{but} \quad \omega 2 > \omega.$$

Exponentiation is also different from exponentiation of cardinals (discussed later in this section):

$$2^\omega = \omega.$$

Students who have not seen ordinal arithmetic before might do:

**Exercise 5.5:8.** Prove the three equalities and two inequalities asserted in the above paragraph.

   You may assume familiar facts about arithmetic of natural numbers, and that the ordinal operations agree with the familiar operations in these cases; but assume nothing about how they behave on infinite ordinals, except the definitions.

**Exercise 5.5:9.** If an ordinal $\alpha$ is the disjoint union of a subset order-isomorphic to the ordinal $\beta$ and a subset order-isomorphic to the ordinal $\gamma$, must we have $\alpha \leq \beta + \gamma$? $\alpha \leq \max(\beta + \gamma, \gamma + \beta)$?

The formulas (5.5.8)–(5.5.10) define *pairwise* arithmetic operations. We can also define arithmetic operations on families of ordinals indexed by (what else?) ordinals. Let us record the case of addition, since we will need this later. Given such a family $(\alpha_\gamma)_{\gamma \in \beta}$, the idea is to define $\sum_{\gamma \in \beta} \alpha_\gamma$ to be the ordinal which, as a well-ordered set, is the union of a chain of disjoint subsets of respective order types $\alpha_\gamma$ $(\gamma \in \beta)$, appearing in that order.

*Definition of infinite ordinal addition:*

(5.5.11)
$$\sum_{\gamma \in 0} \alpha_\gamma = 0, \quad \sum_{\gamma \in \beta+1} \alpha_\gamma = \left(\sum_{\gamma \in \beta} \alpha_\gamma\right) + \alpha_\beta,$$
$$\sum_{\gamma \in \beta} \alpha_\gamma = \bigcup_{\gamma < \beta} \sum_{\delta \in \gamma} \alpha_\delta \quad \text{for } \beta \text{ a limit ordinal} > 0.$$

Taking the $\alpha_\gamma$'s all equal, we see that our recursive definition of $\sum_\beta \alpha_\gamma$ reduces to our definition of multiplication of ordinals; hence

(5.5.12)
$$\sum_{\gamma \in \beta} \alpha = \alpha\beta.$$

**Exercise 5.5:10.**

   (i)   Given a family of ordinals $(\alpha_\gamma)_{\gamma \in \beta}$ indexed by an ordinal, let $\alpha$ denote the ordinal $\bigcup_{\gamma \in \beta} \alpha_\gamma$ (the supremum of the $\alpha_\gamma$'s). Let $P$ be the set $\beta \times \alpha$, lexicographically ordered. Show that the ordinal $\sum_{\gamma \in \beta} \alpha_\gamma$ is isomorphic as a well-ordered set to $\{(\gamma, \delta) \mid \gamma \in \beta, \delta \in \alpha_\gamma\} \subseteq P$.

   (ii)   Deduce from this a description of a well-ordered set isomorphic to the ordinal product $\alpha\beta$ of two arbitrary ordinals.

This description clearly extends inductively to finite products $\prod_{\gamma \in \beta} \alpha_\gamma$ $(\beta < \omega)$, leading, incidentally, to an easy proof of *associativity* of multiplication of ordinals. The extension of these ideas to infinite products will be developed in a later exercise in this section.

We have seen that every well-ordered set is indexed in a canonical way by an ordinal; but we do not yet know whether we can well-order every set. It turns out that we can do so if we assume the Axiom of Choice. This is stated in the second part of the next lemma; the first part gives a key argument (not requiring the Axiom of Choice) used in the proof.

**Lemma 5.5.13.** *Let $X$ be a set. Then*

(i)    *There exists an ordinal $\alpha$ which cannot be put in bijective correspondence with any subset of $X$; equivalently, such that for any well-ordering "$\leq$" of any subset of $Y \subseteq X$, $(Y, \leq)$ is isomorphic to a proper initial segment of $\alpha$.*

(ii)    *Assuming the Axiom of Choice, $X$ itself can be well-ordered.*

*Proof.* The class of well-orderings of subsets of $X$ is easily shown to be a set, hence by the Replacement Axiom, the unique ordinals isomorphic to these various well-ordered sets form a set, hence the union of this set is an ordinal $\beta$. Take $\alpha = \beta + 1$. By construction, any well-ordering of a subset $Y \subseteq X$ induces a bijection of $Y$ with an initial segment of $\beta$, which is a proper initial segment of $\alpha$, yielding the second formulation of (i). To get the first formulation, note that if $\alpha$ could be put in bijective correspondence with a subset of $X$, then the ordering of $\alpha$ would induce a well-ordering of that subset, such that $\alpha$ was the unique ordinal isomorphic to that well-ordered set, giving a contradiction to our preceding conclusion.

Assuming the Axiom of Choice, let us now take a function $c$ which associates to every nonempty subset $Y \subseteq X$ an element $c(y) \in Y$. Let us recursively construct a one-to-one map from some initial subset of the ordinal $\alpha$ of part (i) into $X$ as follows: Suppose we have gotten a function $f_\beta$ from an ordinal $\beta < \alpha$, regarded as a subset of $\alpha$, into $X$. If its image is $X$, we are done. If not, we send the element $\beta$, which is the first element of $\alpha$ on which our map is not yet defined, to $c(X - \text{image}(f_\beta))$. It is easy to verify by induction that each map $f_\beta$ is one-to-one. If this process went on to give a one-to-one map $f_\alpha$ of $\alpha$ into $X$, that would contradict (i). So instead, the construction must terminate at some step, which means we must get a bijection between an initial segment of $\alpha$ and $X$, and hence a well-ordering of $X$, proving (ii). (As in the proof of Lemma 5.5.1, our use of a recursion that terminates before we get through all of $\alpha$ can be formalized by adjoining to $X$ an element DONE.)    □

**Exercise 5.5:11.** Let $P$ be a partially ordered set, and $\text{Ch}(P)$ the set of chains in $P$, partially ordered by writing $A \leq B$ if $A$ is an initial segment of $B$.

(i)   Show that there can be no strict isotone map $f \colon \mathrm{Ch}(P) \to P$. (Suggestion: If there were, show that one could recursively embed any ordinal $\alpha$ in $P$, by sending each element of $\alpha$ to the image under $f$ of the chain of images of all preceding elements.)

(ii)   Deduce from (i) the same statement with $\mathrm{Ch}(P)$ ordered by inclusion. Conclude that $\mathrm{Ch}(P)$, under either ordering, can never be order-isomorphic to $P$.

(iii)   Can you strengthen the result of (i) by replacing $\mathrm{Ch}(P)$ by some natural proper subset thereof?

Let us assume the Axiom of Choice for the rest of this section (though at the beginning of the next section, we will again suspend this assumption).

Recall that two sets are said to have the same *cardinality* if they can be put in bijective correspondence. Statement (ii) of the preceding lemma shows that (assuming the Axiom of Choice), every set has the same cardinality as an ordinal. This means we can use appropriately chosen ordinals as "standard examples" of all cardinalities. In general, there are many ordinals of a given cardinality (e.g., $\omega$, $\omega+1$, $\omega 2$ and $\omega^2$ are all countable), so the ordinal to use is not uniquely determined. The one easily specified choice is the *least* ordinal of the given cardinality; so one makes

**Definition 5.5.14.** *A* cardinal *is an ordinal which cannot be put into bijective correspondence with a proper initial segment of itself.*

*For any set $X$, the least ordinal with which $X$ can be put in bijective correspondence will be called the* cardinality *of $X$, denoted* $\mathrm{card}(X)$. *Thus, this is a cardinal, and is the only cardinal with which $X$ can be put in bijective correspondence.*

There is an arithmetic of cardinals: If $\kappa$ and $\lambda$ are cardinals, $\kappa + \lambda$ is defined as the cardinality of the union of any two disjoint sets one of which has cardinality $\kappa$ and the other cardinality $\lambda$, $\kappa\lambda$ as the cardinality of the direct product of a set of cardinality $\kappa$ and a set of cardinality $\lambda$, and $\kappa^\lambda$ as the cardinality of the set of all functions from a set of cardinality $\lambda$ to a set of cardinality $\kappa$. Unfortunately, if we consider the class of cardinals as a subset of the ordinals, these are different operations from the *ordinal arithmetic* we have just defined! To compare these arithmetics, let us temporarily use the notations $\alpha +_{\mathrm{ord}} \beta$, $\alpha \cdot_{\mathrm{ord}} \beta$ and $\alpha^{\mathrm{ord}\,\beta}$ for ordinal operations, and $\kappa +_{\mathrm{card}} \lambda$, $\kappa \cdot_{\mathrm{card}} \lambda$ and $\kappa^{\mathrm{card}\,\lambda}$ for cardinal operations. A positive statement we can make is that for cardinals $\kappa$ and $\lambda$, the computation of their cardinal sum and product can be reduced to that of their ordinal sum and product, by the formulas

$$(5.5.15) \quad \kappa +_{\mathrm{card}} \lambda = \mathrm{card}(\kappa +_{\mathrm{ord}} \lambda) \quad \text{and} \quad \kappa \cdot_{\mathrm{card}} \lambda = \mathrm{card}(\kappa \cdot_{\mathrm{ord}} \lambda).$$

These are cases of a formula holding for any family of ordinals $(\alpha_\gamma)_{\gamma \in \beta}$ :

(5.5.16)                $\sum_{\gamma \in \beta}^{\mathrm{card}} \mathrm{card}(\alpha_\gamma) = \mathrm{card}(\sum_{\gamma \in \beta}^{\mathrm{ord}} \alpha_\gamma).$

On the other hand, the cardinality of an infinite ordinal product of ordinals is not in general equal to the cardinal product of the cardinalities of these ordinals; in particular, cardinal exponentiation does not in any sense agree with ordinal exponentiation: $2^{\mathrm{card}\omega}$ gives the cardinality of the continuum, which is uncountable, while $2^{\mathrm{ord}\omega} = \omega$. There is no standard notation distinguishing ordinal and cardinal arithmetic; authors either introduce ad hoc notations, or say in words whether cardinal or ordinal arithmetic is meant, or rely on context to show this.

**Exercise 5.5:12.** In this exercise we shall extend the results of Exercise 5.5:10, which characterized the order-types of general sums and finite products of ordinals, to general products. (I have put this off until now so that we would have notation distinguishing the ordinal product $\prod^{\mathrm{ord}}\alpha_\gamma$ from the set-theoretic product.) We will also note a relation with cardinal arithmetic. We need to begin with a generalization of lexicographic ordering.

Suppose $(X_i)_{i \in I}$ is a family of partially ordered sets, indexed by a totally ordered set $I$; and let each $X_i$ have a distinguished element, denoted $0_i$. Define the *support* of $(x_i) \in \prod_I X_i$ as $\{i \in I \mid x_i \neq 0_i\}$, and let $\prod_I^{\mathrm{w.o.supp}} X_i$ denote the set of elements of $\prod_I X_i$ having *well-ordered* support. Similarly, let $\prod_I^{\mathrm{f.supp}} X_i$ denote the set of elements of *finite* support.

(i)   Show that the definition of lexicographic order, which in Definition 5.3.7 was given on $\prod_I X_i$ for $I$ well-ordered, makes sense on $\prod_I^{\mathrm{w.o.supp}} X_i$ for arbitrary totally ordered $I$, and that the resulting ordering is total if each $X_i$ is totally ordered.

(ii)   Show that if $I$ is reverse-well-ordered (has ascending chain condition) then $\prod_I^{\mathrm{w.o.supp}} X_i = \prod_I^{\mathrm{f.supp}} X_i$.

(iii)   Show that if $I$ is reverse-well-ordered and if each $X_i$ has descending chain condition, and has $0_i$ as *least* element, then $\prod_I^{\mathrm{f.supp}} X_i$ has descending chain condition under lexicographic ordering.

(iv)   Let us now be given an ordinal-indexed family of ordinals, $(\alpha_\gamma)_{\gamma \in \beta}$. Write down the definition of $\prod_{\gamma \in \beta}^{\mathrm{ord}} \alpha_\gamma$ analogous to (5.5.11). Verify that if any $\alpha_\gamma$ is 0, your definition gives the ordinal 0. In the contrary case, show that your definition gives an ordinal order-isomorphic to $\prod_{\gamma \in \beta^{\mathrm{op}}}^{\mathrm{f.supp}} \alpha_\gamma$. (Here $\beta^{\mathrm{op}}$ denotes the set $\beta$, but with its ordering—used in defining lexicographic order on our product—reversed. Note that "$\gamma \in \beta^{\mathrm{op}}$" means the same as "$\gamma \in \beta$". For the elements $0_\gamma$ in the definition of $\prod^{\mathrm{f.supp}}$, we take the ordinal $0 \in \alpha_\gamma$, which is why we need to assume all $\alpha_\gamma$ nonzero.)

(v)   Deduce a description of the order-type of $\alpha^{\mathrm{ord}\,\beta}$, and conclude that $\mathrm{card}(\alpha^{\mathrm{ord}\,\beta}) \leq \alpha^{\mathrm{card}\,\beta}$.

You might also want to do

(vi)   Show by examples that (iii) above fails if any of the three hypotheses is deleted.

The set-theoretic concept of cardinality historically antedates the construction of the ordinals, so there is a system of names for cardinals independent of their names as ordinals. The finite cardinals are, of course, denoted by the traditional symbols $0, 1, 2, \ldots$. The least infinite cardinal is denoted $\aleph_0$, the next $\aleph_1$, etc. From our description of the cardinals as a subclass of the ordinals, we see that the class of cardinals is "well-ordered" (written in quotes, as we did for the class of ordinals, because this class is not a set). Hence, now that one has the concept of ordinal, one continues the above set of symbols using ordinal subscripts: The $\alpha$-th cardinal after $\aleph_0$ is written $\aleph_\alpha$.

There is a further notation for cardinals "regarded as ordinals". Each $\aleph_\alpha$, regarded as an ordinal, is written $\omega_\alpha$. Thus one writes $\aleph_0 = \omega_0 = \omega$, $\aleph_1 = \omega_1$, etc.

In the next theorem we set down, without repeating the proofs, some well-known properties of cardinal arithmetic, though we will use them only occasionally.

**Theorem 5.5.17.** *Letting $\kappa$, $\lambda$, etc., denote cardinals, and letting arithmetic notation denote cardinal arithmetic, the following statements are true.*

(i)   *For all $\kappa$, $\lambda$, $\mu$,*

$$\kappa + \lambda = \lambda + \kappa, \qquad \kappa\,\lambda = \lambda\,\kappa, \qquad (\kappa + \lambda)\mu = \kappa\mu + \lambda\mu,$$
$$\kappa^{\lambda+\mu} = \kappa^\lambda \kappa^\mu, \qquad \kappa^{\lambda\mu} = (\kappa^\lambda)^\mu.$$

(ii)   *For sets $X_i$ $(i \in I)$, $\mathrm{card}(\bigcup_I X_i) \leq \sum \mathrm{card}(X_i)$.*

(iii)   *If $\kappa_\beta \leq \lambda_\beta$ for all $\beta \in \alpha$, then*

$$\sum_\alpha \kappa_\beta \leq \sum_\alpha \lambda_\beta, \quad \prod_\alpha \kappa_\beta \leq \prod_\alpha \lambda_\beta, \quad \text{and, if } \lambda_0 > 0, \quad \kappa_0^{\kappa_1} \leq \lambda_0^{\lambda_1}.$$

(iv)   *If $\kappa \leq \lambda$ and $\lambda$ is infinite, then $\kappa + \lambda = \lambda$. If $1 \leq \kappa \leq \lambda$ and $\lambda$ is infinite, then $\kappa\,\lambda = \lambda$. In particular, $\omega\omega = \omega$, hence by (ii) and (iii), a countable union of countable sets is countable.*

(v)   $2^\kappa > \kappa$. *Equivalently, the power set of any set $X$ has strictly larger cardinality than $X$.*

*Proof.* See [32, pp. 17–21], or [34, Appendix 2, §1 and exercises at the end of that appendix].   □

It is interesting that while the statement $\omega\omega = \omega$ is easy to prove without the Axiom of Choice (by describing an explicit bijection), its consequence, "a countable union of countable sets is countable", requires that axiom, to enable us to choose, simultaneously, particular bijections between the set $\omega$ and each of the infinitely many given countable sets.

Incidentally, while the word "countable" is unambiguous when referring to a set that we know is infinite, when that is not the case it may mean either "having cardinality $\aleph_0$" or "having cardinality $\leq \aleph_0$", depending on the author. To make the former meaning unambiguous one can say "countably infinite"; some authors use the word *denumerable* for this.

Turning from arithmetic back to order properties, let me define a concept of interest in the general study of ordered sets, and note a specific application to cardinals.

**Definition 5.5.18.** *If $X$ is a partially ordered set, then the* cofinality *of $X$ means the least cardinality of a cofinal subset $Y \subseteq X$ (Definition 5.1.6).*

*A cardinal $\kappa$ is called* regular *if, as an ordinal, it has cofinality $\kappa$. A cardinal that is not regular is called* singular.

**Exercise 5.5:13.** Show that if a partially ordered set $X$ has cofinality $\kappa$, then every cofinal subset $Y \subseteq X$ also has cofinality $\kappa$.

**Exercise 5.5:14.** Prove:

(i)  Every cardinal of the form $\aleph_{\alpha+1}$ (i.e., every cardinal indexed by a successor ordinal) is regular.

(ii)  The first infinite singular cardinal is $\aleph_\omega$.

The next exercise examines the class of regular cardinals within the class of ordinals.

**Exercise 5.5:15.** Let us call an ordinal $\alpha$ regular if there is no set map from an ordinal $< \alpha$ onto a cofinal subset of $\alpha$.

(i)  Show that regular ordinals are "sparse", by verifying that the only regular ordinals are $0$, $1$, and the regular infinite *cardinals*.

(ii)  On the other hand, part (i) of the preceding exercise shows that within the set of infinite cardinals, the singular cardinals are sparse: They must be limit cardinals, i.e., cardinals $\omega_\alpha$ such that $\alpha$ is a limit ordinal. Prove this if you did not do that exercise.

(iii) Show that among the *limit* cardinals, *regular* cardinals are again sparse, by showing that if $\omega_\alpha$ is regular and $\alpha$ is a limit ordinal, then $\alpha$ must be a cardinal; in fact, $0$ or a cardinal $\kappa$ satisfying

(5.5.19) $$\kappa = \omega_\kappa.$$

Show that the first cardinal $\kappa$ satisfying (5.5.19) is the supremum of the chain $\kappa(i)$ $(i \in \omega)$ defined by $\kappa(0) = 0$, $\kappa(i+1) = \omega_{\kappa(i)}$, but that this cardinal is still *not* regular.

(Regular limit cardinals will come up again in §7.4.)

**Exercise 5.5:16.** Since ordinals are totally ordered sets, they are in particular partially ordered sets, and we can partially order the set-theoretic direct product of two ordinals by componentwise comparison (Definition 5.1.4). Regarding the product-sets $\omega \times \omega$, $\omega \times \omega_1$, $\omega_1 \times \omega_1$, $\omega \times \omega_\omega$, and $\omega_1 \times \omega_{\omega+1}$ as partially ordered in this manner, determine, as far as you can, whether each of these contains a *cofinal chain*. (Partial credit will be given for partial results, and additional credit for general results subsuming some of these particular cases.)

The properties of ordinals allow us to obtain a construction that we wondered about when we considered Stone-Čech compactifications in §4.17:

**Exercise 5.5:17.** Let $S$ be a totally ordered set, and for convenience, let $-\infty$ and $+\infty$ be two elements outside $S$, the former regarded as less than all elements of $S$ and the latter as greater than all elements of $S$. Then we can define the *order topology* on $S$ to have as basis of open sets the intervals $(r, t) = \{s \in S \mid r < s < t\}$, where $r, t \in S \cup \{-\infty, +\infty\}$. (We could do without $-\infty$ and $+\infty$ if we knew that $S$ had no least or greatest element. But if it does, then unless one introduces the above extra elements, one has to define several distinct sorts of basic open sets, instead of just one.)

(i)    Let an ordinal $\alpha$ be given the order topology. Which subsets of $\alpha$ are closed? Which are compact?

(ii)   Show that under the order topology the ordinal $\omega_1$ is not compact, but satisfies condition (b) of Exercise 4.17:9. (Thus, it satisfies condition (a) of that exercise, which was what we were interested in there.)

(iii) If you are familiar with the geometric construction of *the long line*, show that this also satisfies condition (b) of Exercise 4.17:9, and examine its relationship to the ordinal $\omega_1$.

## 5.6. Zorn's Lemma

Ordinals, together with the Axiom of Choice, give a powerful tool for constructing non-uniquely-determined objects in many areas of mathematics.

Let me begin by sketching an example: how those tools can be used to show that every vector space $V$ over a field $k$ has a basis. One constructs *recursively* a chain $(B_\alpha)_{\alpha \in \gamma}$ of linearly independent subsets of $B$ for some cardinal $\gamma$, starting with $B_0 = \emptyset$. For each $\alpha$, as long as $B_\alpha$ does not span $V$, one can use the Axiom of Choice to choose a $v \in |V|$ not in the span of $B_\alpha$, and let $B_{\alpha+1} = B_\alpha \cup \{v\}$, while for $\beta$ a limit ordinal, one can take $B_\beta = \bigcup_{\alpha \in \beta} B_\alpha$. One verifies that each of these steps gives a new linearly independent subset. If we take our indexing ordinal $\gamma$ larger than the cardinality of $V$, this construction cannot continue adding new elements at every non-limit ordinal; so for some $\alpha$, the set $B_\alpha$ must, in fact, span $V$, and hence be a basis.

Abstracting the technique of this example, one should begin a general construction of this nature by deciding what kind of objects one will consider "partial constructions" (above: the linearly independent subsets of $V$), and verifying that these form a set. Hence there exists an ordinal $\gamma$ of greater cardinality than this set. Setting up the recursion over $\gamma$ then involves three tasks:

(i)  Find an "initializing" partial construction to which to map 0. (Above: the set $\emptyset \subseteq |V|$.)

(ii)  If one has built up successive partial constructions through the one associated to an ordinal $\alpha$, and that construction is still not "finished", one must show that it can be extended further, to give an $\alpha+1$-st stage. (Above: if $B_\alpha$ does not span $V$, then for any $v$ outside the span of $B_\alpha$, the set $B_\alpha \cup \{v\}$ is a larger linearly independent set.) Once one has done so, the Axiom of Choice allows one to keep picking such extensions as long as one has not achieved a "finished" construction.

(iii) Specify what to do at a nonzero *limit* ordinal $\alpha$. At such a step, one has a chain of preceding partial constructions, each extending the one before, and it is often easy to verify that their "union", defined in some way, is a partial construction extending all of them. (In the vector space case, as in many others, we literally take the union.)

Since the resulting recursion cannot give a one-to-one map from $\gamma$ into the smaller-cardinality set of all partial constructions, it must, rather, give at some step a "finished" construction, as desired.

The above general technique is a tool that is used repeatedly, so it is natural to seek a lemma whose proof embodies once and for all the set-theoretic side of the argument, and whose statement will show us what we must prove separately for each case. In formulating the statement, one abstracts the set of all "partial constructions" as a partially ordered set $(X, \leq)$, where $\leq$ is intended to be the relation of one construction being a "part of" another. The condition saying that we can initialize our recursion is that $X$ be nonempty. To show that we can extend a partial construction further if it is not yet "finished" is, put in the contrapositive, to show that if $X$ has any *maximal* element, this is a "finished" object, i.e., one of the sort we desire. Finally, the condition we need at steps indexed by limit ordinals, namely that given a chain of partial constructions, we can find one which includes them all, is made the content of a definition.

**Definition 5.6.1.** *A partially ordered set* $X$ *is called* inductive *if for every nonempty chain* $Y \subseteq X$, *there is an element* $z \in X$ *majorizing* $Y$ (*i.e., such that* $z \geq$ *every element of* $Y$).

We can now state the desired result, *Zorn's Lemma* (statement (ii) below), and show that it, and a number of other statements, are each in fact equivalent to the Axiom of Choice.

**Theorem 5.6.2.** *Assuming the axioms of Zermelo–Fraenkel set theory* (*but not the Axiom of Choice*), *the following four statements are equivalent:*

(i)   **The Axiom of Choice:** *If* $X$ *is a set, and* $f$ *is a function associating to every* $x \in X$ *a nonempty set* $f(x)$, *then there exists a function* $g$ *associating to every* $x \in X$ *an element* $g(x) \in f(x)$. (*Equivalently: the direct product of any family of nonempty sets is nonempty.*)

(ii)   **Zorn's Lemma:** *Every nonempty inductive partially ordered set* $(X, \geq)$ *has a maximal element.*

(iii)   **The Well-ordering Principle:** *Every set can be well-ordered.* (*Equivalently: every set can be put in bijective correspondence with an ordinal.*)

(iv)   **Comparability of Cardinalities:** *Given any two sets* $X$ *and* $Y$, *one of these sets can be put in bijective correspondence with a subset of the other.* (*Loosely: the class of cardinalities is totally ordered.*)

*Proof.* The scheme of proof will be (iv) $\iff$ (iii) $\iff$ (i) $\iff$ (ii). That the parenthetical restatement of (iii) is equivalent to the main statement follows from Proposition 5.5.4(iv).

(iv) $\iff$ (iii): Assuming (iv), let $X$ be any set and $\alpha$ an ordinal with the property stated in Lemma 5.5.13(i). By (iv), there is either a bijection between $X$ and a subset of $\alpha$, or between $\alpha$ and a subset of $X$. By choice of $\alpha$, the latter case cannot occur, so there is a bijection between $X$ and a subset $S \subseteq \alpha$. Since $\alpha$ is well-ordered, so is every subset, and the well-ordering of $S$ induces a well-ordering of $X$, proving (iii). Assuming (iii), statement (iv) follows from the comparability of ordinals, Proposition 5.5.3(iii), or more directly, from Lemma 5.5.1.

(iii) $\iff$ (i): We proved (i) $\implies$ (iii) as Lemma 5.5.13(ii). Conversely, assume (iii). Given $X$ and $f$ as in (i), statement (iii) tells us that we can find a well-ordering $\leq$ on the set $\bigcup_{x \in X} f(x)$. We can now define $g$ to take each $x$ to the $\leq$-least element of $f(x)$. (In terms of the axioms, we are using the Replacement Axiom to construct $g$ as $\{(x, y) \mid x \in X$ and $y$ is the least element of $f(x)\}$.)

(i) $\implies$ (ii): Let $(X, \leq)$ be a nonempty inductive partially ordered set, and let us choose as in Lemma 5.5.13(i) an ordinal $\alpha$ which cannot be put in bijective correspondence with any subset of $X$. Note that the combination of conditions "inductive" and "nonempty" is equivalent to saying that for *every* chain $C \subseteq X$, including the empty chain, there is an element $\geq$ all members of $C$.

By (i), we may choose a function $g$ associating to every nonempty subset of $X$ a member of that subset. We will now recursively define an isotone map $h\colon \alpha \to X$. Assuming that for some $\beta \in \alpha$ we have defined an isotone map $h_{<\beta}\colon \beta \to X$, observe that its image will be a chain $C_\beta \subseteq X$. If the set $Y_\beta$ of elements of $X$ greater than all members of $C_\beta$ is nonempty, we define $h(\beta) = g(Y_\beta)$. In the contrary case, the hypothesis that $X$ is inductive still tells us that there is an element $\geq$ all members of $C_\beta$. We conclude that such an element must be equal to some member of $C_\beta$, which means that the chain has a largest element, $c$. In this case, we take $h(\beta) = c$. Note that in this case $c$ must be maximal in $X$, for if not, any element of $X$ greater than it would be greater than all elements of $C_\beta$, contradicting our assumption that $Y_\beta$ was empty.

By choice of $\alpha$, the map $h$ we have constructed cannot be one-to-one, but by the nature of our construction, the only situation in which one-one-ness can fail is if at some point our $h(\beta)$ is a maximal element of $X$. Thus $X$ has a maximal element, as claimed.

(ii) $\implies$ (i): This will be a typical application of Zorn's Lemma. Let $X$ and $f$ be given as in (i). Let $P$ be the set of all maps defined on *subsets* $Y \subseteq X$ and carrying each $x \in Y$ to an element of $f(x)$. Partially order $P$ by setting $g_1 \geq g_0$ if $g_1$ is an *extension* of the map $g_0$. $P$ is nonempty because it contains the empty mapping, and it is easy to see that given any chain $C$ of elements of $P$ under the indicated partial ordering, the union of $C$ will be an element of $P$ that is $\geq$ all elements of $C$; hence $P$ is inductive. Thus $P$ has a maximal element $g$. This element must be a function defined on all of $X$ (otherwise we could extend it further), completing the proof of (i). $\qquad\qquad\square$

At the beginning of this section, when we sketched the situation that is abstracted by Zorn's Lemma, we noted that if one has a chain of partial constructions, then their "union" is usually a partial construction extending them all. So in such cases, in the set of all partial constructions, every chain has not merely an upper bound, but a *least* upper bound. Thus, the weakened form of Zorn's Lemma saying that every partially ordered set with *this* property has at least one maximal element is virtually all one ever uses. Is this equivalent to the full form of Zorn's Lemma? This is answered in

**Exercise 5.6:1.** Show, that the statement "If $P$ is a nonempty partially ordered set such that every nonempty chain in $P$ has a least upper bound, then $P$ has a maximal element", implies the full form of Zorn's Lemma.

(This is not too hard to do by an adaptation of the proof of Theorem 5.6.2. More challenging is the task of finding a proof which obtains the general version of Zorn's Lemma by a direct application of the weakened statement, rather than via one of the other conditions of Theorem 5.6.2.)

Having proved Theorem 5.6.2, we now make

**Convention 5.6.3.** *Throughout the remainder of these notes, we shall assume the Axiom of Choice along with the other axioms of ZFC, and thus may freely use any of the equivalent statements of the preceding theorem.*

Of these equivalent statements, Zorn's Lemma is usually the most convenient.

Note that in the last paragraph of the proof of Theorem 5.6.2 above, our verification that $P$ was nonempty was by the same method used to show that every nonempty chain had an upper bound: To show the latter, we used the union of the chain, while to get an element of $P$ we took the empty function, which is the union of the empty chain. It is my experience that in *most* proofs using Zorn's Lemma, the verification of nonemptiness may be achieved by the same construction that shows every *nonempty* chain has an upper bound; i.e., the assumption "nonempty" is rarely needed in the latter verification. Hence my personal preference would be to use a definition of "inductive" that required *every* chain to have an upper bound, and eliminate "$X$ nonempty" as a separate hypothesis of Zorn's Lemma. (Of course, in some exceptional cases, the verification that all chains have upper bounds may have to treat empty and nonempty chains separately. But curiously, even when the same verification works for both cases, many authors seem embarrassed to use a trivial example to show their $X$ is nonempty, and unnecessarily give a more complicated one instead.) For conformity with common usage, I have stated Zorn's Lemma in terms of the standard definition of "inductive". But we may, at times, skip a separate verification that our inductive set is nonempty, and instead observe that some construction gives an upper bound for any chain, empty or nonempty.

The reader who has not seen proofs by Zorn's Lemma before, and does not see how to begin the next few exercises, might look at a few such proofs in a standard graduate algebra text such as [34], and/or ask his or her instructor for some elementary examples. The steps of identifying the sort of "partial constructions" one wants to use, describing the appropriate partial ordering on that set, verifying that the set is inductive, and verifying that a maximal element corresponds to an entity of the sort one was seeking, take practice to master.

**Exercise 5.6:2.** In Exercise 4.14:2(ii) you were asked to prove that the free Boolean ring on any set could be embedded in the Boolean ring of subsets of some set. Prove now that *every* Boolean ring can be embedded in the Boolean ring of subsets of some set.

**Exercise 5.6:3.** Show that in a commutative ring, every prime ideal contains a *minimal* prime ideal. (Note: though the phrase "maximal ideal" by convention means a maximal element of the set of proper ideals, "minimal prime ideal" means minimal among all prime ideals, without restriction.)

**Exercise 5.6:4.** We saw in Exercise 5.1:10 that the maximal partial orderings on a set $X$ were the total orderings. Deduce now for arbitrary $X$ (as
we were able to deduce there for finite $X$) that

(i)   Every partial ordering on $X$ can be extended to a total ordering.

(ii)  Every partial ordering on $X$ is an intersection of total orderings.

**Exercise 5.6:5.**

(i)   If $X$ is a totally ordered set, show that $X$ has a subset $Y$ well-
ordered under the induced ordering, and cofinal in $X$ (Definition 5.1.6).

(ii)  Show that the $Y$ of (i) can be taken order-isomorphic to a regular
cardinal (Exercise 5.5:14), and that this cardinal is unique. However show
that the set $Y$ itself is not in general unique, and that if the condition of
regularity is dropped, uniqueness of the cardinal can also fail.

(iii) Suppose $(X_i)_{i \in I}$ is a finite family of totally ordered sets, such that
for all $i, j \in I$ the set $X_i \times X_j$, under the product order, contains a cofinal
subchain. Show that the set $\prod_I X_i$ under the product order likewise has
a cofinal subchain.

The final part of this exercise does not depend on the preceding parts;
rather, it is a generalization of part (i).

(iv)  Prove that every *partially* ordered set has a cofinal subset with descending chain condition.

**Exercise 5.6:6.** For a partially ordered set $X$, show that the following conditions are equivalent:

(i)   $X$ has no maximal element.

(ii)  $X$ has two disjoint cofinal subsets.

(ii′) $X$ has an infinite family of disjoint cofinal subsets.

**Exercise 5.6:7.** Suppose $X$ is a partially ordered set which contains a cofinal chain. Show that every cofinal subset of $X$ also contains a cofinal
chain. (I find this harder to prove than I would expect. Perhaps there is
some trick I am missing.)

The next exercise is an example where the "obvious" Zorn's Lemma proof
does not work. The simplest valid proof in this case is by the well-ordering
principle, which is not surprising since it is a result about well-orderability.
However, this can also be turned into a Zorn's Lemma proof, if one is careful.

**Exercise 5.6:8.** Let $X$ be a set, let $P$ be the set of partial order relations
on $X$, partially ordered by inclusion as in Exercise 5.1:10, and let $Q \subseteq P$
consist of those partial orderings having descending chain condition.

(i)   Show that the maximal elements of $Q$ (under the partial ordering
induced from $P$) are the well-orderings of $X$.

(ii)  Show that $Q$ is *not* inductive.

(iii) Prove nonetheless that every element of $Q$ is majorized by a maximal element, and deduce that every partial ordering with DCC on a set $X$ is an intersection of well-orderings. (Hint: Take an appropriate ordinal $\alpha$ and construct an indexing of the elements of $X$ by an initial segment of $\alpha$, in a way "consistent" with the given partial order.)

The next four exercises, though not closely related to Zorn's Lemma, explore further the relation between partially ordered sets and their well-ordered subsets.

**Exercise 5.6:9.** Let $S$ be an infinite set, and $\mathbf{P}(S)$ the set of all subsets of $S$, partially ordered by inclusion. Show by example that $\mathbf{P}(S)$ can contain chains of cardinality $> \operatorname{card}(S)$, but prove that $\mathbf{P}(S)$ cannot contain a *well-ordered* chain of cardinality $> \operatorname{card}(S)$.

**Exercise 5.6:10.**

(i)  Show that every infinite *totally* ordered set has either a subset order-isomorphic to $\omega$ or a subset order-isomorphic to $\omega^{\mathrm{op}}$.

(ii)  Show that every infinite partially ordered set $P$ contains either a subset order-isomorphic to $\omega$, a subset order-isomorphic to $\omega^{\mathrm{op}}$, or a countably infinite antichain (Definition 5.1.6). (Suggestion: If $P$ has no infinite antichain, obtain a finite antichain $B \subseteq P$ maximal for the property that the set $S$ of elements incomparable with all elements of $B$ is infinite; then study the properties this $S$ must have. Alternatively, do the same thing with the roles of comparable and incomparable elements reversed.)

This family of three partially ordered sets is essentially unique for the above property:

(iii) Show that a set $F$ of infinite partially ordered sets has the property that every infinite partially ordered set contains an isomorphic copy of a member of $F$ if and only if $F$ contains a partially ordered set order-isomorphic to $\omega$, a partially ordered set order-isomorphic to $\omega^{\mathrm{op}}$, and a countable antichain.

An application of the preceding exercise is

**Exercise 5.6:11.** Let $P$ be a partially ordered set.
(i)  Show that the following conditions are equivalent:
  (i.a)  $P$ contains no chains order-isomorphic to $\omega^{\mathrm{op}}$.
  (i.b)  Every infinite subset of $P$ contains either a subset order-isomorphic to $\omega$, or an infinite antichain.
  (i.c)  $P$ satisfies the descending chain condition.
(ii)  It is clear from (i) above that conditions (ii.a)–(ii.c) below are equivalent. Show that they are also equivalent to (ii.d):
  (ii.a)  $P$ contains no chains order-isomorphic to $\omega^{\mathrm{op}}$, and no infinite antichains.

(ii.b) Every infinite subset of $P$ contains a subset order-isomorphic to $\omega$.

(ii.c) $P$ has descending chain condition, and contains no infinite antichains.

(ii.d) Every total ordering extending the ordering of $P$ is a well-ordering.

A partially ordered set $P$ with the equivalent properties of (ii) is sometimes called "partially well-ordered".

The first part of the next exercise notes that for uncountable cardinalities, things are more complicated.

**Exercise 5.6:12.**

(i)    Deduce from Exercise 5.6:9 that there exists a totally ordered set $P$ of some cardinality $\kappa$ which contains no well-ordered or reverse-well-ordered subset of cardinality $\kappa$.

(ii)   Suppose $P$ is as in (i), and $\varphi$ is a bijection between $P$ and a well-ordered set $Q$ of cardinality $\kappa$. Consider $\{(p, \varphi(p)) \mid p \in P\}$, under the partial ordering induced by the product ordering on $P \times Q$. Show that this has neither chains nor antichains of cardinality $\kappa$ (in contrast to the result of Exercise 5.1:9 for finite partially ordered sets).

But perhaps one can repair this deficiency. (I have not thought hard about the question asked below.)

(iii)  Exercise 5.1:9 was based on defining the "height" of a partially ordered set as the supremum of the cardinalities of its chains; but a different concept of "height" was introduced for partially ordered sets with descending chain condition in Exercise 5.5:4. Can this definition be extended in some way to general partially ordered sets, or otherwise modified, so as to get an analog of Exercise 5.1:9 for partially ordered sets of arbitrary cardinality? (Or can the definition of "width" be so modified?)

**Exercise 5.6:13.**

(i)    Show that every countable totally ordered set can be embedded in the totally ordered set $(\mathbb{Q}, \leq)$ of rational numbers.

(ii)   Show that $(\mathbb{Q}, \leq)$ is not (up to order-isomorphism) the only countable totally ordered set with the property of part (i); but show that $(\mathbb{Q}, \leq)$ has the property slightly stronger than (i), that for every countable totally ordered set $X$, every embedding of a finite subset $X_0 \supseteq X$ in $(\mathbb{Q}, \leq)$ extends to an embedding of $X$ in $(\mathbb{Q}, \leq)$, and that up to order-isomorphism it is indeed the unique totally ordered set with this property.

**Exercise 5.6:14.** A subset $X$ of a partially ordered set $P$ is called *unbounded* if $X$ is not majorized by any element of $P$.

Note that when we refer below to *set-maps* among partially ordered sets, these are *not* assumed to be isotone.

(i)   Let $P$ be the partially ordered set $\omega \times \omega_1$, with the product ordering, and let $Q$ be either $\omega$ or $\omega_1$. Show that there exist set-maps $f \colon P \to Q$ such that the image under $f$ of every cofinal subset of $P$ is a cofinal subset of $Q$, but no set-maps $g \colon P \to Q$ such that the image under $g$ of every unbounded subset of $P$ is unbounded in $Q$. On the other hand, show that there exist set-maps $h \colon Q \to P$ such that the image under $h$ of every unbounded subset of $Q$ is an unbounded subset of $P$, but no set-maps $i \colon Q \to P$ such that the image under $i$ of every cofinal subset of $Q$ is cofinal in $P$.

A partially ordered set $P$ is called *directed* if any two elements of $P$ have a common upper bound.

(ii)   Show that for any two directed partially ordered sets $P$ and $Q$, the following conditions are equivalent:

(ii.a)   There exists a set-map $f \colon P \to Q$ such that the image under $f$ of every cofinal subset of $P$ is a cofinal subset of $Q$.

(ii.b)   There exist a set-map $h \colon Q \to P$ such that the image under $h$ of every unbounded subset of $Q$ is an unbounded subset of $P$.

(iii)   The equivalent conditions of (ii) above are written $Q \leq_T P$. Show that $\leq_T$ is a preordering on the class of directed partially ordered sets. When $Q \leq_T P$, one says $Q$ is *Tukey reducible* to $P$. Translate the results of (i) into a statement about Tukey reducibility.

(iv)   Show that the following conditions on directed partially ordered sets $P$ and $Q$ are equivalent:

(iv.a) $P \leq_T Q$ and $Q \leq_T P$.

(iv.b) There exist set-maps $f \colon P \to Q$ and $g \colon Q \to P$ such that for all $p \in P$ and $q \in Q$ one has $g\,f(p) \geq p$ and $f\,g(q) \geq q$.

(iv.c) $P$ and $Q$ can be embedded as partially ordered sets (hence, by isotone maps!) in a common directed partially ordered set $R$ so that each is cofinal in $R$. This condition is called *Tukey equivalence*.

(v)   Obtain from the method of proof of Exercise 5.6:7 a result on Tukey equivalence.

For a curious application of the well-ordering principle to the study of abelian groups, see the first section of [45].

## 5.7. Some thoughts on set theory

I have mentioned that when the Axiom of Choice and various equivalent principles were first considered, they were the subject of a heated controversy.

The Axiom of Choice is now known to be *independent* of the other axioms of set theory; i.e., it has been proved that, assuming the consistency of the

Zermelo–Fraenkel axioms without Choice, both the full set of axioms including Choice, and the Zermelo–Fraenkel axioms plus the *negation* of the Axiom of Choice are consistent. And there are further statements (for instance the Continuum Hypothesis, saying that $2^{\aleph_0} = \aleph_1$) which have been shown independent of Zermelo–Fraenkel set theory *with* the Axiom of Choice, and which there do not seem to be any compelling reasons either for accepting or rejecting. This creates the perplexing question of what is the "true" set theory.

Alongside Zermelo–Fraenkel Set Theory with and without Choice, etc., there are still other contenders for the "correct" foundations of mathematics. The Intuitionists objected not only to the Axiom of Choice, but to the "law of the excluded middle", the logical principle that every meaningful statement is either true or false. They claimed (if I understand correctly) that an assertion such as Fermat's Last Theorem (the statement that there are no nontrivial integer solutions to $x^n + y^n = z^n$, $n > 2$, which was unproven at the time) could be said to be false if a counterexample were found, or true if an argument could be found (using forms of reasoning acceptable to them) that proved it, but that it would be neither true nor false if neither a counterexample nor a proof existed. They maintained that the application of the law of the excluded middle to statements which involve infinitely many cases, and which thus cannot be checked case by case, was a fallacious extension to infinite sets of a method correct only for finite sets; in their words, that one cannot reason in this way about an infinite set such as the set of all natural numbers, because it cannot be regarded as a "completed totality".

Although this viewpoint is not current, note that the distinction between sets and proper classes, which got mathematics out of the paradoxes that came from considering "the set of all sets", leaves us wondering whether the *class* of all sets is "a real thing"; and indeed one current textbook on set theory refers to this in terms of the question of whether mathematicians can consider such classes as "completed totalities".

During a painfully protracted correspondence with someone who insisted he could show that Zermelo–Fraenkel set theory was inconsistent, and that the fault lay in accepting infinite sets, which he called "mere phantasms", I was forced to think out my own view of the matter, and the conclusion I came to is that all sets, finite and infinite, are "phantasms"; that none of mathematics is "real", so that there is no true set theory; but that this does not invalidate the practice of mathematics, or the usefulness of choosing a "good" set theory.

To briefly explain this line of thought, let us understand the physical world to be "real". (If your religious or philosophical beliefs say otherwise, you can nevertheless follow the regression to come.)

Our way of perceiving the world and interacting with it leads us to partition it into "objects". This partitioning is useful, but is not a "real thing".

To deal intelligently with objects, we think about families of objects, and, as our thinking gets more sophisticated, families of such families. Though I do not think the families, and families of families are "real things" either, they are useful—as descriptions of the way we classify the world.

Consider in particular our system of numbers, which are themselves not "real things", but which give a model that allows us to use one coherent arithmetic system to deal with the various things in the world that one can count. Note that in spite of this motivation in terms of things one can count, in developing the numbers we use a system that is *not* bounded by the limitations of how high a person could count in a lifetime. A system with such a limitation arbitrarily imposed would be *more* difficult to define, learn, and work with than our system, in which the behavior of arithmetic is uniform for arbitrarily large values! Moreover, our unbounded system turns out to have applications to situations that a system bounded in that way would not be able to deal with: to demographic, geographical, astronomical and other data, which we compute from observations and theoretical models of our world, though no one human being could have counted the numbers involved unit by unit.

Now in thinking about our system of numbers, we are dealing with the concept of "all the numbers in the system"—even those who refuse to call that family a "completed totality" do reason about it!—so, if possible, we want our set theory to be able to handle such concepts. Just as we found it natural to extend the system of numbers beyond the sizes of sets a real person could count, so we may extend our system of "sets" beyond finite sets. This is not as simple as with the number concept. Some plausible approaches turned out to lead to contradictions, e.g., those that allowed one to speak of "the set of all sets". Among the approaches that do not seem to lead to contradictions, some are more convenient than others. I think we are justified in choosing a more convenient system to work in—one in which the "unreal objects" that we are considering are easier to understand and generalize about.

It may seem pointless to work in a set theory which is to some extent "arbitrary", and to which we do not ascribe absolute "truth". But observe that as long as we use a system consistent with the laws of finite arithmetic, any statements we can prove in our system about arithmetic models of aspects of the real world, and which can in principle be confirmed or disproved in each case by a finite calculation, will be correct; i.e., as applicable to the real world as those models are. This is what I see as the "justification" for including the Axiom of Choice and other convenient axioms in our set theory.

(For arguments in favor of adding another axiom, the *Axiom of Projective Determinacy*, to the standard axioms of set theory, see [148].)

Fortunately, making a choice among set theories or systems of reasoning does not consign all others to oblivion. Logicians *do* consider not only which statements hold if the Axiom of Choice is assumed, but also which hold if its negation is assumed. (E.g., [88] shows that in a model of ZF with the negation of Choice, one will have commutative rings with properties contradicting several standard theorems of ZFC ring theory.) Intuitionistic logic is likewise still studied—not, nowadays, as a preferred mode of reasoning, but as a *formal system*, related to objects called Brouwerian lattices (cf. [4]) in the same way standard logic is related to Boolean algebras.

# Chapter 6
# Lattices, Closure Operators, and Galois Connections

## 6.1. Semilattices and lattices

Many of the partially ordered sets $P$ we have seen have a further valuable property: that for any two elements of $P$, there is a least element $\geq$ both of them, and a greatest element $\leq$ both of them, i.e., a *least upper bound* and a *greatest lower bound* for the pair. In this section we shall study partially ordered sets with this property. To get a better understanding of the subject, let us start by looking separately at the properties of having least upper bounds and of having greatest lower bounds.

Recall that an element $x$ is said to be *idempotent* with respect to a binary operation $*$ if $x * x = x$. The binary operation $*$ itself is often called idempotent if $x * x = x$ holds for all $x$.

**Lemma 6.1.1.** *Suppose $X$ is a partially ordered set in which every two elements $x, y \in X$ have a* least *upper bound; that is, such that there exists a least element which majorizes both $x$ and $y$. Then if we write this least upper bound as $x \vee y$, and regard $\vee$ as a binary operation on $X$, this operation satisfies the identities*

$$
\begin{array}{lll}
(\forall\, x \in X) & x \vee x = x & \text{(idempotence)}, \\
(\forall\, x,\, y \in X) & x \vee y = y \vee x & \text{(commutativity)}, \\
(\forall\, x,\, y,\, z \in X) & (x \vee y) \vee z = x \vee (y \vee z) & \text{(associativity)}.
\end{array}
$$

*Conversely, given a set $X$ with a binary operation $\vee$ satisfying the above three identities, there is a unique partial order relation $\leq$ on $X$ for which $\vee$ is the least upper bound function. This relation $\leq$ may be obtained from the operation $\vee$ in two ways: It can be constructed as*

$$\{(x,\, x \vee y) \mid x,\, y \in X\},$$

© Springer International Publishing Switzerland 2015
G.M. Bergman, *An Invitation to General Algebra and Universal Constructions*, Universitext, DOI 10.1007/978-3-319-11478-1_6

*or characterized as the set of elements satisfying an equation:*

$$\{(x, y) \mid x, y \in X \text{ and } y = x \vee y\}. \quad \square$$

**Exercise 6.1:1.** Prove the non-obvious part of the above lemma, namely that every idempotent commutative associative binary operation on a set arises from a partial ordering with least upper bounds. Why is this partial ordering unique?

Hence we make

**Definition 6.1.2.** *An* upper semilattice *means a pair* $S = (|S|, \vee)$, *where* $|S|$ *is a set, and* $\vee$ *(pronounced "join") is an idempotent commutative associative binary operation on* $|S|$. *Informally, the term "upper semilattice" is also used for the equivalent structure of a partially ordered set in which every pair of elements has a least upper bound.*

*Given an upper semilattice* $(|S|, \vee)$, *we shall consider* $|S|$ *as partially ordered by the unique ordering which makes* $\vee$ *the least upper bound operation (characterized in two equivalent ways in the above lemma). The set* $|S|$ *with this partial ordering is sometimes called the "underlying partially ordered set" of the upper semilattice* $S$.

*The join of a finite nonempty family of elements* $x_i$ $(i \in I)$ *in an upper semilattice (which by the associativity and commutativity of the join operation* $\vee$ *makes sense without specification of an order or bracketing for the elements, and which is easily seen to give the least upper bound of* $\{x_i\}$ *in the natural partial ordering) is denoted* $\bigvee_{i \in I} x_i$.

The danger of confusion inherent in the symmetry of the partial order concept is now ready to rear its head! Observe that in a partially ordered set in which every pair of elements $x, y$ has a *greatest lower bound* $x \wedge y$, the operation $\wedge$ will also be idempotent, commutative and associative (it is simply the operation $\vee$ for the opposite partially ordered set), though the partial ordering is recovered from it in the opposite way, by defining $x \leq y$ if and only if $x$ can be written $y \wedge z$, equivalently, if and only if $x = x \wedge y$. We have no choice but to make a formally identical definition for the opposite concept (first half of the first sentence below):

**Definition 6.1.3.** *A* lower semilattice *means a pair* $S = (|S|, \wedge)$, *where* $|S|$ *is a set and* $\wedge$ *(pronounced "meet") is an idempotent commutative associative binary operation on* $|S|$; *or informally, the equivalent structure of a partially ordered set in which every pair of elements has a greatest lower bound. If* $(|S|, \wedge)$ *is such a pair, regarded as a lower semilattice, then* $|S|$ *will be considered partially ordered in the unique way which makes* $\wedge$ *the greatest lower bound operation.*

*The notation for the meet of a finite nonempty family of elements is* $\bigwedge_{i \in I} x_i$.

A partially ordered set $(X, \leq)$ in which every pair of elements $x$ and $y$ has both a least upper bound $x \vee y$ and a greatest lower bound $x \wedge y$ is clearly determined—indeed, redundantly determined—by the 3-tuple $L = (X, \vee, \wedge)$. We see that a 3-tuple consisting of a set, an upper semilattice operation, and a lower semilattice operation arises in this way if and only if these operations are compatible, in the sense that the unique partial ordering for which $\vee$ is the least-upper-bound operation coincides with the unique partial ordering for which $\wedge$ is greatest-lower-bound operation.

Is there a nice formulation for this compatibility condition? The statement that for any two elements $x$ and $y$, the element $y$ can be written $x \vee z$ for some $z$ if and only if the element $x$ can be written $y \wedge w$ for some $w$ would do, but it is awkward. If, instead of using as above the descriptions of how to *construct* all pairs $(x, y)$ with $x \leq y$ with the help of the operations $\vee$ and $\wedge$, we use the formulas that characterize them as solution-sets of equations, we get the condition that for all elements $x$ and $y$, $y = x \vee y \iff x \wedge y = x$. But the best expression for our condition—one that does not use any "can be written"s or " $\iff$ "s—is obtained by playing off one description of $\vee$ against the other description of $\wedge$. This is the fourth pair of equations in

**Definition 6.1.4.** *A lattice will mean a 3-tuple* $L = (|L|, \vee, \wedge)$ *satisfying the following identities for all* $x, y, z \in |L|$ :

| | | |
|---|---|---|
| $x \vee x = x$ | $x \wedge x = x$ | (idempotence), |
| $x \vee y = y \vee x$ | $x \wedge y = y \wedge x$ | (commutativity), |
| $(x \vee y) \vee z = x \vee (y \vee z)$ | $(x \wedge y) \wedge z = x \wedge (y \wedge z)$ | (associativity), |
| $x \wedge (x \vee y) = x$ | $x \vee (x \wedge y) = x$ | (compatibility); |

*in other words, such that* $(|L|, \vee)$ *is an upper semilattice,* $(|L|, \wedge)$ *is a lower semilattice, and the two semilattice structures correspond to the same partial ordering on* $|L|$. *Informally, the term will also be used for the equivalent structure of a partially ordered set in which every pair of elements has both a least upper bound and a greatest lower bound.*

*Given a lattice* $(|L|, \vee, \wedge)$, *we shall consider* $|L|$ *partially ordered by the unique partial ordering (characterizable in four equivalent ways) which makes its join operation the least upper bound and its meet operation the greatest lower bound. The set* $|L|$ *with this partial ordering is sometimes called the "underlying partially ordered set of* $L$. "*

Examples: If $S$ is a set, then the power set $\mathbf{P}(S)$ (the set of all subsets of $S$), partially ordered by the relation of inclusion, has least upper bounds and greatest lower bounds, given by operations of union and intersection of sets; hence $(\mathbf{P}(S), \cup, \cap)$ is a lattice. Since the definition of Boolean algebra was modeled on the structure of the power set of a set, every Boolean algebra $(|B|, \cup, \cap, {}^{c}, 0, 1)$ gives a lattice $(|B|, \cup, \cap)$ on dropping the last three operations; and since we know that Boolean *rings* are equivalent to Boolean algebras, every Boolean ring $(|B|, +, \cdot, -, 0, 1)$ becomes a lattice under the operations $x \vee y = x + y + x\,y$ and $x \wedge y = x\,y$.

Every totally ordered set—for instance, the real numbers—is a lattice, since the larger and the smaller of two elements will respectively be their least upper bound and greatest lower bound. The set of real-valued functions on any set $X$ may be ordered by writing $f \leq g$ if $f(x) \leq g(x)$ for all $x$, and this set is a lattice under *pointwise* maximum and minimum.

Under the partial ordering by divisibility, the set of positive integers has least upper bounds and greatest lower bounds, called "least common multiples" and "greatest common divisors". Note that if we represent a positive integer by its prime factorization, and consider such a factorization as a function associating to each prime a nonnegative integer, then least common multiples and greatest common divisors reduce to pointwise maxima and minima of these functions.

Given a group $G$, if we order the set of subgroups of $G$ by inclusion, then we see that for any two subgroups $H$ and $K$, there is a largest subgroup contained in both, gotten by intersecting their underlying sets, and a smallest subgroup containing both, the subgroup *generated by* the union of their underlying sets. So the set of subgroups of $G$ forms a lattice, called the *subgroup lattice* of $G$. This observation goes over word-for-word with "group" replaced by "monoid", "ring", "vector space", etc.

Some writers use "ring-theoretic" notation for lattices, writing $x + y$ for $x \vee y$, and $x\,y$ for $x \wedge y$. Note, however, that a nontrivial lattice is never a ring (since by idempotence, its join operation cannot be a group structure). We will not use such notation here.

Although one can easily draw pictures of partially ordered sets and semilattices which are not lattices, it takes a bit of thought to find naturally occurring examples. The next exercise notes a couple of these.

**Exercise 6.1:2.**

(i)    If $G$ is a group, show that within the lattice of subgroups of $G$, the finitely generated subgroups form an upper semilattice under the induced order, but not necessarily a lower semilattice, and the finite subgroups form a lower semilattice but not necessarily an upper semilattice. (For partial credit you can verify the positive assertions; for full credit you must find examples establishing the negative assertions as well.)

(ii)    Let us partially order the set of *polynomial* functions on the unit interval $[0, 1]$ by pointwise comparison ($f \leq g$ if and only if $f(x) \leq g(x)$ for all $x \in [0, 1]$). Show that this partially ordered set is neither an upper nor a lower semilattice.

**Exercise 6.1:3.** Give an example of a 3-tuple $(|L|, \vee, \wedge)$ which satisfies all the identities defining a lattice except for *one* of the two compatibility identities. If possible, give a systematic way of constructing such examples. Can you determine for which upper semilattices $(|L|, \vee)$ there will exist operations $\wedge$ such that $(|L|, \vee, \wedge)$ satisfies all the lattice identities except the specified one? (The answer will depend on which identity you leave out; you can try to solve the problem for one or both cases.)

**Exercise 6.1:4.** Show that the two compatibility identities at the end of Definition 6.1.4 together imply the two idempotence identities.

**Exercise 6.1:5.** Show that an element of a lattice is a *maximal* element if and only if it is a *greatest* element. Is this true in every upper semilattice? In every lower semilattice?

A *homomorphism* of lattices, upper semilattices, or lower semilattices means a map of their underlying sets which respects the lattice or semi-lattice operations. If $L_1$ and $L_2$ are lattices, one can speak loosely of an "upper semilattice homomorphism $L_1 \to L_2$," meaning a map of underlying sets which respects joins but not necessarily meets; this is really a homomorphism $(L_1)_\vee \to (L_2)_\vee$, where $(L_i)_\vee$ denotes the upper semilattice $(|L_i|, \vee)$ gotten by forgetting the operation $\wedge$; one may similarly speak of "lower semi-lattice homomorphisms" of lattices. Note that if $f : |L_1| \to |L_2|$ is a lattice homomorphism, or an upper semilattice homomorphism, or a lower semilat-tice homomorphism, it will be an isotone map with respect to the natural order-relations on $|L_1|$ and $|L_2|$, but in general, an isotone map $f$ need not be a homomorphism of any of these sorts.

A *sublattice* of a lattice $L$ is a lattice whose underlying set is a subset of $|L|$ and whose operations are the restrictions to this set of the operations of $L$. A *subsemilattice* of an upper or lower semilattice is defined similarly, and one can speak loosely of an upper or lower subsemilattice of a lattice $L$, meaning a subsemilattice of $L_\vee$ or $L_\wedge$.

**Exercise 6.1:6.**

(i)   Give an example of a subset $S$ of the underlying set of a lattice $L$ such that every pair of elements of $S$ has a least upper bound and a greatest lower bound in $S$ under the induced ordering, but such that $S$ is not the underlying set of either an upper or a lower subsemilattice of $L$.

(ii)   Give an example of an upper semilattice homomorphism between lat-tices that is not a lattice homomorphism.

(iii)   Give an example of a bijective isotone map between lattices which is not an upper or lower semilattice homomorphism.

(iv)   Show that a bijection between lattices *is* a lattice isomorphism if either (a) it is an upper (or lower) semilattice homomorphism, or (b) it and its inverse are both isotone.

**Exercise 6.1:7.** Let $k$ be a field. If $V$ is a $k$-vector space, then the *cosets of subspaces* of $V$, together with the empty set, are called the *affine subspaces* of $V$.

(i)   Show that the affine subspaces of a vector space (ordered by inclusion) form a lattice.

(ii)   Suppose we map the set of affine subspaces of the vector space $k^n$ into the set of *vector subspaces* of $k^{n+1}$ by sending each affine subspace $A \subseteq k^n$ to the vector subspace $s(A) \subseteq k^{n+1}$ spanned by $\{(x_0, \ldots, x_{n-1}, 1) \mid (x_0, \ldots, x_{n-1}) \in A\}$. Show that this map $s$ is one-to-one. One may ask whether $s$ respects meets and/or joins. Show that it respects one of these, and respects the other in "most but not all" cases, in a sense you should make precise.

(The study of the affine subspaces of $k^n$ is called $n$-dimensional *affine geometry*. By the above observations, the geometry of the vector subspaces of $k^{n+1}$ may be regarded as a slight extension of $n$-dimensional affine geometry; this is called $n$-dimensional *projective geometry*. In view of the relation with affine geometry, a one-dimensional subspace of $k^{n+1}$ is called a "point" of projective $n$-space, a two-dimensional subspace, or more precisely, the set of "points" it contains, is called a "line", etc.)

The methods introduced in Chapters 3 and 4 can clearly be used to establish the existence of *free* lattices and semilattices, and of lattices and semilattices presented by *generators and relations*. As in the case of monoids, a "relation" means a statement equating two terms formed from the given generators using the given operations—in this case, the lattice or semilattice operations.

**Exercise 6.1:8.**

(i)   If $P$ is a partially ordered set, show that there exist universal examples of an upper semilattice, a lower semilattice, and a lattice, with isotone maps of $P$ into their underlying partially ordered sets, and that these may be constructed as semilattices or lattices presented by appropriate generators and relations.

(ii)   Show likewise that given any upper or lower semilattice $S$, there is a universal example of a lattice $L$ with an upper, respectively lower semilattice homomorphism of $S$ into it.

(iii)   If the $S$ of part (ii) above "is a lattice" (has both least upper bounds and greatest lower bounds), will this universal semilattice homomorphism be an isomorphism? If the $P$ of part (i) "is a lattice" will the universal isotone maps of that part be isomorphisms of partially ordered sets?

(iv)   Show that the universal maps of (i) and (ii) are in general not surjective, and investigate whether each of them is in general one-to-one.

**Exercise 6.1:9.** Determine a normal form or other description for the free upper semilattice on a set $X$. Show that it will be finite if $X$ is finite.

There exists something like a normal form theorem for free lattices [4, §VI.8], but it is much less trivial than the result for semilattices referred to in the above exercise, and we will not develop it here. However, the next exercise develops a couple of facts about free lattices.

**Exercise 6.1:10.**

(i)  Determine the structures of the free lattices on 0, 1, and 2 generators.

(ii)  Show for some positive integer $n$ that the free lattice on $n$ generators is infinite. (One approach: In the lattice of affine subsets of the plane $\mathbb{R}^2$ (Exercise 6.1:7), consider the sublattice generated by the five lines $x = 0$, $x = 1$, $x = 2$, $y = 0$, $y = 1$.)

**Exercise 6.1:11.**

(i)   Recall (cf. discussion preceding Exercise 5.1:5) that a set map $X \to Y$ induces maps $\mathbf{P}(X) \to \mathbf{P}(Y)$ and $\mathbf{P}(Y) \to \mathbf{P}(X)$. Show that one of these is always, and the other is not always a lattice homomorphism.

(ii)  If $L$ is (a) a lattice, respectively (b) an upper semilattice, (c) a lower semilattice or (d) a partially ordered set, show that there exists a universal example of a set $X$ together with, respectively,

(a) a lattice homomorphism $L \to (\mathbf{P}(X), \cup, \cap)$,

(b) an upper semilattice homomorphism $L \to (\mathbf{P}(X), \cup)$,

(c) a lower semilattice homomorphism $L \to (\mathbf{P}(X), \cap)$, respectively,

(d) an isotone map $L \to (\mathbf{P}(X), \subseteq)$ (unless you did this case in Exercise 5.1:5).

In each case, first formulate the relevant universal properties. These should be based on the construction of part (i) that *does* give lattice homomorphisms. In each case, describe the set $X$ as explicitly as you can.

(iii) In the context of part (i), the map between $\mathbf{P}(X)$ and $\mathbf{P}(Y)$ that does not generally give a lattice homomorphism will nevertheless preserve some of the types of structure named in part (ii). If $L$ is an arbitrary structure of one of those sorts, see whether you can find an example of a set $X$ and a map $|L| \to \mathbf{P}(X)$ respecting that structure, and universal with respect to induced maps in the indicated direction.

(iv) For which of the constructions that you obtained in parts (ii) and/or (iii) can you show the universal map $|L| \to \mathbf{P}(X)$ one-to-one? In the case(s) where you cannot, can you find an example in which it is not one-to-one?

In Exercise 5.6:6 we saw that any partially ordered set without maximal elements has two disjoint cofinal subsets. Let us examine what similar results hold for lattices.

**Exercise 6.1:12.** Let $L$ be a lattice without greatest element.

(i)   If $L$ is *countable*, show that it contains a cofinal chain, that this chain will have two disjoint cofinal subchains, and that these will be disjoint cofinal sublattices of $L$.

(ii)  Show that in general, $L$ need not have a cofinal chain.

(iii) Must $L$ have two disjoint cofinal sublattices? (I don't know the answer.)

(iv) Show that $L$ will always contain two disjoint upper subsemilattices, each cofinal in $L$.

Here is another open question, of a related sort.

**Exercise 6.1:13.**

(i)   (Open question, David Wasserman.) If $L$ is a lattice with more than one element, must $L$ have two proper sublattices $L_1$ and $L_2$ whose union generates $L$?

Parts (ii) and (iv) below, which are fairly easy, give some perspective on this question; parts (iii) and (v) are digressions suggested by (ii) and (iv).

(ii)  Show that if $A$ is a group, monoid, ring or lattice which is finitely generated but cannot be generated by a single element, then $A$ is generated by the union of two proper subgroups, subrings, etc. (You can give one proof that covers all these cases.)

(iii) Determine precisely which finitely generated groups are not generated by the union of any two proper subgroups.

(iv) Let $p$ be a prime and $\mathbb{Z}[p^{-1}]$ the subring of $\mathbb{Q}$ generated by $p^{-1}$, and let $\mathbb{Z}[p^{-1}]_{\mathrm{add}}$ denote its underlying additive group. Show that the abelian group $\mathbb{Z}[p^{-1}]_{\mathrm{add}}/\mathbb{Z}_{\mathrm{add}}$ is non-finitely-generated, and cannot be generated by two proper subgroups.

(v)   Are the groups of parts (iii) and (iv) above the only ones that are not generated by two proper subgroups?

I could not end an introduction to lattices without showing you the concepts introduced in the next two exercises, though this brief mention, and the results developed in the two subsequent exercises, will hardly do them justice. I will refer in these exercises to the following two 5-element lattices:

($M_3$ is sometimes called $M_5$.)

**Exercise 6.1:14.**

(i)   Show that the following conditions on a lattice $L$ are equivalent:

(a) For all $x$, $y$, $z \in |L|$ with $x \le z$, one has $x \vee (y \wedge z) = (x \vee y) \wedge z$.

(b) $L$ has no sublattice isomorphic to $N_5$ (shown above).

(c) For every pair of elements $x$, $y \in |L|$, the intervals $[x \wedge y, y]$ and $[x, x \vee y]$ are isomorphic, the map in one direction being given by $z \mapsto x \vee z$, in the other direction by $z \mapsto z \wedge y$.

(ii)   Show that condition (a) is equivalent to an *identity*, i.e., a statement that a certain equation in $n$ variables and the lattice-operations holds for all $n$-tuples of elements of $L$. (Condition (a) as stated fails to be an identity, because it refers only to 3-tuples satisfying $x \le z$.)

(iii)   Show that the lattice of subgroups of an abelian group satisfies the above equivalent conditions. Deduce that the lattice of submodules of a module over a ring will satisfy the same conditions.

For this reason, a lattice satisfying these conditions is called *modular*.

(iv)   Determine, as far as you can, whether each of the following lattices is in general modular: the lattice of all subsets of a set; the lattice of all subgroups of a group; the lattice of all normal subgroups of a group; the lattice of all ideals of a ring; the lattice of all subrings of a ring; the lattice of all subrings of a Boolean ring; the lattice of elements of a Boolean ring under the operations $x \vee y = x + y + xy$ and $x \wedge y = xy$; the lattice of all sublattices of a lattice; the lattice of all closed subsets of a topological space; the lattices associated with $n$-dimensional affine geometry and with $n$-dimensional projective geometry (Exercise 6.1:7 above).

**Exercise 6.1:15.**

(i)   Show that the following conditions on a lattice $L$ are equivalent:

(a) For all $x$, $y$, $z \in |L|$, one has $x \vee (y \wedge z) = (x \vee y) \wedge (x \vee z)$.

(a*) For all $x$, $y$, $z \in |L|$, one has $x \wedge (y \vee z) = (x \wedge y) \vee (x \wedge z)$.

(b) $L$ has no sublattice isomorphic either to $M_3$ or to $N_5$.

Note that if one thinks of $\vee$ as "addition" and $\wedge$ as "multiplication", then (a*) has the form of the distributive law of ring theory. (Condition (a) is also a distributive law, though that identity does not hold in any nonzero ring.) Hence lattices satisfying the above equivalent conditions are called *distributive*.

(ii)   Show that the lattice of subsets of a set is distributive.

(iii)   Determine, as far as you can, whether lattices of each of the remaining sorts listed in parts (iii) and (iv) of the preceding exercise are always distributive.

(iv)   Show that every finitely generated distributive lattice is finite.

**Exercise 6.1:16.** Let $V$ be a vector space over a field $k$, let $S_1, \ldots, S_n$ be subspaces of $V$, and within the lattice of all subspaces of $V$, let $L$ denote the sublattice generated by $S_1, \ldots, S_n$.

(i)  Show that if $V$ has a basis $B$ such that each $S_i$ is spanned by a subset of $B$, then $L$ is distributive, as defined in the preceding exercise

Below we will prove the converse of (i); so for the remainder of this exercise, *we assume the lattice $L$ generated by the vector subspaces $S_1, \ldots, S_n$ is distributive.*

To prove the existence of a basis as in the hypothesis of (i), it will suffice to prove that $V$ contains a direct sum of subspaces, with the property that each $S_i$ is the sum of some subfamily thereof; so this is what we will aim for. (You'll give the details of why this yields the desired result in the last step.)

You may assume the last result of the preceding exercise, that every finitely generated distributive lattice is finite.

(ii)  Let $T = S_1 + \cdots + S_n$, the largest element of $L$. Assuming $L$ has elements other than $T$, let $W$ be maximal among these. Show that there is a *least* element $U \in L$ not contained in $W$.

(iii)  Let $E$ be a subspace of $V$ such that $U = (U \cap W) \oplus E$. (Why does one exist?) Show that every member of $L$ is either contained in $W$, or is the direct sum of $E$ with a member of $L$ contained in $W$.

(iv)  Writing $L'$ for the sublattice of $L$ consisting of members of $L$ contained in $W$, show that the lattice of subspaces of $V$ generated by $\{S_1, \ldots, S_n, E\}$ is isomorphic to $L' \times \{0, E\}$, and hence is again a finite distributive sublattice of the lattice of subspaces of $V$.

(v)  Conclude by induction (on what?) that there exists a family of subspaces $E_1, \ldots, E_r \subseteq V$ such that every member of $L$, and hence in particular, each of $S_1, \ldots, S_n$, is the direct sum of a subset of this family.

(vi)  Deduce that $V$ has a basis $B$ such that each $S_i$ is spanned by a subset of $B$.

**Exercise 6.1:17.** Let us show that the result of the preceding exercise fails for infinite families $(S_i)_{i \in I}$. Our example will be a chain of subspaces, so

(i)  Verify that every chain in a lattice is a distributive sublattice.

Now let $k$ be a field, and $V$ the $k$-vector-space of all $k$-valued functions on the nonnegative integers. You may assume the standard result that $V$ is uncountable-dimensional. For each nonnegative integer $n$, let $S_n = \{f \in V \mid f(i) = 0 \text{ for } i < n\}$.

(ii)  Show that $V$ does not have a basis $B$ such that each $S_i$ is spanned by a subset of $B$. (One way to start: Verify that for each $n$, $S_{n+1}$ has codimension 1 in $S_n$, and that the intersection of these subspaces is $\{0\}$.)

The preceding exercise suggests

**Exercise 6.1:18.** Can you find necessary and sufficient conditions on a lattice $L$ for it to be true that for every homomorphism $f$ of $L$ into the lattice of subspaces of a vector space $V$, there exists a basis $B$ of $V$ such that every subspace $f(x)$ $(x \in |L|)$ is spanned by a subset of $B$?

We remark that the analog of Exercise 3.3:2 with the finite lattice $N_5$ in place of the finite group $S_3$ is worked out for $n = 3$ in [143].

## 6.2. 0, 1, and completeness

We began this chapter with the observation that many natural examples of partially ordered sets have the property that every *pair* of elements has a least upper bound and a greatest lower bound. But most of these examples in fact have the stronger property that such bounds exist for every *set* of elements. E.g., in the lattice of subgroups of a group, one can take the intersection of, or the subgroup generated by the union of, an arbitrary set of subgroups. The property that every subset $\{x_i \mid i \in I\}$ has a least upper bound (denoted $\bigvee_I x_i$) and a greatest lower bound (denoted $\bigwedge_I x_i$) leads to the class of nonempty *complete* lattices, which we shall consider in this section.

Note that in an ordinary lattice, because every *pair* of elements $x$, $y$ has a least upper bound $x \vee y$, it follows that for every positive integer $n$, every family of $n$ elements $x_0, \ldots, x_{n-1}$ has a least upper bound, namely $\bigvee x_i = x_0 \vee \cdots \vee x_{n-1}$. Hence, to get least upper bounds for *all* families, we need to bring in the additional cases of *infinite* families, and the *empty* family.

Now every element of a lattice $L$ is an upper bound of the empty family, so a *least upper bound* for the empty family means a least *element* in the lattice. Such an element is often written 0, or when there is a possibility of ambiguity, $0_L$. Likewise, a greatest lower bound for the empty family means a greatest element, commonly written 1 or $1_L$.

It is not hard to see that the two conditions that a partially ordered set have pairwise least upper bounds (joins) and that it have a least element (a least upper bound of the empty family) are independent: either, neither, or both may hold. On the other hand, existence of pairwise joins and existence of infinite joins (joins indexed by infinite families, with repetition allowed just as in the case of pairwise joins) are not independent; the latter condition implies the former. However, we may ask whether the property "existence of infinite joins" can somehow be decomposed into the conjunction of existence of pairwise joins, and some natural condition which *is* independent thereof. The next result shows that it can, and more generally, that for any cardinal $\alpha$, the condition "there exist joins of all families of cardinality $< \alpha$" can be so decomposed.

**Lemma 6.2.1.** *Let $P$ be a partially ordered set, and $\alpha$ an infinite cardinal. Then the following conditions are equivalent:*

(i)   *Every nonempty subset of $P$ with $< \alpha$ elements has a least upper bound in $P$.*

(ii)   *Every pair of elements of $P$ has a least upper bound, and every non-empty chain in $P$ with $< \alpha$ elements has a least upper bound.*

*The dual statements concerning greatest lower bounds are likewise equivalent to one another.*

*Proof.* (i) $\Longrightarrow$ (ii) is clear.

Conversely, assuming (ii) let us take any nonempty set $X$ of $< \alpha$ elements of $P$, and index it by an ordinal $\beta < \alpha : X = \{x_\varepsilon \mid \varepsilon < \beta\}$. We shall prove inductively that for $0 < \gamma \leq \beta$, there exists a least upper bound $\bigvee_{\varepsilon<\gamma} x_\varepsilon$. Because we have not assumed a least upper bound for the empty set, this need not be true for $\gamma = 0$, so we start the induction by observing that for $\gamma = 1$, the set $\{x_\varepsilon \mid \varepsilon < 1\} = \{x_0\}$ has least upper bound $x_0$. Now let $1 < \gamma \leq \beta$ and assume our result is true for all positive $\delta < \gamma$. If $\gamma$ is a successor ordinal, $\gamma = \delta + 1$, then we apply the existence of pairwise least upper bounds in $P$ and see that $(\bigvee_{\varepsilon<\delta} x_\varepsilon) \vee x_\delta$ will give the desired least upper bound $\bigvee_{\varepsilon<\gamma} x_\varepsilon$. On the other hand, if $\gamma$ is a limit ordinal, then the elements $\bigvee_{\varepsilon<\delta} x_\varepsilon$ where $\delta$ ranges over all nonzero members of $\gamma$ will form a nonempty chain of $< \alpha$ elements in $P$, which by (ii) has a least upper bound, and this is the desired element $\bigvee_{\varepsilon<\gamma} x_\varepsilon$. So by induction, $\bigvee_{\varepsilon<\beta} x_\varepsilon$ exists, proving (i).

The final statement follows by duality.                                    $\square$

**Definition 6.2.2.** *Let $\alpha$ be a cardinal. Then a lattice or an upper semilattice $L$ in which every nonempty set of $< \alpha$ elements has a least upper bound will be called $<\alpha$-upper semicomplete. A lattice or a lower semilattice satisfying the dual condition is said to be $<\alpha$-lower semicomplete. A lattice satisfying both conditions will be called $<\alpha$-complete.*

*When these conditions hold for all cardinals $\alpha$, one calls $L$ upper semicomplete, respectively lower semicomplete, respectively complete.*

*(We may at times want to refer to the least upper bound or greatest lower bound in a partially ordered set $P$ of a set $X = \{x_i \mid i \in I\}$ indexed by some set other than an ordinal; in such cases we will, of course, write $\bigvee_{i\in I} x_i$ or $\bigwedge_{i\in I} x_i$. If no indexing is given, we can write these as $\bigvee'_{x\in X} x$ or $\bigwedge_{x\in X} x$; which we may abbreviate to $\bigvee X$ or $\bigwedge X$.)*

One can similarly speak of a lattice or semilattice as being *upper* or *lower* $\leq\alpha$-*semicomplete* if all nonempty subsets of cardinality $\leq \alpha$ have least upper bounds, respectively greatest lower bounds. Note, however, that upper or lower $\leq\alpha$-semicompleteness is equivalent to upper or lower $<\alpha^+$-semicompleteness respectively, where $\alpha^+$ is the successor of the cardinal $\alpha$. Since not every cardinal is a successor, the class of conditions named

by the "$<$" properties properly contains the class of conditions named by
the "$\leq$" properties.

Consequently, in the interest of brevity, many authors write "$\alpha$-semicom-
plete" and "$\alpha$-complete" for what we are calling "$<\alpha$-semicomplete" and
"$<\alpha$-complete". I prefer to use a more transparent terminology, however.

We have observed, in effect, that every lattice is $<\aleph_0$-complete; so the
first case of interest among the above completeness conditions is that of
$\leq\aleph_0$-completeness, equivalently, $<\aleph_1$-completeness. This property is commonly called *countable completeness*, even by authors who in their systematic
notation would write it as $\aleph_1$-completeness. Countable upper and lower semicompleteness are defined similarly. We will not, however, use these terms in
these notes, since the cases of greatest interest to us in this section, and the
only cases we will be concerned with after this section, are the full completeness conditions.

Note that in a partially ordered set (e.g., a lattice) with *ascending chain
condition*, all nonempty chains have least upper bounds—since they in fact
have greatest elements. Likewise in a partially ordered set with descending
chain condition, all chains have greatest lower bounds.

**Exercise 6.2:1.** Suppose $\beta$ and $\gamma$ are infinite cardinals, and $X$ a set
having cardinality $\geq \max(\beta, \gamma)$. Let $L = \{S \subseteq X \mid \mathrm{card}(S) < \beta$ or
$\mathrm{card}(X - S) < \gamma\}$. Verify that $L$ is a lattice, and investigate for what
cardinals $\alpha$ this lattice is upper, respectively lower $<\alpha$-semicomplete.

The upper and lower semicompleteness conditions, when not restricted by
a cardinal $\alpha$, have an unexpectedly close relation.

**Proposition 6.2.3.** *Let $L$ be a partially ordered set. Then the following conditions are equivalent:*

(i)  *Every subset of $L$ has a least upper bound; i.e., $L$ is the underlying
partially ordered set of an upper semicomplete upper semilattice with least
element.*

(i\*)  *Every subset of $L$ has a greatest lower bound; i.e., $L$ is the underlying
partially ordered set of a lower semicomplete lower semilattice with greatest
element.*

(ii)  *$L$ is the underlying partially ordered set of a nonempty complete lattice.*

*Proof.* To see the equivalence of the two formulations of (i), recall that a
least upper bound for the empty set is a least element, while the existence
of least upper bounds for all other subsets is what it means to be an upper
semicomplete upper semilattice.

To show (i) $\Longrightarrow$ (ii), observe that the existence of a least element shows
that $L$ is nonempty, and the upper complete upper semilattice condition
gives half the condition to be a complete lattice. It remains to show that any
nonempty subset $X$ of $L$ has a greatest lower bound $u$. In fact, the least
*upper* bound of the set of *all lower* bounds for $X$ will be the desired $u$; the
reader should verify that it has the required property.

Conversely, assuming (ii), we have by definition least upper bounds for all *nonempty* subsets of $L$. A least upper bound for the empty set is easily seen to be given by the greatest lower bound of all of $L$. (How is the nonemptiness condition of (ii) used?)

Since (ii) is self-dual and equivalent to (i), it is also equivalent to (i*).    □

**Exercise 6.2:2.** If $T$ is a topological space, show that the open sets in $T$, partially ordered by inclusion, form a complete lattice. Describe the meet and join operations (finite and arbitrary) of this lattice. Translate these results into statements about the set of closed subsets of $T$.

(General topology buffs may find it interesting to show that, on the other hand, the partially ordered set {open sets} ∪ {closed sets} is not in general a lattice, nor is the partially ordered set of *locally closed* sets.)

**Exercise 6.2:3.** Which ordinals, when considered as ordered sets, form complete lattices?

**Exercise 6.2:4.**

(i)   Show that every *isotone map* from a nonempty complete lattice into itself has a fixed point.

(ii)   Can you prove the same result for a larger class of partially ordered sets?

**Exercise 6.2:5.** Let $L$ be a complete lattice.

(i)   Show that the following conditions are equivalent: (a) $L$ has no chain order-isomorphic to an uncountable cardinal. (b) For every subset $X \subseteq |L|$ there exists a countable subset $Y \subseteq X$ such that $\bigvee Y = \bigvee X$.

(ii)   Let $a$ be any element of $L$. Are the following conditions equivalent? (a) $L$ has no chain which is order-isomorphic to an uncountable cardinal and has join $a$. (b) Every subset $X \subseteq L$ with join $a$ contains a countable subset $Y$ also having join $a$.

Although we write the least upper bound and greatest lower bound of a set $X$ in a complete lattice as $\bigvee X$ or $\bigvee_{x \in X} x$ and $\bigwedge X$ or $\bigwedge_{x \in X} x$, and call these the meet and join of $X$, these "meet" and "join" are not *operations* in quite the sense we have been considering so far. An operation is supposed to be a map $S^n \to S$ for some $n$. One may allow $n$ to be an infinite cardinal (or other set), but when we consider complete lattices, there is no fixed cardinality to use. This suggests that we should consider each of the symbols $\bigvee$ and $\bigwedge$ to stand for a *system* of operations, of varying finite and infinite arities. But how large is this system? In a given complete lattice $L$, all meets and joins reduce (by dropping repeated arguments) to meets and joins of families of cardinalities $\leq \mathrm{card}(|L|)$. But if we want to develop a general theory of complete lattices, then meets and joins of families of arbitrary cardinalities will occur, so this "system of operations" will not be a *set* of operations. We shall eventually see that as a consequence of this, though complete lattices are in many ways like algebras, not all of the results that we prove about algebras will be true for them (Exercise 8.10:6(iii)).

Another sort of complication in the study of complete lattices comes from the equivalence of the various conditions in Proposition 6.2.3: Since these lattices can be characterized in terms of different systems of operations, there are many natural kinds of "maps" among them: maps which respect arbitrary meets, maps which respect arbitrary joins, maps which respect both, maps which respect meets of all *nonempty* sets and joins of all *pairs*, etc. The term "homomorphism of complete lattices" will mean a map respecting meets and joins of all nonempty sets, but the other kinds of maps are also of interest. These distinctions are brought out in:

### Exercise 6.2:6.

(i)    Show that every nonempty complete lattice can be embedded, by a map which respects arbitrary *joins* (including the join of the empty set), in a power set $\mathbf{P}(S)$, for some set $S$, and likewise may be embedded in a power set by a map which respects arbitrary *meets*.

(ii)   On the other hand, show, either using Exercise 6.1:15(ii) or by a direct argument, that the finite lattices $M_3$ and $N_5$ considered there cannot be embedded by any *lattice homomorphism*, i.e., any map respecting both *finite* meets and *finite* joins, in a power set $\mathbf{P}(S)$.

The next few pages contain a large number of exercises which, though I find them of interest, digress from the main point of this section. The reader who so wishes may skim past them and jump to the discussion immediately preceding Definition 6.2.4.

Our proof in Lemma 6.2.1 that the existence of least upper bounds of *chains* made a lattice upper semicomplete really only used well-ordered chains, i.e., chains order-isomorphic to ordinals. In fact, one can do still better:

### Exercise 6.2:7. Recall from Exercise 5.6:5 that every totally ordered set has a cofinal subset order-isomorphic to a regular cardinal.

(i)    Deduce that for $P$ a partially ordered set and $\alpha$ an infinite cardinal, the following two conditions are equivalent:

(a) Every chain in $P$ of cardinality $< \alpha$ has a least upper bound.

(b) Every chain in $P$ which is order-isomorphic to a regular cardinal $\beta < \alpha$ has a least upper bound.

(ii)   With the help of the above result, extend Lemma 6.2.1, adding a third equivalent condition.

There are still more ways than those we have seen to decompose the condition of being a complete lattice, as shown in part (ii) of

**Exercise 6.2:8.**

(i)   Show that following conditions on a partially ordered set $L$ are equivalent:

   (a) Every nonempty subset of $L$ having an upper bound has a least upper bound.

   (b) Every nonempty subset of $L$ having a lower bound has a greatest lower bound.

   (c) $L$ satisfies the *complete interpolation property*: Given two nonempty subsets $X$, $Y$ of $L$, such that every element of $X$ is $\leq$ every element of $Y$, there exists an element $z \in L$ which is $\geq$ every element of $X$ and $\leq$ every element of $Y$.

(ii)   Show that $L$ is a nonempty complete lattice if and only if it has a greatest and a least element, and satisfies the above equivalent conditions.

(iii)   Give an example of a partially ordered set which satisfies (a)–(c) above, but is not a lattice.

(iv)   Give an example of a partially ordered set with greatest and least elements, which has the *finite interpolation property*, i.e., satisfies (c) above for all finite nonempty families $X$ and $Y$, but which is not a lattice.

   This condition-splitting game is carried further in

**Exercise 6.2:9.** If $\sigma$ and $\tau$ are conditions on sets of elements of partially ordered sets, let us say that a partially ordered set $L$ has the $(\sigma, \tau)$-*interpolation property* if for any two subsets $X$ and $Y$ of $L$ such that $X$ satisfies $\sigma$, $Y$ satisfies $\tau$, and all elements of $X$ are $\leq$ all elements of $Y$, there exists an element $z \in L$ which is $\geq$ every element of $X$ and $\leq$ every element of $Y$. Now consider the *nine* conditions on $L$ gotten by taking for $\sigma$ and $\tau$ all combinations of the three properties "is empty", "is a pair" and "is a chain".

(i)   Find simple descriptions for as many of these nine conditions as you can; in particular, note cases that are equivalent to conditions we have already named.

(ii)   Show that $L$ is a nonempty complete lattice if and only if it satisfies all nine of these conditions.

(iii)   How close to independent are these nine conditions? To answer this, determine as well as you can which of the $2^9 = 512$ functions from the set of these conditions to the set {true, false} can be realized by appropriate choices of $L$. (Remark: A large number of these combinations *can* be realized, so to show this, you will have to produce a large number of examples. I therefore suggest that you consider ways that examples with certain *combinations* of properties can be obtained from examples of the separate properties.)

**Exercise 6.2:10.**

(i)    We saw in Exercise 6.1:2(ii) that the set of real polynomial functions on the unit interval $[0, 1]$, partially ordered by the relation ($\forall x \in [0, 1]$) $f(x) \leq g(x)$, does not form a lattice. Show, however, that it has the finite interpolation property. (This gives a solution to Exercise 6.2:8(iii), but far from the easiest solution. The difficulty in proving this result arises from the possibility that some members of $X$ may be tangent to some members of $Y$.)

(ii)    Can you obtain similar results for the partially ordered set of real polynomial functions on a general compact set $K \subseteq \mathbb{R}^n$ ?

An interesting pair of invariants related to Exercise 6.2:6(i) is examined in

**Exercise 6.2:11.**

(i)    For $L$ a nonempty complete lattice and $\alpha$ a cardinal, show that the following conditions are equivalent: (a) $L$ can be embedded, by a map respecting arbitrary *meets*, in the power set $\mathbf{P}(S)$ of a set of cardinality $\alpha$, (b) There exists a subset $T \subseteq |L|$ of cardinality $\leq \alpha$ such that every element of $L$ is the *join* of a (possibly infinite) subset of $T$, (c) $L$ can be written as the image, under a map respecting arbitrary joins, of the power set $\mathbf{P}(U)$ of a set of cardinality $\alpha$.

From condition (b) above we see that for every $L$ there will exist $\alpha$ such that these equivalent conditions hold. Let us call the least cardinal with this property the *upward generating number* of $L$, because of formulation (b). Dually, we have the concept of *downward* generating number.

(ii)    Find a finite lattice $L$ for which these two generating numbers are not equal.

The above exercise concerned complete lattices. On the other hand, if $L$ is a lattice or semilattice with no greatest element, we can't map any power set $\mathbf{P}(X)$ onto it by a homomorphism of semilattices, since a partially ordered set with greatest element can never be taken by an isotone map onto one without greatest element. As a next best possibility, one might ask whether one can map onto any such $L$ some lower complete lower semilattice of the form $\omega^X$, since (unlike $\mathbf{P}(X) = 2^X$) this does not in general have a greatest element, but is nonetheless a full direct product. A nice test case for this idea would be to take for $L$ the lattice of all finite subsets of a set $S$. The first part of the next exercise shows that for this case, the answer to the above question is yes.

**Exercise 6.2:12.**

(i)    Let $S$ be any set, and $\mathbf{P}_{\text{fin}}(S)$ the lower complete lower semilattice of finite subsets of $S$. Let $\omega^S$ denote the lower complete lower semilattice of natural-number-valued functions on $S$ (under pointwise inequality), and $|\omega^S|$ the underlying set of this lattice, so that $\omega^{|\omega^S|}$ is the lower complete lower semilattice of natural-number-valued functions on that set. Show that there exists a surjective homomorphism of lower semicomplete semilattices $\omega^{|\omega^S|} \to \mathbf{P}_{\text{fin}}(S)$.

Suggestion: For each $s \in S$ let $s^* \in |\omega^{|\omega^S|}|$ denote the function sending each element of $\omega^S$ to its value at $s$. Now map each $f \in |\omega^{|\omega^S|}|$ to the set of those $s \in S$ such that $f \geq s^*$. Show that this set is finite, and this map has the desired properties.

(ii)  If $L$ is an arbitrary nonempty lower semicomplete lower semilattice, must there exist a surjective homomorphism $\omega^X \to L$ of such semilattices for some set $X$? If not, can you find necessary and sufficient conditions on $L$ for such a homomorphism to exist?

We noted earlier that the concept of a complete lattice involves meets and joins of arbitrary cardinalities, which form a proper class of operations, and that as a result, it will not quite fit into the concept of an algebra we will develop in Chapter 9 (though for any particular cardinal $\alpha$, the concepts of $(< \alpha\text{-})$complete lattice and semilattice will fall under that concept). The next exercise takes a different approach, and regards possibly infinite meet and join as operators defined on subsets of $|L|$, rather than on tuples of elements. This puts the concept still farther from that of Chapter 9, but the "identities" these operations satisfy have an elegant formulation.

**Exercise 6.2:13.** Let $|L|$ be a set, and suppose that $\bigvee$ and $\bigwedge$ are operators associating to each nonempty subset $X \subseteq |L|$ an element of $|L|$ which will be denoted $\bigvee(X)$, respectively $\bigwedge(X)$. Show that these operators are the greatest lower bound and least upper bound operators arising from a complete lattice structure on $|L|$ if and only if the following three conditions hold.

(a)  For every $x_0 \in |L|$, $\bigvee(\{x_0\}) = x_0 = \bigwedge(\{x_0\})$.

(b)  For every nonempty set $Y$ of nonempty subsets of $|L|$,

$$\bigvee(\{\bigvee(X) \mid X \in Y\}) = \bigvee(\bigcup_{X \in Y}(X)) \text{ and}$$
$$\bigwedge(\{\bigwedge(X) \mid X \in Y\}) = \bigwedge(\bigcup_{X \in Y}(X)).$$

(c)  The pairwise operations defined by $a \vee b = \bigvee(\{a, b\})$ and $a \wedge b = \bigwedge(\{a, b\})$ satisfy the two "compatibility" identities of Definition 6.1.4.

To motivate the next definition, consider the following situation. Let $L$ be the complete lattice of all subgroups of a group $G$, and let $K \in L$ be a *finitely generated* subgroup of $G$, say generated by $g_1, \ldots, g_n$. Suppose this subgroup $K$ is majorized in $L$ by the join of a family of subgroups $H_i$ $(i \in I)$, i.e., is contained in the subgroup generated by the $H_i$. Then each of $g_1, \ldots, g_n$ can be expressed by a group-theoretic term in elements of $\bigcup |H_i|$. But any group-theoretic term involves only finitely many elements; hence $K$ will actually be contained in the subgroup generated by *finitely many* of the $H_i$. The converse also holds, since if $K$ is a *non*-finitely generated subgroup of $G$, then $K$ equals (and hence is contained in) the join of all the cyclic subgroups it contains, but is *not* contained in the join of any finite subfamily thereof.

The property we have just shown to characterize the finitely generated subgroups in the lattice of all subgroups of $G$ has an obvious similarity to the property defining *compact* subsets in a topological space, namely that if such a subset is covered by a family of open subsets, it is covered by some finite subfamily. Hence one makes the definition

**Definition 6.2.4.** *An element $k$ of a complete lattice (or more generally, of a complete upper semilattice) $L$ is called* compact *if every set of elements of $L$ with join $\geq k$ has a finite subset with join $\geq k$.*

By the preceding observations, the compact elements of the subgroup lattice of a group are precisely the finitely generated subgroups. (We will generalize this observation when we have a general theory of algebraic objects.)

We noted in Exercise 6.1:2(i) that the finitely generated subgroups of a group form an upper subsemilattice of the lattice of all subgroups. This suggests

**Exercise 6.2:14.** Do the compact elements of a complete lattice $L$ always form an upper subsemilattice?

**Exercise 6.2:15.** Show that a complete lattice $L$ has ascending chain condition if and only if all elements of $L$ are compact.

There seems to be no standard name for an element of a complete lattice having the dual property to compactness; sometimes such elements are called *co-compact.*

We end this section with some further exercises that, though interesting, are not closely related to material we will use in future sections.

We examined in Exercise 6.2:6 the embedding of lattices in power sets $\mathbf{P}(S)$ (and found that though there were embeddings that respected meets, and embeddings that respected joins, there were not in general embeddings that respected both). Let us look briefly at another fundamental sort of complete lattice, and the problem of embedding arbitrary lattices therein.

If $X$ is a set, and $\approx_0$ and $\approx_1$ are two equivalence relations on $X$, let us say $\approx_1$ *extends* $\approx_0$ if it contains it, as a subset of $X \times X$, and write $\approx_0 \leq \approx_1$ in this situation. Let $\mathbf{E}(X)$ denote the set of equivalence relations on $X$, partially ordered by this relation $\leq$. (One could use the reverse of this order, saying that $\approx_0$ is a *refinement* of $\approx_1$ when the latter extends the former, and justify considering the refinement to be "bigger" by the fact that it gives "more" equivalence classes. So our choice of the sense to give to our ordering is somewhat arbitrary; but let us stick with the ordering based on inclusions of binary relations.)

**Exercise 6.2:16.**

(i)    Verify that the partially ordered set $\mathbf{E}(X)$ defined above forms a complete lattice. Identify the elements $0_{\mathbf{E}(X)}$ and $1_{\mathbf{E}(X)}$.

(ii)   Let $L$ be any nonempty complete lattice, and $f: L \to \mathbf{E}(X)$ a map respecting arbitrary meets (a complete lower semilattice homomorphism respecting greatest elements). Show that for any $x, y \in X$, there is a *least* $d \in L$ such that $(x, y) \in f(d)$. Calling this element $d(x, y)$, verify that the map $d: X \times X \to L$ satisfies the following conditions for all $x, y, z \in X$:

    (a$_0$)  $d(x, x) = 0_L$,

    (b)    $d(x, y) = d(y, x)$,

    (c)    $d(x, z) \leq d(x, y) \vee d(y, z)$.

(iii)  Prove the converse, i.e., that given a nonempty complete lattice $L$ and a set $X$, any function $d: X \times X \to L$ satisfying (a$_0$)–(c) above arises as in (ii) from a unique complete lower semilattice-homomorphism $f: L \to \mathbf{E}(X)$ respecting greatest elements.

    In the remaining parts, we assume that $f: L \to \mathbf{E}(X)$ and $d: X \times X \to L$ are maps related as in (ii) and (iii).

(iv)   Show that the map $f$ respects least elements, i.e., that $f(0_L) = 0_{\mathbf{E}(X)}$, if and only if $d$ satisfies

    (a)    $d(x, y) = 0_L \iff x = y$   (a strengthening of (a$_0$) above).

(v)    Show that $f$ respects joins of finite nonempty families if and only if $d$ satisfies

    (d)    for all $x, y \in X$, $p, q \in |L|$ such that $d(x, y) \leq p \vee q$, there exists a finite sequence $x = z_0, z_1, \ldots, z_n = y$ (think "a path from $x$ to $y$ in $X$") such that for each $i < n$, either $d(z_i, z_{i+1}) \leq p$ or $d(z_i, z_{i+1}) \leq q$.

 

    A function $d$ which satisfies (a)–(c) above might be called an "$L$-valued metric on $X$," and (d) might be called "path sufficiency" of the $L$-valued metric space $X$. Two other properties of importance are noted in

(vi)   Assuming that $f$ respects finite nonempty joins, i.e., satisfies (d) above, show that it respects arbitrary nonempty joins if and only if

    (e)    for all $x, y \in X$, $d(x, y)$ is a compact element of $L$.

(vii)  Show that $f$ is one-to-one if and only if

    (f)    $L$ is generated under (not necessarily finite) joins by the elements $d(x, y)$ $(x, y \in X)$.

    Thus, to get various sorts of embeddings of complete lattices $L$ in complete lattices of the form $\mathbf{E}(X)$, it suffices to construct sets $X$ with appropriate sorts of $L$-valued metrics.

    How can one do this? Note that if we take a tree (in the graph-theoretic sense) with edges labeled in any way by elements of $L$, and define the distance between two vertices to be the join of the labels on the sequence of edges connecting those vertices, then we get an $L$-valued metric, such that the values assumed by this metric generate the same upper semilattice as do the set of labels. This can be used to get a system $(X, d)$ satisfying (a)–(c) and (f). We also want condition (d). This can be achieved by adjoining additional vertices:

**Exercise 6.2:17.** Let $L$ be a nonempty complete lattice.

(i)   Suppose $(X, d)$ is an $L$-valued metric space (i.e., satisfies conditions (a)–(c) of the preceding exercise), $x$ and $y$ are two points of $X$, and $p$ and $q$ are elements of $L$ such that $d(x, y) \leq p \vee q$. Show that we can adjoin new points $z_1$, $z_2$, $z_3$ to $X$ and extend the metric in a consistent way so that the steps of the path $x$, $z_1$, $z_2$, $z_3$, $y$ have lengths $p$, $q$, $p$, $q$ respectively. (The least obvious part is how to define the distance from $z_2$ to a point $w \in X$. To do this, verify that $d(w, x) \vee p \vee q = d(w, y) \vee p \vee q$, and use the common value.)

(ii)   Show that every $L$-valued metric space can be embedded in a path-sufficient one. (This will involve a countable sequence of steps $X = X_0 \subseteq X_1 \subseteq \ldots$ such that each $X_i$ "cures" all failures of path-sufficiency found in $X_{i-1}$, using the idea of part (i). The desired space is then $\bigcup X_i$.)

Now if the given complete lattice $L$ is generated as a complete upper semilattice by the upper subsemilattice $K$ of its compact elements, then one can carry out the above constructions as to get condition (e) above, and hence an embedding of $L$ in $\mathbf{E}(X)$ that respects arbitrary meets and joins. Conversely, one sees that this assumption on $K$ is necessary for such an embedding to exist.

If we don't make this assumption on $K$, we can still use the above construction to embed $L$ in a lattice $\mathbf{E}(X)$ by a map respecting arbitrary meets and *finite* joins.

We shall see in the next section that any lattice can be embedded by a lattice homomorphism in a complete lattice, so the above technique shows that any lattice can be embedded by a lattice homomorphism in a lattice of equivalence relations.

If $L$ is finite, the construction of Exercise 6.2:17 gives, in general, a countable, but not a finite $L$-valued metric space $X$. It was for a long time an open question whether every finite lattice could be embedded in the lattice of equivalence relations of a finite set. This was finally proved in 1980 by P. Pudlák and J. Tůma [126]. However, good estimates for the size of an $X$ such that even a quite small lattice $L$ (e.g., the 15-element lattice $\mathbf{E}(4)^{\mathrm{op}}$) can be embedded in $\mathbf{E}(X)$ remain to be found. The least $m$ such that $\mathbf{E}(n)^{\mathrm{op}}$ embeds in $\mathbf{E}(m)$ has been shown by Pudlák to grow *at least* exponentially in $n$; the earliest *upper* bound obtained for $m$ was $2^{2^{\cdot^{\cdot^{\cdot}}}}$ with $n^2$ exponents! For subsequent better results see [106] and [93, in particular p. 16, top].

## 6.3. Closure operators

We introduced this chapter by noting certain properties common to the partially ordered sets of all subsets of a set, of all subgroups of a group, and similar examples. But so far, we seem to have made a virtue of abstractness, defining semilattice, lattice, etc., without reference to systems of subsets of sets. Neither abstractness nor concreteness is everywhere a virtue; each makes its contribution, and it is time to turn to an important class of concrete lattices.

**Lemma 6.3.1.** *Let $S$ be a set. Then the following data are equivalent:*

(i)   *A lower semicomplete lower subsemilattice of* $\mathbf{P}(S)$ *which contains* $1_{\mathbf{P}(S)} = S$, *in other words, a set $C$ of subsets of $S$ closed under taking arbitrary intersections, including the empty intersection, $S$ itself.*

(ii)   *A function* $\mathrm{cl}\colon \mathbf{P}(S) \to \mathbf{P}(S)$ *with the properties:*

$$
\begin{array}{lll}
(\forall\, X \subseteq S) & \mathrm{cl}(X) \supseteq X & (\text{cl is increasing}), \\
(\forall\, X, Y \subseteq S) & X \subseteq Y \implies \mathrm{cl}(X) \subseteq \mathrm{cl}(Y) & (\text{cl is isotone}), \\
(\forall\, X \subseteq S) & \mathrm{cl}(\mathrm{cl}(X)) = \mathrm{cl}(X) & (\text{cl is idempotent}).
\end{array}
$$

*Namely, given $C$ as in* (i), *one defines* cl *as the operator taking each $X \subseteq S$ to the intersection of all members of $C$ containing $X$, while given* cl *as in* (ii), *one defines $C$ as the set of $X \subseteq S$ satisfying* $\mathrm{cl}(X) = X$, *equivalently, as the set of subsets of $S$ of the form* $\mathrm{cl}(Y)$ $(Y \subseteq S)$.   □

**Exercise 6.3:1.** Verify the above lemma. That is, show that the procedures described do carry families $C$ with the properties of point (i) to operators cl with the properties of point (ii) and vice versa, and are inverse to one another, and also verify the assertion of equivalence in the final clause.

**Definition 6.3.2.** *An operator* cl *on the class of subsets of a set $S$ with the properties described in point* (ii) *of the above lemma is called a* closure operator *on $S$. If* cl *is a closure operator on $S$, the subsets $X \subseteq S$ satisfying* $\mathrm{cl}(X) = X$, *equivalently, the subsets of the form* $\mathrm{cl}(Y)$ $(Y \subseteq S)$, *are called the* closed subsets *of $S$ under* cl.

We see that virtually every mathematical construction commonly referred to as "the ... generated by" (fill in the blank with subgroup, normal subgroup, submonoid, subring, sublattice, ideal, congruence, etc.) is an example of a closure operator on a set. The operation of topological closure on subsets of any topological space is another example. Some cases are called by other names: the *convex hull* of a set of points in Euclidean $n$-space, the *span* of a subset of a vector space (i.e., the vector subspace it generates), the set of *derived operations* of a set of operations on a set (§ 2.6). Incidentally, the constructions of the subgroup and subring generated by a subset of a group or ring illustrate the fact that the closure of the empty set need not be empty.

A very common way of obtaining a closure operator on a set $S$, which includes most of the above examples, can be abstracted as follows: One specifies a certain subset

$$(6.3.3) \qquad\qquad G \subseteq \mathbf{P}(S) \times S,$$

and then calls a subset $X \subseteq S$ *closed* if for all $(A, x) \in G$, $A \subseteq X \implies x \in X$. It is straightforward to verify that the class of closed sets under this definition is closed under arbitrary intersections, and so by Lemma 6.3.1, corresponds to a closure operator cl on $S$.

For example, if $K$ is a group, the operator "subgroup generated by" on subsets of $|K|$ is of this form. One takes for (6.3.3) the set of all pairs of the forms

$$(6.3.4) \qquad\qquad (\{x, y\}, \, x\,y), \quad (\{x\}, \, x^{-1}), \quad (\emptyset, \, e),$$

where $x$ and $y$ range over $|K|$. To get the operator "*normal* subgroup generated by–",we use the above pairs, supplemented by the further pairs $(\{x\}, \, y\,x\,y^{-1})$ $(x, y \in |K|)$. Clearly, the other "... generated by" constructions mentioned above can be characterized similarly. For a non-algebraic example, the operator giving the topological closure of a subset of the real line $\mathbb{R}$ can be obtained by taking $G$ to consist of all pairs $(A, x)$ such that $A$ is the set of points of a convergent sequence, and $x$ is the limit of that sequence.

**Exercise 6.3:2.** Show that for any closure operator cl on a set $S$, there exists a subset $G \subseteq \mathbf{P}(S) \times S$ which determines cl in the sense we have been discussing.

**Exercise 6.3:3.** If $T$ is a set, display a subset $G \subseteq \mathbf{P}(T \times T) \times (T \times T)$ such that the *equivalence relations* on $T$ are precisely the subsets of $T \times T$ closed under the operator cl corresponding to $G$. (The previous exercise gives us a way of doing this "blindly". But what I want here is an explicit set, which one might show to someone who didn't know what "equivalence relation" meant, to provide a characterization of the concept.)

In Chapter 3 we contrasted the approaches of obtaining sets one is interested in "from above" as intersections of systems of larger sets, and of building them up "from below". We have constructed the closure operator associated with a family (6.3.3) by noting that the class of subsets of $S$ we wish to call closed is closed under arbitrary intersections; so we have implicitly obtained these closures "from above". The next exercise constructs them "from below".

**Exercise 6.3:4.** Let $S$ be a set and $G$ a subset of $\mathbf{P}(S) \times S$. For $X$ a subset of $S$ and $\alpha$ any ordinal, let us define $\mathrm{cl}_G^{(\alpha)}(X)$ recursively by:

$$\mathrm{cl}_G^{(0)}(X) = X,$$

$$\mathrm{cl}_G^{(\alpha+1)}(X) = \mathrm{cl}_G^{(\alpha)}(X) \cup \{x \mid (\exists A \subseteq \mathrm{cl}_G^{(\alpha)}(X)) \, (A, x) \in G\}.$$

$$\mathrm{cl}_G^{(\alpha)}(X) = \bigcup_{\beta \in \alpha} \mathrm{cl}_G^{(\beta)}(X) \quad \text{if } \alpha \text{ is a limit ordinal.}$$

(i)   Show (for $S$, $G$ as above) that there exists an ordinal $\alpha$ such that for all $\beta > \alpha$ and all $X \subseteq S$, $\mathrm{cl}_G^{(\beta)}(X) = \mathrm{cl}_G^{(\alpha)}(X)$, and that when this is so, $\mathrm{cl}_G^{(\alpha)}(X)$ is $\mathrm{cl}(X)$ in the sense of the preceding discussion. (Cf. the construction in § 3.2 of the equivalence relation $R$ on group-theoretic terms as the union of a chain of relations $R_i$.)

(ii)   If for all $(A, x) \in G$, $A$ is finite, show that the $\alpha$ of part (i) can be taken to be $\omega$.

(iii)  For each ordinal $\alpha$, can you find an example of a set $S$ and a $G \subseteq \mathbf{P}(S) \times S$ such that $\alpha$ is the least ordinal having the property of part (i)?

We have seen that there are restrictions on the sorts of lattices that can be embedded by lattice homomorphisms into lattices $(\mathbf{P}(S), \cup, \cap)$ (Exercise 6.1:15), or into lattices of submodules of modules (Exercise 6.1:14). In contrast, note statements (ii) and (iii) of

**Lemma 6.3.5.**

(i)   *If* $\mathrm{cl}$ *is a closure operator on a set* $X$, *then the set of* $\mathrm{cl}$-*closed subsets of* $X$, *partially ordered by inclusion, forms a complete lattice, with the meet of an arbitrary family given by its set-theoretic intersection, and the join of such a family given by the* closure *of its union. Conversely,*

(ii)  *Every nonempty complete lattice* $L$ *is isomorphic to the lattice of closed sets of a closure operator* $\mathrm{cl}$ *on some set* $S$; *and*

(iii) *Every lattice* $L$ *is isomorphic to a sublattice of the lattice of closed sets of a closure operator* $\mathrm{cl}$ *on some set* $S$.

*Sketch of proof.* (i): It is straightforward to verify that the indicated operations give a greatest lower bound and a least upper bound to any family of closed subsets.

(ii): Take $S = |L|$, and for each $X \subseteq S$, define $\mathrm{cl}(X) = \{y \mid y \leq \bigvee_{x \in X} x\}$. Then $L$ is isomorphic to the lattice of closed subsets of $S$, by the map $y \mapsto \{x \mid x \leq y\}$.

(iii): Again take $S = |L|$, but since joins of arbitrary families may not be defined in $L$, define $\mathrm{cl}(X)$ to be the set of all elements majorized by joins of finite subsets of $X$. Embed $L$ in the lattice of $\mathrm{cl}$-closed subsets of $S$ by the same map as before.                                                                    $\square$

**Exercise 6.3:5.** Verify that the constructions of (ii) and (iii) above give closure operators on $|L|$, and that the induced maps are respectively an isomorphism of complete lattices and a lattice embedding.

The second of the two closure operators used in the above proof can be thought of as closing a set $X$ in $|L|$ under pairwise joins, and under meets with arbitrary elements of $L$. In the notation that denotes join by $+$ and writes meet as multiplication, this has the same form as the definition of an ideal of a ring. So lattice-theorists often call sets of elements in a lattice closed under these operations "ideals". In particular, $\{y \mid y \leq x\}$ is called the *principal ideal* generated by $x$.

**Exercise 6.3:6.**

(i)    Show that assertion (iii) of the preceding lemma can also be proved by taking the same $S$ and the same map, but taking $\mathrm{cl}(X) \subseteq S$ to be the intersection of all principal ideals of $L$ containing $X$.

(ii)   Will the complete lattices generated by the images of $L$ under these two constructions in general be isomorphic?

**Exercise 6.3:7.** Can the representation of a (complete) lattice $L$ by closed sets of a closure operator given in Lemma 6.3.5(iii) and/or that given in Exercise 6.3:6 be characterized by any universal properties?

**Exercise 6.3:8.** Show that a lattice $L$ is complete and nonempty if and only if every intersection of principal ideals of $L$ (including the intersection of the empty family) is a principal ideal.

The concept of a set with a closure operator is not only general enough to allow representations of all lattices, it is a convenient tool for constructing examples. For example, recall that Exercise 6.1:10(ii), if solved by the hint given, shows that a lattice generated by five elements can be infinite. With more work, that method can be made to give an infinite lattice with four generators, but one can show that any three-generator sublattice of the lattice of affine subspaces of a vector space is finite. However, we shall now give an ad hoc construction of a closure operator whose lattice of closed sets has an infinite three-generator sublattice.

**Exercise 6.3:9.** Let $S = \omega \cup \{x, y\}$, where $\omega$ is regarded as the set of nonnegative integers, and $x$, $y$ are two elements not in $\omega$. Let $G \subseteq \mathbf{P}(S) \times S$ consist of all pairs of the form

$$(\{x, 2m\}, 2m+1), \qquad (\{y, 2m+1\}, 2m+2),$$

where $m$ ranges over $\omega$ in each case. Let $L$ denote the lattice of closed subsets of $S$ under the induced closure operator, and consider the sublattice generated by $\{x\}$, $\{y, 0\}$, and $\omega$. Show by induction that for every $n \geq 0$, this sublattice of $L$ contains the set $\{0, \ldots, n\}$. Thus, this three-generator lattice is infinite.

   (Exercise 8.10:6, which you can do at this point, will show that the same technique, applied to complete lattices, gives three-generator complete lattices of arbitrarily large cardinalities.)

**Exercise 6.3:10.** The lattice of the above exercise contains an infinite chain. Does there exist a three-generator lattice which is infinite but does not contain an infinite chain?

**Exercise 6.3:11.** If $A$ is an abelian group, can a finitely generated sublattice of the lattice of all subgroups of $A$ contain an infinite chain?

We now turn to a property which distinguishes the sort of closure operators commonly occurring in algebra from those arising in topology and analysis.

**Lemma 6.3.6.** *Let* cl *be a closure operator on a set* $S$. *Then the following conditions are equivalent:*

(i)    *For all* $X \subseteq S$, $\mathrm{cl}(X) = \bigcup_{\text{finite subsets } X_0 \subseteq X} \mathrm{cl}(X_0)$.

(ii)   *The union of every chain of closed subsets of* $S$ *is closed.*

(iii)  *The closure of each singleton* $\{s\} \subseteq S$ *is compact in the lattice of closed subsets.*

(iv)   cl *is the closure operator determined by a set* $G \subseteq \mathbf{P}(S) \times S$ *having the property that the first component of each of its members is finite.*    □

**Exercise 6.3:12.** Prove Lemma 6.3.6.

**Definition 6.3.7.** *A closure operator satisfying the equivalent conditions of the above lemma will be called* finitary.

This is because the lattice of subalgebras of an algebra $A$ satisfies condition (iv) of that lemma if the operations of $A$ are all *finitary*, i.e., have finite arity (§ 2.4). (Many authors call such closure operators "algebraic" instead of "finitary", because, as noted, the property is typical of closure operators that come up in algebra.)

**Exercise 6.3:13.** Let cl be a finitary closure operator on a set $S$.

(i)    Show that if $\alpha$ is an infinite cardinal, and $X$ a cl-closed subset of $S$ which is the closure of some subset of cardinality $< \alpha$, then every $Y \subseteq X$ with $\mathrm{cl}(Y) = X$ has a subset $Y'$ with $\mathrm{card}(Y') < \alpha$ such that $\mathrm{cl}(Y') = X$.

(ii)   Show by example that the corresponding statement is not true for finite $\alpha$.

(iii)  Show by example that the result of (i) does not characterize the finitary closure operators; i.e., that not every closure operator satisfying (i) is finitary.

**Exercise 6.3:14.**

(i)    Show that a nonempty complete lattice $L$ is isomorphic to the lattice of all closed sets under a finitary closure operator if and only if every element of $L$ is a (possibly infinite) join of compact elements.

(ii)   For what complete lattices is it true that *every* closure operator cl, on any set, whose lattice of closed sets is isomorphic to $L$, is finitary?

**Exercise 6.3:15.** Show that a closure operator cl is finitary if and only if the compact elements in the lattice of its closed subsets are precisely the closures of finite sets.

For a not necessarily finitary closure operator, prove an inclusion between these two classes of closed subsets, but show that the other inclusion need not hold.

**Exercise 6.3:16.** Consider the following three conditions on a closure operator cl on a set $S$. (a) cl is finitary. (b) The union of any two cl-closed subsets of $S$ is cl-closed. (c) Every singleton subset of $S$ is cl-closed.

For each subset of this set of three properties, find an example of a closure operator that has the properties in that subset, but not any of the others. (Thus, eight examples are asked for.) Where possible, use familiar or important examples.

**Exercise 6.3:17.**

(i)   Show that a closure operator cl on a set $S$ is the operation of topological closure with respect to some topology on $S$ if and only if it satisfies condition (b) of the preceding exercise, and: $(c_0)$ $\emptyset$ is cl-closed in $S$.

(ii)   Assuming $S$ has more than one element, show that cl is closure with respect to a $T_1$ topology if and only if it satisfies conditions (b) and (c) of the preceding exercise.

Since the operation of topological closure determines the topology, this shows that topologies on a space are equivalent to closure operators satisfying the indicated conditions.

**Exercise 6.3:18.** It is well known that if a group $K$ is generated by $\leq \gamma$ elements ($\gamma$ a cardinal), then $\operatorname{card}(|K|) \leq \gamma + \aleph_0$.

(i)   Deduce this fact from simple properties of the set $G \subseteq \mathbf{P}(|K|) \times |K|$ defined in (6.3.4).

(ii)   Try to generalize (i) to a result on the way the cardinalities of sets increase under application of a closure operator cl obtained from a set $G$ as above, in terms of the properties of $G$. Can you show by example that your results are best possible?

When we described how to construct a closure operator cl from a subset $G \subseteq \mathbf{P}(S) \times S$, it would have been tempting to call cl "the closure operator generated by $G$." This would not quite have made sense, because a closure operator is not itself a subset of $\mathbf{P}(S) \times S$. However, we can show what this is "trying to say" by setting up a correspondence between closure operators on $S$ and certain subsets of $\mathbf{P}(S) \times S$:

**Exercise 6.3:19.** Let $S$ be a set.

If cl is a closure operator on $S$, let us write $\sigma(\mathrm{cl}) = \{(A, x) \mid A \subseteq S, x \in \mathrm{cl}(A)\}$ and let us call a subset $H \subseteq \mathbf{P}(S) \times S$ a closure *system* on $S$ if $H = \sigma(\mathrm{cl})$ for some closure operator cl on $S$.

(i)    Show that closure systems on $S$ are precisely the subsets of $\mathbf{P}(S) \times S$ closed under a certain closure operator, $\mathrm{cl}_{\mathrm{sys}}$ (which you should describe).

(ii)   Show that for any subset $G \subseteq \mathbf{P}(S) \times S$, if we write $\mathrm{cl}_G$ for the closure operator determined by $G$ in the sense discussed earlier, then $\sigma(\mathrm{cl}_G) = \mathrm{cl}_{\mathrm{sys}}(G)$.

So though we cannot call $\mathrm{cl}_G$ the closure operator generated by $G$, it is the operator corresponding to the *closure system* generated by $G$.

Of course, I cannot resist adding

(iii)  Describe $\mathrm{cl}_{\mathrm{sys}}$ as the closure operator on $\mathbf{P}(S) \times S$ determined ("generated") by an appropriate set $G_{\mathrm{sys}}$ (of elements of what set?)

We now have three ways of looking at closure data on a set $S$ : as a family of subsets of $S$, as an operator on subsets of $S$, and as a certain kind of subset of $\mathbf{P}(S) \times S$. We take a global look at this data in:

**Exercise 6.3:20.** Let $S$ be a set. Call the set of all families of subsets of $S$ that are closed under arbitrary intersections $\mathrm{Clofam}(S)$, and order this set by inclusion. Call the set of all closure operators on $S$ $\mathrm{Clop}(S)$, and order it by putting $\mathrm{cl}_1 \leq \mathrm{cl}_2$ if for all $X$, $\mathrm{cl}_1(X) \leq \mathrm{cl}_2(X)$. Call the set of closure systems on $S$ in the sense of the preceding exercise $\mathrm{Closys}(S)$, and order it by inclusion.

Verify that $\mathrm{Clofam}(S)$, $\mathrm{Clop}(S)$ and $\mathrm{Closys}(S)$ are all complete lattices. Do the natural correspondences between the three types of data constitute lattice isomorphisms? If not, state precisely the relationships involved. Describe the meet and join operations of $\mathrm{Clop}(S)$ explicitly.

**Exercise 6.3:21.** Investigate the subset of *finitary* closure operators within the set $\mathrm{Clop}(S)$ defined in the preceding exercise. Will it be closed under meets (finite? arbitrary?)—joins (ditto)? Given any $\mathrm{cl} \in \mathrm{Clop}(S)$, will there be a least finitary closure operator containing cl? A greatest finitary closure operator contained in cl?

Descending from the abstruse to the elementary, here is a problem on closure operators that could be explained to a bright High School student, but which has so far defied solution:

**Exercise 6.3:22.** (Péter Frankl's question). Let $S$ be a finite set, and cl a closure operator on $S$ such that $\mathrm{cl}(\emptyset) \neq S$. Must there exist an element $s \in S$ which belongs to *not more than half* of the sets closed under cl?

(I generally state this conjecture to people not in this course in terms of "a system $C$ of subsets of $S$ which is closed under pairwise intersections, and contains at least one proper subset of $S$." There are still other formulations; for instance, as asking whether every nontrivial finite lattice has an element which is join-irreducible (not a join of two smaller elements) and which is majorized by no more than half the elements of the lattice.)

One occasionally encounters the dual of the type of data defining a closure operator—a system $U$ of subsets of a set $S$ closed under forming arbitrary *unions*; equivalently, an operator $f$ on subsets of $S$ which is *decreasing*, idempotent, and isotone. In this situation, the *complements* in $S$ of the sets in $U$ will be the closed sets of a closure operator, namely $X \mapsto {}^c(f({}^cX))$ (where $^c$ denotes complementation). When such an operator $f$ is discovered, it is often convenient to change viewpoints and work with the dual operator $^cf^c$, to which one can apply the theory of closure operators. However, $U$ and $f$ may be more natural in some situations than the dual family and map. In such cases one may refer to $f$ as an *interior operator* (though the term is not widely used), since in a topological space, the complement of the closure of the complement of $X$ is called the interior of $X$. Clearly, every result about closure operators gives a dual result on interior operators.

(Péter Frankl's question, described in the last exercise, is most often stated in dual form, asking whether, given a finite set $C$ of sets which is closed under pairwise unions and contains at least one nonempty set, there must exist an element belonging to at least half the members of $C$. As such, it is called the "union-closed set" question, and papers on the topic can be found by searching for the phrase "union closed".)

## 6.4. Digression: a pattern of threes

It is curious that many basic mathematical definitions involve similar systems of three parts.

A *group structure* on a set is given by (1) a neutral element, (2) an inverse-operation and (3) a multiplication; this family of operations must satisfy (1) the neutral-element laws, (2) the inverse laws and (3) the associative law.

A *partial ordering* on a set is a binary relation that is (1) reflexive, (2) antisymmetric and (3) transitive, while an *equivalence relation* is (1) reflexive, (2) symmetric and (3) transitive.

The operation of a *semilattice* is (1) idempotent, (2) commutative and (3) associative.

A *closure operator* is (1) increasing, (2) isotone and (3) idempotent.

In a *metric space*, the metric satisfies (1) a condition on when distances are 0, (2) symmetry and (3) the triangle inequality.

This parallelism is not just numerical. The general pattern seems to be that the simplest conditions or operations, marked (1) above, have to do with the relation of an element to itself; the intermediate ones, marked (2), tell us, if we know how two elements relate in one order, how they relate in the reverse order; while the strongest, those marked (3), tell us how to use the relation of one element to a second and this second to a third to get a relation between the first and the third.

Let us see this in the examples listed above. We must distinguish in some cases between abstract structures and the "concrete" situations that motivated them.

The concrete situation motivating the concept of a group is that of a group of permutations of a set. For a set of permutations to form a group, (1) it should contain the permutation $e$ that takes every element of the set to itself, (2) if it contains a permutation $x$, it should also contain the permutation $x^{-1}$ which carries $q$ to $p$ whenever $x$ carries $p$ to $q$, and (3) along with any permutations $x$ and $y$, it should contain the permutation $xy$ which carries $p$ to $r$ whenever $y$ carries $p$ to $q$ and $x$ carries $q$ to $r$. So this fits the pattern described.

When we look at the definition of an *abstract* group $G$, the above *closure conditions* are replaced by *operations* of neutral element, inverse, and composition. The conditions on these operations needed to mimic the properties of permutation groups as sets with operations say that when $G$ acts on itself by left or right multiplication, the three operations of $G$ actually behave like the constructions they are modeled on: left or right multiplication by the neutral element leaves all elements of $|G|$ fixed, left or right multiplication by $x$ is "reversed" by the action of $x^{-1}$, and left or right multiplication by $x$ followed by multiplication on the same side by $y$ is equivalent to multiplication by $yx$, respectively $xy$. These are the neutral-element, inverse and associative laws (slightly reformulated). Finally, when we *return* from this abstract concept to its concrete origins via the concept of a $G$-set $X$, we again have three conditions, saying that the actions of the neutral element, of inverses of elements, and of composites of elements of $G$ behave on $X$ in the proper manner. (However, the condition for inverses is a consequence of the other two plus the group identities of $G$, and so is usually omitted from the definition of a $G$-set.)

In the definitions of *partial ordering*, of *equivalence relation*, and of *metric*, we do not have an abstraction of a structure on a set, but such a structure itself. The reader can easily verify that these three-part definitions each have the form we have described.

In the cases of *semilattices* and *closure operators*, one can say roughly that closure operators are the concrete origins and semilattices the abstraction. My general characterization of the three components of these definitions does not, as we shall see, give quite as good a fit in this case. The condition that a closure operator be idempotent, $\mathrm{cl}(\mathrm{cl}(X)) = \mathrm{cl}(X)$, may be considered a "transitivity" type condition, since it says that if you can get some elements from elements of $X$, and some further elements from these, then you get those further elements from $X$. The "reflexivity" type condition is the one saying $\mathrm{cl}(X) \supseteq X$, since it means that what one gets from $X$ includes all of $X$ itself. But I cannot see any way of interpreting the remaining condition,

$X \subseteq Y \implies \mathrm{cl}(X) \subseteq \mathrm{cl}(Y)$, as describing the relation between elements considered in two different orders.

In the abstracted concept, that of a semilattice, the three conditions of idempotence, commutativity, and associativity of the operation $\vee$ do fit the pattern described, but they do not seem to come in a systematic way from the corresponding properties of closure operators.

When one looks at important weakenings of the concepts of group, etc., the middle operation or condition seems to be the one most naturally removed: Monoids are a useful generalization of groups, and preorders are a useful generalization both of partial orders and of equivalence relations.

The folklorist Alan Dundes argued that the number "three" holds a fundamental place in the culture of Western civilization, in ways ranging from traditional stories (three brothers go out to seek their fortune; Goldilocks and the three bears), superstitions ("third time's a charm") and verbal formulas ("Tom, Dick and Harry") to our three-word personal names. (See essay in [75].) He raised the challenge of how many of the "threes" occurring in science (archaeologists' division of each epoch into an "early", a "middle" and a "late" period; the three-stage polio vaccination; the three dimensions of physics, etc.) represent circumstances given to us by nature, and how many we have imposed on nature through cultural prejudice!

In the situation we have been discussing, I would argue that the similarity between the various sets of definitions represents a genuine pattern in "mathematical nature"; that the way the pattern appears, in terms of systems of *three* conditions, in contemporary developments of these topics, is not the only natural way these topics could be developed; but that the fact that they are developed in this way is not a consequence of a prejudice toward the number three, but of chance. As a simple example of how these topics might be differently developed, if basic textbooks regularly first defined "monoid", and then defined a group as a monoid with an inverse operation, and similarly first defined "preorder", then defined partial orders and equivalence relations as preorders satisfying the symmetry or antisymmetry condition, and so on, then, though we would still have a recurring pattern, it would not be a pattern of "threes". More radically, we might define composition in a group or monoid as an operation taking each ordered $n$-tuple of elements $(n \geq 0)$ to its product, and formulate the associative law accordingly, letting the neutral element simply be the empty product, and the neutral-element law a special case of the associative law; and again, no "threes" would be apparent. As to the reason we develop the topics as we do, rather than in one of the above ways, I think this comes out of certain choices regarding pedagogy and notation that have evolved in Western mathematics, for better or worse, without anyone's looking ahead at the number of components this would yield in such definitions. (On the other hand, I freely admit that my choice in § 3.1 to motivate the idea of a free group with the three-generator case was culturally influenced.)

Let me close this discussion by noting that many of the more complicated objects of mathematical study arise by combining one structure that fits, or partially fits, the pattern we have noted, with another. Thus, a lattice is a set with two *semilattice* structures that satisfy compatibility identities; a ring is given by an *abelian group,* together with a bilinear binary operation on this group under which it is a *monoid.*

The reader familiar with the definition of a Lie algebra over a commutative ring $R$ (§ 9.7 below) will note similarly that it is an $R$-module (a concept which fits into the above pattern in the same way as that of $G$-set), with an $R$-bilinear operation, the Lie bracket, which satisfies the alternating identity (which tells *both* the result of bracketing an element with itself, and the relation between bracketings in opposite orders), and the Jacobi identity (which describes how the bracket of an element with the bracket of two others can be described in terms of the operations of bracketing with those elements successively).

Returning to the description of a ring as an abelian group given with a bilinear operation under which it is a monoid, it is interesting to note that various refinements of the concept of ring involve adding one (or more!) conditions that can be thought of as filling in the missing "middle slot" in the monoid structure, concerning how elements relate in opposite orders: A multiplicative inverse operation (on nonzero elements) gives a *division ring* structure; *commutativity* of multiplication determines the favorite class of rings of contemporary algebra; both together give the class of fields. Another important ring-theoretic concept which can be thought of in this way is that of an *involution* on a (not necessarily commutative) ring, that is, an abelian group automorphism $*\colon |R| \to |R|$ satisfying $x^{**} = x$ and $(x\,y)^* = y^*x^*$. The complex numbers have all three structures: multiplicative inverses, commutativity, and the involution of complex conjugation.

The concept of a closure operator has an important special case gotten by imposing an additional condition specifying "how elements relate in opposite orders"; the *exchange axiom*:

$$(6.4.1) \quad y \notin \mathrm{cl}(X), \; y \in \mathrm{cl}(X \cup \{z\}) \implies z \in \mathrm{cl}(X \cup \{y\}) \; (X \subseteq S, \, y, z \in S).$$

This is the condition which, in the theory of vector spaces, allows one to prove that bases have unique cardinalities, and in the theory of transcendental field extensions yields the corresponding result for transcendence bases. (To be precise, in both of these cases (6.4.1) leads to a proof of the uniqueness of the cardinality when the given "bases" are finite. When at least one is infinite, the same result follows from Exercise 6.3:13(i).) Closure operators satisfying (6.4.1) are called (among other names) *matroids.* Cf. [144], and for a ring-theoretic application, [51].

I do not attach great importance to the observations of the above section. But I have noticed them for years, and thought this would be a good place to mention them.

## 6.5. Galois connections

Let me introduce this very general concept using the case from which it gets
its name:

  *Galois theory* deals with the situation where one is given a field $F$ and a
finite group $G$ of automorphisms of $F$. Given any subset $A$ of $F$, let $A^*$
denote the set of elements of the group $G$ fixing all elements of $A$ (where
"$g$ fixes $a$" means $g(a) = a$), and given any subset $B$ of $G$, likewise let
$B^*$ be the set of elements of the field $F$ fixed by all members of $B$. It is not
hard to see that in these situations, $A^*$ is always a subgroup of $G$, and $B^*$
a subfield of $F$. The Fundamental Theorem of Galois Theory says that the
groups $A^*$ give *all* the subgroups of $G$, and similarly that the sets $B^*$ are
*all* the fields between the fixed field of $G$ in $F$ and the whole field $F$, and
gives further information on the relation between corresponding subgroups
and subfields.

  Some parts of the proof of this theorem use arguments specific to fields
and their automorphism groups; but certain other parts can be carried out
without even knowing what the words mean. For instance, the result, "If $A$
is a set of elements of the field $F$, and $A^{**}$ is the set of elements of $F$ fixed
by all automorphisms in $G$ that fix all elements of $A$, then $A^{**} \supseteq A$" is
clearly true independent of what is meant by a "field", an "automorphism",
or "to fix"!

  This suggests that one should look for a general context to which the latter
sort of arguments apply. Replacing the set of elements of our field $F$ by an
arbitrary set $S$, the set of elements of the group $G$ by any set $T$, and the
condition of elements of $F$ being fixed by elements of $G$ by any relation
$R \subseteq S \times T$, we can make the following observations:

**Lemma 6.5.1.** *Let $S$, $T$ be sets, and $R \subseteq S \times T$ a relation. For $A \subseteq S$,
$B \subseteq T$, let us write*

(6.5.2)
$$A^* = \{t \in T \mid (\forall a \in A)\ a\,R\,t\} \subseteq T,$$
$$B^* = \{s \in S \mid (\forall b \in B)\ s\,R\,b\} \subseteq S,$$

*thus defining two operations written* $^*$, *one from* $\mathbf{P}(S)$ *to* $\mathbf{P}(T)$ *and the
other from* $\mathbf{P}(T)$ *to* $\mathbf{P}(S)$. *Then for* $A$, $A' \subseteq S$, $B$, $B' \subseteq T$, *we have*

(i)    $A \subseteq A' \implies A^* \supseteq A'^*$   $B \subseteq B' \implies B^* \supseteq B'^*$   ($^*$ reverses inclusions).
(ii)          $A^{**} \supseteq A$                    $B^{**} \supseteq B$            ($^{**}$ is increasing).
(iii)        $A^{***} = A^*$                  $B^{***} = B^*$          ($^{***} = {}^*$).
(iv)  $^{**}: \mathbf{P}(S) \to \mathbf{P}(S)$ *and* $^{**}: \mathbf{P}(T) \to \mathbf{P}(T)$ *are closure operators on* $S$
*and* $T$ *respectively.*

(v)  *The sets $A^*$ $(A \subseteq S)$ are precisely the closed subsets of $T$, and the sets $B^*$ $(B \subseteq T)$ are precisely the closed subsets of $S$, with respect to these closure operators* **.

(vi)  *The maps* *, *restricted to closed sets, give an antiisomorphism* (*an order-reversing, equivalently,* $\vee$-*and*-$\wedge$-*interchanging, bijection*) *between the complete lattices of* **-*closed subsets of $S$ and of $T$.*

*Proof.* (i) and (ii) are immediate. We shall prove the remaining assertions from those two, without calling again on the definition, (6.5.2).

If we apply * to both sides of (ii), so that the inclusions are reversed by (i), we get $A^{***} \subseteq A^*$, $B^{***} \subseteq B^*$; but if we put $B^*$ for $A$ and $A^*$ for $B$ in (ii) we get $B^{***} \supseteq B^*$, $A^{***} \supseteq A^*$. Together these inclusions give (iii). To get (iv), note that by (i) applied twice, the operators ** are inclusion-preserving, by (ii) they are increasing, and by applying * to both sides of (iii) we find that they are idempotent. To get (v) note that by (iii) every set $B^*$ respectively $A^*$ is closed, and of course every closed set $X$ has the form $Y^*$ for $Y = X^*$. (vi) now follows from (v), (iii) and (i).  $\square$

If for each $t \in T$ we consider the relation $-Rt$ as a condition satisfied by some elements $s \in S$, then for $A \subseteq S$ we can interpret $A^{**}$ as "the set of elements of $S$ which satisfy all conditions (of this sort) that are satisfied by the elements of $A$". From this interpretation, the fact that ** is a closure operator is intuitively understandable.

**Definition 6.5.3.** *If $S$ and $T$ are sets, then a pair of maps* * : $\mathbf{P}(S) \to \mathbf{P}(T)$ *and* * : $\mathbf{P}(T) \to \mathbf{P}(S)$ *satisfying conditions* (i) *and* (ii) *of Lemma 6.5.1* (*and hence the consequences* (iii)–(vi)) *is called a* Galois connection *between the sets $S$ and $T$.*

**Exercise 6.5:1.** Show that every Galois connection between sets $S$ and $T$ arises from a relation $R$ as in Lemma 6.5.1, and that this relation $R$ is in fact unique.

Thus, a Galois connection on a pair of sets $S$, $T$ can be characterized either abstractly, by Definition 6.5.3, or as a structure arising from some relation $R \subseteq S \times T$. In all naturally occurring cases that I know of, the relation $R$ is what we start with, and the Galois connection is obtained from it. On the other hand, the characterization as in Definition 6.5.3 has the advantage that it can be generalized by replacing $\mathbf{P}(S)$ and $\mathbf{P}(T)$ by other partially ordered sets, though we shall not look at this generalization here, except in the second part of the next exercise.

Here is another order-theoretic characterization of Galois connections:

**Exercise 6.5:2.** If $S$ and $T$ are sets, show that a pair of maps * : $\mathbf{P}(S) \to \mathbf{P}(T)$, * : $\mathbf{P}(T) \to \mathbf{P}(S)$ is a Galois connection if and only if for $X \subseteq S$, $Y \subseteq T$, one has

$$X \subseteq Y^* \iff Y \subseteq X^*.$$

More generally, you can show that given two partially ordered sets $(|P|, \leq)$ and $(|Q|, \leq)$, and a pair of maps $*: |P| \to |Q|$, $*: |Q| \to |P|$, these maps will satisfy conditions (i)–(ii) of Lemma 6.5.1 if and only if they satisfy the above condition (with "$\leq$" in place of "$\subseteq$" throughout).

**Exercise 6.5:3.** Show that for every closure operator cl on a set $S$, there exists a set $T$ and a relation $R \subseteq S \times T$ such that the closure operator ** on $S$ induced by $R$ is cl.

Can one in fact take for $T$ any set given with any closure operator whose lattice of closed subsets is antiisomorphic to the lattice of cl-closed subsets of $S$?

A Galois connection between two sets $S$ and $T$ becomes particularly valuable when the **-closed subsets have characterizations of independent interest. Let us give a number of examples, beginning with the one that motivated our definition. (The reader should not worry if he or she is not familiar with all the concepts and results mentioned in these examples.) In describing these examples, I will sometimes, for brevity, ignore the distinction between algebraic objects and their underlying sets.

*Example 6.5.4.* Take for $S$ the underlying set of a field $F$, and for $T$ the underlying set of a finite group $G$ of automorphisms of $F$. For $a \in S$ and $g \in T$ let $a\,R\,g$ mean that $g$ fixes $a$, that is, $g(a) = a$. If we write $K \subseteq F$ for the subfield $G^*$, then, as noted earlier, the Fundamental Theorem of Galois Theory tells us that the closed subsets of $F$ are precisely the subfields of $F$ containing $K$, while the closed subsets of $G$ are all its subgroups. One finds that properties of the field extension $F/K$ are closely related to properties of the group $G$, and can be studied with the help of group theory ([32, Chapter V], [34, Chapter VI]). These further relations between group structure and field structure are not, of course, part of the general theory of Galois connections. That theory gives the underpinnings, over which these further results are built.

*Example 6.5.5.* Let us take for $S$ a vector space over a field $K$, for $T$ the dual space $\mathrm{Hom}_K(S, K)$, and let us take $x\,R\,f$ to mean $f(x) = 0$. In this case, one finds that the closed subsets of $S$ are all its vector subspaces, while those of $T$ are the vector subspaces that are closed in a certain topology. In the finite-dimensional case, this topology is discrete, and so the closed subsets of $T$ are all its subspaces. The resulting correspondence between subspaces of a finite-dimensional vector space and of its dual space is a basic tool which is taught (or should be!) in undergraduate linear algebra. Some details of the infinite-dimensional case are developed in an exercise below.

*Example 6.5.6.* A superficially similar example: Let $S = \mathbb{C}^n$ (complex $n$-space) and $T = \mathbb{Q}[x_0, \ldots, x_{n-1}]$, the polynomial ring in $n$ indeterminates over the rationals, and let $(a_0, \ldots, a_{n-1})R\,f$ mean $f(a_0, \ldots, a_{n-1}) = 0$. This case is the starting-point for classical algebraic geometry, and still the underlying inspiration for much of the modern theory. The closed subsets of $\mathbb{C}^n$ are the solution-sets of systems of polynomial equations, while the Nullstellensatz says that the closed subsets of $T = \mathbb{Q}[x_0, \ldots, x_{n-1}]$ are the "radical ideals".

*Example 6.5.7.* Let $S$ be a finite-dimensional Euclidean space $\mathbb{R}^n$, with inner product $\langle x, y \rangle$, let $T = S \times \mathbb{R}$, and for $x \in S$, $(y, a) \in T$, define $x\,R(y, a)$ to mean $\langle x, y \rangle \leq a$. Then the closed subsets of $S$ turn out to be the closed *convex* sets.

A variant: let us restrict $a$ above to the value 1. Then dropping this constant "1" from our notation, $T$ becomes $S$, and $x\,R\,y$ becomes the condition $\langle x, y \rangle \leq 1$. We thus have a Galois connection between $S$ and itself, under which the closed subsets on each side turn out to be the closed convex subsets containing 0. For instance, in $S = \mathbb{R}^3$, we find that the dual of a cube centered at the origin is a regular octahedron centered at the origin. The regular dodecahedron and icosahedron are similarly dual to one another.

*Example 6.5.8.* Let $S = T$ be a group, semigroup, or ring, and for elements $s$ and $t$ of that object, let $s\,R\,t$ denote the commutativity relation $st = ts$. Then for every subset $X$ of $S$, the set $X^*$ will be a subring, subgroup, or subsemigroup, called the "centralizer" or the "commutant" of $X$, and $X^{**}$ is called the *bicommutant* of $X$.

In particular, if $S = T = $ the ring of endomorphisms of an abelian group $M$ (or more generally, the $k$-algebra of endomorphisms of a $k$-module $M$, for some commutative ring $k$), and if, for $X$ a subring of $S$, we regard $M$ as an $X$-module, then $X^*$ is the ring (respectively, $k$-algebra) of *$X$-module endomorphisms of $M$.*

*Example 6.5.9.* Let $S$ be a set of mathematical objects, $T$ a set of propositions about an object of this sort, and $s\,R\,t$ the relation "the object $s$ satisfies the proposition $t$"; in logician's notation, $s \models t$. Then the closed subsets of $S$ are those sets of objects definable by sets of propositions from $T$, what model theorists call *axiomatic classes*, while the closed subsets of $T$ are what they call *theories*. The theory $B^{**}$ generated by a set $B$ of propositions consists of those members of $T$ that are *consequences* of the propositions in $B$, in the sense that they hold in all members of $S$ satisfying the latter.

(Actually, in the naturally occurring cases of this example, $S$ is often a proper class rather than a *set* of mathematical objects; e.g., the class of all groups. We will see how to deal comfortably with such situations in the next chapter.)

There are cases where it is preferable to use symbols other than "$*$" for the operators of a Galois connection. In Example 6.5.5, it is usual to write the set obtained from a set $A$ as $\mathrm{Ann}(A)$ or $A^\circ$ or $A^\perp$ (the *annihilator* or *null space* of $A$) because "$*$" is commonly used for the dual space. More seriously, whenever $S = T$ but $R$ is not a symmetric relation on $S$, the two constructions $\{s' \mid (\forall s \in A)\ s'\,R\,s\}$ and $\{s' \mid (\forall s \in A)\ s\,R\,s'\}$ will be distinct, so one must denote them by different symbols, such as $A_*$ and $A^*$. An example of such a case is

## Exercise 6.5:4.

(i)  If $S = T = \mathbb{Q}$, the set of rational numbers, and $R$ is the relation $\leq$, characterize the two systems of closed subsets of $\mathbb{Q}$. Describe in as simple a way as possible the structure of the lattices of closed sets.

(ii)  Same question with "$<$" in place of "$\leq$".

## Exercise 6.5:5.

(i)   Let $X$ be a set, $S = T = \mathbf{P}(X)$, the set of all subsets of $X$, and let $R$ be the relation of having nonempty intersection. Since this is a symmetric relation, the two closure operators it induces are the same. Show that this operator ** takes $A \subseteq \mathbf{P}(X)$ to the set of those subsets of $X$ that contain a member of $A$.

Deduce that under this Galois connection, the closed sets which are completely join-irreducible (cannot be written as a finite or infinite join of strictly smaller closed sets) are in natural one-to-one correspondence with the elements of $\mathbf{P}(X)$, and that the general closed sets are precisely the unions of such closed sets.

(ii)  Suppose $X$ is a topological space, $S = T = \{\text{open subsets of } X\}$, and again let $R$ be the relation of having nonempty intersection. Can you characterize the resulting closure operator in this case? Can you get analogs of the remaining statements of part (i)?

The next exercise gives, as promised, some details on the infinite-dimensional case of Example 6.5.5. The one following it is related to Example 6.5.8.

**Exercise 6.5:6.** Let $K$ be a field, $S$ a $K$-vector-space, and $T$ its dual space.

(i)   Show that the subsets of $S$ closed under the Galois connection of Example 6.5.5 are indeed all the vector subspaces of $S$.

To characterize the subsets of $T$ closed under this connection, let us, for each $s \in S$ and $c \in K$, define $U_{s,c} = \{t \in T \mid t(s) = c\}$, and topologize $T$ by making the $U_{s,c}$ a subbasis of open sets.

(ii)  Show that the resulting topology is the weakest such that for each $s \in S$, the evaluation map $t \mapsto t(s)$ is a continuous map from $T$ to the discrete topological space $K$.

(iii)  Show that the subsets of $T$ closed under the Galois connection described above are the vector subspaces of $T$ closed in the above topology.

(There is an elegant characterization of the class of topological vector spaces that arise in this way. They are called *linearly compact* vector spaces. See [108, Chapter II, 27.6 and 32.1], or for a summary, [3, first half of § 24].)

**Exercise 6.5:7.** Let $M$ be the underlying abelian group of the polynomial ring $\mathbb{Q}[t]$ in one indeterminate $t$, let $x\colon M \to M$ be the abelian group endomorphism given by *multiplication* by $t$, and $d\colon M \to M$ the endomorphism given by *differentiation* with respect to $t$. Find the commutant and bicommutant (as defined in Example 6.5.8) of each of the following subrings of $\mathrm{End}(M)$ :

(i)    $\mathbb{Z}[x]$.

(ii)   $\mathbb{Z}[x^2, x^3]$.

(iii)  $\mathbb{Z}[d]$.

(iv)  $\mathbb{Z}\langle x, d\rangle$ (the ring generated by $x$ and $d$. Angle brackets are used to indicate generators of not necessarily commutative rings.)

**Exercise 6.5:8.** If $G$ is a group and $X$ a subset of $G$, then

$$\{g \in G \mid (\forall x \in X)\ g\, x = x\, g\}$$

is called the *centralizer* of $X$ in $G$, often denoted $C_G(X)$. This is easily seen to be a subgroup of $G$.

(i)   Show that if $H$ is a subgroup of a group $G$ then the following conditions are equivalent: (a) $H$ is commutative, and is the centralizer of its centralizer. (b) $H$ is the intersection of some nonempty family of maximal commutative subgroups of $G$.

(ii)  Give a result about Galois connections of which the above is a particular case.

    (You may either state and prove in detail the result of (i), and then for (ii) formulate a general result which can clearly be proved the same way, in which case you need not repeat the argument; or do (ii) in detail, then note briefly how to apply your result to get (i).)

    We recall that for a general closure operator on a set $S$, the union of two closed subsets of $S$ is not in general closed; their join in the lattice of closed sets is the *closure* of this union. However, if we consider the Galois connection between a set of objects and a set of propositions, and if these propositions are the sentences in a language that contains the operator $\vee$ ("or"), then the set of objects satisfying the proposition $s \vee t$ will be precisely the union of the set of objects satisfying $s$ and the set satisfying $t$ :

$$\{s \vee t\}^* \;=\; \{s\}^* \cup \{t\}^*.$$

Likewise, if the language contains the operator $\wedge$ ("and"), then

$$\{s \wedge t\}^* \;=\; \{s\}^* \cap \{t\}^*.$$

In fact, the choice of the symbols $\vee$ and $\wedge$ (modifications of $\cup$ and $\cap$) by logicians to represent these operators was probably suggested by these properties of the sets of objects satisfying such relations. (At least, so I thought when I wrote this. But a student told me he had heard a different explanation: that $\vee$ is an abbreviation of Latin *vel* "or", and $\wedge$ was formed by inverting it. If so, $\cup$ and $\cap$ may have been created as modifications of $\vee$ and $\wedge$, and the fact that $\cup$ looks like an abbreviation of "union" may be a coincidence.)

If we look at closed sets of propositions rather than closed sets of objects, these are, of course, ordered in the reverse fashion: The set of propositions implied by a proposition $s \vee t$ is the *intersection* of those implied by $s$ and those implied by $t$, while the set implied by $s \wedge t$ is the *closure of the union* of the sets implied by $s$ and by $t$. Thus the use of the words "and" (which implies something "bigger") and "or" (which suggests a weakening) is based on the proposition-oriented viewpoint, while the choice of symbols $\wedge$ and $\vee$ corresponds to the object viewpoint.

The conflict between these two viewpoints explains the problem students in precalculus courses have when they are asked, say, to describe by inequalities the set of real numbers $x$ satisfying $x^2 \geq 1$. We want the answer "$x \leq -1$ or $x \geq 1$", meaning $\{x \mid x \leq -1 \text{ or } x \geq 1\}$. But they often put "$x \leq -1$ and $x \geq 1$". What they have in mind could be translated as "$\{x \mid x \leq -1\}$ and $\{x \mid x \geq 1\}$". We can hardly tell them that their difficulty arises from the order-reversing nature of the Galois connection between propositions and objects! But the more thoughtful students might be helped if, without going into the formalism, we pointed out that there is a kind of "reverse relation" between statements and the things they refer to: the larger a set of statements, the smaller the set of things satisfying it; the larger a set of things, the smaller the set of statements they all satisfy; so that "*and*" for sets of real numbers translates to "*or*" among formulas defining them.

I point out this "reverse relation" in a handout on set theory and mathematical notation that I give out in my upper division courses [55, in particular, § 2]. Whether it helps, I don't know.

Logicians often write the propositions $(\forall\, x \in X)\, P(x)$ and $(\exists\, x \in X)\, Q(x)$ as $\bigwedge_{x \in X} P(x)$ and $\bigvee_{x \in X} Q(x)$. Here the universal and existential quantifications are being represented as (generally infinite) conjunctions and disjunctions, corresponding to intersections and unions respectively of the classes of models defined by the given families of conditions $P(x)$ and $Q(x)$, as $x$ ranges over $X$.

We have noted that for many naturally arising types of closure operators cl, the closure of a set $X$ can be constructed both "from above" and "from below"—either by taking the intersection of all closed sets containing $X$, or by "building" elements of $\mathrm{cl}(X)$ from elements of $X$ by iterating some procedure in terms of which cl was defined. Closure operators determined by Galois connections are born with a construction "from above": for $X \subseteq S$, $X^{**}$ is the intersection of those sets $\{t\}^*$ $(t \in T)$ which contain $X$. The definition of a Galois connection does not provide any way of constructing this set "from below"; rather, this is a recurring type of mathematical problem for the particular Galois connections of mathematical interest! Typically, given such a Galois connection, one looks for operations that all the sets $\{t\}^*$

$(t \in Y)$ are closed under, and when one suspects one has found enough of these, one seeks to prove that for every $X$, the set $X^{**}$ is the closure of $X$ under these operations. For instance, the fixed set of an automorphism of a field extension $F/K$ is easily seen to contain all elements of $K$ and to be closed under the field operations of $F$; the Fundamental Theorem of Galois Theory says that under appropriate hypotheses, the closed subsets of $F$ are precisely the subsets closed under these operations. For the case of the Galois connection between mathematical objects and propositions, the problem of finding a way to "build up" the closure of a set of propositions is that of finding an adequate set of *rules of inference* for the type of proposition under consideration, while to construct the closure operator on objects is to characterize intrinsically the axiomatic model classes.

We remark that for a relation $R$ on a pair of sets $X$ and $Y$, there is in general no close connection between the Galois connections determined by $R$ and by its negation $\neg R$ (set-theoretically, the complement of $R$ in $X \times Y$). Typically, when one of the two relations is given by equalities (e.g., the relation $g(a) = a$ on the Galois connection between automorphisms and elements of a field), that relation tends to yield a Galois connection of more mathematical interest than the connection determined by its complement (in the above case, the relation $g(a) \neq a$).

The definition of Galois connection is, unfortunately, seldom presented in courses, and many mathematicians who discover examples of it have not heard of the general concept. Of course, Lemma 6.5.1 is a set of easy observations which can be verified in any particular case without referring to a general result. But it is useful to have the general concept as a guide; and once one has proved Lemma 6.5.1, one can skip those trivial verifications from then on.

# Chapter 7
# Categories and Functors

## 7.1. What is a category?

Let us lead up to the concept of category by first recalling the motivations for some more familiar mathematical concepts:

(a) *Groups.* The definition of a group is motivated by considering the structure on the set $\mathrm{Aut}(X)$ of all automorphisms of a mathematical object $X$. Given $a, b \in \mathrm{Aut}(X)$, the *composite* map $ab$ lies in $\mathrm{Aut}(X)$; for every $a \in \mathrm{Aut}(X)$, its *inverse* $a^{-1}$ is a member of $\mathrm{Aut}(X)$, and, of course, the *identity map* $\mathrm{id}_X$ always belongs to $\mathrm{Aut}(X)$. Thus, $\mathrm{Aut}(X)$ is a set with a binary operation of composition, a unary operation " $^{-1}$ ", and a zeroary operation $\mathrm{id}_X$. When one examines the conditions these operations satisfy, one discovers the associative law, the inverse laws, and the neutral-element laws.

These laws and their consequences turn out to be fundamental to considerations involving automorphisms, so one makes a general definition: A 4-tuple $G = (|G|, \cdot, {}^{-1}, 1)$, where $|G|$ is a set and $\cdot, {}^{-1}, 1$ are operations on $|G|$ satisfying the above laws, is called a *group*.

Let me point out something which is obvious today, but took getting used to for the first generation to see the above definition: The definition does not say that $G$ actually consists of automorphisms of an object $X$—only that it has certain properties we have abstracted from that context. In fact, systems with these properties are also found to arise in other ways:

The additive structures of the sets of integers, rational numbers, and real numbers form groups.

If $(X, x_0)$ is a topological space with basepoint, the set of homotopy classes of closed curves beginning and ending at $x_0$ forms a group, $\pi_1(X, x_0)$.

© Springer International Publishing Switzerland 2015
G.M. Bergman, *An Invitation to General Algebra and Universal Constructions*, Universitext, DOI 10.1007/978-3-319-11478-1_7

And there are groups that are familiar, not because of a particular way they occur, but because of their importance as basic components in the study of other groups. The finite cyclic groups $Z_n$ are the simplest examples.

Despite our abstract definition, and the existence of groups arising in these different ways, the original motivation of the group concept should not be forgotten. A natural question is: *Which* abstract groups can be represented *concretely*, that is, are isomorphic to a family of permutations of a set $X$ under the operations of composition, inverse map, and identity permutation? As we learn in undergraduate algebra, the answer is that *every* group has this property (Cayley's Theorem). Let us rederive the well-known proof.

The idea is to use the simplest nontrivial construction of a $G$-set $X$ : Introduce a single generating element $x \in X$, and let all the elements $g\,x$ ($g \in |G|$) be distinct. Formally we may define $X$ to be the set of symbols "$g\,x$", where $x$ is a fixed symbol and $g$ ranges over $|G|$. We let $G$ act on $X$ in the appropriate way to make this a $G$-action, namely by the law

$$h\,(g\,x) \;=\; (h\,g)\,x \quad (g,\, h \in |G|).$$

The permutations of the set $X$ given by the elements of $G$ are seen to form a "concrete" group isomorphic to $G$. One then observes that the symbol "$x$" is irrelevant to the proof. Stripping it away, we get the textbook proof: "Let $G$ act on $|G|$ by left multiplication ..." ([26, p. 62], [29, p. 121], [30, p. 9], [32, p. 90], [35, p. 52]).

(b) *Monoids.* Suppose we consider not just the *automorphisms* of a mathematical object $X$ but all its *endomorphisms*, that is, homomorphisms into itself. The set $\mathrm{End}(X)$ is closed under composition and contains the identity map, but there is no inverse operation. The operations of composition and identity still satisfy associative and neutral-element laws, and one calls any set with a binary operation and a distinguished element 1 satisfying these laws a *monoid.* Like the definition of a group, this definition does *not* require that a monoid actually consist of endomorphisms of an object $X$.

And indeed, there are again examples which arise in other ways than the one which motivated the definition. The nonnegative integers form a monoid under *multiplication* (with 1 as neutral element), and also under the operation max (with 0 as neutral element). Isomorphism classes of (say) finitely generated abelian groups form a monoid under the operation induced by "$\oplus$", or alternatively under the operation induced by "$\otimes$". (One may remove some set-theoretic difficulties from this example by restricting oneself to a set of finitely generated abelian groups with exactly one member from each isomorphism class.)

One has the precise analog of Cayley's Theorem: Every monoid $S$ is isomorphic to a monoid of maps of a set into itself, and this is proved the same way, by letting $S$ act on $|S|$ by left multiplication.

(c) *Partially ordered sets.* Again let $X$ be any mathematical object, and now let us consider the set $\mathrm{Sub}(X)$ of all *subobjects* of $X$.

In general, we do not have a way of defining interesting *operations* on this set. (There are often operations of "least upper bound" and "greatest lower bound", but not always.) However, $\mathrm{Sub}(X)$ is not structureless; one subobject of $X$ may be *contained in* another, and this inclusion relation is seen to satisfy the conditions of *reflexivity*, *antisymmetry* and *transitivity*.

Again we abstract the situation, calling an arbitrary pair $P = (|P|, \leq)$, where $|P|$ is a set, and $\leq$ is a binary relation on $|P|$ satisfying the above three laws, a *partially ordered set*.

Examples of partial orderings arising in other ways than the above "prototypical" one are the relation "$\leq$" on the integers or the real numbers, and the logical relation "$\Longrightarrow$" on a family of inequivalent propositions. Partially ordered sets are also natural models of various hierarchical and genealogical structures in nature, language, and human society.

Given an arbitrary partially ordered set $P$, will $P$ be isomorphic to a "concrete" partially ordered set—a family of subsets of a set $X$, ordered by inclusion? Again, let us try to build such an $X$ in as simple-minded a way as possible. We want to associate to every $p \in |P|$ a subset $\bar{p}$ of a set $X$, so as to duplicate the order relation among elements of $P$. To make sure all these sets are distinct, let us introduce for each $p \in |P|$ an element $x_p \in X$ belonging to $\bar{p}$, and hence necessarily to every $\bar{q}$ with $q \geq p$, but not to any of the other sets $\bar{q}$ ($q \not\geq p$). It turns out that this works—if we define $X$ to be the set of symbols $\{x_p \mid p \in |P|\}$, and if for $p \in |P|$ we set $\bar{p} = \{x_q \mid q \leq p\} \subseteq X$, we find that $\{\bar{p} \mid p \in |P|\}$, under the relation "$\subseteq$", forms a partially ordered set isomorphic to $P$. Again, the symbol "$x$" is really irrelevant, so we can get a simplified construction by taking $X = |P|$ and $\bar{p} = \{q \mid q \leq p\}$ ($p \in |P|$). Thus we have "Cayley's Theorem for partially ordered sets".

(d) *"Bimonoids."* Let us go back to the idea that led to the definition of a monoid, but make a small change. Suppose that $X$ and $Y$ are two mathematical objects of the same sort (two sets, two rings, etc.), and we consider the family of all homomorphisms among them. What structure does this system have?

First, it is a system of four sets:

$$\mathrm{Hom}(X, X), \qquad \mathrm{Hom}(X, Y), \qquad \mathrm{Hom}(Y, X), \qquad \mathrm{Hom}(Y, Y).$$

Elements of certain of these sets can be composed with elements of others, giving us *eight* composition maps:

$$\mu_{XXX} : \mathrm{Hom}(X, X) \times \mathrm{Hom}(X, X) \to \mathrm{Hom}(X, X),$$
$$\mu_{XXY} : \mathrm{Hom}(X, Y) \times \mathrm{Hom}(X, X) \to \mathrm{Hom}(X, Y),$$

$$\cdot \qquad \cdot \qquad \cdot$$

$$\mu_{YYY} : \mathrm{Hom}(Y, Y) \times \mathrm{Hom}(Y, Y) \to \mathrm{Hom}(Y, Y).$$

(There is no composition on the remaining eight pairs, e.g., $\text{Hom}(X, Y) \times \text{Hom}(X, Y)$.)

These composition operations are associative—we have 16 associative laws; namely, for every 4-tuple $(Z_0, Z_1, Z_2, Z_3)$ of objects from $\{X, Y\}$ (e.g., $(Y, Y, X, Y)$) we get the law

(7.1.1)                                       $(a\,b)\,c \;=\; a\,(b\,c)$

for maps:

$$Z_0 \xrightarrow{\; c \;} Z_1 \xrightarrow{\; b \;} Z_2 \xrightarrow{\; a \;} Z_3.$$

(We could write (7.1.1) more precisely by specifying the four $\mu$'s involved.) We also have two neutral elements, $\text{id}_X \in \text{Hom}(X, X)$ and $\text{id}_Y \in \text{Hom}(Y, Y)$, satisfying eight neutral-element laws, which you can write down.

Cumbersome though this description is, it is clear that we have here a fairly natural mathematical structure, and we might abstract these conditions by defining a *bimonoid* to be any system of sets and operations

$$S \;=\; \left( (|S|_{ij})_{i,\,j \in \{0,1\}},\ (\mu_{ijk})_{i,\,j,\,k \in \{0,1\}},\ (1_i)_{i \in \{0,1\}} \right)$$

such that the $|S|_{ij}$ are sets, the $\mu_{ijk}$ are maps

$$\mu_{ijk}\colon\; |S|_{jk} \times |S|_{ij} \longrightarrow |S|_{ik},$$

satisfying associative laws $(ab)c = a(bc)$ on 3-tuples $(a, b, c) \in |S|_{jk} \times |S|_{ij} \times |S|_{hi}$ for all $h, i, j, k \in \{0,1\}$, and such that the $1_i$ are elements of $|S|_{ii}$ ($i \in \{0,1\}$) satisfying

$$1_j\, a \;=\; a \;=\; a\, 1_i \quad (a \in |S|_{ij}).$$

Again, these objects can arise in ways other than the one just indicated:

We can get an analog of the "$\pi_1$" construction for groups: If $X$ is a topological space and $x_0$, $x_1$ are two points of $X$, then the set of homotopy classes of paths in $X$ whose initial and final points both lie in $\{x_0, x_1\}$ is easily seen to form a "bimonoid" which we might call $\pi_1(X; x_0, x_1)$.

Readers familiar with the ring-theoretic concept of a *Morita context* $(R, S;\ {}_R P_S,\ {}_S Q_R;\ \tau, \tau')$ will see that it also has this form: The underlying sets of the rings $R$ and $S$ play the roles of $|S|_{00}$ and $|S|_{11}$, the underlying sets of the bimodules $P$ and $Q$ give $|S|_{01}$ and $|S|_{10}$, and the required eight multiplication maps are given by the internal multiplication maps of $R$ and $S$, the bimodule structures of $P$ and $Q$, and the bilinear maps $\tau\colon P \times Q \to R$, and $\tau'\colon Q \times P \to S$.

Finally, if $K$ is a field and for any two integers $i$ and $j$ we write $M_{ij}(K)$ for the set of $i \times j$ matrices over $K$, then for any $m$ and $n$, the four systems of matrices $M_{mm}(K)$, $M_{nm}(K)$, $M_{mn}(K)$, $M_{nn}(K)$, form a "bimonoid" under matrix multiplication. (The astute reader will notice that this is really a disguised case of "two mathematical objects and maps among them", since matrix multiplication is designed precisely to encode composition of linear maps between vector spaces $K^m$ and $K^n$. And the ring-theorist will note that this matrix example is a Morita context.)

Is there a "Cayley's Theorem for bimonoids", saying that any bimonoid $S$ is isomorphic to a subbimonoid of the bimonoid of all maps between two sets $X$ and $Y$? Following the models of the preceding cases, our approach should be to introduce a small number of elements in $X$ and/or $Y$, and use them to "generate" the rest of $X$ and $Y$ under the action of elements of $S$. Will it suffice to introduce a single generator $x \in X$, and let $X$ and $Y$ consist of elements obtained from $x$ by application of the elements of the $|S|_{0j}$? In particular, this would mean taking for $Y$ the set $\{t\,x \mid t \in |S|_{01}\}$. For some bimonoids $S$ this will work; but in general it will not. For example, one can define a bimonoid $S$ by taking any two monoids for $|S|_{00}$ and $|S|_{11}$, and using the empty set for both $|S|_{01}$ and $|S|_{10}$. For such an $S$, the above construction gives empty $Y$, though if the monoid $|S|_{11}$ is nontrivial it cannot be represented faithfully by an action on the empty set. Likewise, it will not suffice to take *only* a generator in $Y$.

Let us, therefore, introduce as generators one element $x \in X$ and one element $y \in Y$, and let $X$ be the set of all symbols of either of the forms $s\,x$ or $t\,y$ with $s \in |S|_{00}$, $t \in |S|_{10}$, and $Y$ the set of symbols $u\,x$ or $v\,y$ with $u \in |S|_{01}$, $v \in |S|_{11}$. If we let $S$ "act on" this pair of sets by defining

$$a\,(b\,z) = (a\,b)\,z,$$

whenever $z \in \{x, y\}$, and $a$ and $b$ are members of sets $|S|_{ij}$ such that these symbolic combinations should be meaningful, then we find that this yields an embedding of $S$ in the bimonoid of all maps between $X$ and $Y$, as desired. The interested reader can work out the details.

(e) *Categories.* We could go on in the same vein, looking at maps among three, four, etc., mathematical objects, and define "trimonoids", "quadri-monoids" etc., with larger and larger collections of operations and identities.

But clearly it makes more sense to treat these as cases of one general concept! Let us now, therefore, try to abstract the algebraic structure we find when we look at an arbitrary *family* $\mathbf{X}$ of mathematical objects and the homomorphisms among them.

In the above development of "bimonoids", the index set $\{0, 1\}$ that ran through our considerations was the same for all bimonoids. But in the general situation, the corresponding index set must be specified as part of the object. This is the first component of the 4-tuple described in the next definition.

**Definition 7.1.2** (Provisional). *A category will mean a 4-tuple*

$$\mathbf{C} = (\mathrm{Ob}(\mathbf{C}),\ \mathrm{Ar}(\mathbf{C}),\ \mu(\mathbf{C}),\ \mathrm{id}(\mathbf{C})),$$

*where* $\mathrm{Ob}(\mathbf{C})$ *is any collection of elements,* $\mathrm{Ar}(\mathbf{C})$ *is a family of sets* $\mathbf{C}(X, Y)$ *indexed by the pairs of elements of* $\mathrm{Ob}(\mathbf{C})$:

$$\mathrm{Ar}(\mathbf{C}) = (\mathbf{C}(X, Y))_{X,\,Y \in \mathrm{Ob}(\mathbf{C})},$$

$\mu(\mathbf{C})$ *is a family of operations*

$$\mu(\mathbf{C}) = (\mu_{XYZ})_{X, Y, Z \in \mathrm{Ob}(\mathbf{C})}$$
$$\mu_{XYZ} \colon \ \mathbf{C}(Y, Z) \times \mathbf{C}(X, Y) \to \mathbf{C}(X, Z),$$

*and* $\mathrm{id}(\mathbf{C})$ *is a family of elements*

$$\mathrm{id}(\mathbf{C}) = (\mathrm{id}_X)_{X \in \mathrm{Ob}(\mathbf{C})}$$
$$\mathrm{id}_X \in \mathbf{C}(X, X),$$

*such that, using multiplicative notation for the maps* $\mu_{XYZ}$, *the associative identity*

$$a\,(b\,c) = (a\,b)\,c$$

*is satisfied for all elements*

$$a \in \mathbf{C}(Y, Z), \quad b \in \mathbf{C}(X, Y), \quad c \in \mathbf{C}(W, X) \quad (W, X, Y, Z \in \mathrm{Ob}(\mathbf{C})),$$

*and the identity laws*

$$a\,\mathrm{id}_X = a = \mathrm{id}_Y\,a$$

*are satisfied for all* $a \in \mathbf{C}(X, Y)$ $(X, Y \in \mathrm{Ob}(\mathbf{C}))$.

The above definition is labeled "provisional" because it avoids the question of what we mean by a "*collection* of elements $\mathrm{Ob}(\mathbf{C})$". If we hope to be able to deal with categories within set theory, we should require $\mathrm{Ob}(\mathbf{C})$ to be a *set*. Yet we will find that the most useful applications of category theory are to cases where $\mathrm{Ob}(\mathbf{C})$ consists of *all* algebraic objects of a certain type (e.g., all groups), which calls for larger "collections". We will deal with this dilemma in §7.4. In the next section, where we will give examples of categories, we will interpret "collection" broadly or narrowly as the example requires.

I mentioned that the concept of an "abstract group"—a group given as a set of elements with certain operations on them, rather than as a concrete family of permutations of a set—was confusing to people when it was first introduced. The "abstract" concept of a category still causes many people problems—there is a great temptation for beginning students to imagine that the members of $\mathbf{C}(X, Y)$ must be actual *maps* between *sets* $X$ and $Y$.

One reason for this confusion is that the terminology of category theory is set up to closely mimic the situation which motivated the concept. The word "category" is suggestive to begin with; "$\mathrm{Ob}(\mathbf{C})$" stands for "objects of $\mathbf{C}$", and this is what elements of $\mathrm{Ob}(\mathbf{C})$ are called; elements $f \in \mathbf{C}(X, Y)$ are called "morphisms" from $X$ to $Y$, the objects $X$ and $Y$ are called the "domain" and "codomain" of $f$, these morphisms are often denoted diagrammatically by arrows, $X \xrightarrow{f} Y$, and objects and morphisms are shown together in the sort of diagrams that are used to represent objects and maps in

# Summary of § 7.1

(Read across rows, referring to headings at top, then compare downwards)

| Consider: | Structure: | Properties: | Abstracted concept: | Other examples | Can be represented by: |
|---|---|---|---|---|---|
| All automorphisms of a mathematical object $X$. | Set with composition, inverse operation, and identity element. | $(a\,b)\,c = a\,(b\,c)$, $a^{-1}a=\mathrm{id}=aa^{-1}$, $a\,\mathrm{id} = a = \mathrm{id}\,a$. | group (same properties, but not assumed to arise as at left) | $(\mathbb{Z}, +, -, 0)$, $\pi_1(X, x_0)$, $Z_n$. | permutations of a set (Cayley's Theorem). |
| All endomorphisms of a mathematical object $X$. | Set with composition and identity element. | $(a\,b)\,c = a\,(b\,c)$, $a\,\mathrm{id} = a = \mathrm{id}\,a$. | monoid (same properties, but not assumed to arise as at left) | $(\mathbb{N}, \cdot, 1)$, $(\mathbb{N}, \max, 0)$, $(\{\text{f. g. ab. gps.}\}, \otimes, \mathbb{Z})$. | maps of a set into itself. |
| All subobjects of a mathematical object $X$. | Set with relation $\subseteq$ | transitive, antisymmetric, reflexive. | partially ordered set (same properties, but not assumed to arise as at left) | $(\mathbb{Z}, \leq)$, $\implies$, genealogies. | subsets of a set, under $\subseteq$. |
| All homomorphisms between two mathematical objects $X$ and $Y$. | Four sets, $|S|_{00}, |S|_{01}, |S|_{10}, |S|_{11}$, with composition maps $|S|_{ij} \to |S|_{ik}$ and identity elements $\mathrm{id}_0$, $\mathrm{id}_1$. | $(a\,b)\,c = a\,(b\,c)$ (when defined); $a\,\mathrm{id}_i=a=\mathrm{id}_j\,a$ $(a \in |S|_{ij})$. | "bimonoid" (same properties, but not assumed to arise as at left) | " $\pi_1(X; x_0, x_1)$ ", Morita contexts, matrices. | maps between two sets. |
| All homomorphisms among a family $\mathbf{X}$ of mathematical objects. | Family of sets $\mathrm{Hom}(X,Y)\,(X,Y \in \mathbf{X})$ with composition maps $\mathrm{Hom}(X_1,X_2) \times \mathrm{Hom}(X_0,X_1) \to \mathrm{Hom}(X_0,X_2)$ and identity elements $\mathrm{id}_X \in \mathrm{Hom}(X,X)$. | $(ab)c = a(bc)$ for $X_0 \overset{c}{\to} X_1 \overset{b}{\to} X_2 \overset{a}{\to} X_3$; $a\,\mathrm{id}_X=a=\mathrm{id}_Y\,a$ for $X \overset{a}{\to} Y$. | category (same properties, but not assumed to arise as at left) | coming up, in § 7.2. | family of sets and maps among them (if $\mathrm{Ob}(\mathbf{C})$ is a set). |

other areas of mathematics. In place of $\mathbf{C}(X, Y)$, the notation $\mathrm{Hom}(X, Y)$ is very common. And $\mu_{XYZ}(f, g)$ is generally written $fg$ or $f \cdot g$ or $f \circ g$, and so looks just like a composite of functions.

So I urge you to note carefully the distinction between the situation that motivated our definition, and the definition itself. Within that definition, the collection $\mathrm{Ob}(\mathbf{C})$ is simply an "index set" for the families of elements on which the composition operation is defined. Hence in discussing an abstract category $\mathbf{C}$, one cannot give arguments based on considering "an *element* of the object $X$", "the *image* of the morphism $a$", etc.; any more than in considering a member $g$ of an abstract group $G$ one can refer to such concepts as "the set of points left fixed by $g$". (However, the latter concept is meaningful for concrete groups of permutations, and the former concepts are likewise meaningful for "concrete categories", a concept we will define in §7.5.)

Of course, the motivating situation should not be forgotten, and a natural question is: Is every category isomorphic to a system of maps among some sets? We can give a qualified affirmative answer. The complete answer depends on the set-theoretic matters that we have postponed to §7.4, but if $\mathrm{Ob}(\mathbf{C})$ is actually a *set*, then we can indeed construct sets $(\overline{X})_{X \in \mathrm{Ob}(\mathbf{C})}$, and set maps among these, including the identity map of each of these sets, which form, under composition of maps, a category isomorphic to $\mathbf{C}$. The proof is the analog of the one we sketched for "bimonoids".

**Exercise 7.1:1.** Write out the argument indicated above—"Cayley's Theorem" for a category with only a *set* of objects.

Incidentally, we will now discard the term "bimonoid", since the structure it described was, up to notational adjustment, simply a category having for object-set the two-element set $\{0, 1\}$.

## 7.2. Examples of categories

To describe a category, one should, strictly, specify the class of *objects*, the *morphism-set* associated with any pair of objects, the *composition* operation on morphisms, and the *identity morphism* of each object. In practice, some of this structure is usually clear from context. When one is dealing with the prototype situation—a family of mathematical objects and all homomorphisms among them—the whole structure is usually clear once the class of objects is named. In other cases the morphism-sets must be specified as well; once this is done the intended composition operation is usually (though not always) obvious. As to the identity elements, these are uniquely determined by the remaining structure (just as in groups or monoids), so the only task is to verify that they exist, which is usually easy.

Categories consisting of families of mathematical objects and the homomorphisms among them are generally denoted by boldface or script names for the type of object (often abbreviated. The particular abbreviations may vary from author to author.) Some important examples are:

**Set**, the category of all sets and set maps among them. (Another symbol commonly used for this category is **Ens**, from the French *ensemble*.)

**Group**, the category whose objects are all groups, and whose morphisms are the group homomorphisms; and similarly **Ab**, the category of abelian groups.

**Monoid**, **Semigroup**, **AbMonoid** and **AbSemigroup**, the categories of monoids, semigroups, abelian monoids, and abelian semigroups.

**Ring**[1], and **CommRing**[1], the categories of associative, respectively associative commutative, rings with unity. (One can denote by the same symbols without a superscript "[1]" the corresponding categories of nonunital rings—i.e., 5-tuples $R = (|R|, +, \cdot, -, 0)$, where $|R|$ need not contain an element 1 satisfying the neutral law for multiplication, and where, even if rings happen to possess such elements, morphisms are not required to respect them. But in these notes we will not refer to nonunital rings often enough to need to fix names for these categories.)

If $R$ is an associative unital ring, we will write the category of left $R$-modules $R$-**Mod** and the category of right $R$-modules **Mod**-$R$. (Other common notations are $_R$**Mod** and **Mod**$_R$.) Similarly, for $G$ a group, the category of (left) $G$-sets will be written $G$-**Set**; here the morphisms are the set maps respecting the actions of all elements of $G$.

**Top** denotes the category of all topological spaces and *continuous maps* among them. Topologists often find it useful to work with topological spaces with basepoint, $(X, x_0)$, so we also define the category **Top**$^{\mathrm{pt}}$ of *pointed* topological spaces, the objects of which are such pairs $(X, x_0)$, and the morphisms of which are the continuous maps that send basepoint to basepoint. Much of topology is done under the assumption that the space is Hausdorff; thus one considers the subcategory **HausTop** of **Top** whose objects are the Hausdorff spaces.

We shall write **POSet** for the category of partially ordered sets, with isotone maps for morphisms. If we want to allow only strict isotone maps, i.e., maps preserving the relation "$<$", we can call the resulting category **POSet**$_<$.

We have mentioned that our concept of "bimonoid" is a special case of the concept of category. Let us formalize this idea. The definition of a category requires specification of the object-set, whereas for bimonoids the implicit object-set was always $\{0, 1\}$. So given a bimonoid $S = ((|S|_{ij}), (\mu_{ijk}), (1_i))$, to translate it to a category $\mathbf{C}$, we throw in a formal first component $\mathrm{Ob}(\mathbf{C}) = \{0, 1\}$. We can then define $\mathbf{C}(i, j) = |S|_{ij}$, getting the category $(\{0, 1\}, (|S|_{ij}), (\mu_{ijk}), (1_i))$, which we may denote $S_{\mathbf{cat}}$.

This works because the situation from which we abstracted the concept of a bimonoid was a special case of the situation from which we abstracted the concept of a category. Now in fact, the situations from which we abstracted the concepts of *group*, *monoid*, and *partially ordered set* were also special cases of that situation! Can objects of these types similarly be identified with certain kinds of categories?

The objects most similar to bimonoids are the monoids. Since they are modeled after the algebraic structure on the set of endomorphisms of a single algebraic object, let us associate to an arbitrary monoid $S$ a *one*-object category $S_{\mathbf{cat}}$, with object-set $\{0\}$. The only morphism-set to define is $S_{\mathbf{cat}}(0,0)$; we take this to be $|S|$. For the composition map on pairs of elements of $S_{\mathbf{cat}}(0,0)$, we use the composition operation of $S$, and for the identity morphism, the neutral element of $S$.

Conversely, if $\mathbf{C}$ is any category with only one object, $X$, then the unique morphism set $\mathbf{C}(X,X)$, with its identity element, will form a monoid $S$ under the composition operation of $\mathbf{C}$, such that the category $S_{\mathbf{cat}}$ formed as above is isomorphic to our original category $\mathbf{C}$, the only difference being the name of the one object (originally $X$, now $0$). Thus, a category with exactly one object is "essentially" a monoid.

If we start with a group $G$, we can similarly form a category $G_{\mathbf{cat}}$ with just one object, $0$, whose morphisms are the elements of $G$ and whose composition operation is the composition of $G$. We cannot incorporate the inverse operation of $G$ as an operation of the category; in fact, what we are doing is essentially forgetting the inverse operation, i.e., forming from $G$ the monoid $G_{\mathrm{md}}$, and then applying the previous construction; thus $G_{\mathbf{cat}} = (G_{\mathrm{md}})_{\mathbf{cat}}$. We see that via this construction, a *group* is equivalent to a category which has exactly one object, and in which every morphism is invertible.

Note that for $G$ a group, the one member of $\mathrm{Ob}(G_{\mathbf{cat}})$ should not be thought of as the group $G$; intuitively it is a fictitious mathematical object on which $G$ acts. Thus, morphisms in this category from that one object to itself do not correspond to endomorphisms of $G$, as students sometimes think, but to *elements* of $G$. (One can also define a category with one object whose morphisms comprise the monoid of the endomorphisms of $G$; that is the category $\mathrm{End}(G)_{\mathbf{cat}}$; but $G_{\mathbf{cat}}$ is a more elementary construction.)

The case of partially ordered sets is a little different. In the motivating situation, though we started with a single object $X$, we considered a family of objects obtained from it, namely all its subobjects. Although there might exist many maps among these objects, the structure of partially ordered set only reveals a certain subfamily of these: the inclusion maps. (In fact, since a "homomorphism" means a map which respects the kind of structure being considered, and we are considering these objects as subobjects of $X$, one could say that a homomorphism *as subobjects* should mean a set map which respects the way the objects are embedded in $X$, i.e., an inclusion map; so from this point of view, these really are the only relevant maps.) A composite of inclusion maps is an inclusion map, and identity maps are (trivial) inclusions, so the subobjects of $X$ with the inclusion maps among them form a category. In this category there is a morphism from $A$ to $B$ if and only if $A \subseteq B$, and the morphism is then unique, so the partial ordering of the subobjects determines the structure of the category.

If we start with an abstract partially ordered set $P = (|P|, \leq)$, we can construct from it an abstract category $P_{\mathbf{cat}}$ in the way suggested by this concrete prototype: Take $\mathrm{Ob}(P_{\mathbf{cat}}) = |P|$, and for all $A, B \in |P|$, define there to be one morphism from $A$ to $B$ if $A \leq B$ in $P$, none otherwise. What should we take this one morphism to be? This is like asking in our construction of $G_{\mathbf{cat}}$ what to call the one object. The choice doesn't really matter. Since we want to associate to each ordered pair $(A, B)$ with $A \leq B$ in $P$ some element, the easiest choice is to take for that element the pair $(A, B)$ itself. Thus, we can define $P_{\mathbf{cat}}$ to have object-set $|P|$, and for $A, B \in |P|$, take $P_{\mathbf{cat}}(A, B)$ to be the singleton $\{(A, B)\}$ if $A \leq B$, the empty set otherwise. The reader can easily describe the composition operation and identity elements of $P_{\mathbf{cat}}$.

Incidentally, we see that this construction works equally well if $\leq$ is a preordering rather than a partial ordering.

**Exercise 7.2:1.** Let $\mathbf{C}$ be a category.

(i)   Show that $\mathbf{C}$ is isomorphic to $P_{\mathbf{cat}}$ for some partially ordered set $P$ if and only if "there is at most one morphism between any unordered pair of objects"; in the sense that each hom-set $\mathbf{C}(X, Y)$ has cardinality at most 1, and the hom-sets $\mathbf{C}(X, Y)$ and $\mathbf{C}(Y, X)$ do not *both* have cardinality 1 unless $X = Y$.

(ii)   State a similar condition necessary and sufficient for $\mathbf{C}$ to be isomorphic to $P_{\mathbf{cat}}$ for $P$ a preorder. (No proof required.)

We mentioned that some groups, such as the cyclic groups $Z_n$, are of interest as "pieces" in terms of which we look at general groups. Thus, to give an element of order $n$ in a group $G$ is equivalent to displaying an isomorphic copy of $Z_n$ in $G$, and to give an element satisfying $x^n = e$ is equivalent to displaying a homomorphic image of $Z_n$ in $G$. Various simple categories are of interest for essentially the same reason. For instance a *commutative square* $\begin{smallmatrix} \bullet & \to & \bullet \\ \downarrow & & \downarrow \\ \bullet & \to & \bullet \end{smallmatrix}$ of objects and morphisms in a category $\mathbf{C}$ corresponds to an image in $\mathbf{C}$ of a certain category having four objects, which we can name 0, 1, 2 and 3, and, aside from their four identity morphisms, five arrows, as shown below:

Here the diagonal arrow is *both* the composite of the morphisms from 0 to 1 to 3, and the composite of the morphisms from 0 to 2 to 3. This "diagram category" might be conveniently named "$\begin{smallmatrix} \cdot & \to & \cdot \\ \downarrow & & \downarrow \\ \cdot & \to & \cdot \end{smallmatrix}$".

A simpler example is the diagram category $\cdot \rightrightarrows \cdot$ with two objects and only two nonidentity morphisms, which go in the same direction. Copies of this in a category $\mathbf{C}$ correspond to the type of data one starts with in the definitions of *equalizers* and *coequalizers*. Still simpler is $\cdot \longrightarrow \cdot$, which is often called "**2**"; an image of this in a category corresponds to a choice of two objects and one morphism between them. (So the category **2** takes its place in our vocabulary beside the ordinal 2, the Boolean ring 2, the lattice 2, and the partially ordered set 2!) A larger diagram category is

$$\cdot \longrightarrow \cdot \longrightarrow \cdot \longrightarrow \cdot \longrightarrow \cdot \longrightarrow \cdots ,$$

images of which in $\mathbf{C}$ correspond to right-infinite chains of morphisms. The morphisms of this diagram category are the identity morphisms, the arrows shown in the picture, *and* all composites of these arrows, of which there is exactly one from every object to every object to the right of it. Finally, one might denote by $\circlearrowleft$ a category having one object 0, and, aside from the identity morphism of 0, one other morphism $x$, and all its powers, $x^2$, $x^3$, etc. An image of this in a category $\mathbf{C}$ will correspond to a choice of an object and a morphism from this object to itself.

(In the above discussion I have been vague about what I meant by an "image" of one category in another. In §7.5 we shall introduce the category-theoretic concept analogous to that of homomorphism, in terms of which this can be made precise. At this point, for the sake of giving you some broad classes of examples to think about, I have spoken without having the formal definition at hand.)

The various types of examples we have discussed are by no means disjoint. Three of the above "diagram categories" can be recognized as having the form $P_{\mathbf{cat}}$, where $P$ is respectively, a four-element partially ordered set, the partially ordered set 2, and the partially ordered set of nonnegative integers, while the last example is $S_{\mathbf{cat}}$, for $S$ the free monoid on one generator $x$.

Many of the other "nonprototypical" ways in which we saw that groups, etc., arise also have generalizations to categories:

If $R$ is any ring, we see that multiplication of rectangular matrices over $R$ satisfies precisely the laws for composition of morphisms in a category. Thus, we get a category $\mathbf{Mat}_R$ by defining the objects to be the nonnegative integers, the morphism-set $\mathbf{Mat}_R(m, n)$ to be the set of all $n \times m$ matrices over $R$, the composition $\mu$ to be matrix multiplication, and the morphisms $\mathrm{id}_n$ to be the identity matrices $I_n$. This is not very novel, since as we observed before, matrix multiplication is defined to encode composition of linear maps among free $R$-modules. But it is interesting to note that the abstract system of matrices over $R$ is not limited to serving that function; if $M$ is any left $R$-module, one can use $n \times m$ matrices over $R$ to represent operations which carry $m$-tuples of elements of $M$ to $n$-tuples formed from these elements using linear expressions with coefficients in $R$.

This line of thought suggests similar constructions for other sorts of algebraic objects. For instance, we can define a category $\mathbf{C}$ whose objects are again the nonnegative integers, and such that $\mathbf{C}(m, n)$ represents all ways of getting an $n$-tuple of elements of an arbitrary *group* from an $m$-tuple using *group operations*. Precisely, we can define $\mathbf{C}(m, n)$ to be the set of all $n$-tuples of *derived group-theoretic operations* in $m$ variables. The composition maps

$$(7.2.1) \qquad \mathbf{C}(n, p) \times \mathbf{C}(m, n) \longrightarrow \mathbf{C}(m, p)$$

can be described in terms of substitution of derived operations into one another.

Generalizing the construction of the fundamental group of a topological space with basepoint $(X, p)$, one can associate to any topological space $X$ a category $\pi_1(X)$ whose objects are all points of $X$, and where a morphism from $x_0$ to $x_1$ means a homotopy class of paths from one point to the other.

We can also define categories which have familiar mathematical entities for their objects, but put unexpected twists into the definitions of the morphismsets. Recall that in the category **Set**, the morphisms from the set $X$ to the set $Y$ are all functions from $X$ to $Y$. Now formally, a function is a relation $f \subseteq X \times Y$ such that for every $x \in X$ there exists a unique $y \in Y$ such that $(x, y) \in f$. Suppose we drop this restriction, and consider arbitrary relations $R \subseteq X \times Y$. One can compose these using the same formula by which one composes functions: If $R \subseteq X \times Y$ and $S \subseteq Y \times Z$, one defines

$$S \cdot R = \{(x, z) \in X \times Z \mid (\exists y \in Y)\ (x, y) \in R,\ (y, z) \in S\}.$$

This operation of composing relations is associative, and the identity relations satisfy the identity laws; hence one can define a category **RelSet**, whose objects are ordinary sets, but such that **RelSet**$(X, Y)$ is the set of all relations in $X \times Y$.

Algebraic topologists work with topological spaces, but instead of individual maps among them, they are concerned with *homotopy classes* of maps. Thus, they use the category **HtpTop** whose objects are topological spaces, and whose morphisms are such homotopy classes. Composition of continuous maps respects homotopy, allowing one to define the composition operation of this category.

In complex variable theory, one often fixes a point $z$ of the complex plane and considers all analytic functions defined on neighborhoods of $z$. Different functions in this set are defined on different neighborhoods of $z$, so these functions do not all have any domain of definition in common. Further, functions which are the same in a neighborhood of $z$ may not agree on the full intersection of their domains, if this intersection is not connected. For example, the natural logarithm function $\ln(z)$ with value zero at $z = 1$ extends to some connected regions of the plane so as to assume the value $\pi i$ at the point $-1$, and to other such regions so as to assume the value $-\pi i$ at that point. To eliminate distinctions which are not relevant to the behavior of functions in the vicinity of the specified point $z$, one introduces the concept of a *germ of a function* at $z$. This is an equivalence class of functions defined on neighborhoods of $z$, under the relation making two functions equivalent if they agree on some common neighborhood of $z$.

An apparent inconvenience of this concept is that for germs of functions at $z$, one does not have a well-defined operation of composition. For instance, if $f$ and $g$ are germs of analytic functions at $z = 0$, one cannot generally attach a meaning to $g\,f$ unless $f(0) = 0$, because $g$ does not have a well-defined "value" at $f(0)$. (This is the analog of the algebraic problem that given formal power series $f(z) = a_0 + a_1 z + \dots$ and $g(z) = b_0 + b_1 z + \dots$, one cannot in general "substitute $f$ into $g$" to get another formal power series in $z$, unless $a_0 = 0$.) But this ceases to be a problem if we define a category **GermAnal**, whose objects are the points of the complex plane, and where a morphism from $z$ to $w$ means a germ of an analytic function at $z$ whose value at $z$ is $w$. Then for any three points $z_0$, $z_1$, $z_2$, one sees that one does indeed have a well-defined composition operation

$$\mathbf{GermAnal}(z_1,\ z_2) \times \mathbf{GermAnal}(z_0,\ z_1) \longrightarrow \mathbf{GermAnal}(z_0,\ z_2).$$

That is, the partial operation of composition of germs of analytic functions is defined in exactly those cases needed to make these germs the morphisms of a category.

These examples allow endless modification as needed. A topologist may impose the restriction that the topological spaces considered in a given context be Hausdorff, be locally compact, be given with basepoint, etc., and modify the category he or she uses accordingly. The definition of a *germ of a function* is not limited to complex variable theory, so analogs of **GermAnal** can be set up wherever needed. Here is an interesting case:

**Exercise 7.2:2.** If $G$ and $H$ are groups, let us define an *almost-homomorphism* from $G$ to $H$ to mean a homomorphism $f : G_f \to H$, whose domain $G_f$ is a subgroup of *finite index* in $G$. .Given two almost-homomorphisms $f$ and $g$ from $G$ to $H$, with domains $G_f$ and $G_g$, let us write $f \approx g$ if the subgroup $\{x \in |G_f| \cap |G_g| \mid f(x) = g(x)\}$ also has finite index in $G$.

(i) Show that $\approx$ is an equivalence relation on the set of almost-homomorphisms from $G$ to $H$.

(ii) Show how one may define a category **C** whose objects are all groups, and whose morphisms are the equivalence classes of almost-homomorphisms, under $\approx$.

(iii) Describe the endomorphism-monoid $\mathbf{C}(\mathbb{Z}, \mathbb{Z})$, where $\mathbf{C}$ is the category described above, and $\mathbb{Z}$ is the additive group of integers.

We noted earlier that isomorphism classes of abelian groups formed a monoid under $\otimes$. The reader familiar with bimodules and the tensor operation on these might like the following generalization of this monoid to a category.

**Exercise 7.2:3.** Show that one can define a category $\mathbf{C}$ such that $\mathrm{Ob}(\mathbf{C})$ is the class of all rings, for each $R, S \in \mathrm{Ob}(\mathbf{C})$, $\mathbf{C}(R, S)$ is the family of all isomorphism-classes $[P]$ of $(S, R)$-bimodules $P$, and for each $[P] \in \mathbf{C}(S, T)$, $[Q] \in \mathbf{C}(R, S)$, the composite $[P][Q]$ is the isomorphism class of the tensor product, $[P \otimes_S Q]$. (Either ignore the problem that the classes involved in this definition are not sets, or modify the statement in some reasonable way to avoid this problem.)

(If you are familiar with Morita equivalence, you will find that two objects are isomorphic in this category if and only if they are Morita equivalent as rings.)

The following example shows that not every plausible definition works:

**Exercise 7.2:4.** Suppose one attempts to define a category $\mathbf{C}$ by taking all sets for the objects, and letting $\mathbf{C}(X, Y)$ consist of all equivalence classes of set maps $X \to Y$, under the relation that makes $f \approx g$ if $\{x \in X \mid f(x) \neq g(x)\}$ is finite.

(i)   Show that this does not work, i.e., that composition of set maps does not induce a composition operation on equivalence classes of set maps.

On the other hand

(ii)   Find the least equivalence relation $\sim$ on set maps which contains the above equivalence relation $\approx$, and has the property that composition of set maps does induce a composition operation on equivalence classes of set maps under $\sim$.

(Precisely, $\sim$ will be a family of equivalence relations: a relation $\sim_{X, Y}$ on $\mathbf{C}(X, Y)$ for each pair of sets $X$ and $Y$. So what you should show is that among such families of equivalence relations, there is a least $\sim$ such that composition of set maps induces composition operations on the factor sets $\mathbf{C}(X, Y)/\sim_{X, Y}$, and such that $f \approx g \implies f \sim g$ for all $f$ and $g$; and describe these relations $\sim_{X, Y}$ .)

Here is an interesting variant of the construction $S_{\mathbf{cat}}$, for $S$ a monoid. (For an application, see [50].)

**Exercise 7.2:5.** Let $S$ be a monoid, and $X$ an $S$-set. One can define a category whose objects are the elements of $X$, and such that a morphism $x \to y$ $(x, y \in |X|)$ is an element $s \in |S|$ such that $s\,x = y$. However, to help remind us of the intended domain and codomain of each morphism, let us, rather, take the morphisms $x \to y$ to be all 3-tuples $(y, s, x)$ such that $s \in |S|$ and $s\,x = y$. We define composition by $(z, t, y)(y, s, x) = (z, t\,s, x)$; the definition of the identity morphisms should be clear.

(i)   Show that the construction $S_{\mathbf{cat}}$ is a special case of this construction.

(ii)   In general, can one reconstruct the monoid $S$ and the $S$-set $X$ from the structure of the category $X_{\mathbf{cat}}$?

I don't know the answer to the first part of

(iii)   Given a category $\mathbf{C}$, is there a nice necessary and sufficient condition for there to exist a monoid $S$ and an $S$-set $X$ such that $\mathbf{C} \cong X_{\mathbf{cat}}$? For there to exist a *group* $G$ and a $G$-set $X$ such that this isomorphism holds?

## 7.3. Other notations and viewpoints

The language and notation of category theory are still far from uniform. Let me note some of the commonest variations on the conventions I have presented.

I have mentioned that what we are writing $\mathbf{C}(X, Y)$ is often written $\mathrm{Hom}(X, Y)$; this may be made more explicit as $\mathrm{Hom}_{\mathbf{C}}(X, Y)$; there is also the shorter notation $(X, Y)$. Even though we shall not use the notation $\mathrm{Hom}(X, Y)$, we shall often call these sets "hom sets".

More problematically, some authors reverse the order in which the objects are written; i.e., they write the set of morphisms from $X$ to $Y$ as $\mathbf{C}(Y, X)$, $\mathrm{Hom}(Y, X)$, etc. There are advantages to each choice: The order we are using matches the order of words when we speak of going "from $X$ to $Y$", and the use of arrows drawn from left to right, $X \longrightarrow Y$, but has the disadvantage that composition of morphisms $X \to Y \to Z$ must be described as a map $\mathbf{C}(Y, Z) \times \mathbf{C}(X, Y) \to \mathbf{C}(X, Z)$, while under the reversed notation, it goes more nicely, $\mathbf{C}(Z, Y) \times \mathbf{C}(Y, X) \to \mathbf{C}(Z, X)$. A different cure for the same problem is to continue to think of elements of $\mathbf{C}(X, Y)$ as morphisms from $X$ to $Y$ (as we are doing), but reverse the way composition is written, letting the composite of $X \xrightarrow{a} Y \xrightarrow{b} Y$ be denoted $a\,b \in \mathbf{C}(X, Z)$, rather than $b\,a$. However if one does this, then when writing functions on sets, one is more or less forced to abandon the conventional notation $f(x)$, which leads to the usual order of composition, and write $x\,f$ instead.

Note that the above difficulties in category-theoretic notation simply mirror conflicts of notation already existing within mathematics! (Cf. [56].)

The elements of $\mathbf{C}(X, Y)$, which we call "morphisms", are called "arrows" by some. Our notation $\mathrm{Ar}(\mathbf{C})$ for the family of morphism-sets is based on that word; some authors write $\mathrm{Fl}(\mathbf{C})$, based on the French *flèche* (arrow). Colloquially they are also called "maps" from $X$ to $Y$, and I may allow myself to fall into this easy usage at times, hoping that you understand by now that they are *not* maps in the literal sense, i.e., functions.

The identity element in $\mathbf{C}(X, X)$ which we are writing $\mathrm{id}_X$ is also written $I_X$ (like an identity matrix) or $1_X$ (just as the identity element of a group is often written $1$).

The student has probably noticed at some point in his or her study of mathematics the petty but vexing question: If $X$ is a subset of $Y$, is the inclusion map of $X$ into $Y$ the "same" as the identity map of $X$? If we follow the convenient formalization of a function as a set of ordered pairs $(x, f(x))$, then they are indeed the same. But this means that a question like "Is $f$ surjective?" is meaningless; one can only ask whether $f$ is surjective as a map from $X$ to $Y$, as a map from $X$ to $X$, etc. A formalization more in accord with the way we think about these things might be to define a function $f : X \to Y$ as a 3-tuple $(X, Y, |f|)$, where $|f|$ is the set of ordered pairs used in the usual definition. Then $f$ is surjective if and only if the set of second components of members of $|f|$ equals the whole set $Y$. (Since $X$ is determined by $|f|$, our making $X$ a component of the 3-tuple is, strictly, unnecessary; but it seems worth doing for symmetry. Note that if one wants to use a similar notation for general *relations* $|R| \subseteq X \times Y$, then neither $X$ nor $Y$ will be determined by $|R|$, so one needs both of these in the tuple describing the relation. Having both in the tuple describing a function then allows one to continue to regard the functions from $X$ to $Y$ as a subset of the relations between these sets.)

The same problem arises when we abstract our functions in the definition of a category: Can an element be a member of two different morphism-sets, $\mathbf{C}(X, Y)$ and $\mathbf{C}(X', Y')$, with $(X, Y) \neq (X', Y')$? Under our definitions, yes. However, some authors add to the definition of a category the condition that the sets of morphisms between distinct pairs of objects be disjoint.

Let us note what such a condition would entail. In the category **Group**, as an example, a group homomorphism $f : G \to H$ would have to determine not merely its set-theoretic domain and codomain $|G|$ and $|H|$, but the full group structures $G = (|G|, \mu_G, \iota_G, e_G)$ and $H = (|H|, \mu_H, \iota_H, e_H)$. When one thinks about it, this makes good sense, not only from the point of view of category theory but from that of group theory; for without knowing the group structures on $|G|$ and $|H|$, one cannot say whether a map $f : |G| \to |H|$ is a homomorphism, let alone answer such group-theoretic questions as, say, whether its kernel contains all elements of order $2$.

In set theory, when one defines a function as a set of ordered pairs, though its codomain is not uniquely determined, most other things one would want to know about it are; for example, the composite $f g$ of two composable functions can be constructed from those functions. But there is nothing in the definition we have given of a category that says that if $g$ lies in both $\mathbf{C}(X, Y)$ and $\mathbf{C}(X', Y')$, while $f$ lies in both $\mathbf{C}(Y, Z)$ and $\mathbf{C}(Y', Z')$, then the composites $\mu_{XYZ}(f, g)$ and $\mu_{X'Y'Z'}(f, g)$ must be the same; so even the symbol "$f g$" is formally ambiguous.

On the whole, I think it desirable to include in the definition of a category the condition that morphism-sets be disjoint. However, we shall not do so in these notes, largely because it would increase the gap between our category theory and ordinary mathematical usage. So the difficulties mentioned above mean that we have to be careful, understanding for instance that in a given context, we are using $f\,g$ as a shorthand for $\mu_{XYZ}(f,\,g)$, which is the only really unambiguous expression. Note that given any structure which is a category $\mathbf{C}$ under our definition, we can form a new category $\mathbf{C}^{\mathrm{disj}}$ with disjoint morphism-sets, by using the same objects, and letting $\mathbf{C}^{\mathrm{disj}}(X,\,Y)$ consist of all 3-tuples $f = (X,\,Y,\,|f|)$ with $|f| \in \mathbf{C}(X,\,Y)$, and composition operations obtained in the obvious way from those of $\mathbf{C}$.

Authors who require morphism-sets to be disjoint can play some interesting variations on the definition of category. Instead of defining $\mathrm{Ar}(\mathbf{C})$ to be a family of sets, $\mathrm{Ar}(\mathbf{C}) = (\mathbf{C}(X,\,Y))_{X,\,Y \in \mathrm{Ob}(\mathbf{C})}$, they can take it to be a single set (or class), the union of all the $\mathbf{C}(X,\,Y)$'s. To recover *domains* and *codomains* of morphisms, they then add to the definition of a category two operations, dom, cod: $\mathrm{Ar}(\mathbf{C}) \to \mathrm{Ob}(\mathbf{C})$. They can then make composition of morphisms a single map

$$\mu: \{(f,\,g) \in \mathrm{Ar}(\mathbf{C})^2 \mid \mathrm{cod}(g) = \mathrm{dom}(f)\} \longrightarrow \mathrm{Ar}(\mathbf{C}).$$

One can be even more radical and eliminate all reference to objects, as sketched in the next exercise.

**Exercise 7.3:1.**

(i)    Let $\mathbf{C}$ be a category such that distinct ordered pairs of objects $(X,\,Y)$ have disjoint morphism-sets. Let $A = \bigcup_{X,\,Y} \mathbf{C}(X,\,Y)$, and let $\mu$ denote the composition operation in $A$, considered now as a partial map from $A \times A$ to $A$, i.e., a function from a subset of $A \times A$ to $A$. Show that the pair $(A,\,\mu)$ determines $\mathbf{C}$ up to isomorphism.

(ii)   Find conditions on a pair $(A,\,\mu)$, where $A$ is a set and $\mu$ a partial binary operation on $A$, which are necessary and sufficient for it to arise, as above, from a category $\mathbf{C}$ with disjoint morphism sets. (Try to formulate these conditions so that they give a nice self-contained characterization of the sort of structure in question.)

One gets a still nicer structure by combining the above approach with the idea of giving functions specifying the domain and codomain of each morphism. Namely, given a category $\mathbf{C}$ with disjoint morphism-sets, let $A$ be defined as in (i), let dom: $A \to A$ be the map associating to each morphism $f$ the *identity* morphism of its domain, and similarly let cod: $A \to A$ associate to each morphism the identity morphism of its codomain. Since the pair $(A,\,\mu)$ determines $\mathbf{C}$ up to isomorphism, the same will be true, a fortiori, of the 4-tuple $(A,\,\mu,\,\mathrm{dom},\,\mathrm{cod})$.

(iii)  Find simple necessary and sufficient conditions on a 4-tuple $(A,\,\mu,\,\mathrm{dom},\,\mathrm{cod})$ for it to arise as above from a category $\mathbf{C}$ with disjoint morphism-sets.

So one could redefine a category as an ordered pair $(A, \mu)$ or 4-tuple $(A, \mu, \text{dom}, \text{cod})$ satisfying appropriate conditions.

These differences in definition do not make a great difference in how one actually works with categories. If, for instance, one defines a category as a 5-tuple $\mathbf{C} = (\text{Ob}(\mathbf{C}), \text{Ar}(\mathbf{C}), \text{dom}_{\mathbf{C}}, \text{cod}_{\mathbf{C}}, \text{id}_{\mathbf{C}})$, one then immediately makes the definition

$$\mathbf{C}(X, Y) = \{f \in \text{Ar}(\mathbf{C}) \mid (\text{dom}(f) = X) \wedge (\text{cod}(f) = Y)\},$$

and works with these morphism sets as other category-theorists do. (But I will mention one notational consequence of the morphisms-only approach that can be confusing to the uninitiated: the use, by some categorists, of the name of an object as the name for its identity morphism as well.)

Changing the topic from technical details to attitudes, category theory has been seen by some as the new approach that would revolutionize, unify, and absorb all of mathematics; by others as a pointless abstraction whose content is trivial where it is not incomprehensible.

Neither of these characterizations is justified, but each has a grain of truth. The subject matter of essentially every branch of mathematics can be viewed as forming a category (or a family of categories); but this does not say how much value the category-theoretic viewpoint will have for workers in a given area. The actual role of category theory in mathematics is like that of group theory: Groups come up in all fields of mathematics because for every sort of mathematical object, we can look at its symmetries, and generally make use of them. In some situations the contribution of group theory is limited to a few trivial observations, and to providing a language consistent with that used for similar considerations in other fields. In others, deep group-theoretic results are applicable. Finally, group theory is a branch of algebra in its own right, with its own intrinsically interesting questions. All the corresponding observations are true of category theory.

As with the concept of "abstract group" for an earlier generation, many people are troubled by that of an "abstract category", whose "objects" are structureless primitives, not mathematical objects with "underlying sets", so that in particular, one cannot reason by "chasing elements" around diagrams. I think the difficulty is pedagogic. The problem comes from *expecting* to be able to "chase elements". As one learns category theory (or a given branch thereof), one learns the techniques one *can* use, which is, after all, what one needs to do before one can feel at home in any area of mathematics. These include some reasonable approximations of element-chasing when one needs them.

And there is no objection to sometimes using a mental image in which objects are sets and morphisms are certain maps among them, since this is an important class of cases. One must merely bear in mind that, like all the mental images we use to understand mathematics, it is imperfect.

(Of course, strictly speaking, the objects of a category *are* sets, since in ZFC there are no "primitive objects". But the morphisms of a category are not in general maps between these sets, and the set-theoretic structure of these objects is of no more relevance to the concept of category than the set-theoretic structure of "1/2" is to the functional analyst.)

When one thinks of categories as *algebraic entities* themselves, one should note that the item in the definition of a category analogous to the *element* in the definition of a group, monoid etc., is the *morphism*. It is on these that the composition operation, corresponding to the multiplication in a group or monoid, is defined. The *object*-set of **C**, which has no analog in groups or monoids, is essentially an index set, used to classify these elements.

While on the subject of terminology, I will mention one distinction among words (relevant to, but not limited to category theory) which many mathematicians are sloppy about, but which I try to maintain: the distinction between *composite* and *composition*. If $f$ and $g$ are maps of sets, or morphisms in a category, such that $g\,f$ makes sense, it is their *composite*. The operation carrying the pair $(f, g)$ to this element $g\,f$ is *composition*. This is analogous to the distinction between the *sum* of two integers, $a + b$, and the operation of *addition*.

## 7.4. Universes

Let us now confront the problem we postponed, of how we can both encompass category theory within set theory, and have category theory include structures like "the category of sets".

One approach is the following. One formulates the general definition of a category **C** so that Ob(**C**), and even the families **C**$(X, Y)$, are *classes*. One does as much as one can in that context—the resulting animals are called *large* categories. One then goes on to consider those categories in which, at least, every morphism-class **C**$(X, Y)$ is a set, and proves better results about these—they are called *legitimate* categories; note that most of the examples of §7.2 were of this sort. Finally, one considers categories such that both the class Ob(**C**) and the classes **C**$(X, Y)$ are sets. One calls these *small* categories, and in studying them one can use the full power of set theory.

Unfortunately, in conventional set theory one has one's hands tied behind one's back when trying to work with large, or even legitimate categories, for there is no concept of a *collection of classes*. To get around this, one might try extending set theory. One could remove the assumption that every member of a class must be a set, so as to allow certain classes of proper classes, and extend the axioms to apply to such classes as well as sets—and one would find essentially no difficulty—except that what one had been calling "classes" are looking more and more like sets!

So suppose we changed their names, and called our old sets *small sets*, and introduce the term *large set* to describe the things we have been calling classes (our original sets and classes, collections of those, etc. The word "class" itself we would then restore to the function of referring to arbitrary collections of the sets, large and small, in our new set theory. This includes the class of *all* sets, which again would not itself be a member of that theory.) We would assume the axioms of ZFC for arbitrary sets, large or small. Thus, no distinction between large and small sets would appear in our axioms.

Note, however, that not only would those axioms be satisfied by the collection of *all sets*; they would also be satisfied by the subcollection consisting of the *small* sets. It is not hard to see that this is equivalent to saying that the *set* $\mathbb{U}$ of all small sets would have the properties listed in the following definition.

**Definition 7.4.1.** *A* universe *is a set $\mathbb{U}$ satisfying*

(i)   $X \in Y \in \mathbb{U} \implies X \in \mathbb{U}$.

(ii)  $X, Y \in \mathbb{U} \implies \{X, Y\} \in \mathbb{U}$.

(iii) $X \in \mathbb{U} \implies \mathbf{P}(X) \in \mathbb{U}$.

(iv)  $X \in \mathbb{U} \implies (\bigcup_{A \in X} A) \in \mathbb{U}$.

(v)   $\omega \in \mathbb{U}$.

(vi)  *If $X \in \mathbb{U}$ and $f \colon X \to \mathbb{U}$ is a function, then $\{f(x) \mid x \in X\} \in \mathbb{U}$.*

The axioms of ZFC introduced in §5.4 do not guarantee the existence of a set with the above properties. However, if a universe $\mathbb{U}$ is assumed given, the above discussion suggests that we give members of $\mathbb{U}$ the name "small sets", call arbitrary sets "large sets", and, in terms of these kinds of sets, define "small category", "legitimate category" and "large category" as above. We would define "group", "ring", "lattice", "topological space", etc., as we always did; we would further define one of these objects to be "small" if it is a member of $\mathbb{U}$. Although *all* groups would still not form a set, all *small* groups would (though not a small set!) We would then define **Set**, **Group**, etc., to mean the categories of all small sets, all small groups, etc., make the tacit assumption that small objects are all that "ordinary mathematics" cares about, and use large categories to study them! All that needs to be added to ZFC is an axiom saying that there exists a universe $\mathbb{U}$; and such an axiom is considered reasonable by set-theorists. (Note that the operations of power set, direct product, etc., will be the same within the "sub-set-theory" of members of $\mathbb{U}$ as in the total set theory.)

The above is the approach used by Mac Lane [20, pp. 21–24]. However one can go a little further, and, following A. Grothendieck [78, §1.1], use ZFC plus an assumption that seems no less reasonable than the existence of a single universe, and more elegant. Namely,

**Axiom of Universes.** Every set is a member of a universe.

So in particular, under this assumption every universe is a member of a larger universe.

**Convention 7.4.2.** *We shall assume ZFC with the Axiom of Universes from now on.*

Given this set of axioms, we no longer have to think in terms of a two-tiered set theory such that "ordinary" mathematicians work in the lower tier of small sets, and category theorists have access to the higher tier of large sets. Rather, categories, just like other mathematical objects, can exist "at any level". But when we want to use categories to study a given sort of mathematical object, we study the category of these objects at a fixed level (i.e., belonging to a fixed universe $\mathbb{U}$), while that category itself lies at every higher level (in every universe having $\mathbb{U}$ as a member).

Let us make this formal.

**Definition 7.4.3.** *The concept of* category *will be defined as in the provisional Definition 7.1.2, but with the "system" of objects* $\mathrm{Ob}(\mathbf{C})$ *explicitly meaning a set.*

**Definition 7.4.4.** *If* $\mathbb{U}$ *is a universe, a set* $X$ *will be called* $\mathbb{U}$-*small if* $X \in \mathbb{U}$. *A mathematical object (e.g., a group, a ring, a topological space, a category) will be called* $\mathbb{U}$-*small if it is so as a set. In addition, a category* $\mathbf{C}$ *will be called* $\mathbb{U}$-*legitimate if* $\mathrm{Ob}(\mathbf{C}) \subseteq \mathbb{U}$ *and for all* $X, Y \in \mathrm{Ob}(\mathbf{C})$ *one has* $\mathbf{C}(X, Y) \in \mathbb{U}$.

*The categories of* $\mathbb{U}$-*small sets,* $\mathbb{U}$-*small groups, etc., will be denoted* $\mathbf{Set}_{(\mathbb{U})}$, $\mathbf{Group}_{(\mathbb{U})}$, *etc.*

Thus, $\mathbf{Set}_{(\mathbb{U})}$, $\mathbf{Group}_{(\mathbb{U})}$ etc., are $\mathbb{U}$-legitimate categories. Note that this implies that for every universe $\mathbb{U}'$ having $\mathbb{U}$ as a member, they are $\mathbb{U}'$-small categories.

But we don't want to encumber our notation with these subscripts "$_{(\mathbb{U})}$", so we agree to suppress them most of the time:

**Definition 7.4.5.** *When we are not discussing universes, some chosen universe* $\mathbb{U}$ *will be understood to be fixed, and the terms "small" and "legitimate" will mean "*$\mathbb{U}$-*small" and "*$\mathbb{U}$-*legitimate". When we speak of mathematical objects (sets, groups, rings, topological spaces etc.), these will be assumed small if the contrary is not stated. As an exception, "category" will mean legitimate category if the contrary is not stated; and sets used to index the objects of categories will merely be assumed to be subsets of* $\mathbb{U}$ *unless otherwise specified. In particular, symbols such as* $\mathbf{Set}$, $\mathbf{Group}$, $\mathbf{Top}$ *etc., will denote the legitimate categories of all small sets, groups, topological spaces, etc.*

*Large (referring to sets or categories) will mean "not necessarily small or legitimate".*

Thus, the term "large" does not specify any conditions on a set; it simply *removes* the assumption of smallness.

Things now look more or less as they did before, except that we know what we are doing!

The distinctions between small and large objects will come into our considerations from time to time. For instance, when we generalize the construction of free groups and other universal objects as *subobjects of direct products*, we will see that the key condition we need is that we be able to choose an appropriate *small* set of objects over which to take the direct product.

**Exercise 7.4:1.** Assume a universe $\mathbb{U}$ is given.

(Results such as those asked for below will in general be taken for granted in this course; but in doing this exercise, you are asked to show detailed deductions. For this purpose, recall that *functions* are understood to be appropriate sets of ordered pairs, where such pairs are taken to be sets of the form $\{\{X\}, \{X, Y\}\}$ as discussed in the paragraph following (5.4.1); but that aside from that case, $n$-tuples of elements of a set $S$ are defined to be functions $n \to S$.)

(i)   Let $S$ be a large set (a set not necessarily in $\mathbb{U}$), and let $f: S \to \mathbb{U}$ be a function. Show that $f \in \mathbb{U} \iff S \in \mathbb{U}$. (So in particular, any map from a member of $\mathbb{U}$ to $\mathbb{U}$ is a member of $\mathbb{U}$, but no map $\mathbb{U} \to \mathbb{U}$ is.)

(ii)   Show that a large group $G$ is small if and only if $|G|$ is small.

(iii)  Show that a large category $\mathbf{C}$ is small if and only if $\mathrm{Ob}(\mathbf{C})$ is small, and for all $X, Y \in \mathrm{Ob}(\mathbf{C})$, the set $\mathbf{C}(X, Y)$ is small.

Although, as we have seen, one uses *non-small* categories to study *small* mathematical objects of other sorts, the tables can be turned. For instance, we may consider closure operators on classes of small (or legitimate) categories, and the lattice of closed sets of such an operator will then be a *large* lattice.

The next few exercises show some properties of the class of universes. (The Axiom of Universes is, of course, to be assumed if the contrary is not stated.)

**Exercise 7.4:2.**

(i)   Show that the class of universes is not a set.

(ii)   Will this same result hold if we weaken the Axiom of Universes to the statement that there is at least one universe (as in Mac Lane)? What if we use the intermediate statement that there is a universe, and that every universe is a member of a larger universe? (In answering these questions, you may assume that there *exists* a model of set theory satisfying the full Axiom of Universes.)

**Exercise 7.4:3.** Let us, one the one hand, recursively define the *rank* of a set by the condition that $\mathrm{rank}(X)$ is the least ordinal greater than all the ordinals $\mathrm{rank}(Y)$ for $Y \in X$, and on the other hand, define the *hereditary cardinality* of a set by the condition that $\mathrm{her.card}(X)$ is the least cardinal that is both $\geq \mathrm{card}(X)$ and $\geq \mathrm{her.card}(Y)$ for all $Y \in X$.

(i)    Explain why we can make these definitions. (Cf. Exercise 5.4:1.)

(ii)   Show that for every universe $\mathbb{U}$ there exists a cardinal $\alpha$ such that $\mathbb{U}$ consists of all sets of hereditary cardinality $< \alpha$, and/or show that for every universe $\mathbb{U}$ there exists a cardinal $\alpha$ such that $\mathbb{U}$ consists of all sets of rank $< \alpha$.

(iii)  Obtain bounds for the hereditary cardinality of a set in terms of its rank, and vice versa, and if you only did one part of the preceding point, deduce the other part.

(iv)   Characterize the cardinals $\alpha$ which determine universes as in (ii).

Do the arguments you have used require the Axiom of Universes?

(Incidentally, the term "rank" is used as above by set theorists, who write $V_\alpha$ for the class of sets of rank $< \alpha$; but my use of "hereditary cardinality" above is a modification of their usage, which speaks of sets as being "*hereditarily* of cardinality $< \alpha$". The class of sets with this property is denoted $H_\alpha$.)

**Exercise 7.4:4.** Show that if $\mathbb{U} \neq \mathbb{U}'$ are universes, then either $\mathbb{U} \in \mathbb{U}'$ or $\mathbb{U}' \in \mathbb{U}$. Deduce that the relation "$\in$ or $=$" is a well-ordering on the class of universes. (You may wish to use some results from the preceding exercise.)

**Exercise 7.4:5.** Suppose that we drop from our axioms for set theory the Axiom of Infinity, and in our definition of "universe" replace the condition that a universe contain $\omega$ by the condition that it contain $\emptyset$. Show that under the new axiom-system, one can recover the Axiom of Infinity using the Axiom of Universes. Show that all but one of the sets which are "universes" under the new definition will be universes under our existing definition, and characterize the one exception.

**Exercise 7.4:6.** In Exercise 5.5:15 we found that "most" infinite cardinals were regular, namely, that all singular cardinals were limit cardinals; but we also saw that among limit cardinals, regular cardinals were rare, and we found no example but $\aleph_0$. Show now that the cardinality of any *universe* is a regular limit cardinal.

Remarks: Set-theorists call a regular limit cardinal a *weakly inaccessible cardinal*, because it cannot be "reached" from lower cardinals using just the cardinal successor operation and the operation of taking the union of a chain of cardinals indexed by a lower cardinal. The *inaccessible* cardinals, which are the cardinalities of universes, are the cardinals which cannot be reached from lower cardinals using *all* of the constructions of ZFC; i.e., the above two constructions together with the *power set* construction, and the Axioms of the Empty Set and Infinity, which hand us 0 and $\aleph_0$. (Inaccessible cardinals are sometimes called *strongly inaccessible* cardinals.) Whether every weakly inaccessible cardinal is inaccessible depends on the assumptions one makes on one's set theory. The student familiar with the Generalized Continuum Hypothesis will see that this assumption implies that these two concepts do coincide. Discussions of inaccessible cardinals can be found in basic texts on set theory. (For their relation to universes, cf. [5, 105]; for some alternative proposals for set-theoretic foundations of category theory, [77, 112]; and for a proposal in the opposite direction, [18].)

Notice that introducing "large sets" has not eliminated the need for the concept of a "class". In discussing set theory, one wants to refer to the collection of *all* sets; and one of the above exercises refers to the class of all universes. However, the need to work with classes, and the difficulties arising from not being able to use set-theoretic techniques in doing so, is greatly reduced, because for many purposes, references to large sets will now do.

We cannot be sure that the axiomatization we have adopted will be satisfactory for all the needs of category theory. It is based on the assumption that "ordinary mathematics" can be done within any universe $\mathbb{U}$, so that the set of all $\mathbb{U}$-small objects is a reasonable substitute for what was previously treated as the class of *all* objects. If some area of mathematics studied using category theory should itself require the full strength of the Axiom of Universes, then to get an adequate version of the category of "all" objects in that area, one might want to define a "second-order universe" to mean a universe $\mathbb{U}'$ such that every set $X \in \mathbb{U}'$ is a member of a universe $\mathbb{U} \in \mathbb{U}'$, and introduce a Second Axiom of Universes, saying that every set belongs to a second-order universe! However, the fact that for pre-category-theoretic mathematics, ZFC seemed an adequate foundation suggests that the set theory we have adopted here should be good for a while.

Concerning the basic idea of what we have done, namely to assume a set theory that contains "sub-set-theories" which themselves look like traditional set theory, let us note that these are "sub-set-theories" in the strongest sense: They involve the same membership relation, the same power set operation, etc. Set theorists often work with "sub-set-theories" in weaker senses; for example, allowing certain sets $X$ to belong to the sub-set-theory without making all subsets of $X$ members of the sub-set-theory. (For example, they may allow only those that are "constructible" in some way.) The resulting model may still satisfy general axioms such as ZFC, but have other properties significantly different from those of the set theory one started with. This technique is used in proving results of the sort, "If a certain set of axioms is consistent, so is a modified set of axioms". The distinction in question can be compared with the difference between considering a sublattice of a lattice, which by assumption has the same meet and join operations, and considering a subset which also has least upper bounds and greatest lower bounds, and hence can again be regarded as a lattice, but where these least upper and greatest lower bounds are not the same as in the original lattice, so that the object is not a sublattice.

We will find the following concept useful at times.

**Definition 7.4.6.** *A mathematical object will be called* quasi-small *if it is isomorphic to a small object.*

Here "isomorphic" is to be understood in the sense of the sort of object in question. Thus, a quasi-small set means a set with the same cardinality as a small set. A quasi-small group is easily seen to be a group whose underlying set is a quasi-small set.

We shall now return to category theory proper. Our language will in general be, superficially, as before; but as stated in Definition 7.4.5, there is now a fixed universe $\mathbb{U}$ in the background, and when the contrary is not stated, words such as "group" etc. now mean "$\mathbb{U}$-small group", etc., while "category" means "$\mathbb{U}$-legitimate category".

## 7.5. Functors

Since categories are themselves a sort of mathematical object, we should have a concept of "subcategory", and some sort of concept of "homomorphism" between categories. The first of these concepts is described in

**Definition 7.5.1.** *If* **C** *is a category, a* subcategory *of* **C** *means a category* **S** *such that* (i) Ob(**S**) *is a subset of* Ob(**C**), (ii) *for each* $X, Y \in$ Ob(**S**), **S**$(X, Y)$ *is a subset of* **C**$(X, Y)$, *and* (iii) *the composition and identity operations of* **S** *are the restrictions of those of* **C**.

Examples are clear: The category **Ab** of abelian groups is a subcategory of **Group**. Within **Monoid**, we can look at the subcategory whose objects are monoids all of whose elements are invertible (and whose morphisms are still all monoid-homomorphisms between these); this will be isomorphic to **Group**. **Lattice** is likewise isomorphic to a subcategory of **POSet**; here the lattice homomorphisms form a *proper* subset of the isotone maps. A subcategory of **POSet** with the same objects as the whole category, but a smaller set of morphisms, is the one we called **POSet**$_<$. Similarly, **Set** is a subcategory of **RelSet** with the same set of objects, but a more restricted set of morphisms. The *empty category* (no objects, and hence no morphisms) is a subcategory of every category.

The analog of homomorphism for categories is defined in

**Definition 7.5.2.** *If* **C** *and* **D** *are categories, then a* functor $F: \mathbf{C} \to \mathbf{D}$ *means a pair* $(F_{\mathrm{Ob}}, F_{\mathrm{Ar}})$, *where* $F_{\mathrm{Ob}}$ *is a map* Ob(**C**) $\to$ Ob(**D**), *and* $F_{\mathrm{Ar}}$ *is a family* $F_{\mathrm{Ar}} = (F(X, Y))_{X, Y \in \mathrm{Ob}(\mathbf{C})}$ *of maps*

$$F(X, Y): \mathbf{C}(X, Y) \longrightarrow \mathbf{D}(F_{\mathrm{Ob}}(X), F_{\mathrm{Ob}}(Y)) \quad (X, Y \in \mathrm{Ob}(\mathbf{C})),$$

*such that*

(i)   *for any two composable morphisms* $X \xrightarrow{g} Y \xrightarrow{f} Z$ *in* **C**, *one has*

$$F(X, Z)(f\,g) = F(Y, Z)(f)\, F(X, Y)(g),$$

*and*

(ii)   *for every* $X \in \mathrm{Ob}(\mathbf{C})$,

$$F(X, X)(\mathrm{id}_X) = \mathrm{id}_{F_{\mathrm{Ob}}(X)}.$$

When there is no danger of ambiguity, $F_{\mathrm{Ob}}$, $F_{\mathrm{Ar}}$, and $F(X, Y)$ are generally all abbreviated to $F$. Thus, in this notation, the last three displays become (*more readably*)

$$F\colon\ \mathbf{C}(X, Y) \longrightarrow \mathbf{D}(F(X), F(Y))\quad (X, Y \in \mathrm{Ob}(\mathbf{C})),$$
$$F(f\,g) = F(f)\, F(g),$$
$$F(\mathrm{id}_X) = \mathrm{id}_{F(X)}.$$

How do functors arise in the prototypical situation where **C** and **D** consist of mathematical objects and homomorphisms among them? Since we must first specify the object of **D** to which each object of **C** is carried, such a functor typically starts with a *construction* which gives us for each object of **C** an object of **D**. And in fact, most mathematical constructions, though often discussed as merely associating to each object of one sort an object of another, *also* have the property that to every morphism of objects of the first sort there corresponds naturally a morphism between the constructed objects, in a manner which satisfies the conditions of the above definition.

Consider, for example the construction of the free group, with which we began this course. To every $X \in \mathrm{Ob}(\mathbf{Set})$ this associates a group $F(X)$, together with a map $u_X \colon X \to |F(X)|$ having a certain universal property. Now if $f\colon X \to Y$ is a set map, it is easy to see how to get a homomorphism $F(f)\colon F(X) \to F(Y)$. Intuitively, $F(f)$ acts by "substituting $f(x)$ for $x$" in elements of $F(X)$ and evaluating the results in $F(Y)$. Recall that in terms of the universal property of $F(X)$, "substituting values in a group $G$ for the generators of $F(X)$" means determining a group homomorphism $F(X) \to G$ by specifying its composite with the set map $u_X \colon X \to |F(X)|$. In particular, for $f\colon X \to Y$, our above description of $F(f)$ translates to say that it is the unique group homomorphism $F(X) \to F(Y)$ such that $F(f) \cdot u_X = u_Y \cdot f$ :

$$
\begin{array}{ccc}
X & \xrightarrow{\ f\ } & Y \\[4pt]
{\scriptstyle u_X}\big\downarrow & & \big\downarrow{\scriptstyle u_Y} \\[4pt]
|F(X)| & \xrightarrow{\ F(f)\ } & |F(Y)|
\end{array}
$$

It is easy to check that if we define $F(f)$ in this way for each set-map $f$, we get $F(f g) = F(f) F(g)$ and $F(\mathrm{id}_X) = \mathrm{id}_{F(X)}$. Hence the free group construction gives a functor $F \colon \mathbf{Set} \to \mathbf{Group}$.

Looking in the same way at the construction of *abelianization*, associating to each group $G$ the abelian group $G^{\mathrm{ab}} = G/[G, G]$, we see that every group homomorphism $f \colon G \to H$ yields a homomorphism of abelian groups $f^{\mathrm{ab}} \colon G^{\mathrm{ab}} \to H^{\mathrm{ab}}$ (Exercise 4.4:3), describable either concretely in terms of cosets, or by a commutative diagram construction using the universal property of the canonical homomorphism $G \to G^{\mathrm{ab}}$. The constructions of free semilattices, universal abelianizations of rings, etc., give similar examples.

Like most mathematical concepts, the concept of functor also has "trivial" examples, that by themselves would not justify the general definition, yet which play important roles in the theory. The "construction" associating to every group $G$ its underlying set $|G|$ is a functor $\mathbf{Group} \to \mathbf{Set}$, since homomorphisms of groups certainly give maps of underlying sets. One similarly has underlying-set functors from $\mathbf{Ring}^1$, $\mathbf{Lattice}$, $\mathbf{Top}$, $\mathbf{POSet}$, etc., to $\mathbf{Set}$. These all belong to the class of constructions called "forgetful functors". Those listed above "forget" all structure on the object, and so give functors to $\mathbf{Set}$; other forgetful functors we have seen are the construction $G \mapsto G_{\mathrm{md}}$ of §4.11, taking a group $(|G|, \cdot, {}^{-1}, e)$ to the monoid $(|G|, \cdot, e)$, which "forgets" the inverse operation, and the construction taking a ring to its underlying additive group, or to its underlying multiplicative monoid.

The term "forgetful functor" is not a technical one, so one cannot say precisely whether it should be applied to constructions like the one taking a lattice to its "underlying" partially ordered set ("underlying" in quotes because the partial ordering is not part of the 3-tuple formally defining the lattice); but in any case, this is another example of a functor. I likewise don't know whether one would apply the term "forgetful" to the inclusion of the subcategory $\mathbf{Ab}$ in the category $\mathbf{Group}$, which might be said to "forget" that the groups are abelian, but this too, and indeed, the inclusion of any subcategory in any category, is easily seen to be a functor. In particular, the inclusion of any category $\mathbf{C}$ in itself is the identity functor, $\mathrm{Id}_{\mathbf{C}}$, which takes each object and each morphism to itself.

If, instead of looking at the whole underlying set of a group, we consider the set of its elements of exponent 2, we get another example of a functor $\mathbf{Group} \to \mathbf{Set}$. (Clearly every group homomorphism gives a map between the corresponding sets.)

If $R$ is a ring, the *opposite* ring $R^{\mathrm{op}}$ is defined to have the same underlying set, and the same operations $+, -, 0, 1$ as $R$, but reversed multiplication: $x * y = y x$. A ring homomorphism $f \colon R \to S$ will also be a homomorphism $R^{\mathrm{op}} \to S^{\mathrm{op}}$, and we see that this makes $(\ )^{\mathrm{op}}$ a functor $\mathbf{Ring}^1 \to \mathbf{Ring}^1$; one which, composed with itself, gives the identity functor. One has similar opposite-multiplication constructions for monoids and groups. The definitions of the opposite (or dual) of a partially ordered set or lattice give functors with similar properties.

Recall that **HtpTop** is defined to have the same objects as **Top**, but has for morphisms *equivalence classes* of continuous maps under homotopy. Thus we have a functor **Top** $\to$ **HtpTop** which preserves objects, and sends every morphism to its homotopy class.

We have mentioned *diagram categories*, such as the "commuting square diagram" $\begin{smallmatrix}\bullet\to\bullet\\\downarrow\quad\downarrow\\\bullet\to\bullet\end{smallmatrix}$ which is useful because "images" of it in any category **C** correspond to commuting squares of objects and arrows in **C**. We can now say this more precisely: Commuting squares in **C** correspond to *functors* from this diagram-category into **C**.

Let us also note a few examples of mathematical constructions that are *not* functors. These tend to be of two sorts: those in which morphisms from one object to another may not preserve the properties used by the construction, and those that involve arbitrary choices. We have noted that the construction associating to every group $G$ the set of elements of exponent 2, $\{x \in |G| \mid x^2 = e\}$, is a functor **Group** $\to$ **Set**. However, if we define $T(G)$ to be the set of elements of *order* 2, $\{x \in |G| \mid x^2 = e, x \neq e\}$, we find that a group homomorphism $f$ may take some of these elements to the identity element, so there is no natural way to define "$T(f)$". Similarly, the important group-theoretic construction of the *center* $Z(G)$ of a group $G$ (the subgroup of elements $a \in |G|$ that commute with all elements of $G$) is not functorial, because if $a$ is in the center of $G$ and we apply a homomorphism $f : G \to H$, some elements of $H$ outside the image of $G$ may fail to commute with $f(a)$. The construction Aut, taking a group $G$ to its automorphism group, is also not a functor, roughly because when we map $G$ into another group $H$, there is no guarantee that $H$ will have all the "symmetries" that $G$ does.

Some constructions of these sorts can be "made into" functors by modifying the choice of domain category so as to restrict the morphisms thereof to maps that don't "disturb" the structure involved. Thus, the construction associating to every group its set of elements of order 2 does give a functor **Group**$_{\text{inj}}$ $\to$ **Set**, if we define **Group**$_{\text{inj}}$ to be the category whose objects are groups and whose morphisms are *injective* (one-to-one) group homomorphisms. The construction of the center likewise gives a functor **Group**$_{\text{surj}}$ $\to$ **Group**, where the morphisms of **Group**$_{\text{surj}}$ are the *surjective* group homomorphisms. One may make Aut a functor by restricting morphisms to *isomorphisms* of groups.

An example of the other sort, where a construction is not a functor because it involves choices that cannot be made in a canonical way, is that of finding a basis for a vector space. Even limiting ourselves to finite-dimensional vector spaces, so that bases may be constructed without the Axiom of Choice, the finitely many choices one must make are still arbitrary, so that if one chooses a basis $B_V$ for a vector space $V$, and a basis $B_W$ for a vector space $W$, there is no natural way to associate to every linear map $V \to W$ a set map $B_V \to B_W$.

In the above discussion, we have merely indicated where straightforward attempts to make these constructions into functors go wrong. In the next four exercises you are asked to prove more precise negative results.

**Exercise 7.5:1.**

(i)   Show that there can be no functor $F$: **Group** → **Set** taking each group to the set of its elements of order 2, no matter how $F$ is made to act on morphisms.

   On the other hand,

(ii)  Show how to define a functor **Group** → **RelSet** taking every group to its set of elements of order 2. (Since **RelSet** is an unfamiliar category, verify explicitly all parts of the definition of functor.)

**Exercise 7.5:2.**

(i)   Show that there can be no functor $F$: **Group** → **Group** taking each group to its center.

(ii)  Can one construct a functor **Group** → **RelSet** taking every group to the set of its central elements?

**Exercise 7.5:3.**

(i)   Give an example of a group homomorphism $f: G \to H$ and an automorphism $a$ of $G$ such that there does not exist a unique automorphism $a'$ of $H$ such that $a'f = fa$. In fact, find such examples with $f$ one-to-one but not onto, and with $f$ onto but not one-to-one, and in each of these cases, if possible, an example where such $a'$ does not exist, and an example where such $a'$ exists but is not unique. (If you cannot get an example of one of the above combinations, can you show that it does not occur?)

(ii)  Find similar examples involving partially ordered sets in place of groups.

(iii) Prove that there is no functor from **Group** (alternatively, from **POSet**) to **Set** (or even to **RelSet**) taking each object to its set of automorphisms.

**Exercise 7.5:4.** If $K$ is a field, let $\overline{K}$ denote the algebraic closure of $K$. We recall that any field homomorphism $f: K \to L$ can be extended to a homomorphism of algebraic closures, $\overline{f}: \overline{K} \to \overline{L}$.

(i)   Show, however, that in general there is no way to choose an extension $\overline{f}$ of each field homomorphism $f$ so as to make the algebraic closure construction a functor.

(ii)  If we remove the restriction that $\overline{f}$ be an extension of $f$, can we make algebraic closure a functor?

The exercise below is instructive and entertaining. A full solution to the second part is difficult, but one can get many interesting partial results.

**Exercise 7.5:5.** Let **FSet** denote the subcategory of **Set** having for objects the finite sets, and for morphisms all set maps among these.

(i)   Show that every functor $F$ from **FSet** to **FSet** determines a function $f$ from the nonnegative integers to the nonnegative integers, such that for every finite set $X$, $\mathrm{card}(F(X)) = f(\mathrm{card}(X))$.

(ii)   Investigate *which* integer-valued functions $f$ can occur as the functions associated to such functors. If possible, determine necessary and sufficient conditions on $f$ for such an $F$ to exist.

Note that given functors $\mathbf{C} \overset{G}{\to} \mathbf{D} \overset{F}{\to} \mathbf{E}$ between any three categories, we can form the *composite* functor $\mathbf{C} \overset{FG}{\longrightarrow} \mathbf{E}$ taking each object $X$ to $F(G(X))$ and each morphism $f$ to $F(G(f))$. Composition of functors is clearly associative, and identity functors satisfy the identity laws, so we have a "category of categories"! This is named in

**Definition 7.5.3. Cat** *will denote the (legitimate) category whose objects are all* small *categories, and where for two small categories* **C** *and* **D**, **Cat**(**C**, **D**) *is the set of all functors* **C** → **D**, *with composition of functors defined as above.*

You might be disappointed with this definition, since only a few of the categories we have mentioned have been small (the diagram-categories, and the categories $S_{\mathbf{cat}}$ and $P_{\mathbf{cat}}$ constructed from monoids $S$ and partially ordered sets $P$). Thus, **Cat** would appear to be of limited importance. But here the Axiom of Universes comes to our aid. The universe $\mathbb{U}$ relative to which we have defined "small category" is arbitrary. If we want to study the categories of all groups, rings, etc., belonging to a universe $\mathbb{U}$, and functors among these categories, we may choose a universe $\mathbb{U}'$ having $\mathbb{U}$ as a member, and note that the abovementioned categories, and indeed, all $\mathbb{U}$-legitimate categories, are $\mathbb{U}'$-small, hence are objects of $\mathbf{Cat}_{(\mathbb{U}')}$. Thus we can apply general results about the construction **Cat** to this situation.

(For some purposes, it might also be useful to have a symbol for the category of all $\mathbb{U}$-legitimate categories, which lies strictly between $\mathbf{Cat}_{(\mathbb{U})}$ and $\mathbf{Cat}_{(\mathbb{U}')}$, but we shall not introduce one here.)

Considering functors as "homomorphisms" among categories, we should like to define properties of functors analogous to "one-to-one-ness" and "onto-ness". The complication is that a functor acts both on objects and on morphisms. We have observed that it is the *morphisms* in a category that are like the *elements* of a group or monoid; this leads to the pair of concepts named below. They are not the only analogs of one-one-ness and onto-ness that one ever uses, but they are the most important:

**Definition 7.5.4.** *Let* $F \colon \mathbf{C} \to \mathbf{D}$ *be a functor.*

  $F$ *is called* faithful *if for all* $X, Y \in \mathrm{Ob}(\mathbf{C})$, *the map* $F(X, Y) \colon$ $\mathbf{C}(X, Y) \to \mathbf{D}(F(X), F(Y))$ *is one-to-one.*

$F$ *is called* full *if for all* $X, Y \in \mathrm{Ob}(\mathbf{C})$, *the map* $F(X, Y) \colon \mathbf{C}(X, Y) \to \mathbf{D}(F(X), F(Y))$ *is onto.*

*A subcategory of* $\mathbf{C}$ *is said to be* full *if the corresponding* inclusion functor *is full.*

Thus, a full subcategory of $\mathbf{C}$ is determined by specifying a subset of the object-set; the morphisms of the subcategory are then *all* the morphisms among these objects. The subcategory $\mathbf{Ab}$ of $\mathbf{Group}$ is an example. Some examples of nonfull subcategories are $\mathbf{Set} \subseteq \mathbf{RelSet}$ and $\mathbf{POSet}_< \subseteq \mathbf{POSet}$. The inclusion of a full subcategory in a category is a full and faithful functor, while the inclusion of a nonfull subcategory is a faithful functor, but is not full. The reader should verify that most of our examples of forgetful functors are faithful but not full, as is, also, the free-group functor $\mathbf{Set} \to \mathbf{Group}$. The functor $\mathbf{Top} \to \mathbf{HtpTop}$ which takes every object (topological space) to itself, and each morphism to its *homotopy class*, is an example of a functor that is full but not faithful. The functor associating to every group the set of its elements of exponent 2 is neither full nor faithful.

**Exercise 7.5:6.** Show that the abelianization construction, $\mathbf{Group} \to \mathbf{Ab}$ is neither full nor faithful.

**Exercise 7.5:7.** Is the functor $\mathbf{Monoid} \to \mathbf{Group}$ associating to every monoid its group of invertible elements full? Faithful?

**Exercise 7.5:8.**

(i)   Show that the construction associating to each partially ordered set $P$ the category $P_{\mathbf{cat}}$ can be made in a natural way into a functor $F \colon \mathbf{POSet} \to \mathbf{Cat}$, and that as such it is full and faithful. This says that the concept of functor, when restricted to the class of categories that correspond to partially ordered sets, just gives the concept of isotone map between these sets!

(ii)   Which isotone maps between partially ordered sets correspond under $F$ to full functors? To faithful functors?

(iii)  Show similarly that the construction associating to each monoid $S$ the category $S_{\mathbf{cat}}$ is a full and faithful functor $E \colon \mathbf{Monoid} \to \mathbf{Cat}$. Which monoid homomorphisms are sent by $E$ to full, respectively faithful functors?

**Exercise 7.5:9.** Show that for $F \colon \mathbf{C} \to \mathbf{D}$ a functor, neither of the following conditions implies the other: (a) $F$ is full, (b) for all $X, Y \in \mathrm{Ob}(\mathbf{D})$ and $f \in \mathbf{D}(X, Y)$ there exist $X_0, Y_0 \in \mathrm{Ob}(\mathbf{C})$ and $f_0 \in \mathbf{C}(X_0, Y_0)$ such that $F(X_0) = X,\ F(Y_0) = Y,$ and $F(f_0) = f$.

In §7.1 we sketched a way of "concretizing" any small category $\mathbf{C}$ (Exercise 7.1:1 and preceding discussion). Let us make the details precise now.

**Definition 7.5.5.** *A* concrete category *means a category* **C** *given with a* faithful *functor* $U: \mathbf{C} \to \mathbf{Set}$ *(a "concretization functor"). (More formally, one would say that the concrete category is the ordered pair* $(\mathbf{C}, U)$.)

So the result in question was that given any small category **C**, there exists a faithful functor $U: \mathbf{C} \to \mathbf{Set}$. The idea was to let the family of representing sets – in our present notation, the system of sets $U(X)$ $(X \in \mathrm{Ob}(\mathbf{C}))$ – be "generated" by a family of elements $z_Y \in U(Y)$, one for each $Y \in \mathrm{Ob}(\mathbf{C})$, so that the general element of $U(X)$ would look like $U(a)(z_Y)$ for $Y \in \mathrm{Ob}(\mathbf{C})$ and $a \in \mathbf{C}(Y, X)$; and to impose no additional relations on these elements, so that they are all distinct.

Let us use the ordered pair $(Y, a)$ for the element that is to become $U(a)(z_Y)$. Then we should define $U$ to take $X \in \mathrm{Ob}(\mathbf{C})$ to $\{(Y, a) \mid Y \in \mathrm{Ob}(\mathbf{C}), a \in \mathbf{C}(Y, X)\}$. Given $b \in \mathbf{C}(X, W)$, we see that $U(b)$ should take $(Y, a) \in U(X)$ to $(Y, ba) \in U(W)$. It is easy to verify that this defines a faithful functor $U: \mathbf{C} \to \mathbf{Set}$, proving

**Theorem 7.5.6** (Cayley's Theorem for small categories). *Every small category admits a concretization, i.e., a faithful functor to the category of small sets.* ☐

**Exercise 7.5:10.** Verify that the above construction $U$ is a functor, and is faithful. Which element of each set $U(Y)$ corresponds to the $z_Y$ of our motivating discussion?

Incidentally, if we had required that categories have disjoint morphism-sets, we could have dropped the $Y$'s from the pairs $(Y, a)$, since each $a$ would determine its domain. Then we could simply have taken $U(X) = \bigcup_{Y \in \mathrm{Ob}(\mathbf{C})} \mathbf{C}(Y, X)$.

It is natural to hope for stronger results, so you can try

**Exercise 7.5:11.**

(i) Does every legitimate category admit a concretization—a faithful functor to the (legitimate) category of small sets? (Obviously, most of those we are familiar with do.)

Since this question involves "big" cardinalities, you might prefer to examine a mini-version of the same problem:

(ii) Suppose **C** is a category with countably many objects, and such that for all $X, Y \in \mathrm{Ob}(\mathbf{C})$, the set $\mathbf{C}(X, Y)$ is finite. Must **C** admit a faithful functor into the category of finite sets?

(iii) If the answer to either question is negative, can you find necessary and sufficient conditions on **C** for such concretizations to exist?

Of course, a given concretizable category will admit many concretizations, just as a given group has many faithful representations by permutations.

Recall that the proof of Theorem 7.5.6 sketched above came out of our proof of the corresponding result for "bimonoids", and that in trying to prove that result, we first wondered whether it would suffice to adjoin a generator in just one of the two representing sets $X$ and $Y$, but saw that the resulting representation of our bimonoid might not be faithful. Given a category $\mathbf{C}$ and an object $Y$ of $\mathbf{C}$, we can similarly construct a functor $U : \mathbf{C} \to \mathbf{Set}$ by introducing only one generator $z_Y \in U(Y)$, again with no relations imposed among the elements $U(a)(z_Y)$. Though these functors also generally fail to be faithful, they will play an important role in our subsequent work. Note that each such functor is the "part" of the construction we used in Theorem 7.5.6 consisting of the elements $U(a)(z_Y)$ for one fixed $Y$. With $Y$ fixed, each such element is determined by $a \in \mathbf{C}(Y, X)$, so $U$ may be described as taking each object $X$ to the hom-set $\mathbf{C}(Y, X)$; hence its name:

**Definition 7.5.7.** *For* $Y \in \mathrm{Ob}(\mathbf{C})$, *the* hom *functor* induced by $Y$, $h_Y : \mathbf{C} \to \mathbf{Set}$, *is defined on objects by*

$$h_Y(X) = \mathbf{C}(Y, X) \quad (X \in \mathrm{Ob}(\mathbf{C})),$$

*while for a morphism* $b \in \mathbf{C}(X, W)$, $h_Y(b)$ *is defined to carry* $a \in \mathbf{C}(Y, X)$ *to* $ba \in \mathbf{C}(Y, W)$.

(Some authors denote the above construction $h^Y$, with $h_Y$ used for a dual construction. I will address this point when we introduce that dual construction, in Definition 7.6.3.)

Examples: Let $Z$ denote the infinite cyclic group, with generator $x$. Then on the category **Group**, the functor $h_Z$ takes each group $G$ to **Group**$(Z, G)$. But a homomorphism from $Z$ to $G$ is determined by what it does on the generator $x \in |Z|$, so the elements of $h_Z(G)$ correspond to the elements of the underlying set of $G$; i.e., $h_Z$ is essentially the underlying set functor. You should verify that its behavior on morphisms also agrees with that functor. Similarly, writing $Z_2$ for the cyclic group of order 2, the functor $h_{Z_2}$ may be identified with the functor taking each group to the set of its elements of exponent 2.

Recalling that $2 \in \mathrm{Ob}(\mathbf{Set})$ is a two-element set, we see that $h_2 : \mathbf{Set} \to \mathbf{Set}$ is essentially the construction $X \mapsto X^2$.

For a topological example, consider the category of topological spaces with basepoint, and homotopy classes of basepoint-preserving maps, and let $(S^1, 0)$ denote the circle with a basepoint chosen. Then $h_{(S^1, 0)}(X, x_0) = |\pi_1(X, x_0)|$. (Of course, the interesting thing about $\pi_1(X, x_0)$ is its group structure. How *this* can be described category-theoretically we shall discover in Chapter 10!)

In the last few paragraphs, I have said a couple of times that a certain functor is "essentially" a certain construction. What was meant should be intuitively clear. We will see how to make these statements precise in §7.9.

## 7.6. Contravariant functors, and functors of several variables

Consider the construction associating to every set $X$ the additive group $\mathbb{Z}^X$ of integer-valued functions on $X$, with pointwise operations. This takes objects of **Set** to objects of **Ab**, but given a set map $f\colon X \to Y$, there is not a natural map $\mathbb{Z}^X \to \mathbb{Z}^Y$ – rather, there is a homomorphism $\mathbb{Z}^Y \to \mathbb{Z}^X$ carrying each integer-valued function $a$ on $Y$ to the function $af$ on $X$.

There are many similar examples – the construction associating to any set $X$ the Boolean algebra $(\mathbf{P}(X), \cup, \cap, {}^{\mathrm{c}}, \emptyset, X)$ of its subsets, the construction associating to a set $X$ the lower semilattice $(\mathbf{E}(X), \cap)$ of equivalence relations on $X$, the construction associating to a vector space $V$ its dual $V^*$, the construction associating to a commutative ring the partially ordered set of its prime ideals. All have the property that a map going one way among the given objects yields a map going the *other* way among constructed objects. It is clear that these constructions take identity maps to identity maps and composite maps to composite maps (though the order of composition must be reversed because of the reversal of the direction of the maps). These properties look like the definition of a functor turned backwards. Let us set up a definition to cover this:

**Definition 7.6.1.** *If* **C** *and* **D** *are categories, then a* contravariant functor $F\colon \mathbf{C} \to \mathbf{D}$ *means a pair* $(F_{\mathrm{Ob}}, F_{\mathrm{Ar}})$, *where* $F_{\mathrm{Ob}}$ *(written* $F$ *when there is no danger of ambiguity) is a map* $\mathrm{Ob}(\mathbf{C}) \to \mathrm{Ob}(\mathbf{D})$, *and* $F_{\mathrm{Ar}}$ *is a family of maps*

$$F(X, Y)\colon \mathbf{C}(X, Y) \to \mathbf{D}(F(Y), F(X)) \quad (X, Y \in \mathrm{Ob}(\mathbf{C})),$$

*such that (abbreviating these maps* $F(X, Y)$ *to* $F$),

(i)  *for any two composable morphisms* $X \xrightarrow{g} Y \xrightarrow{f} Z$ *in* **C**, *one has*

$$F(f\,g) \;=\; F(g)\,F(f) \;\; in \;\; \mathbf{D},$$

*and*

(ii)  *for every* $X \in \mathrm{Ob}(\mathbf{C})$, *one has*

$$F(\mathrm{id}_X) \;=\; \mathrm{id}_{F(X)}.$$

*Functors of the sort defined in the preceding section are called* covariant *functors when one wants to contrast them with contravariant functors. But when the contrary is not indicated, "functor" (unmodified) will still mean covariant functor.*

It is easy to see that a composite of two contravariant functors is a covariant functor, while a composite of a covariant and a contravariant functor, in either order, is a contravariant functor.

Contravariant functors can in fact be expressed in terms of covariant functors, thus eliminating the need to prove results separately for them. We shall do this with the help of

**Definition 7.6.2.** *If* $\mathbf{C}$ *is a category, then* $\mathbf{C}^{\mathrm{op}}$ *will denote the category defined by*

$$\mathrm{Ob}(\mathbf{C}^{\mathrm{op}}) = \mathrm{Ob}(\mathbf{C}) \qquad \mathbf{C}^{\mathrm{op}}(X, Y) = \mathbf{C}(Y, X),$$

$$\mu(\mathbf{C}^{\mathrm{op}})(f, g) = \mu(\mathbf{C})(g, f), \qquad \mathrm{id}(\mathbf{C}^{\mathrm{op}})_X = \mathrm{id}(\mathbf{C})_X.$$

Thus, a *contravariant* functor $\mathbf{C} \to \mathbf{D}$ is equivalent to a covariant functor $\mathbf{C}^{\mathrm{op}} \to \mathbf{D}$. Of course, one could also describe it as equivalent to a covariant functor $\mathbf{C} \to \mathbf{D}^{\mathrm{op}}$, and at this point we have no way of deciding which reduction is preferable. However, we shall see soon that putting the " $^{\mathrm{op}}$ " on the domain category is more convenient.

As in the theory of partially ordered sets, the "opposite" construction allows us to dualize results. Whenever we have proved a result about a general category $\mathbf{C}$, the statement obtained by reversing the directions of all morphisms and the orders of all compositions is also a theorem, which may be proved by applying the original theorem to $\mathbf{C}^{\mathrm{op}}$.

There is a slight notational difficulty in dealing with a category $\mathbf{C}^{\mathrm{op}}$, while referring also to the original category $\mathbf{C}$. Though in the formal definition given above we could distinguish the two composition operations as $\mu(\mathbf{C})$ and $\mu(\mathbf{C}^{\mathrm{op}})$, the usual notation for composition, $f \cdot g$ or $f g$, does not allow such a distinction. There are various ways of getting around this. One can use a modified symbol, such as $\cdot^{\mathrm{op}}$ or $*$, for the composition of $\mathbf{C}^{\mathrm{op}}$. Or one can continue to denote composition by juxtaposition, but use different symbols for the same objects and morphisms when considered as elements of $\mathbf{C}$ and of $\mathbf{C}^{\mathrm{op}}$; e.g., one can let the morphism written $f \in \mathbf{C}(X, Y)$ also be written $\widetilde{f} \in \mathbf{C}^{\mathrm{op}}(\widetilde{Y}, \widetilde{X})$, so that one would have $\widetilde{f g} = \widetilde{g}\, \widetilde{f}$, relying on the convention that the meaning of juxtaposition is determined by context – specifically, by the structure to which the elements being juxtaposed belong. Still other solutions are possible. For example, one could be daring, and denote the same composite by $f g$ in both $\mathbf{C}$ and $\mathbf{C}^{\mathrm{op}}$, using different conventions, $f g = \mu(f, g)$ in $\mathbf{C}$ and $f g = \mu(g, f)$ in $\mathbf{C}^{\mathrm{op}}$; i.e., writing morphisms with domains "on the right" in one category and "on the left" in the other. (Cf. [56].)

Most often, one avoids the problem by not writing formulas in $\mathbf{C}^{\mathrm{op}}$. One uses this category as an auxiliary concept in discussing contravariant functors and in dualizing results, but avoids talking explicitly about objects and morphisms inside it.

In these notes, we shall regularly write a contravariant functor from $\mathbf{C}$ to $\mathbf{D}$ as $F \colon \mathbf{C}^{\mathrm{op}} \to \mathbf{D}$, where $F$ is a covariant functor on $\mathbf{C}^{\mathrm{op}}$, and shall take advantage of the principle of duality mentioned. These are the main uses we shall make of the $^{\mathrm{op}}$ construction; in the rare cases where we have to work explicitly inside $\mathbf{C}^{\mathrm{op}}$, we will generally use modified symbols such as $\widetilde{X}$, $\widetilde{f}$ (or $X^{\mathrm{op}}$, $f^{\mathrm{op}}$) for objects and morphisms in $\mathbf{C}^{\mathrm{op}}$.

Note that in the category of categories, **Cat**, the morphisms are the *covariant* functors.

**Exercise 7.6:1.**

(i)   Show how to make $^{\mathrm{op}}$ a functor $R$ from **Cat** to **Cat**. Is $R$ a covariant or a contravariant functor?

(ii)   Let $R\colon \mathbf{Cat} \to \mathbf{Cat}$ be as in part (i), let $R'\colon \mathbf{POSet} \to \mathbf{POSet}$ be the functor taking every partially ordered set $P$ to the opposite partially ordered set $P^{\mathrm{op}}$, and let $C\colon \mathbf{POSet} \to \mathbf{Cat}$ denote the functor taking each partially ordered set $P$ to the category $P_{\mathbf{cat}}$ (§7.2). Show that $RC \cong CR'$.

Thus, the construction of the opposite of a partially ordered set is essentially a case of the construction of the opposite of a category!

(iii)   State the analogous result with *monoids* in place of partially ordered sets.

We noted in earlier chapters that given a set map $X \to Y$, there are ways of getting both a map $\mathbf{P}(X) \to \mathbf{P}(Y)$ and a map $\mathbf{P}(Y) \to \mathbf{P}(X)$ (where $\mathbf{P}$ denotes the power-set construction). The next few exercises look at this and some similar situations.

**Exercise 7.6:2.**

(i)   Write down explicitly how to get from a set map $f\colon X \to Y$ a set map $P_1(f)\colon \mathbf{P}(X) \to \mathbf{P}(Y)$ and also a set map $P_2(f)\colon \mathbf{P}(Y) \to \mathbf{P}(X)$. Show that these constructions make the power set construction a functor $P_1\colon \mathbf{Set} \to \mathbf{Set}$ and a functor $P_2\colon \mathbf{Set}^{\mathrm{op}} \to \mathbf{Set}$ respectively. (These are called the covariant and contravariant power set functors.)

(ii)   Examine what structure on $\mathbf{P}(X)$ is *respected* by maps of the form $P_1(f)$ and $P_2(f)$ defined as above. In particular, determine whether each sort of map always respects the operations of finite meets, finite joins, empty meet, empty join, unions of chains, intersections of chains, complements, and the relations "$\subseteq$" and "$\subset$" in power-sets $\mathbf{P}(X)$. (Cf. Exercise 6.1:11. If you are familiar with the standard topologization of $\mathbf{P}(X)$, you can also investigate whether maps of the form $P_1(f)$ and $P_2(f)$ are continuous.) Accordingly, determine whether the constructions $P_1$ and $P_2$ which we referred to above as functors from **Set**, respectively **Set**$^{\mathrm{op}}$, to **Set**, can in fact be made into functors from **Set** and/or **Set**$^{\mathrm{op}}$ to $\vee$-**Semilat**, to **Bool**[1] (the category of Boolean rings), etc. In each case, note only the strongest structure that you are showing. (For example, there is no need to note that you can make a functor $\vee$-semilattice-valued if you will in fact make it lattice-valued.)

**Exercise 7.6:3.** Investigate similarly the construction associating to every set $X$ the set $\mathbf{E}(X)$ of *equivalence relations* on $X$. That is, for a set map $f\colon X \to Y$, look for functorial ways of inducing maps in one or both directions between the sets $\mathbf{E}(X)$, $\mathbf{E}(Y)$, and determine what structure on these sets is respected by each such construction.

**Exercise 7.6:4.**

(i)   Do the same for the construction associating to every group $G$ the set of subgroups of $G$.

(ii)   Do the same for the construction associating to every group $G$ the set of *normal* subgroups of $G$.

As with covariant functors, there is an important class of contravariant functors which one can define on every category:

**Definition 7.6.3.** *For any category $\mathbf{C}$ and any object $Y \in \mathrm{Ob}(\mathbf{C})$, the contravariant hom functor induced by $Y$, $h^Y : \mathbf{C}^{\mathrm{op}} \to \mathbf{Set}$, is defined on* objects *by*

$$h^Y(X) = \mathbf{C}(X, Y) \quad (X \in \mathrm{Ob}(\mathbf{C})).$$

*For a* morphism *$b \in \mathbf{C}(W, X)$ the morphism $h^Y(b) : \mathbf{C}(X, Y) \to \mathbf{C}(W, Y)$ is defined to carry $a \in \mathbf{C}(X, Y)$ to $ab \in \mathbf{C}(W, Y)$. (The functor $h_Y$ which we previously named "the hom functor induced by $Y$" will henceforth be called "the covariant hom functor induced by $Y$".)*

*(Note: Many authors, in particular, algebraic geometers, use $h^Y$ for what we call $h_Y$, and vice versa. The usage is divided; I have chosen the usage given above because it matches the convention in homology and homotopy theory, where subscripts appear on the covariant functors of* homology *and* homotopy, *and superscripts on the contravariant* cohomology *and cohomotopy* functors.*)*

Examples: Let $\mathbf{C} = \mathbf{Set}$, and let $Y$ be the set $2 = \{0, 1\}$. Recall that every map from a set $X$ into $2$ is the characteristic function of a unique subset $S \subseteq X$. Hence $\mathbf{Set}(X, 2)$ can be identified with $\mathbf{P}(X)$. The reader should verify that the behavior of $h^2 : \mathbf{Set} \to \mathbf{Set}$ on morphisms is exactly that of the contravariant power-set functor.

Let $k$ be a field, and in the category $k$-$\mathbf{Mod}$ of $k$-vector spaces, let $k$ denote this field considered as a one-dimensional vector space. Then for any vector space $V$, $h^k(V)$ is the underlying set of the *dual* vector space, and for any linear map $b : V \to W$, $h^k(b)$ is the induced map from the dual of the space $W$ to the dual of the space $V$.

Let $\mathbb{R} \in \mathrm{Ob}(\mathbf{Top})$ denote the real line. Then $h^{\mathbb{R}}$ is the construction associating to every topological space $X$ the set of continuous real-valued functions on $X$. One can vary this example using categories of differentiable manifolds and differentiable maps, etc., in place of $\mathbf{Top}$.

Here are three further examples for students familiar with the areas in question.

In the category of commutative algebras over the rational numbers, if $\mathbb{C}$ denotes the algebra of complex numbers, then $h^{\mathbb{C}}$ is the functor associating to every algebra $R$ the set of its "complex-valued points", its *classical spectrum*. In particular, if $R$ is presented by generators $x_0, \ldots, x_{n-1}$ and relations $p_0 = 0, \ldots, p_{m-1} = 0$, then $h^{\mathbb{C}}(R)$ can be identified with the solution-set, in complex $n$-space $\mathbb{C}^n$, of the system of polynomial equations $p_0 = 0, \ldots, p_{m-1} = 0$.

If **LocCpAb** is the category of locally compact topological abelian groups, and $S = \mathbb{R}/\mathbb{Z}$ is the circle group, then $h^S(A)$ is the underlying set of the *Pontryagin dual* of the group $A$ [130, §1.7]. (In the study of non-topological abelian groups, $\mathbb{Q}/\mathbb{Z}$ plays a somewhat similar role [34, p. 145, Remark 2].)

Finally, in the category **HtpTop**, the set $h^{S^n}(X)$ (where $S^n$ denotes the $n$-sphere) gives the underlying set of the $n$-th *cohomotopy* group, $\pi^n(X)$.

**Exercise 7.6:5.** Let $2 \in \mathrm{Ob}(\mathbf{POSet})$ denote the set $2 = \{0, 1\}$, ordered in the usual way.

(i)  Show that $h^2 \colon \mathbf{POSet}^{\mathrm{op}} \to \mathbf{Set}$ is faithful.

(ii)  Show that for $P \in \mathrm{Ob}(\mathbf{POSet})$, the set $h^2(P)$ can be made a *lattice* with a greatest and a least element, under pointwise operations. Show that in this way $h^2$ induces a functor $A \colon \mathbf{POSet}^{\mathrm{op}} \to \mathbf{Lattice}^{0,1}$, where **Lattice**$^{0,1}$ denotes the category of lattices with greatest and least elements, and lattice homomorphisms respecting these elements.

(iii)  Let us also write $2 \in \mathrm{Ob}(\mathbf{Lattice}^{0,1})$ for the two-element lattice! Thus we get a functor $h^2 \colon (\mathbf{Lattice}^{0,1})^{\mathrm{op}} \to \mathbf{Set}$. Show that this functor is *not* faithful.

(iv)  Show that for $L \in \mathrm{Ob}(\mathbf{Lattice}^{0,1})$, the set $h^2(L)$ is not in general closed under pointwise meet or join, and may not contain a greatest or least element, but that if we partially order lattice homomorphisms by pointwise comparison, $h^2$ yields a functor $B \colon (\mathbf{Lattice}^{0,1})^{\mathrm{op}} \to \mathbf{POSet}$.

(v)  Show that for $P$ a *finite* partially ordered set, $B(A(P)) \cong P$.

The above is just a teaser. The interested student might examine this pair of functors further, and see what more he or she can prove; or wait till we return to the topic in §10.12 with general tools at our disposal.

**Exercise 7.6:6.** Following up on the idea of Exercise 7.5:5, observe that every *contravariant* functor from the category **FSet** of finite sets into itself also determines a nonnegative integer-valued function on the nonnegative integers. Investigate which functions on the nonnegative integers arise as functions associated with contravariant functors.

**Exercise 7.6:7.** Let **RelFSet** denote the full subcategory of **RelSet** whose objects are finite sets. Investigate similarly the integer-valued functions associated with functors **RelFSet** $\to$ **FSet**, **FSet** $\to$ **RelFSet**, and **RelFSet** $\to$ **RelFSet**. In these cases, it does not matter whether we look at covariant or contravariant functors—why not?

**Exercise 7.6:8.** We have noted that a composite of two contravariant functors is a covariant functor, etc. But in terms of the description of contravariant functors as covariant functors $\mathbf{C}^{\mathrm{op}} \to \mathbf{D}$, it is not clear how to formally describe the composite of two contravariant functors (or a composite of the form (contravariant functor)·(covariant functor)). Show how to reduce these cases to composition of covariant functors, with the help of Exercise 7.6:1(i).

There are still some types of well-behaved mathematical constructions which we have not yet fitted into our functorial scheme: (a) Given a *pair* of sets $(A, B)$, we can form the *product* set $A \times B$. We likewise have product constructions for groups, rings, topological spaces, etc., *coproducts* for most of the same types of objects, and the *tensor product* construction on abelian groups. (b) From two objects $A$ and $B$ of any category $\mathbf{C}$, one gets $\mathbf{C}(A, B) \in \mathrm{Ob}(\mathbf{Set})$. (c) There are also constructions that combine objects of different categories. For instance, from a commutative ring $R$ and a set $X$, one can form the *polynomial ring* over $R$ in an $X$-tuple of indeterminates, $R[X]$.

In each of these cases, maps on the given objects yield maps on the constructed objects. In cases (a) and (c), the maps of constructed objects go the same way as the maps of the given objects, while in case (b) the direction depends on which argument one varies: A morphism $Y \to Y'$ yields a map $\mathbf{C}(X, Y) \to \mathbf{C}(X, Y')$, but a morphism $X \to X'$ yields a map $\mathbf{C}(X', Y) \to \mathbf{C}(X, Y)$.

It is natural to call these constructions *functors of two variables*. Like the concept of contravariant functor, that of a functor of more than one variable can be reduced to our original definition of functor via an appropriate construction on categories.

**Definition 7.6.4.** *Let* $(\mathbf{C}_i)_{i \in I}$ *be a family of categories. Then the* product category $\prod_{i \in I} \mathbf{C}_i$ *will mean the category* $\mathbf{C}$ *defined by*

$$\mathrm{Ob}(\mathbf{C}) = \prod_{i \in I} \mathrm{Ob}(\mathbf{C}_i) \qquad \mathbf{C}((X_i)_{i \in I}, (Y_i)_{i \in I}) = \prod_{i \in I} \mathbf{C}_i(X_i, Y_i),$$

$$\mu((f_i)_I, (g_i)_I) = (\mu(f_i, g_i))_I, \qquad\qquad \mathrm{id}_{(X_i)_{i \in I}} = (\mathrm{id}_{X_i})_{i \in I}.$$

*The product of a finite family of categories is often written* $\mathbf{C} \times \cdots \times \mathbf{E}$.

*A functor* $F$ *on a product category is called a* functor of several variables; *a functor of two variables is often called a* bifunctor.

Thus, a functor on a category of the form $\mathbf{C} \times \mathbf{D}^{\mathrm{op}}$ may be described as a "bifunctor covariant in a $\mathbf{C}$-valued variable and contravariant in a $\mathbf{D}$-valued variable". Note that if we tried to express contravariance by putting "$^{\mathrm{op}}$" onto the *codomains* instead of the domains of functors, we would not be able to express this mixed type of functor; hence the preference for putting $^{\mathrm{op}}$ on domains.

A product category $\prod_{i \in I} \mathbf{C}_i$ has a *projection functor* onto each of the categories $\mathbf{C}_i$ $(i \in I)$, taking each object and each morphism to its $i$-th component, and as we might expect from our experience with products of other sorts, this is characterizable by the following universal property:

**Theorem 7.6.5.** *Let* $(\mathbf{C}_i)_{i \in I}$ *be a family of categories,* $\mathbf{C} = \prod \mathbf{C}_i$ *their product, and* $P_i \colon \mathbf{C} \to \mathbf{C}_i$ *the projection functors. Then for every category* $\mathbf{D}$ *and family of functors* $F_i \colon \mathbf{D} \to \mathbf{C}_i$, *there exists a unique functor* $F \colon \mathbf{D} \to \mathbf{C}$ *such that for each* $i \in I$, $F_i = P_i F$. □

**Exercise 7.6:9.** Prove the above theorem.

**Exercise 7.6:10.** Show that a family of categories also has a *coproduct.* (First state the universal property desired.)

I claim now that the two sorts of hom-functors, $h_X$ and $h^Y$, are pieces of a single bifunctor. In the definition of this functor below, we use "$\widetilde{X}$"-notation for objects and morphisms in opposite categories, though in presentations elsewhere, you are likely to see no distinctions made.

**Definition 7.6.6.** *The* bivariant hom-functor *of a category* $\mathbf{C}$ *means the functor*

$$h \colon \ \mathbf{C}^{\mathrm{op}} \times \mathbf{C} \ \longrightarrow \ \mathbf{Set}$$

*which is defined on objects by*

$$h(\widetilde{X}, Y) \ = \ \mathbf{C}(X, Y) \quad (X, Y \in \mathrm{Ob}(\mathbf{C})),$$

*while for a morphism* $(\widetilde{p}, q) \in \mathbf{C}^{\mathrm{op}}(\widetilde{X}, \widetilde{W}) \times \mathbf{C}(Y, Z)$ *(formed from morphisms* $p \in \mathbf{C}(W, X)$, $q \in \mathbf{C}(Y, Z)$*) we define* $h(\widetilde{p}, q)$ *to carry* $a \in \mathbf{C}(X, Y)$ *to* $q \, a \, p \in \mathbf{C}(W, Z)$.

Thus, each *covariant* hom-functor $h_X \colon \mathbf{C} \to \mathbf{Set}$ can be described as taking objects $Y$ to the objects $h(\widetilde{X}, Y)$, and morphisms $q$ to the morphisms $h(\widetilde{\mathrm{id}_X}, q)$; and the *contravariant* hom-functors $h^Y \colon \mathbf{C}^{\mathrm{op}} \to \mathbf{Set}$ are similarly obtained by putting $Y$ and $\mathrm{id}_Y$ in the right-hand slot of the bifunctor $h$.

**Exercise 7.6:11.** Extend further the ideas of Exercises 7.5:5–7.6:7, by investigating functions in two nonnegative-integer-valued variables induced by bifunctors $\mathbf{FSet} \times \mathbf{FSet} \to \mathbf{FSet}$, $\mathbf{FSet}^{\mathrm{op}} \times \mathbf{FSet} \to \mathbf{FSet}$, etc.

## 7.7. Category-theoretic versions of some common mathematical notions: properties of morphisms

We have mentioned that in an abstract category, one cannot speak of "elements" of an object, hence one cannot meaningfully ask whether a given morphism is one-to-one or onto. However, we have occasionally spoken of two objects of a category $\mathbf{C}$ being "isomorphic". What we meant was, I hope, clear: An *isomorphism* between $X$ and $Y$ means an element $f \in \mathbf{C}(X, Y)$ for which there exists a two-sided inverse, that is, a morphism $g \in \mathbf{C}(Y, X)$ such that $fg = \mathrm{id}_Y$, $gf = \mathrm{id}_X$. It is clear that in virtually any naturally occurring category, the invertible morphisms are the things one wants to think of as the isomorphisms. (However, for some mathematical objects other words are traditionally used: In set theory the term is *bijection*, an invertible morphism in **Top** is called a *homeomorphism*, and differential geometers call their invertible maps *diffeomorphisms*.) If $X$ and $Y$ are isomorphic, we will as usual write $X \cong Y$. An isomorphism of an object $X$ with itself is called an *automorphism* of $X$; these together comprise the *automorphism group* of $X$.

**Exercise 7.7:1.** Let $\mathbf{C}$ be a category.

(i)   Show that if a morphism $f \in \mathbf{C}(X, Y)$ has both a right inverse $g$ and a left inverse $g'$, then these are equal. (Hence if $h$ and $h'$ are both two-sided inverses of $f$, then $h = h'$.)

(ii)   Show that the relation $X \cong Y$ is an equivalence relation on $\mathrm{Ob}(\mathbf{C})$.

(iii)  Show that isomorphic objects in a category have isomorphic automorphism groups.

Our aim in this and the next section will be to look at various other concepts occurring in "concrete mathematics" and ask, in each case, whether we can define a concept for abstract categories which will yield the given concept in *many* concrete cases. We cannot expect that there will always be as perfect a fit as there was for the concept of isomorphism! But lack of perfect fit with existing concepts will not necessarily detract from the usefulness of the concepts we find.

Let us start with the concepts of "one-to-one map" and "onto map". The next exercise shows that no condition can give a perfect fit in these cases.

**Exercise 7.7:2.** Show that a category $\mathbf{C}$ can have concretizations $T$, $U$, $V$, $W: \mathbf{C} \to \mathbf{Set}$ such that for a particular morphism $f$ in $\mathbf{C}$,

$T(f)$ is one-to-one and onto,

$U(f)$ is one-to-one but not onto,

$V(f)$ is onto but not one-to-one, and

$W(f)$ is neither one-to-one nor onto.

(Suggestion: Take $\mathbf{C} = S_{\mathbf{cat}}$, where $S$ is the free monoid on one generator, or $\mathbf{C} = 2_{\mathbf{cat}}$, where $2$ is the two-element totally ordered set.)

Nevertheless, there is a category-theoretic property which in the vast majority of naturally occurring concrete categories does correspond to one-one-ness.

**Definition 7.7.1.** *A morphism* $f : X \to Y$ *in a category* **C** *is called a* monomorphism *if for all* $W \in \mathrm{Ob}(\mathbf{C})$ *and all pairs of morphisms* $g, h \in \mathbf{C}(W, X)$, *one has* $f g = f h \implies g = h$; *equivalently, if every covariant hom-functor* $h_W : \mathbf{C} \to \mathbf{Set}$ $(W \in \mathrm{Ob}(\mathbf{C}))$ *carries* $f$ *to a one-to-one set map.*

**Exercise 7.7:3.**

(i)  Show that if $(\mathbf{C}, U)$ is a concrete category (i.e., **C** is a category and $U : \mathbf{C} \to \mathbf{Set}$ a faithful functor) and $f$ is a morphism in **C** such that $U(f)$ is one-to-one, then $f$ is a monomorphism in **C**.

(ii)  Show that if **C** is a *small* category and $f$ a morphism in **C**, then $f$ is a monomorphism if and only if there exists a concretization functor $U : \mathbf{C} \to \mathbf{Set}$ such that $U(f)$ is one-to-one.

**Exercise 7.7:4.** Show that in the categories **Set**, **Group**, **Monoid**, **Ring**[1], **POSet** and **Lattice**, a morphism is one-to-one on underlying sets if and only if it is a monomorphism. (Suggestion: look for one method that works in all six cases.) If you are familiar with the basic definitions of general topology, also verify this for **Top**.

Naturally occurring concrete categories where monomorphisms are not the one-to-one maps are rare, but here is an example:

**Exercise 7.7:5.** A group $G$ is called *divisible* if for every $x \in |G|$ and every positive integer $n$, there exists $y \in |G|$ such that $x = y^n$.

(i)  Show that in the category of divisible groups (a full subcategory of **Group**), the quotient map $\mathbb{Q} \to \mathbb{Q}/\mathbb{Z}$ (where $\mathbb{Q}$ is the additive group of rational numbers and $\mathbb{Z}$ the subgroup of integers) is a monomorphism, though it is not a one-to-one map.

(ii)  Can you characterize group-theoretically the homomorphisms that are monomorphisms in the category of divisible abelian groups? Of all divisible groups?

(iii)  Can you find a category-theoretic property equivalent in either of these categories to being one-to-one?

If you are familiar with topological group theory, you may in the above questions consider the category of connected abelian Lie groups and the quotient map $\mathbb{R} \to \mathbb{R}/\mathbb{Z}$, instead of or in addition to divisible groups and $\mathbb{Q} \to \mathbb{Q}/\mathbb{Z}$.

It is natural to dualize the concept of monomorphism.

**Definition 7.7.2.** *A morphism* $f : X \to Y$ *in a category* **C** *is called an* epimorphism *if for all* $Z \in \mathrm{Ob}(\mathbf{C})$ *and all pairs of morphisms* $g, h \in \mathbf{C}(Y, Z)$ *one has* $g f = h f \implies g = h$; *equivalently, if all the contravariant hom-functors* $h^Z : \mathbf{C} \to \mathbf{Set}$ $(Z \in \mathrm{Ob}(\mathbf{C}))$ *carry* $f$ *to one-to-one set maps; equivalently, if in* $\mathbf{C}^{\mathrm{op}}$ *the morphism* $\tilde{f}$ *is a monomorphism.*

This concept coincides with that of a surjective map in many naturally occurring concrete categories; but in about equally many, it does not:

**Exercise 7.7:6.**

(i)   Show that if $(\mathbf{C}, U)$ is a concrete category, and $f$ a morphism in $\mathbf{C}$ such that $U(f)$ is surjective, then $f$ is an epimorphism in $\mathbf{C}$.

(ii)   Show that in the categories **Set** and **Ab**, the epimorphisms are precisely the surjective morphisms.

(iii) Show that in the category **Monoid**, the inclusion of the free monoid on one generator in the free group on one generator is an epimorphism, though not surjective with respect to the underlying-set concretization. (Hint: uniqueness of inverses.) Show similarly that in $\mathbf{Ring}^1$, the inclusion of any integral domain in its field of fractions is an epimorphism.

(iv) If you are familiar with elementary point-set topology, show that in the category **HausTop** of Hausdorff topological spaces, the epimorphisms are precisely the continuous maps with *dense* image.

**Exercise 7.7:7.**

(i)   Determine the epimorphisms in **Group**.

(ii)   Show the relation between this problem and Exercise 4.10:10.

(iii)   Does the method you used in (i) also yield a description of the epimorphisms in the category of *finite* groups? If not, can you nevertheless determine these?

**Exercise 7.7:8.** Let $\mathbf{C} = \mathbf{Ring}^1$, or, if you prefer, $\mathbf{CommRing}^1$.

(i)   Show that for an object $A$ of $\mathbf{C}$, the following conditions are equivalent: (a) The unique morphism $\mathbb{Z} \to A$ is an epimorphism. (b) For each object $R$ of $\mathbf{C}$, there is at most one morphism $A \to R$.

(ii)   Investigate the class of rings $A$ with the above property. (Cf. Exercise 4.10:10, and last sentence of Exercise 7.7:6(iii).)

**Exercise 7.7:9.**

(i)   Show that if $R$ is a commutative ring, and $f : R \to S$ is an epimorphism in $\mathbf{Ring}^1$, then $S$ is also commutative.

(Hint: Given a ring $A$, construct a ring $A'$ of formal sums $a + b\varepsilon$ $(a, b \in A)$ with multiplication given by $(a + b\varepsilon)(c + d\varepsilon) = ac + (ad + bc)\varepsilon$. For fixed $r \in A$, on what elements of $A$ do the two homomorphisms $A \to A'$ given by $x \mapsto x$ and $x \mapsto (1 + r\varepsilon)^{-1}x(1 + r\varepsilon)$ agree?)

(ii)   Show that if $f : R \to S$ is an epimorphism in $\mathbf{CommRing}^1$, then it is also an epimorphism in $\mathbf{Ring}^1$.

(Hint: Given homomorphisms $g, h : S \to T$ agreeing on $f(R)$, reduce to the situation where the image of $R$ in $T$ is in the center. Then look at the ring of endomorphisms of the additive group of $T$ generated by left multiplications by elements of $g(S)$ and right multiplications by elements of $h(S)$.)

(iii) Prove the converse of (ii), i.e., that if a homomorphism of commutative rings is an epimorphism in $\mathbf{Ring}^1$, then it is also an epimorphism in $\mathbf{CommRing}^1$. In fact, show that this is an instance of a general property of epimorphisms in a category and a subcategory.

Unlike the result of (iii), the results of (i) and (ii) are rather exceptional, as indicated by

(iv) Show that for a commutative ring $k$, the inclusion of the ring of upper triangular $2 \times 2$ matrices over $k$ (matrices $(a_{ij})$ such that $a_{21} = 0$) in the ring of all $2 \times 2$ matrices over $k$ is an epimorphism in $\mathbf{Ring}^1$. Show, however, that the identity $(x\,y - y\,x)^2 = 0$ holds in the former ring but not the latter.

Thus, although the result of (i) can be formulated as saying "If $f : R \to S$ is an epimorphism in $\mathbf{Ring}^1$, and $R$ satisfies the identity $x\,y - y\,x = 0$, then so does $S$", the corresponding statement with $x\,y - y\,x$ replaced by $(x\,y - y\,x)^2$ is false.

(v) Similarly, give an example showing that the analog of (ii) does not remain true if $\mathbf{Ring}^1$ and $\mathbf{CommRing}^1$ are replaced by an arbitrary category and any full subcategory thereof.

(vi) Does the analog of (i) and/or (ii) hold for the category $\mathbf{Monoid}$ and its subcategory $\mathbf{AbMonoid}$?

As some of the above exercises show, the property of being an epimorphism is not a reliable equivalent of surjectivity; but they also show that it is an interesting concept in its own right. In concrete categories, the statement that $f : A \to B$ is an epimorphism means intuitively that the image $f(A)$ "controls" $B$, in terms of behavior under morphisms.

There is an unfortunate tendency for some categorical enthusiasts to consider epimorphism to be the "category-theoretically correct" translation of surjective map, even in cases when the concepts do not agree. For instance, a standard definition in module theory calls a module $P$ *projective* if for every surjective module homomorphism $f : M \to N$, every homomorphism $P \to N$ factors through $f$. (If you haven't seen this concept, draw a diagram, and verify that every *free* module is projective.) I have heard it claimed that one should therefore define an object $P$ of a general category $\mathbf{C}$ to be projective if and only if for every *epimorphism* $f : M \to N$ of $\mathbf{C}$, every morphism $P \to N$ factors through $f$. This property is certainly of interest, but there is no reason to consider it to the exclusion of others. In particular, if $\mathbf{C}$ is a category having some natural concretization functor $U : \mathbf{C} \to \mathbf{Set}$, there is no reason to reject the concept of projective object defined in terms of factorization through "surjective" maps, i.e., maps $f : M \to N$ such that $U(f)$ is surjective. The fact that a property can be defined purely category-theoretically does not make it automatically superior to another property.

(The right context for developing a theory of "projective objects" is probably that of a category $\mathbf{C}$ given with a subfamily of morphisms $S$, which we wish to put in the role of surjections. To make things behave nicely, one will presumably want to put certain restrictions on $S$; for instance that it be *contained* in the class of epimorphisms, as the surjective maps in concrete categories always are by Exercise 7.7:6(i); probably also that it contain all invertible morphisms, and be closed under composition. We would then say that an object $P$ is "projective with respect to the class $S$" if for every morphism $f : M \to N$ belonging to $S$, every morphism $P \to N$ factors through $f$. This relative approach is taken in [102], where a large number of properties are defined relative to a *pair* of classes of morphisms, one in the role of the surjections and the other in the role of the injections.)

The use of the words "monomorphism" and "epimorphism" is itself unsettled. In the days before category theory, the words were introduced by Bourbaki with the meanings "injective (i.e., one-to-one) homomorphism" and "surjective (i.e., onto) homomorphism". The early category-theorists brazenly gave these words the abstract category-theoretic meanings we have been discussing. This made the terms ambiguous in situations where the category-theoretic definition did not agree with the old meaning. Mac Lane [20] tries to remedy the situation by restoring "monomorphism" and "epimorphism" to their old meanings (applicable in concrete categories) and calling the general category-theoretic concepts that we have been discussing "monic" and "epic" morphisms, or "monos" and "epis" for short. However, the category-theoretic meanings are already well-established in many areas; e.g., there have been many published papers dealing with epimorphisms in categories of rings. (A concept which includes the construction of the field of fractions of a commutative domain is bound to be of interest!) My feeling is that "epimorphism" and "epic morphism" sound too similar to usefully carry Mac Lane's distinction; and that we should now stick with the category-theoretic meanings of "epimorphism" and "monomorphism". The phrases "surjective (or onto) homomorphism" and "injective (or one-to-one) homomorphism" give us more than enough ways of referring to the concrete concepts.

In any case, when you see these words used by other authors, make sure which meaning they are giving them.

**Exercise 7.7:10.** Suppose $f \in \mathbf{C}(Y, Z)$, $g \in \mathbf{C}(X, Y)$. Investigate implications holding among the conditions "$f$ is a monomorphism", "$g$ is a monomorphism", "$fg$ is a monomorphism" "$f$ is an epimorphism", "$g$ is an epimorphism" and "$fg$ is an epimorphism".

A full answer would be an exact determination of which among the 64 possible combinations of truth-values for these six statements can hold for a pair of morphisms, and which cannot! As a partial answer, you might determine which of the eight possible combinations of truth-values of the first three conditions can hold. Then see whether duality allows you to deduce which combinations of the last three can hold, and whether, by examining when morphisms in a *product* of categories are monomorphisms or epimorphisms, you can use the results you have found to get a complete or nearly complete answer to the full 64-case question.

**Exercise 7.7:11.** Although in most natural categories of mathematical objects the two obvious questions about a morphism are whether it is one-to-one and whether it is onto, in the category **RelSet** we can ask additional questions such as whether a given relation is a *function*.

(i) Can you find a general condition on a morphism in an arbitrary category, which, for a morphisms $f: X \to Y$ in **RelSet**, is equivalent to being a set-theoretic function $X \to Y$?

(ii) Examine other properties of relations, and whether they can be characterized by category-theoretic properties in **RelSet**. For instance, which members of **RelSet**$(X, X)$ represent partial orderings on $X$? Given $f, g \in$ **RelSet**$(X, Y)$, how can one determine whether $f \subseteq g$ as relations? Can one construct from the category-structure of **RelSet** the contravariant functor $R:$ **RelSet**$^{op} \to$ **RelSet** taking each relation $f \in$ **RelSet**$(X, Y)$ to the opposite relation, $R(f) \in$ **RelSet**$(Y, X)$?

(iii) Can you find a necessary and sufficient condition on a subset $f \subseteq X \times Y$ for it to be, when regarded as morphism in **RelSet**, a monomorphism, or an epimorphism in that category? Left or right invertible?

Because of the way we used duality in getting from the concept of monomorphism to that of epimorphism, both of them refer to *one-one-ness* of the images of a morphism under certain hom-functors. Let us look at the conditions that these same images be *onto*:

**Exercise 7.7:12.**

(i) Given $f \in$ **C**$(X, Y)$, show that the following conditions are equivalent:

(a) For all $Z \in \text{Ob}(\mathbf{C})$, $h_Z(f)$ is surjective.

(b) $f$ is right invertible; i.e., there exists $g \in$ **C**$(Y, X)$ such that $f g = \text{id}_Y$.

(c) For every covariant functor $F: \mathbf{C} \to \mathbf{Set}$, $F(f)$ is surjective.

(d) For every contravariant functor $F: \mathbf{C} \to \mathbf{Set}$, $F(f)$ is injective.

(e) For every category **D** and covariant functor $F: \mathbf{C} \to \mathbf{D}$, $F(f)$ is an epimorphism.

(f) For every category **D** and contravariant functor $F: \mathbf{C} \to \mathbf{D}$, $F(f)$ is a monomorphism.

(For partial credit, simply establish the equivalence of (a) and (b). Hint: $\text{id}_Y \in h_Y(Y)$.)

(ii) State the result which follows from the result of (i) by duality, indicating briefly how one deduces this dual result from that of (i).

Let us look at what condition (b) of the above exercise means in familiar categories; in other words, what it means to have morphisms $f$ and $g$ satisfying a one-sided inverse relation,

(7.7.3)                    $f g = \mathrm{id}_Y \quad (f \in \mathbf{C}(X, Y),\, g \in \mathbf{C}(Y, X))$.

First take $\mathbf{C} = \mathbf{Set}$. Then we see that if (7.7.3) holds, $g$ must be one-to-one (if two elements of $Y$ fell together under $g$, there would be no way for $f$ to "separate" them); so let us think of $g$ as embedding a copy of $Y$ in $X$. The map $f$ sends $X$ to $Y$ so as to take each element $g(y)$ back to $y$, while acting in an unspecified way on elements of $X$ that are not in the image of $g$. Thus the composite $g f \in \mathbf{C}(X, X)$ leaves elements of the image of $g$ fixed, and carries all elements not in that image into that image; i.e., it "retracts" $X$ onto the embedded copy of $Y$. Hence in an arbitrary category, a pair of morphisms satisfying (7.7.3) is called a *retraction* of the object $X$ onto the object $Y$. In this situation $Y$ is said to be a *retract* of $X$ (via the morphisms $f$ and $g$).

**Exercise 7.7:13.**

(i)   Show that a morphism in **Set** is left invertible if and only if it is one-to-one, with the exception of certain cases involving $\emptyset$ (which you should show are indeed exceptions) and right invertible if and only it is onto (without exceptions).

(ii)   Show that $X$ is a retract of $Y$ in the category **Ab** of abelian groups (or more generally, the category $R$-**Mod** of left $R$-modules) if and only if $X$ is isomorphic to a direct summand in $Y$.

(iii)   Give examples of a morphism in **Ab** that is surjective, but not right invertible, and a morphism that is one-to-one, but not left invertible.

(iv)   Characterize retractions in **Group** in terms of group-theoretic constructions. Do they all arise from direct-product decompositions, as in **Ab**?

Combining part (i) of the above exercise with Exercises 7.7:3(i) and 7.7:6(iii), we see that for morphisms in any concrete category, one has

left invertible $\implies$ one-to-one $\implies$ monomorphism,
right invertible $\implies$ onto $\implies$ epimorphism.

On the other hand, part (iii) of the above exercise and similar examples given in earlier exercises show that none of these implications are reversible.

**Exercise 7.7:14.**  Give an example of a morphism in some category which is both an epimorphism and a monomorphism, but not an isomorphism. Investigate what combinations of the properties "epimorphism", "monomorphism", "left invertible" and "right invertible" force a morphism to be an isomorphism.

(Warning in connection with the above discussion and exercises: The meanings of the terms "left" and "right" invertible become reversed when category-theorists—or other mathematicians—compose their maps in the opposite sense to the one we are using!)

We have noted that in the situation of (7.7.3) the composite $e = gf$ is an idempotent endomorphism of the object $X$, whose image, in concrete situations, is a copy of the retract $Y$. The next exercise establishes two category-theoretic versions of the idea that this idempotent morphism "determines" the structure of the retract $Y$ of $X$.

**Exercise 7.7:15.**

(i) Let $X, Y, Y' \in \mathrm{Ob}(\mathbf{C})$, and suppose that $f \in \mathbf{C}(X, Y)$, $f' \in \mathbf{C}(X, Y')$ have right inverses $g$, $g'$ respectively. Show that $gf = g'f' \implies Y \cong Y'$.

(ii) Let $\mathbf{C}$ be a category, and $e \in \mathbf{C}(X, X)$ be an idempotent morphism: $e^2 = e$. Show that $\mathbf{C}$ may be embedded as a full subcategory in a category $\mathbf{D}$, *unique up to isomorphism*, with one additional object $Y$ (i.e., with $\mathrm{Ob}(\mathbf{D}) = \mathrm{Ob}(\mathbf{C}) \cup \{Y\}$) and such that there exist morphisms $f \in \mathbf{D}(X, Y)$, $g \in \mathbf{D}(Y, X)$ satisfying

$$fg = \mathrm{id}_Y \ (\text{in } \mathbf{D}(Y, Y)), \quad gf = e \ (\text{in } \mathbf{D}(X, X) = \mathbf{C}(X, X)).$$

Returning to our search for conditions which correspond to familiar mathematical concepts in many cases, let us ask whether we can define a concept of a *subobject* of an object $X$ in a category $\mathbf{C}$.

If by this we mean a criterion telling *which* objects of a category such as **Set** or **Group** are actually *contained* in which other objects, the answer is "certainly not": There can be no way to distinguish an object that is a subobject of another from one that is simply *isomorphic* to such a subobject. However, in particular categories of mathematical objects, we may well be able to say when a given morphism is an *embedding*, i.e., corresponds to an isomorphism of its domain object with a subobject of its codomain. For instance, in **Set**, **Group**, **Monoid**, **Ring**[1], **Lattice** and similar categories, the embeddings are the monomorphisms. In these cases, and more generally, whenever we know which morphisms we want to regard as embeddings, we can recover the partially ordered set of subobjects of $X$ as equivalence classes of such morphisms:

**Exercise 7.7:16.** Let $\mathbf{C}$ be a category, and suppose we are given a subcategory $\mathbf{C}_{\mathrm{emb}}$ of $\mathbf{C}$ whose object-set consists of all the objects of $\mathbf{C}$, and whose set of morphisms is *contained in* the set of *monomorphisms* of $\mathbf{C}$. The morphisms of $\mathbf{C}_{\mathrm{emb}}$ are the morphisms of $\mathbf{C}$ that we intend to *think of* as embeddings. (But you may not assume anything about $\mathbf{C}_{\mathrm{emb}}$ except the conditions stated above.) For any object $X$ of $\mathbf{C}$, let $\mathbf{Emb}_X$ denote the category whose objects are pairs $(Y, f)$, where $Y \in \mathrm{Ob}(\mathbf{C})$ and

$f \in \mathbf{C}_{\mathrm{emb}}(Y, X)$, and where a morphism from $(Y, f)$ to $(Z, g)$ means a morphism $a : Y \to Z$ of $\mathbf{C}$ such that $f = g\,a$.

(i)    Show that each hom-set $\mathbf{Emb}_X(U, V)$ has at most one element. Deduce that $\mathbf{Emb}_X$ is of the form $\mathrm{Emb}(X)_{\mathbf{cat}}$ for some (possibly large) preorder $\mathrm{Emb}(X)$.

(ii)    Let us call the partially ordered set constructed from the preorder $\mathrm{Emb}(X)$ as in Proposition 5.2.2 "$\mathrm{Sub}(X)$". Show that if $\mathbf{C}$ is one of $\mathbf{Set}$, $\mathbf{Group}$, $\mathbf{Ring}$ or $\mathbf{Lattice}$, and we take $\mathbf{C}_{\mathrm{emb}}$ to have for its morphisms all the monomorphisms of $\mathbf{C}$, then $\mathrm{Sub}(X)$ is isomorphic to the partially ordered set of subsets, subgroups, etc., of $X$.

(iii)    Let $X$ be a set, in general infinite, and $S$ the monoid of set maps of $X$ into itself. Form the category $S_{\mathbf{cat}}$, and take $(S_{\mathbf{cat}})_{\mathrm{emb}}$ to have the monomorphisms of $S_{\mathbf{cat}}$ for its morphisms. Calling the one object of $S_{\mathbf{cat}}$ "$0$", describe the partially ordered set $\mathrm{Sub}(0)$.

The categories of algebraic objects mentioned so far in discussing one-one-ness have had the property that every one-to-one morphism gives an isomorphism of its domain with a subobject of its codomain. An example of a category for which this is not true is $\mathbf{POSet}$. For instance if $P$ and $Q$ are finite partially ordered sets having the same underlying set, but the order-relation on $Q$ is stronger than that of $P$, then the identity map of the underlying set is a one-to-one isotone map from $P$ to $Q$, but some elements of $Q$ will satisfy order-relations that they don't satisfy in $P$; so we cannot regard $P$ as a subobject of $Q$ with the induced ordering. This leads to the questions:

**Exercise 7.7:17.**

(i)    Suppose the construction of the preceding exercise is applied with $\mathbf{C}$ the category $\mathbf{POSet}$, and $\mathbf{C}_{\mathrm{emb}}(X, Y)$ the set of all monomorphisms in $\mathbf{C}(X, Y)$. For $X \in \mathrm{Ob}(\mathbf{POSet})$, describe the partially ordered set $\mathrm{Sub}(X)$.

(ii)    Can you find a category-theoretic property characterizing those morphisms of $\mathbf{POSet}$ which are "genuine" embeddings, i.e., correspond to isomorphisms of their domain with subsets of their codomain, partially ordered under the induced ordering?

## 7.8. More categorical versions of common mathematical notions: special objects

I shall start this section with some "trivialities".

In many of the classes of structures we have dealt with, there were one, or sometimes two objects that one would call the "trivial" objects: the one-element group; the one-element set and also the empty set; the one-element lattice and likewise the empty lattice. The following definition abstracts the common properties of these objects.

**Definition 7.8.1.** *An* initial object *in a category* **C** *means an object* $I$ *such that for every* $X \in \mathrm{Ob}(\mathbf{C})$, $\mathbf{C}(I, X)$ *has exactly one element.*

*A* terminal object *in a category* **C** *means an object* $T$ *such that for every* $X \in \mathrm{Ob}(\mathbf{C})$, $\mathbf{C}(X, T)$ *has exactly one element.*

*An object that is both initial and terminal is often called a* zero object.

Thus, in **Set**, the empty set is the unique initial object, while any one-element set is a terminal object. In **Group**, a one-element group is both initial and terminal, hence is a zero object. The categories **Lattice**, **POSet**, **Top** and **Semigroup** are like **Set** in this respect, while **Top**$^{\mathrm{pt}}$ and **Monoid** are like **Group**. In **Ring**[1], the initial object is $\mathbb{Z}$, though we would usually not call it "trivial"; the terminal object is the one-element ring with $1 = 0$ (which some people do not call a ring).

A category need not have an initial or terminal object: The category of nonempty sets, or nonempty partially ordered sets, or nonempty lattices, or finite rings, has no initial object; **POSet**$_<$ has no terminal object, nor does the category of nonzero rings (rings in which $1 \neq 0$). If $P$ is the partially ordered set of the integers, then $P_{\mathbf{cat}}$ has neither an initial nor a terminal object. Terminal objects are also called "final" objects, and I may sometimes slip and use that word in class.

**Lemma 7.8.2.** *If* $I$, $I'$ *are two initial objects in a category* **C**, *then they are isomorphic, via a unique isomorphism. Similarly, any two terminal objects are isomorphic via a unique isomorphism.* □

**Exercise 7.8:1.** Prove Lemma 7.8.2.

**Exercise 7.8:2.** Suppose **C** is a category with an initial object $I$ and a terminal object $T$, and suppose $f$ is a morphism with domain $I$ or $T$. We would like to know whether $f$ will *always, sometimes,* or *almost never* be an epimorphism or a monomorphism. Here by "almost never" I mean "only if it is an isomorphism", while by "sometimes", I mean "not for all choices of **C** and $f$, but in at least some cases other than when $f$ is an isomorphism".

(i)     For each of the four combinations of one of the two distinguished objects $I$ and $T$, and one of the two conditions "epimorphism" or "monomorphism", answer the above question, i.e., prove the answer if it is "always" or "almost never", while if it is "sometimes", give the two examples needed to establish this.

(ii)     State the corresponding results for morphisms with *codomain* $I$ and $T$, noting briefly how the results of part (i) can be used to get these.

**Exercise 7.8:3.** Consider the following conditions on a category $\mathbf{C}$:

(a) $\mathbf{C}$ has a zero object (an object that is both initial and terminal).

(b) It is possible to choose in each hom-set $\mathbf{C}(X, Y)$ a morphism $0_{X, Y}$ in such a way that for all $X, Y, Z \in \mathrm{Ob}(\mathbf{C})$ and $f \in \mathbf{C}(X, Y)$, $g \in \mathbf{C}(Y, Z)$ one has $0_{Y, Z} f = 0_{X, Z} = g \, 0_{X, Y}$.

(c) It is possible to choose in each hom-set $\mathbf{C}(X, Y)$ a morphism $0_{X, Y}$ such that for all $X, Y, Z \in \mathrm{Ob}(\mathbf{C})$ one has $0_{Y, Z} \, 0_{X, Y} = 0_{X, Z}$.

(d) For all $X, Y \in \mathrm{Ob}(\mathbf{C})$, $\mathbf{C}(X, Y) \neq \emptyset$.

(i)   Show that (a) $\implies$ (b) $\implies$ (c) $\implies$ (d), but that none of these implications is reversible.

(ii)  Show that if $\mathbf{C}$ has either an initial or a terminal object, then the first and third implications are reversible, but not, in general, the second.

(iii) Show that if $\mathbf{C}$ has an initial object *and* a terminal object (as the majority of naturally occurring categories do), then (d) $\implies$ (a), so that all four conditions are equivalent.

**Exercise 7.8:4.** If $\mathbf{C}$ is a category with a terminal object $T$, let $\mathbf{C}^{\mathrm{pt}}$ denote the category whose objects are pairs $(X, p)$, where $X \in \mathrm{Ob}(\mathbf{C})$, $p \in \mathbf{C}(T, X)$, and where $\mathbf{C}^{\mathrm{pt}}((X, p), (Y, q)) = \{f \in \mathbf{C}(X, Y) \mid f p = q\}$.

(i)   Verify that this defines a category, and that $\mathbf{C}^{\mathrm{pt}}$ will have a zero object.

(ii)  Show that if $\mathbf{C} = \mathbf{Top}$, this gives the category we earlier named $\mathbf{Top}^{\mathrm{pt}}$.

(iii) Show that if $\mathbf{C}$ already had a zero object, then $\mathbf{C}^{\mathrm{pt}}$ is isomorphic to $\mathbf{C}$.

**Exercise 7.8:5.** If $\mathbf{C}$ is a category, call an object $A$ of $\mathbf{C}$ *quasi-initial* if it satisfies the condition of Exercise 7.7:8(i)(b). Generalize the result "(a) $\iff$ (b)" of that exercise to a characterization of quasi-initial objects in categories with initial objects.

What about the concept of *free* object? The definition of a free group $F$ on a set $X$ refers to *elements* of groups, hence the generalization should apply to a *concrete* category $(\mathbf{C}, U)$. You should verify that when $\mathbf{C} = \mathbf{Group}$ and $U$ is the underlying set functor, the following definition reduces to the usual definition of free group.

**Definition 7.8.3.** *If $\mathbf{C}$ is a category, $U: \mathbf{C} \to \mathbf{Set}$ a faithful functor, and $X$ a set, then a free object of $\mathbf{C}$ on $X$ with respect to the concretization $U$ will mean a pair $(F_X, u)$, where $F_X \in \mathrm{Ob}(\mathbf{C})$, $u \in \mathbf{Set}(X, U(F_X))$, and this pair has the universal property that for any pair $(G, v)$ with $G \in \mathrm{Ob}(\mathbf{C})$, $v \in \mathbf{Set}(X, U(G))$, there is a unique morphism $h \in \mathbf{C}(F_X, G)$ such that $v = U(h) \, u$.*

Loosely, we often call the object $F_X$ the free object, and $u$ the associated universal map.

**Exercise 7.8:6.** Let $V$ denote the functor associating to every group $G$ the set $|G|^2$ of ordered pairs $(x, y)$ of elements of $G$, and $W$ the functor associating to $G$ the set of "unordered pairs" $\{x, y\}$ of elements of $G$ (where $x = y$ is allowed).

(i)    State how these functors should be defined on morphisms. (I don't ask you to verify the fairly obvious fact that these descriptions do make $V$ and $W$ functors.) Show that they are both faithful.

(ii)   Show that for any set $X$, there exists a free group with respect to the functor $V$, and describe this group.

(iii)  Show that there do not in general exist free groups with respect to $W$.

**Exercise 7.8:7.** Let $U \colon \mathbf{Ring}^1 \to \mathbf{Set}$ be the functor associating to every ring $R$ the set of $2 \times 2$ invertible matrices over $R$. Show that $U$ is faithful. Does there exist for every set $X$ a free ring $R_X$ on $X$ with respect to $U$?

The next exercise shows why the property of being a monomorphism characterizes the one-to-one maps in most of the concrete categories we know—or more precisely, shows that this characterization follows from another property we have noted in these categories.

**Exercise 7.8:8.** Let $(\mathbf{C}, U)$ be a concrete category. Show that if there exists a free object on a one-element set with respect to $U$, then a morphism $f$ of $\mathbf{C}$ is a monomorphism if and only if $U(f)$ is one-to-one.

We could go further into the study of free objects, proving, for instance, that they are unique up to isomorphism when they exist, and that when $\mathbf{C}$ has free objects on all sets, the free-object construction gives a functor $\mathbf{Set} \to \mathbf{C}$. Some of this will be done in Exercise 7.9:9 later in this chapter, but for the most part we shall get such results in the next chapter, as part of a theory embracing wide classes of universal constructions.

Let us turn to another pair of constructions that we have seen in many categories (including $\mathbf{Cat}$ itself), those of *product* and *coproduct*. No concretization or other additional structure is needed to translate these concepts into category-theoretic terms.

**Definition 7.8.4.** *Let $\mathbf{C}$ be a category, $I$ a set, and $(X_i)_{i \in I}$ a family of objects of $\mathbf{C}$.*

*A* product *of this family in $\mathbf{C}$ means a pair $(P, (p_i)_{i \in I})$, where $P \in \mathrm{Ob}(\mathbf{C})$ and for each $i \in I$, $p_i \in \mathbf{C}(P, X_i)$, having the universal property that for any pair $(Y, (y_i)_{i \in I})$ $(Y \in \mathrm{Ob}(\mathbf{C}), y_i \in \mathbf{C}(Y, X_i))$ there exists a unique morphism $r \in \mathbf{C}(Y, P)$ such that $y_i = p_i r$ $(i \in I)$.*

*Likewise, a* coproduct *of the family $(X_i)_{i \in I}$ means a pair $(Q, (q_i)_{i \in I})$, where $Q \in \mathrm{Ob}(\mathbf{C})$ and for each $i \in I$, $q_i \in \mathbf{C}(X_i, Q)$, having the universal*

*property that for any pair* $(Y, (y_i)_{i\in I})$ $(Y \in \mathrm{Ob}(\mathbf{C}),\ y_i \in \mathbf{C}(X_i, Y))$ *there
exists a unique morphism* $r \in \mathbf{C}(Q, Y)$ *such that* $y_i = r\, q_i$ $(i \in I)$.

Loosely, we call $P$ and $Q$ the product and coproduct of the objects $X_i$,
the $p_i : P \to X_i$ the projection *maps, and the* $q_i : X_i \to Q$ *the coprojection
maps. (The term* injection *is used by some authors instead of* coprojection.)

The category $\mathbf{C}$ *is said to have finite products if every finite family of
objects of* $\mathbf{C}$ *has a product in* $\mathbf{C}$, *and to have small products (often simply
"to have products") if every family of objects of* $\mathbf{C}$ *indexed by a small set
has a product; and similarly for finite and small coproducts.*

Standard notations for product and coproduct objects are $P = \prod_{i\in I} X_i$
and $Q = \coprod_{i\in I} X_i$. For a product of finitely many objects one also writes
$X_0 \times \cdots \times X_{n-1}$. There is no analogous standard notation for coproducts
of finitely many objects; we used "$*$" as the operation-symbol in Chapter 4,
following group-theorists' notation for "free products". One sometimes sees
$+$ or $\oplus$, based on module-theoretic notation. Still another notation that is
used, and which I will follow from now on in these notes, is $X_0 \amalg \ldots \amalg X_{n-1}$.

Observe that a product of the empty family is equivalent to a terminal
object, while a coproduct of the empty family is equivalent to an initial
object.

**Exercise 7.8:9.** If $P$ is a partially ordered set, what does it mean for a
family of objects of $P_{\mathbf{cat}}$ to have a product? A coproduct?

**Exercise 7.8:10.**

(i)   Suppose we are given a *family of families of objects* in a category $\mathbf{C}$,
$((X_{ij})_{i\in I_j})_{j\in J}$, such that for each $j$, $\prod_{I_j} X_{ij}$ exists, and such that we can
also find a product of these product objects, $P = \prod_{j\in J}(\prod_{i\in I_j} X_{ij})$. Show
that $P$ will be a product of the family $(X_{ij})_{i\in I_j,\, j\in J}$.

(ii)   Deduce that if a category has products of pairs of objects, it has
products of all finite nonempty families of objects.

(iii)  Consider the case of (i) where $J = \{0, 1\}$, $I_0 = \emptyset$, and $I_1 = \{0\}$.
What form do the products described there take, and what does the conclusion tell us? Also state the dual of the result you get.

**Exercise 7.8:11.**

(i)   Let $X$ be a set (in general infinite) and $S$ the monoid of maps of $X$
into itself. When, if ever, does the category $S_{\mathbf{cat}}$ have products of pairs of
objects? (Of course, there is only one ordered pair of objects, and only one
object to serve as their product, so the question comes down to whether
two morphisms $p_1$ and $p_2$ can be found having appropriate properties.)

(ii)   Is there, in some sense, a "universal" example of a monoid $S$ such
that $S_{\mathbf{cat}}$ has products of pairs of objects?

**Exercise 7.8:12.** Let $k$ be a field. Show that one can define a category $\mathbf{C}$ whose objects are the $k$-vector-spaces, and such that for vector spaces $U$ and $V$, $\mathbf{C}(U, V)$ is the set of equivalence classes of linear maps $U \to V$ under the equivalence relation that makes $f \sim g$ if and only if the linear map $f - g$ has finite rank. Show that in this category, finite families of objects have products and coproducts, but infinite families in general have neither.

We saw in Exercise 7.7:13(i) that in $\mathbf{Ab}$ and $R\text{-}\mathbf{Mod}$, any retraction of an object arises from a decomposition as a direct sum, which in those categories is both a product and coproduct. The next exercise examines the relation between retractions, products and coproducts in general.

**Exercise 7.8:13.**

(i) Show that if $\mathbf{C}$ is a category with a zero object, then for any objects $A$ and $B$ of $\mathbf{C}$, if the product $A \times B$ exists, then $A$ can be identified with a retract of this product, and if the coproduct $A \amalg B$ exists, then $A$ can be identified with a retract of this coproduct.

(ii) Can you find a condition weaker than the existence of a zero object under which these conclusions hold?

Though we saw in (i) that in a category with a zero object, a decomposition of an object as a product or a coproduct leads to a retraction, it is not in general true that every retraction comes from a product decomposition, nor that every retraction comes from a coproduct decomposition. Indeed,

(iii) Let $A$ and $B$ be nontrivial objects of **Group**. Thus, by part (i) above, the subgroup $A \subseteq A \amalg B$ is a retract. Show, however, that it is not a factor in any *product* decomposition of that group. Likewise, show that $A \subseteq A \times B$, though a retract, is not a factor in any *coproduct* decomposition of that group.

Some related facts are noted in the next exercise. (Part (iii) thereof requires some group-theoretic expertise, or some ingenuity.)

**Exercise 7.8:14.**

(i) Show that if $A$ is the free group or free abelian group on a generating set $X$, and $Y$ is a subset of $X$, then the subgroup of $A$ generated by $Y$ is a retract of $A$.

(ii) For the case of a free abelian group $A$, show, conversely, that if $B$ a retract of $A$, then $A$ has a basis $X$ such that $B$ is the subgroup generated by a subset of $X$.

(iii) On the other hand, show that if $A$ is the free group on two generators $x$ and $y$, then $A$ has cyclic subgroups which are retracts, but are not generated by any subset of any free generating set for $A$. (Suggestion: try the cyclic subgroup generated by $x^2 y^3$, or by $x^2 y x^{-1} y^{-1}$.)

The next exercise shows that when one requires even *large* families of objects to have products, one's categories tend to become degenerate.

**Exercise 7.8:15.** Let $\mathbf{C}$ be a category and $\alpha$ a cardinal such that $\mathrm{Ob}(\mathbf{C})$ and all morphism sets $\mathbf{C}(X, Y)$ have cardinality $\leq \alpha$ (e.g., the cardinality of a universe with respect to which $\mathbf{C}$ is legitimate).

(i)     Show that if every family of objects of $\mathbf{C}$ indexed by a set of cardinality $\leq \alpha$ has a product in $\mathbf{C}$, then $\mathbf{C}$ has the form $P_{\mathbf{cat}}$, where $P$ is a preorder whose associated partially ordered set $P/\approx$ is a complete lattice.

(ii)    Deduce that in this case *every* family of objects of $\mathbf{C}$ (indexed by any set whatsoever) has a product and a coproduct.

It is an easy fallacy to say, "since product is a category-theoretic notion, functors must respect products." Rather

**Exercise 7.8:16.** Find an example of categories $\mathbf{C}$ and $\mathbf{D}$ having finite products, and a functor $\mathbf{C} \to \mathbf{D}$ which does not respect such products.

On the other hand:

**Exercise 7.8:17.** Show that if $(\mathbf{C}, U)$ is a concrete category, and there exists a free object on one generator with respect to $U$, then $U$ respects all products which exist in $\mathbf{C}$. (Cf. Exercise 7.8:8.)

Thus, in most of the concrete categories we have been interested in, the underlying set of a product object is the direct product of the underlying sets of the given objects. However, there is a well-known example for which this fails:

**Exercise 7.8:18.** A *torsion* group (also called a "periodic group") is a group all of whose elements are of finite order. Let $\mathbf{TorAb}$ be the category of all torsion abelian groups.

(i)     Show that a product in $\mathbf{Ab}$ of an infinite family of torsion abelian groups is not in general a torsion group.

(ii)    Show, however, that the category $\mathbf{TorAb}$ has small products.

(iii) Deduce that the underlying set functor $\mathbf{TorAb} \to \mathbf{Set}$ does not respect products.

**Exercise 7.8:19.** Does the category $\mathbf{TorGroup}$ of *all* torsion groups have small products?

**Exercise 7.8:20.** Consider a category $\mathbf{C}$ having finite products. When we spoke of making the product construction into a functor (in motivating the concept of a functor of two variables), the domain category was to be the set of *pairs* of objects of $\mathbf{C}$. Clearly we can do the same using $I$-tuples for any *fixed* finite set $I$. But what if we look at the product construction as simultaneously applying to $I$-tuples of objects as $I$ ranges over *all* finite index sets?

To make this question precise, let $\mathrm{Ob}(\mathbf{C})^+$ denote the class of all families $(X_i)_{i \in I}$ such that $I$ is a finite set (varying from family to family) and the $X_i$ are objects of $\mathbf{C}$. Can you make this the object-set of a category $\mathbf{C}^+$ in a natural way (which allows morphisms between families indexed by sets of possibly different sizes), so that the product construction becomes a functor $\mathbf{C}^+ \to \mathbf{C}$? If so, will the same category $\mathbf{C}^+$ serve as domain for the *coproduct* construction, assuming $\mathbf{C}$ has finite coproducts?

For future reference, let us make

**Definition 7.8.5.** *Let $I$ be a set (for instance, a natural number or other cardinal), and $\mathbf{C}$ a category having $I$-fold products. If $X$ is an object of $\mathbf{C}$, then when the contrary is not stated, $X^I$ will denote the $I$-fold product of $X$ with itself, which we may call the "$I$-th power of $X$". Likewise, if $F$ is a functor from another category $\mathbf{D}$ to $\mathbf{C}$, then when the contrary is not stated, $F^I$ will denote the functor taking each object $Y$ of $\mathbf{D}$ to the object $F(Y)^I$ of $\mathbf{C}$, and behaving in the obvious way on morphisms.*

(Note that if $F \colon \mathbf{C} \to \mathbf{C}$ is an endofunctor of a category $\mathbf{C}$, we might want to write $F^n$ for the $n$-fold composite of $F$ with itself. In such a case we would have to make an explicit exception to the above convention.)

What about category-theoretic versions of the constructions of *kernel* and *cokernel*?

We saw that these constructions were specific to fairly limited kinds of mathematical objects, such as groups and rings, but that a pair of concepts which embrace them but are much more versatile are those of *equalizer* and *coequalizer*. The latter concepts are abstracted in

**Definition 7.8.6.** *Let $\mathbf{C}$ be a category, $X, Y \in \mathrm{Ob}(\mathbf{C})$, and $f, g \in \mathbf{C}(X, Y)$.*

*Then an* equalizer *of $f$ and $g$ means a pair $(K, k)$, where $K$ is an object, and $k \colon K \to X$ a morphism which satisfies $f k = g k$, and is universal for this property, in the sense that for any pair $(W, w)$ with $W$ an object and $w \colon W \to X$ a morphism such that $f w = g w$, there exists a unique morphism $h \colon W \to K$ such that $w = k h$.*

*Likewise, a* coequalizer *of $f$ and $g$ means a pair $(C, c)$ where $C$ is an object, and $c \colon Y \to C$ a morphism which satisfies $c f = c g$, and is universal for this property, in the sense that for any pair $(Z, z)$ with $Z$ an object and $z \colon Y \to Z$ a morphism such that $z f = z g$, there exists a unique morphism $h \colon C \to Z$ such that $z = h c$.*

*Loosely, $K$ and $C$ are called the* equalizer *and* coequalizer *objects, and $k$, $c$ the* equalizer *and* coequalizer *morphisms, or the* canonical morphisms *associated with the equalizer and coequalizer objects. We say that $\mathbf{C}$ has* equalizers *(respectively coequalizers) if all pairs of morphisms between pairs of objects of $\mathbf{C}$ have equalizers (coequalizers).*

It turns out that in familiar categories, the concept of coequalizer yields a better approximation to that of *surjective map* than does the concept of epimorphism:

**Exercise 7.8:21.**

(i)   Show that in each of the categories **Group, Ring**[1], **Set, Monoid,** a morphism out of an object $Y$ is surjective on underlying sets if and only if it is a coequalizer morphism of some pair of morphisms from an object $X$ into $Y$.

(ii)   Is the same true in **POSet**? In the category of *finite* groups?

(iii)   In the categories considered in (i) (and optionally, those considered in (ii)) investigate whether, likewise, the condition of being an *equalizer* is equivalent to one-one-ness.

(iv)   Investigate what implications hold in a general category between the conditions of being an epimorphism, being right invertible, and being a coequalizer map.

**Exercise 7.8:22.** Let $f, g \in \mathbf{Set}(X, Y)$ be morphisms, and $(C, c)$ their coequalizer.

(i)   Show that $\mathrm{card}(X) + \mathrm{card}(C) \geq \mathrm{card}(Y)$. If you wish, assume $X$ and $Y$ are finite.

(ii)   Can one establish some similar relation between the cardinalities of $X$, of $Y$, and of the *equalizer* of $f$ and $g$ in **Set**?

(iii)   What can be said of the corresponding questions in **Ab**? In **Group**?

In categories such as **Group, Ab** and **Monoid** which have a zero object, concepts of *kernel* and *cokernel* of a morphism $f : X \to Y$ may also be defined, namely as the equalizer and coequalizer of $f$ with the zero morphism $X \to Y$ (see Exercise 7.8:3).

We turn next to a pair of constructions which we have not discussed before, but which are related both to products and coproducts and to equalizers and coequalizers.

**Definition 7.8.7.** *Given objects $X_1$, $X_2$, $X_3$ of a category $\mathbf{C}$, and morphisms $f_1 : X_1 \to X_3$, $f_2 : X_2 \to X_3$ (diagram below), a pullback of the pair of morphisms $f_1$, $f_2$ means a 3-tuple $(P, p_1, p_2)$, where $P$ is an object, and $p_1 : P \to X_1$, $p_2 : P \to X_2$ are morphisms satisfying $f_1 p_1 = f_2 p_2$, and which is universal for this property, in the sense that any 3-tuple $(Y, y_1, y_2)$, with $y_1 : Y \to X_1$, $y_2 : Y \to X_2$ satisfying $f_1 y_1 = f_2 y_2$, is induced by a unique morphism $h : Y \to P$.*

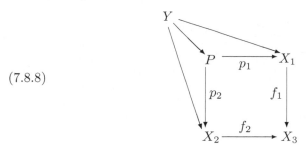

(7.8.8)

*Dually, for objects $X_0$, $X_1$, $X_2$ and morphisms $g_1 : X_0 \to X_1$, $g_2 : X_0 \to X_2$, a pushout of $g_1$ and $g_2$ means a 3-tuple $(Q, q_1, q_2)$, where $q_1 : X_1 \to Q$, $q_2 : X_2 \to Q$ satisfy $q_1\,g_1 = q_2\,g_2$, and which is universal for this property in the sense shown below:*

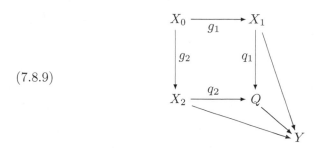

(7.8.9)

*As in the case of products and coproducts, the universal morphisms $p_1$, $p_2$ from a pullback object $P$ are called its* projection morphisms *to the $X_i$, and the universal morphisms $q_1$, $q_2$ to a pushout object $Q$ are called its* coprojection *morphisms.*

*A commuting square in* **C** *is called a pullback diagram (respectively, a pushout diagram) if the upper left-hand (lower right-hand) object is a pullback (pushout) of the remainder of the diagram. We say that a category* **C** *has* pullbacks *if every diagram of objects and morphisms $X_1$, $X_2$, $X_3$, $f_1$, $f_2$ as in (7.8.8) has a pullback $P$, and that* **C** *has* pushouts *if every diagram of objects and morphisms $X_0$, $X_1$, $X_2$, $g_1$, $g_2$ as in (7.8.9) has a pushout $Q$.*

The next exercise shows how to construct these creatures.

**Exercise 7.8:23.**

(i)  Show that if a category **C** has finite products and has equalizers, then it has pullbacks. Namely, for every system of objects and morphisms, $X_1$, $X_2$, $X_3$, $f_1$, $f_2$ as in the first part of the above definition, construct a pullback as the equalizer of a certain pair of morphisms $X_1 \times X_2 \to X_3$.

(ii)  State the dual result for pushouts (including the statement of how the pushout may be obtained).

To get a picture of pullbacks in **Set**, note that any set map $f : X \to Y$ can be regarded as a decomposition of the set $X$ into subsets $f^{-1}(y)$, indexed by the elements $y \in Y$. When one looks at $f$ this way, one calls $X$ a set *fibered* by $Y$, and calls $f^{-1}(y)$ the *fiber* of $X$ at $y \in Y$. Now in a pullback situation (7.8.8) in **Set**, we see that from two sets $X_1$ and $X_2$, each fibered by $X_3$, we obtain a third set $P$ fibered by $X_3$, with maps into the first two. From the preceding exercise one can verify that the fiber of $P$ at each $y \in X_3$ is the direct product of the fibers of $X_1$ and of $X_2$ at $y$. Consequently, pullbacks are sometimes called *fibered products*, whether or not one is working in a concrete category. The next exercise shows that "fibered products" can be regarded as products in an appropriate category of "fibered objects".

**Exercise 7.8:24.** Given a category $\mathbf{C}$ and any $Z \in \mathrm{Ob}(\mathbf{C})$, let $\mathbf{C}_Z$ denote the category of "objects of $\mathbf{C}$ fibered by $Z$", that is, the category having for objects all pairs $(X, f)$ where $X \in \mathrm{Ob}(\mathbf{C})$ and $f \in \mathbf{C}(X, Z)$, and having for morphisms $(X, f) \to (Y, g)$ all members of $\mathbf{C}(X, Y)$ making commuting triangles with the morphisms $f$ and $g$ into $Z$.

Show that a *pullback* (7.8.8) in $\mathbf{C}$ is equivalent to a *product* of the objects $(X_1, f_1)$, $(X_2, f_2)$ in $\mathbf{C}_{X_3}$.

The *pushout* $Q$ of a diagram (7.8.9) is also often called suggestively the "coproduct of $X_1$ and $X_2$ with *amalgamation* of $X_0$", especially in concrete situations where the morphisms $f_1$ and $f_2$ are embeddings. It also has names specific to particular fields: In topology, the $Q$ of (7.8.9) is the space gotten by "gluing together" the spaces $X_1$ and $X_2$ along a common image of $X_0$. In commutative ring theory, where the $X_0$, $X_1$ and $X_2$ of (7.8.9) might be denoted $K$, $R$ and $S$, the pushout $Q$ is written $R \otimes_K S$, and called the tensor product of $R$ and $S$ over $K$ as $K$-algebras.

In the spirit of Chapter 4, you might do

**Exercise 7.8:25.**

(i)   Show by a generators-and-relations argument that the category **Group** has pushouts.

(ii)   Obtain a normal form or other explicit description for the pushout, in the category of groups, of one-to-one group homomorphisms $f_1 : G_0 \to G_1$ and $f_2 : G_0 \to G_2$. Assume for notational convenience that these maps are inclusions, and that the underlying sets of $G_1$ and $G_2$ are disjoint except for the common subgroup $G_0$.

This is a classical construction, called by group theorists "the free product of $G_1$ and $G_2$ with amalgamation of the common subgroup $G_0$". (If you are already familiar with this construction, and the proof of its normal form by van der Waerden's trick, skip to the next part.)

(iii) Describe how to reduce the construction of an arbitrary pushout of groups to the case where the given maps $f_1$ and $f_2$ are one-to-one, as above.

**Exercise 7.8:26.** Show by example that given a pair of one-to-one monoid homomorphisms $g_1 : M_0 \to M_1$ and $g_2 : M_0 \to M_2$, if we construct their pushout $M$, the universal maps $M_1 \to M$ and $M_2 \to M$ need not be one-to-one.

Must the common map $M_0 \to M$ be one-to-one?

**Exercise 7.8:27.** This exercise, for students who have done (or are familiar with the result of) Exercise 7.8:25(ii), will construct an infinite, finitely presented group with no nontrivial finite homomorphic images.

The group is

$$(7.8.10) \qquad \begin{aligned} G = \langle w, x, y, z \mid & w^{-1} x w = x^2, \; x^{-1} y x = y^2, \\ & y^{-1} z y = z^2, \; z^{-1} w z = w^2 \rangle. \end{aligned}$$

Exercise 7.8:25 will not be needed until step (vi); in particular, it is not needed in proving that $G$ has no nontrivial finite homomorphic images (steps (i)–(ii)).

(i)   Show that if a group $H$ has nonidentity elements $x$, $y$, both of finite order, satisfying $x^{-1} y x = y^2$, then the least prime dividing the order of $x$ is strictly less than the least prime dividing the order of $y$.

(ii)   Deduce that the group $G$ of (7.8.10) has no nontrivial finite homomorphic images.

In proving the nontriviality of $G$, let us make the convention that for any subset $I$ of our set of generators $\{w, x, y, z\}$, we shall write $G_I$ for the group presented by the generators in the subset $I$, and those of the four relations in (7.8.10) that involve only terms in that subset. Moreover, in cases where $I$ is an explicit list of generators, we shall drop the brackets and commas from this notation. So, for instance, $G_{xy}$ will denote $\langle x, y \mid x^{-1} y x = y^2 \rangle$, with two generators and one relation, while $G_{xz}$ is presented by the two generators $x$, $z$ and no relations, i.e., it denotes the free group on those generators.

(iii)   For subsets $I \subseteq J \subseteq \{w, x, y, z\}$, explain why there will exist a homomorphism $f_{I,J} : G_I \to G_J$ taking each generator of $G_I$ to the generator of $G_J$ denoted by the same symbol.

In the symbol $f_{I,J}$, as in $G_I$, we will drop set-brackets and commas within such brackets in the symbols for $I$ and $J$ when these sets are shown explicitly.

(iv)   Obtain a normal form or other computationally convenient description of $G_{xy}$ (or if you have done Exercise 4.3:3, quote your result from that exercise to get such a description). Verify from this description that the elements $x$ and $y$ of this group each generate an infinite cyclic subgroup, i.e., that the maps $f_{x,xy}$ and $f_{y,xy}$ are both one-to-one. Also verify that their images have trivial intersection.

In view of the symmetry of the presentation (7.8.10), you may, after proving a result such as the preceding, use any statement obtained from it via cyclic permutation of the generator-symbols $w$, $x$, $y$, $z$, merely noting that this is what you are doing.

(v)  Show that $G_{wxy}$ is the pushout of the diagram formed from the two maps $f_{x,\,wx}$ and $f_{x,\,xy}$.

(vi)  Deduce from part (v) above and Exercise 7.8:25(ii) that the subgroup of $G_{wxy}$ generated by $w$ and $y$ is free on those generators. Translate this into a statement about the map $f_{wy,\,wxy}$.

(vii) Show that $G = G_{wxyz}$ is the pushout of the diagram formed from the two maps $f_{wy,\,wxy}$ and $f_{wy,\,yzw}$. Conclude that $G$ is infinite.

This example shows, in particular, that a finitely presented group need not be *residually finite* (as defined just before Exercise 4.5:2).

**Exercise 7.8:28.** Let $\mathbf{C}$ be a category having pullbacks and pushouts, and let $X_1$, $X_2$, $X_3$, $f_1$, $f_2$ be as in (7.8.8). Suppose we form their pullback $P$, then form the pushout of the system $P$, $X_2$, $X_3$, $p_1$, $p_2$, and so on, going back and forth between pullbacks and pushouts. Will this process ever "stabilize"?

(Suggestion: Given the two objects $X_1$ and $X_2$, consider the set $A$ of all objects $W$ given with morphisms into $X_1$ and $X_2$, and the set $B$ of all objects $Y$ given with morphisms into them from $X_1$ and $X_2$, and let $R \subseteq A \times B$ denote the relation "the four morphisms form a commuting square". Examine the resulting Galois connection between $A$ and $B$.)

We note

**Lemma 7.8.11.** *A morphism* $f \colon X \to Y$ *of a category* $\mathbf{C}$ *is a monomorphism if and only if the diagram*

$$
\begin{array}{ccc}
X & \xrightarrow{\ \mathrm{id}_X\ } & X \\
\Big\downarrow{\scriptstyle \mathrm{id}_X} & & \Big\downarrow{\scriptstyle f} \\
X & \xrightarrow{\ f\ } & Y
\end{array}
$$

*is a pullback diagram. Similarly* $f$ *is an epimorphism if and only if*

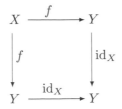

$$
\begin{array}{ccc}
X & \xrightarrow{\ f\ } & Y \\
\Big\downarrow{\scriptstyle f} & & \Big\downarrow{\scriptstyle \mathrm{id}_X} \\
Y & \xrightarrow{\ \mathrm{id}_X\ } & Y
\end{array}
$$

*is a pushout diagram.* $\qquad\qquad \square$

**Exercise 7.8:29.** Prove Lemma 7.8.11.

The category of "objects of $\mathbf{C}$ fibered over $Z$" used in Exercise 7.8:24 has a far-reaching generalization:

**Definition 7.8.12.** *Given three categories and two functors,* $\mathbf{D} \xrightarrow{S} \mathbf{C} \xleftarrow{T}$ $\mathbf{E}$, *we shall denote by* $(S \downarrow T)$ *the category having for objects all 3-tuples* $(D, f, E)$, *where* $D \in \mathrm{Ob}(\mathbf{D})$, $E \in \mathrm{Ob}(\mathbf{E})$, *and* $f \in \mathbf{C}(S(D), T(E))$, *and where a morphism* $(D, f, E) \to (D', f', E')$ *means a pair consisting of morphisms* $d\colon D \to D'$, $e\colon E \to E'$, *such that* $S(d)$ *and* $T(e)$ *make a commuting square with* $f$ *and* $f'$.

This construction is sometimes written $(S, T)$. We follow Mac Lane [20] in writing it $(S \downarrow T)$, because, as he observes, "the comma is already overworked". However, the older notation is the source of its name, the *comma category* construction. The most frequently used cases of this construction are those noted in (ii) and (iii) of the next exercise.

**Exercise 7.8:30.**

(i)   Verify that Definition 7.8.12 makes sense. Namely, write out the indicated commutativity condition, say how composition should be defined in $(S \downarrow T)$, and verify that the result is a category.

(ii)   Given a category $\mathbf{C}$, suppose we let $\mathbf{D} = \mathbf{C}$, with $S\colon \mathbf{C} \to \mathbf{C}$ the identity functor, while we take for $\mathbf{E}$ the trivial category, with only one object, denoted $0$, and its identity morphism. Let $T\colon \mathbf{E} \to \mathbf{C}$ be any functor; thus $T$ will be determined by the choice of one object, $T(0)$, which we shall call $Z$.

Show that the category $(S \downarrow T)$ can then be identified with the category we called $\mathbf{C}_Z$ in Exercise 7.8:24. This is often denoted $(\mathbf{C} \downarrow Z)$.

(iii)   For $S$ and $T$ as in (ii) above, also describe the category $(T \downarrow S)$ (often denoted $(Z \downarrow \mathbf{C})$).

(In the symbols for particular comma categories mentioned at the ends of (ii) and (iii) above, note that the object-name "$Z$" is used as an abbreviation for the functor on the trivial category taking its one object to $Z$, while $\mathbf{C}$ is used as an abbreviation for the identity functor of $\mathbf{C}$. The latter is an instance of the use, mentioned in §7.3, of the symbol for an object to denote that object's identity morphism. Though we are not adopting that usage in general, it is convenient in this case, where we know the two slots in the comma category symbol must be filled with names of functors, so there is no danger of confusion.)

If $\mathbf{C}$ is a category with a terminal object $T$, the construction $\mathbf{C}^{\mathrm{pt}}$ of Exercise 7.8:4 can clearly be described as $(T \downarrow \mathbf{C})$. However, there is a different comma category construction that is also sometimes called the category of "pointed objects" of $\mathbf{C}$ :

**Exercise 7.8:31.** Suppose $(\mathbf{C}, U)$ is a concrete category having a free object $F(1)$ on the one-element set $1 = \{0\}$. Show that the following categories are isomorphic:

(i)   $(F(1) \downarrow \mathbf{C})$.

(ii)   The category whose objects are pairs $(X, x_0)$, where $X$ is an object of $\mathbf{C}$ and $x_0$ an element of $U(X)$, and whose morphisms are the morphisms (in $\mathbf{C}$) of first components that respect second components.

(iii)   $(1 \downarrow U)$, where "1" denotes the functor from the one-object one-morphism category to **Set** which takes the unique object to the one-element set 1.

Since the one-point topological space is both the terminal object $T$ of **Top** and the free object $F(1)$ on one generator in that category (under the concretization by underlying sets), the constructions $(T \downarrow \mathbf{C})$ and $(F(1) \downarrow \mathbf{C})$ agree in this case, leading to the above situation of one concept from topology with two equally natural but inequivalent generalizations to category theory. As mentioned, each of these constructions is sometimes called the category of "pointed objects of $\mathbf{C}$," though they can be quite different from one another. The term "pointed" presumably comes from the term "pointed topological space" for an object of **Top**$^{\mathrm{pt}}$, regarded as a space with a distinguished basepoint.

Note that since the terminal object of **Group** is also initial (i.e., is a zero object, as defined in Definition 7.8.1), every object of **Group** admits a unique homomorphism of this terminal object into it. Hence such a homomorphism contains no new information, and **Group**$^{\mathrm{pt}}$ is isomorphic to **Group**. Therefore when an author speaks of the category of "pointed groups" one can guess that he or she does not mean **Group**$^{\mathrm{pt}}$, but $(F_{\mathbf{Group}}(1) \downarrow \mathbf{Group})$, equivalently $(1 \downarrow U_{\mathbf{Group}})$, the category of groups with a distinguished element.

We end this section with a particularly simple example of a category-theoretic translation of a familiar concept. Let $G$ be a group, and recall that a $G$-set is a set with an *action* of $G$ on it by permutations. More generally, one can consider an action of $G$ by automorphisms on any object $X$ of a category $\mathbf{C}$; one defines such an action as a homomorphism $f$ of $G$ into the monoid $\mathbf{C}(X, X)$. Now observe that the pair consisting of such an object $X$ and such a homomorphism $f : G \to \mathbf{C}(X, X)$ is equivalent to a *functor* $G_{\mathbf{cat}} \to \mathbf{C}$; the object $X$ gives the image of the one object of $G_{\mathbf{cat}}$, and $f$ determines the images of the morphisms. Thus, group actions are examples of functors!

## 7.9. Morphisms of functors (or "natural transformations")

We have seen that various sorts of mathematical structures can be regarded as functors from "diagram" categories to categories of simpler objects: As just noted, $G$-sets are equivalent to functors from $G_{\mathbf{cat}}$ to **Set**; another example is the type of structure which is the input of the equalizer and coequalizer constructions, consisting of two objects of a category **C** and a pair of morphisms from the first object to the second, $(X, Y, f, g)$. If we call such a 4-tuple a "parallel pair" of morphisms in **C**, then as observed in §7.2, parallel pairs correspond to functors from the two-object diagram category $\cdot \rightrightarrows \cdot$ to **C**.

Now if we regard such functors as new sorts of mathematical "objects", it is natural to ask whether we can define *morphisms* among these objects.

There is a standard concept of a morphism of $G$-sets—a set map which "respects" the action of $G$. Is there a similar concept of "morphism of parallel pairs"? Given two parallel pairs $S = (X, Y, f, g)$ and $S' = (X', Y', f', g')$, it seems reasonable to define a morphism $S \to S'$ to be a pair of morphisms $x \in \mathbf{C}(X, X'), y \in \mathbf{C}(Y, Y')$ which respects the structure of parallel pairs, in the sense that $y f = f'x$ and $y g = g'x$ :

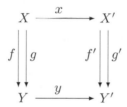

It is clear how to compose such morphisms, and immediate to verify that this composition makes the class of parallel pairs in **C** into a category.

We find that with this definition, equalizers and coequalizers join the ranks of constructions which, though originally thought of only as defined on objects, can also be applied to morphisms. Indeed, if the two parallel pairs of the above diagram each have an equalizer, then it is not hard to check that the morphism $(x, y)$ induces a morphism $z$ of the equalizer objects, and if *every* parallel pair in **C** has an equalizer, then this way of associating to every morphism of parallel pairs a morphism of their equalizer objects makes the equalizer construction a functor. Likewise, if each parallel pair has a coequalizer, the coequalizer construction becomes a functor.

**Exercise 7.9:1.** Prove the assertions about equalizers in the above paragraph.

Exactly similar considerations apply to the configurations in a category $\mathbf{C}$ for which we defined the concepts of *pullbacks* and *pushouts*. Such configurations can be regarded as functors from diagram categories $\overset{\bullet}{\underset{\bullet}{\downarrow}}\,$, respectively $\bullet \rightarrow\overset{\bullet}{\underset{\bullet}{\downarrow}}$ , into $\mathbf{C}$, and the set of all configurations of one or the other of these kinds can be made into a category, by letting a morphism from one such configuration to another mean a system of maps between corresponding objects, which respect the given morphisms among these. One can verify that this makes the pullback and pushout constructions, when they exist, into functors on these categories of configurations.

In each of these cases, we have had a diagram category $\mathbf{D}$ and a general category $\mathbf{C}$, and we have discovered a concept of "morphism" between functors from $\mathbf{D}$ to $\mathbf{C}$. So, although we have seen that we can regard functors as the morphisms of $\mathbf{Cat}$, it seems that there is also a concept of morphisms among functors! We formalize this as

**Definition 7.9.1.** *Let* $\mathbf{C}$ *and* $\mathbf{D}$ *be categories and* $F, G\colon \mathbf{D} \to \mathbf{C}$ *functors. Then a* morphism of functors $a\colon F \to G$ *means a family* $(a(X))_{X\in\mathrm{Ob}(\mathbf{D})}$ *such that for each* $X \in \mathrm{Ob}(\mathbf{D})$, *one has* $a(X) \in \mathbf{C}(F(X), G(X))$, *and for each morphism* $f\colon X \to Y$ *in* $\mathbf{D}$, *one has*

$$(7.9.2) \qquad\qquad a(Y)\,F(f) \;=\; G(f)\,a(X) \quad \text{in } \mathbf{C}.$$

*Pictorially, this means that for each arrow* $f$ *of* $\mathbf{D}$ *as at left below, we have commutativity of the square at right:*

$$
\begin{array}{ccc}
X & \qquad\qquad F(X) \xrightarrow{\;a(X)\;} G(X) \\[2pt]
\big\downarrow f & \qquad\qquad \big\downarrow F(f) \qquad\qquad \big\downarrow G(f) \\[2pt]
Y & \qquad\qquad F(Y) \xrightarrow{\;a(Y)\;} G(Y)
\end{array}
$$

*Given functors* $F, G, H\colon \mathbf{D} \to \mathbf{C}$ *and morphisms* $F \xrightarrow{\;a\;} G \xrightarrow{\;b\;} H$, *the composite morphism* $b\,a\colon F \to H$ *is defined by*

$$b\,a(X) \;=\; b(X)\,a(X) \quad (X \in \mathrm{Ob}(\mathbf{D})).$$

*Likewise, the identity morphism* $\mathrm{id}_F$ *of a functor* $F\colon \mathbf{D} \to \mathbf{C}$ *is defined by*

$$\mathrm{id}_F(X) \;=\; \mathrm{id}_{F(X)} \quad (X \in \mathrm{Ob}(\mathbf{D})).$$

*The category whose objects are all the functors from* $\mathbf{D}$ *to* $\mathbf{C}$, *with morphisms, composition, and identity defined as above, will be denoted* $\mathbf{C}^{\mathbf{D}}$.

Note that if $\mathbf{D}$ is small, then $\mathbf{C}^{\mathbf{D}}$ will be small or legitimate according as $\mathbf{C}$ is, but that if $\mathbf{D}$ is legitimate, then even if $\mathbf{C}$ is small, $\mathbf{C}^{\mathbf{D}}$ need not be legitimate. (Its hom-sets may not be small.) But the Axiom of Universes shows us that we may consider these large functor categories as small categories with respect to a larger universe.

We see that if $G$ is a group, the above definition of a morphism between functors $G_{\mathbf{cat}} \to \mathbf{Set}$ indeed agrees with the concept of a morphism between $G$-sets, hence the category $G$-$\mathbf{Set}$ can be identified with $\mathbf{Set}^{G_{\mathbf{cat}}}$. Since $G_{\mathbf{cat}}$ is a small category, $\mathbf{Set}^{G_{\mathbf{cat}}}$ is a legitimate category.

Let us note some examples (where $\mathbf{D}$ will not be a small category) of morphisms between functors we have seen before. Let $F, A \colon \mathbf{Set} \to \mathbf{Group}$ be the functors taking a set $X$ to the free group and the free abelian group on $X$ respectively. For every set $X$ there is a homomorphism $a(X) \colon F(X) \to A(X)$ taking each generator of $F(X)$ to the corresponding generator of $A(X)$. It is easy to see that these form commuting squares with group homomorphisms induced by set maps, hence they constitute a morphism of functors $a \colon F \to A$.

Let $F$ again be the free group construction, and let $U \colon \mathbf{Group} \to \mathbf{Set}$ be the underlying set functor. Recall that for each $X \in \mathrm{Ob}(\mathbf{Set})$, the universal property of $F(X)$ involves a set map $u(X) \colon X \to U(F(X))$. It is easy to check that these maps $u(X)$, taken together, give a morphism $u \colon \mathrm{Id}_{\mathbf{Set}} \to UF$ of functors $\mathbf{Set} \to \mathbf{Set}$, where $\mathrm{Id}_{\mathbf{Set}}$ denotes the identity functor of the category $\mathbf{Set}$.

**Exercise 7.9:2.** Verify the above claim that $u$ is a morphism of functors.

We note

**Lemma 7.9.3.** *Let $\mathbf{C}$ and $\mathbf{D}$ be categories, $F, G \colon \mathbf{D} \to \mathbf{C}$ functors, and $a \colon F \to G$ a morphism of functors. Then $a$ is an* isomorphism *of functors (i.e., an isomorphism in the category $\mathbf{C}^{\mathbf{D}}$) if and only if for every $X \in \mathrm{Ob}(\mathbf{C})$, the morphism $a(X)$ is an isomorphism of objects of $\mathbf{D}$.*

*Proof.* If $a$ has a two-sided inverse $b$, then it is immediate from the definition of composition of functors that for each $X \in \mathrm{Ob}(\mathbf{C})$, $b(X)$ is a two-sided inverse to $a(X)$.

Conversely, if for each $X \in \mathrm{Ob}(\mathbf{C})$, $a(X)$ has a two-sided inverse $b(X)$, then we want to show that these maps together give a morphism of functors $G \to F$; once we have this, that morphism will clearly be a two-sided inverse to $a$. So let $f \colon X \to Y$ be any morphism in $\mathbf{C}$. The fact that $a$ is a morphism of functors tells us that $a(Y)F(f) = G(f)a(X)$ (7.9.2). Composing on the left with $b(Y)$ and on the right with $b(X)$, we get $F(f)b(X) = b(Y)G(f)$. Since this is true for all $f$, $b$ is a morphism of functors, as required.    $\square$

Statements that two different constructions are "essentially the same" can usually be formulated precisely as saying that they are isomorphic as functors. For instance

**Exercise 7.9:3.**

(i)   Let $F : \mathbf{Set} \to \mathbf{Group}$ denote the free group construction, $A : \mathbf{Set} \to \mathbf{Ab}$ the free abelian group construction, and $C : \mathbf{Group} \to \mathbf{Ab}$ the abelianization construction. Show that $CF \cong A$. (In what functor category?)

(ii)   When we gave examples of covariant hom-functors $h_X : \mathbf{C} \to \mathbf{Set}$ at the end of §7.5, we observed that for $\mathbf{C} = \mathbf{Group}$, the functor $h_Z$ was "essentially" the underlying set functor, and that for $\mathbf{C} = \mathbf{Set}$ and $2 = \{0, 1\} \in \mathrm{Ob}(\mathbf{Set})$, $h_2$ was "essentially" the construction $X \mapsto X \times X$. Similarly, in §7.6, we noted that the contravariant hom-functor $h^2$ on $\mathbf{Set}$ "could be identified with" the contravariant power-set functor. Verify that in each of these cases, we have an *isomorphism* of functors.

(iii) Let $T : \mathbf{Ab} \times \mathbf{Ab} \to \mathbf{Ab}$ be the tensor product construction, and $R : \mathbf{Ab} \times \mathbf{Ab} \to \mathbf{Ab} \times \mathbf{Ab}$ the construction taking each pair of abelian groups $(A, B)$ to the pair $(B, A)$, and acting similarly on morphisms. Show that $T \cong TR$.

(iv)  Show that the isomorphisms of Exercise 7.6:5(v) give an isomorphism of functors $\mathrm{Id}_{\mathbf{FPOSet}} \cong BA$.

A venerable example of an isomorphism of functors arises in considering duality of finite-dimensional vector spaces. We know that a finite-dimensional vector space $V$, its dual $V^*$, and its double dual $V^{**}$ are all isomorphic. Now the isomorphism $V \cong V^*$ is not "natural" – these spaces are isomorphic simply because they have the same dimension. But there *is* a natural way to construct an isomorphism $V \cong V^{**}$, by taking each vector $v$ to the operator $\bar{v}$ defined by $\bar{v}(f) = f(v)$ $(f \in V^*)$. What this natural construction shows is that for $\mathbf{C}$ the category of finite-dimensional $k$-vector spaces, the functors $\mathrm{Id}_{\mathbf{C}}$ and $^{**}$ are isomorphic. (One cannot even attempt to construct an isomorphism between $\mathrm{Id}_{\mathbf{C}}$ and $^*$, since one functor is covariant and the other contravariant.)

Examples such as this had long been referred to as "natural isomorphisms", and people had gradually noticed that these and other sorts of "natural" constructions respected maps among objects. When Eilenberg and Mac Lane introduced category theory in [9], they therefore gave the name *natural transformation* to what we are calling a *morphism of functors*. The former term is still widely used, though we shall not use it here. One can also call such an entity a *functorial map*, to emphasize that it is not merely a system of maps between individual objects $F(X)$ and $G(X)$, but that these respect the morphisms $F(f)$ and $G(f)$ that make the constructions $F$ and $G$ *functors*.

In fact, we used this term "functorial"—deferring explanation—in Exercises 3.3:6 and 3.3:7. What we called there a "functorial group-theoretic operation in $n$ variables" is in our new language a morphism $U^n \to U$, where $U$ is the underlying-set functor $\mathbf{Group} \to \mathbf{Set}$, and $U^n$ is (as indicated in Definition 7.8.5) the functor associating to every group $G$ the direct product of $n$ copies of $U(G)$, i.e., the set of $n$-tuples of elements of $U(G)$. Some cases of those exercises reappear, along with other problems, in the following exercises, which should give you practice thinking about morphisms of functors. Note, incidentally, that the set-theoretic difficulties referred to when we introduced Exercises 3.3:6 and 3.3:7 have been overcome by our adoption of the Axiom of Universes, and our convention that "group" now means "$\mathbb{U}$-small group".

**Exercise 7.9:4.** In each part below, attempt to describe *all* morphisms among the functors listed, including morphisms from functors to themselves. (I describe functors below in terms of their behavior on objects. The definitions of their behavior on morphisms should be clear. If you are at all in doubt, begin your answer by saying how you think these functors should act on morphisms.)

(i)  The functors Id, $A$ and $B \colon \mathbf{Set} \to \mathbf{Set}$ given by $\mathrm{Id}(S) = S$, $A(S) = S \times S$, $B(S) = \{\{x, y\} \mid x, y \in S\}$. (Note that a member of $B(S)$ may have either one or two elements.)

(ii)  The functors $U$, $V$ and $W \colon \mathbf{Group} \to \mathbf{Set}$ given by $U(G) = |G|$, $V(G) = |G| \times |G|$, $W(G) = \{x \in |G| \mid x^2 = e\}$.

(iii)  The underlying set functor $U \colon \mathbf{FGroup} \to \mathbf{Set}$, where $\mathbf{FGroup}$ is the category of finite groups.

**Exercise 7.9:5.**

(i)  Show that for any category $\mathbf{C}$, the monoid $\mathbf{C}^{\mathbf{C}}(\mathrm{Id}_{\mathbf{C}}, \mathrm{Id}_{\mathbf{C}})$ of endomorphisms of the identity functor of $\mathbf{C}$ is commutative.

(ii)  Attempt to determine this monoid for the following categories $\mathbf{C} \colon \mathbf{Set}$, $\mathbf{Group}$, $\mathbf{Ab}$, and $\mathbf{FAb}$, the last being the category of finite abelian groups.

(iii)  Do the same for $\mathbf{C} = S_{\mathbf{cat}}$ where $S$ is an arbitrary monoid.

(iv)  Is the endomorphism monoid of a full and faithful functor $F \colon \mathbf{C} \to \mathbf{D}$ in general isomorphic to the endomorphism monoid of the identity functor of the full subcategory of $\mathbf{D}$ that is its image? If not, is it at least abelian? If you get such a result, can either "full" or "faithful" be deleted from the hypothesis?

**Exercise 7.9:6.** (i)  Let $F \colon \mathbf{Set} \to \mathbf{Set}$ be the functor associating to every set $S$ the set $S^{\omega}$ of all sequences $(s_0, s_1, \dots)$ of elements of $S$. Determine all morphisms from $F$ to the identity functor of $\mathbf{Set}$.

(ii)  Let $G \colon \mathbf{FSet} \to \mathbf{Set}$ be the restriction of the above functor to the category of finite sets; i.e., the functor taking every finite set $S$ to the (generally infinite) set of all sequences of members of $S$. Determine all morphisms from $G$ to the inclusion functor $\mathbf{FSet} \to \mathbf{Set}$.

**Exercise 7.9:7.** Give an example of two functors $F, G : \mathbf{D} \to \mathbf{C}$ such that for every object $X$ of $\mathbf{D}$, $F(X) \cong G(X)$ in $\mathbf{C}$, but such that $F$ and $G$ are not isomorphic as functors. In fact, if possible give one example in which $\mathbf{C}$ and $\mathbf{D}$ are both one-object categories, and another in which they are naturally occurring categories of mathematical objects.

**Exercise 7.9:8.** Suppose $F, G : \mathbf{C} \to \mathbf{D}$ are functors, and $a : F \to G$ a morphism of functors. In Lemma 7.9.3, we saw that $a$ will be invertible if and only if each of the morphisms $a(X)$ is invertible. Let us look at some related questions.

(i)   What implications if any hold between the conditions: (a) for all $X \in \mathrm{Ob}(\mathbf{C})$, $a(X) \in \mathbf{D}(F(X), G(X))$ is left invertible in $\mathbf{D}$, and (b) $a \in \mathbf{D}^{\mathbf{C}}(F, G)$ is left invertible in $\mathbf{D}^{\mathbf{C}}$?

(Suggestion: Consider the example where $\mathbf{C}$ is the category whose objects are pairs $(S, T)$, for $T$ a nonempty set and $S$ a nonempty subset of $T$, and where the morphisms $(S, T) \to (S', T')$ are the set-maps $T \to T'$ which carry $S$ into $S'$. Let $F, G : \mathbf{C} \to \mathbf{Set}$ carry $(S, T)$ to $S$ and $T$ respectively. Examine the obvious morphism $a : F \to G$.)

(ii)   What implications if any hold between the conditions: (a) for all $X \in \mathrm{Ob}(\mathbf{C})$, $a(X) \in \mathbf{D}(F(X), G(X))$ is a monomorphism in $\mathbf{D}$, and (b) $a \in \mathbf{D}^{\mathbf{C}}(F, G)$ is a monomorphism in $\mathbf{D}^{\mathbf{C}}$?

In the same spirit:

(iii)  Suppose $F_1, F_2, P : \mathbf{C} \to \mathbf{D}$ are functors, and $p_1 : P \to F_1$, $p_2 : P \to F_2$ are morphisms. What implications if any hold between the conditions (a) for all $X \in \mathrm{Ob}(\mathbf{C})$, $P(X)$ is a product of $F_1(X)$ and $F_2(X)$ in $\mathbf{D}$, with projection morphisms $p_1(X)$ and $p_2(X)$, and (b) $P$ is a product of $F_1$ and $F_2$ in $\mathbf{D}^{\mathbf{C}}$, with projection morphisms $p_1$ and $p_2$?

We have mentioned that constructions such as that of free groups, product objects, etc., could be made into functors by using the universal properties to get the required morphisms between the constructed objects. Since then, we have talked about *the* free group functor, *the* product functor on a category, etc. Part (ii) of the next exercise justifies this use of the definite article.

**Exercise 7.9:9.**

(i)   Let $(\mathbf{C}, U)$ be a concrete category having free objects, and let $\Phi$ be a function associating to every $X \in \mathrm{Ob}(\mathbf{Set})$ a free object on $X$ in $\mathbf{C}$, $\Phi(X) = (F(X), u(X))$. Show that there is a unique way of extending $F$ (the first component of $\Phi$) to a functor (i.e., defining $F(f)$ for each morphism $f$ of $\mathbf{Set}$ in a functorial manner) so that $u$ becomes a morphism of functors $\mathrm{Id}_{\mathbf{Set}} \to UF$.

(ii)   Suppose $\Phi : X \mapsto (F(X), u(X))$ and $\Psi : X \mapsto (G(X), v(X))$ are two constructions *each* assigning to every set $X$ a free object in $\mathbf{C}$ with respect to $U$. Show that the functors $F$ and $G$ obtained from $\Phi$ and $\Psi$ as in part (i) above (so that the second components $u$ and $v$ become morphisms of functors) are isomorphic; in fact, that there is a *unique* isomorphism making an appropriate diagram commute.

(iii) Write up the analogs of (i) and (ii) for one other functor associated with a universal construction, e.g., products, equalizers, tensor products of abelian groups, etc. You may abbreviate steps that parallel the free-object case closely.

To motivate what comes next, let us consider the following three pairs of constructions: (a) To every group $G$, we may associate the set of its elements of exponent 2, and also its set of elements of exponent 4; this gives two functors $V_2$ and $V_4$ from **Group** to **Set** such that for every $G$, $V_2(G) \subseteq V_4(G)$. (b) To every set $X$ we can associate the set $\mathbf{P}(X)$ of its subsets, and also the set $\mathbf{P}_f(X)$ of its finite subsets. If we regard the power-set construction as a covariant functor $P \colon \mathbf{Set} \to \mathbf{Set}$, this gives a second covariant functor $P_f \colon \mathbf{Set} \to \mathbf{Set}$ such that for all $X$, $P_f(X) \subseteq P(X)$. (We used the covariant power-set functor here because the inverse image of a finite set under a set map may not be finite, so there is no natural way to make a *contravariant* functor out of $\mathbf{P}_f$.) (c) If Inv: **Monoid** $\to$ **Monoid** denotes the functor associating to every monoid its submonoid of invertible elements, then for each monoid $S$, $\mathrm{Inv}(S)$ is a submonoid of $S = \mathrm{Id}_{\mathbf{Monoid}}(S)$.

These examples suggest that we want a concept of a "subfunctor" of a functor. Of course, the examples were based on having the concept of a "subobject" of an object, and as we have observed, there is no unique way to define this in a category. However, if we assume a concept of subobject *given*, we can define the concept of subfunctor relative to it:

**Lemma 7.9.4.** *Let* $\mathbf{C}$ *be a category, and* $\mathbf{C}_{\mathrm{incl}}$ *be a subcategory having for objects all the objects of* $\mathbf{C}$, *and having for morphisms a subclass of the monomorphisms of* $\mathbf{C}$, *called the* inclusions, *such that there is at most one inclusion morphism between any unordered pair of objects (i.e., such that* $\mathbf{C}_{\mathrm{incl}}$ *is a (large) partially ordered set). For* $X_0, X \in \mathrm{Ob}(\mathbf{C})$, *let us call* $X_0$ *a subobject of* $X$ *(or when there is a possibility of ambiguity, a "subobject with respect to the distinguished subcategory* $\mathbf{C}_{\mathrm{incl}}$ *") if there exists an inclusion morphism* $X_0 \to X$. *If* $X_0$ *and* $Y_0$ *are subobjects of* $X$ *and* $Y$ *respectively, and* $f \in \mathbf{C}(X, Y)$, *let us say* $f$ *carries* $X_0$ *into* $Y_0$ *if there exists a (necessarily unique!) morphism* $f_0 \in \mathbf{C}(X_0, Y_0)$ *making a commuting square with* $f$ *and the inclusions of* $X_0$ *and* $Y_0$ *in* $X$ *and* $Y$.

*Then for* $\mathbf{C}$ *and* $\mathbf{C}_{\mathrm{incl}}$ *as above, and* $F$ *any functor from another category* $\mathbf{D}$ *into* $\mathbf{C}$, *the following data are equivalent:*

(a) *A choice for each* $X \in \mathrm{Ob}(\mathbf{D})$ *of a subobject* $G(X)$ *of* $F(X)$ *such that for each* $f \in \mathbf{D}(X, Y)$, $F(f)$ *carries* $G(X)$ *into* $G(Y)$.

(b) *A functor* $G \colon \mathbf{D} \to \mathbf{C}$ *such that each* $G(X)$ *is a subobject of* $F(X)$, *and such that the inclusion maps give a morphism of functors* $G \to F$.

(c) *A subobject* $G$ *of* $F$ *as objects of* $\mathbf{C}^{\mathbf{D}}$ *with respect to the subcategory of thereof having for objects all the objects of that category (all functors*

**D** → **C**), *and for morphisms those morphisms of functors whose values at
all objects of* **D** *are inclusion morphisms (morphisms in* **C**$_{\text{incl}}$*). We may
call such an* $G$ a subfunctor *of* $F$.                                    □

**Exercise 7.9:10.** Prove the above lemma, including the assertion of unicity
noted parenthetically near the end of the first paragraph, and the implicit
assertion that the subcategory referred to in (c) has the same properties as-
sumed for **C**$_{\text{incl}}$. (Can that subcategory of **C**$^{\text{D}}$ be described as (**C**$_{\text{incl}}$)$^{\text{D}}$ ?)

In considering categories **C** of familiar algebraic objects, when we speak
of subobjects and subfunctors, the distinguished subcategory **C**$_{\text{incl}}$ will be
understood to have for morphisms the "ordinary" inclusions, unless the contrary is stated.

**Exercise 7.9:11.** Let $G$ be a group.

(i)   Show that if $S$ is a subfunctor of the identity functor of **Group**,
then $S(G)$ will be a subgroup of $G$ which is carried into itself by every
endomorphism of $G$. (Group theorists call such a subgroup *fully invariant.*)

(ii)   Is it true, conversely, that if $H$ is any fully invariant subgroup of $G$,
then there exists a subfunctor $S$ of Id$_{\text{Group}}$ such that $S(G) = H$ ?

(iii)   Given a subgroup $H$ of $G$ such that *some* subfunctor $S$ of Id$_{\text{Group}}$
exists for which $S(G) = H$, will there exist a *least* $S$ with this property?
A greatest?

(iv)   Generalize your answers to (i)–(iii) in one way or another.

**Exercise 7.9:12.** Let $k$ be a field of characteristic 0, and $k$-**Mod** the
category of $k$-vector-spaces. For each positive integer $n$ let $\otimes^n : k$-**Mod** $\rightarrow$
$k$-**Mod** denote the $n$-fold tensor product functor, $V \mapsto V^{\otimes n} =_{\text{def}} V \otimes \cdots \otimes$
$V$ ($n$ factors).

(i)   Determine all subfunctors of the functors $\otimes^1$ and $\otimes^2$.

(ii)   Investigate subfunctors of higher $\otimes^n$ 's.

(iii)   Are the results you obtained in (i) and/or (ii) valid over fields $k$ of
arbitrary characteristic?

We have observed that the idea that two constructions of some sort of
mathematical object are "equivalent" can often be made precise as a statement that two functors are isomorphic. A different type of statement is that
two *sorts* of mathematical object are "equivalent". In some cases, this can be
formalized by giving an *isomorphism* (invertible functor) between the categories of the two sorts of objects. For example, the category of Boolean rings
is isomorphic to the category of Boolean algebras, and **Group** is isomorphic
to the category of those monoids all of whose elements are invertible. But
there are times when this does not work, because the two sorts of objects
differ in certain "irrelevant" structure which makes it impossible, or unnatural, to set up such an isomorphism. For instance, groups with underlying

set contained in $\omega$ are "essentially" the same as arbitrary countable groups, although there cannot be an isomorphism between these two categories of groups, because one is small while the object-set of the other has the cardinality of the universe in which we are working. Monoids are "essentially the same" as categories with just one object, but the natural construction taking one-object categories to monoids is not one-to-one, because it forgets what element was the one object; and the way we found to go in the other direction (inserting " 1 " as the object) is likewise not onto. For these purposes, a concept weaker than isomorphism is useful.

**Definition 7.9.5.** *A functor* $F : \mathbf{C} \to \mathbf{D}$ *is called an* equivalence *between the categories* $\mathbf{C}$ *and* $\mathbf{D}$ *if there exists a functor* $G : \mathbf{D} \to \mathbf{C}$ *such that* $GF \cong \mathrm{Id}_{\mathbf{C}}$ *and* $FG \cong \mathrm{Id}_{\mathbf{D}}$ *(isomorphisms of functors). If such an equivalence exists, one says "*$\mathbf{C}$ *is equivalent to* $\mathbf{D}$ *", often written* $\mathbf{C} \approx \mathbf{D}$.

**Lemma 7.9.6.** *A functor* $F : \mathbf{C} \to \mathbf{D}$ *is an equivalence if and only if it is full and faithful, and every object of* $\mathbf{D}$ *is isomorphic to* $F(X)$ *for some* $X \in \mathrm{Ob}(\mathbf{C})$.

*Idea of proof.* "$\Longrightarrow$" is straightforward. To show "$\Longleftarrow$", choose for each object $Y$ of $\mathbf{D}$ an object $G(Y)$ of $\mathbf{C}$ and an isomorphism $i(Y) : Y \to FG(Y)$. One finds that there is a unique way to make $G$ a functor so that $i$ becomes an isomorphism $\mathrm{Id}_{\mathbf{D}} \cong FG$, and that there is a straightforward way to construct an isomorphism $\mathrm{Id}_{\mathbf{C}} \cong GF$. □

**Exercise 7.9:13.** Give the details of the proof of the above lemma (including the part described as straightforward).

Note that it is clear from Definition 7.9.5 that the relation $\approx$ is symmetric and reflexive, but it is not entirely clear whether a composite of equivalences is an equivalence, hence whether $\approx$ is transitive. That condition, however, is easily seen from Lemma 7.9.6. So the relation $\approx$ of equivalence between categories is, as one would hope, an equivalence relation.

If one merely assumes a functor $F : \mathbf{C} \to \mathbf{D}$ is full and faithful, but not the final condition of the above lemma, then it is not hard to deduce that this is equivalent to saying that it is an equivalence of $\mathbf{C}$ with a full subcategory of $\mathbf{D}$.

**Exercise 7.9:14.** Let $k$ be a field and $k$-**fgMod** the category of finite-dimensional vector spaces over $k$. Let $\mathbf{Mat}_k$ denote the category whose objects are the nonnegative integers, and such that a morphism from $m$ to $n$ is an $n \times m$ matrix over $k$, with composition of morphisms given by matrix multiplication. Show that $\mathbf{Mat}_k \approx k$-**fgMod**.

**Exercise 7.9:15.** Let $k$ and $k$-**fgMod** be as in the preceding exercise. Show that duality of vector spaces gives a *contravariant equivalence* of $k$-**fgMod** with itself, i.e., an equivalence between $k$-**fgMod**$^{\mathrm{op}}$ and $k$-**fgMod**.

**Exercise 7.9:16.** Let $\mathbf{FBool}^1$ denote the category of finite Boolean rings, and $\mathbf{FSet}$ the category of finite sets. In $\mathbf{FBool}^1$, 2 will denote the two-element Boolean ring with underlying set $\{0, 1\}$, while in $\mathbf{FSet}$, 2 will as usual denote the set $\{0, 1\}$.

For each $B \in \mathrm{Ob}(\mathbf{FBool}^1)$, if we define $B^* = \mathbf{FBool}^1(B, 2)$, we get a natural homomorphism $m_B : B \to 2^{B^*}$ which takes $x \in B$ to $(h(x))_{h \in B^*}$.

(i)   Show that $m_B$ is always an isomorphism. In particular, this says that every finite Boolean ring is a finite product of copies of the ring 2.

(ii)   Show with the help of (i) that the category $\mathbf{FBool}^1$ is equivalent to $\mathbf{FSet}^{\mathrm{op}}$.

**Exercise 7.9:17.** Let $R$ be a ring, $n$ a positive integer, and $M_n(R)$ the ring of $n \times n$ matrices over $R$. For any left $R$-module $M$, let $\mathrm{Col}_n(M)$ denote the set of column vectors of height $n$ of elements of $M$, and let this be made a left $M_n(R)$-module in the obvious way. This gives a functor $\mathrm{Col}_n : R\text{-}\mathbf{Mod} \to M_n(R)\text{-}\mathbf{Mod}$.

Show that $\mathrm{Col}_n$ is an equivalence of categories.

(Pairs of rings which, like $R$ and $M_n(R)$, have equivalent module categories are said to be *Morita equivalent*; cf. [107, §§18–19].)

The following definition and lemma reduce the question of whether two categories are *equivalent* to the question of whether two other categories are *isomorphic*.

**Definition 7.9.7.** *If* $\mathbf{C}$ *is a category, then a* skeleton *of* $\mathbf{C}$ *means a full subcategory having exactly one representative of each isomorphism class of objects of* $\mathbf{C}$; *i.e., by Lemma 7.9.6, a minimal full subcategory* $\mathbf{C}_0$ *such that the inclusion of* $\mathbf{C}_0$ *in* $\mathbf{C}$ *is an equivalence.*

The Axiom of Choice clearly allows us to construct a skeleton for every category.

**Lemma 7.9.8.** *Let* $\mathbf{C}$ *and* $\mathbf{D}$ *be categories, with skeleta* $\mathbf{C}_0$ *and* $\mathbf{D}_0$. *Then* $\mathbf{C}$ *and* $\mathbf{D}$ *are equivalent if and only if* $\mathbf{C}_0$ *and* $\mathbf{D}_0$ *are isomorphic.*   $\square$

**Exercise 7.9:18.** Write out the proof of Lemma 7.9.8.

Lemma 7.9.8 shows that equivalent categories agree in all properties that respect isomorphism of categories and "don't depend on how many isomorphic copies each object has"; that is, intuitively speaking, in all "genuinely category-theoretic" properties.

**Exercise 7.9:19.** Show that $\mathbf{Set}$ is not equivalent to $\mathbf{Set}^{\mathrm{op}}$ by finding a category-theoretic property possessed by one of these categories but not the other, and proving that equivalent categories must agree with respect to whether this property holds. For additional credit, demonstrate the non-equivalence of a few other pairs of familiar categories, e.g., show that **Set** is not equivalent to **Group**.

**Exercise 7.9:20.** Let $X$ be a pathwise connected topological space. Recall that one can define a category $\pi_1(X)$ whose objects are the points of $X$, and in which a morphism from $x$ to $y$ means a homotopy class of paths from $x$ to $y$. What does a skeleton of this category look like?

**Exercise 7.9:21.** Suppose **C** and **D** are equivalent categories, and $\mathbf{C}_0$ is a subcategory of **C**. Must **D** have a subcategory $\mathbf{D}_0$ equivalent to $\mathbf{C}_0$?

## 7.10. Properties of functor categories

In the preceding section we defined morphisms of functors, and saw some applications of the resulting category structure of $\mathbf{C}^{\mathbf{D}}$. Let us now set down a few basic properties of such categories.

First, consider any bifunctor

$$F \colon \mathbf{D} \times \mathbf{E} \longrightarrow \mathbf{C},$$

in other words, any object of $\mathbf{C}^{\mathbf{D} \times \mathbf{E}}$. If we fix an object $Y \in \mathrm{Ob}(\mathbf{E})$, it is easy to verify that $F$ induces a functor $F(-, Y) \colon \mathbf{D} \to \mathbf{C}$, i.e., an object of $\mathbf{C}^{\mathbf{D}}$, sending each object $X$ of **D** to $F(X, Y)$ and each morphism $f$ of **D** to $F(f, \mathrm{id}_Y)$.

Having made this observation for each *object* of **E**, let us now note that for each *morphism* between such objects, $g \in \mathbf{E}(Y, Y')$, the morphisms $F(\mathrm{id}_X, g)$ $(X \in \mathrm{Ob}(\mathbf{D}))$ yield a morphism of functors $F(-, g) \colon F(-, Y) \to F(-, Y')$. Thus our system of objects $F(-, Y)$ $(Y \in \mathrm{Ob}(\mathbf{E}))$ of $\mathbf{C}^{\mathbf{D}}$ has become a functor $F' \colon \mathbf{E} \to \mathbf{C}^{\mathbf{D}}$. That is, from our object $F$ of $\mathbf{C}^{\mathbf{D} \times \mathbf{E}}$ we have gotten an object $F'$ of $(\mathbf{C}^{\mathbf{D}})^{\mathbf{E}}$.

In constructing $F'$, we have not used the values of $F$ at all the morphisms of $\mathbf{D} \times \mathbf{E}$, but only at morphisms of the forms $(\mathrm{id}_X, g)$ and $(f, \mathrm{id}_Y)$; so we might wonder whether $F'$ embodies all the information contained in $F$. But in fact, an arbitrary morphism of $\mathbf{D} \times \mathbf{E}$, $(f, g) \colon (X, Y) \to (X', Y')$, can be written $(f, \mathrm{id}_{Y'})(\mathrm{id}_X, g)$, so the images under $F$ of morphisms of those two sorts do indeed determine the images of all morphisms of $\mathbf{D} \times \mathbf{E}$. In fact, we have

**Lemma 7.10.1** (Law of exponents for categories). *For any categories* **C**, **D**, **E** *one has* $\mathbf{C}^{\mathbf{D} \times \mathbf{E}} \cong (\mathbf{C}^{\mathbf{D}})^{\mathbf{E}}$, *via the construction sketched above.*    □

**Exercise 7.10:1.** Prove the above lemma. In particular, describe how to map morphisms of $\mathbf{C}^{\mathbf{D} \times \mathbf{E}}$ to morphisms of $(\mathbf{C}^{\mathbf{D}})^{\mathbf{E}}$.

**Exercise 7.10:2.** Does one have other laws of exponents for functor categories? In particular, is $(\mathbf{C} \times \mathbf{D})^{\mathbf{E}} \cong (\mathbf{C}^{\mathbf{E}}) \times (\mathbf{D}^{\mathbf{E}})$, and is $\mathbf{C}^{\mathbf{D} \amalg \mathbf{E}} \cong (\mathbf{C}^{\mathbf{D}}) \times (\mathbf{C}^{\mathbf{E}})$? (For the meaning of $\mathbf{D} \amalg \mathbf{E}$, cf. Exercise 7.6:10.)

Next, suppose that $G_1, G_2 \colon \mathbf{D} \to \mathbf{C} \cdot$ are functors, and $a \colon G_1 \to G_2$ is a morphism between them. If $H$ is a functor from any other category into $\mathbf{D}$, we can form the composite functors $G_1 H$ and $G_2 H$, and we find, not surprisingly, that the morphism $a \colon G_1 \to G_2$ induces a morphism $G_1 H \to G_2 H$. Likewise, given a functor $F$ out of $\mathbf{C}$, $a$ induces a morphism $F G_1 \to F G_2$. These induced morphisms of functors are written $a \circ H \colon G_1 H \to G_2 H$ and $F \circ a \colon F G_1 \to F G_2$ respectively.

For example, let $a$ be the canonical morphism from the free group functor $F$ to the free abelian group functor $A$. If we compose on the right with, say, the functor $U$ taking every lattice to its underlying set,

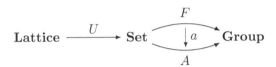

we get a morphism of functors $a \circ U$ mapping free groups on the underlying sets of lattices $L$ homomorphically to the free abelian groups on the same underlying sets. If instead we compose on the left with the underlying-set functor $V$ out of the category of groups,

we get a morphism of functors $V \circ a$ mapping the underlying set of the free group on each set $X$ to the underlying set of the free abelian group on $X$.

We record below the above constructions and note the basic laws that they satisfy. The reader is advised to draw (or visualize) pictures like those above for the various situations described.

**Lemma 7.10.2.** *Let* $\mathbf{C}$, $\mathbf{D}$ *and* $\mathbf{E}$ *be categories.*
(i)    *Given a morphism* $a \colon F_1 \to F_2$ *of functors* $\mathbf{D} \to \mathbf{C}$, *and any functor* $G \colon \mathbf{E} \to \mathbf{D}$, *a morphism* $a \circ G \colon F_1 G \to F_2 G$ *is defined by setting* $(a \circ G)(X) = a(G(X))$ $(X \in \mathrm{Ob}(\mathbf{E}))$.

(ii)  *Given any functor* $F\colon \mathbf{D} \to \mathbf{C}$, *and a morphism* $b\colon G_1 \to G_2$ *of functors* $\mathbf{E} \to \mathbf{D}$, *a morphism* $F \circ b\colon FG_1 \to FG_2$ *is defined by setting* $(F \circ b)(X) = F(b(X))$  $(X \in \mathrm{Ob}(\mathbf{E}))$.

(iii)  *Given morphisms* $F_1 \xrightarrow{a_1} F_2 \xrightarrow{a_2} F_3$ *of functors* $\mathbf{D} \to \mathbf{C}$, *and any functor* $G\colon \mathbf{E} \to \mathbf{D}$, *one has*

$$(a_2\, a_1 \circ G) = (a_2 \circ G)(a_1 \circ G).$$

(iv)  *Given any functor* $F\colon \mathbf{D} \to \mathbf{C}$, *and morphisms* $G_1 \xrightarrow{b_1} G_2 \xrightarrow{b_2} G_3$ *of functors* $\mathbf{E} \to \mathbf{D}$, *one has*

$$(F \circ b_2\, b_1) = (F \circ b_2)(F \circ b_1).$$

(v)  *Given functors* $F\colon \mathbf{D} \to \mathbf{C}$ *and* $G\colon \mathbf{E} \to \mathbf{D}$, *one has*

$$\mathrm{id}_F \circ G \;=\; \mathrm{id}_{FG} \;=\; F \circ \mathrm{id}_G.$$

(vi)  *Given both a morphism* $a\colon F_1 \to F_2$ *of functors* $\mathbf{D} \to \mathbf{C}$, *and a morphism* $b\colon G_1 \to G_2$ *of functors* $\mathbf{E} \to \mathbf{D}$, *one has*

(7.10.3) $$(a \circ G_2)(F_1 \circ b) \;=\; (F_2 \circ b)(a \circ G_1)$$

*as morphisms* $F_1\, G_1 \to F_2\, G_2$.

(vii) *Hence, the operation of composing functors, which a priori is a set map*

(7.10.4) $$\mathbf{Cat}(\mathbf{D},\, \mathbf{C}) \times \mathbf{Cat}(\mathbf{E},\, \mathbf{D}) \;\longrightarrow\; \mathbf{Cat}(\mathbf{E},\, \mathbf{C}),$$

*actually gives a functor*

(7.10.5) $$\mathbf{C}^{\mathbf{D}} \times \mathbf{D}^{\mathbf{E}} \;\longrightarrow\; \mathbf{C}^{\mathbf{E}},$$

*which acts on morphisms by taking each* $(a, b) : (F_1, G_1) \to (F_2, G_2)$ *as in* (vi) *to the common value of the two sides of* (7.10.3), *with details of functoriality established by* (i)–(v).                                      $\square$

### Exercise 7.10:3.

(i)  Prove statements (i)–(vi) of the above lemma.

(ii)  Verify that statement (vii) summarizes all of statements (i)–(v), except for the descriptions of how $F \circ a$ and $b \circ G$ are defined.

   In making **Cat** a category, we had to verify that the set map (7.10.4) satisfied the associativity and identity laws; we now ought to check that these laws hold, not merely as equalities of set maps, but as equalities of functors! The case of the identity laws is easy, but as part (ii) of the next exercise shows, is still useful:

**Exercise 7.10:4.**

(i)   Given a morphism $a \colon G_1 \to G_2$ of functors $G_1, G_2 \colon \mathbf{D} \to \mathbf{C}$, show that

$$a \circ \mathrm{Id}_{\mathbf{D}} = a = \mathrm{Id}_{\mathbf{C}} \circ a.$$

(ii)  Show that the above result, together with Lemma 7.10.2(vi), immediately gives the result of Exercise 7.9:5(i).

It is more work to write out the details of

**Exercise 7.10:5.** For categories $\mathbf{B}$, $\mathbf{C}$, $\mathbf{D}$ and $\mathbf{E}$ establish identities (like those of Lemma 7.10.2) showing that the two iterated-composition functors $\mathbf{B}^{\mathbf{C}} \times \mathbf{C}^{\mathbf{D}} \times \mathbf{D}^{\mathbf{E}} \to \mathbf{B}^{\mathbf{E}}$ are equal as functors.

In doing the above exercises, you may wish to use the notation which represents the common value of the two sides of (7.10.3) as $a \circ b$. In this connection, note part (i) of

**Exercise 7.10:6.**

(i)   Show that if the above notation is adopted, there are situations where $a \circ b$ and $ab$ are both defined, but are unequal.

(ii)  Can you find any important class of cases where they must be equal?

**Exercise 7.10:7.** Suppose we have an equivalence of categories, given by functors $F \colon \mathbf{C} \to \mathbf{D}$, $G \colon \mathbf{D} \to \mathbf{C}$ with $\mathrm{Id}_{\mathbf{C}} \cong GF$, $\mathrm{Id}_{\mathbf{D}} \cong FG$. Given a particular isomorphism of functors $i \colon \mathrm{Id}_{\mathbf{C}} \to GF$, can one in general choose an isomorphism $j \colon \mathrm{Id}_{\mathbf{D}} \to FG$ such that the two isomorphisms of functors, $i \circ G$, $G \circ j \colon G \to GFG$ are equal, and likewise the two isomorphisms $j \circ F$, $F \circ i \colon F \to FGF$?

How are we to look at a functor category $\mathbf{C}^{\mathbf{D}}$? Should we think of the functors which are its objects as "maps" or as "things"? As a category, is it "like" $\mathbf{C}$, "like" $\mathbf{D}$, or "like" neither?

My general advice is to think of its objects as "things" and its morphisms as "maps". More precisely, its objects are "things" which consist of *systems* of objects of $\mathbf{C}$, linked together by morphisms of $\mathbf{C}$, in a way *parametrized* by $\mathbf{D}$; its morphisms are further systems of morphisms of $\mathbf{C}$ uniting parallel structures of this sort. With respect to basic properties, such a functor category usually behaves more like $\mathbf{C}$ than like $\mathbf{D}$. For example, if $\mathbf{C}$ has finite products, so does $\mathbf{C}^{\mathbf{D}}$: one can construct the product $F \times G$ of two functors $F, G \in \mathbf{C}^{\mathbf{D}}$ "objectwise", by taking $(F \times G)(X)$ to be the product $F(X) \times G(X)$ for each $X \in \mathrm{Ob}(\mathbf{D})$ (cf. Exercise 7.9:8(iii)). On the other hand, so far as I know, existence of products in $\mathbf{D}$ tells us nothing about $\mathbf{C}^{\mathbf{D}}$.

## 7.11. Enriched categories (a sketch)

A recurring theme in category theory is that one characterizes some type of mathematical entity as a certain kind of structure in a particular category, such as **Set**, analyzes what properties of **Set** are needed for the concept to make sense, and then creates a generalized definition, which is like the original one except that **Set** is replaced by a general category having the required properties.

There is in fact an important application of this idea to the concept of *category* itself! I shall sketch this briefly below. I will begin with a few examples to motivate the idea, and then discuss what is involved in the general case, though I shall not give formal definitions.

Recall that a category, as we have defined it, is given by a *set* of objects, and a *set* of morphisms between any two objects, with composition operations given by *set maps*, $\mu \colon \mathbf{C}(Y, Z) \times \mathbf{C}(X, Y) \to \mathbf{C}(X, Z)$. But now consider the category **Cat**. Though we still have a *set* of objects, we have seen that for each pair of objects **C**, **D** we can speak of a *category* of morphisms $\mathbf{C}^{\mathbf{D}}$, and composition in **Cat** is given by *bifunctors* $\mu_{\mathbf{C}\,\mathbf{D}\,\mathbf{E}} \colon \mathbf{E}^{\mathbf{D}} \times \mathbf{D}^{\mathbf{C}} \to \mathbf{E}^{\mathbf{C}}$.

Likewise, for any ring $R$, it is well known that the homomorphisms from one $R$-module to another form an additive group, so the category $R$-**Mod** can be described as having, for each pair of objects, an *abelian group* of morphisms $X \to Y$. Here composition is given by *bilinear maps* $(R\text{-}\mathbf{Mod}(Y, Z), R\text{-}\mathbf{Mod}(X, Y)) \to R\text{-}\mathbf{Mod}(X, Z)$ among these abelian groups.

One expresses these facts by saying that **Cat** can be regarded as a **Cat**-*based* category, or a **Cat**-category for short, and $R$-**Mod** as an **Ab**-category. Similarly, in situations where one has a natural topological structure on sets of morphisms, and the composition maps are continuous, one can say one has a **Top**-based category.

These generalized categories are called *enriched* categories.

Note that when we referred to $R$-**Mod** as being an **Ab**-category, this included the observation that the composition maps $\mu_{XYZ}$ are *bilinear*. Thus, they correspond to abelian group homomorphisms (which by abuse of notation we shall denote by the same symbols):

$$\mu_{XYZ} \colon \ \mathbf{C}(Y, Z) \otimes \mathbf{C}(X, Y) \longrightarrow \mathbf{C}(X, Z).$$

The general definition of enriched category requires that the "base category" (the category in which the hom-objects are taken to lie; i.e., **Set**, **Cat**, **Ab**, **Top**, etc.) be given with a bifunctor into itself having certain properties, which is used, as above, in describing the composition maps. In the case where the base category is **Ab**, this is the *tensor-product* bifunctor, while in the cases of **Set**, **Cat**, and **Top**, the corresponding role is filled by the *product* bifunctor. See [20, §VII.7] for a few more details, and [100] for a thorough development of the subject.

One should, strictly, distinguish between $R$-**Mod** as an ordinary (i.e., **Set**-based) category and as an **Ab**-category, writing these two entities as, say, $R$-**Mod** and $R$-**Mod**$_{(\mathbf{Ab})}$, and similarly distinguish **Cat** and **Cat**$_{(\mathbf{Cat})}$ — just as one ought to distinguish between the set of integers, the additive group of integers, the lattice of integers, the ring of integers, etc. This notational problem will not concern us, however, since we will not formalize the concept of enriched category in these notes. Outside this section, when we occasionally refer to the special properties of categories such as $R$-**Mod** or **Cat**, we shall not assume any general theory of enriched categories, but simply use in an ad hoc fashion what we know about the extra structure.

We remark that the concept of **Ab**-based categories, and more generally, of $k$-**Mod**-based categories for $k$ a commutative ring (called "$k$-linear categories"), are probably more widely used than all the other sorts of enriched categories together. See [11] for a lively development of the subject.

The **Cat**-based category **Cat** contains a vast number of interesting sub-**Cat**-categories. Here is one:

**Exercise 7.11:1.** Consider the full subcategory of **Cat** whose objects are the categories $G_{\mathbf{cat}}$, for groups $G$.

(i)   Characterize in group-theoretic terms the morphisms and morphisms of morphisms, and the operation $\circ$ of Lemma 7.10.2, in this **Cat**-category.

(ii)   Translate your answer into a description of a **Cat**-category structure one can put on the category **Group**.

(iii)   Describe the structure of the full **Cat**-subcategory of the above category having for objects the *abelian* groups.

The student interested in ring theory might note that the category of Exercise 7.2:3 (with rings as objects, and bimodules $_RB_S$ as morphisms) can be made a **Cat**-category, by using bimodule homomorphisms as the morphisms-among-morphisms; moreover, each morphism-category $\mathbf{C}(R, S)$ (for $R$ and $S$ rings) is in fact an **Ab**-category! What this says is that this category is an **AbCat**-category, where **AbCat** is the category of **Ab**-categories. There is an explanation: This category is equivalent to the subcategory of **Cat** whose objects are the **Ab**-categories $R$-**Mod** for rings $R$, and whose morphisms are the functors $_RB \otimes_S - : S$-**Mod** $\to R$-**Mod** induced by bimodules $_RB_S$. So this observation is really a special case of the fact that **AbCat** is an **AbCat**-category, just as **Ab** is an **Ab**-category and **Cat** is a **Cat**-category.

We have mentioned (in Exercise 7.3:1 and the preceding discussion) that there is a version of the definition of category which eliminates reference to objects, and thus involves only one kind of element, the morphism (the objects being hidden under the guise of their identity morphisms). If we apply this idea twice to the concept of a **Cat**-category, we likewise get a structure with only one type of element – what we have been calling the morphisms of morphisms – but with two partial composition operations on these elements, *ab* and *a* ∘ *b* (Exercise 7.10:6). Described in this way, **Cat**-categories have been called "2-categories" [20, p. 44]. (The relation between the two types of composition is slightly asymmetric. If one drops the asymmetric condition— that every identity element with respect to the *first* composition is also an identity element with respect to the second—one gets a slightly more general concept, also defined in [20], and called a "double category".)

Having begun by considering **Cat** as an ordinary, i.e., **Set**-based category, with objects and morphisms (i.e., functors), and then having found that there was an important concept of *morphisms between morphisms* (morphisms of functors), we may ask whether one can define, further, *morphisms between morphisms between morphisms* in this category. The answer is "yes and no". On the yes side, observe that one can set up a concept of "morphisms between morphisms" in *any* category **C**! For a morphism in **C** is the same as an object of $\mathbf{C^2}$, where **2** denotes the diagram category $\cdot \longrightarrow \cdot$, and we know how to make $\mathbf{C^2}$ a category. So in particular, given categories **C** and **D** we can define a "morphism of morphisms in $\mathbf{C^D}$", which is thus a "morphism of morphisms of morphisms" in **Cat**.

However, this construction does not constitute a nontrivial enrichment of structure, since the concept of morphism of morphisms we have just described in an arbitrary category **C** is defined in terms of its existing category structure. (Indeed, when applied to **Cat**, it does not give the concept of "morphism of functors", but that of "commuting square of functors".) So we come to the "no" side of the answer—so far as I know, the category **Cat** has no enriched structure beyond that of a **Cat**-category.

However, if one turns from the category **Cat** of all **Set**-based categories, to the category **CatCat** of all **Cat**-based categories, one finds that here one has a natural and nontrivial concept of morphisms between morphisms between morphisms—in other words, **CatCat** is a **CatCat**-based category. And this process can be iterated ad infinitum.

But it is time to return from these vertiginous heights to the main stream of our subject.

# Chapter 8
# Universal Constructions
# in Category-Theoretic Terms

The language of category theory has enabled us to give general definitions of "free object", "product", "coproduct", "equalizer" and various other universal constructions. It is clear that these different constructions have many properties in common. Let us now look for ways to unify them, so that we will be able to prove results about them by general arguments, rather than piecemeal.

## 8.1. Universality in terms of initial and terminal objects

In all the above constructions, we deal with mathematical entities with certain "extra" structure, and seek one entity $E$ with such structure that is "universal". This suggests that we make the class of entities with such extra structure into a category, and examine the universal property of $E$ there.

For instance, the free group on three generators is universal among systems $(G, a, b, c)$ where $G$ is a group, and $a, b, c \in |G|$. If we define a category whose objects are these systems $(G, a, b, c)$, and where a morphism $(G, a, b, c) \to (G', a', b', c')$ means a group homomorphism $f : G \to G'$ such that $f(a) = a'$, $f(b) = b'$, $f(c) = c'$, we see that the universal property of the free group $(F, x, y, z)$ says that it has a unique morphism into every object of the category—in other words, that it is an initial object.

Similarly, given a group $G$, the *abelianization* of $G$ is universal among pairs $(A, f)$ where $A$ is an abelian group, and $f$ a group homomorphism $G \to A$. If we define a morphism from one such pair $(A, f)$ to another such pair $(B, g)$ to mean a group homomorphism $m : A \to B$ such that $m f = g$, we see that the definition of the abelianization of $G$ says that it is initial in *this* category.

G.M. Bergman, *An Invitation to General Algebra and Universal Constructions*, Universitext, DOI 10.1007/978-3-319-11478-1_8

Likewise, a group, a ring, a lattice, etc., with a presentation $\langle X \mid R \rangle$ clearly means an initial object in the category whose objects are groups, etc., with specified $X$-tuples of elements satisfying the system of equations $R$, and whose morphisms are homomorphisms respecting these distinguished $X$-tuples of elements.

The above were examples of what we named "left universal" properties in §4.8. Let us look at one "right universal" property, that of a *product* of two objects $A$ and $B$ in a category $\mathbf{C}$. We see that the relevant auxiliary category should have for objects all 3-tuples $(X, a, b)$, where $X \in \mathrm{Ob}(\mathbf{C})$, $a \in \mathbf{C}(X, A)$  $b \in \mathbf{C}(X, B)$, and for morphisms $(X, a, b) \to (Y, a', b')$ all morphisms $X \to Y$ in $\mathbf{C}$ making commuting triangles with the maps into $A$ and $B$. A direct product of $A$ and $B$ in $\mathbf{C}$ is seen to be a *terminal* object $(P, p_1, p_2)$ in this category.

You can likewise easily translate the universal properties of *pushouts, pullbacks* and *coproducts* in arbitrary categories to those of initial or terminal objects in appropriately defined auxiliary categories.

So all the universal properties we have considered reduce to those of being initial or terminal objects in appropriate categories. This view of universal constructions is emphasized by Lang [34, p. 57 et seq.], who gives these two types of objects the poetic designations "universally repelling" and "universally attracting". Since a terminal object in $\mathbf{C}$ is an initial object in $\mathbf{C}^{\mathrm{op}}$, all these universal properties ultimately reduce to that of initial objects!

Lemma 7.8.2 tells us that initial (and hence terminal) objects are *unique* up to unique isomorphism. This gives us, in one fell swoop, uniqueness up to canonical isomorphism for free groups, abelianizations of groups, products, coproducts, pushouts, pullbacks, objects presented by generators and relations, and all the other universal constructions we have considered. The canonical isomorphisms which these constructions are "unique up to" correspond to the unique morphisms between any two initial objects of a category. That is, given two realizations of one of our universal constructions, these isomorphisms will be the unique morphisms from each to the other that preserve the extra structure.

We will look at questions of *existence* of initial objects in §8.10.

## 8.2. Representable functors, and Yoneda's Lemma

The above approach to universal constructions is impressive for its simplicity; but we would also like to relate these universal objects to the original categories in question: Though the free group on an $S$-tuple of generators is initial in the category of groups given with $S$-tuples of elements, and the kernel of a group homomorphism $f \colon G \to H$ is terminal in the category of groups $L$ given with homomorphisms $L \to G$ having trivial composite with $f$, we also want to understand these constructions in relation to the category **Group**.

Note that the objects of the various auxiliary categories we have used can be written as pairs $(X, a)$, where $X$ is an object of the original category $\mathbf{C}$, and $a$ is some additional structure on $X$. If for each object $X$ of $\mathbf{C}$ we write $F(X)$ for the set of *all possible values* of this additional structure (e.g., in the case that leads to the free group on a set $S$, the set of all $S$-tuples of elements of $X$), we find that $F$ is in general a functor, covariant or contravariant, from $\mathbf{C}$ to $\mathbf{Set}$. The condition characterizing a *left* universal pair $(R, u)$ is that for every $X \in \mathrm{Ob}(\mathbf{C})$ and $x \in F(X)$, there should be a unique morphism $f : R \to X$ such that $F(f)(u) = x$. This condition—which we see requires a covariant $F$ for the latter equation to make sense—is equivalent to saying that for each object $X$, the set of morphisms $f \in \mathbf{C}(R, X)$ is sent bijectively to the set of elements of $F(X)$ by the map $f \mapsto F(f)(u)$. The bijectivity of this correspondence for each $X$ leads to an isomorphism between the functor $\mathbf{C}(R, -)$, i.e., $h_R : \mathbf{C} \to \mathbf{Set}$, and the given functor $F : \mathbf{C} \to \mathbf{Set}$. Thus, the universal property of $R$ can be formulated as a statement of this isomorphism:

**Theorem 8.2.1.** *Let $\mathbf{C}$ be a category, and $F : \mathbf{C} \to \mathbf{Set}$ a functor. Then the following data are equivalent:*

(i)    *An object $R \in \mathrm{Ob}(\mathbf{C})$ and an element $u \in F(R)$ having the* universal property *that for all $X \in \mathrm{Ob}(\mathbf{C})$ and all $x \in F(X)$, there exists a unique $f \in \mathbf{C}(R, X)$ such that $F(f)(u) = x$.*

(ii)    *An* initial object *$(R, u)$ in the category whose objects are all ordered pairs $(X, x)$ with $X \in \mathrm{Ob}(\mathbf{C})$ and $x \in F(X)$, and whose morphisms are morphisms among the first components of these pairs which respect the second components.*

(iii)    *An object $R \in \mathrm{Ob}(\mathbf{C})$ and an isomorphism of functors $i : h_R \cong F$ in $\mathbf{Set}^{\mathbf{C}}$.*

*Namely, given $(R, u)$ as in* (i) *or* (ii), *one obtains the isomorphism $i$ of* (iii) *by letting $i(X)$ take $f \in h_R(X)$ to $F(f)(u) \in F(X)$, while in the reverse direction, one obtains $u$ from $i$ as $i(R)(\mathrm{id}_R)$.*

*Sketch of proof.* The equivalence of the structures described in (i) and (ii) is immediate.

Concerning our description of how to pass from these structures to that of (iii), it is straightforward to verify that for any $u \in F(R)$, the map $i$ described there gives a *morphism of functors* $h_R \to F$. That this is an isomorphism is then the content of the universal property of (i). In the opposite direction, given an isomorphism $i$ as in (iii), if $u$ is defined as indicated, then the universal property of (i) is just a restatement of the bijectivity of the maps $i(X) : h_R(X) \to F(X)$.

Finally, it is easy to check that if one goes as above from universal element to isomorphism of functors and back, one recovers the original element, and if one goes from isomorphism to universal element and back, one recovers the original isomorphism.                                                                  □

**Exercise 8.2:1.** Write out the "straightforward verification" referred to in the second sentence of the above proof, and those implied in the phrases "is then the content of" and "is just a restatement of" in the next two sentences.

Dualizing (i.e., applying Theorem 8.2.1 to $\mathbf{C}^{\mathrm{op}}$ and stating the resulting assertion in terms of contravariant functors on $\mathbf{C}$), we get

**Theorem 8.2.2.** *Let $\mathbf{C}$ be a category, and $F$ a contravariant functor from $\mathbf{C}$ to $\mathbf{Set}$ (i.e., a functor $\mathbf{C}^{\mathrm{op}} \to \mathbf{Set}$). Then the following data are equivalent:*

(i)  *An object $R \in \mathrm{Ob}(\mathbf{C})$ and an element $u \in F(R)$ with the universal property that for any $X \in \mathrm{Ob}(\mathbf{C})$ and $x \in F(X)$, there exists a unique $f \in \mathbf{C}(X, R)$ such that $F(f)(u) = x$.*

(ii)  *A terminal object $(R, u)$ in the category whose objects are all ordered pairs $(X, x)$ with $X \in \mathrm{Ob}(\mathbf{C})$ and $x \in F(X)$, and whose morphisms are morphisms among the first components of these pairs which respect the second components.*

(iii)  *An object $R \in \mathrm{Ob}(\mathbf{C})$ and an isomorphism of contravariant functors $i \colon h^R \cong F$ in $\mathbf{Set}^{\mathbf{C}^{\mathrm{op}}}$.*

*Namely, given $(R, u)$ as in (i) or (ii), one obtains an isomorphism $i$ as in (iii) by letting $i(X)$ take $f \in h^R(X)$ to $F(f)(u) \in F(X)$, while in the reverse direction, one obtains $u$ from $i$ as $i(R)(\mathrm{id}_R)$.*                                      □

Note that in Theorem 8.2.1(ii), the last phrase, "which respect second components", meant that for a morphism $f \colon X \to Y$ to be considered a morphism $(X, x) \to (Y, y)$, we required $F(f)(x) = y$, while in Theorem 8.2.2(ii), the corresponding condition is $F(f)(y) = x$.

We remark that the auxiliary categories used in point (ii) of the above two theorems are comma categories, $(1 \downarrow F)$ (see Exercise 7.8:30(iii)).

The properties described above have names:

**Definition 8.2.3.** *Let $\mathbf{C}$ be a category.*

*A covariant functor $F \colon \mathbf{C} \to \mathbf{Set}$ is said to be* representable *if it is isomorphic to a covariant hom-functor $h_R$ for some $R \in \mathrm{Ob}(\mathbf{C})$.*

*A contravariant functor $F \colon \mathbf{C}^{\mathrm{op}} \to \mathbf{Set}$ is likewise said to be* representable *if it is isomorphic to a contravariant hom-functor $h^R$ for some $R \in \mathrm{Ob}(\mathbf{C})$.*

*In each case, $R$ is called the* representing object *for $F$, and if $i$ is the given isomorphism from the indicated hom functor to $F$, then $i(R)(\mathrm{id}_R)$ is called the* associated *universal element of $F(R)$.*

So from this point of view, universal problems of the sort considered above in a category $\mathbf{C}$ are questions of the *representability* of certain set-valued functors on $\mathbf{C}$. Let us examine a few set-valued functors, and see which of them are representable.

If $U$ is the underlying-set functor on **Group**, a representing object for $U$ should be a group with a universal element of its underlying set. The object with this property is the free group on one generator. More generally, if a category has free objects with respect to a concretization $U$, then $U$ will be represented by the free object on one generator, while the free object on a general set $I$ can be characterized as representing the functor $U^I$ (Definition 7.8.5).

The functor associating to every group the set of its elements of exponent 2 is represented by the group $Z_2$. More generally, the group with presentation by generators and relations $\langle X \mid R \rangle$ represents the functor associating to every group $G$ the set of $X$-tuples of members of $G$ which satisfy the relations $R$.

Is the functor associating to every commutative ring $K$ the set $|K[t]|$ of all polynomials over $K$ in one indeterminate $t$ representable? A representing object would be a ring $R$ with a universal polynomial $u(t) \in |R[t]|$. The universal property would say that given any polynomial $p(t)$ over any ring $K$, there should exist a unique homomorphism $R \to K$ which, applied coefficient-wise to polynomials, carries $u(t)$ to $p(t)$. But clearly there is a problem here: The polynomial $u$ will have some degree $n$, and if we choose a polynomial $p$ of degree $> n$, it cannot be obtained from $u$ in this way. So the set-of-polynomials functor is not representable.

However, there is a concept close to that of polynomial but not subject to the restriction that only finitely many of the coefficients be nonzero, that of a *formal power series* $a_0 + a_1 t + a_2 t^2 + \cdots$. If $K$ is a ring, then the ring of formal power series over $K$ is denoted $K[[t]]$; its underlying set $|K[[t]]| = \{a_0 + a_1 t + a_2 t^2 + \ldots\}$ can be identified with the set of all sequences $(a_0, a_1, \ldots)$ of elements of $K$, i.e., with $|K|^\omega = U^\omega(K)$. We know that the functor $U^\omega$ is represented by the free commutative ring on an $\omega$-tuple of generators, that is, the polynomial ring $\mathbb{Z}[A_0, A_1, \ldots]$. And indeed, the formal power series ring over this polynomial ring contains the element $A_0 + A_1 t + A_2 t^2 + \cdots$, which clearly has the property of a universal power series.

**Exercise 8.2:2.**

(i)   Show that the functor associating to every monoid $S$ the set of its invertible elements is representable, but that the functor associating to every monoid $S$ the set of its right-invertible elements is not.

(ii)   What about the functor associating to every monoid $S$ the set of pairs $(x, y)$ such that $xy = e$ and $yx = e$? The set of pairs $(x, y)$ merely satisfying $xy = e$? The set of 3-tuples $(x, y, z)$ such that $xy = xz = e$?

(iii)   Determine which, if any, of the functors mentioned in (i) and (ii) are isomorphic to one another.

**Exercise 8.2:3.** Let $P$ denote the contravariant power-set functor, associating to every set $X$ the set $\mathbf{P}(X)$ of its subsets, and $E$ the contravariant functor associating to every set $X$ the set $\mathbf{E}(X)$ of equivalence relations on $X$. Determine whether each of these is representable.

**Exercise 8.2:4.** Let $A$, $B$ be objects of a category $\mathbf{C}$. Describe a set-valued functor $F$ on $\mathbf{C}$ such that a *product* of $A$ and $B$, if it exists in $\mathbf{C}$, means a representing object for $F$, and likewise a functor $G$ such that a *coproduct* of $A$ and $B$ in $\mathbf{C}$ means a representing object for $G$. (One of these will be covariant and the other contravariant.)

**Exercise 8.2:5.** Let $(\mathbf{C}, U)$ be a concrete category. Show that the following conditions are equivalent. (a) The concretization functor $U$ is representable. (b) $\mathbf{C}$ has a free object on one generator with respect to $U$.

Moreover, show that if $\mathbf{C}$ has coproducts, then these are also equivalent to: (b') $\mathbf{C}$ has free objects with respect to $U$ on all sets.

Students who know some Lie group theory might try

**Exercise 8.2:6.** Let **LieGp** denote the category whose objects are all Lie groups (understood to be finite-dimensional) and whose morphisms are the continuous (equivalently, analytic) group homomorphisms among these. Let $T\colon \mathbf{LieGp} \to \mathbf{Set}$ denote the functor associating to a Lie group $L$ the set of tangent vectors to $L$ at the neutral element.

Which of the following covariant functors $\mathbf{LieGp} \to \mathbf{Set}$ are representable? (a) the functor $T$, (b) the functor $T^2\colon L \mapsto T(L) \times T(L)$, (c) the functor $L \mapsto \{(x, y) \in T(L) \times T(L) \mid [x, y] = x\}$.

The equivalence, in each of Theorems 8.2.1 and 8.2.2, of parts (ii) and (iii) shows that the concept of representable functor can be characterized in terms of initial and terminal objects. The reverse is also true:

**Exercise 8.2:7.** Let $\mathbf{C}$ be any category. Display a covariant functor $F$ and a contravariant functor $G$ from $\mathbf{C}$ to $\mathbf{Set}$, such that an initial, respectively a terminal object of $\mathbf{C}$ is equivalent to a representing object for $F$, respectively $G$.

The bijection (i) $\longleftrightarrow$ (iii) of Theorem 8.2.1 shows that an isomorphism between the hom-functor $h_R$ associated with an object $R$, and an arbitrary functor $F$, is equivalent to a specification of an element of $F(R)$ with the universal property given in (i). In fact, *every* morphism, invertible or not, from a hom-functor $h_R\colon \mathbf{C} \to \mathbf{Set}$ to another functor $F\colon \mathbf{C} \to \mathbf{Set}$ corresponds to a choice of *some* element of $F(R)$. Though utterly simple to prove, this is an important tool. We give both this result and its contravariant dual in

**Theorem 8.2.4** (Yoneda's Lemma). *Let $\mathbf{C}$ be a category, and $R$ an object of $\mathbf{C}$.*

*If $F\colon \mathbf{C} \to \mathbf{Set}$ is a covariant functor, then morphisms $f\colon h_R \to F$ are in one-to-one correspondence with elements of $F(R)$, under the map*

(8.2.5) $$f \longmapsto f(\mathrm{id}_R)$$

*(where on the right-hand side, $f$ is short for $f(R)\colon h_R(R) \to F(R)$).*

*Likewise, if* $F\colon \mathbf{C}^{\mathrm{op}} \to \mathbf{Set}$ *is a contravariant functor, then morphisms* $f\colon h^R \to F$ *are in one-to-one correspondence with elements of* $F(R)$, *under the map described by the same formula* (8.2.5).

*Proof.* In the covariant case, to show that (8.2.5) is a bijection we must describe how to get from an element $x \in F(R)$ an appropriate morphism $f_x\colon h_R \to F$. We define $f_x(A)$ to carry $a \in h_R(A) = \mathbf{C}(R, A)$ to $F(a)(x) \in F(A)$. The verification that this is a morphism of functors, and that this construction is inverse to the indicated map from morphisms of functors to elements of $F(R)$, is immediate.

The contravariant case follows by duality (or by the dualized argument).

□

**Exercise 8.2:8.** Show the verifications omitted in the first paragraph of the proof of the above result.

Rather than starting as above with the functor $h_R$, and proving that it has the property of Theorem 8.2.4, one can approach things the other way around. Given a category $\mathbf{C}$, observe that a functor $H : \mathbf{C} \to \mathbf{Set}$ is a family of sets $H(X)$ (one for each $X \in \mathrm{Ob}(\mathbf{C})$) given with maps among them satisfying certain composition relations. For any $R \in \mathrm{Ob}(\mathbf{C})$, we can ask whether there is such a system $H$ of sets and maps "freely generated" by an element $x$ in the $R$-th set, $H(R)$. It is not hard to see that for our system to be "generated" by $x$ means that for each object $X$ of $\mathbf{C}$, each element of $H(X)$ is obtained from $x$ by applying $H(a)$, for some $a \in \mathbf{C}(R, X)$. If we write $H(a)(x)$ as $a\,x$, it turns out that to make our system "free" on $x$, we should let $a\,x = b\,x$ only when $a = b$. As in the development of Cayley's Theorem for groups at the beginning of §7.1, we can now drop the symbol $x$, and regard each $H(X)$ as consisting of the appropriate elements $a$; i.e., take $H(X) = \mathbf{C}(R, X)$. So our system $H$ free on one element of $H(R)$ is $h_R$. The free generator that we began by calling $x$ can be written $\mathrm{id}_R\,x$, so when we drop the symbol $x$, it takes the form $\mathrm{id}_R$. The next result makes explicit this view of Yoneda's Lemma as expressing a universal property.

**Corollary 8.2.6.** *Let $\mathbf{C}$ be a category and $R$ an object of $\mathbf{C}$.*

*In the (large) category whose objects are pairs $(F, x)$ where $F$ is a functor $\mathbf{C} \to \mathbf{Set}$ and $x$ an element of $F(R)$, the pair $(h_R, \mathrm{id}_R)$ is the initial object. Equivalently, the object $h_R \in \mathrm{Ob}(\mathbf{Set}^{\mathbf{C}})$ is a representing object for the "evaluation at $R$" functor $\mathbf{Set}^{\mathbf{C}} \to \mathbf{Set}$, its universal element being $\mathrm{id}_R \in h_R(R)$.*

*Likewise, in the category whose objects are pairs $(F, x)$ where $F$ is a functor $\mathbf{C}^{\mathrm{op}} \to \mathbf{Set}$ and $x$ an element of $F(R)$, the pair $(h^R, \mathrm{id}_R)$ is the initial object; equivalently, the object $h^R \in \mathrm{Ob}(\mathbf{Set}^{\mathbf{C}^{\mathrm{op}}})$ represents the (again covariant!) "evaluation at $R$" functor $\mathbf{Set}^{\mathbf{C}^{\mathrm{op}}} \to \mathbf{Set}$.* □

This points to a general principle worth keeping in mind: when dealing with a morphism from a hom-functor to an arbitrary set-valued functor, look at its value on the identity map!

What if we apply Yoneda's Lemma (covariant or contravariant) to the case where the arbitrary functor $F$ is another hom-functor, $h_S$ or $h^S$? We get

**Corollary 8.2.7.** *Let* **C** *be a category.*

*Then for any two objects* $R, S \in \mathrm{Ob}(\mathbf{C})$, *the morphisms from* $h_R$ *to* $h_S$ *as functors* $\mathbf{C} \to \mathbf{Set}$ *are in one-to-one correspondence with morphisms* $S \to R$. *Thus, the mapping* $R \mapsto h_R$ *gives a* contravariant *full embedding of* **C** *in* $\mathbf{Set}^{\mathbf{C}}$, *i.e., a full and faithful functor* $\mathbf{C}^{\mathrm{op}} \to \mathbf{Set}^{\mathbf{C}}$, *the "Yoneda embedding".*

*Likewise, morphisms from* $h^R$ *to* $h^S$ *as functors* $\mathbf{C}^{\mathrm{op}} \to \mathbf{Set}$ *correspond to morphisms* $R \to S$, *giving a* covariant *"Yoneda embedding"* $\mathbf{C} \to \mathbf{Set}^{\mathbf{C}^{\mathrm{op}}}$.

*These two embeddings may both be obtained from the bivariant hom-functor* $\mathbf{C}^{\mathrm{op}} \times \mathbf{C} \to \mathbf{Set}$ *by distinguishing one or the other argument, i.e., regarding this bifunctor in one case as a functor* $\mathbf{C}^{\mathrm{op}} \to \mathbf{Set}^{\mathbf{C}}$, *and in the other as a functor* $\mathbf{C} \to \mathbf{Set}^{\mathbf{C}^{\mathrm{op}}}$.

*Sketch of proof.* By Lemma 7.10.1 the bivariant hom functor does indeed yield functors $\mathbf{C}^{\mathrm{op}} \to \mathbf{Set}^{\mathbf{C}}$ and $\mathbf{C} \to \mathbf{Set}^{\mathbf{C}^{\mathrm{op}}}$ on distinguishing one or the other argument, and we see that the object $R$ is sent to $h_R$, respectively $h^R$. Given a morphism $f: S \to R$ in **C**, one verifies that the induced morphism of functors $h_f: h_R \to h_S$ takes $\mathrm{id}_R$ to $f \in h_S(R)$. Yoneda's Lemma with $F = h_S$ tells us that the map $f \mapsto h_f$ is one-to-one and onto, so our functor $\mathbf{C}^{\mathrm{op}} \to \mathbf{Set}^{\mathbf{C}}$ is full and faithful. The contravariant case follows by duality.
$\square$

**Exercise 8.2:9.** Verify the property used above of the morphism $h_f: h_R \to h_S$ induced by a morphism $f: S \to R$.

**Exercise 8.2:10.** Show how to answer most of the parts of Exercise 7.9:4, and also Exercise 7.9:6(i), using Yoneda's Lemma.

*Remark 8.2.8.* It may seem paradoxical that we get the *contravariant* Yoneda embedding using *covariant* hom-functors, and the *covariant* Yoneda embedding using *contravariant* hom-functors, but there is a simple explanation. When we write the hom bifunctor $\mathbf{C}^{\mathrm{op}} \times \mathbf{C} \to \mathbf{Set}$ as a functor to a functor category $\mathbf{C} \to \mathbf{Set}^{\mathbf{C}^{\mathrm{op}}}$ or $\mathbf{C}^{\mathrm{op}} \to \mathbf{Set}^{\mathbf{C}}$ by distinguishing one variable, the variance in that variable determines the variance of the resulting Yoneda embedding, while the variance in the other variable determines the variance of the hom-functors that the embedding takes as its values. Whichever way we slice it, we get covariance in one, and contravariance in the other.

What is the value of the Yoneda embedding? First, categories of the form $\mathbf{Set}^{\mathbf{C}}$ are very nicely behaved; e.g., they have small products and coproducts, pushouts, pullbacks, equalizers and coequalizers, all of which can be constructed "objectwise". So Yoneda embeddings embed arbitrary categories into "good" categories. Moreover, if one wishes to extend a category $\mathbf{C}$ by adjoining additional objects with particular properties, one can often do this by identifying $\mathbf{C}$ with the category of representable contravariant functors on $\mathbf{C}$, or with the opposite of the category of representable covariant functors, and then taking for the additional objects certain other functors that are not representable, constructed in one way or another from the representable ones.

In §7.5 (discussion following Definition 7.5.2), we saw that systems of universal constructions could frequently be linked together, by natural morphisms among the constructed objects, to give functors. From the above corollary, we see that this should happen in situations where the functors that these universal objects are constructed to *represent* are linked by a corresponding system of morphisms of functors, in other words (by Lemma 7.10.1) where they form the components of a *bifunctor*. There is a slight complication in formulating this precisely, because the given representable functors are not themselves assumed to be hom-functors $h_R$ or $h^R$, but only isomorphic to these, and the choice of representing objects $R$ is likewise determined only up to isomorphism. To prepare ourselves for this complication, let us prove that a system of objects given with isomorphisms to the values of a functor in fact form the values of an isomorphic functor.

**Lemma 8.2.9.** *Let $F : \mathbf{C} \to \mathbf{D}$ be a functor, and for each $X \in \mathrm{Ob}(\mathbf{C})$, let $i(X)$ be an isomorphism of $F(X)$ with another object $G(X) \in \mathrm{Ob}(\mathbf{D})$.*

*Then there is a unique way to assign to each morphism of $\mathbf{C}$, $f \in \mathbf{C}(X, Y)$ a morphism $G(f) \in \mathbf{D}(G(X), G(Y))$ so that the objects $G(X)$ and morphisms $G(f)$ constitute a functor $G : \mathbf{C} \to \mathbf{D}$, and $i$ constitutes an isomorphism of functors, $F \cong G$.*

*Proof.* If $G$ is to be a functor and $i$ a morphism of functors, then for each $f \in \mathbf{C}(X, Y)$ we must have $G(f) i(X) = i(Y) F(f)$. Since $i(X)$ is an isomorphism, we can rewrite this as $G(f) = i(Y) F(f) i(X)^{-1}$. It is straightforward to verify that $G$, so defined on morphisms, is indeed a functor. This definition of $G(f)$ insures that $i$ is a morphism of functors $F \to G$, and it clearly has an inverse, defined by $i^{-1}(X) = i(X)^{-1}$.                                  □

**Exercise 8.2:11.** Write out the verification that $G$, constructed as above, is a functor.

We can now get our desired result about tying representing objects together into a functor. In thinking about results such as the next lemma, I find it useful to keep in mind the case where $\mathbf{C} = \mathbf{Set}$, $\mathbf{D} = \mathbf{Group}$, and $A$ is the bifunctor associating to every set $X$ and group $G$ the set $|G|^X$ of $X$-tuples of elements of $G$.

**Lemma 8.2.10.** *Suppose that* **C** *and* **D** *are categories, and that for each* $X \in \mathrm{Ob}(\mathbf{C})$ *we are given a functor* $A(X, -)\colon \mathbf{D} \to \mathbf{Set}$ *and an object* $F(X) \in \mathrm{Ob}(\mathbf{D})$ *representing this functor, via an isomorphism*

$$(8.2.11) \qquad\qquad i(X)\colon\; A(X, -) \cong h_{F(X)}.$$

*Then*

(i)   *If the given functors* $A(X, -)$ *are in fact the values of a bifunctor* $A\colon \mathbf{C}^{\mathrm{op}} \times \mathbf{D} \to \mathbf{Set}$ *at the objects of* **C**, *then the objects* $F(X)$ *of* **D** *can be made the values of a functor* $F\colon \mathbf{C} \to \mathbf{D}$ *in a unique way so that the isomorphisms* (8.2.11) *comprise an isomorphism of bifunctors*

$$(8.2.12) \qquad\qquad i\colon\; A(-, -) \cong \mathbf{D}(F(-), -).$$

(ii)   *Conversely, if the objects* $F(X)$ *are the values, at the objects* $X$, *of a functor* $F\colon \mathbf{C} \to \mathbf{D}$, *we can make the family of functors* $A(X, -)$ *into a bifunctor* $A\colon \mathbf{C}^{\mathrm{op}} \times \mathbf{D} \to \mathbf{Set}$ *in a unique way so that the isomorphisms* (8.2.11) *again give an isomorphism* (8.2.12) *of bifunctors.*

*Proof.* On the one hand, if $A\colon \mathbf{C}^{\mathrm{op}} \times \mathbf{D} \to \mathbf{Set}$ is a bifunctor, the induced system of functors $A(X, -)\colon \mathbf{D} \to \mathbf{Set}$ will together constitute a single functor which we may call $B\colon\; \mathbf{C}^{\mathrm{op}} \to \mathbf{Set}^{\mathbf{D}}$ (Lemma 7.10.1). For each $X \in \mathrm{Ob}(\mathbf{C})$ we have an isomorphism $i(X)$ of $B(X)$ with a hom-functor $h_{F(X)}$, so by the preceding lemma we get an isomorphic functor $C\colon \mathbf{C}^{\mathrm{op}} \to \mathbf{Set}^{\mathbf{D}}$, such that $C(X) = h_{F(X)}$, and the isomorphism $i\colon B \cong C$ is made up of the $i(X)$'s. Now by Corollary 8.2.7, the covariant hom-functors $h_Y$ $(Y \in \mathrm{Ob}(\mathbf{D}))$ form a full subcategory of $\mathbf{Set}^{\mathbf{D}}$ isomorphic to $\mathbf{D}^{\mathrm{op}}$ via the Yoneda embedding $Y \mapsto h_Y$. Hence the functor $C\colon\; \mathbf{C}^{\mathrm{op}} \to \mathbf{Set}^{\mathbf{D}}$ is induced by precomposing this embedding $\mathbf{D}^{\mathrm{op}} \to \mathbf{Set}^{\mathbf{D}}$ with a unique functor $\mathbf{C}^{\mathrm{op}} \to \mathbf{D}^{\mathrm{op}}$, which is equivalent to a functor $F\colon \mathbf{C} \to \mathbf{D}$, and this $F$ is the functor of the statement of the lemma.

Inversely, if $F$ is given as a functor, let us consider each functor $A(X, -)$ as an object $B(X)$ of $\mathbf{Set}^{\mathbf{D}}$. Then for each $X$ we have an isomorphism $i(X)\colon B(X) \cong h_{F(X)}$, and applying the preceding lemma to the isomorphisms $i(X)^{-1}$, we conclude that the objects $B(X)$ are the values of a functor $B\colon \mathbf{C}^{\mathrm{op}} \to \mathbf{Set}^{\mathbf{D}}$, which we may regard as a bifunctor $A\colon \mathbf{C}^{\mathrm{op}} \times \mathbf{D} \to \mathbf{Set}$, and again the values of $i$ become an isomorphism of bifunctors.   □

The above lemma concerns systems of objects representing covariant hom-functors; let us state the corresponding result for contravariant hom-functors. A priori, this means replacing **D** by $\mathbf{D}^{\mathrm{op}}$. But it is then natural to replace the "parametrizing" category $\mathbf{C}^{\mathrm{op}}$ by **C** so as to keep the parametrization of the constructed objects of **D** covariant. And having done that much, why not interchange the names of **C** and **D** so as to get a set-up parallel to that of the preceding case? Doing so, we get

**Lemma 8.2.13.** *Suppose that* **C** *and* **D** *are categories, and that for each* $Y \in \mathrm{Ob}(\mathbf{D})$ *we are given a functor* $A(-, Y) : \mathbf{C}^{\mathrm{op}} \to \mathbf{Set}$ *and an object* $U(Y) \in \mathrm{Ob}(\mathbf{C})$ *representing this contravariant functor, via an isomorphism*

$$(8.2.14) \qquad\qquad j(Y): \ A(-, Y) \cong h^{U(Y)}.$$

*Then*

(i)  *If the given functors* $A(-, Y)$ *are the values of a bifunctor* $A: \mathbf{C}^{\mathrm{op}} \times \mathbf{D} \to \mathbf{Set}$ *at the objects of* **D**, *the family of objects* $U(Y)$ *of* **C** *can be made the values of a functor* $U: \mathbf{D} \to \mathbf{C}$ *in a unique way so that the isomorphisms* (8.2.14) *constitute an isomorphism of bifunctors*

$$(8.2.15) \qquad\qquad j: \ A(-, -) \cong \mathbf{C}(-, U(-)).$$

(ii)  *Conversely, if the objects* $U(Y)$ *are the values at the objects* $Y$ *of a functor* $U: \mathbf{C} \to \mathbf{D}$, *we can make the family of functors* $A(-, Y)$ *into a bifunctor* $A: \mathbf{C}^{\mathrm{op}} \times \mathbf{D} \to \mathbf{Set}$ *in a unique way so that the isomorphisms* (8.2.14) *together give an isomorphism* (8.2.15) *of bifunctors.*  $\square$

## 8.3. Adjoint functors

Let us look at some examples of the situation of the two preceding lemmas—families of objects that we characterized individually as the representing objects for certain naturally occurring functors, but that turned out, themselves, to fit together into a functor. By the above lemmas, this means that the system of functors that these objects represented fit together into a bifunctor. We shall see that in each of these cases, this structure of bifunctor was actually present in the original situation, providing an explanation of why our constructions yielded functors.

The free group on each set $X$ is the object of **Group** representing the functor $G \mapsto |G|^X = \mathbf{Set}(X, U(G))$. So the free group *functor* arises by representing the family of functors **Group** $\to$ **Set** obtained by inserting all sets $X$ as the first argument of the *bifunctor*

$$\mathbf{Set}(-, U(-)): \ \mathbf{Set}^{\mathrm{op}} \times \mathbf{Group} \longrightarrow \mathbf{Set}.$$

The analogous description obviously applies in any category **C** having free objects with respect to a concretization $U: \mathbf{C} \to \mathbf{Set}$.

If $G$ is a group, the *abelianization* of $G$ is the object of **Ab** representing the functor **Ab** $\to$ **Set** given by $A \mapsto \mathbf{Group}(G, A)$. The symbol $\mathbf{Group}(G, A)$ makes sense because **Ab** is a subcategory of **Group**; to put this example in the context of the general pattern, let us write $V$ for the inclusion functor of **Ab** in **Group**. We then see that the abelianization functor arises by representing the family of set-valued functors obtained by inserting values in the first argument of the bifunctor

$$\mathbf{Group}(-, V(-)): \ \mathbf{Group}^{\mathrm{op}} \times \mathbf{Ab} \longrightarrow \mathbf{Set}.$$

In the same way, if $W$ denotes the forgetful functor $\mathbf{Group} \to \mathbf{Monoid}$, then the functor taking a monoid to its universal enveloping group arises by representing the family of set-valued functors obtained by inserting values in the first argument of the bifunctor

$$\mathbf{Monoid}(-, W(-)): \ \mathbf{Monoid}^{\mathrm{op}} \times \mathbf{Group} \longrightarrow \mathbf{Set}.$$

The above were left universal examples, that is, constructions $F : \mathbf{C} \to \mathbf{D}$ such that each object $F(X)$ represented a covariant functor $\mathbf{D} \to \mathbf{Set}$. We see that in each such case, the bifunctor from which these covariant functors were extracted had the form

(8.3.1)                     $\mathbf{C}(-, U(-)): \ \mathbf{C}^{\mathrm{op}} \times \mathbf{D} \longrightarrow \mathbf{Set},$

for some functor $U : \mathbf{D} \to \mathbf{C}$. Taking (8.3.1) to be the $A$ in Lemma 8.2.10, we see that the universal properties of the objects $F(X)$ in terms of $U$ can be formulated in each of these cases as

$$\mathbf{C}(-, U(-)) \ \cong \ \mathbf{D}(F(-), -)$$

—a strikingly symmetrical condition!

Let us consider one right universal example. Given a monoid $S$, we considered above the construction of the universal group $G$ with a homomorphism of $S$ into $G_{\mathrm{md}}$; but there is also a universal group $G$ with a homomorphism of $G_{\mathrm{md}}$ into $S$, namely the group $G = S_{\mathrm{inv}}$ of invertible elements ("units") of $S$. If we write $F : \mathbf{Group} \to \mathbf{Monoid}$ for the forgetful functor $G \mapsto G_{\mathrm{md}}$, and call the above group-of-units functor $U : \mathbf{Monoid} \to \mathbf{Group}$, we see that $U(S)$ represents the contravariant functor associating to each group $G$ the set $\mathbf{Monoid}(F(G), S)$. If we write $\mathbf{C}$ and $\mathbf{D}$ for $\mathbf{Group}$ and $\mathbf{Monoid}$, then on taking $\mathbf{D}(F(-), -)$ for the bifunctor $A$ in the second paragraph of Lemma 8.2.13, we get an isomorphism characterizing this right universal construction $U$ :

$$\mathbf{D}(F(-), -) \ \cong \ \mathbf{C}(-, U(-)).$$

This is exactly the same as the isomorphism characterizing our examples of left universal constructions—but written in reverse order, and looked at as characterizing $U$ in terms of $F$, rather than $F$ in terms of $U$! The fact that these two situations are characterized by the same isomorphism means that a functor $F$ gives objects representing the covariant functors $\mathbf{C}(X, U(-))$ if and only if $U$ gives objects representing the contravariant functors $\mathbf{D}(F(-), Y)$.

Let us test this conclusion, by turning our characterization of the free group construction upside down. Since the free group $F(X)$ on a set $X$ is left universal among groups $G$ with set maps of $X$ into their underlying sets $U(G)$, the *underlying set* $U(G)$ of a group $G$ should be right-universal among all sets $X$ with group homomorphisms from the free group $F(X)$ into $G$. And indeed, though it may seem bizarre to treat the free-group construction as something given and the underlying-set construction as something to be characterized, the universal property certainly holds: For any group $G$, $U(G)$ is a set with a homomorphism $u\colon F(U(G)) \to G$, such that given any homomorphism $f$ from a free group $F(X)$ on a set into $G$, there is a unique set map $h\colon X \to U(G)$ (which you should be able to describe) such that $f = uF(h)$. This property of underlying sets is sometimes even useful. For instance, in showing that every group can be presented by generators and relations, one wishes to write an arbitrary group $G$ as a homomorphic image of a free group on some set $X$. The above property says that there is a universal choice of such $X$, namely the underlying set $U(G)$ of $G$.

Before setting out to tie together all our ways of describing these universal constructions, let us prove a lemma that will allow us to relate isomorphisms of bifunctors as above to systems of maps $X \to U(F(X))$ and $F(U(Y)) \to Y$. (The lemma is an instance of the general principle noted following Corollary 8.2.6.)

**Lemma 8.3.2.** *Let $\mathbf{C}$ and $\mathbf{D}$ be categories and $U\colon \mathbf{D} \to \mathbf{C}$, $F\colon \mathbf{C} \to \mathbf{D}$ functors, and consider the two bifunctors $\mathbf{C}^{\mathrm{op}} \times \mathbf{D} \to \mathbf{Set}$,*

$$\mathbf{C}(-, U(-)), \quad \mathbf{D}(F(-), -).$$

*Then a morphism of bifunctors*

(8.3.3) $$a\colon \ \mathbf{C}(-, U(-)) \longrightarrow \mathbf{D}(F(-), -)$$

*is determined by its values on identity morphisms $\mathrm{id}_{U(D)} \in \mathbf{C}(U(D), U(D))$ $(D \in \mathrm{Ob}(\mathbf{D}))$. In fact, given $a$ as above, if we write*

$$\alpha(D) = a(U(D), D)(\mathrm{id}_{U(D)}) \in \mathbf{D}(F(U(D)), D),$$

*then this family of morphisms comprises a morphism of functors,*

(8.3.4) $$\alpha\colon \ FU \longrightarrow \mathrm{Id}_{\mathbf{D}}$$

*and this construction yields a bijection between morphisms (8.3.3) and morphisms (8.3.4). Given a morphism (8.3.4), the corresponding morphism (8.3.3) can be described as acting on $f \in \mathbf{C}(C, U(D))$ by first applying $F$ to get $F(f)\colon F(C) \to FU(D)$, then composing this with $\alpha(D)\colon FU(D) \to D$, getting*

$$a(f) = \alpha(D)\,F(f)\colon \ F(C) \longrightarrow D.$$

*Likewise, a morphism of bifunctors in the opposite direction to* (8.3.3),

$$(8.3.5) \qquad\qquad b\colon \mathbf{D}(F(-), -) \longrightarrow \mathbf{C}(-, U(-))$$

*is determined by its values on identity morphisms, in this case morphisms*
$\mathrm{id}_{F(C)} \in \mathbf{D}(F(C), F(C))$ $(C \in \mathrm{Ob}(\mathbf{C}))$, *and writing*

$$\beta(C) = b(C, F(C))(\mathrm{id}_{F(C)}) \in \mathbf{C}(C, U(F(C))),$$

*we get a bijection between morphisms* (8.3.5) *and morphisms*

$$(8.3.6) \qquad\qquad \beta\colon \mathrm{Id}_{\mathbf{C}} \longrightarrow UF.$$

*Given* $\beta$, *the corresponding morphism* $b$ *can be described as taking* $f \in$
$\mathbf{D}(F(C), D)$ *to*

$$b(f) = U(f)\beta(C)\colon\ C \longrightarrow U(D).$$

*Sketch of proof.* Consider a morphism $a$ as in (8.3.3). For each $D \in \mathrm{Ob}(\mathbf{D})$
this gives a morphism of functors $\mathbf{C}(-, U(D)) \to \mathbf{D}(F(-), D)$. Since the
first of these functors is $h^{U(D)}$, the Yoneda Lemma says this morphism is
determined by its value on the identity morphism of $U(D)$. That value will
be a member of $\mathbf{D}(FU(D), D)$; let us call it $\alpha(D)\colon FU(D) \to D$. I claim
that these maps constitute a morphism (8.3.4).

To see this, let $f\colon D_1 \to D_2$ be any morphism in $\mathbf{D}$, and consider the
diagram

$$(8.3.7)$$

The left square commutes by functoriality of $a$ in its second argument; the
right square by functoriality in its first argument. Note that the identity morphisms in the upper left and upper right hom-sets have the same image in
the upper middle hom-set, namely $U(f)$, and hence they must have equal
images in the lower middle hom-set. But those same images, computed by going first downward and then to the center, are the composites whose equality
is need to prove the functoriality of $\alpha$. The reader can easily check that the
description of how to recover (8.3.3) from (8.3.4) also leads to a morphism of
functors, and that this construction is inverse to the first.

The second paragraph follows by duality.                              □

**Exercise 8.3:1.** Verify the assertion that the equality proved using (8.3.7) is the condition for $\alpha$ to be a morphism of functors, and show the "easy check" mentioned after that.

To get a feel for the above construction, you might start with the morphism of bifunctors $a$ that associates to every set map from a set $X$ to the underlying set $U(G)$ of a group $G$ the induced group homomorphism from the free group $F(X)$ into $G$. Determine the morphism of functors $\alpha$ that the above construction yields, and check explicitly that the "inverse" construction described does indeed recover $a$. In this example, one finds that $a$ is invertible; calling its inverse $b$, you might similarly work out for this $b$ the constructions of the second assertion of the lemma.

With the help of Lemmas 8.2.10, 8.2.13, and 8.3.2, we can now give several descriptions of the type of universal construction discussed at the beginning of this section.

**Theorem 8.3.8.** *Let $\mathbf{C}$ and $\mathbf{D}$ be categories. Then the following data are equivalent:*

(i)  *A pair of functors $U: \mathbf{D} \to \mathbf{C}$, $F: \mathbf{C} \to \mathbf{D}$, and an isomorphism*

$$i: \ \mathbf{C}(-, U(-)) \cong \mathbf{D}(F(-), -)$$

*of functors $\mathbf{C}^{\mathrm{op}} \times \mathbf{D} \to \mathbf{Set}$.*

(ii)  *A functor $U: \mathbf{D} \to \mathbf{C}$, and for every $C \in \mathrm{Ob}(\mathbf{C})$, an object $R_C \in \mathrm{Ob}(\mathbf{D})$ and an element $u_C \in \mathbf{C}(C, U(R_C))$ which are universal among such object-element pairs, i.e., which represent the covariant functor $\mathbf{C}(C, U(-)): \mathbf{D} \to \mathbf{Set}$ (cf. Theorem 8.2.1 and Definition 8.2.3).*

(ii\*)  *A functor $F: \mathbf{C} \to \mathbf{D}$, and for every $D \in \mathrm{Ob}(\mathbf{D})$, an object $R_D \in \mathrm{Ob}(\mathbf{C})$ and an element $v_D \in \mathbf{D}(F(R_D), D)$ which are universal among such object-element pairs, i.e., which represent the contravariant functor $\mathbf{D}(F(-), D): \mathbf{C}^{\mathrm{op}} \to \mathbf{Set}$.*

(iii)  *A pair of functors $U: \mathbf{D} \to \mathbf{C}$, $F: \mathbf{C} \to \mathbf{D}$, and a pair of morphisms of functors*

$$\eta: \mathrm{Id}_{\mathbf{C}} \longrightarrow UF, \qquad \varepsilon: FU \longrightarrow \mathrm{Id}_{\mathbf{D}},$$

*such that the two composites*

$$U \xrightarrow{\eta \circ U} UFU \xrightarrow{U \circ \varepsilon} U, \qquad F \xrightarrow{F \circ \eta} FUF \xrightarrow{\varepsilon \circ F} F,$$

*are the identity morphisms of $U$ and $F$ respectively. (For the "$\circ$" notation see Lemma 7.10.2.)*

*Sketch of proof.* The equivalence (i) $\Longleftrightarrow$ (ii) is given by Lemma 8.2.10 with $A(-, -) = \mathbf{C}(-, U(-))$; similarly, (i) $\Longleftrightarrow$ (ii\*) is given by Lemma 8.2.13 with $A(-, -) = \mathbf{D}(F(-), -)$. By Lemma 8.3.2, an isomorphism of bifunctors as in (i) must correspond to a pair of morphisms of functors $\eta: \mathrm{Id}_{\mathbf{C}} \to UF$,

$\varepsilon : FU \to \mathrm{Id}_{\mathbf{D}}$ which induce mutually inverse morphisms of bifunctors. I claim that the conditions needed for these induced morphisms to be mutually inverse are those shown diagrammatically in (iii).

In the verification of this statement (made an exercise below), one assumes $\alpha$ and $\beta$ given as in Lemma 8.3.2, and uses the formulas for $a$ and $b$ in terms of these to express the composites $ab$ and $ba$. One must then prove that these composites are the identity morphisms. By Yoneda's Lemma, it suffices to check these equalities on appropriate identity morphisms. (With what objects of $\mathbf{C}$ and $\mathbf{D}$ in the slots of $\mathbf{D}(F(-), -)$, respectively $\mathbf{C}(-, U(-))$?) This approach quickly gives the desired statements. However, if one prefers to see directly that these statements are equivalent to $ab$ and $ba$ fixing *all* morphisms $f \in \mathbf{D}(F(C), D)$, respectively $g \in \mathbf{C}(C, U(D))$, then one may combine the equations saying that the latter conclusions hold with the commutativity of the diagram expressing the functoriality of $a$, respectively $b$, applied to the morphism $f$, respectively $g$.                                        □

### Exercise 8.3:2.

(i)   Write out the verification sketched in the last paragraph of the above proof (including, of course, the verification that the two constructions are mutually inverse).

(ii)  Show that $\eta$ will be composed of the "universal morphisms" $u_C$ of point (ii) of the theorem, and $\varepsilon$ will be composed of the universal morphisms $v_D$ of point (ii*).

(iii) Take one universal construction, e.g., that of free groups, write down the equalities expressed diagrammatically in part (iii) of the above theorem for this construction in terms of maps of set- and group-elements, and explain why they hold in *this case*.

**Definition 8.3.9.** *Given categories* $\mathbf{C}$ *and* $\mathbf{D}$ *and functors* $U : \mathbf{D} \to \mathbf{C}$, $F : \mathbf{C} \to \mathbf{D}$, *an isomorphism*

$$i : \quad \mathbf{C}(-, U(-)) \cong \mathbf{D}(F(-), -)$$

*of bifunctors* $\mathbf{C}^{\mathrm{op}} \times \mathbf{D} \to \mathbf{Set}$, *or equivalently, a pair of morphisms of functors* $\varepsilon$, $\eta$ *satisfying the condition of point* (iii) *of the above theorem, is called an* adjunction *between* $U$ *and* $F$.

*In this situation,* $U$ *is called the "right adjoint" of the functor* $F$, *and* $F$ *the "left adjoint" of* $U$ *(referring to their occurrence in the right and left slots of the hom-bifunctors in the above isomorphism). The morphisms of functors* $\eta$ *and* $\varepsilon$ *are called, respectively, the* unit *and* counit *of the adjunction.*

Historical note: The term "adjoint" was borrowed from analysis, where the adjoint of a bounded operator between Hilbert spaces, $A : X \to Y$, is the operator $B : Y \to X$ characterized by the condition on inner products $(x, By) = (Ax, y)$.

The student who finds condition (iii) of Theorem 8.3.8 hard to grasp will be happy to know that we will not make much use of it in the next few chapters. (I have trouble with it myself.) But we *will* use the morphisms $\eta$ and $\varepsilon$ named in that condition, so you should get a clear idea of how these act. (What we will seldom use is the fact that the indicated compositional condition on a pair of morphisms $\eta$, $\varepsilon$ is equivalent to their being the unit and counit of an adjunction. Nevertheless, I recommend working Exercise 8.3:2 this once.)

We can now characterize as right or left adjoints many of the universal constructions we are familiar with. The three diagrams below show the cases we used above to motivate the concept. In each of these, a pair of successive vertical arrows between two categories represents a pair of mutually adjoint functors, the right adjoint being shown on the right and the left adjoint on the left.

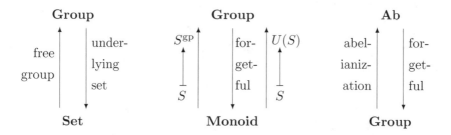

The middle diagram is interesting in that the forgetful functor there (in the notation of §4.11, $G \mapsto G_{\mathrm{md}}$) has both a left and a right adjoint. In the first diagram above, we can, as mentioned, replace **Group** with any category **C** having free objects with respect to a concretization $U$. A slight generalization is noted in the next exercise.

**Exercise 8.3:3.** If you did not do Exercise 8.2:5, prove that if **C** is a category with small coproducts and $U : \mathbf{C} \to \mathbf{Set}$ a functor, then $U$ has a left adjoint if and only if it is *representable*.

  (Exercise 8.2:5 was essentially the case of this result where $U$ was faithful, so that it could be called a "concretization" and its left adjoint a "free object" construction; but faithfulness played no part in the proof. In Chapter 10 we shall extend the concept of "representable functor" from set-valued functors to algebra-valued functors, and generalize the above result to the resulting much wider context.)

**Exercise 8.3:4.** Show that (i) the left (or right) adjoint of a functor, if one exists, is unique up to canonical isomorphism, and conversely, that (ii) if $A$ and $B$ are isomorphic functors, then any functor which can be made a left (or right) adjoint of $A$ can also be made a left (or right) adjoint of $B$.

**Exercise 8.3:5.** Show that if $A\colon \mathbf{C} \to \mathbf{D}$, $B\colon \mathbf{D} \to \mathbf{C}$ give an equivalence of categories (Definition 7.9.5), then $B$ is both a right and a left adjoint to $A$.

The next exercise is a familiar example in disguise.

**Exercise 8.3:6.** Let $\mathbf{C}$ be the category with $\mathrm{Ob}(\mathbf{C}) = \mathrm{Ob}(\mathbf{Group})$, but with morphisms defined so that for groups $G$ and $H$, $\mathbf{C}(G, H) = \mathbf{Set}(|G|, |H|)$. Thus $\mathbf{Group}$ is a subcategory of $\mathbf{C}$, with the same object set but smaller morphism sets. Does the inclusion functor $\mathbf{Group} \to \mathbf{C}$ have a left and/or a right adjoint?

There are many other constructions whose universal properties translate into adjointness statements: The forgetful functor $\mathbf{Ring}^1 \to \mathbf{Monoid}$ that remembers only the multiplicative structure has as left adjoint the *monoid ring* construction. The forgetful functor $\mathbf{Ring}^1 \to \mathbf{Ab}$ that remembers only the additive structure has for left adjoint the *tensor ring* construction. (These two ring constructions were discussed briefly toward the end of §4.12.) The inclusion of the category of compact Hausdorff spaces in that of arbitrary topological spaces has for left adjoint the Stone-Čech compactification functor (§4.17). The functor associating to every commutative ring its Boolean ring of idempotent elements has as left adjoint the construction asked for in Exercise 4.14:3(iv). The forgetful functors going from $\mathbf{Lattice}$ to $\vee$-$\mathbf{Semilattice}$ and $\wedge$-$\mathbf{Semilattice}$, and from these in turn to $\mathbf{POSet}$, have left adjoints which you were asked to construct in Exercise 6.1:8.

The student familiar with Lie algebras (§9.7 below) will note that the functor associating to an associative algebra $A$ the Lie algebra $A_{\mathrm{Lie}}$ with the same underlying vector space as $A$, and with the commutator operation of $A$ for Lie bracket, has for left adjoint the *universal enveloping algebra* construction. (The Poincaré-Birkhoff-Witt Theorem gives a normal form for this universal object; cf. [46, §3].)

Suppose $\mathbf{C}$ is a category having *products* and *coproducts* of all pairs of objects. We know that each of these constructions will give a functor $\mathbf{C} \times \mathbf{C} \to \mathbf{C}$. We may ask whether these functors can be characterized as adjoints of some functors $\mathbf{C} \to \mathbf{C} \times \mathbf{C}$. Similarly, we may ask whether the *tensor product* functor $\mathbf{Ab} \times \mathbf{Ab} \to \mathbf{Ab}$ can be characterized as an adjoint of some functor $\mathbf{Ab} \to \mathbf{Ab} \times \mathbf{Ab}$.

The universal property of the product functor $\mathbf{C} \times \mathbf{C} \to \mathbf{C}$ is a right universal one, so if it arises as an adjoint, it should be a right adjoint to some functor $A\colon \mathbf{C} \to \mathbf{C} \times \mathbf{C}$. No such functor was evident in our definition of products. However, we can search for such an $A$ by posing the universal problem whose solution would be a *left* adjoint to the product functor: Given $X \in \mathrm{Ob}(\mathbf{C})$, does there exist $(Y, Z) \in \mathrm{Ob}(\mathbf{C} \times \mathbf{C})$ with a universal example of a morphism $X \to Y \times Z$? Since a morphism $X \to Y \times Z$ corresponds to a morphism $X \to Y$ and a morphism $X \to Z$, this asks whether there exists a pair $(Y, Z)$ of objects of $\mathbf{C}$ universal for having a morphism from $X$ to each member of this pair. In fact, the pair $(X, X)$ is easily seen to have the desired universal property. This leads us to define the "diagonal functor" $\Delta: \mathbf{C} \to \mathbf{C} \times \mathbf{C}$ taking each object $X$ to $(X, X)$, and each morphism $f$ to $(f, f)$. It is now easy to check that the universal property of the direct product construction is that of a right adjoint to $\Delta$. Similar reasoning shows that the universal property of the coproduct is that of a left adjoint to $\Delta$; so in a category $\mathbf{C}$ having both products and coproducts, we have the diagram of adjoint functors,

We recall that if $\mathbf{C}$ is $\mathbf{Ab}$, the constructions of pairwise products and coproducts ("direct products and direct sums") coincide. So in that case we get an infinite periodic chain of adjoints.

**Exercise 8.3:7.** Does the direct product construction on $\mathbf{Set}$ have a *right* adjoint? Does the coproduct construction have a *left* adjoint?

The next exercise is one of my favorites:

**Exercise 8.3:8.** Recall that $\mathbf{2}$ denotes the category with two objects, 0 and 1, and exactly one nonidentity morphism, $0 \to 1$, so that for any category $\mathbf{C}$, an object of $\mathbf{C}^{\mathbf{2}}$ corresponds to a choice of two objects $A_0, A_1 \in \mathrm{Ob}(\mathbf{C})$ and a morphism $f: A_0 \to A_1$.

Let $p_0: \mathbf{Group}^{\mathbf{2}} \to \mathbf{Group}$ denote the functor taking each object $(A_0, A_1, f)$ to its first component $A_0$, and likewise every morphism $(a_0, a_1): (A_0, A_1, f) \to (B_0, B_1, g)$ of $\mathbf{Group}^{\mathbf{2}}$ to its first component $a_0$.

Investigate whether $p_0$ has a left adjoint, and whether it has a right adjoint. If a left adjoint is found, investigate whether this in turn has a left adjoint (clearly it has a right adjoint – namely $p_0$); likewise if $p_0$ has a right adjoint, investigate whether this in turn has a right adjoint; and so on, as long as further adjoints on either side can be found.

**Exercise 8.3:9.** Let $G$ be a group, and $G$-$\mathbf{Set}$ the category of all $G$-sets.

You can probably think of one or more very easily described functors from $\mathbf{Set}$ to $G$-$\mathbf{Set}$, or vice versa. Choose one of them, and apply the idea of the preceding exercise; i.e., look for a left adjoint and/or a right adjoint, and for further adjoints of these, as long as you can find any.

When you are finished, does the chain of functors you have gotten contain all the "easily described functors" between these two categories that you were able to think of? If not, take one that was missed, and do the same with it.

**Exercise 8.3:10.** Translate the idea indicated in observation (a) following Exercise 4.8:1 into questions of the existence of adjoints to certain functors between categories $G_1$-**Set** and $G_2$-**Set**, determine whether such adjoints do in fact exist, and if they do, describe them as concretely as you can.

Let us now consider the case of the tensor product construction, $\otimes \colon \mathbf{Ab} \times \mathbf{Ab} \to \mathbf{Ab}$. It is the solution to a left universal problem, and we can characterize this problem as arising, as described in Lemma 8.2.10, from the bifunctor $\mathrm{Bil} \colon (\mathbf{Ab} \times \mathbf{Ab})^{\mathrm{op}} \times \mathbf{Ab} \to \mathbf{Set}$, where for abelian groups $A$, $B$, $C$ we let $\mathrm{Bil}((A, B), C)$ denote the set of bilinear maps $(A, B) \to C$. From the preceding examples, we might expect $\mathrm{Bil}((A, B), C)$ to be expressible in the form $(\mathbf{Ab} \times \mathbf{Ab})((A, B), U(C))$ for some functor $U \colon \mathbf{Ab} \to \mathbf{Ab} \times \mathbf{Ab}$.

But, in fact, it cannot be so expressed; in other words, the tensor product construction $\mathbf{Ab} \times \mathbf{Ab} \to \mathbf{Ab}$, though it is a left universal construction, is *not* a left adjoint. The details (and a different sense in which the tensor product *is* a left adjoint) are something you can work out:

**Exercise 8.3:11.**

(i)  Show that the functor $\otimes \colon \mathbf{Ab} \times \mathbf{Ab} \to \mathbf{Ab}$ has no left or right adjoint.

(ii)  On the other hand, show that for any fixed abelian group $A$, the functor $A \otimes - \colon \mathbf{Ab} \to \mathbf{Ab}$ is left adjoint to the functor $\mathrm{Hom}(A, -) \colon \mathbf{Ab} \to \mathbf{Ab}$. (I am writing $\mathrm{Hom}(A, B)$ for the *abelian group* of homomorphisms from $A$ to $B$, in contrast to $\mathbf{Ab}(A, B)$ the *set* of such homomorphisms – a temporary ad hoc notational choice.)

(iii)  Investigate whether the functor $A \otimes - \colon \mathbf{Ab} \to \mathbf{Ab}$ has a left adjoint, and whether $\mathrm{Hom}(A, -) \colon \mathbf{Ab} \to \mathbf{Ab}$ has a right adjoint. If such adjoints do not *always* exist, do they exist for *some* choices of $A$?

If you are familiar enough with ring theory, generalize the above problems to modules over a fixed commutative ring $k$, or to bimodules over pairs of noncommutative rings.

**Exercise 8.3:12.** For a fixed set $A$, does the functor $\mathbf{Set} \to \mathbf{Set}$ given by $S \mapsto S \times A$ have a left or right adjoint?

A situation which is similar to that of the tensor product in that the question of whether a construction is an adjoint depends on what we take as the variable, though quite different in terms of which versions have and which don't have adjoints, is considered in

**Exercise 8.3:13.** In this exercise "ring" will mean commutative ring with 1; recall that we denote the category of such rings $\mathbf{CommRing}$[1].

If $R$ is a ring and $X$ any set, $R[X]$ will denote the polynomial ring over $R$ in an $X$-tuple of indeterminates.

(i)   Show that for $X$ a nonempty set, the functor $P_X \colon \mathbf{CommRing}^1 \to \mathbf{CommRing}^1$ taking each ring $R$ to $R[X]$ has neither a right nor a left adjoint, and similarly that for $R$ a ring, the functor $Q_R \colon \mathbf{Set} \to \mathbf{CommRing}^1$ taking each set $X$ to $R[X]$ has neither a right nor a left adjoint.

(ii)   On the other hand, show that the functor $\mathbf{CommRing}^1 \times \mathbf{Set} \to \mathbf{CommRing}^1$ taking a pair $(R, X)$ to $R[X]$ is an adjoint (on the appropriate side) of an easily described functor.

(iii)   For any ring $R$, let $\mathbf{CommRing}^1_R$ denote the category of commutative $R$-algebras (rings $S$ given with homomorphisms $R \to S$), and $R$-algebra homomorphisms (ring homomorphisms making commuting triangles with $R$. In the notation of Exercise 7.8:30(ii), this is the comma category $(R \downarrow \mathbf{CommRing}^1)$.)

Similarly, for any set $X$, let $\mathbf{CommRing}^1_X$ denote the category of rings $S$ given with set maps $X \to |S|$, and again having for morphisms the ring homomorphisms making commuting triangles. (This is the comma category $(X \downarrow U)$, where $U$ is the underlying set functor of $\mathbf{CommRing}^1$. Note: to keep the symbols $\mathbf{CommRing}^1_R$ and $\mathbf{CommRing}^1_X$ unambiguous, we must remember to use distinct symbols for rings and sets.)

Show that for any $R$, the functor $\mathbf{Set} \to \mathbf{CommRing}^1_R$ taking $X$ to $R[X]$ can be characterized as an adjoint, and that for any $X$, the functor $\mathbf{CommRing}^1 \to \mathbf{CommRing}^1_X$ taking $R$ to $R[X]$ can also be characterized as an adjoint.

(iv)   Investigate similar questions for the formal power series construction, $R[[X]]$; in particular, whether the analog of (i) is true.

Here is still another way to make the tensor product construction into an adjoint functor:

**Exercise 8.3:14.**

(i)   Let $\mathbf{Bil}$ be the category whose objects are all 4-tuples $(A, B, \beta, C)$, where $A, B, C$ are abelian groups and $\beta \colon (A, B) \to C$ is a bilinear map, and where morphisms are defined in the natural way. (Say what this natural way is!) Show that the forgetful functor $\mathbf{Bil} \to \mathbf{Ab} \times \mathbf{Ab}$, taking each such 4-tuple to its first two components, has a left adjoint, which is "essentially" the tensor-product construction.

(ii)   Show that an analogous trick can be used to convert any isomorphism of bifunctors as in the Lemma 8.2.10 into an adjunction. (Between what categories?) Do the same for the situation of Lemma 8.2.13.

**Exercise 8.3:15.** Describe all pairs of adjoint functors at least one member of which is a *constant* functor, i.e., a functor taking all objects of its domain category to a single object $X$ of its codomain category, and all morphisms of its domain category to $\mathrm{id}_X$.

What happens when we compose two functors arising from adjunctions?

Note that the *abelianization* of the *free* group on a set $X$ is a *free abelian group* on $X$. That is, when we compose these two functors, each of which is a left adjoint, we get another functor with that property. The general statement is simple, and is delightfully easy to prove.

**Theorem 8.3.10.** *Suppose* $\mathbf{E} \underset{F}{\overset{U}{\rightleftarrows}} \mathbf{D} \underset{G}{\overset{V}{\rightleftarrows}} \mathbf{C}$ *are pairs of adjoint functors, with* $U$ *and* $V$ *the right adjoints,* $F$ *and* $G$ *the left adjoints. Then* $\mathbf{E} \underset{FG}{\overset{VU}{\rightleftarrows}} \mathbf{C}$ *are also adjoint, with* $VU$ *the right adjoint and* $FG$ *the left adjoint.*

*Proof.* $\mathbf{C}(-, VU(-)) \cong \mathbf{D}(G(-), U(-)) \cong \mathbf{E}(FG(-), -)$. $\qquad\square$

**Exercise 8.3:16.** Suppose $U$, $V$, $F$ and $G$ are as above, $\eta$ and $\varepsilon$ are the unit and counit of the adjoint pair $U$, $F$, and $\eta'$ and $\varepsilon'$ are the unit and counit of the adjoint pair $V$, $G$. Describe the unit and counit of the adjoint pair $VU$, $FG$.

For further examples of the above theorem, consider two ways we can factor the forgetful functor from $\mathbf{Ring}^1$ to $\mathbf{Set}$. We can first pass from a ring to its multiplicative monoid, then go to the underlying set thereof, or we can first pass from the ring to its additive group, and then to the underlying set:

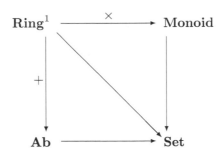

Taking left adjoints, we get the two decompositions of the *free ring* construction noted in §4.12: as the free-monoid functor followed by the monoid-ring functor, and as the free abelian group functor followed by the tensor ring functor:

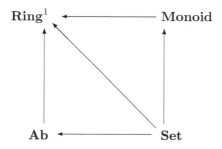

## 8.4. Number-theoretic interlude: the $p$-adic numbers and related constructions

While you digest the concept of adjunction (fundamentally simple, yet daunting in its multiple facets), let us look at some constructions of a different sort, which we did not meet in the "Cook's tour" of Chapter 4. In this section we will develop a particular case important in number theory; the general category-theoretic concept will be defined in the next section. A broader generalization, which embraces this and several constructions we *have* studied, will be developed in the section after that.

Suppose we are interested in solving the equation

$$(8.4.1) \qquad\qquad x^2 = -1$$

in the integers, $\mathbb{Z}$. Of course, we know it has no solution in the real numbers, let alone the integers, but we will ignore that dreary fact for the moment. ("If the fool would persist in his folly, he would become wise," William Blake [62].)

We may observe that the above equation does have a solution in the finite ring $\mathbb{Z}/5\,\mathbb{Z}$, in fact, two solutions, $2$ and $3$. Up to sign, these are the same, so let us look for a solution of (8.4.1) in $\mathbb{Z}$ satisfying

$$x \equiv 2 \pmod 5.$$

An integer $x$ which is $\equiv 2 \pmod 5$ has the form

$$(8.4.2) \qquad\qquad x = 5y + 2,$$

so we may rewrite (8.4.1) as

$$(5y + 2)^2 = -1$$

and expand, to see what information we can learn about $y$. We get $25y^2 + 20y = -5$. Hence $20y \equiv -5 \pmod{25}$, and dividing by $5$, we get $4y \equiv -1 \pmod 5$. This has the unique solution

$$y \equiv 1 \pmod{5},$$

which, substituted into (8.4.2), determines $x$ modulo $25$ :

$$x = 5y + 2 \equiv 5 \cdot 1 + 2 = 7 \pmod{25}.$$

We continue in the same fashion: At the next stage, putting $x = 25z + 7$ we have $(25z + 7)^2 = -1$. You should verify that this implies

$$z \equiv 2 \pmod{5},$$

which leads to

$$x \equiv 57 \pmod{125}.$$

Can we go on indefinitely? This is answered in

**Exercise 8.4:1.**

(i)   Show that given $i > 0$, and $c \in \mathbb{Z}$ such that $c^2 \equiv -1 \pmod{5^i}$, there exists $c' \in \mathbb{Z}$ such that $c'^2 \equiv -1 \pmod{5^{i+1}}$, and $c' \equiv c \pmod{5^i}$.

(ii)   Show that any integer is uniquely determined by its residue classes modulo 5, $5^2$, $5^3$, ..., $5^i$, ...; i.e., its images in $\mathbb{Z}/5\,\mathbb{Z}$, $\mathbb{Z}/5^2\,\mathbb{Z}$, $\mathbb{Z}/5^3\,\mathbb{Z}$, ..., $\mathbb{Z}/5^i\mathbb{Z}$, ....

(Aside: I find that some students have been taught to consider the underlying set of $\mathbb{Z}/n\mathbb{Z}$ to be the set of $n$ integers $\{0, \ldots, n-1\}$. Under the standard definition it is, rather, the set of *equivalence classes* of integers (positive and negative) under the relation of congruence modulo $n$. Though the representatives $0, \ldots, n-1$ of these equivalence classes are easier to talk about and compute with, the description as a set of equivalence classes is better for understanding the construction and its relation to factor-rings in general. Because two positive integers are in the same equivalence class modulo $n$ if and only if they give the same residue (remainder) in $\{0, \ldots, n-1\}$ on dividing by $n$, these equivalence classes have been named *residue classes*; and by extension, that phrase is used for the equivalence classes that comprise any factor-ring $R/I$, even when no natural concept of division with remainder is applicable. Once one understands that the $n$ elements of $\mathbb{Z}/n\mathbb{Z}$ are equivalence classes, one can, of course, use $0, \ldots, n-1$ as abbreviated names for those classes.)

Part (ii) of the above exercise shows that if there *were* an integer satisfying (8.4.1), the sequence of residues arising by repeated application of part (i) of that exercise would determine it. But now let us return to our senses, and remember that (8.4.1) has no real solution, and ask what, if anything, we *have* found.

Clearly, we have shown that there exists a sequence of residues, $x_1 \in |\mathbb{Z}/5\,\mathbb{Z}|$, $x_2 \in |\mathbb{Z}/5^2\,\mathbb{Z}|$, ..., $x_i \in |\mathbb{Z}/5^i\mathbb{Z}|$, ..., each of which satisfies (8.4.1) in the ring in which it lives, and which are "consistent", in the sense that each $x_{i+1}$ is a "lifting" of $x_i$, under the series of natural ring homomorphisms

$$\cdots \longrightarrow \mathbb{Z}/5^{i+1}\mathbb{Z} \longrightarrow \mathbb{Z}/5^{i}\mathbb{Z} \longrightarrow \cdots \longrightarrow \mathbb{Z}/5^{2}\mathbb{Z} \longrightarrow \mathbb{Z}/5\mathbb{Z}.$$

Let us name the $i$-th homomorphism in the above sequence $f_i \colon \mathbb{Z}/5^{i+1}\mathbb{Z} \to \mathbb{Z}/5^{i}\mathbb{Z}$; thus, $f_i$ is the map taking the residue of any integer $n$ modulo $5^{i+1}$ to the residue of $n$ modulo $5^{i}$. Now note that the set of all infinite strings

(8.4.3)
$$(\ldots, x_i, \ldots, x_2, x_1) \text{ such that}$$
$$x_i \in |\mathbb{Z}/5^{i}\mathbb{Z}| \text{ and } f_i(x_{i+1}) = x_i \quad (i = 1, 2, \ldots)$$

forms a ring under componentwise operations. What we have shown is that *this ring* contains a square root of $-1$. Since, as we have noted, an integer $n$ is determined by its residues modulo the powers of 5, the ring $\mathbb{Z}$ is *embedded* in this ring by the map taking each integer to its residues modulo the powers of 5, though of course the square root, in this ring, of $-1 \in |\mathbb{Z}|$ does not lie in the embedded copy of the ring $\mathbb{Z}$.

The ring of sequences (8.4.3) is called the *ring of 5-adic integers*. The corresponding object constructed for any prime $p$, using the system of maps

(8.4.4)     $\cdots \longrightarrow \mathbb{Z}/p^{i+1}\mathbb{Z} \longrightarrow \mathbb{Z}/p^{i}\mathbb{Z} \longrightarrow \cdots \longrightarrow \mathbb{Z}/p^{2}\mathbb{Z} \longrightarrow \mathbb{Z}/p\mathbb{Z},$

is called the ring of *p-adic* integers. These rings are of fundamental importance in modern number theory, and come up in many other areas as well. The notation for them is not uniform; the symbol we will use here is $\widehat{\mathbb{Z}}_{(p)}$. (The $(p)$ in parenthesis denotes the *ideal* of the ring $\mathbb{Z}$ generated by the element $p$. What is meant by putting it as a subscript of $\mathbb{Z}$ and adding a hat will be seen a little later. Most number-theorists simply write $\mathbb{Z}_p$ for the $p$-adic integers, denoting the field of $p$ elements by $\mathbb{Z}/p\mathbb{Z}$ or $\mathbb{F}_p$; cf. [27, p. 272], [34, p. 162, Example].)

The construction of $\widehat{\mathbb{Z}}_{(p)}$ can be described in another way, analogous to the construction of the real numbers from the rationals. Real numbers are entities that can be approximated by rational numbers under the *distance metric*; $p$-adic integers are entities that can be approximated by integers via *congruence* modulo increasing powers of $p$. This analogy is made stronger in

**Exercise 8.4:2.** Let $p$ be a fixed prime number. If $n$ is any integer, let $v_p(n)$ denote the greatest integer $e$ such that $p^e$ divides $n$, or the symbol $+\infty$ if $n = 0$. The $p$-adic metric on $\mathbb{Z}$ is defined by $d_p(m, n) = p^{-v_p(m-n)}$. Thus, it makes $m$ and $n$ "close" if they are congruent modulo a high power of $p$.

(i)    Verify that $d_p$ is a metric on $\mathbb{Z}$, and that the ring operations are uniformly continuous in this metric. Deduce that the *completion* of $\mathbb{Z}$ with respect to this metric (the set of Cauchy sequences modulo the usual equivalence relation) can be made a ring containing $\mathbb{Z}$.

(In fact, you will find that $d_p$ satisfies a stronger condition than the triangle inequality, namely $d_p(x, z) \leq \max(d_p(x, y), d_p(y, z))$. This is called the *ultrametric inequality*.)

(ii)  Show that this completion is isomorphic to $\widehat{\mathbb{Z}}_{(p)}$.

(iii) Show that the function $v_p$ can be extended in a natural way to the completion $\widehat{\mathbb{Z}}_{(p)}$, and that both the original function and this extension satisfy the identities $v_p(x\,y) = v_p(x) + v_p(y)$ and $v_p(x+y) \geq \min(v_p(x), v_p(y))$ $(x, y \in \mathbb{Z})$.

(iv)  Deduce that $\widehat{\mathbb{Z}}_{(p)}$ is an integral domain.

## Exercise 8.4:3.

(i)   Show that the function $v_p$ on $\mathbb{Z}$ of the preceding exercise can be extended in a unique manner to a $\mathbb{Z} \cup \{+\infty\}$-valued function on $\mathbb{Q}$, again satisfying the conditions of part (iii) of that exercise.

(ii)  Show that the completion of $\mathbb{Q}$ with respect to the metric $d_p$ induced by the above extended function $v_p$ is the field of fractions of $\widehat{\mathbb{Z}}_{(p)}$.

This field is called the field of *p-adic rationals*, and denoted $\widehat{\mathbb{Q}}_{(p)}$ (or $\mathbb{Q}_p$).

An interesting way of representing elements of these rings is noted in

## Exercise 8.4:4.

(i)   Show that every element $x$ of $\widehat{\mathbb{Z}}_{(p)}$ has a unique "left-facing base-$p$ expression" $x = \sum_{0 \leq i < \infty} c_i\, p^i$, where each $c_i \in \{0, 1, \ldots, p-1\}$. In particular, show that any such infinite sum is convergent in the $p$-adic metric. What is the expression for $-1$ in this form?

(ii)  Show likewise that elements of the field $\widehat{\mathbb{Q}}_{(p)}$ have expansions $x = \sum_i c_i p^i$, where again $c_i \in \{0, 1, \ldots, p-1\}$, and where $i$ now ranges over all integer values (not necessarily positive), but subject to the condition that the set of $i$ such that $c_i$ is nonzero is bounded below.

We showed earlier that one could find a solution to the equation $x^2 = -1$ in $\widehat{\mathbb{Z}}_{(5)}$. Let us note some simpler equations one can also solve.

## Exercise 8.4:5.

(i)   Show that every integer $n$ not divisible by $p$ is invertible in $\widehat{\mathbb{Z}}_{(p)}$.

(ii)  Are the "base-$p$ expressions" (in the sense of Exercise 8.4:4(i)) for the elements $n^{-1}$ eventually periodic?

It follows from part (i) of the above exercise that we can embed into the $p$-adic integers not only $\mathbb{Z}$, but the subring of $\mathbb{Q}$ consisting of all fractions with denominators not divisible by $p$. Now when one adjoins to a commutative ring $R$ inverses of all elements not lying in some prime ideal $P$, the resulting ring (which, if $R$ is an integral domain, is a subring of the field of fractions of $R$) is denoted $R_P$, so what we have embedded in the $p$-adic integers is the ring $\mathbb{Z}_{(p)}$. In $\mathbb{Z}_{(p)}$, every nonzero element is clearly an invertible element times a power of $p$, from which it follows that the nonzero ideals are precisely the ideals $(p^i)$. It is easy to verify that the factor-ring $\mathbb{Z}_{(p)}/(p^i)$ is isomorphic to $\mathbb{Z}/p^i\mathbb{Z}$; hence the system of finite rings and homomorphisms (8.4.4) can be described as consisting of *all* the proper factor-rings of $\mathbb{Z}_{(p)}$, together with the canonical maps among them. Hence the $p$-adic integers can be thought of as elements which can be approximated by members of $\mathbb{Z}_{(p)}$ modulo all *nonzero ideals* of that ring. Ring-theorists call the ring of such elements the *completion* of $\mathbb{Z}_{(p)}$ with respect to the system of its nonzero ideals, hence the symbol $\widehat{\mathbb{Z}}_{(p)}$.

We will not go into a general study of what algebraic equations have solutions in the ring of $p$-adic integers. A result applicable to a large class of rings including the $p$-adics is *Hensel's Lemma*; see [27, Theorem 8.5.6] or [25, §III.4.3] for the statement.

Let us now characterize abstractly the relation between the diagram (8.4.4) and the ring of $p$-adic integers which we have constructed from it. Since a $p$-adic integer is by definition a sequence $(\ldots, x_i, \ldots, x_2, x_1)$ with each $x_i \in \mathbb{Z}/p^i\mathbb{Z}$, the ring of $p$-adic integers has *projection* homomorphisms $p_i$ onto each ring $\mathbb{Z}/p^i\mathbb{Z}$. (Apologies for the double use of the letter "$p$"!) Since the components $x_i$ of each element satisfy the compatibility conditions $f_i(x_{i+1}) = x_i$, these projection maps satisfy

$$f_i \, p_{i+1} = p_i,$$

i.e., they make a commuting diagram

(8.4.5)

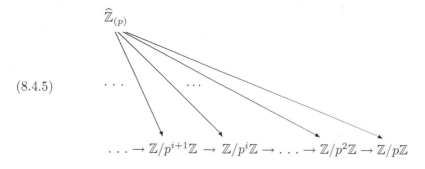

$$\ldots \to \mathbb{Z}/p^{i+1}\mathbb{Z} \to \mathbb{Z}/p^i\mathbb{Z} \to \ldots \to \mathbb{Z}/p^2\mathbb{Z} \to \mathbb{Z}/p\mathbb{Z}$$

I claim that $\widehat{\mathbb{Z}}_{(p)}$ is right universal for these properties. Indeed, given any ring $R$ with homomorphisms $r_i \colon R \to \mathbb{Z}/p^i\mathbb{Z}$ which are "compatible", i.e., satisfy $f_i \, r_{i+1} = r_i$, we see that for any $a \in R$, the system of images $(\ldots, r_i(a), \ldots, r_2(a), r_1(a))$ defines an element $r(a) \in \widehat{\mathbb{Z}}_{(p)}$. The resulting map $r \colon R \to \widehat{\mathbb{Z}}_{(p)}$ will be a homomorphism such that $r_i = p_i r$ for each $i$, and will be uniquely determined by these equations.

This universal property is expressed by saying that $\widehat{\mathbb{Z}}_{(p)}$ is the *inverse limit* of the system (8.4.4); one writes

$$\widehat{\mathbb{Z}}_{(p)} = \varprojlim_i \mathbb{Z}/p^i\mathbb{Z}.$$

We will give the formal definition of this concept in the next section.

A very similar example of an inverse limit is that of the system of homomorphic images of the ring $k[x]$ of polynomials in $x$ over a field $k$,

$$\ldots \longrightarrow k[x]/(x^{i+1}) \longrightarrow k[x]/(x^i) \longrightarrow \ldots \longrightarrow k[x]/(x^2) \longrightarrow k[x]/(x),$$

where $(x^i)$ denotes the ideal of all multiples of $x^i$. A member of $k[x]/(x^i)$ can be thought of as a polynomial in $x$ specified modulo terms of degree $\geq i$. If we take a sequence of such partially specified polynomials, each extending the next, these determine a *formal power series* in $x$. So the inverse limit of the above system is the formal power series ring $k[[x]]$. This ring is well known as a place where one can solve various sorts of equations. Some of these results are instances of Hensel's Lemma, referred to above; others, such as the existence of formal-power-series solutions to differential equations, fall outside the scope of that lemma.

We constructed the $p$-adic integers using the canonical *surjections* $\mathbb{Z}/p^{i+1}\mathbb{Z} \to \mathbb{Z}/p^i\mathbb{Z}$. Now there are also canonical *embeddings* $\mathbb{Z}/p^i\mathbb{Z} \to \mathbb{Z}/p^{i+1}\mathbb{Z}$, sending the residue of $n$ modulo $p^i$ to the residue of $pn$ modulo $p^{i+1}$. These respect addition but not multiplication, i.e., they are homomorphisms of abelian groups but not of rings. If we write out this system of groups and embeddings,

$$(8.4.6) \qquad \mathbb{Z}/p\mathbb{Z} \longrightarrow \mathbb{Z}/p^2\mathbb{Z} \longrightarrow \ldots \longrightarrow \mathbb{Z}/p^i\mathbb{Z} \longrightarrow \mathbb{Z}/p^{i+1}\mathbb{Z} \longrightarrow \ldots$$

it is natural to think of each group as a subgroup of the next, and to try to take their "union" $G$. But they are not literally subgroups of one another, so we need to think further about what we want this $G$ to be.

Clearly, for every element $x$ of each group in the above system, we want there to be an element of $G$ representing the image of $x$. Furthermore, if an element $x$ of one of the above groups is mapped to an element $y$ of another by some composite of the maps shown in (8.4.6), then these two elements should have the same image in $G$. Hence to get our $G$, let us form a disjoint union of the underlying sets of the given groups, and divide out by the equivalence relation that equates two elements if the image of one under a composite of the given maps is the other. It is straightforward to verify that this *is* an equivalence relation on the disjoint union, and that because the maps in the above diagram are group homomorphisms, the quotient by this relation inherits a group structure. If we call the maps in (8.4.6) $e_i : \mathbb{Z}/p^i\mathbb{Z} \to \mathbb{Z}/p^{i+1}\mathbb{Z}$, and the maps to the group we have constructed $q_i : \mathbb{Z}/p^i\mathbb{Z} \to G$, then the identifications we have made have the effect that for each $i$,

$$q_{i+1}\, e_i = q_i,$$

i.e., that the diagram

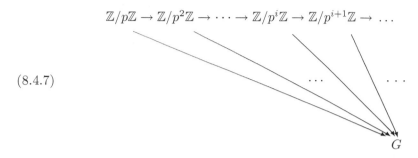

$$\mathbb{Z}/p\mathbb{Z} \to \mathbb{Z}/p^2\mathbb{Z} \to \cdots \to \mathbb{Z}/p^i\mathbb{Z} \to \mathbb{Z}/p^{i+1}\mathbb{Z} \to \ldots$$

(8.4.7)

$G$

commutes. Since we have made *only* these identifications, $G$ will have the universal property that given any group $H$ and family of homomorphisms $r_i \colon \mathbb{Z}/p^i\mathbb{Z} \to H$ satisfying $r_{i+1}e_i = r_i$ for each $i$, there will exist a unique homomorphism $r \colon G \to H$ such that $r_i = r\, q_i$ for all $i$. This universal property is expressed by saying that the group $G$ is the *direct limit*, $\varinjlim_i \mathbb{Z}/p^i\mathbb{Z}$, of the given system of groups.

Group theorists denote the direct limit $G$ of the system (8.4.6) by the suggestive symbol $Z_{p^\infty}$. It is often called the "Prüfer $p$-group".

**Exercise 8.4:6.**

(i)   Show that $Z_{p^\infty}$ is isomorphic to the subgroup of $\mathbb{Q}/\mathbb{Z}$ generated by the elements $[p^{-1}]$, $[p^{-2}]$, ....

(ii)   Show that the ring of endomorphisms of the abelian group $Z_{p^\infty}$ is isomorphic to $\widehat{\mathbb{Z}}_{(p)}$.

(Incidentally, in these notes I generally write $\mathbb{Z}/n\mathbb{Z}$ only for the *ring* of integers modulo $n$, but use the group-theorists' notation $Z_n$ for the cyclic group of order $n$, for which the construction as a quotient of the integers is only one of many realizations. However, above, to emphasize the relationship between (8.4.4) and (8.4.6), I have used the former notation in both cases. But for the direct limit of (8.4.6), the group-theorists' symbol $Z_{p^\infty}$ has no ring-theoretic analog.)

## 8.5.  Direct and inverse limits

Before we give abstract definitions of our two types of limits, let us give an example showing that one may want to consider limits of systems indexed by more general partially ordered sets than the natural numbers. Consider the concept of a *germ of a function* at a point $z$ of the complex plane or any other topological space $X$. This arises by considering, for every neighborhood $S$ of $z$, the set $F(S)$ of functions of the desired sort on the set $S$ (for instance, analytic functions, if $X$ is the complex plane), and observing that when one goes from a neighborhood $S$ to a smaller neighborhood $T$, one gets a restriction map $F(S) \to F(T)$ (not in general one-to-one, since distinct functions on the set $S$ may have the same restriction to the subset $T$, and not necessarily onto, since not every admissible function on $T$ need extend to an admissible function on $S$). To get *germs* of functions at $z$, one intuitively wants to consider this system of sets of functions for smaller and smaller neighborhoods of $z$, and "take the limit". To do this formally, one takes a disjoint union of all the sets $F(S)$, and divides out by the equivalence relation that makes two functions $a \in F(S_1)$, $b \in F(S_2)$ equivalent if and only if they have the same image in $F(T)$ for some neighborhood $T \subseteq S_1 \cap S_2$ of $z$.

If the sets of functions $F(S)$ are given with some algebraic structure (structures of groups, rings, etc.) for which the above restriction maps are homomorphisms, we find that an algebraic structure of the same sort is induced on the direct limit set. The key point is that given functions $a$, $b$ defined on different neighborhoods $S$ and $T$ of $z$, both will have images in the neighborhood $S \cap T$ of $z$, and these images can be added, multiplied, etc. there, allowing us to define the sum, product, etc., of the images of $a$ and $b$ in the limit set.

If we look for the conditions on a general partially ordered index set that allow us to reason in this way, we get

**Definition 8.5.1.** *Let $P$ be a partially ordered set.*

*$P$ is said to be* directed *(or upward* directed*) if it is nonempty, and for any two elements $x, y$ of $P$, there exists an element $z$ majorizing both $x$ and $y$.*

*$P$ is said to be* inversely directed *(or downward* directed*) if it is nonempty and for any two elements $x, y$ of $P$, there exists an element $z$ which is $\leq$ both $x$ and $y$; equivalently, if $P^{\mathrm{op}}$ is upward directed.*

*(The word "filtered" is sometimes used instead of "directed" in these definitions.)*

(If you did Exercise 6.2:9, you will find that these conditions are certain of the "interpolation" properties of that exercise.)

We can now give the general definitions of direct and inverse limits. The formulations we give below assume that the morphisms of our given systems go in the "upward" direction with respect to the ordering on the indexing set. It happens that in our initial example of $\widehat{\mathbb{Z}}_{(p)}$, the standard ordering on the positive integers is such that the morphisms went the *opposite* way; in our construction of $Z_{p^\infty}$ they went the "right" way; while in the case of germs of analytic functions, if one orders neighborhoods of $z$ by inclusion, the morphisms again go the "wrong" way (namely, from the set of functions on a larger neighborhood to the set of functions on a smaller neighborhood). This can be corrected formally by using, when necessary, the opposite partial ordering on the index set. Informally, in discussing direct and inverse limits one often just specifies the system of *objects and maps*, and understands that to apply the formal definition, one should partially order the set indexing the objects so as to make maps among them go "upward".

**Definition 8.5.2.** *Let* $\mathbf{C}$ *be a category, and suppose we are given a family of objects* $X_i \in \mathrm{Ob}(\mathbf{C})$ $(i \in I)$, *a partial ordering on the index set* $I$, *and a system* $(f_{ij})$ *of morphisms,* $f_{ij} \in \mathbf{C}(X_i, X_j)$ $(i{<}j,\ i, j \in I)$ *such that for* $i < j < k$, *one has* $f_{jk}\, f_{ij} = f_{ik}$. *(In brief, suppose we are given a partially ordered set* $I$, *and a functor* $F : I_{\mathbf{cat}} \to \mathbf{C}$.)

*If* $I$ *is inversely directed, then* $(X_i, f_{ij})_I$ *is called an* inversely directed system *of objects and maps in* $\mathbf{C}$. *An* inverse limit *of this system means an object* $L$ *of* $\mathbf{C}$ *given with morphisms* $p_i : L \to X_i$ *which are compatible, in the sense that for all* $i < j \in I$, $p_j = f_{ij}\, p_i$, *and which is* universal *for this property, in the sense that given any object* $W$ *and morphisms* $w_i : W \to X_i$ *such that for all* $i < j \in I$, $w_j = f_{ij}\, w_i$, *there exists a unique morphism* $w : W \to L$ *such that* $w_i = p_i w$ *for all* $i \in I$.

*Likewise, if* $I$ *is directed, then* $(X_i, f_{ij})_I$ *is called a* directed system *in* $\mathbf{C}$; *and a* direct limit *of this system means an object* $L$ *of* $\mathbf{C}$ *given with morphisms* $q_i : X_i \to L$ *such that for all* $i < j \in I$, $q_i = q_j\, f_{ij}$, *and which is* universal *in the sense that given any object* $Y$ *and morphisms* $y_i : X_i \to Y$ *such that for all* $i < j \in I$, $y_i = y_j\, f_{ij}$, *there exists a unique morphism* $y : L \to Y$ *such that* $y_i = y\, q_i$ *for all* $i \in I$.

*(Synonyms sometimes used for inverse and direct limit are* projective *and* inductive *limit, respectively.)*

*Loosely, one often writes the inverse limit object* $\varprojlim_i X_i$, *and the direct limit object* $\varinjlim_i X_i$. *More precisely, letting* $F$ *denote the functor* $I_{\mathbf{cat}} \to \mathbf{C}$ *corresponding to the inversely directed or directed system* $(X_i, f_{ij})$, *one writes these objects as* $\varprojlim F$ *and* $\varinjlim F$ *respectively.*

*The morphisms* $p_j : \varprojlim_i X_i \to X_j$ *are called the* projection *maps associated with this inverse limit, and the* $q_j : X_j \to \varinjlim_i X_i$ *the* coprojection *maps associated with the direct limit.*

In the next-to-last paragraph of the above definition, by the "functor ... corresponding to the ... system $(X_i, f_{ij})$" we understand the functor which takes on the value $X_i$ at the object $i$, the value $f_{ij}$ at the morphism $(i, j)$ $(i < j$ in $I)$, and the value $\mathrm{id}_{X_i}$ at the morphism $(i, i)$. In the case where the indexing partially ordered set consists of the positive or negative integers, note that the full system of morphisms is determined by the morphisms $f_{i,\,i+1}$ (which can be arbitrary), hence in such cases one generally specifies only these morphisms in describing the system.

Direct and inverse limits in **Set** may be constructed by the techniques we illustrated earlier:

**Lemma 8.5.3.** *Every inversely directed system* $(X_i, f_{ij})$ *of sets and set maps has an inverse limit, given by*

$$(8.5.4) \qquad \varprojlim X_i = \{(x_i) \in \textstyle\prod_I X_i \mid x_j = f_{ij}(x_i) \text{ for } i < j \in I\},$$

*with the* $p_j$ *given by projection maps,* $\varprojlim X_i \subseteq \prod X_i \to X_j$.

*Likewise, every directed system $(X_i, f_{ij})$ of sets and set maps has a direct limit, gotten by forming the disjoint union of the $X_i$ and dividing out by the equivalence relation under which $x \in X_i$ and $x' \in X_{i'}$ are equivalent if and only if they have the same image in some $X_j$ $(j \geq i, i')$. Each map $q_i$ acts as the inclusion of $X_i$ in the disjoint union, followed by the map of that union into its factor-set by the above equivalence relation.* ☐

One may ask what the point is, in our definitions of direct and inverse limit, of requiring that the partially ordered set $I$ be directed or inversely directed. One could set up the definitions without that restriction, and in most familiar categories one can, in fact, construct objects which satisfy the resulting conditions. But the behavior of these constructions tends to be quite different from those we have discussed *unless* these directedness assumptions are made. (For instance, the explicit description in Lemma 8.5.3 of the equivalence relation in the construction of a direct limit of sets is no longer correct.) In any case, such generalized definitions would be subsumed by the still more general definitions to be made in the next section! So the value of the definitions in the form given above is that they single out a situation in which the limit objects can be studied by certain techniques.

**Exercise 8.5:1.**

(i)   If $(X_i, f_{ij})_I$ is a directed system in a category $\mathbf{C}$, and $J$ a *cofinal* subset of $I$, show that $\varinjlim_J X_j \cong \varinjlim_I X_i$. Precisely, show that $J$ will also be a directed partially ordered set, and that any object with the universal property of the direct limit of the given system can be made into a direct limit of the subsystem in a natural way, and vice versa.

(ii)  Show that the isomorphism of (i) is an instance of a morphism (in one direction or the other) between $\varinjlim_J X_j$ and $\varinjlim_I X_i$ which can be defined whenever $J \subseteq I$ are both directed and both limits exist, whether or not $J$ is cofinal.

(iii) State the result corresponding to (i) for inverse limits. (For this we need a term for a subset of a partially ordered set which has the property of being cofinal under the opposite ordering; let us use "downward cofinal". When speaking of inverse systems, one sometimes just says "cofinal", with the understanding that this is meant in the sense that is relevant to such systems.)

(iv)  What can you deduce from (i) and (iii) about direct limits over directed partially ordered sets having a greatest element, and inverse limits over inversely directed partially ordered sets having a least element?

(v)   Given any directed partially ordered set $I$ and any *non*cofinal directed subset $J$ of $I$, show that there exists a directed system of sets, $(X_i, f_{ij})$, indexed by $I$, such that $\varinjlim_I X_i \not\cong \varinjlim_J X_j$.

**Exercise 8.5:2.** Suppose $(X_i, f_{ij})_I$ is a directed system in a category **C**, and $f\colon J \to I$ a surjective isotone map, such that $J$, like $I$, is directed. Show that $\varinjlim_{j \in J} X_{f(j)} \cong \varinjlim_{i \in I} X_i$.

The next few exercises concern direct and inverse limits of *sets*. We shall see in the next chapter that direct and inverse limits of algebras have as their underlying sets the direct or inverse limits of the objects' underlying sets (assuming, in the case of *direct* limits, that the algebras have only finitary operations); hence the results obtained for sets in the exercises below will be applicable to algebras.

The construction of the $p$-adic integers was based on a system of *surjective* homomorphisms. The first point of the next exercise looks at inverse systems with the opposite property, and the second part considers the dual situation for direct limits.

**Exercise 8.5:3.**

(i) Let $(S_i, f_{ij})$ be an inversely directed system in **Set** such that all the morphisms $f_{ij}$ are one-to-one, and let us choose any element $i_0 \in I$. Show that $\varprojlim_i S_i$ can be identified with the intersection, in $S_{i_0}$, of the sets $f_{i i_0}(S_i)$ $(i < i_0)$.

(ii) Let $(S_i, f_{ij})$ be a directed system in **Set** such that all the morphisms $f_{ij}$ are onto, and let us choose any element $i_0 \in I$. Show that $\varinjlim_i S_i$ can be identified with the quotient set of $S_{i_0}$ by the union of the equivalence relations induced by the maps $f_{i_0 i}\colon S_{i_0} \to S_i$ $(i > i_0)$.

**Exercise 8.5:4.**

(i) Show that the inverse limit in **Set** of any inverse system of *finite nonempty* sets is nonempty.

(Suggestions: Either build the description of an element of the inverse limit up "from below", by looking at partial assignments satisfying appropriate extendibility conditions, and apply Zorn's Lemma to get a maximal such assignment, or else "narrow down on an element from above", by looking at "subsystems" of the given inverse system, i.e., systems of nonempty subsets of the given sets carried into one another by the given mappings, and using Zorn's Lemma to get a minimal such subsystem. You might find it instructive to work out both of these proofs.)

(ii) Show that (i) can fail if the condition "finite" is removed, even for inverse limits over the totally ordered set of negative integers.

(iii) If you have some familiarity with general topology, see whether you can generalize statement (i) to a result on topological spaces, with "compact Hausdorff" replacing "finite".

As an application of part (i) of the above exercise, suppose we are given a subdivision of the plane into regions, possibly infinitely many, and a positive integer $n$, and we wish to study the problem of coloring these regions with $n$ colors so that no two adjacent regions are the same color. Let the set of all our regions be denoted $R$, the adjacency relation $A \subseteq R \times R$ (i.e., $(r_1, r_2) \in A$ if and only if $r_1$ and $r_2$ are adjacent regions), and the set of colors $C$. For any subset $S \subseteq R$, let $X_S$ denote the set of all colorings of $S$ (maps $S \to C$) under which no two adjacent regions have the same color; let us call these "permissible colorings of $S$". Given subsets $S \subseteq T$, the restriction to $S$ of a permissible coloring of $T$ is a permissible coloring of $S$; thus we have a restriction map $X_T \to X_S$. Now–

**Exercise 8.5:5.**

(i)   Show that in the above situation, the sets $X_S$, as $S$ ranges over the *finite* subsets of $R$, form an inversely directed system, and that $X_R$ may be identified with the inverse limit of this system in **Set**.

(ii)   Deduce using Exercise 8.5:4(i) that if each finite family $S \subseteq R$ can be colored, then the whole picture $R$ can be colored.

(Note: the assumption that every finite family $S$ can be colored does *not* say that *every* permissible coloring of a finite family $S$ can be extended to a permissible coloring of every larger finite family $T$!)

**Exercise 8.5:6.**

(i)   Show that if $(X_i, f_{ij})$ is a directed system of sets, and each $f_{ij}$ is one-to-one, then the canonical maps $q_j : X_j \to \varinjlim X_i$ are all one-to-one.

(ii)   Let $(X_i, f_{ij})$ be an inversely directed system of sets such that each $f_{ij}$ is surjective. Show that if $I$ is *countable*, then the canonical maps $p_j : \varprojlim X_i \to X_j$ are surjective. (Suggestion: First prove this in the case where $I$ is the set of negative integers. Then show that any countable inversely directed partially ordered set either has a least element, or has a downward-cofinal subset order-isomorphic to the negative integers, and apply Exercise 8.5:1(iii).)

(iii)   Does this result remain true for uncountable $I$? In particular, what if $I$ is the opposite of an uncountable cardinal?

**Exercise 8.5:7.**   Show that every group is a direct limit of finitely presented groups.

(This result is not specific to groups. We shall be able to extend it to more general algebras when we have developed the necessary language in the next chapter.)

The remaining exercises in this section develop some particular examples and applications of direct and inverse limits, including some further results on the $p$-adic integers. In these exercises you may assume the result which, as noted earlier, will be proved in the next chapter, that a direct or inverse limit of algebras whose operations are finitary can be constructed by forming the corresponding limit of underlying sets and giving this an induced algebra structure. None of these exercises, or the remarks connecting them, is needed for the subsequent sections of these notes.

One can sometimes achieve interesting constructions by taking direct limits of systems in which all objects are the same; this is illustrated in the next three exercises. The first shows a sophisticated way to get a familiar construction; in the next two, direct limits are used to get curious examples.

**Exercise 8.5:8.** Consider the directed system $(X_i,\ f_{ij})_{i,j\in I}$ in **Ab**, where $I$ is the set of positive integers, partially ordered by divisibility ($i$ considered less than or equal to $j$ if and only if $i$ divides $j$), each object $X_i$ is the additive group $\mathbb{Z}$, and for $j = ni$, $f_{ij}\colon \mathbb{Z} \to \mathbb{Z}$ is given by multiplication by $n$.

(i) Show that $\varinjlim X_i$ may be identified with the additive group of the rational numbers.

(ii) Show that if you perform the same construction starting with an arbitrary abelian group $A$ in place of $\mathbb{Z}$, the result is a $\mathbb{Q}$-vector-space which can be characterized by a universal property relative to $A$.

**Exercise 8.5:9.** For this exercise, assume as known the facts that every subgroup of a free group is free, and in particular, that in the free group on two generators $x$, $y$, the subgroup generated by the two commutators $x^{-1}y^{-1}xy$ and $x^{-2}y^{-1}x^2y$ is free on those two elements.

Let $F$ denote the free group on $x$ and $y$, and $f$ the endomorphism of $F$ taking $x$ to $x^{-1}y^{-1}xy$ and $y$ to $x^{-2}y^{-1}x^2y$. Let $G$ denote the direct limit of the system $F \to F \to F \to \dots$, where all the arrows between successive objects are the above morphism $f$.

Show that $G$ is a nontrivial group such that every finitely generated subgroup of $G$ is free, but that $G$ is equal to its own commutator, $G = [G, G]$; i.e., that the abelianization of $G$ is the trivial group. Deduce that though $G$ is "locally free", it is not free.

**Exercise 8.5:10.** Let $k$ be a field. Let $R$ denote the direct limit of the system of $k$-algebras $k[x] \to k[x] \to k[x] \to \dots$, where each arrow is the homomorphism sending $x$ to $x^2$. Show that $R$ is an integral domain in which every finitely generated ideal is principal, but not every ideal is finitely generated. (Thus, for each ideal, the minimum cardinality of a generating set is either 0, 1, or infinite.)

For the student familiar with the Galois theory of finite-degree field extensions, the next exercise shows how the Galois groups of *infinite* extensions can be characterized in terms of the finite-dimensional case.

**Exercise 8.5:11.** Suppose $E/K$ is a normal algebraic field extension, possibly of infinite degree. Let $I$ be the set of subfields of $E$ normal and of *finite* degree over $K$. If $F_2 \subseteq F_1$ in $I$, let $f_{F_1,F_2}\colon \mathrm{Aut}_K F_1 \to \mathrm{Aut}_K F_2$ denote the map which acts by restricting automorphisms of $F_1$ to the subfield $F_2$.

(i)   Show that the definition of $f_{F_1,F_2}$ makes sense, and gives a group homomorphism.

(ii)   Show that if we order $I$ by reverse inclusion of fields, then the groups $\mathrm{Aut}_K F$ $(F \in I)$ and homomorphisms $f_{F_1,F_2}$ $(F_1 \leq F_2)$ form an inversely directed system of groups.

(iii)   Show that $\mathrm{Aut}_K E$ is the inverse limit of this system in **Group**.

(iv)   Can you find a normal algebraic field extension whose automorphism group is isomorphic to the additive group of the $p$-adic integers?

**Exercise 8.5:12.**

(i)   (Open question.) Suppose a group $G$ is the inverse limit of a system of finite groups. If $G$ is a torsion group (i.e., if all elements of $G$ are of finite order), must $G$ have finite exponent (i.e., must there exist an integer $n$ such that $x^n = e$ is an identity of $G$)?

Though the above question is very difficult, the next two parts are reasonable exercises, and may help render that question more tractable.

(ii)   Show that (i) is equivalent to the corresponding question in which we assume that $G$ is the inverse limit of a system of finite groups indexed by the negative integers (under the natural ordering), with all connecting morphisms surjective.

(iii)   Translate (i) (possibly with the help of (ii)) into a question on finite groups which you could pose to a person not familiar with the concept of inverse limit. (The more natural-sounding, the better.)

Back to the $p$-adic integers. A curious property of the additive group of $\widehat{\mathbb{Z}}_{(p)}$ is noted in

**Exercise 8.5:13.** Let us call an element $x$ of a group $G$ *completely divisible* if for every positive integer $n$ there is a $y \in |G|$ such that $y^n = x$ (or if $G$ is written additively, $n y = x$).

(i)   Show that no nonzero element of the additive group of $\widehat{\mathbb{Z}}_{(p)}$ is completely divisible.

On the other hand

(ii)   Show that if $A$ is any nonzero subgroup of $\widehat{\mathbb{Z}}_{(p)}$ such that $\widehat{\mathbb{Z}}_{(p)}/A$ is torsion-free, then *every* element of $\widehat{\mathbb{Z}}_{(p)}/A$ is completely divisible; in fact, that $\widehat{\mathbb{Z}}_{(p)}/A$ is the underlying additive group of a $\mathbb{Q}$-vector-space.

Part (i) of the next exercise seemed to me too simple to be true when I saw it described (in a footnote in a Ph.D. thesis, as "well-known"). But it is, in fact, not hard to verify

**Exercise 8.5:14.**

(i)   Show that $\mathbb{Z}[[x]]/(x-p) \cong \widehat{\mathbb{Z}}_{(p)}$, where $\mathbb{Z}[[x]]$, we recall, denotes the ring of formal power series over $\mathbb{Z}$ in one indeterminate $x$, and $(x-p)$ denotes the ideal of that ring generated by $x - p$.

(ii)   Examine other factor-rings of formal power series rings. For instance, can you describe $\mathbb{Z}[[x]]/(x - p^2)$? $\mathbb{Z}[[x]]/(x^2 - p)$? $\mathbb{Z}[[x]]/(px^2 - 1)$? $R[[x]]/(f(x))$ for a general commutative ring $R$ and a polynomial or power series $f(x)$, perhaps subject to some additional conditions? $R[[x, y]]/I$ for some fairly general class of ideals $I$?

(If you consider $\mathbb{Z}[[x]]/(x - n)$ for $n$ not a prime power, you might first look at Exercise 8.5:15 below.)

Is the "adic" construction limited to primes $p$, or can one construct, say, a ring of "10-adic integers", $\widehat{\mathbb{Z}}_{(10)}$? One encounters a trivial difficulty in that there are two ways of interpreting this symbol. But we shall see below that they lead to the same ring; so there is a well-defined object to which we can give this name. However, its properties will not be as nice as those of the $p$-adic integers for prime $p$.

**Exercise 8.5:15.** Let $\mathbb{Z}_{(10)}$ denote the ring of all rational numbers which can be written with denominators relatively prime to 10.

(i)   Determine all nonzero ideals $I \subseteq \mathbb{Z}_{(10)}$ and the structures of the factor-rings $\mathbb{Z}_{(10)}/I$. Show that each of these factor-rings is isomorphic to a ring $\mathbb{Z}/n\mathbb{Z}$. Writing them in this way, sketch the diagram of the inverse system of these factor-rings and the canonical maps among them.

(ii)   Show that the inverse system $\cdots \to \mathbb{Z}/10^i\,\mathbb{Z} \to \cdots \to \mathbb{Z}/100\,\mathbb{Z} \to \mathbb{Z}/10\,\mathbb{Z}$ constitutes a downward *cofinal* subsystem of the above inverse system.

Hence by Exercise 8.5:1 the inverse limits of these two systems are isomorphic, and we shall denote their common value $\widehat{\mathbb{Z}}_{(10)}$. It is clear from the form of the second inverse system that elements of $\widehat{\mathbb{Z}}_{(10)}$ can be described by "infinite decimal expressions to the left of the decimal point".

(iii)   Show that the relation $[2] \cdot [5] = [0]$ in $\mathbb{Z}/10\,\mathbb{Z}$ can be lifted to get a pair of nonzero elements which have product 0 in $\mathbb{Z}/100\,\mathbb{Z}$, that these can be lifted to such elements in $\mathbb{Z}/1000\,\mathbb{Z}$, and so on, and deduce that $\widehat{\mathbb{Z}}_{(10)}$ is not an integral domain.

(iv)   Prove, in fact, that $\widehat{\mathbb{Z}}_{(10)} \cong \widehat{\mathbb{Z}}_{(2)} \times \widehat{\mathbb{Z}}_{(5)}$.

A construction often used in number theory is characterized in

**Exercise 8.5:16.** Show that the inverse limit of the system of all factor-rings of $\mathbb{Z}$ by nonzero ideals is isomorphic to $\prod_p \widehat{\mathbb{Z}}_{(p)}$, where the direct product is taken over all primes $p$. (This ring is denoted $\widehat{\mathbb{Z}}$.)

A feature we have not yet mentioned, which is important in the study of inverse limits, is topological structure. Recall that the inverse limit of a system of sets and set maps $(X_i, f_{ij})$ was constructed as a subset of $\prod X_i$. Let us regard each $X_i$ as a discrete topological space, and give $\prod X_i$ the product topology. In general, a product of discrete spaces is not discrete; however, a product of compact spaces *is* compact, so if our discrete spaces

$X_i$ are *finite*, their product will be compact. It is not hard to show that the subset $\varprojlim X_i \subseteq \prod X_i$ will be closed in the product topology, and hence, if the $X_i$ are finite, will be compact in the induced topology.

**Exercise 8.5:17.**

(i)   Verify the assertion that if the $X_i$ are discrete, then $\varprojlim X_i \subseteq \prod X_i$ is always closed in the product topology, and is therefore compact if all $X_i$ are finite.

(ii)   Show that the result of Exercise 8.5:4(i) (and hence that of Exercise 8.5:5(ii)) can be deduced using the compactness of $\varprojlim X_i$.

(iii)   Show that the compact topology described above agrees in the case of $\widehat{\mathbb{Z}}_{(p)}$ with the topology arising from the metric $d_p$ of Exercise 8.4:2.

In fact, results like Exercise 8.5:5(ii), saying that a family of conditions can be satisfied simultaneously if all finite subfamilies of these conditions can be so satisfied, are called by logicians "compactness" results, because the statements can generally be formulated in terms of the compactness of some topological space.

I can now say that the usual formulation of the open question of Exercise 8.5:12(i) is, "If a compact topological group is torsion, must it have finite exponent?" (Note that a topological group is by definition required to have a Hausdorff topology.) The equivalence of this with the question of that exercise follows from a deep result, that any compact group is an inverse limit of surjective maps of compact Lie groups (see [118, Theorem IV.4.6, p. 175]), combined with the observation that if any of these Lie groups had positive dimension, we would get elements of infinite order. Thus, any compact torsion group is an inverse limit of 0-dimensional compact Lie groups, i.e., finite discrete groups, under the product topology.

An inverse limit of finite structures is called *profinite* (based on the synonym "projective limit" for "inverse limit"). Let us look briefly at the condition of pro-finite-dimensionality in

**Exercise 8.5:18.** Let $V$ be a vector space over a field $k$.

(i)   Show that the dual space $V^*$ is the inverse limit, over all finite-dimensional subspaces $V_0 \subseteq V$, of the spaces $V_0^*$.

(ii)   Can you get the result of (i) as an instance of a general result describing *duals* of *direct limits* of vector spaces?

(iii)   If you did Exercise 6.5:6(ii)–(iii), show that the topology described there is that of the inverse limit of the finite-dimensional discrete spaces $V_0^*$ referred to above. Show moreover that the only linear functionals $V^* \to k$ continuous in this topology are those induced by the elements of $V$.

For some interesting results on profinite structures, especially groups, see [57, §§3–5].

To motivate the final exercise in this section, suppose $L$ is the direct limit in a category $\mathbf{C}$ of a system of objects and morphisms indexed by a certain directed partially ordered set $I$. Thinking about Exercises 8.5:1 and 8.5:2 can lead us to wonder: what other partially ordered index sets we can use to get $L$ as a direct limit of (some of) these same objects, using (some of) these same morphisms?

The objects used in the original direct limit construction, and the morphisms used, together with their composites, form a subcategory $\mathbf{D}$ of $\mathbf{C}$; so the question can be posed as below.

**Exercise 8.5:19.** Let $\mathbf{C}$ be a category, $\mathbf{D}$ a subcategory, $I$ a directed partially ordered set, and $L$ an object of $\mathbf{C}$ which can be written as the direct limit in $\mathbf{C}$ of a system of objects and morphisms of $\mathbf{D}$, indexed by $I$. We would like to know for what other partially ordered sets $J$ there will exist $J$-directed systems in $\mathbf{D}$ having $L$ as their direct limit.

(i)    Show with the help of Exercise 8.5:2 that $L$ can be written as a direct limit in $\mathbf{C}$ of a system of objects and morphisms in $\mathbf{D}$ indexed by a directed partially ordered set of the form $\mathbf{P}_{\mathrm{fin}}(S)$, where $S$ is a set, and $\mathbf{P}_{\mathrm{fin}}(S)$ denotes the partially ordered set of all finite subsets of $S$, ordered by inclusion.

If we think of the above result as showing that the class of partially ordered sets of the form $\mathbf{P}_{\mathrm{fin}}(S)$ is quite "strong", the next result shows that a different class of partially ordered sets is less strong.

(ii)   Letting $\mathbf{C} = \mathbf{Set}$ and $\mathbf{D} = \mathbf{FinSet}$, show that every object of the former category can be written as a direct limit of a directed system of objects in the latter subcategory, but that if we restrict attention to directed systems indexed by *chains* (totally ordered sets), then only the finite and countable sets can be so expressed.

The next statement may appear to contradict (ii), but does not.

(iii) Show that the closure of $\mathbf{FinSet}$ in $\mathbf{Set}$ under taking direct limits indexed by chains is all of $\mathbf{Set}$.

The idea of looking at families of partially ordered sets in terms of their relative "strength" in achieving various direct limits suggests some sort of Galois connection. This is made explicit in

(iv) Let $X$ denote the large set of all partially ordered sets, and $Y$ the large set of all 3-tuples $(\mathbf{C}, \mathbf{D}, L)$ where $\mathbf{C}$ is a small category, $\mathbf{D}$ is a subcategory of $\mathbf{C}$, and $L$ is an object of $\mathbf{C}$. Let $R \subseteq X \times Y$ be the binary relation consisting of those pairs $(P, (\mathbf{C}, \mathbf{D}, L))$ such that $L$ cannot be written as a direct limit in $\mathbf{C}$ of a system of objects and maps from $\mathbf{D}$ indexed by $P$. Express the results of (i) and (ii) above as statements about the Galois connection between $X$ and $Y$ induced by $R$. (So far as I can see, those results cannot be expressed as statements about the simpler-looking relation $\neg R$, i.e., "$L$ *can* be written as a direct limit in $\mathbf{C}$ of a system of objects and maps from $\mathbf{D}$ indexed by $P$".)

## 8.6. Limits and colimits

The universal properties defining direct and inverse limits are similar to those defining several other constructions we have seen. Let us recall these.

Given two objects $X_1$, $X_2$ of a category $\mathbf{C}$, a *product* of $X_1$ and $X_2$ in $\mathbf{C}$ is an object $P$ given with morphisms $p_1$ and $p_2$ into $X_1$ and $X_2$, and universal for this property.

Given a pair of parallel morphisms $X_1 \rightrightarrows X_2$ in $\mathbf{C}$, an *equalizer* of this system is an object $K$ given with a morphism $k$ into $X_1$ having equal composites with those two morphisms, and again universal. To improve the parallelism with similar constructions, let us rename the morphism $k$ as $k_1$, and let $k_2 : K \to X_2$ denote the common value of the composites of $k_1$ with the two morphisms $X_1 \rightrightarrows X_2$. Then we can describe $K$ as having a morphism into *each* of $X_1$, $X_2$, such that the composite of $k_1 : K \to X_1$ with each of the two given morphisms $X_1 \to X_2$ is $k_2 : K \to X_2$, and such that $(K, k_1, k_2)$ is universal for these properties. We see that this is exactly like the universal property of an inverse limit, except that the indexing category $\cdot \rightrightarrows \cdot$ is not of the form $I_{\mathbf{cat}}$ for a partially ordered set $I$.

In the same way, a *pullback* of a pair of morphisms $f_1 : X_1 \to X_3$, $f_2 : X_2 \to X_3$ can be redefined as an object $P$ given with morphisms $p_1$, $p_2$, $p_3$ into $X_1$, $X_2$, $X_3$ respectively, satisfying $f_1 p_1 = p_3$ and $f_2 p_2 = p_3$, and universal for this property.

Let us look at a case we haven't discussed yet. If $G$ is a group and $S$ a $G$-set, then the *fixed-point set* of the action of $G$ on $S$ means $\{x \in |S| \mid (\forall g \in |G|)\ g\,x = x\}$. If we denote the action of each $g \in |G|$ on $S$ by $g_S : |S| \to |S|$, then the fixed-point set is universal among sets $A$ with maps $i : A \to |S|$ such that for all $g \in |G|$, $i = g_S\,i$. Given an object $X$ of any category $\mathbf{C}$, and an action of a group $G$ on $X$, we can look for an object with the same universal property, and, if it exists, call it the "fixed object" of the action.

We have seen constructions dual to those of product, equalizer and pullback. A construction dual to that of fixed object should take an object $X$ of $\mathbf{C}$ with an action of $G$ on it to an object $B$ of $\mathbf{C}$ with a map $j : X \to B$ unchanged under composition on the right with the actions on $X$ of elements of $G$, and universal for this property. Examples of this concept are examined in

**Exercise 8.6:1.** Let $G$ be a group.

(i)   If $X$ is a set on which $G$ acts by permutations, and $x$ an element of $X$, one defines the *orbit* of $x$ under $G$ to be the set $G\,x = \{g\,x \mid g \in |G|\}$. Let $B$ be the set of such orbits $G\,x$, called the *orbit space* of $X$. Show that this set $B$, together with the map $X \to B$ taking $x$ to $G\,x$, has the universal property discussed above.

(ii)   Show that if $G$ acts by automorphisms on (say) a ring $R$, then there is an object $S$ in the category of rings with this same universal property, but that its underlying set will not in general be the orbit space of the action of $G$ on the underlying set of $R$.

(iii) If $G$ acts by automorphisms on an object $X$ of **POSet**, again show the existence of an object $B$ with the above universal property. Show moreover that if $G$ is finite, the underlying set of $B$ will be the orbit space of the underlying set of $X$, and the universal map $X \to B$ will be *strictly* isotone; but that if $G$ is infinite, neither statement need be true.

(iv) Do the assertions of (iii) about the case where $G$ is finite remain true if we replace **POSet** by **Lattice**?

As noted above, the universal properties we have been examining have statements formally identical with those of direct and inverse limits, except that the partially ordered set $I$ of that definition is replaced by other categories **D** (for example the two-object category $\cdot \rightrightarrows \cdot$, or the one-object category $G_{\mathbf{cat}}$). As names for the general concepts embracing such cases, one uses modified versions of the terms "inverse limit" and "direct limit".

**Definition 8.6.1.** *Let* **C** *and* **D** *be categories, and* $F\colon \mathbf{D} \to \mathbf{C}$ *a functor.*

*Then a* limit *of* $F$*, written* $\varprojlim F$ *or* $\varprojlim_{\mathbf{D}} F(X)$*, means an object* $L \in \mathrm{Ob}(\mathbf{C})$ *given with a morphism* $p(X)\colon L \to F(X)$ *for each* $X \in \mathrm{Ob}(\mathbf{D})$*, such that for* $f \in \mathbf{D}(X, Y)$ *one has* $p(Y) = F(f)p(X)$*, and universal for this property, in the sense that given any object* $M \in \mathrm{Ob}(\mathbf{C})$ *and family of morphisms* $m(X)\colon M \to F(X)$ $(X \in \mathrm{Ob}(\mathbf{D}))$ *which similarly make commuting triangles with the morphisms* $F(f)$*, there exists a unique morphism* $h\colon M \to L$ *such that for all* $X$*,* $m(X) = p(X)\,h$*.*

*Dually, a* colimit *of* $F$*, written* $\varinjlim F$ *or* $\varinjlim_{\mathbf{D}} F(X)$*, means an object* $L \in \mathrm{Ob}(\mathbf{C})$ *given with morphisms* $q(X)\colon F(X) \to L$ *for all* $X \in \mathrm{Ob}(\mathbf{D})$ *such that for* $f \in \mathbf{D}(X, Y)$ *one has* $q(X) = q(Y)F(f)$*, and universal for this property, in the sense that given* $M \in \mathrm{Ob}(\mathbf{C})$ *and morphisms* $m(X)\colon F(X) \to M$ $(X \in \mathrm{Ob}(\mathbf{D}))$ *making commuting triangles with the morphisms* $F(f)$*, there exists a unique morphism* $h\colon L \to M$ *such that for all* $X$*,* $m(X) = h\,q(X)$*.*

*The morphisms* $p(X)$ *in the definition of a limit may be called the associated* projection *morphisms, and the* $q(X)$ *in the definition of colimit the associated* coprojection *morphisms.*

*One says that a category* **C** *"has small limits" if all functors from small categories* **D** *into* **C** *have limits, and that* **C** *"has small colimits" if all functors from small categories into* **C** *have colimits.*

*Remarks on terminology.* Since the above concepts generalize not only direct and inverse limits, but also a large number of other pairs of constructions, they might just as well have been given names suggestive of one of the other pairs. I think that the reason "limit" and "colimit" were chosen is that each of the other relevant universal constructions involves a more or less fixed diagram, while the diagrams involved in direct and inverse limits are varied.

Hence in developing the latter concepts, people were forced to formulate a fairly general definition, and just a little more generality gave the concepts noted above.

But though the choice is historically explainable, I think it is unfortunate. As we can see from the examples of products and coproducts, or of kernels and cokernels, the objects given by limit and colimit constructions over diagram categories other than directed partially ordered sets are not "approximated arbitrarily closely" by the objects from which they are constructed, as the term "limit" would suggest. The particular cases that best exemplify the general concepts are not, I think, inverse and direct limits, but pullbacks and pushouts, so it would be preferable if the limit and colimit of $F : \mathbf{D} \to \mathbf{C}$ were renamed the *pullback* and the *pushout* of $F$ (regarded as a system of objects and maps in $\mathbf{C}$). But it seems too late to turn the tide of usage.

Note also the initially confusing fact that *limits* generalize *inverse* limits, while *colimits* generalize *direct* limits. The explanation is that the words "direct" and "inverse" refer to forward and backward orientation with respect to the arrows in the diagram, while the terms "limit" and "colimit" are related by the principle of using a simple term for a right universal construction, and adding "co-" to it to get the name of the dual left universal construction. That principle arose from such cases as "products and coproducts", and "kernels and cokernels", where the right universal constructions, being more elementary, were named first. There is no reason why two such principles of naming should agree as to which concept gets the "plain" and which the "modified" name, and in this case, they do not.

There is another pair of words for the same constructions: Freyd has named them "roots" and "coroots", probably because if one pictures a system of objects and morphisms as a graph, the addition of the universal object makes it a *rooted* graph, with the universal object at the root. However there is no evident connection with roots of equations etc., and this terminology has not caught on.

Following the associations of the word "limit", Mac Lane [20] calls a category $\mathbf{C}$ *small-complete* if it has small limits, *small-cocomplete* if it has small colimits.

**Exercise 8.6:2.** If $S$ is a monoid, then as for groups, an $S$-set is equivalent to a functor $F : S_{\mathbf{cat}} \to \mathbf{Set}$. Show how to construct the limit (easy) and the colimit (not so easy) of such a functor.

A useful observation is

**Lemma 8.6.2.** *Let* $\mathbf{D}$ *be a category and* $X_0$ *an object of* $\mathbf{D}$ *such that there are morphisms from* $X_0$ *to every object of* $\mathbf{D}$*. Let* $F : \mathbf{D} \to \mathbf{C}$ *be a functor having a limit* $L$*. Then the projection morphism* $p(X_0) : L \to F(X_0)$ *is a monomorphism. In particular, all equalizer maps are monomorphisms.*

*Likewise, if* $\mathbf{D}$ *is a category having an object* $X_0$ *such that there are morphisms from every object of* $\mathbf{D}$ *to* $X_0$*, and* $F : \mathbf{D} \to \mathbf{C}$ *is a functor* *having a colimit $L$, then the coprojection morphism $q(X_0)\colon F(X_0) \to L$ is an epimorphism. In particular, coequalizer maps are epimorphisms.*

*Proof.* Assume the first situation. The universal property of $L$ implies that a morphism $h\colon M \to L$ in $\mathbf{C}$ is uniquely determined by the system of morphisms $p(X)\,h\colon M \to F(X)$ $(X \in \mathrm{Ob}(\mathbf{D}))$. But for any $X \in \mathrm{Ob}(\mathbf{D})$, we can find a morphism $f\colon X_0 \to X$ in $\mathbf{D}$, and we then have $p(X) = F(f)\,p(X_0)$. Thus any $h\colon M \to L$ in $\mathbf{C}$ is uniquely determined by the single morphism $p(X_0)\,h$. This is equivalent to saying $p(X_0)$ is a monomorphism. The result for colimits follows by duality. $\qquad\square$

For example, when $\mathbf{D}$ has the form $G_{\mathbf{cat}}$ for $G$ a group or a monoid, so that a functor $\mathbf{D} \to \mathbf{C}$ is an action of $G$ on an object $X$ of $\mathbf{C}$, the above result tells us that the projection from the limit of that action to $X$ is a monomorphism; a fact that is clear from Exercise 8.6:2 when $\mathbf{C} = \mathbf{Set}$. One similarly sees that equalizer maps are monomorphisms; and maps arising in the duals of these two ways are epimorphisms. However,

**Exercise 8.6:3.** Show that there exist categories $\mathbf{C}$ having monomorphisms $f$ which cannot, for any choice of $F$, $\mathbf{D}$, $X_0$ as in Lemma 8.6.2, be represented as projection morphisms $p(X_0)$. (Suggestion: What about a morphism that is both an epimorphism and a monomorphism?)

We have seen that the constructions of pairwise product and coproduct, when they exist for all pairs of objects of a category $\mathbf{C}$, give right and left adjoints to the "diagonal" functor $\Delta\colon \mathbf{C} \to \mathbf{C} \times \mathbf{C}$. These statements generalize to limits and colimits.

**Proposition 8.6.3.** *Let $\mathbf{C}$ and $\mathbf{D}$ be categories, with $\mathbf{D}$ small, and denote by $\Delta\colon \mathbf{C} \to \mathbf{C}^{\mathbf{D}}$ the "diagonal" functor, taking every object $X \in \mathrm{Ob}(\mathbf{C})$ to the "constant" functor $\Delta(X) \in \mathrm{Ob}(\mathbf{C}^{\mathbf{D}})$ with value $X$ at all objects of $\mathbf{D}$ and value $\mathrm{id}_X$ at all morphisms of $\mathbf{D}$, and likewise taking each morphism $f \in \mathbf{C}(X, Y)$ to the morphism of functors $\Delta(f)\colon \Delta(X) \to \Delta(Y)$ with value $f$ at all objects of $\mathbf{D}$.*

*Then a limit of a functor $F\colon \mathbf{D} \to \mathbf{C}$ is the same as an object $L$ representing the contravariant functor $\mathbf{C}^{\mathbf{D}}(\Delta(-), F)\colon \mathbf{C}^{\mathrm{op}} \to \mathbf{Set}$. In particular, if $\mathbf{C}$ and $\mathbf{D}$ are such that all functors $\mathbf{D} \to \mathbf{C}$ have limits, then the construction $\varprojlim_{\mathbf{D}}\colon \mathbf{C}^{\mathbf{D}} \to \mathbf{C}$ is a right adjoint to the diagonal functor $\Delta\colon \mathbf{C} \to \mathbf{C}^{\mathbf{D}}$. Likewise, a colimit of $F\colon \mathbf{D} \to \mathbf{C}$ is an object $L$ representing the covariant functor $\mathbf{C}^{\mathbf{D}}(F, \Delta(-))\colon \mathbf{C} \to \mathbf{Set}$. Thus, when all functors $\mathbf{D} \to \mathbf{C}$ have colimits, the construction $\varinjlim_{\mathbf{D}}\colon \mathbf{C}^{\mathbf{D}} \to \mathbf{C}$ is a left adjoint to the diagonal functor $\Delta\colon \mathbf{C} \to \mathbf{C}^{\mathbf{D}}$.* $\qquad\square$

(Above, we assumed $\mathbf{D}$ small so that $\mathbf{C}^{\mathbf{D}}$ would be a legitimate category, so that $\mathbf{C}^{\mathbf{D}}(\Delta(-), F)$ would take values in the category $\mathbf{Set}$ of small sets. If we are interested in the case where $\mathbf{D}$ is merely assumed legitimate, we can apply the above result in any universe $\mathbb{U}'$ larger than $\mathbb{U}$, replacing $\mathbf{Set}$ in the statement with $\mathbf{Set}_{(\mathbb{U}')}$.)

These adjointness relationships are shown below.

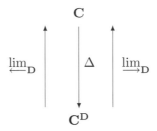

Note that if, as above, $\mathbf{C}$ has colimits of all functors $F \in \mathbf{C}^{\mathbf{D}}$, then our observation that $\varinjlim_{\mathbf{D}} : \mathbf{C}^{\mathbf{D}} \to \mathbf{C}$ is left adjoint to $\Delta$ tells us, in particular, that it is a *functor*. Thus, given a morphism

$$f : F \longrightarrow G$$

in $\mathbf{C}^{\mathbf{D}}$, we get an induced morphism

$$\varinjlim_{\mathbf{D}} f : \varinjlim_{\mathbf{D}} F \longrightarrow \varinjlim_{\mathbf{D}} G$$

in $\mathbf{C}$. This will be characterized by the equations

(8.6.4)        $(\varinjlim_{\mathbf{D}} f)\, q_F(X) = q_G(X)\, f(X) \quad (X \in \mathrm{Ob}(\mathbf{D}))$

where $q_F(X) : F(X) \to \varinjlim_{\mathbf{D}} F$ and $q_G(X) : G(X) \to \varinjlim_{\mathbf{D}} G$ are the coprojection maps for these objects and colimits.

Similarly, if functors in $\mathbf{C}^{\mathbf{D}}$ have limits, then $\varprojlim_{\mathbf{D}} : \mathbf{C}^{\mathbf{D}} \to \mathbf{C}$ becomes a functor, with

$$\varprojlim_{\mathbf{D}} f : \varprojlim_{\mathbf{D}} F \longrightarrow \varprojlim_{\mathbf{D}} G$$

characterized by

(8.6.5)        $p_G(X)(\varprojlim_{\mathbf{D}} f) = f(X)\, p_F(X) \quad (X \in \mathrm{Ob}(\mathbf{D}))$

where $p_F(X) : \varprojlim_{\mathbf{D}} F \to F(X)$ and $p_G(X) : \varprojlim_{\mathbf{D}} G \to G(X)$ are projection maps.

In drawing a picture of a morphism $\Delta(M) \to F$ or $F \to \Delta(M)$ ($M \in \mathrm{Ob}(\mathbf{C})$), we can for convenience collapse the copies of the object $M$ and the identity arrows among them which constitute $\Delta(M)$ into a single "$M$". (E.g., we can collapse the picture representing Proposition 8.6.3 into the picture representing Definition 8.6.1.) What we have then looks like a "cone" of maps, with $M$ at the apex; cf. (8.4.5), and for the dual picture, (8.4.7). Hence a morphism of functors $\Delta(M) \to F$ or $F \to \Delta(M)$ is often called a "cone" from the object $M$ to the functor $F$, or from the functor $F$ to the object $M$; and the limit or colimit of a functor $F$ may be described as an object with a "universal cone" to or from $F$.

**Exercise 8.6:4.** Let $\mathbf{C}$ and $\mathbf{D}$ be categories. By Lemma 7.10.1 ("Law of Exponents for Functors"), the functor $\Delta\colon \mathbf{C} \to \mathbf{C}^{\mathbf{D}}$ corresponds to some functor $\mathbf{D} \times \mathbf{C} \to \mathbf{C}$. Describe this functor.

Our construction in Lemma 8.5.3 of the inverse limit of an inverse system of sets $(X_i,\ f_{ij})$ as the subset of $\prod X_i$ determined by "compatibility" conditions can be generalized to give a construction of general limits in any category having appropriate products and equalizers, and it dualizes to a construction of colimits in categories with appropriate coproducts and coequalizers. (The latter construction may be thought of as generalizing our construction of the direct limit of a directed system of sets as the quotient of a disjoint union by an equivalence relation, though the simple way that equivalence relation could be described when $\mathbf{D}$ was a directed partially ordered set and $\mathbf{C}$ was **Set** does not go over to the general situation.) In the case of inverse limits of sets, the compatibility conditions say that for all $i < j$ in $I$, the pair of maps $p_j$, $f_{ij}\,p_i$ must agree on elements of our subset of $\prod X_i$. This family of conditions can in fact be translated to a condition saying that a single pair of maps into an appropriate product object should agree. Using this idea, we get

**Proposition 8.6.6.** *Let $\mathbf{C}$ be a category and $\mathbf{D}$ a small category, and let $\alpha$ be an infinite cardinal such that $\mathbf{D}$ has $< \alpha$ objects and $< \alpha$ morphisms.*
*Then if $\mathbf{C}$ has products of all families of $< \alpha$ objects, and has equalizers, then every functor $F\colon \mathbf{D} \to \mathbf{C}$ has a limit.*
*Dually, if $\mathbf{C}$ has coproducts of all families of $< \alpha$ objects, and has coequalizers, then every functor $F\colon \mathbf{D} \to \mathbf{C}$ has a colimit.*

*Proof.* Under the hypotheses of the first assertion, let

$$P = \prod_{X \in \mathrm{Ob}(\mathbf{D})} F(X),$$

$$P' = \prod_{X,\,Y \in \mathrm{Ob}(\mathbf{D}),\, f \in \mathbf{D}(X,Y)} F(Y).$$

(If we required categories to have disjoint hom-sets, we could write the latter definition more simply as $P' = \prod_{f \in \mathrm{Ar}(\mathbf{D})} F(\mathrm{cod}(f))$.) Denote the projection morphisms associated with these two product objects by $p_X\colon P \to F(X)$ $(X \in \mathrm{Ob}(\mathbf{D}))$ and $p'_{X,Y,f}\colon P' \to F(Y)$ $(X,Y \in \mathrm{Ob}(\mathbf{D}),\ f \in \mathbf{D}(X,Y))$. We shall construct $L$ as the equalizer of two maps $a, b\colon P \to P'$. Since $a$ and $b$ are to be morphisms into the direct product object $P'$, they may be defined by specifying their composites with the projection morphisms $p'_{X,Y,f}\colon P' \to F(Y)$. Define them so that

$$p'_{X,Y,f}\, a = p_Y, \quad p'_{X,Y,f}\, b = F(f)\, p_X.$$

If $L$ is the equalizer of $a$ and $b$, and $k\colon L \to P$ the canonical morphism, we see that the universal property of $L$ as an equalizer is equivalent to the statement that the morphisms $p_X\,k\colon L \to F(X)$ form commuting triangles with the morphisms $F(f)$ and are universal for this property. Thus, the object $L$ together with the morphisms $p_X\,k$ has the property characterizing $\varprojlim F$.

The result for colimits follows by duality.                                   $\square$

**Exercise 8.6:5.** Verify the assertion following the phrase "we see that", near the end of the above proof.

Of course, some limits or colimits may exist even if the category does not have enough (co)products and (co)equalizers to obtain them by the above lemma. Such a case is noted in part (iv) of the next exercise. (But the most useful part of this exercise is (i), and the most difficult, surprisingly, is (ii).)

**Exercise 8.6:6.** Let $\mathbf{C}$ be a category.

(i)  Show that an initial object of $\mathbf{C}$ is equivalent to a *colimit* of the unique functor from the empty category into $\mathbf{C}$.

(ii)  Show that such an initial object is also equivalent to a *limit* of the identity functor of $\mathbf{C}$.

(iii)  State the corresponding results for a terminal object.

(iv)  Give an example where the limit of (ii) exists, but $\mathbf{C}$ does not satisfy the hypotheses needed to get this from Proposition 8.6.6.

Here is another degenerate case of the concept of limit:

**Exercise 8.6:7.** Characterize the categories $\mathbf{D}$ with the property that every constant functor from $\mathbf{D}$ to any category $\mathbf{C}$, i.e., any functor of the form $\Delta(C)\colon \mathbf{D} \to \mathbf{C}$ $(C \in \mathrm{Ob}(\mathbf{C}))$ has a limit given by the object $C$ itself, with universal cone consisting of identity morphisms of $C$. State the corresponding result for colimits.

We have seen that a product or coproduct of objects in a category may or may not coincide with their product or coproduct in a subcategory to which they also belong. E.g., the coproduct of two abelian groups in the category of *all* groups and their coproduct in the category of all *abelian* groups are different, since the former is generally nonabelian. We note below that for full subcategories, such phenomena occur if and only if the constructed object in the larger category fails to lie in the subcategory.

**Lemma 8.6.7.** *Let $\mathbf{C}$ be a category, $\mathbf{B}$ a full subcategory of $\mathbf{C}$, $I\colon \mathbf{B} \to \mathbf{C}$ the inclusion functor, and $F\colon \mathbf{D} \to \mathbf{B}$ a functor from an arbitrary category into $\mathbf{B}$.*

*If $\varprojlim IF$ exists (loosely, if there exists "a limit of the system of objects $F(X)$ in the larger category $\mathbf{C}$"), and if as an object it belongs to*

**B**, *then this same object, with the same cone to the objects* $F(X)$, *constitutes a limit* $\varprojlim F$ *(loosely, it is also "a limit of the given system within the subcategory* **B**"*).

*The same is true for colimits* $\varinjlim IF$ *and* $\varinjlim F$. □

### Exercise 8.6:8.

(i) Prove the above lemma.

(ii) Does the above result remain true if the hypothesis that the subcategory **B** is *full* in **C** is deleted? If not, does it help to add the hypothesis that not only the object $\varinjlim IF$, but also the projection maps from this object to the values of $F$ belong to **B**?

The example of the coproduct of two abelian groups in **Ab** and in **Group** shows that if we merely assume in the above lemma that $\varinjlim IF$ exists, this does not guarantee that it belongs to **B**. By duality, one can get from this colimit example an analogous example for limits. (Using what category and subcategory?)

Another way that the same object can be a limit of two related functors is examined in

### Exercise 8.6:9. Given functors $\mathbf{D_0} \xrightarrow{D} \mathbf{D_1} \xrightarrow{F} \mathbf{C}$, note that if $F$ has a limit $L = \varprojlim F$, then the cone from $L$ to $F$ induces a cone from $L$ to $FD$, and we can look for conditions under which $L$, with this cone, is also a limit of $FD$. In particular, we can ask which functors $D \colon \mathbf{D_0} \to \mathbf{D_1}$ have the property that this is true for all functors $F$ with domain $\mathbf{D_1}$ which have limits.

Exercise 8.5:1 answered this question for functors $D \colon (P_0)_{\mathbf{cat}} \to (P_1)_{\mathbf{cat}}$ induced by inclusions of partially ordered sets $P_0 \subseteq P_1$. Investigate the same question for functors $D$ between general (i.e., arbitrary small, or perhaps legitimate) categories; that is, look for necessary and/or sufficient conditions on a functor $D$ for this property to hold.

We indicated in the last two paragraphs of §7.10 that if a category **C** has finite products, then any functor category $\mathbf{C^E}$ will also have such products, which can be computed "objectwise". To formulate the analogous result for general limits and colimits, suppose **C** and **E** are categories and $E$ an object of **E**; then let us write $c_E \colon \mathbf{C^E} \to \mathbf{C}$ for the "$E$-th coordinate functor", taking functors and morphisms of functors to their values at the object $E$. Likewise, if $f \colon E_1 \to E_2$ is a morphism in **E**, then $c_f \colon c_{E_1} \to c_{E_2}$ will denote the induced morphism of coordinate functors. You should find it easy to verify

**Lemma 8.6.8.** *Let* **C**, **D** *and* **E** *be categories. Then if all functors* $\mathbf{D} \to \mathbf{C}$ *have limits, so do all functors* $\mathbf{D} \to \mathbf{C^E}$. *Namely, given* $F \colon \mathbf{D} \to \mathbf{C^E}$, *the object* $L = \varprojlim_{\mathbf{D}} F$ *of* $\mathbf{C^E}$ *can be described as the functor taking each* $E \in \mathrm{Ob}(\mathbf{E})$ *to* $\varprojlim_{\mathbf{D}} c_E F$, *and each* $f \in \mathbf{E}(E_1, E_2)$ *to* $\varprojlim_{\mathbf{D}} c_f \circ F$.

*Likewise, if all functors* $\mathbf{D} \to \mathbf{C}$ *have colimits, then all functors* $\mathbf{D} \to \mathbf{C^E}$ *have colimits, which are similarly constructed "object- and morphism-wise".*

$\square$

**Exercise 8.6:10.** Prove Lemma 8.6.8 for the case of limits. .

## 8.7. What respects what?

It is natural to ask what one can say about *limits* and *colimits* of systems of objects constructed by *adjoint functors*, about the values of *adjoint functors* on objects constructed by *limits* and *colimits*, and similar questions for other sorts of universal constructions.

Some quick examples: It is not hard to see that the free group on a disjoint union of sets, $X \sqcup Y$, is the coproduct of the free groups on $X$ and $Y$. If we look similarly at the free group on the coequalizer of a pair of set maps, $f, g : X \rightrightarrows Y$ we find that it is the coequalizer of the induced maps of free groups, $F(f), F(g) : F(X) \rightrightarrows F(Y)$. On the other hand, a direct product of free groups is in general not a free group, in particular not the free group on the direct product set. So the *free group* construction seems to respect *colimits*, but not limits.

If we look at its right adjoint, the underlying set functor, we find the opposite: The underlying set of a product or equalizer of groups is the product or equalizer of the underlying sets of the groups (that is how we constructed products and equalizers of groups), but the underlying set of a coproduct of groups is not the coproduct (disjoint union) of their underlying sets, both because the group operation within this coproduct generally produces new elements from the elements of the two given groups, and because the two identity elements fall together in this coproduct. Similarly, when we take a coequalizer of two group homomorphisms $f, g : G \rightrightarrows H$, more identifications of elements are forced than in the set-theoretic coequalizer: not only must pairs of elements $f(a)$ and $g(a)$ $(a \in |G|)$ fall together, but also pairs such as $f(a)\,b$ and $g(a)\,b$ $(a \in |G|, b \in |H|)$.

These examples suggest the general principle that "left universal constructions respect left universal constructions, and right universal constructions respect right universal constructions". We shall prove a series of theorems of that form in this and the next section.

We have seen left universal constructions in four guises: initial objects, representing objects for covariant set-valued functors, left adjoint functors, and colimits. Since an initial object of a category may be described as the object representing a certain trivial set-valued functor (Exercise 8.2:7) or as the colimit of a functor from a certain trivial category (Exercise 8.6:6(i)), let us focus on relations among the remaining three types of constructions. These give us six unordered pairs of constructions to consider. The first of these pairs would correspond to the question of whether the construction of an object representing one covariant set-valued functor $U$ "respects" the construction of an object representing another such set-valued functor $V$; but I see no meaning to give this question. However, for the next case, concerning representing objects for covariant set-valued functors on the one hand, and left adjoint functors on the other, there is a nice sense in which these respect one another. We give this, along with its dual, as

**Theorem 8.7.1.** *Suppose* $\mathbf{D} \xrightarrow[F]{U} \mathbf{C}$ *are adjoint functors, with* $U$ *the right adjoint and* $F$ *the left adjoint, and with unit* $\eta$ *and counit* $\varepsilon$. *Then*

(i) *If* $A\colon \mathbf{C} \to \mathbf{Set}$ *is a representable functor, with representing object* $R \in \mathrm{Ob}(\mathbf{C})$ *and universal element* $u \in A(R)$, *then* $AU\colon \mathbf{D} \to \mathbf{Set}$ *is also representable, with representing object* $F(R)$ *and universal element* $A(\eta(R))(u) \in A(U(F(R)))$.

*Likewise,*

(ii) *If* $B\colon \mathbf{D}^{\mathrm{op}} \to \mathbf{Set}$ *is representable, with representing object* $R \in \mathrm{Ob}(\mathbf{D})$ *and universal element* $u \in B(R)$, *then* $BF\colon \mathbf{C}^{\mathrm{op}} \to \mathbf{Set}$ *is representable, with representing object* $U(R)$ *and universal element* $B(\varepsilon(R))(u) \in B(F(U(R)))$.

*Proof.* In the first situation, $AU(-) \cong \mathbf{C}(R, U(-)) \cong \mathbf{D}(F(R), -)$, showing that $AU$ is represented by $F(R)$. The identification of the universal element, corresponding to the identity morphism in $\mathbf{D}(F(R), F(R))$ is straightforward. The second situation is the dual of the first. $\qquad\square$

As an example, suppose we wish to construct the ring with a universal pair of elements $x$, $y$ satisfying the relation $x\,y = y\,x^2$. We notice that this ring-theoretic relation is "actually a monoid relation"; the formal statement is that the functor we want to represent can be written $AU$, where $U$ is the forgetful functor from $\mathbf{Ring}^1$ to $\mathbf{Monoid}$, and $A$ the functor associating to any monoid $S$ the set of pairs $(x, y)$ of elements of $S$ satisfying $x\,y = y\,x^2$. It is not hard to see that we can construct our ring by first forming the monoid $R$ presented by these generators and relation, and then passing to the monoid ring $\mathbb{Z}R$, i.e., applying the left adjoint to $U$, and that this is an instance of the above theorem. Note that the universal ring elements $x$ and $y$ satisfying the given equation are the images of the corresponding universal monoid elements, under the canonical map $\eta(R)\colon R \to U(F(R))$ (informally, the inclusion map $R \to \mathbb{Z}R$). Applying $\eta(R)$ to this pair of elements of $R$ corresponds to applying $A(\eta(R))$ to the element $(x, y) \in A(R)$, as in the statement of the theorem.

The above example makes it clear that Theorem 8.7.1 is a powerful tool, and that it indeed deserves to be described as saying that "left adjoint functors respect the construction of objects representing covariant set-valued functors".

Note, however, that, the sense in which the latter statement is true is rather idiosyncratic; the formulation involves both the left adjoint functor and its right adjoint, and it does not appear to be a special case of any natural concept of a left adjoint functor respecting a general construction, or of a general functor respecting the construction of representing objects. There is a similarly idiosyncratic sense in which "left adjoint functors respect other left adjoint functors"; this is Theorem 8.3.10, already proved, which says that the composite of the left adjoints of two functors is the left adjoint of their composite (in the opposite order).

In contrast, when one looks at how our three sorts of left universal construction interact with colimits (these three cases being all we have left to consider of our six possible sorts of interaction), one finds that there *is* a natural definition of an arbitrary functor's respecting a colimit. We will examine that concept in the next section, and verify the remaining cases of our observation that left universal constructions respect left universal constructions, and hence, dually, that right universal constructions respect right universal constructions.

**Exercise 8.7:1.** Prove the following converse to the first assertion of Theorem 8.7.1: If $U : \mathbf{D} \to \mathbf{C}$ is a functor such that for every representable functor $A : \mathbf{C} \to \mathbf{Set}$, the composite functor $AU : \mathbf{D} \to \mathbf{Set}$ is representable, then $U$ has a left adjoint. Also state the dual result.

## 8.8. Functors respecting limits and colimits

Here is the definition of a functor "respecting" a limit or colimit.

**Definition 8.8.1.** *Let* $\mathbf{C}$, $\mathbf{C}'$ *be categories, and* $F : \mathbf{C} \to \mathbf{C}'$ *a functor.*

*Then if* $S : \mathbf{E} \to \mathbf{C}$ *is a functor into* $\mathbf{C}$, *having a limit* $\varprojlim S$, *with cone of projection maps* $p_E : \varprojlim S \to S(E)$ $(E \in \mathrm{Ob}(\mathbf{E}))$, *one says that* $F$ *respects the limit of* $S$ *if the object* $F(\varprojlim S)$, *together with the cone of morphisms* $F(p_E) : F(\varprojlim S) \to F(S(E))$ *from this object to the functor* $FS : \mathbf{E} \to \mathbf{C}'$, *is a limit of the functor* $FS$.

*We shall say that* $F$ *respects small limits if for every functor* $S$ *from a small category* $\mathbf{E}$ *to* $\mathbf{C}$ *which has a limit,* $F$ *respects the limit of* $S$. *We shall say that* $F$ *respects large limits if this is true without the restriction that* $\mathbf{E}$ *be small. Likewise, we shall say that* $F$ *respects pullbacks, terminal objects, small products, large products, small inverse limits, large inverse limits, etc., if it respects all instances of the sort of limit named.*

*Dually, if* $S : \mathbf{E} \to \mathbf{C}$ *is a functor having a colimit* $\varinjlim S$, *with cone of coprojection maps* $q_E : S(E) \to \varinjlim S$, *then we shall say that* $F$ *respects the colimit of* $S$ *if the object* $F(\varinjlim S)$, *with the cone from* $FS$ *to it given by the morphisms* $F(q_E) : F(S(E)) \to F(\varinjlim S)$, *is a colimit of* $FS$; *and we will say* *that $F$ respects small colimits, large colimits, pushouts, initial objects, small or large direct limits, etc., if it respects all colimits having these respective descriptions.*

*In all of these situations, we may use "commutes with" as a synonym for "respects".*

(Many authors, e.g., Mac Lane [20], again following the topological associations of the word "limit", call a functor respecting limits "continuous", and one respecting colimits "cocontinuous". But we will not use these terms here.)

The distinctions between the "small" and "large" cases of the above definition are technically necessary, but there are situations where they can be ignored:

**Observation 8.8.2.** *Suppose that for each universe $\mathbb{U}$ we are given a condition $P(\mathbb{U})$ on functors between $\mathbb{U}$-legitimate categories, such that*

*(a) For every universe $\mathbb{U}$, all functors $F$ between $\mathbb{U}$-legitimate categories that satisfy $P(\mathbb{U})$ respect $\mathbb{U}$-small limits (respectively, $\mathbb{U}$-small colimits; or $\mathbb{U}$-small limits or colimits of a particular sort, such as products or coproducts), and*

*(b) Functors $F$ satisfying $P(\mathbb{U})$ also satisfy $P(\mathbb{U}')$ for all universes $\mathbb{U}' \supseteq \mathbb{U}$. (The commonest case will be where $P(\mathbb{U})$ does not refer to $\mathbb{U}$; e.g., where it is a condition such as "$F$ is a right adjoint functor".)*

*Then all functors $F$ satisfying $P(\mathbb{U})$ in fact respect $\mathbb{U}$-large limits (respectively, $\mathbb{U}$-large colimits, products, coproducts, etc.).*

*Hence, in discussing properties $P$ which are preserved, in the sense of (b) above, under enlarging the (usually unnamed) universe, if we make assertions that all functors $F$ satisfying $P$ "respect limits" etc., we need not specify "small" or "large".*

*Proof.* Regard any $\mathbb{U}$-large limit as a $\mathbb{U}'$-small limit for a universe $\mathbb{U}' \supseteq \mathbb{U}$, and apply (a) with $\mathbb{U}'$ in the role of $\mathbb{U}$. $\qquad\qquad\square$

The above observation will allow us to ignore the small / large distinction in formulating the results of this section. For instance, since the properties of being left and right adjoints do not depend on the universe, we do not need to worry about smallness in stating

**Theorem 8.8.3.** *Left adjoint functors respect colimits, and right adjoint functors respect limits.*

*Proof.* Let $\mathbf{D} \underset{F}{\overset{U}{\rightleftarrows}} \mathbf{C}$ be adjoint functors, with $U$ the right and $F$ the left adjoint, and suppose $S : \mathbf{E} \to \mathbf{C}$ has a colimit $L$, with coprojection maps $q_E : S(E) \to L$ ($E \in \mathrm{Ob}(\mathbf{E})$). Recall that $L$ represents the functor

$$(8.8.4) \qquad\qquad \mathbf{C}^{\mathbf{E}}(S, \Delta(-)) : \ \mathbf{C} \longrightarrow \mathbf{Set},$$

i.e., the construction taking each object $C \in \mathrm{Ob}(\mathbf{C})$ to the set of cones $\mathbf{C}^{\mathbf{E}}(S, \Delta(C))$ and acting correspondingly on morphisms, and that the cone $(q_E)_{E \in \mathrm{Ob}(\mathbf{E})}$ is the universal element for this representing object.

Applying Theorem 8.7.1 to the statement that $L$ represents (8.8.4), we see that $F(L)$ will represent the functor $\mathbf{D} \to \mathbf{Set}$ given by

$$(8.8.5) \qquad \mathbf{C}^{\mathbf{E}}(S, \Delta(U(-))) = \mathbf{C}^{\mathbf{E}}(S, U\Delta(-)) \cong \mathbf{D}^{\mathbf{E}}(FS, \Delta(-));$$

in other words, it will be a colimit of $FS$.

The universal cone could hardly be anything but $(F(q_E))_{E \in \mathrm{Ob}(\mathbf{E})}$; but we need to check this formally. By Theorem 8.7.1, to get this universal element we apply to $L$ the unit $\eta$ of our adjunction, getting a morphism $\eta(L) \colon L \to UF(L)$, apply the functor $\mathbf{C}^{\mathbf{E}}(S, \Delta(-))$ to it, getting a set map

$$\mathbf{C}^{\mathbf{E}}(S, \Delta(\eta(L))) \colon \ \mathbf{C}^{\mathbf{E}}(S, \Delta(L)) \longrightarrow \mathbf{C}^{\mathbf{E}}(S, \Delta(UF(L))),$$

and apply this set map to our original universal cone. Now the above set map is given by left composition with $\eta(L)$, so it transforms our original cone $(q_E)$ from $S$ to $L$ into the cone $(\eta(L) q_E)$ from $S$ to $UF(L)$. Following (8.8.5), we identify cones from the functor $S$ to objects $U(D)$ $(D \in \mathrm{Ob}(\mathbf{D}))$ with cones from $FS$ to the objects $D$ by use of the given adjunction. This identification works by applying $F$ to the given morphisms, then applying the counit of the adjunction to the codomains of the resulting morphisms. So the morphisms $\eta(L) q_E$ of our cone are first transformed to $F(\eta(L) q_E) = F(\eta(L)) F(q_E)$, then composed on the left with $\varepsilon(F(L))$. By Theorem 8.3.8(iii), the latter morphism is left inverse to $F(\eta(L))$, so the composite is $F(q_E)$, as claimed.

The assertion about right adjoint functors and limits follows by duality.

$$\square$$

For example, suppose $(\mathbf{C}, U)$ is a concrete category having free objects on all sets, i.e., such that $U$ has a left adjoint $F$. Then we see by applying the above theorem to appropriate colimits in $\mathbf{Set}$ that a free object in $\mathbf{C}$ on a disjoint union of sets is a coproduct of the free objects on the given sets, and that a free object on the empty set is an initial object. (These facts were noted for particular cases in Chapter 4.) The fact that right adjoints respect limits tells us, likewise, that for $\mathbf{C}$ and $U$ as above, if we call $U(X)$ the "underlying set" of $X \in \mathrm{Ob}(\mathbf{C})$, then underlying sets of product objects, terminal objects, equalizers, and inverse limits are, respectively, direct products of underlying sets, the one-element set, equalizers of underlying sets, and inverse limits of underlying sets. This explains why, in so many familiar cases, the construction of the latter objects begins by applying the corresponding construction to underlying sets. (The perceptive reader may note that what this actually does is reduce these many facts to the one unexplained fact that the underlying set functors of the categories arising in algebra tend to have left adjoints – though they rarely have right adjoints.)

**Exercise 8.8:1.**

(i)   Combining the above theorem with Lemma 7.8.11, obtain results on how left and right adjoint functors behave with respect to epimorphisms and monomorphisms.

(ii)   The results you get will not say that *both* left and right adjoint functors preserve *both* epimorphisms and monomorphisms. Find examples showing that the implications *not* proved in part (i) do not, in general, hold.

(For some related observations, positive and negative, cf. Exercise 7.7:9(v)–(vi) and Exercise 7.8:8.)

Let us look next at how limits and colimits interact with objects that represent functors. In this form, there is not an obvious question to ask; but we can ask whether *representable functors* respect limits and colimits. Our definition of a functor $F$ respecting a limit or colimit assumed $F$ covariant, so to cover contravariant as well as covariant representable functors, we also need a name for the dual property:

**Definition 8.8.6.** *Let* **C**, **C**$'$ *be categories, and* $F$ *a* contravariant *functor from* **C** *to* **C**$'$, *i.e., a functor* **C**$^{\mathrm{op}} \to$ **C**$'$.

*Then if* $S : \mathbf{E} \to \mathbf{C}$ *is a functor having a limit* $\varprojlim S$, *with projection maps* $p_E : \varprojlim S \to S(E)$, *one says that* $F$ *turns the limit of* $S$ *into a colimit if the object* $F(\varprojlim S)$, *together with the cone from the functor* $FS : \mathbf{E}^{\mathrm{op}} \to \mathbf{C}'$ *to this object given by the morphisms* $F(p_E) : F(S(E)) \to F(\varprojlim S)$, *is a colimit of that functor* (*equivalently, if, viewing* $\varprojlim S$ *and* $(p_E)$ *as an object and a cone of morphisms in* **C**$^{\mathrm{op}}$ *which comprise a colimit of the functor* $S^{\mathrm{op}} : \mathbf{E}^{\mathrm{op}} \to \mathbf{C}^{\mathrm{op}}$, *the functor* $F$ *respects this colimit*).

*This yields the obvious definitions of statements such as that* $F$ *"turns small limits into colimits", "turns pullbacks into pushouts", "turns terminal objects into initial objects", etc..*

*We define analogously the concept of* $F$ *turning the* colimit *of a functor* $S$ *into a* limit (*and thus the concepts of turning coproducts into products, pushouts into pullbacks, etc.*).

We can now state

**Theorem 8.8.7.** *Let* **C** *be a category. Then covariant representable functors* $V : \mathbf{C} \to \mathbf{Set}$ *respect limits, while contravariant representable functors on* **C**, $W : \mathbf{C}^{\mathrm{op}} \to \mathbf{Set}$, *turn colimits* (*in* **C**) *into limits* (*in* **Set**).

*Sketch of proof.* The second statement is equivalent to the first applied to the category **C**$^{\mathrm{op}}$, so it suffices to prove the first assertion.

Without loss of generality we may take $V = h_R$ where $R \in \mathrm{Ob}(\mathbf{C})$. Let $L$ be the limit of a functor $S : \mathbf{E} \to \mathbf{C}$. Thus, $L$ is given with a universal cone to $S$, i.e., a universal $\mathrm{Ob}(\mathbf{E})$-tuple of morphisms $p_E : L \to S(E)$ making commuting diagrams with the morphisms $S(f)$ arising from morphisms $f$ in **E**. Applying $h_R$, we get a cone of set maps from the set $h_R(L) = \mathbf{C}(R, L)$

to the sets $\mathbf{C}(R, S(E))$. The fact that it is a cone tells us that each element of $\mathbf{C}(R, L)$ determines, under these maps, a family of elements, one for each of the sets $\mathbf{C}(R, S(E))$, that is respected by the maps $h_R(S(f))$.

Moreover, the universal property of $L$ tells us that each system of elements of the $\mathbf{C}(R, S(E))$ that is respected by the maps $h_R(S(f))$ arises in this way from a unique element of $\mathbf{C}(R, L)$. Now limits over $\mathbf{E}$ in $\mathbf{Set}$ are given by $\mathrm{Ob}(\mathbf{E})$-tuples of elements satisfying just these compatibility conditions; so we see that $\varprojlim_{\mathbf{E}} h_R(S(E))$ and its universal cone to the sets $h_R(S(E))$ can be identified with $h_R(L)$ and its universal cone to these same sets. Thus, the functors $\varprojlim_{\mathbf{E}}(h_R \circ -)$ and $h_R \varprojlim_{\mathbf{E}}$ agree on objects. Their behavior on morphisms is determined by compatibility with maps among cones, and so also agrees.                                                                      $\square$

## Exercise 8.8:2.

(i)   Show by example that covariant representable functors $\mathbf{Ab} \to \mathbf{Set}$ need not respect *colimits*. In fact, give examples of failure to respect coproducts, failure to respect coequalizers, and failure to respect direct limits.

(ii)   Similarly show by examples that contravariant representable functors on $\mathbf{Ab}$ in general fail to turn products, equalizers, and inverse limits into coproducts, coequalizers and direct limits respectively.

Finally, we come to the interaction of colimits with colimits, and of limits with limits. Suppose $B: \mathbf{D} \times \mathbf{E} \to \mathbf{C}$ is a bifunctor. Then each object $D$ of $\mathbf{D}$ induces a functor $B(D, -): \mathbf{E} \to \mathbf{C}$, and each morphism $f: D \to D'$ in $\mathbf{D}$ yields a morphism of functors, $B(f, -) : B(D, -) \to B(D', -)$. (Cf. Lemma 7.10.1 and preceding discussion.) If for each $D$ the functor $B(D, -)$ has a colimit, let us write these objects $\varinjlim_{\mathbf{E}} B(D, E) \in \mathrm{Ob}(\mathbf{C})$. The morphisms between functors $B(D, -)$ induce morphisms among these colimit objects (cf. (8.6.4) and preceding display), so that the construction of $\varinjlim_{\mathbf{E}} B(D, E)$ from $D$ becomes a functor $\varinjlim_{\mathbf{E}} B(-, E) : \mathbf{D} \to \mathbf{C}$. Suppose this functor in turn has a colimit, which we write $\varinjlim_{\mathbf{D}}(\varinjlim_{\mathbf{E}} B(D, E))$. Then the composites of coprojections

$$
(8.8.8) \qquad \begin{array}{c} B(D_0, E_0) \longrightarrow \varinjlim_{\mathbf{E}} B(D_0, E) \longrightarrow \varinjlim_{\mathbf{D}}(\varinjlim_{\mathbf{E}} B(D, E)) \\ (D_0 \in \mathrm{Ob}(\mathbf{D}), \, E_0 \in \mathrm{Ob}(\mathbf{E})) \end{array}
$$

constitute a cone of morphisms from the $B(D_0, E_0)$ to our iterated colimit, and it is straightforward to verify that the latter object, together with this cone, has the universal property of $\varinjlim_{\mathbf{D} \times \mathbf{E}} B(D, E)$.

## Exercise 8.8:3.

(i)   Prove the above claim, that if $\varinjlim_{\mathbf{D}}(\varinjlim_{\mathbf{E}} B(D, E))$ exists, the morphisms (8.8.8) form a cone with respect to which the right hand object satisfies the universal property of $\varinjlim_{\mathbf{D} \times \mathbf{E}} B(D, E)$.

(ii)   On the other hand, give an example where $\varinjlim_{\mathbf{D}\times\mathbf{E}} B(D,\,E)$ exists, but $\varinjlim_{\mathbf{D}} (\varinjlim_{\mathbf{E}} B(D,\,E))$ does not.

This gives us the first isomorphism of (8.8.10) in the next theorem. By symmetry, we likewise have the second isomorphism if the rightmost colimit exists. The isomorphisms of the second display similarly hold under the dual hypotheses.

**Theorem 8.8.9.** *Colimits commute with colimits, and limits commute with limits.*

*Precisely, let* $B: \mathbf{D}\times\mathbf{E}\to\mathbf{C}$ *be a bifunctor. Then*

$$(8.8.10) \quad \varinjlim_{\mathbf{D}}(\varinjlim_{\mathbf{E}} B(D,\,E)) \cong \varinjlim_{\mathbf{D}\times\mathbf{E}} B(D,\,E) \cong \varinjlim_{\mathbf{E}}(\varinjlim_{\mathbf{D}} B(D,\,E)),$$

*in the sense that if the left side of the above display is defined, then this object also has the universal property of the middle object, via the cone of morphisms (8.8.8), and similarly, if the right side is defined, it has the property of the middle object via the analogous cone. Hence, if both sides are defined, they are isomorphic.*

*Likewise*

$$(8.8.11) \quad \varprojlim_{\mathbf{D}}(\varprojlim_{\mathbf{E}} B(D,\,E)) \cong \varprojlim_{\mathbf{D}\times\mathbf{E}} B(D,\,E) \cong \varprojlim_{\mathbf{E}}(\varprojlim_{\mathbf{D}} B(D,\,E))$$

*in the same sense.* □

As formulated, (8.8.10) is not an instance of a functor "respecting" colimits in the precise sense of Definition 8.8.1, because the minimalist hypotheses we assumed in the above theorem do not make $\varinjlim_{\mathbf{D}}$ a functor on all of $\mathbf{C}^{\mathbf{D}}$. If we in fact assume that all functors from $\mathbf{D}$ to $\mathbf{C}$ have colimits (e.g., if $\mathbf{C}$ has small colimits and $\mathbf{D}$ is small), then the isomorphism $\varinjlim_{\mathbf{D}}(\varinjlim_{\mathbf{E}}(B(D,\,E))) \cong \varinjlim_{\mathbf{E}}(\varinjlim_{\mathbf{D}}(B(D,\,E)))$ becomes a case of Theorem 8.8.3, since $\varinjlim_{\mathbf{D}}$ becomes a left adjoint functor $\mathbf{C}^{\mathbf{D}}\to\mathbf{C}$. However, the identification of the common value of the two iterated colimits as $\varinjlim_{\mathbf{D}\times\mathbf{E}} B(D,\,E)$ must still be stated and proved separately. (In the same spirit, if $\mathbf{C}$ has small coproducts, the covariant case of Theorem 8.8.7 follows from Theorem 8.8.3 and Exercise 8.3:3.)

The case of (8.8.10) where $\mathbf{E}$ is the empty category says that colimits respect initial objects; i.e., that if $I$ is an initial object of $\mathbf{C}$, then for any $\mathbf{D}$, the diagram $\Delta(I) \in \mathbf{C}^{\mathbf{D}}$ has colimit $I$. For instance, the coproduct in $\mathbf{Ring}^1$ of two copies of $\mathbb{Z}$ is again $\mathbb{Z}$. The next exercise examines variants of this result.

**Exercise 8.8:4.**

(i)   Show, conversely, that if an object $I$ of a category $\mathbf{C}$ has the property that for all small categories $\mathbf{D}$, the functor $\Delta(I) \in \mathbf{C}^{\mathbf{D}}$ has a colimit isomorphic to $I$, then $I$ is an initial object of $\mathbf{C}$.

(Contrast Exercise 8.6:7, which asks for a description of those *categories* **D** such that this property holds for all *objects* of **C**.)

(ii)  Can you characterize those objects $I$ of a category **C** for which the hypothesis of (i) holds for all *nonempty* small categories **D**?

(iii)  Show that in $\mathbf{Ring}^1$ (or if you prefer, $\mathbf{CommRing}^1$), every ring of the form $\mathbb{Z}/n\mathbb{Z}$ has the property of (ii).

Despite the similar nomenclature, category-theoretic double limits behave quite differently from double limits in topology. The contrast is explored in

**Exercise 8.8:5.**

(i)   For nonnegative integers $i, j$, define $b_{ij}$ to be 1 if $i > j$, 2 if $i \le j$. Show that as limits of real-valued functions, $\lim_{i \to \infty} (\lim_{j \to \infty} b_{ij})$ and $\lim_{j \to \infty} (\lim_{i \to \infty} b_{ij})$ exist and are unequal.

(ii)  Let the set $\omega \times \omega$ be partially ordered by setting $(i, j) \le (i', j')$ if and only if $i \le i'$ and $j \le j'$. Show that there exist functors (directed systems) $B : (\omega \times \omega)_{\mathbf{cat}} \to \mathbf{Set}$ satisfying $\mathrm{card}(B(i, j)) = b_{ij}$, for the function $b_{ij}$ defined in (i).

(iii)  Deduce from Theorem 8.8.9 that a functor as in (ii) can never have the property that for each $i$, the given morphisms $B(i, j) \to B(i, j+1)$ and $B(i, j) \to B(i+1, j)$ are isomorphisms for all sufficiently large $j$.

(iv)  Establish the result of (iii) directly, without using the concept of category-theoretic colimit.

In earlier sections, there were several exercises asking you to determine whether functors $F$ were representable or had right or left adjoints. If you go back over the cases where $F$ turned out *not* to be representable, or not to have an adjoint, you will find that, whatever ad hoc arguments you may have used at the time, each of these negative results can be deduced from Theorem 8.8.7 or 8.8.3 by noting that the $F$ in question fails to respect some limit or colimit.

Since limits and colimits come in many shapes and sizes, it is useful to note that to test whether a functor $F$ respects these constructions, it suffices to check two basic cases.

**Corollary 8.8.12** (to proof of Proposition 8.6.6). *Let* **C**, **D** *be categories and* $F : \mathbf{C} \to \mathbf{D}$ *a functor.*

*If* **C** *has small colimits, then* $F$ *respects such colimits if and only if it respects coequalizers and respects coproducts of small families of objects.*

*Likewise, if* **C** *has small limits, then* $F$ *respects these if and only if it respects equalizers, and products of small families.*                                                       □

One can break things down further, if one wishes:

**Exercise 8.8:6.**

(i)   Let **C** be a category having coproducts of pairs of objects, and hence of finite nonempty families of objects. Show that the universal property of a coproduct of an arbitrary family $\coprod_I X_i$ is equivalent to that of a direct limit, over the directed partially ordered set of finite nonempty subsets $I_0 \subseteq I$, of the finite coproducts $\coprod_{I_0} X_i$.

(ii) Deduce that a category has small colimits if and only if it has co-equalizers, finite coproducts, and colimits over directed partially ordered sets; and that a functor on such a category will respect small colimits if and only if it respects those three constructions.

State the corresponding result for *limits*.

(iii) For every two of the three conditions "respects equalizers", "respects finite products", "respects inverse limits over inversely directed partially ordered sets" (the conditions occurring in the dual to the result of (ii)), try to find an example of a functor among categories having small limits which satisfies those two conditions but not the third. As far as possible, use naturally occurring examples.

You might look at further similar questions; e.g., whether you can find an example respecting both finite and infinite products, but not inverse limits; or whether you can still get a full set of examples if you break the condition of respecting finite products into the two conditions of respecting pairwise products and respecting the terminal object (the product of the empty family).

One can go into this more deeply. I do not know the answers to most of the questions raised in

**Exercise 8.8:7.** Let $A$ denote the (large) set of all small categories, and $B$ the (large) set of all legitimate categories. Define a relation $R \subseteq A \times B$ by putting $(\mathbf{E}, \mathbf{C}) \in R$ if all functors $\mathbf{E} \to \mathbf{C}$ have colimits.

(i) The above relation $R$ induces a Galois connection between $A$ and $B$. Translate results proved about existence of colimits in Proposition 8.6.6 and part (ii) of the preceding exercise into statements about the closure operator $^{**}$ on $A$.

(ii) Investigate further the properties of the lattice of closed subsets of $A$. Is it finite, or infinite? Can you characterize the induced closure operator on the subclass of $A$ or of $B$ consisting of categories $P_{\mathbf{cat}}$ for partially ordered sets $P$?

The above questions concerned *existence* of colimits. To study preservation of colimits, let $C$ denote the class of functors $F$ whose domain and codomain are legitimate categories having small colimits, and let us define a relation $S \subseteq A \times C$ by putting $(\mathbf{E}, F) \in S$ if $F : \mathbf{C} \to \mathbf{D}$ respects the colimits of all functors $\mathbf{E} \to \mathbf{C}$. This relation induces a Galois connection between $A$ and $C$; so let us ask

(iii) Can you obtain results relating the lattice of closed subsets of $A$ under this new Galois connection and the lattice of subsets of $A$ closed under the Galois connection of part (i)? If they are not identical, investigate the structure of the new lattice. (You will have to use a notation that distinguishes between the two Galois connections.)

In studying situations where we do not know that one functor respects the (co)limit of another, but where the two (co)limits in question both exist, there is a natural way to compare them:

**Definition 8.8.13.** *If* $\mathbf{E} \xrightarrow{S} \mathbf{C} \xrightarrow{F} \mathbf{D}$ *are functors such that* $\varinjlim S$ *and* $\varinjlim FS$ *both exist, then by the* comparison morphism

$$\varinjlim FS \longrightarrow F(\varinjlim S)$$

*we shall mean the unique morphism from the former object to the latter which makes a commuting diagram with the natural cones of maps from the functor* $FS$ *to these two objects (namely, the universal cone from* $FS$ *to* $\varinjlim FS$, *and the cone obtained by applying* $F$ *to the universal cone from* $S$ *to* $\varinjlim S$. *The existence and uniqueness of this map follow from the universal property of the former cone.)*

*Likewise, if* $\varprojlim S$ *and* $\varprojlim FS$ *both exist, then by the* comparison mor-phism

$$F(\varprojlim S) \longrightarrow \varprojlim FS$$

*we shall mean the unique morphism which makes a commuting diagram with the obvious cones from these two objects to the functor* $FS$.

*In particular, we may use the term "comparison morphism" in connection with coproducts, products, coequalizers, equalizers, etc., regarding these as colimits and limits.*

These comparison morphisms measure whether the functor $F$ respects these colimits and limits. I.e., comparing Definitions 8.8.13 and 8.8.1 we have

**Lemma 8.8.14.** *Given* $S$ *and* $F$ *as in the first paragraph of Defini-tion 8.8.13, the functor* $F$ *respects the colimit of* $S$ *if and only if the com-parison morphism* $\varinjlim FS \to F(\varinjlim S)$ *is an isomorphism. Likewise, under the assumptions of the second paragraph of that definition,* $F$ *respects the limit of* $S$ *if and only if the comparison morphism* $F(\varprojlim S) \to \varprojlim FS$ *is an isomorphism.* $\qquad\square$

**Exercise 8.8:8.** Suppose $\mathbf{C}$, $\mathbf{D}$ and $\mathbf{E}$ are categories such that $\mathbf{C}$ has colimits of all functors $\mathbf{D} \to \mathbf{C}$, and also of all functors $\mathbf{E} \to \mathbf{C}$, so that $\varinjlim_{\mathbf{D}}$ becomes a functor $\mathbf{C}^{\mathbf{D}} \to \mathbf{C}$ and $\varinjlim_{\mathbf{E}}$ a functor $\mathbf{C}^{\mathbf{E}} \to \mathbf{C}$. Show that for any bifunctor $B \colon \mathbf{D} \times \mathbf{E} \to \mathbf{C}$, the above definition yields com-parison morphisms $\varinjlim_{\mathbf{D}}(\varinjlim_{\mathbf{E}} B(D, E)) \to \varinjlim_{\mathbf{E}}(\varinjlim_{\mathbf{D}} B(D, E))$ and also $\varinjlim_{\mathbf{E}}(\varinjlim_{\mathbf{D}} B(D, E)) \to \varinjlim_{\mathbf{D}}(\varinjlim_{\mathbf{E}} B(D, E))$, and that these are *inverse* to one another. This gives another proof of the isomorphism between the two sides of (8.8.10) under these hypotheses.

Earlier in this section, I said that there was no obvious way to talk about limits or colimits "respecting" the construction of objects representing functors, and we looked instead at the subject of representable functors respecting limits and colimits. But there are actually some not-so-obvious results one can get on limits and colimits of objects that represent functors. Conveniently, these reduce to statements that certain functors respect limits and colimits. You can develop these in

**Exercise 8.8:9.**

(i) Show that the covariant Yoneda embedding $\mathbf{C} \to \mathbf{Set}^{\mathbf{C}^{\mathrm{op}}}$ respects small limits, and that the contravariant Yoneda embedding $\mathbf{C}^{\mathrm{op}} \to \mathbf{Set}^{\mathbf{C}}$ turns small colimits into limits. (Idea: combine Lemma 8.6.8 and Theorem 8.8.7.)

(ii) Turn the above results into statements on the representability of set-valued functors which are limits or colimits of other representable functors, and characterizations of the objects that represent these.

(iii) Deduce the characterization, noted near the beginning of §4.6, of pairwise coproducts of groups defined by presentations, and the assertion of Exercise 8.5:7, that every group is a direct limit of finitely presented groups.

(iv) Show by example that the covariant Yoneda embedding of a category need not respect small colimits, and that the contravariant Yoneda embedding need not turn small limits into colimits.

(v) Suppose $\mathbf{C}$, $\mathbf{D}$, $\mathbf{E}$ are categories, with $\mathbf{E}$ small, and $U : \mathbf{E} \to \mathbf{C}^{\mathbf{D}}$ a functor such that each of the functors $U(E) : \mathbf{D} \to \mathbf{C}$ has a left adjoint $F(E)$. Under appropriate assumptions on existence of small limits and/or colimits in one or more of these categories, deduce from preceding parts of this exercise that $\varprojlim_{\mathbf{E}} U(E)$ exists (as an object of $\mathbf{C}^{\mathbf{D}}$), and that (as a functor $\mathbf{D} \to \mathbf{C}$) it has a left adjoint, constructible from the $F(E)$.

(vi) Show by example that the analogous statement about *colimits* of functors which have left adjoints is false.

## 8.9. Interaction between limits and colimits

Since limits are right universal constructions and colimits are left universal, these two sorts of constructions cannot be expected to respect one another in general. However, there are important special cases where they do. We observed in §8.5 (and will prove formally in the next chapter) that one can form the direct limit of any directed system of algebras with finitary operations by taking the direct limit of their underlying sets, and putting operations on this set in a natural manner. The essential reason for this is that algebra structures are given by operations $|A| \times \cdots \times |A| \to |A|$ on sets, and that in **Set**, direct limits commute with finite products – although, generally, colimits do not.

When we ask whether a given limit and a given colimit commute, there are potentially two comparison morphisms to consider, one a case of the comparison morphism that measures whether a limit is respected by a general functor, the other of the comparison morphism for a colimit and a general functor. A priori, one of these might be an isomorphism and the other not, or they might give different isomorphisms between the same objects. Fortunately, these anomalies cannot occur; as we shall now prove, the two comparison morphisms coincide. (Note that these morphisms go in the same direction, because the comparison morphism for limits goes into the limit object, while the comparison morphism for colimits comes out of the co-limit object. Contrast the interaction between limits and limits, or between colimits and colimits, where the two comparison morphisms go in opposite directions, and, as shown in Exercise 8.8:8, are inverse to one another.)

**Lemma 8.9.1.** *Suppose* **C**, **D** *and* **E** *are categories, such that* **C** *has co-limits of all functors with domain* **D**, *and has limits of all functors with domain* **E**; *and let* $B : \mathbf{D} \times \mathbf{E} \to \mathbf{C}$ *be a bifunctor. Then the two comparison morphisms*

$$\varinjlim_{\mathbf{D}} \varprojlim_{\mathbf{E}} B(D, E) \longrightarrow \varprojlim_{\mathbf{E}} \varinjlim_{\mathbf{D}} B(D, E)$$

*coincide, their common value being characterizable as the unique morphism* $c_B$ *such that for every* $D_0 \in \mathrm{Ob}(\mathbf{D})$ *and* $E_0 \in \mathrm{Ob}(\mathbf{E})$, *the following diagram commutes:*

(8.9.2)

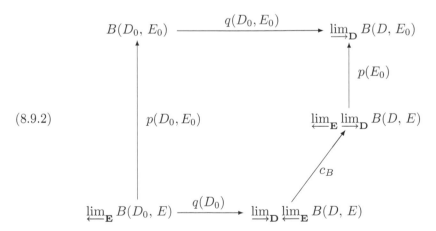

*Here* $p(D_0, E_0)$ *and* $p(E_0)$ *denote the* $E_0$-*th projection maps out of the respective limits* $\varprojlim_{\mathbf{E}} B(D_0, E)$ *and* $\varprojlim_{\mathbf{E}} \varinjlim_{\mathbf{D}} B(D, E)$, *and* $q(D_0, E_0)$ *and* $q(D_0)$, *the* $D_0$-*th coprojection morphisms into the colimits* $\varinjlim_{\mathbf{D}} B(D, E_0)$ *and* $\varinjlim_{\mathbf{D}} \varprojlim_{\mathbf{E}} B(D, E)$.

*Proof.* Let us first take $c_B$ to be the comparison map between the two indicated objects of (8.9.2) which tests whether the construction $\varprojlim_{\mathbf{E}} : \mathbf{C}^{\mathbf{E}} \to \mathbf{C}$, regarded simply as a functor (and not specifically as a limit) respects $\varinjlim_{\mathbf{D}}$. We shall verify that this is the unique morphism making the family of diagrams (8.9.2) commute, i.e., giving the same morphisms via the two routes from the lower left to the upper right corners. The dual argument will then clearly show the same for the other comparison map, proving the lemma.

The defining property of the colimit-comparison morphism $c_B$ is that it respects the cones from the family of objects $\varprojlim_{\mathbf{E}} B(D_0, E)$ ($D_0 \in \mathrm{Ob}(\mathbf{D})$) to the two objects it connects, where the cone to its domain is the universal one for the colimit over $\mathbf{D}$, and consists of the bottom arrows of the diagrams, while we get the cone to the codomain by applying $\varprojlim_{\mathbf{E}}(-, E)$ to the family of coprojection maps $(q(D_0, E_0))_{E_0 \in \mathrm{Ob}(\mathbf{E})}$ (top arrow in (8.9.2); cf. Lemma 8.6.8). Now when we apply $\varprojlim_{\mathbf{E}}(-, E)$ to such a family, the resulting morphism is characterized by the condition that for each $E_0$, it form a commuting square with the projection maps to the objects indexed by $E_0$ (cf. (8.6.5)). In our case, those projection maps are the vertical arrows in (8.9.2); thus the condition is that for all $E_0$, our map $\varprojlim_{\mathbf{E}} B(D_0, E) \to \varprojlim_{\mathbf{E}} \varinjlim_{\mathbf{D}} B(D, E)$ should form a commuting square with the three arrows above it. Since by choice of $c_B$, it also equals the composite of the two arrows below it, (8.9.2) will commute for all $D_0$ and $E_0$; and we see from the universal properties involved that $c_B$ will be the unique morphism making this happen. □

Before proving that in certain cases the above comparison morphism is an isomorphism, let us note some elementary cases where it is not.

**Exercise 8.9:1.** Let $\mathbf{D}$ and $\mathbf{E}$ each be the category with object-set $\{0, 1\}$, and no morphisms other than identity morphisms.

(i)  Suppose $L$ is a lattice, $L_{\mathrm{pos}}$ its underlying partially ordered set, and $\mathbf{C} = (L_{\mathrm{pos}})_{\mathbf{cat}}$. For these choices of $\mathbf{C}$, $\mathbf{D}$ and $\mathbf{E}$, say what it means to give a bifunctor $B$ as in Lemma 8.9.1, verify that the indicated limits and colimits exist, and identify the morphism $c_B$ of the lemma. Show that even if $\mathbf{C}$ is the two-element lattice, this morphism can fail to be an isomorphism.

(ii)  Analyze similarly the case where $\mathbf{C} = \mathbf{Set}$, and $\mathbf{D}$, $\mathbf{E}$ are again as above.

We shall now prove our positive result, which includes the claim in the first paragraph of this section about direct limits of finite products in $\mathbf{Set}$. The proof will involve chasing elements in objects of $\mathbf{Set}$, and we shall see subsequently that the corresponding statement with $\mathbf{Set}$ replaced by a general category $\mathbf{C}$ is false. In thinking about what the result says, you might begin with the case where $\mathbf{D} = \omega_{\mathbf{cat}}$ ($\omega$ the partially ordered set of natural numbers) and $\mathbf{E}$ is either the two-object category such that limits over $\mathbf{E}$ are equalizers, or the three-object category such that limits over $\mathbf{E}$ are pullbacks, or the one-object category $G_{\mathbf{cat}}$ for $G$ a finitely generated group;

and convince yourself that the assertion of the proposition is true in one or more of these cases, before reading the proof for the general case.

Let us note a piece of notation that will be used in the proof. If $B : \mathbf{D} \times \mathbf{E} \to \mathbf{C}$ is a bifunctor, $D$ an object of $\mathbf{D}$, and $f : E_1 \to E_2$ a morphism of $\mathbf{E}$, then one often writes $B(D, f)$ for the induced morphism $B(D, E_1) \to B(D, E_2)$, which is, strictly, $B(\mathrm{id}_D, f)$. Similarly, given a morphism $g$ of $\mathbf{D}$ and an object $E$ of $\mathbf{E}$, one may write $B(g, E)$ for $B(g, \mathrm{id}_E)$.

**Proposition 8.9.3.** *If* $\mathbf{D}$ *is a category of the form* $P_{\mathbf{cat}}$, *for* $P$ *a directed partially ordered set, and* $\mathbf{E}$ *is a nonempty category which has only finitely many objects, and whose morphism-set is finitely generated under composition, then for any bifunctor* $B : \mathbf{D} \times \mathbf{E} \to \mathbf{Set}$, *the morphism* $c_B$ *of Lemma 8.9.1 is an isomorphism.* (*Briefly:* "*In* $\mathbf{Set}$, *direct limits commute with finite limits.*")

*Proof.* Let $E_0, \ldots, E_{m-1}$ be the objects of $\mathbf{E}$, and $f_0, \ldots, f_{n-1}$ a generating set for the morphisms of $\mathbf{E}$, with

$$(8.9.4) \qquad\qquad f_j \in \mathbf{E}(E_{u(j)}, E_{v(j)}).$$

Given elements $D \le D'$ in the partially ordered set $P$, let us write $g_{D, D'}$ for the unique morphism $D \to D'$ in $P_{\mathbf{cat}} = \mathbf{D}$. Projection and coprojection morphisms associated to limits and colimits of our system will be named as in (8.9.2).

To show surjectivity of $c_B$, let $x$ be any element of $\varprojlim_{\mathbf{E}} \varinjlim_{\mathbf{D}} B(D, E)$. For each of the finitely many objects $E_i$ of $\mathbf{E}$, consider $p(E_i)(x) \in \varinjlim_{\mathbf{D}} B(D, E_i)$. By the construction of direct limits in $\mathbf{Set}$ (second paragraph of Lemma 8.5.3), there must exist for each $i$ a $D(i) \in P = \mathrm{Ob}(\mathbf{D})$, such that this element arises from some $x_i \in B(D(i), E_i)$, i.e.,

$$(8.9.5) \qquad p(E_i)(x) = q(D(i), E_i)(x_i) \quad (i = 0, \ldots, m-1).$$

Since the partially ordered set $P$ is directed, we can find $D_0 \in P$ majorizing all the $D(i)$. Thus we have images of all the $x_i$ at the "$D_0$ level"; let us denote these

$$(8.9.6) \qquad x_i' = B(g_{D(i), D_0}, E_i)(x_i) \in B(D_0, E_i) \quad (i = 0, \ldots, m-1).$$

In view of the commutativity relations in the description of $\varinjlim_{\mathbf{D}} B(D, E_i)$, (8.9.5) and (8.9.6) imply

$$(8.9.7) \qquad p(E_i)(x) = q(D_0, E_i)(x_i') \quad (i = 0, \ldots, m-1).$$

Now the definition of $\varprojlim_{\mathbf{E}} \varinjlim_{\mathbf{D}} B(D, E)$ as a limit tells us that the system of elements on the left-hand side of (8.9.7) is "respected" by all morphisms of $\mathbf{E}$, equivalently, by the generating family of morphisms $f_j$. That is, recalling that $f_j$ has domain $E_{u(j)}$ and codomain $E_{v(j)}$, the map

$$
\text{(8.9.8)} \quad
\begin{aligned}
&\varinjlim_{\mathbf{D}} B(D, f_j)\colon\ \varinjlim_{\mathbf{D}} B(D, E_{u(j)}) \to \varinjlim_{\mathbf{D}} B(D, E_{v(j)}) \\
&\text{carries } p(E_{u(j)})(x) \text{ to } p(E_{v(j)})(x) \quad (j = 0, \dots, n-1).
\end{aligned}
$$

It is not necessarily true that the system of preimages $x_i' \in B(D_0, E_i)$ that we have found for these elements satisfy the corresponding relations, i.e., that $B(D_0, f_j)$ carries $x'_{u(j)}$ to $x'_{v(j)}$; but by the construction of direct limits in **Set** referred to earlier, applied to the direct limit objects of (8.9.8), we see that for each $j$, there is some $D'(j) \geq D_0$ such that the corresponding relation holds, namely

$$
\begin{aligned}
B(D'(j), f_j)\, & (B(g_{D_0, D'(j)}, E_{u(j)})(x'_{u(j)})) = \\
& B(g_{D_0, D'(j)}, E_{v(j)})(x'_{v(j)}) \quad (j = 0, \dots, n-1).
\end{aligned}
$$

Hence taking $D_1$ majorizing all the $D'(j)$'s, and letting

$$
x_i'' = B(g_{D_0, D_1}, E_i)(x_i') \in B(D_1, E_i) \quad (i = 0, \dots, m-1)
$$

we have the desired "lifting" of the system of equations (8.9.8):

$$
B(D_1, f_j)(x''_{u(j)}) = x''_{v(j)} \quad (j = 0, \dots, n-1).
$$

That is, the $f$'s do respect the $x_i''$. Now since every morphism of $\mathbf{E}$ is a composite of the $f_j$, every morphism of $\mathbf{E}$ similarly respects the $x_i''$; so the $x_i''$ define an element $x'' \in \varprojlim_{\mathbf{E}} B(D_1, E)$. The element $q(D_1)(x'') \in \varinjlim_{\mathbf{D}} \varprojlim_{\mathbf{E}} B(D, E)$ is the required inverse image of $x$ under $c_B$. (Cf. (8.9.2).)

The proof that $c_B$ is one-to-one is similar, but easier; indeed, it does not need any hypothesis on the morphisms of $\mathbf{E}$, but only the finiteness of the object-set. Suppose $x, y \in \varinjlim_{\mathbf{D}} \varprojlim_{\mathbf{E}} B(D, E)$ with $c_B(x) = c_B(y)$. Since $\mathbf{D}$ is directed, there will exist $D_0 \in \mathrm{Ob}(\mathbf{D})$ such that we can write $x$ and $y$ as the images of some $x_0, y_0 \in \varprojlim_{\mathbf{E}} B(D_0, E)$. By assumption, these elements fall together when mapped into $\varprojlim_{\mathbf{E}} \varinjlim_{\mathbf{D}} B(D, E)$, which means that for each $i$, the projections $p(D_0, E_i)(x_0)$ and $p(D_0, E_i)(y_0)$ fall together in $\varinjlim_{\mathbf{D}} B(D, E_i)$. By the construction of direct limits in **Set**, this means that for each $i$ there is some $D(i) \geq D_0$ such that the images of these elements already agree in $B(D(i), E_i)$. Let $D_1 \in \mathrm{Ob}(\mathbf{D})$ majorize all these $D(i)$. Thus the images of $x_0$ and $y_0$ fall together in all the $B(D_1, E_i)$, hence in $\varprojlim_{\mathbf{E}} B(D_1, E)$. Hence in $\varinjlim_{\mathbf{D}} \varprojlim_{\mathbf{E}} B(D, E)$, $x = y$. $\qquad\square$

**Exercise 8.9:2.** Show that the above proposition remains true if the condition that $\mathbf{E}$ be nonempty is replaced by the condition that $\mathbf{D}$ be nonempty, but fails when both are empty. The proof we gave for the proposition does not explicitly refer to the nonemptiness of $\mathbf{E}$; where is it used implicitly? (Note that a statement that something is true "for all $E_i$" does not require that the set of $E_i$ be nonempty—it is vacuously true if the set is empty. So you need to find something less obvious than that.)

The next exercise shows that in the above proposition, neither the assumption that **E** has finite object-set nor the assumption that its morphism-set is finitely generated can be dropped.

**Exercise 8.9:3.**

(i)    Show that direct limits in **Set** do not commute with infinite products. In fact, give both an example where the comparison map fails to be one-to-one, and an example where it fails to be onto.

Now, a product over a set $X$ is a limit over the category $X_{\mathbf{cat}}$ having object-set $X$ and only identity morphisms; thus, the morphism-set of that category may be regarded as generated by the empty set. Hence in the examples you have just constructed, **E** has infinite object-set, but finitely generated morphism-set.

(ii)    To show that finite generation of the morphism-set cannot be dropped either, let **E** $= G_{\mathbf{cat}}$ for $G$ a non-finitely-generated group, and let $P$ be the partially ordered set of all finitely generated subgroups $H \subseteq G$. Take the direct limit over $P$ of the $G$-sets $G/H$, examine the action of $\varprojlim_{\mathbf{E}}$ on this direct limit, and show that this gives the desired counterexample.

The above examples show the need for our hypotheses on **E**. What about the condition that **D** have the form $P_{\mathbf{cat}}$ for $P$ a directed partially ordered set? A simple example of a partially ordered set that is not directed is $\bigvee$, while some examples of categories not having the form $P_{\mathbf{cat}}$ for any partially ordered set are the two-object category $\cdot \rightrightarrows \cdot$, and the one-object category $Z_{\mathbf{cat}}$ where $Z$ is the infinite cyclic group. So

**Exercise 8.9:4.** Give examples showing that the fixed-point-set construction on $Z$-sets (which takes limits over a one-object category with finitely generated morphism set) respects neither pushouts, nor coequalizers, nor orbit-sets of actions of $Z$ commuting with the given action.

Part (i) of the next exercise shows that we cannot interchange the hypotheses on **D** and **E** in Proposition 8.9.3. As one can see from part (iii), this is equivalent to saying that Proposition 8.9.3 does not remain true if we replace the category **Set** by **Set**$^{\mathrm{op}}$; in particular, we cannot replace **Set** in that proposition by a general category having small limits and colimits.

**Exercise 8.9:5.**

(i)    Show that inverse limits in **Set** do *not* commute with coequalizers.

(ii)    Show, on the other hand, that inverse limits in **Set** *do* commute with small coproducts.

(iii)   Translate the results of (i) and (ii) into statements about constructions in **Set**$^{\mathrm{op}}$.

Though by Exercise 8.9:3(ii), the finite generation hypothesis of Proposition 8.9.3 cannot be dropped, the next exercise shows that it can sometimes be weakened. (Cf. [52].)

**Exercise 8.9:6.** If $S$ is a monoid, then a *left congruence* on $S$ means an equivalence relation $\sim$ on $|S|$ such that $y \sim z \implies x\,y \sim x\,z$. It is easy to verify that an equivalence relation $\sim$ is a left congruence if and only if the natural structure of left $S$-set on $|S|$ induces a structure of left $S$-set on $|S|/\!\sim$. Given any subset $R \subseteq |S| \times |S|$, there is a least left congruence on $|S|$ containing $R$, the left congruence "generated by" $R$. We shall call the set $|S| \times |S|$ itself the *improper* left congruence on $S$.

(i)    Show that the following conditions on a monoid $S$ are equivalent: (a) The improper left congruence on $S$ is finitely generated. (b) The one-element $S$-set is finitely presented. (c) The fixed-point-set functor $S$-**Set** $\to$ **Set** respects direct limits.

(ii)    Show that the monoids satisfying the equivalent conditions of (i) include all finitely generated monoids, and all monoids having a right zero element (an element $z$ such that $x\,z = z$ for all $x$).

(iii)   Find a monoid $S$ which satisfies the equivalent conditions of (i), but such that $S^{\mathrm{op}}$ does not.

(iv)    Deduce from (ii) or (iii), or, preferably, from each of them, that the class of categories **E** with finitely many objects such that limits over **E** respect direct limits of sets is strictly larger than the class of such categories with finitely generated morphism sets.

(v)    Can you generalize the result of (i) to get a necessary and sufficient condition on a category **E** (perhaps under the assumption that it has only finitely many objects, or some weaker condition) for colimits over **E** to respect direct limits of sets?

We noted in the last paragraph of the proof of Proposition 8.9.3 that the one-one-ness part of the conclusion did not require finite generation of the morphism set of **E**. It also does not require the non-emptiness assumption on the object-set; moreover, even the assumption that the object-set be finite can be weakened, using the idea of Lemma 8.6.2, to say that it contains a "good" finite subset. Thus, you can easily verify

**Corollary 8.9.9** (to proofs of Proposition 8.9.3 and Lemma 8.6.2). *Let **D** be a category of the form $P_{\mathbf{cat}}$, for $P$ a directed partially ordered set, and let **E** be a category with only finitely many objects, or more generally, having a finite family of objects $E_0, \ldots, E_{m-1}$ such that every object $E$ admits a morphism $E_i \to E$ for some $i$. Then for any bifunctor $B \colon \mathbf{D} \times \mathbf{E} \to \mathbf{Set}$, the comparison morphism $c_B$ of Lemma 8.9.1 is one-to-one.*    □

Let us note next that the role of finiteness in all the above considerations is easily generalized. The interested reader will find that under the next definition, the proofs of our proposition and corollary yield the proposition stated below.

**Definition 8.9.10.** *If $\alpha$ is a cardinal and $P$ a partially ordered set, then $P$ will be called $<\alpha$-directed if every subset of $P$ of cardinality $<\alpha$ has an upper bound in $P$.*

**Proposition 8.9.11.** *Let* $\alpha$ *be an infinite cardinal. If* $\mathbf{D}$ *is a category of the form* $P_{\mathbf{cat}}$, *for* $P$ *a* $<\alpha$-*directed partially ordered set, and* $\mathbf{E}$ *is a non-empty category which has* $<\alpha$ *objects, and whose morphism-set is generated under composition by a set of* $<\alpha$ *morphisms (which, except in the case* $\alpha = \omega$, *is equivalent to saying that* $\mathbf{E}$ *has* $<\alpha$ *morphisms), then for any bifunctor* $B \colon \mathbf{D} \times \mathbf{E} \to \mathbf{Set}$, *the morphism* $c_B$ *of Lemma 8.9.1 is an isomorphism. (Briefly: "In* $\mathbf{Set}$, $<\alpha$-*directed direct limits commute with limits over* $<\alpha$-*generated categories.")*

*Further, the one-one-ness of* $c_B$ *continues to hold if we weaken the hypotheses on* $\mathbf{E}$ *to say merely that there is a set* $S$ *of* $<\alpha$ *objects of* $\mathbf{E}$ *such that every object of* $\mathbf{E}$ *admits a morphism from a member of* $S$. $\qquad\square$

Here are a few more exercises on commuting limits and colimits, some of them open-ended.

**Exercise 8.9:7.** Generalizing part (ii) of Exercise 8.9:5, determine the class of all small categories $\mathbf{E}$ such that limits over $\mathbf{E}$ in $\mathbf{Set}$ commute with coproducts.

**Exercise 8.9:8.** Let $G$ be a group or monoid, let $(X_i)_{i \in P}$ be an inverse system of $G$-sets, and let $c_X : \varinjlim_{G_{\mathbf{cat}}} \varprojlim_{i \in P} X_i \to \varprojlim_{i \in P} \varinjlim_{G_{\mathbf{cat}}} X_i$ be the associated comparison morphism. (In the case where $G$ is a group, recall that $\varinjlim_{G_{\mathbf{cat}}}$ is the *orbit-set* construction of Exercise 8.6:1.)

(i)   Show that if $G$ is a group and $P$ is countable, then $c_X$ is surjective. (Hint: Use Exercise 8.5:6(ii).)

(ii)   Does the result of (i) remain true for $G$ a monoid? For $P$ not necessarily countable? If either of these generalizations fails, can you find any additional conditions under which it again becomes true?

(iii)   What can you say (positive or negative) about conditions under which $c_X$ will be one-to-one?

**Exercise 8.9:9.**

(i)   In the spirit of Exercise 8.8:7, investigate the Galois connection between the set of small categories $\mathbf{D}$ and the set of small categories $\mathbf{E}$ determined by the relation "colimits over $\mathbf{D}$ commute with limits over $\mathbf{E}$ in $\mathbf{Set}$."

(ii)   Investigate the Galois connections (still on the class of all small categories) obtained by replacing "$\mathbf{Set}$" in (i) with one or more other natural categories; e.g., $\mathbf{Ab}$.

We have been considering the interaction between limits and colimits. One can also look at the interaction between *limits* and *left adjoint* functors, and between right adjoint functors and colimits. For example

**Exercise 8.9:10.** Does the abelianization functor $(\ )^{\mathrm{ab}} \colon \mathbf{Group} \to \mathbf{Ab}$ respect inverse limits? Products? Equalizers? In each case where the answer is negative, is it one-one-ness, surjectivity, or both properties of the comparison morphism that can fail? (Cf. Exercise 4.7:1.)

A different sort of "comparison morphism" is considered in

**Exercise 8.9:11.** Given functors $\mathbf{D} \xrightarrow{F} \mathbf{E} \xrightarrow{S} \mathbf{C}$ such that $S$ and $SF$ both have colimits in $\mathbf{C}$, describe a natural morphism (in one direction or the other) between these colimit objects, and examine conditions under which these morphisms will or will not be invertible. (Cf. Exercise 8.5:1.) Also state the corresponding results for limits.

## 8.10. Some existence theorems

Having defined several sorts of universal objects, and established facts about them, it would be nice to have some general results on when such objects exist.

Basic results on the existence of *algebras* with universal properties must wait for the next chapter, where we will set up a general theory of algebras. What we can prove before then are relative results, to the effect that if in a category one can perform certain constructions, then one can perform others; for instance, Proposition 8.6.6 was of this sort. With this limitation in mind, can we abstract any of the methods by which we proved the existence of *free groups* in Chapter 3?

The construction by terms modulo consequences of the identities depends on the fact that one is considering algebras; generalizing this will be one of the first things we do in Chapter 9.

The *normal form* description is still more specialized. As mentioned toward the end of §3.4, different sorts of algebras vary widely as to whether such results hold.

But the *subobject of a big direct product* approach of §3.3 seems amenable to a category-theoretic development, and we shall in fact obtain below several results that have evolved from that construction. The approach is due to Peter Freyd.

We know how to translate the concept of direct product into category theoretic terms. There were two other key ideas in the construction of §3.3: a cardinality estimate, which allowed us to find a *small* set of groups to take the direct product of, and the passage to "the subgroup of the product generated by the given family". The first of these will simply be made a hypothesis – that there exists a small set of objects with an appropriate property. What about the concept of "subalgebra generated"? We know that there is not a canonical concept of "subobject" in category theory, but is there one that is appropriate to this proof?

We saw at various points in Chapters 3 and 4 that if we had an object satisfying one of our left universal properties, except possibly for the *uniqueness* of the factoring maps, then the added condition of uniqueness was equivalent to the object being generated by the appropriate set (e.g., Exercise 3.1:2, and end of proof of Proposition 4.3.3). To put things negatively, in the case of the universal property of a free group on $X$, we saw in Exercise 3.1:1 that if our candidate $F$ for a free group was not generated by the image of $X$, then we could get a pair of group homomorphisms from $F$ into some group which agreed on the elements of $X$, but were not equal on all of $F$. This suggests that the subgroup generated by $X$ may be obtainable as an *equalizer*, using pairs of morphisms having equal composites with the image of $X$. That is the idea which we shall abstract below.

Repeating the progression in the first half of this chapter, let us start with an existence result for initial objects. In reading the next lemma and its proof, you might think of the case where $\mathbf{C}$ is the category of 4-tuples $(G, a, b, c)$ with $G$ a group and $a, b, c \in |G|$, and of the principle that guided us to the subgroup-of-a-product construction, that if one such object $(G, a, b, c)$ is mappable to another such object $(H, a', b', c')$, then the set of relations satisfied by $a, b, c$ in $G$ is contained in the set of relations satisfied by $a', b', c'$ in $H$.

**Lemma 8.10.1.** *Let $\mathbf{C}$ be a (legitimate) category having small limits. Suppose there exists a small set of objects $S \subseteq \mathrm{Ob}(\mathbf{C})$, such that for every $X \in \mathrm{Ob}(\mathbf{C})$ there is a $Y \in S$ with $\mathbf{C}(Y, X)$ nonempty. Then $\mathbf{C}$ has an initial object.*

*Proof.* Let $J = \prod_{Y \in S} Y \in \mathrm{Ob}(\mathbf{C})$. For every $X \in \mathrm{Ob}(\mathbf{C})$ there is at least one morphism from $J$ to $X$, since we can compose the projection of $J$ to some $Y \in S$ with a morphism $Y \to X$. Hence our hypothesis on the set of objects $S$ has been concentrated in this one object $J$, and we may henceforth forget $S$ and work with $J$.

We wish to form the "intersection of the equalizers of all pairs of maps from $J$ into objects of $\mathbf{C}$". If we were working in a category of algebras, this would make sense, for even though all such pairs of maps do not form a small set, the underlying set of $J$ would be small, and hence the set of subalgebras that are equalizers of such pairs of maps would be small, and we could take its intersection. That argument is not available here; but it turns out that, just as we were able to use the family $S$ as a substitute for the class of all objects in forming our product $J$, so it can serve as a substitute for the class of all objects in this second capacity, though a less obvious argument will be needed. Moreover, since the hypothesis on $S$ has been concentrated in the object $J$, we may use $J$ in place of $S$ in this function.

So let us form a product $\prod_{(u,\, v)} J$ of copies of $J$ indexed by the set of all pairs of morphisms $u, v \in \mathbf{C}(J, J)$. (Since $\mathbf{C}$ is legitimate, such pairs form a small set.) Let $a, b : J \rightrightarrows \prod_{(u,\, v)} J$ be defined by the conditions that for all $u, v \in \mathbf{C}(J, J)$, $a$ followed by the projection of the product onto the $(u, v)$ component gives $u$, and $b$ followed by that projection gives $v$. Let us form the equalizer $i \colon I \to J$ of this pair of morphisms. Note that by the universal property of $I$,

(8.10.2)         $u\,i = v\,i$ for any two endomorphisms $u, v$ of $J$.

Since $J$ can be mapped to every object of $\mathbf{C}$, we can find a morphism going the other way, $x \colon J \to I$. Now suppose $c$ is any endomorphism of $I$. By (8.10.2), the morphisms $I \to J$ given by $i$, $i\,x\,i$, and $i\,c\,x\,i$ are equal. But by Lemma 8.6.2 (second sentence), $i$ is a monomorphism; hence we can cancel it on the left and conclude that $\mathrm{id}_I$, $x\,i$, and $c\,x\,i$ are equal. Substituting the equation $x\,i = \mathrm{id}_I$ into $c\,x\,i = x\,i$, we get $c = \mathrm{id}_I$; so $I$ has no nonidentity endomorphisms.

$I$ also inherits from $J$ the property of having morphisms into every object of $\mathbf{C}$, so we can now forget $J$ and work with $I$ only.

We claim that $I$ is an initial object of $\mathbf{C}$. We know it has morphisms into every $X \in \mathrm{Ob}(\mathbf{C})$; consider two such morphisms $u, v \in \mathbf{C}(I, X)$. We may form their equalizer, $k \colon K \to I$, and take an arbitrary morphism the other way, $d \colon I \to K$. Then $k\,d$ is an endomorphism of $I$, hence $k\,d = \mathrm{id}_I$. By choice of $k$, $u\,k = v\,k$, hence $u\,k\,d = v\,k\,d$, i.e., $u = v$; so $I$ has exactly one morphism into each object of $\mathbf{C}$, as claimed.                    $\square$

**Exercise 8.10:1.** The final part of the proof of the above lemma uses the facts that (a) the object $I$ of $\mathbf{C}$ has morphisms into all objects, (b) $I$ has no nonidentity endomorphism, and (c) $\mathbf{C}$ has equalizers. Do (a) and (b) alone imply that $I$ is initial in $\mathbf{C}$?

For some perspective on the above result, recall Exercise 8.6:6, which showed that an initial object of a category $\mathbf{C}$ is equivalent to a *colimit* of the unique functor from the empty category to $\mathbf{C}$, and also to a *limit* of the *identity functor* of $\mathbf{C}$. Now in the study of categories of algebraic objects (for instance, the category of groups with 3-tuples of distinguished elements), one does not have, to begin with, any easy way of constructing colimits, even for as trivial a functor as the one from the empty category! One can, however, construct products and equalizers using the corresponding constructions on the underlying sets of one's algebras; hence one can get all small limits. This suggests trying to construct an initial object as a limit of the identity functor of the whole category. The difficulty is, of course, that the domain of that functor is not small. Hence one looks for a small set $S$ of objects of $\mathbf{C}$ which "get around enough" to serve in place of the set of *all* objects.

In fact, if this had been used as our motivation for the above lemma, we would have gotten a proof in which the initial object $I$ was constructed in one step, as the limit of the inclusion functor of the full subcategory with object-set $S$ into $\mathbf{C}$. But I preferred the present approach because the characterization of limits of identity functors is itself not easy to prove. In [20] you can find both versions of the proof, as Theorem 1 on p. 116 and Theorem 1 on p. 231 respectively.

In results such as the above, the assumption that there exists a small set $S$ which, for the purposes in question, is "as good as" the set of all objects, is known as the "solution-set condition".

On, now, to our next result in this family. Since a representing object for a functor $U\colon \mathbf{C} \to \mathbf{Set}$ is equivalent to an initial object in an appropriate auxiliary category $\mathbf{C}'$, let us see under what conditions we can apply Lemma 8.10.1 to such an auxiliary category to get a representability result. By Theorem 8.8.7, if $U$ is representable it must respect limits, so the condition of respecting limits must somehow be a precondition for the application of Lemma 8.10.1 in this way. The next result shows that, indeed, for the auxiliary category $\mathbf{C}'$ to *have* small limits is equivalent to $U$ respecting such limits.

**Lemma 8.10.3.** *Let $\mathbf{C}$ be a category, $U\colon \mathbf{C} \to \mathbf{Set}$ any functor, and $\mathbf{C}'$ the category whose objects are pairs $(X, \dot{x})$ with $X \in \mathrm{Ob}(\mathbf{C})$ and $x \in U(X)$, and whose morphisms are morphisms of first components respecting second components (in the notation of Exercise 7.8:30, the comma category $(1 \downarrow U)$). Let $V\colon \mathbf{C}' \to \mathbf{C}$ denote the forgetful functor taking $(X, x)$ to $X$. Then*

(i)   *If $\mathbf{D}$ is a small category and $G\colon \mathbf{D} \to \mathbf{C}$ a functor having a limit in $\mathbf{C}$, the following conditions are equivalent:*

  (a)  *$U$ respects the limit of $G$; in other words, the comparison morphism $c\colon U(\varprojlim_{\mathbf{D}} G) \to \varprojlim_{\mathbf{D}} UG$ is a bijection of sets.*

  (b)  *Every functor $F\colon \mathbf{D} \to \mathbf{C}'$ satisfying $VF = G$ has a limit in $\mathbf{C}'$.*

*Hence,*

(ii)  *If $\mathbf{C}$ has small limits, then $\mathbf{C}'$ will have small limits if and only if $U$ respects small limits.*

*Sketch of proof.* We shall prove (i), from which (ii) will clearly follow.

Note that a functor $F$ that "lifts $G$" as in (b) is essentially a compatible way of choosing for each $X \in \mathrm{Ob}(\mathbf{D})$ an element $x \in UG(X)$; hence it corresponds to an element $y \in \varprojlim_{\mathbf{D}} UG$. Now assuming (a), such an element $y$ has the form $c(z)$ for a unique $z \in U(\varprojlim_{\mathbf{D}} G)$, and it is immediate that the pair $(\varprojlim_{\mathbf{D}} G, z)$ is a limit of $F$ in $\mathbf{C}'$, giving (b).

Conversely, assuming (b), let $y$ be any element of $\varprojlim_{\mathbf{D}} UG$. As noted, this corresponds to a functor $F\colon \mathbf{D} \to \mathbf{C}'$, and by (b) $F$ has a limit $(Z, z)$ in $\mathbf{C}'$. The cone from this limit object to $F$, applied to first components, gives a cone from $Z$ to the objects $G(X)$, under which the second component, $z$ is carried to the components of $y$; hence the map $Z \to \varprojlim_{\mathbf{D}} G$ induced by this cone carries $z \in U(Z)$ to an element $w \in U(\varprojlim_{\mathbf{D}} G)$, which is taken by $c$ to $y \in \varprojlim_{\mathbf{D}} UG$. This establishes the surjectivity of $c$.

Suppose now that $c$ also takes another element $w' \in U(\varprojlim_{\mathbf{D}} G)$ to $y$. By the universal property of $(Z, z)$, there is a morphism $\varprojlim_{\mathbf{D}} G \to Z$ carrying $w'$ to $z$; composing this with our morphism $Z \to \varprojlim_{\mathbf{D}} G$ we get an endomorphism of $\varprojlim_{\mathbf{D}} G$ carrying $w'$ to $w$. But all these morphisms, and hence this endomorphism in particular, respect cones to $G$ in $\mathbf{C}$, hence by the universal property of $\varprojlim_{\mathbf{D}} G$, this endomorphism must be the identity morphism of $\varprojlim_{\mathbf{D}} G$. This shows that $w' = w$, proving one-one-ness of $c$.

$\square$

**Exercise 8.10:2.** Give the details of the proof of (i) $\implies$ (ii) above.

**Exercise 8.10:3.** In part (i) of the above lemma, we assumed that the functor $G$ had a limit. We may ask whether this assumption is needed in proving (b) $\implies$ (a), or whether the existence of the limits assumed in (b) implies this.

To answer this question, let $\mathbf{C}$ be the category whose objects are pairs $(G, S)$ where $G$ is a group and $S$ a *cyclic* subgroup of $G$ (a subgroup generated by one element), and where a morphism $(G, S) \to (H, T)$ means a homomorphism $G \to H$ which carries the subgroup $S$ *onto* the subgroup $T$. Let $U : \mathbf{C} \to \mathbf{Set}$ be the functor which carries each pair $(G, S)$ to the set of cyclic generators of $S$, i.e., $U(G, S) = \{s \in |S| \mid S = \langle s \rangle\}$.

Show how to define $U$ on morphisms. Show that $\mathbf{C}$ does not, in general, have products of pairs of objects, but that the category $\mathbf{C}'$, defined as in the above lemma, has all small limits, hence, in particular, pairwise products. Apply this example to answer the original question.

The reader should verify that Lemmas 8.10.1 and 8.10.3 now give the desired criterion for representability, namely

**Proposition 8.10.4.** *Let $\mathbf{C}$ be a category with small limits, and $U : \mathbf{C} \to \mathbf{Set}$ a functor. Then $U$ is representable if and only if*

(a) *$U$ respects small limits, and*

(b) *there exists a small set $S$ of objects of $\mathbf{C}$ such that for every object $Y$ of $\mathbf{C}$ and $y \in U(Y)$, there exist $X \in S$, $x \in U(X)$, and $f \in \mathbf{C}(X, Y)$ such that $y = U(f)(x)$.* $\square$

Finally, let us get from this a condition for the existence of *adjoints*. (In reading the next result, observe that for $\mathbf{D} = \mathbf{Group}$ and $U$ its underlying-set functor, condition (b) below was precisely what we had to come up with in showing the existence of free groups on arbitrary sets $Z$.)

**Theorem 8.10.5** (Freyd's Adjoint Functor Theorem). *Let $\mathbf{C}$ and $\mathbf{D}$ be categories such that $\mathbf{D}$ has small limits. Then a functor $U : \mathbf{D} \to \mathbf{C}$ has a left adjoint $F : \mathbf{C} \to \mathbf{D}$ if and only if*

(a) *$U$ respects small limits, and*

(b) *for every $Z \in \mathrm{Ob}(\mathbf{C})$ there exists a small set $S \subseteq \mathrm{Ob}(\mathbf{D})$ such that for every $Y \in \mathrm{Ob}(\mathbf{D})$ and $y \in \mathbf{C}(Z, U(Y))$, there exist $X \in S$, $x \in \mathbf{C}(Z, U(X))$ and $f \in \mathbf{D}(X, Y)$ such that $y = U(f)(x)$.*

*Proof.* The existence of a left adjoint to $U$ is equivalent by Theorem 8.3.8(ii) to the representability, for every $Z \in \mathrm{Ob}(\mathbf{C})$, of the functor $\mathbf{C}(Z, U(-)) \colon \mathbf{D} \to \mathbf{Set}$. Condition (b) is clearly the form that condition (b) of the preceding proposition takes for this class of functors. As for condition (a), we know by Theorem 8.8.3 that it, too, is necessary for the existence of a left adjoint, so it suffices to show that it implies that each set-valued functor $\mathbf{C}(Z, U(-))$ respects limits. If we write this functor as $h_Z U$, and recall that covariant representable functors $h_Z$ respect limits, this implication is immediate.                                                                           □

**Exercise 8.10:4.** Show the converse of the observation used in the last step of the above proof: If $\mathbf{C}$ and $\mathbf{D}$ are categories, and $U \colon \mathbf{D} \to \mathbf{C}$ a functor such that for every $Z \in \mathrm{Ob}(\mathbf{C})$, $h_Z U$ respects limits, then $U$ respects limits.

I remarked in §8.8 that for every example *we had seen* of a functor that was not representable or did not have a left adjoint, the nonrepresentability or nonexistence of an adjoint could be proved by showing that the functor did not respect some limit. We can now understand this better. On a category having small limits, the only way a functor *respecting* these limits can fail to have a left adjoint or a representing object is if the solution-set condition fails. Since the solution-set condition says "a *small* set is sufficient", its failure must involve uncircumventable set-theoretic difficulties, which are rare in algebraic contexts. However, knowing now what to look for, we can find examples. The next exercise gives a simple, if somewhat artificial example. The example in the exercise after that is more complicated, but more relevant to constructions mathematicians are interested in.

**Exercise 8.10:5.** Let $\mathbf{D}$ be the subcategory of $\mathbf{Set}$ whose objects are all sets (or if you prefer, all ordinals; in either case, "small" is understood, since by definition $\mathbf{Set}$ is the category of all small sets), and whose morphisms are the *inclusion* maps among these. Show that $\mathbf{D}$ has small colimits (and has limits over all nonempty categories, though this will not be needed), but has no terminal object.

   Hence, letting $\mathbf{C} = \mathbf{D}^{\mathrm{op}}$, the category $\mathbf{C}$ has small limits (and colimits over nonempty categories) but no initial object. Translate the nonexistence of an initial object for $\mathbf{C}$ to the nonrepresentability of a certain functor $U \colon \mathbf{C} \to \mathbf{Set}$ which respects limits (cf. Exercise 8.2:7).

   The results of this section would imply the existence of an initial object of $\mathbf{C}$, and of a representing object for $U$, if a certain solution-set condition held. State this condition, and note why it does not hold.

The next exercise validates the comment made in §6.2, that because the class of *complete lattices* is not defined by a small set of operations, it fails in some ways to behave like classes of "ordinary" algebras. The exercise shows that the solution-set condition required for the existence of the free complete lattice on three generators fails, and indeed, that there is no such free object.

**Exercise 8.10:6.** First, a preparatory observation:

(i) Show that every ordinal has a unique decomposition $\alpha = \beta + n$, where $\beta$ is a limit ordinal (possibly 0) and $n \in \omega$. Let us call $\alpha$ *even* or *odd* respectively according as the summand $n$ in this decomposition is even or odd.

Now let $\alpha$ be an arbitrary ordinal, let $S = \alpha \cup \{x, y\}$ where $x$, $y$ are two elements that are not ordinals, and let $L$ be the lattice of all subsets $T \subseteq S$ such that (a) if $T$ contains $x$ and all ordinals less than an odd ordinal $\beta \in \alpha$, then it contains $\beta$, and (b) if $T$ contains $y$ and all ordinals less than an even ordinal $\beta \in \alpha$, then it contains $\beta$.

(ii) Show that the complete sublattice of $L$ generated by the three elements $\{x\}$, $\{0, y\}$ and $\alpha$ (i.e., the closure of this set of three elements under arbitrary meets and joins within $L$) has cardinality $\geq \mathrm{card}(\alpha)$. (This is an extension of the trick of Exercise 6.3:9.)

(iii) Deduce that there can be no free complete lattice on three generators.

(This was first proved in [84], by a different construction. Three proofs of the similar result that there is no free complete Boolean algebra on countably many generators are given in [79, 84, 132].)

This is not to say that a class of algebras having a large set of primitive operations *cannot* have free objects on all sets. The next exercise gives an example of one that does.

**Exercise 8.10:7.** Complete ∨-semilattices with least elements, like complete lattices, have an $\alpha$-fold join operation for every cardinal $\alpha$. Nevertheless:

(i) Show that a complete ∨-semilattice with least element generated by an $X$-tuple of elements has at most $\mathrm{card}(\mathbf{P}(X))$ elements.

(ii) Deduce from Freyd's Adjoint Functor Theorem that there exist free complete ∨-semilattices with least elements on all sets. (This despite the fact that complete ∨-semilattices with least elements are, as partially ordered sets, the same objects as nonempty complete lattices!)

(iii) Does the category of ∨-complete *lattices* with least element behave, in this respect, like that of complete ∨-semilattices with least element, or like that of complete lattices? I.e., does it have free objects on all sets or not?

In §4.17, where we constructed the Stone-Čech compactification of a topological space, we found that one way to obtain the solution-set condition was via the fact that in a compact Hausdorff space, continuous maps to the unit interval $[0, 1]$ separate points. (See the discussion beginning in the paragraph just before the one containing (4.17.7).) Freyd [11, Exercise 3-M, p. 89] cf. [20, §V.8] abstracts this observation to give a variant of Theorem 8.10.5, called the "special adjoint functor theorem", in which the solution-set hypothesis is replaced by an assumption that there exists such an object, called a "cogenerator" of the category, together with a smallness assumption on sets of monomorphisms. However, since the existence of cogenerators is not as common in algebra as the direct verifiability of the solution set condition, we will not develop that result here.

You may have noticed that in this section, I have not followed my usual practice of stating every result both for left and for right universal constructions. That practice is, of course, logically unnecessary, since one result can always be deduced immediately from the other by putting $\mathbf{C}^{\mathrm{op}}$ for $\mathbf{C}$ and making appropriate notational translations. In earlier sections I nonetheless gave dual pairs of formulations, because both statements were generally of comparable importance. However, when one studies categories of algebras, objects characterized by right universal properties are usually easier to construct directly than those characterized by left universal properties, so we have little need for results obtaining the former from the latter; hence my one-sided presentation. (It is also true that short-term generalizations about what cases are important may fail in the longer run! However, we can always call on the duals of the results of this section if we find we need them.)

Here is a somewhat vague question, to which I don't know an answer.

**Exercise 8.10:8.** Suppose a functor $U$ has a left adjoint $F$, which in turn has a left adjoint $G$. Can one conclude more about $U$ itself than the results that we have shown to follow from the existence of $F$? In other words, are there any nice necessary conditions for existence of *double* left adjoints, comparable to the property of respecting limits as a condition for existence of a single left adjoint?

## 8.11. Morphisms involving adjunctions

I am not planning on using the results of this section in subsequent chapters, so the reader may excuse a little sketchiness. (However, the material in the *next* section *will* be referred to in subsequent chapters, and should be read with your usual vigilance.)

Let $\mathbf{C}$ and $\mathbf{D}$ be categories, and $\mathbf{D} \mathrel{\substack{U \\ \longrightarrow \\ \longleftarrow \\ F}} \mathbf{C}$ adjoint functors. We recall the isomorphism which characterizes their adjointness:

$$(8.11.1) \qquad\qquad \mathbf{C}(-, U(-)) \cong \mathbf{D}(F(-), -).$$

Suppose now that we have functors from a third category into each of these categories, $P\colon \mathbf{E} \to \mathbf{C}$ and $Q\colon \mathbf{E} \to \mathbf{D}$. It is not hard to verify that if we formally "substitute $P$ and $Q$ into the blanks" in (8.11.1), we get a bijection between sets of morphisms of functors:

$$\mathbf{C}^{\mathbf{E}}(P, UQ) \longleftrightarrow \mathbf{D}^{\mathbf{E}}(FP, Q).$$

As one would expect, this bijection is functorial in $P$ and $Q$, i.e., respects morphisms $P \to P'$, $Q \to Q'$; in other words, writing $F\circ$ and $U\circ$ for the operations of composing on the left with $F$ and $U$ respectively, the above bijection gives an isomorphism of bifunctors $\mathbf{C}^{\mathbf{E}} \times \mathbf{D}^{\mathbf{E}} \to \mathbf{Set}$ :

$$\mathbf{C}^{\mathbf{E}}(-, U \circ -) \cong \mathbf{D}^{\mathbf{E}}(F \circ -, -).$$

This means that we have an adjoint pair of functors on functor categories, $\mathbf{D}^{\mathbf{E}} \underset{F\circ}{\overset{U\circ}{\rightleftarrows}} \mathbf{C}^{\mathbf{E}}$. We can also describe this adjunction in terms of its unit and counit; these will be $\eta \circ : \mathrm{Id}_{(\mathbf{C}^{\mathbf{E}})} \to (UF)\circ$ and $\varepsilon \circ : (FU)\circ \to \mathrm{Id}_{(\mathbf{D}^{\mathbf{E}})}$, where $\eta$ and $\varepsilon$ are the unit and counit of the adjunction between $U$ and $F$. In fact, the quickest way to prove that $U\circ$ and $F\circ$ are adjoint is to note that the equations in $\eta$ and $\varepsilon$ which establish the adjointness of $U$ and $F$ (Theorem 8.3.8(iii)) give equations in $\eta \circ$ and $\varepsilon \circ$ establishing the adjointness of $U\circ$ and $F \circ$.

The above fits with our comment at the end of §7.9 that a functor category such as $\mathbf{C}^{\mathbf{E}}$ or $\mathbf{D}^{\mathbf{E}}$ behaves very much like its codomain category, $\mathbf{C}$ or $\mathbf{D}$. What that observation does not prepare us for is that analogous results hold for composition on the *right* with adjoint functors. Given adjoint functors $U$ and $F$, still as in (8.11.1) above, let us take a category $\mathbf{B}$ and functors $R : \mathbf{D} \to \mathbf{B}$, $S : \mathbf{C} \to \mathbf{B}$. I claim we get a bijection

$$\mathbf{B}^{\mathbf{D}}(SU, R) \longleftrightarrow \mathbf{B}^{\mathbf{C}}(S, RF)$$

and thus an isomorphism

$$\mathbf{B}^{\mathbf{D}}(- \circ U, -) \cong \mathbf{B}^{\mathbf{C}}(-, - \circ F),$$

i.e., a pair of adjoint functors, $\mathbf{B}^{\mathbf{D}} \underset{\circ U}{\overset{\circ F}{\rightleftarrows}} \mathbf{B}^{\mathbf{C}}$, where this time $\circ U$ is the left adjoint and $\circ F$ the right adjoint. I don't know a way of seeing this directly from (8.11.1), but it comes out easily if we check the formal properties of the unit and counit $\circ \eta$ and $\circ \varepsilon$.

Let us cook up a random example. We shall take for $U$ and $F$ the familiar case of the underlying set functor on groups and the free group functor; so $\mathbf{C} = \mathbf{Set}$, $\mathbf{D} = \mathbf{Group}$. To avoid overlap with the result we proved earlier about composition of adjoints (Theorem 8.3.10), let us take for $R$ and $S$ functors which are not adjoints on either side: Let $\mathbf{B} = \vee\text{-}\mathbf{Semilattice}^0$, the category of upper semilattices with least elements $0$, i.e., with arbitrary finite joins, including the empty join, and let $R : \mathbf{Group} \to \vee\text{-}\mathbf{Semilattice}^0$ take a group $G$ to the upper semilattice of *subgroups* of $G$, and $S : \mathbf{Set} \to \vee\text{-}\mathbf{Semilattice}^0$ take a set $X$ to the upper semilattice of *equivalence relations* on $X$. We make these behave in the obvious way on morphisms: given a group homomorphism $h : G \to H$,

$R(h)$ takes each subgroup of $G$ to its image under $h$, while given a set-map
$f \colon X \to Y$, $S(f)$ takes each equivalence relation $e$ on $X$ to the equivalence
relation on $Y$ generated by $\{(f(x), f(y)) \mid (x, y) \in e\}$. A morphism from
$SU$ to $R$ thus means a way of associating to every equivalence relation on
the underlying set of a group a subgroup of that group, in a way that respects
joins (including the empty join), and respects maps induced by group homomorphisms. Though I truly chose the functors without specific examples of
such morphisms in mind, there turn out to exist several constructions with
these properties: Given an equivalence relation $E$ on the underlying set of
a group $G$, one can form (a) the subgroup of $G$ generated by the elements
$x y^{-1}$ for $(x, y) \in E$, (b) the subgroup generated by the elements $y^{-1}x$, as
well as the subgroups generated by (c) both types of elements and (d) neither
(the trivial subgroup); and each of these constructions can easily be seen to
have the properties required to be a morphism of functors.

On the other hand, a morphism from $S$ to $RF$ means a way of associating to every equivalence relation on a set $X$ a subgroup of the free group
$F(X)$, again respecting joins and morphisms. The adjointness result stated
above implies that there should be such a morphism $S \to RF$ corresponding
to each of the morphisms $SU \to R$ just listed; and indeed, these can be
described as associating to an equivalence relation $E$ on $X$ the subgroup of
$F(X)$ generated by the elements $x y^{-1}$ respectively $y^{-1}x$, respectively both,
respectively neither, for $(x, y) \in E$. To get these morphisms formally from
the morphisms (a)-(d) above, we look at any equivalence relation $E \in S(X)$,
use it and the natural map $X \to U(F(X))$ to induce an equivalence relation on $U(F(X))$, i.e., a member of $S(U(F(X)))$, then apply the chosen
morphism $SU \to R$.

You can look further into the above example in

**Exercise 8.11:1.** Let $U$, $F$, $S$ and $R$ be as in the above example. Given
any set of nonzero integers, $I \subseteq \mathbb{Z}-\{0\}$, let $m_I \colon SU \to R$ associate to each
equivalence relation $E$ on the underlying set of a group $G$ the subgroup
of $G$ generated by all the elements $x^i y^{-i}$ $((x, y) \in E, i \in I)$.

(i)    Show that the $m_I$ are morphisms of functors, and are all distinct.

(ii)    Try to determine whether these are all the morphisms $SU \to R$. Are
there any morphisms which respect finite joins (including empty joins) but
not infinite joins?

Returning to the question of why adjointness is preserved not only by
the construction $(-)^{\mathbf{E}}$ but also (with roles of right and left reversed) by
the construction $\mathbf{B}^{(-)}$, the explanation seems to be that the definition of
adjointness can be expressed as the condition that certain equations hold
among given functors and morphisms in the **Cat**-enriched structure (§7.11)
of **Cat**, namely those of Theorem 8.3.8(iii), and that these equations will
be preserved by any functor preserving **Cat**-enriched structure. And $(-)^{\mathbf{E}}$
and $\mathbf{B}^{(-)}$ both do so, one covariantly and the other contravariantly. (For
an analogous but simpler situation, observe that, although conditions on a

morphism $a$ in a category such as being an epimorphism or a monomorphism are not preserved by arbitrary functors, the conditions of left, right and two-sided invertibility are preserved, because they come down to the existence of another morphism $b$ satisfying one or both of the equations $a\,b = \mathrm{id}_X$, $b\,a = \mathrm{id}_Y$, and these conditions are preserved by functors. The formulation of adjointness in terms of unit and counit morphisms in Theorem 8.3.8(iii), is similarly "robust".)

To complicate things a bit further, consider next any two functors $P$ and $Q$ (the vertical arrows below), any adjoint pair of functors between their domain categories, and any adjoint pair of functors between their codomain categories:

(8.11.2)

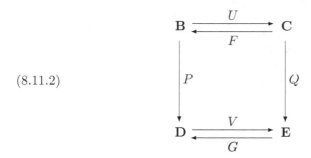

(No commutativity conditions are assumed in this diagram!) Now we may apply on the one hand our isomorphisms involving composition on the right with adjoint pairs of functors, and on the other hand our isomorphisms involving composition on the left with such pairs, getting four bijections of morphism-sets

(8.11.3)

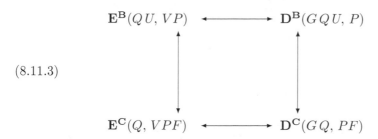

Because composition with functors on the left commutes with composition with other functors on the right, the above diagram of bijections commutes. This result is statement (iii) of the next proposition; the preceding observations of this section comprise statements (i) and (ii).

**Proposition 8.11.4.** *Suppose* $\mathbf{D} \underset{F}{\overset{U}{\rightleftarrows}} \mathbf{C}$ *are adjoint functors, with* $F$ *the left adjoint and* $U$ *the right adjoint, and with unit* $\eta\colon \mathrm{Id}_{\mathbf{C}} \to UF$ *and counit* $\varepsilon\colon FU \to \mathrm{Id}_{\mathbf{D}}$. *Then*

(i)   *For any category* $\mathbf{E}$, *the functors* $\mathbf{D}^{\mathbf{E}} \underset{F\circ}{\overset{U\circ}{\rightleftarrows}} \mathbf{C}^{\mathbf{E}}$ *are adjoint, with* $F\circ$ *the left adjoint,* $U\circ$ *the right adjoint, unit* $\eta \circ \colon \mathrm{Id}_{\mathbf{CE}} \to UF\circ$ *and counit* $\varepsilon \circ \colon FU\circ \to \mathrm{Id}_{\mathbf{DE}}$.

(ii)   *For any category* $\mathbf{B}$, *the functors* $\mathbf{B}^{\mathbf{D}} \underset{\circ U}{\overset{\circ F}{\rightleftarrows}} \mathbf{B}^{\mathbf{C}}$ *are adjoint, with* $\circ U$ *the left adjoint,* $\circ F$ *the right adjoint, unit* $\circ \eta \colon \mathrm{Id}_{\mathbf{BC}} \to \circ UF$ *and counit* $\circ \varepsilon \colon \circ FU \to \mathrm{Id}_{\mathbf{BD}}$.

(iii)   *Given two pairs of adjoint functors as in* (8.11.2), *the square of isomorphisms of bifunctors* $\mathbf{E}^{\mathbf{C}} \times \mathbf{D}^{\mathbf{B}} \to \mathbf{Set}$

$$\mathbf{E}^{\mathbf{B}}(- \circ U, V \circ -) \cong \mathbf{D}^{\mathbf{B}}(G \circ - \circ U, -)$$

(8.11.5)                  $\|\wr$                                   $\|\wr$

$$\mathbf{E}^{\mathbf{C}}(-, V \circ - \circ F) \cong \mathbf{D}^{\mathbf{C}}(G \circ -, - \circ F)$$

*commutes.*                                                                        $\square$

**Exercise 8.11:2.** Give the details of the proof of parts (i) and/or (ii) of the above proposition.

My reason for setting down the above observations is to help understand a better known result, which we can get from (8.11.3) by taking $\mathbf{B} = \mathbf{D}$, $\mathbf{C} = \mathbf{E}$, and for $P$, $Q$ the identity functors of these categories.

**Corollary 8.11.6.** *Suppose* $\mathbf{D} \underset{F}{\overset{U}{\rightleftarrows}} \mathbf{C}$ *and* $\mathbf{D} \underset{G}{\overset{V}{\rightleftarrows}} \mathbf{C}$ *are two pairs of adjoint functors between a common pair of categories* $\mathbf{C}$ *and* $\mathbf{D}$ ($F$ *and* $G$ *the left adjoints,* $U$ *and* $V$ *the right adjoints). Then there is a natural bijection* $i\colon \mathbf{D}^{\mathbf{C}}(G, F) \longleftrightarrow \mathbf{C}^{\mathbf{D}}(U, V)$ (*an instance of the diagonal bijection of* (8.11.5) *above, described explicitly below). In other words, morphisms in one direction between left adjoints correspond to morphisms in the other direction between right adjoints.*

*Description of the bijection.* Given $f \in \mathbf{D}^{\mathbf{C}}(G, F)$, one may apply $U\colon \mathbf{D} \to \mathbf{C}$ on the right to get

$$f \circ U \in \mathbf{D}^{\mathbf{D}}(GU, FU).$$

Composing with the counit morphism $\varepsilon_{U,F}\colon FU \to \mathrm{Id}_{\mathbf{D}}$ we get $(\varepsilon_{U,F})(f \circ U) \in \mathbf{D}^{\mathbf{D}}(GU, \mathrm{Id}_{\mathbf{D}})$. Finally, using the adjunction between $G$ and $V$ in a manner analogous to our above use of the adjunction between $F$ and $U$, we turn this into the desired member of $\mathbf{C}^{\mathbf{D}}(U, V)$, namely

$$i(f) = (V \circ \varepsilon_{U, F})(V \circ f \circ U)(\eta_{V, G} \circ U). \quad \square$$

As an example, let $U$ and $V$ both be the underlying set functor $\mathbf{Group} \to \mathbf{Set}$, so that $F$ and $G$ are both the free group functor $\mathbf{Set} \to \mathbf{Group}$. Then the above result says that there is a natural bijection between endomorphisms of these adjoint functors. We have already looked at endomorphisms of $U$; in the language of Exercises 3.3:6 they are the "functorial generalized group-theoretic operations in one variable", which we found were just the derived group-theoretic operations in one variable, i.e., the operations of exponentiation by arbitrary integers $n$. (Cf. also Exercises 7.9:4(ii), 8.2:10.)

As for endomorphisms of $F$, it is not hard to see that such an endomorphism is determined by the endomorphism it induces on the free group on one generator. That endomorphism will send the generator $x$ to $x^n$ for some integer $n$; conversely, we easily verify that for each $n$, an endomorphism of the whole functor $F$ with this behavior on the free group on one generator exists; hence endomorphisms of $F$ also correspond to exponentiation by arbitrary integers $n$.

In the above example, because $U = V$ and $F = G$, it is hard to see that the direction of the morphisms has been reversed. So for another example, let $\mathbf{C} = \mathbf{Group}$ and $\mathbf{D} = \mathbf{CommRing}^1$. Fix a positive integer $n$, and let $U$ be the functor taking each commutative ring with 1, $R$, to the group $\mathrm{GL}(n, R)$ of $n \times n$ invertible matrices over $R$, and $V$ the functor taking the same $R$ to its group of invertible elements (units). Clearly there is an important morphism $a : U \to V$, the map taking each invertible $n \times n$ matrix to its *determinant*. The left adjoint $F$ of $U$ takes every group $A$ to the commutative ring $F(A)$ presented by generators and relations that create a universal image of $A$ in the group of $n \times n$ invertible matrices over $F(A)$, and likewise the left adjoint $G$ of $V$ takes a group $A$ to the commutative ring $G(A)$ with a universal image of $A$ in its group of units. (The latter is easily seen to be the group ring of the abelianization of $A$.) If we look at the *determinants* of the matrices over $F(A)$ comprising the universal $n \times n$ matrix representation of $A$, we see that these give a homomorphism of $A$ into the group of units of $F(A)$, which by the universal property of $G(A)$ is equivalent to a ring homomorphism $G(A) \to F(A)$. This gives the morphism of functors $G \to F$ in $(\mathbf{CommRing}^1)^{\mathbf{Group}}$ corresponding to our determinant morphism $U \to V$ in $\mathbf{Group}^{\mathbf{CommRing}^1}$.

Mac Lane [20, p. 98, top] calls a pair of morphisms of functors related under the bijection of Corollary 8.11.6 *conjugate*.

Of course, we should have proved more about this phenomenon than we have stated in Corollary 8.11.6; in particular, that the conjugate of the composite of two morphisms between three right adjoint functors $\mathbf{D} \rightrightarrows \mathbf{C}$ is the composite of their conjugates in reversed order, i.e., that conjugation constitutes a contravariant equivalence between the category of all functors $\mathbf{D} \to \mathbf{C}$ having left adjoints and the category of all functors $\mathbf{C} \to \mathbf{D}$ which have right adjoints; and likewise that conjugacy behaves properly with respect to composition of functors. Once one verifies these statement, one can look at the situation as follows: Within the **Cat**-based category **Cat**, suppose we define the subcategory **RightAdj** to have the same objects as **Cat**, to have for morphisms those functors which are right adjoints (equivalently, have left adjoints) in **Cat**, while its morphisms-of-morphisms are again unrestricted (that is, if functors $F$, $G \colon \mathbf{C} \rightrightarrows \mathbf{D}$ both lie in **RightAdj**, we let the morphisms $F \to G$ be the same in **RightAdj** as in **Cat**). Suppose we likewise form the subcategory **LeftAdj**, as above except that the functors are those that are left adjoints in **Cat**. Then we get an equivalence of **Cat**-based categories **RightAdj** $\approx$ **LeftAdj**$^{\mathrm{op}}$. (Actually, one needs a notation to show that there is a "double $^{\mathrm{op}}$" here, applying both to composition of functors and to composition of morphisms of functors!) One might most elegantly consider a third **Cat**-category, **Adj**, isomorphic to these two and defined to have *adjoint pairs* of functors for its morphisms, and conjugate pairs of morphisms of functors for its morphisms of morphisms. For more details, see [20, pp. 97–102].

We could also have brought into the statement of Corollary 8.11.6 the upper right-hand and lower left-hand corners of (8.11.3). For instance, in the case involving groups and commutative rings discussed above, the reader can easily describe a morphism $\mathrm{Id}_{\mathbf{Group}} \to VF$, i.e., a functorial way of mapping each group $A$ into the group of units of the commutative ring with a universal $n \times n$ representation of $A$, again based on the determinant function, and a morphism $GU \to \mathrm{Id}_{\mathbf{CommRing}^1}$, i.e., a functorial way of mapping the group ring on the abelianization of the group of invertible $n \times n$ matrices over a ring $R$ into $R$, yet again based on the determinant.

Here is another twist on the ideas we have discussed.

**Exercise 8.11:3.** Suppose $\mathbf{C}$, $\mathbf{D}$, $\mathbf{E}$ are categories and $A \colon \mathbf{E} \times \mathbf{D} \to \mathbf{C}$ a functor. For each $Z \in \mathrm{Ob}(\mathbf{E})$, let us write $U_Z$ for the functor $A(Z, -) \colon \mathbf{D} \to \mathbf{C}$, and suppose each of these functors $U_Z$ has a left adjoint $F_Z$. In the spirit of §8.8, examine the relation between the condition that the system of functors $U_Z \colon \mathbf{D} \to \mathbf{C}$ indexed by $\mathbf{E}$ have a limit or colimit in $\mathbf{C}^{\mathbf{D}}$, and that the system of functors $F_Z \colon \mathbf{C} \to \mathbf{D}$ have a limit or colimit in $\mathbf{D}^{\mathbf{C}}$.

## 8.12. Contravariant adjunctions

The concept of an adjoint pair of functors is *self-dual*, in the sense that if we write down the definition of adjointness of $\mathbf{D} \xrightarrow[F]{U} \mathbf{C}$, put $\mathbf{C}^{\mathrm{op}}$ and $\mathbf{D}^{\mathrm{op}}$

in place of $\mathbf{C}$ and $\mathbf{D}$, and describe the resulting structure in language natural for our new $\mathbf{C}$ and $\mathbf{D}$, the result has the same form as the original definition, though with the roles of $\mathbf{C}$ and $\mathbf{D}$ interchanged, and hence likewise $U$ and $F$, and $\eta$ and $\varepsilon$.

But since the concept of adjunction involves more than one category, it also has "partial dualizations". For instance, if in the definition of adjunction we only replace $\mathbf{C}$ by $\mathbf{C}^{\mathrm{op}}$, we get a condition on a pair of functors $\mathbf{C}^{\mathrm{op}} \rightleftarrows \mathbf{D}$. Note that the one going to the right is a contravariant functor from $\mathbf{C}$ to $\mathbf{D}$, and the other is *equivalent to* a contravariant functor from $\mathbf{D}$ to $\mathbf{C}$, i.e., a functor $\mathbf{D}^{\mathrm{op}} \to \mathbf{C}$. Writing it in the latter form, we arrive at a setup which is symmetric, that is, in which the two categories and the two functors play equivalent roles – but which is *not* self-dual. We describe this construction and its dual in the definition below.

When we defined ordinary adjunctions, we wrote the isomorphism of bi-functors "$\mathbf{C}(-, U(-)) \cong \mathbf{D}(F(-), -)$", with the tacit understanding that the first argument "$-$" on the left matched the first argument on the right, and similarly for second arguments. But below, the first argument on one side of our isomorphism will represent the same variable as the second argument on the other side. To make this clear, I will use distinct place-holders, "$-$" and "$\sim$", for the two arguments.

**Definition 8.12.1.** *Let* $U \colon \mathbf{C}^{\mathrm{op}} \to \mathbf{D}$ *and* $V \colon \mathbf{D}^{\mathrm{op}} \to \mathbf{C}$ *be contravariant functors between categories* $\mathbf{C}$ *and* $\mathbf{D}$.

*Then a* contravariant right *adjunction between* $U$ *and* $V$ *means an iso-morphism*

$$(8.12.2) \qquad\qquad \mathbf{C}(-, V(\sim)) \cong \mathbf{D}(\sim, U(-))$$

*of bifunctors* $\mathbf{C}^{\mathrm{op}} \times \mathbf{D}^{\mathrm{op}} \to \mathbf{Set}$, *where* "$-$" *denotes the* $\mathbf{C}$-*valued argu-ment and* "$\sim$" *the* $\mathbf{D}$-*valued argument; equivalently, an adjunction between* $U \colon \mathbf{C}^{\mathrm{op}} \to \mathbf{D}$ *and the functor* $V^{\mathrm{op}} \colon \mathbf{D} \to \mathbf{C}^{\mathrm{op}}$ *corresponding to* $V$, *with* $U$ *the right and* $V^{\mathrm{op}}$ *the left adjoint; equivalently, an adjunction between* $V \colon \mathbf{D}^{\mathrm{op}} \to \mathbf{C}$ *and* $U^{\mathrm{op}} \colon \mathbf{C} \to \mathbf{D}^{\mathrm{op}}$, *with* $V$ *the right and* $U^{\mathrm{op}}$ *the left adjoint.*

*Likewise, a contravariant* left *adjunction between* $U$ *and* $V$ *means an isomorphism*

$$(8.12.3) \qquad\qquad \mathbf{C}(V(\sim), -) \cong \mathbf{D}(U(-), \sim)$$

*of bifunctors* $\mathbf{C} \times \mathbf{D} \to \mathbf{Set}$, *equivalently, an adjunction between* $V$ *(left) and* $U^{\mathrm{op}}$ *(right); equivalently, an adjunction between* $U$ *(left) and* $V^{\mathrm{op}}$ *(right).*

Of course, these two new kinds of adjointness also have descriptions cor-responding to the other ways of describing adjoint functors noted in Theo-rem 8.3.8. For instance, given $U \colon \mathbf{C}^{\mathrm{op}} \to \mathbf{D}$, to find a contravariant right adjoint to $U$ is equivalent to finding, for each object $D$ of $\mathbf{D}$, a represent-ing object for the contravariant functor $\mathbf{D}(D, U(-)) \colon \mathbf{C}^{\mathrm{op}} \to \mathbf{Set}$; in other words, an object $R_D$ of **C** with a map $D \to U(R_D)$, which is universal among objects of **C** with such maps.

As an example of a contravariant right adjunction, let **C** = **Set** and **D** = **Bool**$^1$, the category of Boolean rings, and let $U$ be the contravariant functor taking every set $S$ to the Boolean ring $\mathbf{P}(S)$ of its subsets. Then given any Boolean ring $B$, there is a universal set $S_B$ with a homomorphism $B \to \mathbf{P}(S_B)$. Namely, $S_B$ is the set of all homomorphisms $g : B \to 2$ (where 2 denotes the two-element Boolean ring), and the universal map $B \to \mathbf{P}(S_B)$ takes each $x \in B$ to the set of $g$ such that $g(x) = 1$. It is easy to verify that this is a homomorphism, and that any homomorphism $f : B \to \mathbf{P}(T)$ for any set $T$ factors through the above map, via the map $T \to S_B$ taking each $t \in T$ to the map $g_t : B \to 2$ having value 0 at those $b \in B$ with $t \notin f(b)$ and 1 at those with $t \in f(b)$. We will see further examples, and some of their interesting properties, in §10.12.

When we have a contravariant right adjunction, it is of interest to look for a natural interpretation of the common value of the two sides of (8.12.2). For instance, in the example just discussed, for $T$ a set and $B$ a Boolean ring, how can we describe the common value of $\mathbf{Set}(T, \mathbf{Bool}^1(B, 2)) \cong \mathbf{Bool}^1(B, \mathbf{P}(T))$? It is not hard to see that a member of either set can be thought of as a function $T \times |B| \to 2$ such that, when one fixes the $|B|$-coordinate and varies the $T$-coordinate, one simply has a set-map, while when one fixes the $T$-coordinate and varies the $|B|$-coordinate, one has a homomorphism of Boolean rings.

In another direction,

**Exercise 8.12:1.** Show that if $P$ and $Q$ are partially ordered sets, then a contravariant right adjunction between $P_{\mathbf{cat}}$ and $Q_{\mathbf{cat}}$ is equivalent to a *Galois connection* between $P$ and $Q$, in the generalized sense characterized in the last sentence of Exercise 6.5:2.

Contravariant *left* adjunctions rarely come up in algebra. In fact, it is shown in [49] that all such adjunctions among the kind of categories of algebras we will be studying in this course (*varieties of algebras*, to be defined in §9.4) must be very degenerate.

It may seem peculiar that we get *three* phenomena—covariant adjointness, contravariant right adjointness, and contravariant left adjointness—as the orbit of one phenomenon (the first of these) under a group of symmetries (interchanging **C** and **C**$^{\mathrm{op}}$ and interchanging **D** and **D**$^{\mathrm{op}}$) that seems to have the structure $Z_2 \times Z_2$. In fact, there is another sort of symmetry that we have implicitly used: interchanging the roles of **C** and **D**. Together with the symmetries just noted, this gives an action of $D_8$, the eight-element symmetry group of the square. The set of distinct phenomena that we find depends on our choice of what constructions to consider "essentially the same". When the set of choices we have made is formalized, the set of "distinct" constructions turns out to take the form of a double-coset space $H \backslash D_8 / K$, where

$H$ and $K$ are certain two-element subgroups of $D_8$; and that double coset space indeed has three elements [59].

Incidentally, there is yet another sort of symmetry one can consider: that given by reversing the direction (and hence order of composition) of the *functors* in our statements. In general, results of category theory are *not* preserved by this transformation, because **Cat** is not equivalent to **Cat**$^{\mathrm{op}}$. But concepts and results which are not specific to **Cat**, but can be formulated or proved for arbitrary **Cat**-based categories, may be dualized in this way. We noted in the preceding section that the concept of adjointness is meaningful in an arbitrary **Cat**-based category; hence we can apply this duality to it. It turns out to take each of the three kinds of adjointness to itself, leaving the roles of **C**, **D**, $\varepsilon$ and $\eta$ unchanged, but interchanging $U$ and $F$. Indeed, the invariance of adjointness under this symmetry is the reason for the unexpected result, Proposition 8.11.4(ii).

**Exercise 8.12:2.** Prove the claim made above that **Cat** is not equivalent to **Cat**$^{\mathrm{op}}$. (You can do this by finding an appropriate statement which holds for **Cat** but whose dual does not.)

One might ask why, if **Cat** is not equivalent to **Cat**$^{\mathrm{op}}$, the concept of **Cat**-based category should be invariant under reversing order of composition. Briefly, this is because in applying that reversal to a statement about a **Cat**-based category **X**, one does not replace **Cat** by **Cat**$^{\mathrm{op}}$ in the definition of the categories $\mathbf{X}(\mathbf{C}, \mathbf{D})$ occurring in the statement. Rather, one replaces composition maps $\mathbf{X}(\mathbf{D}, \mathbf{E}) \times \mathbf{X}(\mathbf{C}, \mathbf{D}) \to \mathbf{X}(\mathbf{C}, \mathbf{E})$ by maps in which the order of the product is reversed; in other words, one uses the fact that the product bifunctor on **Cat** *is* symmetric. (Replacing **Cat** by **Cat**$^{\mathrm{op}}$ would instead redefine composition as being given by functors $\mathbf{X}(\mathbf{C}, \mathbf{E}) \to \mathbf{X}(\mathbf{D}, \mathbf{E}) \amalg \mathbf{X}(\mathbf{C}, \mathbf{D})$.)

This is similar to the fact that though **Set** is non-self-dual, the symmetry of its product bifunctor allows us to define a functor $(-)^{\mathrm{op}} : \mathbf{Cat} \to \mathbf{Cat}$, and use this in ordinary (i.e., **Set**-based) category theory to prove the dual of any true result.

# Chapter 9
# Varieties of Algebras

We are at last ready to set up a general theory of algebras!

We recall our convention that a fixed universe $\mathbb{U}$ is assumed chosen, and that when the contrary is not stated, a "set" (or for emphasis, "small set") means a set which is a member of $\mathbb{U}$, while a "category" means a $\mathbb{U}$-legitimate category.

We will begin by formalizing some of the ideas we sketched in §§ 2.4–2.7. (The reader who was not previously familiar with them might review those sections before beginning this formal development.)

## 9.1. The category $\Omega$-Alg

In studying structures consisting of a set $|A|$ given with some operations, we will want to say that two such structures are of the *same type* if we have indexed their operations in the same way, with corresponding operations having the same arities (cf. § 2.4). Hence, below, we shall define a "type" to be an index set for the operations, with an arity associated to each of its members.

Without loss of generality one could index the operations by a cardinal, and also take the arities to be cardinals; and indeed, one or both of these assumptions is usually made. But allowing more general index sets and arities in our definition involves no complication, so let us do so.

**Definition 9.1.1.** *A type will mean a pair* $\Omega = (|\Omega|, \mathrm{ari}_\Omega)$, *where* $|\Omega|$ *is a set, and* $\mathrm{ari}_\Omega$ *(written* $\mathrm{ari}$ *when there is no danger of ambiguity) is a map from* $|\Omega|$ *to sets. The elements* $s \in |\Omega|$ *are called the* operation-symbols *of* $\Omega$, *and for each such* $s$, *the set* $\mathrm{ari}(s)$ *is called the* arity *of the operation-symbol* $s$. *(As mentioned in § 2.4, a more common notation in the literature for the arity of* $s$ *is* $n(s)$.)

© Springer International Publishing Switzerland 2015
G.M. Bergman, *An Invitation to General Algebra and Universal Constructions*, Universitext, DOI 10.1007/978-3-319-11478-1_9

$\Omega$ is called finitary if all of its operation-symbols have finite arity, i.e., if for all $s \in |\Omega|$, $\mathrm{card}(\mathrm{ari}(s)) < \omega$.

We will call a type $\Omega$ conventional if $|\Omega|$ is a cardinal, and for each $s \in |\Omega|$, $\mathrm{ari}(s)$ is a cardinal. In this situation, $\Omega$ may be expressed by giving the arity function as a tuple of cardinals, $(\mathrm{ari}(0), \mathrm{ari}(1), \dots)$.

**Definition 9.1.2.** If $\Omega$ is a type, then an algebra of type $\Omega$, or $\Omega$-algebra, will mean a pair $A = (|A|, (s_A)_{s \in |\Omega|})$, where $|A|$ is a set, and for each $s \in |\Omega|$, $s_A$ is an $\mathrm{ari}(s)$-ary operation on $|A|$, i.e., a map $|A|^{\mathrm{ari}(s)} \to |A|$.

For example, the type $\Omega$ which indexes the operations of *groups* has three operation-symbols, which we may write $\mu$, $\iota$, $\varepsilon$, with $\mathrm{ari}(\mu) = 2$, $\mathrm{ari}(\iota) = 1$, $\mathrm{ari}(\varepsilon) = 0$. Every group is an algebra of this type, but not every algebra of this type is a group, since there are algebras of this type not satisfying the associative, inverse and identity laws. If we replaced this by a "conventional type" and followed the usage that represents a type by its arity function, we would say that groups are certain algebras "of type $(2, 1, 0)$."

If $R$ is a ring, then right or left $R$-modules can be described as certain algebras of type $\Omega$, where $|\Omega| = \{+, -, 0\} \sqcup |R|$, and all these operation-symbols are unary except $+$, which is binary, and $0$, which is zeroary. Here the first three operations specify an additive group structure, while the remaining, generally infinite family give the scalar multiplications by members of $R$. To translate this type into conventional notation, one would index $|R|$ by a cardinal $\alpha$, and let $|\Omega|$ be the cardinal $3 + \alpha$; here the notational convenience of allowing more general sets for $|\Omega|$ is clear!

For an example in which it is natural to regard some operations as having for their *arities* sets other than cardinals, let $n$ be a fixed positive integer, and for every commutative ring $R$, let $d$ denote the determinant function taking $n \times n$ matrices over $R$ to elements of $R$. Suppose one wishes to construct from each commutative ring $R = (|R|, +, -, 0, \cdot, 1)$ the object $(|R|, +, -, 0, d)$, i.e., to study the set of elements of $R$ as an additive group with an $n \times n$ "determinant" operation. One would conventionally consider $d$ as $n^2$-ary, which would mean writing a typical value as $d(x_0, \dots, x_{n^2-1})$. But it is more natural to treat $d$ as an $(n \times n)$-ary operation, and write $d(x_{00}, x_{01}, \dots, x_{n-1\,n-1})$, i.e., to call the typical argument of $d$ the $(i, j)$ argument where $0 \le i, j < n$, rather than the $m$-th argument where $m = ni + j$.

If there were a significant advantage in restricting ourselves to conventional algebra-types, then we might say, "Let us use conventional types in our formal development. We can always *translate* our results into the form appropriate to a particular area of mathematics when we make our applications." But I see no advantage in such a restriction. At some points we will indeed find it convenient to restrict attention to cardinal-valued arities, but we will still put no restriction on the set of operation-symbols.

Let us note in passing the unfortunate ambiguity of the word "algebra". There is the ring-theoretic concept of "an algebra over a commutative ring", and the present much broader concept used in General Algebra. It would be desirable if a new word could be coined to replace one of them; but there is a large literature in both fields, so it would be hard to get such a change accepted. Since the literature in ring theory is the more enormous of the two, I suppose it is the General Algebra definition that would have to change.

In situations where there is a danger of misunderstanding, authors generally specify "an algebra over a commutative ring $k$" on the one hand, or "an algebra in the sense of Universal Algebra" on the other. The Russians shorten the latter phrase to "a universal algebra", which is easier to say, but somewhat inappropriate, since it suggests an object with a universal property. (The term "algebra in the sense of Universal Algebra" should now presumably be changed to "algebra in the sense of General Algebra", for the reasons mentioned in § 1.5.)

Incidentally, what is the original source of the word "algebra"? It goes back to a ninth century Arabic text, *Al-jabr w'al-muqābalah*. This title is composed of two technical terms concerning the solving of equations, whose literal meanings are something like "restoration and comparison". This title was transliterated, rather than translated, into medieval Latin, so that the book became known as *Algebra*, which eventually became the name of the subject. Not only this work but also its author, abu-Ja'far Muḥammed ibn-Mūsā, has entered mathematical language: He was known as Al-Khuwārizmi, "the person from Khuwarizm". This name was rendered as *algorism*, and, further distorted in English, has become the word *algorithm*.

Of course, we want to make the set of $\Omega$-algebras into a category, so:

**Definition 9.1.3.** *A* homomorphism *between algebras of the same type means a map of underlying sets which respects operations.*

*Precisely, if $A$ and $B$ are algebras of type $\Omega$, a homomorphism $A \to B$ means a set map $f : |A| \to |B|$ such that for each $s \in |\Omega|$ and $(x_i)_{i \in \mathrm{ari}(s)} \in |A|^{\mathrm{ari}(s)}$, one has*

$$f(s_A((x_i)_{i \in \mathrm{ari}(s)})) \;=\; s_B((f(x_i))_{i \in \mathrm{ari}(s)}).$$

*For each type $\Omega$, the category of all $\Omega$-algebras, with these homomorphisms as the morphisms, will be denoted $\Omega$-**Alg**.*

Note that when applying a set map to a tuple of elements, one generally drops one pair of parentheses, e.g., shortens $f((x_1, x_2, x_3))$ to $f(x_1, x_2, x_3)$, or $f((a_i)_{i \in I})$ to $f(a_i)$. So the above equation saying that $f$ respects $s$ can be written $f(s(x_i)) = s(f(x_i))$. If one abbreviates the ari($s$)-tuple $(x_i)$ to $x$ and uses parenthesis-free notation for functions, one can still further shorten this to $f s x = s f x$, or, distinguishing between $f$, which acts on elements of $|A|$, and the induced map which acts componentwise on ari($s$)-tuples of such elements, $f s x = s f^{\mathrm{ari}(s)} x$.

**Definition 9.1.4.** *Let $A$ be an $\Omega$-algebra.*

*Then a* subalgebra *of $A$ means an $\Omega$-algebra $B$ such that $|B| \subseteq |A|$, and such that the operations of $B$ are restrictions of the corresponding operations of $A$; equivalently, such that the inclusion map $|B| \to |A|$ is a homomorphism $B \to A$. Thus, the subalgebras of $A$ correspond to the subsets of $|A|$ closed under the operations of $A$. If $B$ is a subalgebra of $A$ we will, by a slight abuse of notation, write "$B \subseteq A$". We shall consider the set of subalgebras of $A$ to be partially ordered by inclusion (of underlying sets).*

*A* homomorphic image *of $A$ means an algebra $B$ given with a homomorphism $f : A \to B$ which is surjective on underlying sets.*

Another notational problem: If $A$ is an algebra, and if we have shown that some subset $S \subseteq |A|$ is closed under the operations of $A$, we have no simple notation for the subalgebra of $A$ whose underlying set is $S$. We shall give such algebras ad hoc names when we refer to them, though it would be tempting to fall back on the sloppy usage which does not distinguish between an algebra and its underlying set.

**Lemma 9.1.5.** *If $A$ is any $\Omega$-algebra, the set of subalgebras of $A$ is "closed under intersections"; i.e., for every set of subalgebras $B_i$ of $A$ $(i \in I)$, the intersection of the underlying sets, $\bigcap_I |B_i|$, is the underlying set of a subalgebra, which we may loosely call $\bigcap_I B_i$. Hence the subalgebras of $A$ form a complete lattice, with meets given by intersections of underlying sets.*

*If $X$ is any subset of $|A|$, the intersection of the underlying sets of all subalgebras of $A$ containing $X$ will be the underlying set of the least subalgebra containing $X$, called the subalgebra generated by $X$. We shall say that $A$ is generated by a subset $X \subseteq |A|$ if the subalgebra of $A$ generated by $X$ is all of $A$.* $\square$

As we observed in Chapter 2, a *zeroary* operation on a set is equivalent to a choice of a distinguished element of that set. Note that if $\Omega$ is a type without zeroary operation-symbols, then the empty set can be made an $\Omega$-algebra in a unique way. On the other hand, the empty set does not admit any zeroary operations, so if $\Omega$ has any operation-symbols of arity $0$, all $\Omega$-algebras are nonempty. The least element of the subalgebra lattice of an algebra $A$ of *any* type $\Omega$ will be the subalgebra generated by the empty set; this can also be described as the subalgebra generated, under the operations of *positive* arity, by the values of the *zeroary* operations. So if the type has zeroary operations, this least subalgebra is nonempty, while if it does not, it is empty.

Empty algebras sometimes constitute special cases in algebraic considerations, and many general algebraists avoid this "problem" by requiring, in their *definitions*, that all algebras have nonempty underlying sets. But the problem gets back at them: For instance, they can no longer define subalgebra lattices as above, since when an algebra has no zeroary operations, an intersection of nonempty subalgebras can be empty. Thus they make definitions such as "the subalgebra lattice of an algebra $A$ consists of all subalgebras of

$A$, and also the empty set if $A$ has no zeroary operations." I feel strongly that it is best *not* to exclude empty algebras, but to allow them when dealing with a type without zeroary operations, and accept the need to occasionally give special arguments for them, or mention them as exceptions to some statements.

Let us note that in the category $\Omega$-**Alg** we can construct products in the manner to which we have become accustomed: If $(A_i)_{i \in I}$ is a family of $\Omega$-algebras, then the set $\prod_I |A_i|$ becomes an $\Omega$-algebra $P$ under componentwise operations; that is, for each $s \in |\Omega|$, and each $\mathrm{ari}(s)$-tuple of elements of $\prod_I |A_i|$, say

$$(a_j)_{j \in \mathrm{ari}(s)} \; = \; ((a_{ij})_{i \in I})_{j \in \mathrm{ari}(s)} \in |P|^{\mathrm{ari}(s)} \; = \; (\textstyle\prod_I |A_i|)^{\mathrm{ari}(s)},$$

we define

$$s_P(a_j) \; = \; (s_{A_i}((a_{ij})_{j \in \mathrm{ari}(s)}))_{i \in I}.$$

The resulting algebra $P$ is easily seen to have the universal property of the product $\prod_I A_i$ in $\Omega$-**Alg**. Products in $\Omega$-**Alg** are often called by the traditional name, *direct products*.

Similarly, given a pair of homomorphisms of $\Omega$-algebras $f, g \colon A \to B$, their equalizer as set maps will be the underlying set of a subalgebra of $A$, and that subalgebra will constitute an equalizer of $f$ and $g$ in $\Omega$-**Alg**.

Since general *limits* can be constructed from products and equalizers (Proposition 8.6.6), we have

**Proposition 9.1.6.** *Let $\Omega$ be any type. Then the category $\Omega$-**Alg** has small limits, which can be constructed by taking the limits of the underlying sets and making them $\Omega$-algebras under pointwise operations.*

*Explicitly, if $\mathbf{D}$ is a small category and $F \colon \mathbf{D} \to \Omega$-**Alg** a functor, then the set*

$$\varprojlim_{\mathbf{D}} |F(D)| \; = \; \{(a_D) \in \textstyle\prod_{D \in \mathrm{Ob}(\mathbf{D})} |F(D)| \; | \\ (\forall\, D_1, D_2 \in \mathrm{Ob}(\mathbf{D}), \, \forall\, f \in \mathbf{D}(D_1, D_2))\; a_{D_2} = F(f)(a_{D_1})\}$$

*is the underlying set of a subalgebra of $\prod_{\mathbf{D}} F(D)$, which constitutes a limit of $F$ in $\Omega$-**Alg**.* $\qquad\square$

**Exercise 9.1:1.** Show that if empty algebras are excluded from $\Omega$-**Alg**, the resulting category can fail to have small limits.

On the other hand, *colimits* and other *left-universal* constructions are not, in general, the same in $\Omega$-**Alg** as in **Set**. We will construct general colimits in § 9.3; but there are two cases that we can obtain now. We first need to note

**Lemma 9.1.7.** *Let $A$ be an $\Omega$-algebra and $E \subseteq |A| \times |A|$ an equivalence relation on $|A|$. Then the following conditions are equivalent:*

(a) *The set* $|A|/E$ *can be made the underlying set of an* $\Omega$-*algebra* $A/E$ *in such a way that the canonical map* $|A| \to |A|/E$ *is a homomorphism* $A \to A/E$.

(b) $E$ *is the equivalence relation on* $|A|$ *induced by a homomorphism of* $\Omega$-*algebras with domain* $A$. (*I.e., there exists an* $\Omega$-*algebra* $B$ *and a homo-morphism* $f : A \to B$ *such that* $E = \{(x, y) \in |A| \times |A| \mid f(x) = f(y)\}$.)

(c) $E$ *is the underlying set of a subalgebra of* $A \times A$.

*Further, if* $R$ *is any subset of* $|A| \times |A|$, *and* $E$ *the intersection of all underlying sets of subalgebras of* $A \times A$ *which contain* $R$, *and which form equivalence relations on* $|A|$, *then* $E$ *will again be the underlying subset of a subalgebra, and will form an equivalence relation; and* $A/E$ *will be universal (initial) among algebras* $B$ *given with homomorphisms* $f : A \to B$ *such that for all* $(r, s) \in R$, $f(r) = f(s)$. $\square$

**Definition 9.1.8.** *If* $A$ *is an* $\Omega$-*algebra, then an equivalence relation* $E$ *on* $|A|$ *which is the underlying set of a subalgebra of* $A \times A$ (*as in condition* (c) *of the above lemma*) *will be called a* congruence *on the algebra* $A$, *and* $A/E$ (*defined as in condition* (a) *thereof*) *will be called the* quotient algebra (*or factor-algebra*) *of* $A$ *by the congruence* $E$.

*The complete lattice of all congruences on* $A$ *is called the* congruence lattice *of* $A$. *The least congruence containing a given subset* $R \subseteq |A| \times |A|$ *is called the congruence on* $A$ generated *by* $R$, *and the quotient of* $A$ *by this congruence is often called the algebra obtained by* imposing *on* $A$ *the family of relations* $R$, *or loosely, the family of relations* $(x = y)_{(x, y) \in R}$.

I say "loosely" in the last sentence because (as we noted in passing in §4.3), there is an abuse of notation in writing such a relation as "$x = y$". The symbol $x = y$ usually denotes a proposition, i.e., an assertion about elements of $A$, and this proposition is generally false in the case where the relation is one we wish to *impose* on $A$! What is true is that in our quotient algebra the *images* of $x$ and $y$ satisfy the corresponding relation; and when there is no danger of ambiguity, one may denote these images by the same symbols $x$ and $y$ as the original elements of $A$, so that $x = y$ becomes a true statement in that quotient algebra. But in more precise notation, the statement which is true in the latter algebra must be written using modified symbols, e.g., $\bar{x} = \bar{y}$ or $[x] = [y]$. We will be precise about this here; but in informal algebraic use, the language of "imposing the relation $x = y$ on $A$" is very convenient.

Many workers in General Algebra and Logic make a convention half-way between these extremes, defining "relations" or "identities" to be symbols of the form "$x \approx y$". (E.g., [21, p. 234].) These are essentially just our ordered pairs $(x, y)$, written in a more suggestive form. A notation that allows one to avoid ambiguity while using the same symbols for elements of different algebras is that of Model Theory, where one writes $A/E \models x = y$ to mean "$x = y$ holds in $A/E$", so that this is distinguishable from $A \models x = y$.

**Exercise 9.1:2.**

(i)   Show that in the context of Lemma 9.1.7, if the type $\Omega$ is finitary, then the three equivalent conditions of that lemma are also equivalent to the condition

(c′)  For every $s \in |\Omega|$ and every ari($s$)-tuple
$$((a_0, b_0), \ldots, (a_{\mathrm{ari}(s)-1}, b_{\mathrm{ari}(s)-1})) \in E^{\mathrm{ari}(s)} \subseteq (A \times A)^{\mathrm{ari}(s)}$$
such that $a_i \neq b_i$ for *at most one* $i \in \mathrm{ari}(s)$, one has
$$s_A((a_0, b_0), \ldots, (a_{\mathrm{ari}(s)-1}, b_{\mathrm{ari}(s)-1})) \in E.$$

(ii)  Show that the analog of (i) is not in general true if the type $\Omega$ is not finitary.

Using the quotient algebra construction, we immediately get

**Lemma 9.1.9.** *For any type $\Omega$, the category $\Omega$-**Alg** has coequalizers. Namely, the coequalizer of a pair of maps $f, g : A \rightrightarrows B$ may be constructed as $(B/E, q)$, where $E$ is the congruence on $B$ generated by $\{(f(x), g(x)) \mid x \in |A|\}$, and $q \colon B \to B/E$ is the canonical map.*   □

The other left universal construction that we can get easily is that of direct limit, assuming appropriate restrictions on the arities of our operations.

**Lemma 9.1.10.** *If $\Omega$ is a finitary type (Definition 9.1.1) then $\Omega$-**Alg** has direct limits, i.e., colimits over directed partially ordered sets. Namely, suppose $J$ is a directed partially ordered set and $A \colon J_{\mathbf{cat}} \to \Omega$-**Alg** a functor, whose values at objects and morphisms of $J_{\mathbf{cat}}$ will be written $A_j$ $(j \in J)$ and $A(j, j')$ $(j \leq j' \in J)$ respectively. Then the $\Omega$-algebra structures of the algebras $A_j$ induce an $\Omega$-algebra structure on the set-theoretic direct limit $\varinjlim_J |A_j|$ which makes it a direct limit algebra, $\varinjlim_J A_j$, with the same coprojection maps as for the set-theoretic direct limit.*

*More generally, if $\alpha$ is an infinite cardinal, and $\Omega$ a type in which all arities have cardinality $< \alpha$, then the category $\Omega$-**Alg** has direct limits over all $<\alpha$-directed partially ordered sets (Definition 8.9.10), again constructed by giving an $\Omega$-algebra structure to the direct limit of the underlying sets.*

*Proof.* We will prove the general case. Let $|L| = \varinjlim_J |A_j|$, and let $q_j \colon |A_j| \to |L|$ $(j \in J)$ be the coprojection maps. We wish to define an $\Omega$-algebra structure on $|L|$. Given $s \in |\Omega|$ and an ari($s$)-tuple $(x_i)_{i \in \mathrm{ari}(s)}$ of elements of $|L|$, let us write each $x_i$ as $q_{j(i)}(y_i)$ for some $j(i) \in J$ and $y_i \in |A_{j(i)}|$. Because $J$ is $<\alpha$-directed and $\mathrm{card}(\mathrm{ari}(s)) < \alpha$, we can choose $j \in J$ majorizing all the $j(i)$. Taking such a $j$, and letting $z_i = A(j(i), j)(y_i) \in A(j)$ for each $i$, we have

(9.1.11)                    $x_i = q_j(z_i)$    for all $i \in \mathrm{ari}(s)$.

To define $s_L$, let us say that whenever we have a family $(x_i) \in |L|^{\mathrm{ari}(s)}$ expressed as in (9.1.11) for some $j \in J$, we will let

$$s_L(x_i) \; = \; q_j(s_{A_j}(z_i)) \in |L|.$$

The verification that these operations $s_L$ are well-defined, and that the resulting $\Omega$-algebra $L$ has the universal property of $\varinjlim A$, are straightforward, again by the method of "going far enough out along the $<\alpha$-directed set $J$".                                                                                    □

**Exercise 9.1:3.** Write out these final verifications.

As noted at the beginning of § 8.9, the "reason" the above lemma holds is that in **Set**, direct limits respect finite products (a case of Proposition 8.9.3) and more generally, that direct limits over $<\alpha$-directed partially ordered sets respect $\alpha$-fold products (Proposition 8.9.11). Similarly, Proposition 9.1.6 holds because arbitrary products in **Set** (indeed, in any category) respect arbitrary limits (Theorem 8.8.9).

We shall prove in § 9.3 that $\Omega$-**Alg** has general colimits, so the arity-restrictions of the above lemma are not needed for the existence statements to hold. But they are needed for the direct limits in question to have the description given. Indeed

**Exercise 9.1:4.** Show by example that the last sentence of the first paragraph of Lemma 9.1.10 fails if the assumption that $\Omega$ is finitary is dropped. Specifically, show that there may not exist an algebra with underlying set the direct limit of the $|A_j|$, and having the universal property of $\varinjlim A_i$.

**Exercise 9.1:5.** Let $M$ be a monoid. As mentioned in Exercise 8.9:6, a *left congruence* on $M$ means an equivalence relation $\sim$ on $|M|$ such that for all $a, b, c \in M$ one has $a \sim b \implies ca \sim cb$. (We will not call on that exercise here, though part (i) below asks you to prove a slight strengthening of a result referred to there as "easy to verify".)

(i)   Show that a binary relation $\sim$ on $|M|$ is a left congruence if and only if there exists a left action of $M$ on a set $X$, and an element $x \in X$, such that $\sim$ is $\{(a, b) \in |M| \times |M| \mid a\,x = b\,x\}$.

(ii)   If $G$ is a group and $M = G_{\mathrm{md}}$ (the monoid obtained by "forgetting" the inverse operation of $G$, as in § 4.11), show that the left congruences on $M$ are in natural bijective correspondence with the subgroups of $G$. (Not, as might seem more natural, with the submonoids of $M$.)

(iii)   (Open question of Hotzel [89].) If a monoid $M$ has ascending chain condition on left congruences, must $M$ be finitely generated?

It has been shown [103] that if $M$ has ascending chain condition on both left and right congruences, then it is indeed finitely generated. (Students wishing to read that paper might want to have the standard reference work [68] in hand, since semigroup theorists use some rather idiosyncratic terminology.)

Incidentally, the converse to the statement asked for in (iii) is not true. This can be seen from (ii), and the fact that a finitely generated group need not have ascending chain condition on subgroups.

**Exercise 9.1:6.** Let us call an $\Omega$-algebra "just infinite" if it is infinite, but every proper homomorphic image (i.e., every image under a non-one-to-one homomorphism) is finite.

(i)   Show that if $\Omega$ has only finitely many operations, and all are finitary, then every infinite finitely generated $\Omega$-algebra has a just-infinite homomorphic image.

(ii)   Can any of the above hypotheses be dropped? Or can you find any variant results in which one or more occurrences of "finite" and/or "infinite" are replaced by "$< \alpha$" and/or "$\geq \alpha$" for a more general choice of cardinal $\alpha$?

**Exercise 9.1:7.**

(i)   Suppose $\Omega$ is a type having only finitely many operations, all of finite arities, and let $A$, $B$ be finitely generated $\Omega$-algebras. Must $A \times B$ be finitely generated?

(ii)   If the answer to part (i) is yes, can the assumptions on $\Omega$ be dropped or weakened? If, on the other hand, the answer is no, can you find additional assumptions on $\Omega$ and/or $A$ and $B$ that will make the assertion true?

## 9.2. Generating algebras from below

We want to construct other left universal objects in $\Omega$-**Alg**—free algebras, coproducts, arbitrary small colimits, etc. In general, these will contain elements created by applying operations of $\Omega$ to tuples of the elements we start with, further elements obtained by applying the operations to elements we get in this way, and so on. Whatever methods we use to justify these constructions must involve showing that this iteration process "eventually ends".

"Eventually" does not mean in a finite number of steps, of course—even in constructing algebras with operations of finite arity such as groups, we needed countably many iterations to get the full set of such elements. When we have infinitary operations, we may have to continue longer than that.

To see how long, let us examine the process by which a subset of an algebra generates a subalgebra. Let $\Omega$ be an arbitrary type, and $A$ an $\Omega$-algebra. Given a subset $X \subseteq |A|$, define a sequence of subsets of $|A|$ indexed by the ordinals,

$$S^{(0)} = X,$$

(9.2.1)
$$S^{(\alpha+1)} = S^{(\alpha)} \cup \{s_A(x_i) \mid s \in |\Omega| \text{ and } x_i \in S^{(\alpha)}$$
$$\text{for all } i \in \mathrm{ari}(s)\},$$

$$S^{(\alpha)} = \bigcup_{\beta < \alpha} S^{(\beta)} \text{ if } \alpha \text{ is a limit ordinal } > 0.$$

We see by induction that the $S^{(\alpha)}$'s increase monotonically. Since $|A|$ is a small set, $S^{(\alpha)}$ and $S^{(\alpha+1)}$ cannot be distinct for all cardinals $\alpha$, and clearly as soon as one such pair is equal, the chain becomes constant. The constant value $S$ that it assumes will contain $S^{(0)} = X$ and be closed under the operations $s_A$; moreover, by induction on $\alpha$, each $S^{(\alpha)}$, and so in particular, $S$, is contained in every subalgebra of $A$ containing $X$. Hence $S$ is the underlying set of the least subalgebra of $A$ containing $X$, i.e., the subalgebra generated by $X$.

We now want to bound in terms of properties of $\Omega$ the least value of $\alpha$ for which $S^{(\alpha)} = S$. (Above, we implicitly bounded it in terms of card $|A|$.)

We know how to show that if $\Omega$ is finitary, $S = S^{(\omega)}$. Namely, given finitely many elements $s_0, \ldots, s_{n-1} \in S^{(\omega)}$, all of the $s_i$ will have been reached by some finite step $S^{(N)}$, hence the value of any operation of $A$ on this family lies in $S^{(N+1)}$, and hence in $S^{(\omega)}$; so $S^{(\omega+1)} = S^{(\omega)}$,

The next case is that of a type $\Omega$ in which all operations have arities of cardinality $\leq \omega$, equivalently, $< \omega_1$. (Recall that $\omega_1$ denotes the first uncountable ordinal.) It is then no longer true that the above process converges by the $\omega$-th step: If $s \in |\Omega|$ is $\omega$-ary, and we can find for each nonnegative $n$ an element $x_n$ which first appears in $S^{(n)}$, then $S^{(\omega)}$ will not in general contain $s_A(x_0, x_1, \ldots, x_n, \ldots)$. Rather, this will appear in $S^{(\omega+1)}$, and further elements obtained from *it* under the operations of $A$ will in general appear at still later steps. However, I claim that this process stabilizes by the $\omega_1$ th step. Indeed, given a countable (possibly finite) family of elements $x_i \in S^{(\omega_1)}$, each occurs in some $S^{(\alpha_i)}$ for a countable ordinal $\alpha_i \in \omega_1$, hence all the $x_i$ will occur in $S^{(\alpha)}$ where $\alpha = \sup(\alpha_i)$, and this ordinal $\alpha$ is still $< \omega_1$, since $\sup(\alpha_i)$ is $\leq$ the ordinal sum of the $\alpha_i$ (defined as in (5.5.11)), which has cardinality equal to the cardinal sum of the $\mathrm{card}(\alpha_i)$, which is a countable sum of countable cardinals, hence countable. For $\alpha$ so chosen, the value at $(x_0, x_1, \ldots, x_n, \ldots)$ of any operation of countable arity lies in $S^{(\alpha+1)} \subseteq S^{(\omega_1)}$, showing that $S^{(\omega_1)}$ is closed under the operations of $A$, and hence that (9.2.1) stabilizes by the $\omega_1$ th step. The next exercise has you show that in this statement, we cannot replace the estimate $\omega_1$ by any smaller ordinal (such as $\omega^2$ or $\omega^\omega$).

**Exercise 9.2:1.** Let $\gamma$ be any uncountable ordinal, and let $A$ be an algebra with underlying set $\gamma$ and three operations: the zeroary operation taking the value $0 \in \gamma$, the unary operation taking $\alpha \in \gamma$ to $\alpha+1$ if $\alpha+1 < \gamma$, or to $0$ if $\alpha+1 = \gamma$, and the $\omega$-ary operation taking $(\alpha_0, \alpha_1, \ldots)$ to $\bigcup \alpha_i$ if this is $< \gamma$, to $0$ otherwise. Taking $X = \emptyset \subseteq |A|$, determine explicitly the sequence of subsets $S^{(\alpha)}$, and show that this sequence does not become constant until $S^{(\omega_1)}$.

The same argument will show that if all members of $|\Omega|$ have arity $\leq \omega_1$, then we get our desired algebra as $S^{(\omega_2)}$, that if all arities are $\leq \omega_2$, we get it as $S^{(\omega_3)}$, etc.; and it might appear that the proper general statement is that if $\alpha$ is any infinite ordinal of cardinality *greater* than the arities of all members of $|\Omega|$, then $S^{(\alpha)}$ is closed under the operations of $\Omega$.

But this is not quite correct. The first value of $\alpha$ for which it fails is $\omega_\omega$. If $A$ has operations of arities $\omega$, $\omega_1$, $\omega_2$, etc. (all the infinite cardinals $< \omega_\omega$), then the chain of subalgebras $S^{(\omega)} \subseteq S^{(\omega_1)} \subseteq S^{(\omega_2)} \subseteq \ldots$ may be strictly increasing. If we now choose an element $x_i \in S^{(\omega_{i+1})} - S^{(\omega_i)}$ for each $i$, we get a countable family of elements of $S^{(\omega_\omega)}$, and the value of an operation of (merely!) countable arity on this family cannot be expected to lie in $S^{(\omega_\omega)}$.

**Exercise 9.2:2.** Construct an explicit example with the properties sketched above, i.e., an algebra $A$ all of whose operations have arities $< \omega_\omega$, and a subset $X \subseteq |A|$, such that the chain of subsets $S^{(\alpha)}$ does not reach its maximum value till $S^{(\omega_\omega+1)}$. (Suggestion: Adapt the idea of the preceding exercise.)

To state the right choice of $\alpha$, we recall from Definition 5.5.18 that an infinite cardinal $\alpha$ is called *regular* if, as a partially ordered set, $\alpha$ has no cofinal subset of cardinality $< \alpha$, and that a cardinal that is not regular is called *singular*. What we have run into is the first singular infinite cardinal, $\omega_\omega$. Fortunately, *regular* cardinals are quite abundant: as shown in Exercise 5.5:14, the cardinal $\omega$ is regular, and every infinite *successor* cardinal, i.e., every cardinal of the form $\omega_{\alpha+1}$ for $\alpha$ an ordinal, is also regular. We can now show

**Lemma 9.2.2.** *Let $\Omega$ be a type, and $\gamma$ a regular infinite cardinal such that $\mathrm{card}(\mathrm{ari}(s)) < \gamma$ for all $s \in |\Omega|$ (e.g., the least such regular cardinal). Then for any $\Omega$-algebra $A$, and any subset $X \subseteq |A|$, if we define the chain of sets $S^{(\alpha)}$ by (9.2.1), then $S^{(\gamma)}$ is closed under the operations of $A$, hence is the underlying set of the subalgebra of $A$ generated by $X$.*

*Proof.* Consider any $s \in |\Omega|$ and elements $x_i \in S^{(\gamma)}$ ($i \in \mathrm{ari}(s)$). Since $\gamma$ is a limit ordinal, $S^{(\gamma)} = \bigcup_{\beta < \gamma} S^{(\beta)}$, hence each $x_i$ lies in some $S^{(\beta_i)}$ ($\beta_i \in \gamma$). Since $\mathrm{card}(\mathrm{ari}(s)) < \gamma$ and $\gamma$ is regular, the set $\{\beta_i \mid i \in \mathrm{ari}(s)\}$ is not cofinal in $\gamma$, hence that set is majorized by some $\beta < \gamma$. For this choice of $\beta$, all $x_i$ lie in $S^{(\beta)}$, hence $s(x_i) \in S^{(\beta+1)} \subseteq S^{(\gamma)}$, as required.                □

In the next section we will apply the above result to the construction of left universal objects.

For later use, we record the following generalization of the familiar observation that if an algebra with finitary operations is generated by a set $X$, each element of the algebra can be expressed in terms of finitely many elements of $X$.

**Lemma 9.2.3.** *Let $\Omega$ be a type, and $\gamma$ a regular infinite cardinal such that* card(ari($s$)) $< \gamma$ *for all* $s \in |\Omega|$. *Let $A$ be any $\Omega$-algebra, and $X$ any generating set for $A$. Then each element of $|A|$ belongs to the subalgebra of $A$ generated by a subset $X_0 \subseteq X$ of cardinality $< \gamma$.*

*Sketch of proof.* It is easy to verify that under the given hypothesis the set of elements of $|A|$ belonging to subalgebras generated by $< \gamma$ elements of $X$ forms a subalgebra. As it contains $X$, it must be all of $|A|$.                □

**Exercise 9.2:3.** Write out the easy verification referred to. Show that the result becomes false if the regularity assumption on $\gamma$ is deleted.

It may now seem anomalous that in our results on direct limits over $<\alpha$-directed partially ordered sets, Proposition 8.9.11 and Lemma 9.1.10, we did *not* have to assume $\alpha$ regular! This is explained by

**Exercise 9.2:4.** Show that if $\alpha$ is a singular infinite cardinal and $J$ a $<\alpha$-directed partially ordered set, then $J$ is also $<\alpha^+$-directed, where $\alpha^+$ is the successor cardinal to $\alpha$.

   Thus, if $J$ is $<\alpha$-directed for a cardinal $\alpha$ greater than the arities of all operations of $\Omega$, it is in fact $<\alpha'$-directed for a *regular* cardinal $\alpha'$ greater than the arities of those operations.

We could have avoided using the concept of regular cardinal in this section by taking $\gamma$ in our results to be "the successor cardinal of the least infinite upper bound of the arities of the operation-symbols of $\Omega$". However, in the case where $\Omega$ is finitary, this would have given $\gamma = \omega_1$, whereas the development we have used shows that $\omega$ suffices in that important case.

**Exercise 9.2:5.**

(i)   Let $\Omega$ be a finitary type, and $A$ an $\Omega$-algebra. Show that a subalgebra of $A$ is finitely generated if and only if it is *compact* as an element of the lattice of subalgebras of $A$. (Cf. Lemma 6.3.6.)

(ii)   Deduce that a congruence on $A$ is finitely generated if and only if it is a compact element of the lattice of congruences.

(iii)   Deduce, in turn, that the subalgebra lattice, respectively the congruence lattice, has ascending chain condition if and only if every subalgebra of $A$, respectively every congruence on $A$, is finitely generated.

(iv)   Show that in each of the preceding "if and only if" statements, one direction can fail if $\Omega$ is not finitary, but the other will continue to hold.

## 9.3. Terms and left universal constructions

Given a type $\Omega$ and a set $X$, Lemma 9.2.2 can be used to obtain a bound on the size of an $\Omega$-algebra generated by an $X$-tuple of elements, and hence to establish the *solution set* hypotheses needed by the existence results for left universal constructions developed in § 8.10. Now such a bound can be thought of as an estimate of the number of "$\Omega$-algebra terms in an $X$-tuple of variable-symbols", and rather than just giving the existence proof suggested above, we can, with little additional work, construct such a set of terms, thus laying the groundwork for the more explicit approach to universal constructions that was sketched in § 3.2.

Let us first define precisely the concept of a "term". At the beginning of this course (Definition 2.5.1) we described the set of "group-theoretic terms in the elements of $X$" as a set $T$ given with certain structure: a map of $X$ into it, and a family of "formal group-theoretic operations" satisfying some further conditions. If we make the corresponding definition for $\Omega$-algebras, we see that the "formal operations" in fact make the set $T$ into an $\Omega$-algebra. (We could not similarly say that formal operations made the set of group-theoretic terms into a group, because they did not satisfy the group identities. But in the present development, we are studying algebras of type $\Omega$ in general, before introducing identities.) So we state the definition accordingly:

**Definition 9.3.1.** *Let $\Omega$ be any type, and $X$ any set. Then an "$\Omega$-term algebra on $X$" will mean a pair $(F, u)$, where $F$ is an $\Omega$-algebra, and $u \colon X \to |F|$ a set map, such that*

(i)   *the map $u \colon X \to |F|$, and all the maps $s_F \colon |F|^{\mathrm{ari}(s)} \to |F|$ are one-to-one,*

(ii)   *the images in $|F|$ of the above maps are disjoint,*

(iii)   *$F$ is generated as an $\Omega$-algebra by $u(X)$.*

Note that the first two parts of the above definition can be stated as a single condition: If we write $\sqcup$ for disjoint union of sets, and consider the map $u$ and the operations $s_F$ as defining a single map $X \sqcup \bigsqcup_{s \in |\Omega|} |F|^{\mathrm{ari}(s)} \to |F|$, then (i) and (ii) say that this map is one-to-one.

Since the concept of $\Omega$-algebra involves no identities, the idea of constructing free objects by taking "terms modulo identities" simplifies in this case to

**Lemma 9.3.2.** *Let $\Omega$ be any type, and $X$ any set. Suppose there exists an $\Omega$-term algebra $(F, u)$ on $X$. Then $(F, u)$ is a free $\Omega$-algebra on $X$.*

*Proof.* To prove that $(F, u)$ has the universal property of a free $\Omega$-algebra on $X$, suppose $A$ is an $\Omega$-algebra and $v \colon X \to |A|$ any set map. We wish to construct a homomorphism $f \colon F \to A$ such that $v = f\, u$. Intuitively $f$

should represent "substitution of the elements $v(x)$ for the variable-symbols $u(x)$ in our terms".

Let us write $|F|$ as the union of a chain of subsets $S^{(\alpha)}$ as in (9.2.1), starting with the generating set $S^{(0)} = u(X)$. Assume recursively that $f$ has been defined on all the sets $S^{(\beta)}$ with $\beta < \alpha$; we wish to extend $f$ to $S^{(\alpha)}$. If $\alpha = 0$, $S^{(\alpha)}$ consists of elements $u(x)$ $(x \in X)$, all distinct, and we let $f(u(x)) = v(x) \in |A|$. If $\alpha$ is a successor ordinal $\beta + 1$, then an element which first appears in $S^{(\alpha)}$ will have the form $s_F(t_i)$, where $s \in |\Omega|$ and each $t_i \in |S^{(\beta)}|$. Thus the $f(t_i)$ have already been defined, and we define $f(s_F(t_i)) = s_A(f(t_i))$. If $\alpha$ is a nonzero limit ordinal, then $S^{(\alpha)} = \bigcup_{\beta<\alpha} S^{(\beta)}$, and having defined $f$ consistently on $S^{(\beta)}$ for all $\beta < \alpha$, we have defined it on $S^{(\alpha)}$.

In each case, the one-one-ness condition (i) and the disjointness condition (ii) of Definition 9.3.1 insure that if an element of $F$ occurs at some stage as $u(x)$ or $s_F(t_i)$, it cannot occur (at the same or another stage) in a different way as $u(x')$ or $s'_F(t'_i)$. Hence our definition of $f$ is unambiguous. By construction, $f$ is a homomorphism of $\Omega$-algebras and satisfies $f\,u = v$; and by (iii) it is unique for this property. $\qquad\square$

We have not proved the converse statement, that if a free $\Omega$-algebra on $X$ exists, it will be an $\Omega$-term algebra on $X$. We would want this if we planned to prove the existence of free algebras first and deduce from this the existence of term algebras, but we shall be going the other way. However, this implication is not hard to prove; I will make it

**Exercise 9.3:1.** Show (without assuming the existence of $\Omega$-term algebras) that if $(F, u)$ is a free $\Omega$-algebra on $X$, then it is an $\Omega$-term algebra on $X$.

(Hint: If $F$ fails to satisfy one of conditions (i)–(iii) of Definition 9.3.1, you want to find a pair $(A, v)$ for which the universal property of $(F, u)$ fails. If condition (iii) fails, make $A$ a subalgebra of $F$; if (i) or (ii) fails, obtain $A$ by replacing one element $p$ of $F$ by two elements $p_1$ and $p_2$, and defining the operations appropriately on $|F| - \{p\} \cup \{p_1, p_2\}$. Since the operations of $\Omega$-algebras are not required to satisfy any identities, any definition of these operations yields an $\Omega$-algebra.)

Let us now prove

**Theorem 9.3.3.** *Let $\Omega$ be any type, and $X$ any set. Then there exists an $\Omega$-term algebra on $X$; equivalently, a free $\Omega$-algebra on $X$.*

*Proof.* Let $*$ be any element not in $|\Omega|$, and $\gamma$ an infinite regular cardinal which is $> \mathrm{card}(\mathrm{ari}(s))$ for all $s \in |\Omega|$. We define recursively a chain $(S^{(\alpha)})_{\alpha\le\gamma}$ of sets of ordered pairs, by taking

$$S^{(0)} = \{(*, x) \mid x \in X\},$$

$$S^{(\alpha+1)} = S^{(\alpha)} \cup \{(s, (x_i)) \mid s \in |\Omega|,\ (x_i) \in (S^{(\alpha)})^{\mathrm{ari}(s)}\},$$

$$S^{(\alpha)} = \bigcup_{\beta < \alpha} S^{(\beta)} \text{ if } \alpha \text{ is a limit ordinal with } 0 < \alpha \leq \gamma.$$

Let $|F| = S^{(\gamma)}$, and define $u \colon X \to |F|$, and maps $s_F \colon |F|^{\mathrm{ari}(s)} \to |F|$ ($s \in |\Omega|$), by

$$u(x) = (*, x) \quad (x \in X),$$

$$s_F(x_i) = (s, (x_i)) \quad (s \in |\Omega|, \ (x_i) \in |F|^{\mathrm{ari}(s)}).$$

That the operations $s_F$ carry $|F| = S^{(\gamma)}$ into itself follows from our choice of $\gamma$, by the same argument we used in proving Lemma 9.2.2. Thus these operations make $|F|$ an $\Omega$-algebra $F$. That $F$ satisfies conditions (i) and (ii) of Definition 9.3.1 follows from the set-theoretic fact that an ordered tuple uniquely determines its components. To get (iii), one verifies by induction that any subalgebra containing $X$ must contain each $S^{(\alpha)}$. $\qquad\Box$

I should mention that the technique of explicit induction or recursion on the forms of elements, which we have used in proving Lemma 9.3.2 and Theorem 9.3.3, will hardly ever have to be used after this point. Beginners often assume that they need to use such methods when proving results to the effect that if an algebra $A$ is generated by a set $X$ of elements having an appropriate sort of property $P$, then all elements of $A$ satisfy $P$ (e.g., Exercise 9.2:3 above). But in fact, such results can generally be obtained more simply by verifying that the set of elements of $A$ satisfying $P$ is closed under the algebra operations, hence forms a subalgebra containing $X$, hence is all of $|A|$. Likewise, if we want to construct a homomorphism on the free algebra $A$ on a set $X$ given its values on elements of $X$, we can do this using the universal property of $A$ as a free object. In the case of free objects of $\Omega$-**Alg**, we have just proved by *recursion on elements* that a certain object has that universal property, but this one application of that method will free us from having to repeat the same argument in similar situations.

Since we have free $\Omega$-algebras on all sets $X$, these give a left adjoint to the underlying-set functor from $\Omega$-**Alg** to **Set**.

**Exercise 9.3:2.** Show how we could, alternatively, have gotten the existence of such an adjoint using Freyd's Adjoint Functor Theorem (Theorem 8.10.5) and Lemma 9.2.2.

Let us fix a notation for these functors.

**Definition 9.3.4.** *The underlying-set functor of $\Omega$-**Alg** and its left adjoint, the free algebra functor, will be denoted $U_\Omega \colon \Omega$-**Alg** $\to$ **Set** and $F_\Omega \colon$ **Set** $\to$ $\Omega$-**Alg** respectively.*

*When there is no danger of misunderstanding, we may abbreviate a symbol such as $F_\Omega(\{x_0, \ldots, x_{n-1}\})$ to $F_\Omega(x_0, \ldots, x_{n-1})$.*

The "danger of misunderstanding" referred to is that the symbol $F_\Omega(X)$ for the free $\Omega$-algebra on a set $X$ might be misinterpreted, under the above convention, as meaning the one-generator free algebra $F_\Omega(\{X\})$. But in context, there is almost never any doubt as to whether a given entity is meant to be treated as a free generator, or as a set of free generators.

There is another sort of looseness in our usage, which we noted in Chapter 3. Although we have formally defined free algebras to be pairs $(F, u)$, we also sometimes use the term for the first components of such pairs, thought of as algebras "given with" the set-maps $u$. (E.g., when we spoke of the free-algebra functor above, the values of the functor were algebras $F$, not ordered pairs $(F, u)$; the maps $u$ are the values of the unit of the adjunction, $u = \eta(X)\colon X \to U_\Omega(F_\Omega(X))$.) At other times, we speak of an algebra $F$ as being free on a given set of its elements, without specifying an indexing of this set by any external set (though we can always index it by its identity map to itself). Finally, we may speak of an algebra as being "free", meaning that there *exists* a generating set on which it is free, without choosing a particular such set, as when we say that "any subgroup of a free abelian group is free abelian". So we need to be sure it is always clear which version of the concept we are using.

The next exercise shows that in a category of the form $\Omega$-**Alg**, and in certain others, the last two of the above senses of "free algebra" essentially coincide.

**Exercise 9.3:3.**

(i)   Show that a free $\Omega$-algebra is free on a unique set of generators. That is, if $(F, u)$ is a free $\Omega$-algebra, then the image in $|F|$ of the set map $u$ (and hence in particular, the cardinality of the domain of $u$) is determined by the $\Omega$-algebra structure of $F$. (Hint: Definition 9.3.1.)

(ii)   Is the analogous statement true for free groups? Free monoids? Free rings?

(iii)   Same question for free upper (or lower) semilattices.

(iv)   Same question for free lattices. (If you know the structure theorem for free lattices this is not hard. Even if you do not, a little ingenuity will yield the answer by a direct argument.)

**Exercise 9.3:4.**

(i)   Show that every subalgebra $A$ of a free $\Omega$-algebra $F$ is free.

We mentioned above the fact (proved in the standard beginning graduate algebra course) that the corresponding statement holds for free abelian groups. It is also a basic (though harder to prove) result of group theory that it holds for free groups. But

(ii)   Is the analogous statement true for free monoids? Free rings? Free upper semilattices? Free lattices?

**Exercise 9.3:5.**

(i)   Let $\Omega$ be a finitary type without zeroary operation symbols, and $F_\Omega(x)$ the free $\Omega$-algebra on a single generator $x$. Show that the monoid of endomorphisms $\mathrm{End}(F_\Omega(x))$ (under composition) is a free monoid. If you wish, you may for simplicity assume that $|\Omega|$ consists of a single binary operation-symbol (since even in this case, the description of the free generating set for the monoid $\mathrm{End}(F_\Omega(x))$ is nontrivial).

(ii)   Does the result of (i) remain true if the assumption that $\Omega$ is finitary is removed?

(iii) Show that the corresponding result is never true if $\Omega$ has zeroary operations. Can you describe the monoid in this case?

(iv) If all operation-symbols of $\Omega$ have arity 1, describe the monoid $\mathrm{End}(F_\Omega(x))$ precisely in terms of $|\Omega|$.

The next result is easily seen from the explicit description of free $\Omega$-algebras in our proof of Theorem 9.3.3.

**Corollary 9.3.5** (to proof of Theorem 9.3.3). *If $a\colon X \to Y$ is an injective (respectively surjective) map of sets, then the induced map of free $\Omega$-algebras $F_\Omega(a)\colon F_\Omega(X) \to F_\Omega(Y)$ is likewise injective (surjective) on underlying sets.*

$\square$

We can also get the above result from a very general observation, though in that case we need a special argument to handle the free algebra on the empty set:

**Exercise 9.3:6.**

(i)   Show that *every* functor $A\colon \mathbf{Set} \to \mathbf{Set}$ carries surjective maps to surjective maps, and carries injective maps with nonempty domains to injective maps. (Hint: Use right and left invertibility.)

(ii)   Show that (i) becomes false if the qualification about nonempty domains is dropped.

(iii) Show, however, that if $A$ has the form $UF$, where $U$ is a functor from some category to $\mathbf{Set}$, and $F$ is a left adjoint to $U$, then $A$ carries maps with empty domain to injective maps.

(iv) Deduce Corollary 9.3.5 from the above result without calling on an explicit description of free $\Omega$-algebras.

Using free algebras, we can obtain other left universal constructions. A basic tool will be

**Definition 9.3.6.** *Let $\Omega$ be a type, $X$ a set, and $(F_\Omega(X), u_X)$ a free $\Omega$-algebra on $X$. An $\Omega$-algebra relation in an $X$-tuple of variables will mean an element $(s, t) \in |F_\Omega(X)| \times |F_\Omega(X)|$ (often informally written "$s = t$"). An $X$-tuple $v$ of elements of an $\Omega$-algebra $A$ is said to* satisfy *the relation*

$(s, t)$ *if the unique homomorphism* $f\colon F_\Omega(X) \to A$ *such that* $f u_X = v$ *has the property* $f(s) = f(t)$.

*If* $Y \subseteq |F_\Omega(X)| \times |F_\Omega(X)|$ *is a set of relations, then an* $\Omega$-*algebra presented by generators* $X$ *and relations* $Y$ *will mean an initial object* $(B, w)$ *in the category whose objects are pairs* $(A, v)$ *with* $A$ *an* $\Omega$-*algebra and* $v$ *an* $X$-*tuple of elements of* $|A|$ *satisfying all the relations in* $Y$, *and whose morphisms are homomorphisms of first components respecting second components; equivalently, a representing object for the functor* $\Omega$-**Alg** $\to$ **Set** *associating to every* $\Omega$-*algebra* $A$ *the set of* $X$-*tuples* $v$ *satisfying all the relations in* $Y$. *Such an algebra* $B$ *will be denoted* $\langle X \mid Y \rangle_{\Omega\text{-}\mathbf{Alg}}$, *or, when there is no danger of ambiguity,* $\langle X \mid Y \rangle$.

(If we wanted to be more precise, we might write our relations as $(s, t, (F_\Omega(X), u_X))$, since formally, a given pair of elements $s$ and $t$ can belong to underlying sets of various free algebras. But to avoid messy notation, we will assume that there is no ambiguity as to which free algebra is meant. Also, strictly speaking, the object presented by the generating set $X$ and relation set $Y$ should be given as a pair $(\langle X \mid Y \rangle, w)$, where $w$ is the canonical map $X \to |\langle X \mid Y \rangle|$. But again we will speak of it as $\langle X \mid Y \rangle$, and leave it understood that $w$ is there if we need to refer to it.)

**Theorem 9.3.7.** *Let* $\Omega$ *be a type. Then* $\Omega$-**Alg** *has algebras* $\langle X \mid Y \rangle$ *presented by arbitrary sets of generators* $X$ *and relations* $Y$.

*Proof.* $\langle X \mid Y \rangle$ can be constructed as the quotient of $F_\Omega(X)$ by the congruence generated by $Y$ (Definition 9.1.8). ☐

**Exercise 9.3:7.** Give an alternative proof of the above theorem using the results of § 8.10.

**Exercise 9.3:8.** At the end of § 4.5 we introduced the term *residually finite* to describe a group $G$ with the property that for any two elements $x \neq y \in |G|$, there exists a homomorphism $f$ of $G$ into a *finite* group such that $f(x) \neq f(y)$. The same definition applies to $\Omega$-algebras, for arbitrary $\Omega$.

Show that if $\Omega$ is a finitary type, then every finitely related $\Omega$-algebra (i.e., every $\Omega$-algebra that has a presentation $\langle X \mid Y \rangle$ with $Y$ finite) is residually finite.

(In contrast, we constructed in Exercise 7.8:27 a finitely related group $G$ that was not residually finite. There is no contradiction between these results, because that $G$, though finitely related as a group, is not finitely related as an $\Omega$-algebra for $\Omega$ the type to which groups belong; it is presented by the finitely many group-theoretic relations that we used, and infinitely many instances of the group identities.)

**Exercise 9.3:9.** This exercise will show that finitely related $\Omega$-algebras tend to have large free subalgebras.

Let $\Omega$ be a finitary type, $X$ a finite set, and $(F, u_F)$ a free $\Omega$-algebra on $X$.

(i)    Show that for every element $a \in |F|$ not belonging to the subalgebra generated by the empty set, there exists a subalgebra $A \subseteq F$ not containing $a$, such that $|F| - |A|$ is finite.

(ii)   Let $Y = \{(s_0, t_0), \ldots, (s_{n-1}, t_{n-1})\} \subseteq |F|^2$. Show that an algebra $A$ isomorphic to $\langle X \mid Y \rangle_\Omega$ can be obtained from $F$ by (roughly—you fill in the details) the following two steps.

(a) Define a certain algebra $B$ with underlying set $|F|$, and with operations and distinguished $X$-tuple of elements $u_B$ that differ from those of $F$ in a total of at most $n$ places. (I.e., such that the number of elements of $X$ at which $u_B$ differs from $u_F$, and the number of ari(s)-tuples at which the various $s_B$ differ from the $s_F$, add up to at most $n$. The idea is to slightly modify the structure of $F$ so as to make the relations of $Y$ hold in $B$.) And then,

(b) Let $A$ be the subalgebra of $B$ generated by $u_B(X)$ under the operations $s_B$.

(iii) Show from (i) and (ii) that if $\Omega$ has no zeroary operations, and $Y$ is as in part (ii) above, then $\langle X \mid Y \rangle_\Omega$ has a subalgebra $C$ such that $|\langle X \mid Y \rangle_\Omega| - |C|$ is finite and $C$ is isomorphic to a subalgebra of $A$, hence, by Exercise 9.3:4, is free.

After working out your proof, you might see whether you can weaken the assumption that $\Omega$ have no zeroary operations, replacing it either with a weaker condition on $\Omega$, or with a condition on $Y$.

Returning to our task of obtaining basic universal constructions in $\Omega$-**Alg**, let us prove

**Theorem 9.3.8.** *The category* $\Omega$-**Alg** *has all small colimits.*

*Proof.* By Proposition 8.6.6 (last statement), it is enough to show that $\Omega$-**Alg** has coequalizers of pairs of morphisms, and has small coproducts. We obtained coequalizers in Lemma 9.1.9; we shall now construct the coproduct of a small family of $\Omega$-algebras $(A_i)_{i \in I}$.

We assume without loss of generality that the $A_i$ have disjoint underlying sets (since we can replace them with disjoint isomorphic algebras if they do not). Let $A$ be the algebra presented by the generating set $\bigcup |A_i|$ and, for relations, all the relations satisfied within the separate $A_i$'s. (Precisely, we take for relations the images in $|F_\Omega(\bigcup_I |A_i|)| \times |F_\Omega(\bigcup_I |A_i|)|$, under the canonical maps $F_\Omega(|A_j|) \to F_\Omega(\bigcup_I |A_i|)$, of all the relations $(s, t) \in |F_\Omega(|A_j|)| \times |F_\Omega(|A_j|)|$ holding in the given algebras $A_j$.) It is easy to verify that $A$ is the desired coproduct.                                                                   $\square$

We end this section with two exercises which assume familiarity with point-set topology, and which concern certain algebras with a single binary operation. The first exercise sets up a general construction and establishes some

of its properties, to give you the feel of things. The second restricts attention to a particular instance of this construction, and asks you to establish a
seemingly bizarre universal property of that object.

**Exercise 9.3:10.** Let the set $2^\omega$ of all sequences $(\iota_0, \iota_1, \dots)$ of 0's and 1's
  be given the product topology induced by the discrete topology on $\{0, 1\}$.
  (The resulting space can be naturally identified with the Cantor set.) Let
  us define two continuous maps $\alpha$, $\beta \colon 2^\omega \to 2^\omega$, by letting

  $$(9.3.9) \quad \alpha(\iota_0, \iota_1, \dots) = (0, \iota_0, \iota_1, \dots), \text{ and } \beta(\iota_0, \iota_1, \dots) = (1, \iota_0, \iota_1, \dots).$$

  Thus, $2^\omega$ is the disjoint union of the two copies of itself, $\alpha(2^\omega)$ and $\beta(2^\omega)$.
     Now,

  (9.3.10)  let $\Omega$ be the type determined by a single binary operation $*$,

and let us define a covariant functor $H$ from the category **HausTop**
of Hausdorff topological spaces to $\Omega$-**Alg**. For every space $S$, the set
$|H(S)|$ will be **HausTop**$(2^\omega, S)$, i.e., the space of continuous $S$-valued
functions on $2^\omega$. Thus, these sets are given by the covariant hom-functor
$h_{2^\omega} \colon$ **HausTop** $\to$ **Set**. To describe the binary operation, let $u, v \in$
$|H(S)|$. Then we define $u * v$ to be the function $2^\omega \to S$ such that

$$(9.3.11) \quad (u * v)(\alpha(x)) = u(x), \quad (u * v)(\beta(x)) = v(x), \quad (x \in 2^\omega).$$

Thus, if we identify $2^\omega$ with the Cantor set, $u * v$ is the map whose
graph on the first half of that set looks like the graph of $u$ compressed
horizontally, and whose graph on the second half of the Cantor set is a
similarly compressed copy of the graph of $v$. Let $H(S) = (|H(S)|, *)$.

(i)   Show that for every $S$, the map $* \colon |H(S)| \times |H(S)| \to |H(S)|$ is
bijective.

(ii)  Let $S$ be any Hausdorff topological space and $X$ any finite subset of
$|H(S)|$. Let $X_0$ be the set of those $x \in X$ which, as maps $2^\omega \to S$, are
constant, and $X_1$ the set of $x \in X$ which are not constant, and such that
$x$ does not belong to the $\Omega$-subalgebra of $H(S)$ generated by $X - \{x\}$.
Show that the $\Omega$-subalgebra of $H(S)$ generated by $X$ can be presented
by the generating set $X_0 \cup X_1$, and the relations $x * x = x$ for $x \in X_0$.

(iii) Deduce that the set of nonconstant elements of $H(S)$ forms a subalgebra $N$ every finitely generated subalgebra of which is free. Show, however,
that if $S$ contains a homeomorphic copy of $2^\omega$, then $N$ itself is not free.
     (Can you find necessary and sufficient conditions on $S$ for $N$ to be
free?)

  Our definition above of the element $u * v$ involved its composites on the
right with $\alpha$ and $\beta$. We shall now let our construction take its tail in its

mouth, by applying it with $S = 2^\omega$. Since elements of the resulting algebra also have $2^\omega$ as *codomain*, we can also compose them on the *left* with $\alpha$ and $\beta$.

**Exercise 9.3:11.** Let $\alpha, \beta : 2^\omega \to 2^\omega$ and $H : \mathbf{HausTop} \to \Omega\text{-}\mathbf{Alg}$ be defined as in the first two paragraphs of the preceding exercise, and let $A = H(2^\omega)$, an $\Omega$-algebra with underlying set $\mathbf{HausTop}(2^\omega, 2^\omega)$.

(i)   Show that each of the $\Omega$-algebra homomorphisms $H(\alpha), H(\beta) : A \to A$ is an embedding, and that $A$ is the coproduct in $\Omega\text{-}\mathbf{Alg}$ of the images of these homomorphisms.

This is equivalent to saying that $A$ is a coproduct of two copies of itself, with coprojection maps $H(\alpha)$ and $H(\beta)$; or, fixing an arbitrary coproduct of two copies of $A$ and calling it $A \amalg A$, and calling its coprojection maps $q_0$ and $q_1$, it is equivalent to saying that the unique homomorphism $f : A \amalg A \to A$ satisfying $f q_0 = \alpha$ and $f q_1 = \beta$ is an isomorphism.

We now come to the strange universal property. Let $\mathbf{m}_A : A \to A \amalg A$ be the *inverse* of the above map $f$.

(ii)   Show that if $B$ is *any* $\Omega$-algebra given with a homomorphism $\mathbf{m}_B : B \to B \amalg B$, there exists a unique homomorphism $\theta : B \to A$ such that the following diagram commutes:

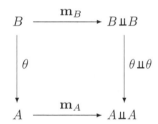

(Since our construction of $A$ uses topology, the same will necessarily be true of the proofs of (i) and (ii). Note, however, that the statements of these properties of $A$ are purely algebraic. We will be able to make sense of the above universal property in Chapter 10.)

## 9.4. Identities and varieties

Here is a definition that needs no introduction!

**Definition 9.4.1.** *Let* $\Omega$ *be a type,* $X$ *a set, and* $(F_\Omega(X), u_X)$ *a free* $\Omega$*-algebra on* $X$. *An* identity *in an* $X$*-tuple of variables will mean an element* $(s, t) \in |F_\Omega(X)| \times |F_\Omega(X)|$, *i.e., formally the same thing as a relation, and likewise often informally written "$s = t$". However an* $\Omega$*-algebra* $A$ *will be* *said to "satisfy" the identity* $(s, t)$ *if and only if* every $X$-*tuple* $v$ *of elements of* $|A|$ *satisfies* $(s, t)$ *as a relation; that is, if and only if for* every *homomorphism* $f : F_\Omega(X) \to A$, *one has* $f(s) = f(t)$.

The next result relates identities in different sets of variables.

**Lemma 9.4.2.** *Let* $\Omega$ *be a type,* $X$ *a set, and* $(s, t) \in |F_\Omega(X)| \times |F_\Omega(X)|$ *an identity in an* $X$-*tuple of variables. Then if* $f : X \to Y$ *is a one-to-one set map, an* $\Omega$-*algebra* $A$ *satisfies the identity* $(s, t)$ *if and only if it satisfies the identity in a* $Y$-*tuple of variables,* $(F_\Omega(f)(s), F_\Omega(f)(t))$.

*Hence if* $\gamma$ *is an infinite cardinal such that* $\gamma \geq \operatorname{card}(\operatorname{ari}(s))$ *for all* $s \in |\Omega|$, *every identity* $(s, t)$ *in any set* $X$ *of variables is equivalent to an identity* $(s', t')$ *in a* $\gamma$-*tuple of variables (i.e., there is an identity* $(s', t') \in |F_\Omega(\gamma)| \times |F_\Omega(\gamma)|$ *which is satisfied by an* $\Omega$-*algebra* $A$ *if and only if* $A$ *satisfies* $(s, t)$).

*Proof.* First statement: It is easy to see (without assuming the given map $f$ one-to-one) that for any map $f : X \to Y$ and any $Y$-tuple of elements of $|A|$, $v : Y \to |A|$, the induced $X$-tuple $v f : X \to |A|$ will satisfy the *relation* $(s, t)$ if and only if $v$ satisfies the relation $(F_\Omega(f)(s), F_\Omega(f)(t))$. Hence if $A$ satisfies $(s, t)$ as an *identity* it will likewise satisfy $(F_\Omega(f)(s), F_\Omega(f)(t))$ as an identity. The converse will hold for $f$ a one-to-one map if we can show that *every* map $w : X \to |A|$ can be written $v f$ for some $v : Y \to |A|$. It is clear how to define $v$ on elements of the one-to-one image of $X$ in $Y$ under $f$. If $|A|$ is nonempty, we can extend this map by giving $v$ arbitrary values on other elements of $Y$. If $|A|$ is empty, on the other hand, then there can be no homomorphisms to $A$ from the algebra $F_\Omega(X)$ (which is nonempty because it contains $s$ and $t$) so this case is vacuous. (An empty algebra satisfies every identity $(s, t)$, because the hypothesis of the implication defining "satisfaction" can never hold.)

Now let $\gamma$ be as in the second statement, and let $\gamma'$ denote the successor cardinal to $\gamma$. Then $\gamma'$ will be a regular cardinal (by Exercise 5.5:14(i)) which is greater than the arity of every operation of $\Omega$; hence given any set $X$ and any $s, t \in |F_\Omega(X)|$, Lemma 9.2.3 tells us that $s$ and $t$ lie in the subalgebra generated by some subset $X_0 \subseteq X$ of cardinality $< \gamma'$, hence $\leq \gamma$; hence the set $X_0$ can be mapped injectively into $\gamma$. Hence applying the first statement of this lemma to the inclusion of $X_0$ in $X$ on the one hand, and to an embedding of $X_0$ in $\gamma$ on the other, we see that $(s, t)$ is equivalent to some identity in a $\gamma$-tuple of variables. $\square$

Thus, for the purpose of studying families of identities satisfied by $\Omega$-algebras, and classes of algebras determined by identities, we can restrict ourselves to identities in a $\gamma$-tuple of variables for $\gamma$ as above. In particular, identities in a countable set of variables suffice for finitary algebras, and even for algebras all of whose operations have countable arity.

In making the second assertion of the above lemma, why have we looked at a cardinal such that all operations have cardinalities $\leq \gamma$, rather than following the pattern recommended earlier, of looking at cardinals that strictly bound the quantities we are interested in? Although generally speaking, the latter pattern gives one greater flexibility in stating conditions, in this case the conclusion we wanted was that there was a single free algebra in terms of which all our identities could be expressed, so we wanted a particular value for the cardinality of a generating set; and the non-strict inequalities used above yielded the smallest such cardinal.

Unfortunately, to study *direct limits* later in this section, we will also want a cardinal satisfying strict inequalities. Hence

**Convention 9.4.3.** *For the remainder of this section, $\Omega$ will denote a fixed type, $\gamma_0$ will denote an infinite cardinal that is $\geq \mathrm{ari}(s)$ for all $s \in |\Omega|$, and $\gamma_1$ will denote a regular infinite cardinal that is $> \mathrm{ari}(s)$ for all $s \in |\Omega|$. An identity will mean an $\Omega$-algebra identity in a $\gamma_0$-tuple of variables.*

*In writing identities, we shall often write $x_\alpha$ for the image $u(\alpha) \in |F_\Omega(\gamma_0)|$ of $\alpha \in \gamma_0$. We may also at times write $x$, $y$, etc., for $x_0$, $x_1$, etc.*

Note that if the smallest possible choice for $\gamma_1$ is a successor cardinal, then the smallest possible choice for $\gamma_0$ is its predecessor, while if the smallest choice for $\gamma_1$ is a limit cardinal, the smallest choice for $\gamma_0$ is the same cardinal. In particular, in the classical case, where all operations are finitary, $\aleph_0$ can be used for both $\gamma_0$ and $\gamma_1$.

The next exercise shows that when all arities are 0 or 1, one can do better than described above (though we will not use this result in what follows).

**Exercise 9.4:1.** Show that if all operation-symbols of $\Omega$ are of arity $\leq 1$, then the statement of Lemma 9.2.3 holds with $\gamma = 2$ (even though 2 is not a regular cardinal), and deduce that the final statement of Lemma 9.4.2 also holds for $\gamma = 2$. On the other hand, show by example that it does not hold for $\gamma = 1$.

Let us denote the set of all $\Omega$-algebra identities by

$$(9.4.4) \qquad\qquad I_\Omega = |F_\Omega(\gamma_0)| \times |F_\Omega(\gamma_0)|.$$

Thus we have a relation of *satisfaction* (Definition 9.4.1) defined between elements of the large set $\mathrm{Ob}(\Omega\text{-}\mathbf{Alg})$ of all $\Omega$-algebras and elements of the small set $I_\Omega$ of all identities. If $C$ is a (not necessarily small) set of $\Omega$-algebras, let us for the moment write $C^*$ for the set of identities satisfied by all members of $C$, and if $J$ is a set of identities, let us write $J^*$ for the (large) set of $\Omega$-algebras that satisfy all identities in $J$. The theory of Galois connections (§ 6.5) tells us that the two composite operators ** will be closure operators, that every set $J^*$ or $C^*$ will be closed under the appropriate closure operator **, and that the operators * give an antiisomorphism between the complete lattice of all closed sets of algebras and the complete lattice of all closed sets of identities.

In talking about this Galois connection, it is obviously not convenient to apply to sets of algebras our convention that sets are small if the contrary is not stated; so we make

**Convention 9.4.5.** *For the remainder of this chapter, we suspend for sets of algebras (as we have done from the start for object-sets of categories) the assumption that sets are small if the contrary is not stated.*

*(However, we still assume that any set of algebras is a subset of our universe $\mathbb{U}$ if the contrary is not stated; i.e., the smallness convention still applies to the underlying set of each algebra.)*

**Definition 9.4.6.** *A* variety *of $\Omega$-algebras means a full subcategory* **V** *of $\Omega$-***Alg** *having for object-set the set $J^*$ of algebras determined by some set $J$ of identities. The variety with object-set $J^*$ will be written* **V**$(J)$. *A category is called a* variety of algebras *if it is a variety of $\Omega$-algebras for some type $\Omega$.*

*If* **V** *is a variety, an algebra belonging to* **V** *will be called a* **V**-*algebra. The least variety of $\Omega$-algebras whose object-set contains a given set $C$ of algebras, that is, the full subcategory of $\Omega$-***Alg** *with object-set $C^{**}$, is called the variety* generated by $C$, *written* **Var**$(C)$.

*An* equational theory *for $\Omega$-algebras means a subset of $I_\Omega$ (i.e., a set of identities for $\Omega$-algebras) which can be written $C^*$ for some set $C$ of $\Omega$-algebras; $C^*$ is called the equational theory of the set of algebras $C$. If* **C** *is a full subcategory of $\Omega$-***Alg**, *then we may also call the equational theory of* Ob(**C**) *"the equational theory of* **C**." *The least equational theory containing a set $J$ of identities, namely, $J^{**}$, is called the equational theory* generated by $J$.

Examples: The categories we have named **Group**, **Ab**, **Monoid**, **Semigroup**, **Ring**[1], **CommRing**[1], ∨-**Semilattice**, ∧-**Semilattice** and **Lattice** are all varieties of algebras (up to trivial notational adjustment; e.g., we originally defined an object of **Group** as a 4-tuple $(|G|, \mu, \iota, \varepsilon)$; under Definition 9.1.2 it must be described as a pair $(|G|, (\mu, \iota, \varepsilon))$). For every group $G$, the category $G$-**Set** is a variety; for every ring $R$ the category $R$-**Mod** is a variety, and for every commutative ring $k$ the category of all associative $k$-algebras is a variety. For every type $\Omega$, the whole category $\Omega$-**Alg** is a variety (the greatest element in the complete lattice of varieties of $\Omega$-algebras, definable by the empty set of identities. Its equational theory consists of the tautological identities $(s, s)$.) Taking for $\Omega$ the trivial type, with no operation-symbols, we see that **Set** is (up to trivial notational adjustment) a variety.

If **C** is the full subcategory of **Monoid** consisting of those monoids all of whose elements are invertible, then **C** is not a variety of algebras, since invertibility is not an identity; nevertheless, this category is *equivalent* (Definition 7.9.5) to the variety **Group**.

Finally, some categories we have looked at which are not varieties, and are not in any obvious way equivalent to varieties, are **POSet**, **Top**,

**Set**$^{op}$, **RelSet**, the category of *complete* lattices, the full subcategory of **CommRing**[1] consisting of the *integral domains*, and the category of *torsion-free* groups (groups without elements of finite order other than $e$). How to determine whether or not any of these is nonetheless equivalent to some variety of algebras is a question we are not yet ready to tackle.

*Remark 9.4.7.* An algebra $A$ satisfies the identity $x_0 = x_1$ if and only if all its elements are equal. Hence an algebra satisfying this identity satisfies all identities; i.e., $\{(x_0, x_1)\}^{**} = I_\Omega$, the greatest element of the lattice of equational theories of $\Omega$-algebras. The corresponding variety of $\Omega$-algebras is the least element of the lattice of such varieties, and consists of algebras with *at most one* element. If $\Omega$ has any zeroary operation-symbols, then this variety consists only of one-element algebras, which are all isomorphic; thus the variety is equivalent to the category **1** with only one object and its identity morphism. If $\Omega$ has no zeroary operations, then this least variety contains both the empty algebra and all one-element algebras, and is equivalent to the 2-object category **2**.

Let us prove some easy results about varieties.

**Proposition 9.4.8.** *Let* $\mathbf{V} \subseteq \Omega\text{-}\mathbf{Alg}$ *be a variety. Then:*

(i)   *Any subalgebra of an algebra in* $\mathbf{V}$ *again lies in* $\mathbf{V}$.

(ii)   *The limit* $\varprojlim_{\mathbf{D}} A(D)$*, taken in* $\Omega\text{-}\mathbf{Alg}$*, of any functor* $A$ *from a small category* $\mathbf{D}$ *to* $\mathbf{V} \subseteq \Omega\text{-}\mathbf{Alg}$ *again lies in* $\mathbf{V}$.

(iii)   *Any homomorphic image of an algebra in* $\mathbf{V}$ *again lies in* $\mathbf{V}$.

(iv)   *The direct limit (colimit)* $\varinjlim A_j$*, taken in* $\Omega\text{-}\mathbf{Alg}$*, of any* $<\gamma_1$*-directed system of* $\mathbf{V}$*-algebras again lies in* $\mathbf{V}$*. (For* $\gamma_1$ *see Convention 9.4.3. Lemma 9.1.10 describes this direct limit.)*

   *In particular, the category* $\mathbf{V}$ *has small limits, has coequalizers, and has colimits of* $<\gamma_1$*-directed systems, and all of these constructions are the same in* $\mathbf{V}$ *as in* $\Omega\text{-}\mathbf{Alg}$.

*Proof.* It is straightforward that if an algebra satisfies an identity, any subalgebra or homomorphic image satisfies the same identity, giving (i) and (iii), and that a direct product of algebras satisfying an identity again satisfies that identity. Since arbitrary limits can be constructed using products and equalizers, and in $\Omega\text{-}\mathbf{Alg}$ equalizers are certain subalgebras, we get (ii).

To show (iv), let $L$ be the direct limit in $\Omega\text{-}\mathbf{Alg}$ of a $<\gamma_1$-directed system of algebras $(A_i)_I$ of $\mathbf{V}$, let $(s, t)$ be an identity of $\mathbf{V}$, say involving the first $\alpha < \gamma_1$ variables, which we regard as an identity $(s', t')$ in an $\alpha$-tuple of variables, and let $v$ be an $\alpha$-tuple of elements of $L$. By Lemma 9.1.10 (second paragraph) $|L|$ is the direct limit of the sets $|A_i|$, hence by $<\gamma_1$-directedness of $I$, we can find an $i \in I$ such that $A_i$ contains an inverse image of each component of $v$. The $\alpha$-tuple formed from these inverse images will satisfy the relation $(s', t')$, hence so does $v$, its image. Hence $L$ satisfies the identity $(s', t')$, and hence the equivalent identity $(s, t)$. Knowing that $L \in \mathrm{Ob}(\mathbf{V})$, its universal property as a direct limit in $\Omega$-**Alg** implies the corresponding property in **V**. (This was Lemma 8.6.7.)                                                     □

*Remark 9.4.9.* If we consider classes of algebras defined by sorts of propositions more general than identities, involving logical operators such as $\exists$, $\Longrightarrow$ and $\vee$ (for instance, torsion-free groups, and integral domains, both mentioned above, divisible groups, which were considered in Exercise 7.7:5, and fields, considered in Exercises 3.3:3 and 3.3:4), we find, in general, that one or more of the statements of Proposition 9.4.8 fail. This is why we stated in Chapter 2 that it was "better" to define the concept of group using three operations and the *identities* (2.2.1), than using just one operation, and the more complicated conditions (2.2.2). Of course, it is also worth studying what results *are* true of classes of algebras defined by other sorts of propositions, and these are also studied in General Algebra. But this course cannot cover everything, and varieties form a very broad, important, and well-behaved class; so we will focus on these here.

The parts of the preceding proposition saying that certain constructions have the same form in **V** as in $\Omega$-**Alg**, together with some earlier results yield:

**Corollary 9.4.10.** *Let* **V** *be a variety of* $\Omega$-*algebras. Then*

(i)   *The forgetful functor from* **V** *to* **Set** *respects limits, and respects co-limits over* $< \gamma_1$-*directed partially ordered sets.*

(ii)   *The inclusion functor into* **V** *of any subvariety* **W** *respects these constructions, and also respects coequalizers.*

(iii)   *Direct limits in* **V** *over* $< \gamma_1$-*directed partially ordered sets respect limits in* **V** *over categories* **D** *having* $< \gamma_1$ *objects and whose morphism-sets are generated by* $< \gamma_1$ *morphisms.*                                                     □

**Exercise 9.4:2.** Verify that the above corollary indeed follows from results we have proved in this and earlier sections.

We saw in Lemma 7.9.4 that if a category **C** is given with a concept of a *subobject* of an object, then one likewise gets a concept of a *subfunctor* of a **C**-valued functor. Let us make, for future reference

**Definition 9.4.11.** *If* **V** *is a variety of algebras, then unless the contrary is stated, references to subfunctors of* **V**-*valued functors* $F$ *are to be interpreted with "subobject" meaning "subalgebra".*

*Thus, for any category* **C** *and functor* $F : \mathbf{C} \to \mathbf{V}$, *a subfunctor* $G$ *of* $F$ *is (essentially) a construction associating to every* $X \in \mathrm{Ob}(\mathbf{C})$ *a subalgebra* $G(X) \subseteq F(X)$, *in such a way that for every morphism* $f : X \to Y$ *of* **C**, *the* **V**-*algebra homomorphism* $F(f)$ *carries* $G(X) \subseteq F(X)$ *into* $G(Y) \subseteq F(Y)$.

The subfunctors of group- and vector-space-valued functors considered in the exercises following Lemma 7.9.4 are examples of this concept. (If you didn't do the last part of Exercise 7.9:11, this might be a good time to look at it again.)

We can now prove for general varieties a pair of facts that we noted earlier in many special cases.

**Proposition 9.4.12.** *A morphism $f \colon A \to B$ in a variety $\mathbf{V}$ is one-to-one if and only if it is a monomorphism, and is surjective if and only if it is a coequalizer map.*

*Proof.* One direction of each statement is immediate: As we saw in Exercise 7.8:8, if $f$ is one-to-one on underlying sets, then it is a monomorphism, while we have observed that coequalizers in $\mathbf{V}$ are coequalizers in $\Omega$-**Alg**, and that these are surjective.

To get the converse statements, consider the congruence $E$ associated to $f$. This is the underlying set of a subalgebra $C$ of $A \times A$, hence it is an object of $\mathbf{V}$, and the projections of this object onto the two factors are morphisms $C \rightrightarrows A$ having the same composite with $f$. The coequalizer of these projections is $A/E \cong f(A)$, hence if $f$ is surjective, $f$ is indeed a coequalizer map. On the other hand, if $f$ has the cancellation property defining a monomorphism, these two projections must be equal, which means that $E$ can contain no nondiagonal elements of $|A| \times |A|$, which says that $f$ is one-to-one.

(Remark: These arguments are not valid in an arbitrary full subcategory of $\mathbf{V}$, since such a subcategory may contain $f \colon A \to B$ without containing $C$; cf. Exercise 7.7:5(i). On the other hand, they can be extended to a wide class of full subcategories, for instance, the subcategory of all finite objects of $\mathbf{V}$, where Exercise 7.8:8 is not applicable.) $\qquad\square$

The above result does not discuss the relation between onto-ness and being an *epimorphism*, nor between one-one-ness and the condition of being an *equalizer* map. If $f$ is onto, then by the above proposition it is a coequalizer map, hence by Lemma 8.6.2, it is an epimorphism; likewise, by that lemma an equalizer map is a monomorphism, hence by the above proposition, is one-to-one. Neither of the converse statements is true, but they have a surprising relationship to one another:

**Exercise 9.4:3.**

(i)   Show that if a variety $\mathbf{V}$ of algebras has an epimorphism which is not surjective (cf. Exercise 7.7:6(iii)) then it also has a one-to-one map which is not an equalizer.

(ii)   Is the converse to (i) true?

**Exercise 9.4:4.** The proof of Proposition 9.4.12 used the facts that $\mathbf{V}$ is closed in $\Omega$-**Alg** under products and subalgebras. Which of these two conditions is missing in the example from Exercise 7.7:5 mentioned in that proof? Can you also find an example in which only the other condition is missing?

We turn now to constructions which are not the same in a variety $\mathbf{V}$ and the larger category $\Omega$-$\mathbf{Alg}$. We will get these via the next lemma. Let us give both the proof of that result based on the "big direct product" idea (Freyd's Adjoint Functor Theorem), and the one based on "terms modulo consequences of identities".

**Lemma 9.4.13.** *If* $\mathbf{V}$ *is a variety of* $\Omega$-*algebras, the inclusion functor of* $\mathbf{V}$ *into* $\Omega$-$\mathbf{Alg}$ *has a left adjoint.*

*First proof.* We have seen that $\mathbf{V}$ has small limits and that these are respected by the inclusion functor into $\Omega$-$\mathbf{Alg}$, so by Freyd's Adjoint Functor Theorem, it suffices to verify the solution-set condition. If $A \in \mathrm{Ob}(\Omega\text{-}\mathbf{Alg})$, then every $\Omega$-algebra homomorphism $f$ of $A$ into a $\mathbf{V}$-algebra $B$ factors through the quotient of $A$ by the congruence $E$ associated to $f$. Since the factor-algebra $A/E$ is isomorphic to a subalgebra of $B$, it belongs to $\mathbf{V}$. Hence the set of all factor-algebras of the given $\Omega$-algebra $A$ which belong to $\mathbf{V}$, with the canonical morphisms $A \to A/E$, is the desired solution-set.

*Second proof.* Let $\mathbf{V}$ be $\mathbf{V}(J)$, the variety determined by the set of identities $J \subseteq |F_\Omega(\gamma_0)| \times |F_\Omega(\gamma_0)|$. Given $A \in \mathrm{Ob}(\Omega\text{-}\mathbf{Alg})$, let $E \subseteq |A| \times |A|$ be the congruence on $A$ generated by all pairs $(f(s),\, f(t))$ with $(s,\, t) \in J$, and $f \colon F_\Omega(\gamma_0) \to A$ a homomorphism. Then it is straightforward to verify that $A/E$ belongs to $\mathbf{V}$, and is universal among homomorphic images of $A$ belonging to $\mathbf{V}$.                                                                               $\square$

We shall call the above left adjoint functor the construction of *imposing the identities of* $\mathbf{V}$ *on an* $\Omega$-*algebra* $A$. Note that if we impose the identities of $\mathbf{V}$ on an algebra already in $\mathbf{V}$, we get the same algebra.

We can now get the rest of the constructions we want:

**Theorem 9.4.14.** *Let* $\mathbf{V}$ *be a variety of* $\Omega$-*algebras. Then* $\mathbf{V}$ *has small colimits, objects presented by generators and relations, and free objects on all small sets. All of these constructions can be achieved by performing the corresponding constructions in* $\Omega$-$\mathbf{Alg}$, *and then imposing the identities of* $\mathbf{V}$ *on the resulting algebras (i.e., applying the left adjoint obtained in the preceding lemma).*

*Proof.* The existence of these constructions in $\Omega$-$\mathbf{Alg}$ was shown in Theorems 9.3.8, 9.3.7 and 9.3.3. That left adjoints respect such constructions was proved in Theorems 8.8.3, 8.7.1, and 8.3.10.                                                     $\square$

We note the                                                         .

**Corollary 9.4.15** (to proof of Lemma 9.4.13). *If* $\mathbf{V} \subseteq \mathbf{W}$ *are varieties of* $\Omega$-*algebras, then the inclusion functor of* $\mathbf{V}$ *in* $\mathbf{W}$ *has a left adjoint, given by the composite of the inclusion of* $\mathbf{W}$ *in* $\Omega$-$\mathbf{Alg}$ *with the left adjoint of the inclusion of* $\mathbf{V}$ *in* $\Omega$-$\mathbf{Alg}$.

*Proof.* Given an algebra $A$ in **W**, the assertion to be proved is that if we regard $A$ as an object of $\Omega$-**Alg** and as such impose on it the identities of **V**, the resulting **V**-algebra $B$ will be universal as a **V**-algebra with a **W**-algebra homomorphism of $A$ into it. This is immediate because a **W**-algebra homomorphism $A \to B$ is the same as an $\Omega$-algebra homomorphism $A \to B$. □

We remark that the above corollary does not follow from Lemma 9.4.13 by Theorem 8.3.10 (on composites of left adjunctions). Rather, it can be regarded as a special case of some important results not yet discussed. One, which will come in the next chapter, gives general conditions for a functor among varieties to have a left adjoint. It can also be gotten from Lemma 9.4.13 using a result on subcategories whose inclusion functors have left adjoints ("reflective subcategories" [20, §IV.3]).

**Exercise 9.4:5.** Give an example of a variety **V** in which the free algebra on one generator $x$ is also generated *non-freely* by some element $y$.

The next exercise, which leads up to an open question in part (iii), is about the left adjoint to the inclusion functor $\mathbf{Bool}^1 \to \mathbf{CommRing}^1$, though its statement does not use the word "functor".

**Exercise 9.4:6.** Recall that $\mathbf{Bool}^1$, the variety of Boolean rings, is the subvariety of $\mathbf{CommRing}^1$ determined by the one additional identity $x^2 - x = 0$.

(i)  Show that the following conditions on a commutative ring $R$ are equivalent: (a) $R$ admits no homomorphism to the field $\mathbb{Z}/2\mathbb{Z}$. (b) $R$ admits no homomorphism to a nontrivial Boolean ring. (c) The ideal of $R$ generated by all elements $r^2 - r$ $(r \in R)$ is all of $R$. (Hint: from the fact that every commutative ring admits a homomorphism onto a field, deduce that every Boolean ring admits a homomorphism to $\mathbb{Z}/2\mathbb{Z}$.)

Let us call a ring $R$ satisfying the above equivalent conditions "Boolean-trivial".

(ii)  For every positive integer $n$, let $T_n$ denote the commutative ring presented by $2n$ generators $a_0, \ldots, a_{n-1}, b_0, \ldots, b_{n-1}$ and one relation $\sum a_i (b_i^2 - b_i) = 1$. Show that a commutative ring $R$ is Boolean-trivial if and only if for some $n$ there exists a ring homomorphism $T_n \to R$. Show, moreover, that there exist certain homomorphisms among the rings $T_n$, which allow one to deduce that the above family of conditions forms a chain under implication.

Ralph McKenzie (unpublished) has raised the question

(iii)  Are the implications from some point on in the above chain all reversible?

The above question is equivalent to asking whether beyond some point there exist homomorphisms in the "nonobvious" direction among the $T_n$.

This seems implausible, but to my knowledge no one has found a way to prove that such homomorphisms do not exist, and the question is open.

For those familiar with the language of logic, what McKenzie actually asked was whether Boolean-triviality was a first-order condition. Since, as noted in the exercise, that condition is the disjunction of a countable chain of first-order conditions, his question is equivalent to asking whether it is given by one member of the chain.

His version of the question also differed from the above in that he on the one hand restricted attention to rings $R$ of characteristic 2, and on the other hand did not restrict attention to commutative $R$. However, the general-characteristic case is equivalent to the characteristic 2 case, since given a ring $R$, one can translate a first-order sentence about $R/2R$ into a sentence about $R$ by replacing relations "$x = y$" with "$(\exists z)\, x = y + 2\,z$". Concerning commutativity, I felt that since the expected answer is negative, and since a negative answer in the commutative case would imply a negative answer in the general case, and since people are more familiar with commutative rings than with general rings, this modification of the question would be for the better.

One can, however, raise the same question about the relation between commutative and noncommutative rings, and it seems equally difficult to answer:

**Exercise 9.4:7.** Prove results analogous to (i)(b) $\Longleftrightarrow$ (c) and to (ii) of the preceding exercise with the varieties **CommRing**$^1$ and **Bool**$^1$ replaced by **Ring**$^1$ and **CommRing**$^1$ respectively, and see whether you can make any progress on the question analogous to (iii) for this case.

We remark that one does not have a result analogous to Exercise 9.4:6(ii) for every pair consisting of a variety and a subvariety; e.g., **Group** and **Ab**. What is special about the varieties of the above two exercises is that triviality of an object is equivalent to a single relation, $0 = 1$. (Another variety with this property is **Lattice**$^{0,1}$, the variety of lattices with greatest and least element made into zeroary operations.)

Returning to the general theory of varieties of algebras, let us extend some notation that we had set up for the categories $\Omega$-**Alg** :

**Definition 9.4.16.** *The free-object functor and the underlying-set functor associated with a variety* **V** *will be denoted* $F_{\mathbf{V}} \colon$ **Set** $\to$ **V** *and* $U_{\mathbf{V}} \colon$ **V** $\to$ **Set**. *The* **V**-*algebra presented by a generating set* $X$ *and relation set* $R$ *will be denoted* $\langle X \mid R \rangle_{\mathbf{V}}$, *or* $\langle X \mid R \rangle$ *when there is no danger of confusion.*

In presenting a **V**-algebra, it is often convenient to take a "relation" in an $X$-tuple of variables to mean a pair of elements of $F_{\mathbf{V}}(X)$ rather than of $F_{\Omega}(X)$. If we write $q \colon F_{\Omega}(X) \to F_{\mathbf{V}}(X)$ for the canonical homomorphism, it is clear that given $(s,\, t) \in |F_{\Omega}(X)| \times |F_{\Omega}(X)|$ and an $X$-tuple $v$ of elements of a **V**-algebra $A$, the elements $s$ and $t$ will fall together under the homomorphism $F_\Omega(X) \to A$ determined by $v$ if and only if $q(s)$ and $q(t)$ fall together under the homomorphism $F_{\mathbf{V}}(X) \to A$ determined by $v$; so the same condition is expressed by the original relation $(s, t) \in |F_\Omega(X)| \times |F_\Omega(X)|$, and by the induced pair $(q(s), q(t)) \in |F_{\mathbf{V}}(X)| \times |F_{\mathbf{V}}(X)|$. Thus, if $Y$ is a subset of $|F_{\mathbf{V}}(X)| \times |F_{\mathbf{V}}(X)|$, we will often denote by $\langle X \mid Y \rangle_{\mathbf{V}}$ the quotient of $F_{\mathbf{V}}(X)$ by the congruence generated by $Y$.

When we considered the concept of *representable functors* $\mathbf{C} \to \mathbf{Set}$, we noted (in the third paragraph following Definition 8.2.3) that for $\mathbf{C} = \mathbf{Group}$, a presentation of the group representing such a functor yielded a nice concrete description of the functor. This observation is true in arbitrary varieties of algebras, and can be made into a characterization of representable functors.

**Lemma 9.4.17.** *Let* $\mathbf{W}$ *be a variety of algebras, and* $U \colon \mathbf{W} \to \mathbf{Set}$ *a functor. Then the following conditions are equivalent:*

(i)   *$U$ is representable; i.e., there exists an object $R$ of $\mathbf{W}$ such that $U$ is isomorphic to the functor $h_R = \mathbf{W}(R, -)$.*

(ii)   *There exists a set $X$, and a set of relations in an $X$-tuple of variables, $Y \subseteq |F_\Omega(X)| \times |F_\Omega(X)|$, such that $U$ is isomorphic to the functor associating to every object $A$ of $\mathbf{W}$ the set $\{\xi \in |A|^X \mid (\forall (s, t) \in Y) \; s_A(\xi) = t_A(\xi)\}$, and to every morphism $f \colon A \to B$ in $\mathbf{W}$ the map that applies $f$ componentwise to $X$-tuples of elements of $|A|$.*

*Proof.* If $U$ is represented by $R$, take a presentation $R = \langle X \mid Y \rangle_{\mathbf{W}}$; then $U$ will have the form shown in (ii). Conversely, if $U$ is as in (ii), it is represented by the algebra with presentation $\langle X \mid Y \rangle_{\mathbf{W}}$.   □

Thus, we immediately see that such functors on **Group** as $G \mapsto \{x \in |G| \mid x^2 = e\}$ and $G \mapsto \{(x, y) \in |G|^2 \mid x\,y = y\,x\}$ are representable. A less obvious case is the "set of invertible elements" functor on monoids. If we try to use the criterion of the above lemma, taking $X$ to be a singleton, it does not work, because the condition of invertibility is not an equation in $x$ alone. However, because inverses are *unique* when they exist, we see that this construction is isomorphic to the functor $S \mapsto \{(x, y) \in |S|^2 \mid x\,y = e = y\,x\}$, which is of the required form.

In condition (ii) of the above lemma, $X$ and/or $Y$ may, of course, be empty. If $Y$ is empty, then $U$ is the $X$-th power of the underlying-set functor (Definition 7.8.5), and is represented by $F_{\mathbf{V}}(X)$. An example with $X$ but not $Y$ empty is the functor $\mathbf{Ring}^1 \to \mathbf{Set}$ represented by $\mathbb{Z}/n\mathbb{Z}$ for an integer $n$. We recall that this ring is presented by the empty set of generators, and the one relation $n = 0$ (where "$n$" as a ring element means the $n$-fold sum $1 + \ldots + 1$). This ring admits no homomorphism to a ring $A$ unless $n = 0$ in $A$; when $A$ does satisfy that equation, there is a unique ring homomorphism $\mathbb{Z}/n\mathbb{Z} \to A$ (namely, the additive group map taking $1_{\mathbb{Z}/n\mathbb{Z}}$ to $1_A$). Thus,

$h_{\mathbb{Z}/n\mathbb{Z}}$ takes $A$ to the empty set if the characteristic of $A$ does not divide $n$, and to a one-element set if it does. In terms of point (ii) of the above lemma, this functor must be described as sending $A$ to "the set of 0-tuples of elements of $A$ such that $n = 0$". This sounds peculiar because the words following "such that" do not refer to what is named in the preceding phrase; but it is logically correct: we get the unique 0-tuple if $n = 0$ in $A$, and nothing otherwise.

**Exercise 9.4:8.** Determine which of the following covariant set-valued functors are representable. In each case where the answer is affirmative, give an "$X$" and "$Y$" as in Lemma 9.4.17. In (i)–(v), $n$ is a fixed integer $> 1$.

(i)   The functor on **Ring**$^1$ taking $A$ to a singleton if $n$ is invertible in $A$, and to the empty set otherwise.

(ii)  The functor on **Ring**$^1$ taking $A$ to its underlying set if $n$ is invertible in $A$, and to the empty set otherwise.

(iii) The functor on **Ab** taking $A$ to the kernel of the endomorphism "multiplication by $n$".

(iv)  The functor on **Ab** taking $A$ to the image of this endomorphism.

(v)   The functor on **Ab** taking $A$ to the cokernel of this endomorphism.

(vi)  The functor on **Lattice** taking $A$ to the set of pairs $(x, y)$ such that $x \leq y$.

(vii) For $P$ a fixed partially ordered set, the functor on **Lattice** taking $A$ to the set of isotone maps from $P$ to the "underlying" partially ordered set of $|A|$.

We mentioned in §8.12 that it was shown in [49] that contravariant *left* adjunctions among "the kind of categories of algebras we will be studying in this course" were degenerate. We now have the language in which to state the restriction on such functors (which will not be proved here!): If $W : \mathbf{V}^{\mathrm{op}} \to \mathbf{W}$ and $V : \mathbf{W}^{\mathrm{op}} \to \mathbf{V}$ are mutually left adjoint contravariant functors between *varieties of algebras*, then all objects $W(A)$ are epimorphs of the initial object of $\mathbf{W}$ (i.e., codomains of epimorphisms with the initial object as domain), and all objects $V(B)$ are epimorphs of the initial object of $\mathbf{V}$. The next exercise shows how to get trivial examples of such adjunctions, and then gives an example which, though rather unnatural, is nontrivial.

**Exercise 9.4:9.**

(i)   Show that if $\mathbf{C}$ and $\mathbf{D}$ are any two categories having initial objects, then the contravariant functors between $\mathbf{C}$ and $\mathbf{D}$ each of which takes every object of its domain to the initial object of its codomain, and takes all morphisms to the identity morphism of that object, are mutually left adjoint.

(ii)  Let $n$ be a positive integer, and **CommRing**$^1_{\mathbb{Z}/n\mathbb{Z}}$ the variety of commutative $\mathbb{Z}/n\mathbb{Z}$-algebras; equivalently, commutative rings satisfying the identity $n = 0$. Show how to define a functor $F\colon (\mathbf{CommRing}^1_{\mathbb{Z}/n\mathbb{Z}})^{\mathrm{op}} \to$ $\mathbf{CommRing}^1$ such that if $R$ has characteristic $m \mid n$, then $F(R) = \mathbb{Z}[m^{-1}]$, and verify that $F$ has a left adjoint.

(iii) Verify that the above functor and its adjoint both take arbitrary objects to epimorphs of the initial object of their codomain categories.

## 9.5. Derived operations

Having identified $\Omega$-algebra terms $s$ with elements of free $\Omega$-algebras $F_{\Omega\text{-}\mathbf{Alg}}(X)$, our viewpoint in "evaluating" these terms has been, "a choice of an $X$-tuple $v$ of elements in an $\Omega$-algebra $A$ induces an evaluation homomorphism $F_{\Omega\text{-}\mathbf{Alg}}(X) \to A$". But as in §2.6, we can modify which variable(s)— the $X$-tuple $v$, the term $s$, or both—we foreground. We do this in the next definition, again replacing $\Omega$-$\mathbf{Alg}$ with a general variety $\mathbf{V}$.

**Definition 9.5.1.** *Let $\mathbf{V}$ be a variety of algebras, $X$ a set, and $(F_{\mathbf{V}}(X), u)$ the free $\mathbf{V}$-algebra on $X$. For every element $s \in |F_{\mathbf{V}}(X)|$, every $\mathbf{V}$-algebra $A$, and every $X$-tuple $v$ of elements of $|A|$, let*

$$\mathrm{eval}(s, A, v) \in |A|$$

*denote the image of the element $s$ under the unique homomorphism $f\colon F_{\mathbf{V}}(X) \to A$ such that $f\,u = v$. (Intuitively, the result of substituting into the term $s$ the $X$-tuple $v$ of elements of $A$.)*

*For fixed $s$ and $A$, let us define*

$$s_A\colon\ |A|^X \longrightarrow |A|$$

*by*

$$s_A(v) = \mathrm{eval}(s, A, v).$$

*Then a* derived $X$-ary operation *on $A$ will mean a map $|A|^X \to |A|$ which is equal to $s_A$ for some $s \in |F_{\mathbf{V}}(X)|$.*

*More generally, given $s \in |F_{\mathbf{V}}(X)|$ and any full subcategory $\mathbf{C}$ of $\mathbf{V}$ (e.g., a one-object subcategory, or all of $\mathbf{V}$), if we write $U_{\mathbf{C}}\colon \mathbf{C} \to \mathbf{Set}$ for the restriction to $\mathbf{C}$ of the underlying-set functor of $\mathbf{V}$, and $U_{\mathbf{C}}^X\colon \mathbf{C} \to \mathbf{Set}$ for the functor carrying an object $A$ to the set $U_{\mathbf{C}}(A)^X$ (cf. Definition 7.8.5), then $s_{\mathbf{C}}\colon U_{\mathbf{C}}^X \to U_{\mathbf{C}}$ will denote the morphism between these functors $\mathbf{C} \to \mathbf{Set}$ which on each object $A$ of $\mathbf{C}$ acts by $s_A$. A morphism $U_{\mathbf{C}}^X \to U_{\mathbf{C}}$ which can be written $s_{\mathbf{C}}$ for some $s \in |F_{\mathbf{V}}(X)|$ will be called a* derived $X$-ary operation of $\mathbf{C}$.

Note that the derived operations will in particular include the *primitive operations* $s_A\colon |A|^{\mathrm{ari}(s)} \to |A|$ (respectively, $s_{\mathbf{C}}\colon U_{\mathbf{C}}^{\mathrm{ari}(s)} \to U_{\mathbf{C}}$) induced by the operation symbols $s \in |\Omega|$, and the *projection* operations $p_{X,\,x} \colon |A|^X \to |A|$ (respectively $U_{\mathbf{C}}^X \to U_{\mathbf{C}}$), induced by the free generators $u(x) \in |F_{\mathbf{V}}(X)|$.

Let us now follow up on some ideas that we toyed with at the end of §3.3. Given any full subcategory $\mathbf{C}$ of our variety $\mathbf{V}$, consider the large set of all "generalized operations on $\mathbf{C}$ in an $X$-tuple of variables", i.e., functions $f$ associating to each object $A$ of $\mathbf{C}$ a map $f_A \colon |A|^X \to |A|$ in an arbitrary way. If we look at the set of all these generalized operations as a direct product, $\prod_{A \in \mathrm{Ob}(\mathbf{C})} |A|^{|A|^X}$ (living in the next larger universe, since $\mathbf{C}$ need not be small), we see that it can be made the underlying set of a large $\mathbf{V}$-algebra, namely the product, $\prod_{A \in \mathrm{Ob}(\mathbf{C})} A^{|A|^X}$; let us denote this algebra by $\mathrm{GenOp}_{\mathbf{C}}(X)$. We are not interested in this bloated monster for itself, but for the observation that the (still generally large) set of those elements thereof which constitute morphisms of functors $U_{\mathbf{C}}^X \to U_{\mathbf{C}}$, i.e., $\mathbf{Set}^{\mathbf{C}}(U_{\mathbf{C}}^X, U_{\mathbf{C}})$, forms a subalgebra thereof. (A description of the $\mathbf{V}$-algebra structure on this hom set might have seemed unnatural without the context of the algebra structure on $\mathrm{GenOp}(\mathbf{C})$, which is why we began with the latter. Incidentally, when we first discussed this in §3.3, we were not sure it made sense to talk about large sets. Having adopted the Axiom of Universes, and the associated interpretation of large sets, we can deal with these safely!) We shall call this the algebra of *functorial $X$-ary operations* on $\mathbf{C}$. The *derived* $X$-ary operations of $\mathbf{C}$ form a subalgebra of this subalgebra:

$$(9.5.2) \qquad \mathrm{DerOp}_{\mathbf{C}}(X) \subseteq \mathbf{Set}^{\mathbf{C}}(U_{\mathbf{C}}^X, U_{\mathbf{C}}) \subseteq \mathrm{GenOp}_{\mathbf{C}}(X).$$

Note that the algebra of derived operations is quasi-small, i.e., isomorphic to a small algebra, since it is a homomorphic image of $F_{\mathbf{V}}(X)$. The image of each generator $x \in X$ will be the function carrying every $X$-tuple to its $x$-th coordinate. Clearly, these "coordinate functions" generate $\mathrm{DerOp}_{\mathbf{C}}(X)$ as an $\Omega$-algebra. We can describe the resulting algebra nicely, and, under appropriate hypotheses, the algebra of functorial operations as well:

**Lemma 9.5.3.** *Let $\mathbf{C}$ be a full subcategory of a variety $\mathbf{V}$, and $X$ a (small) set. Then the (large) algebra of derived $X$-ary operations on $\mathbf{C}$ is isomorphic to the (small) algebra $F_{\mathbf{Var}(\mathrm{Ob}(\mathbf{C}))}(X)$.*

*Moreover, if $\mathbf{C}$ contains the free $\mathbf{V}$-algebra on $X$, then every functorial $X$-ary operation on $\mathbf{C}$ is a derived operation; i.e., $\mathbf{Set}^{\mathbf{C}}(U_{\mathbf{C}}^X, U_{\mathbf{C}}) = \mathrm{DerOp}_{\mathbf{C}}(X) \cong F_{\mathbf{V}}(X)$.*

*In particular, if $\mathbf{C} = \mathbf{V}$, the above equality holds for all sets $X$.*

*Sketch of proof.* The first assertion is straightforward, since for two terms $s$ and $t$, we have $s_{\mathbf{C}} = t_{\mathbf{C}}$ if and only if $(s, t)$ is an identity of every algebra in $\mathbf{C}$. To prove the second, assume $\mathbf{C}$ contains the free algebra $F_{\mathbf{V}}(X)$, and verify that each functorial $X$-ary operation on $\mathbf{C}$ is determined by its value on the universal $X$-tuple $u$ of elements of $F_{\mathbf{V}}(X)$; equivalently, apply Yoneda's Lemma to the pair of functors $\mathbf{C} \to \mathbf{Set}$ given by $U_{\mathbf{V}}^X \cong h_{F_{\mathbf{V}}(X)}$ and $U_{\mathbf{V}} \cong h_{F_{\mathbf{V}}(1)}$. The final assertion clearly follows. $\qquad\square$

**Exercise 9.5:1.** Give the details of the above proof.

**Exercise 9.5:2.**

(i)   Show that if $\mathbf{C}$ is the full subcategory of all *finite algebras* in $\mathbf{V}$, then the algebra of functorial $X$-ary operations on $\mathbf{C}$ can be described as the inverse limit of all finite factor algebras of $F_{\mathbf{V}}(X)$. (Make this statement precise.)

(ii)   Show that if $\mathbf{V} = \mathbf{Group}$ and $\mathbf{C}$ is as in (i), and $X = 1$, then the group of functorial unary operations on $\mathbf{C}$ is uncountable. Give an explicit example of an operation in this group that is not a derived group-theoretic operation.

(iii)  Interpret Exercise 7.9:6, especially part (ii) thereof, in terms of part (i) above, and if you had not yet successfully done that exercise, see whether you can make further progress on it.

In part (ii) above, the map from functorial operations on general groups to functorial operations on finite groups failed to be surjective. There are also situations where such maps fail to be one-to-one:

**Exercise 9.5:3.**

(i)   Give an example of a variety $\mathbf{V}$ not generated by its finite algebras. (If possible, get such an example in which the variety is defined by finitely many operation-symbols, all of finite arities, and finitely many identities.)

(ii)   Show that the property asked for in the first sentence of (i) is equivalent to saying that the restriction map from functorial operations on $\mathbf{V}$ to functorial operations on the full subcategory of finite objects of $\mathbf{V}$ is not one-to-one.

Since Exercise 9.5:2(ii) above shows that, though the variety of all groups has only countably many functorial operations of any finite arity, its full subcategory of *finite* groups has uncountably many such operations, one may ask whether, for $\mathbf{C}$ a full subcategory of a variety $\mathbf{V}$, the class of finitary functorial operations of $\mathbf{C}$ need even be quasi-small!

The answer depends on one's foundational assumptions; I will briefly summarize the situation. Logicians have asked the question,

(9.5.4)   Does there exist a proper class (in our language, a non-small set) of (small) models of some first-order theory, none of which is embeddable in another?

The answer to (9.5.4) turns out to depend on one's choice of universe. If $\mathbb{U}$ is the smallest universe, or is a successor element in the well-ordered set of universes, the answer is yes. The negative answer, on the other hand, is called "Vopěnka's principle"; the *existence* of a universe for which this holds is equivalent to the existence of a cardinal with some special properties (which force it to be enormous) but which are thought likely to be consistent with ZFC.

Now the positive answer to (9.5.4), which, as just noted, is true in "most" universes, is known to imply the existence of a non-small set $C$ of small *algebras* of some finitary type $\Omega$ such that there are no homomorphisms between distinct members of $C$. Given such a $C$, let $\mathbf{C}$ be the full subcategory of $\Omega$-**Alg** with $C$ as object-set. Then we see that the definition of a functorial operation $f$ on $\mathbf{C}$ involves no conditions relating the behavior of $f$ on *different* objects. So, for instance, for every subset $B \subseteq C$, there is a functorial binary operation on $\mathbf{C}$ which acts as the first-coordinate function on algebras in $B$, and as the second-coordinate function on algebras not in $B$. The set of operations so defined has cardinality $2^{\mathrm{card}(C)}$, which is certainly not small. Thus, in "most" universes we have a class of algebras with a non-quasi-small set of functorial binary operations. (Cf. [124, 139].) The converse result, that in a universe where Vopěnka's principle does hold, the class of functorial operations on every class of algebras of a given type is quasi-small, is obtained in [42, Theorem 10.1].

Let me end this section with some questions about operations on the real and rational numbers which, so far as I know, are open.

**Exercise 9.5:4.** (Harvey Friedman).

(i)   If we make the set of real numbers an algebra under the single binary operation $a(x, y) = x^2 + y^3$, does this algebra satisfy any nontrivial identities?

(ii)   If we make the set of nonnegative real numbers an algebra under the single binary operation $c(x, y) = x^{1/2} + y^{1/3}$, does this algebra satisfy any nontrivial identities?

(iii)   Does there exist a derived binary operation on the ring $\mathbb{Q}$ of rational numbers which is one-to-one as a map $|\mathbb{Q}| \times |\mathbb{Q}| \to |\mathbb{Q}|$?

If you cannot answer this last question, you might hand in proofs that the answer to the corresponding question for the ring of integers is "yes", and for the ring of real numbers, "no".

Another question posed by Friedman along the lines of (i) and (ii) above was whether the group of bijective maps $\mathbb{R} \to \mathbb{R}$ generated by the two maps $p(x) = x + 1$ and $q(x) = x^3$ is free on those two generators. This was answered affirmatively in [145], with 3 replaced by any odd prime. (See [70] for a simplified proof.) The result has subsequently been generalized to show, essentially, that the group of maps generated by exponentiation by all positive rational numbers and addition of all real constants is the coproduct of the two groups generated by these two sorts of maps [69], and, in another direction [36], to show that the group generated by exponentiation by positive rationals with odd numerator and denominator, addition of real algebraic numbers, and *multiplication* by nonzero real algebraic numbers, is the coproduct of the group generated by the above addition and multiplication operations and the group generated by the multiplication and exponentiation operations, with amalgamation of the subgroup of multiplication operations. (For the meaning and structure of a coproduct of groups with amalgamation of a common subgroup, cf. Exercise 7.8:25(ii) and paragraph preceding that exercise.)

Another open question, somewhat similar to the above, though more a question in number theory than general algebra, is

**Exercise 9.5:5** (B. Poonen). Does there exist a polynomial $f \in \mathbb{Q}(x, y)$ such that $f(\mathbb{Z} \times \mathbb{Z}) = \mathbb{N}$; or even such that $f(\mathbb{Z} \times \mathbb{Z})$ is a subset of $\mathbb{N}$ which has positive upper density (i.e., does *not* satisfy $\lim_{n \to \infty} \mathrm{card}(f(\mathbb{Z} \times \mathbb{Z}) \cap n)/n = 0$)?

## 9.6. Characterizing varieties and equational theories

We observed at the end of §6.5 that when one obtains a Galois connection from a relation on a pair of sets, $R \subseteq S \times T$, the closure $X^{**}$ or $Y^{**}$ of a subset $X \subseteq S$ or $Y \subseteq T$ is constructed "from above", namely as the set of members of $S$ or $T$ that satisfy certain conditions determined by members of the other set; and that a recurring type of mathematical question is how to describe these closures "from below", as all elements obtainable by starting with the elements of $X$ or $Y$ and iterating some constructions. In the case of the Galois connection between $\Omega$-algebras and identities, these questions are: Given a set $C$ of $\Omega$-algebras, how can we construct from these algebras all the algebras in the set $C^{**} = \mathrm{Ob}(\mathbf{Var}(C))$ that they generate; and given a set $J$ of identities, how can we construct from these all members of the equational theory $J^{**}$ that they generate? Answers to these questions should, in particular, give internal criteria for when a set of algebras is a variety, and for when a set of identities is an equational theory.

I said near the end of §6.5 that a general approach to this kind of question is to look for operations which carry every set $Y^*$ or $X^*$ into itself, and having found as many as one can, to try to show that closure under these operations (or better, under some nice subset of them) is sufficient, as well as necessary, for a set to be closed.

Now we have shown that a variety of algebras is closed under forming subalgebras, homomorphic images, products, and $<\gamma_1$-directed direct limits. (Closure under general limits need not be mentioned, since it is implied by closure under products and subalgebras. On the other hand, the existence of free objects, coproducts, etc., are not closure conditions, since these are defined relative to the variety we are trying to construct.) The next result shows that three of the above four closure conditions suffice to characterize varieties.

In reading that result, recall that by Convention 9.4.5, sets $C$ of algebras are not assumed small.

**Theorem 9.6.1** (Birkhoff's Theorem). *Let $\Omega$ be a type. Then a set of $\Omega$-algebras forms a variety if and only if it is closed under forming homomorphic images, subalgebras, and products (of small families).*

*In fact, if $C$ is a set of $\Omega$-algebras, then any object of $\mathbf{Var}(C)$ can be written as a homomorphic image of a subalgebra of a product of a small set of members of $C$.*

*Proof.* Clearly, it suffices to prove the final assertion. Let $\mathbf{V} = \mathbf{Var}(C)$. Then an algebra belonging to $\mathbf{V}$ can be written as a *homomorphic image* of the free $\mathbf{V}$-algebra $F_{\mathbf{V}}(X)$ for some set $X$, hence it suffices to show that $F_{\mathbf{V}}(X)$ can be obtained as a *subalgebra* of a *product* of objects in $C$. To show this, let $N \subseteq |F_\Omega(X)| \times |F_\Omega(X)|$ denote the set of all pairs $(s, t)$ that are *not* identities of $\mathbf{V}$; equivalently, which are not identities of all members of $C$. For each $(s, t) \in N$, choose an $X$-tuple $v_{(s,t)}$ of elements of an algebra $A_{(s,t)} \in C$ such that $v_{(s,t)}$ fails to satisfy the relation $(s, t)$. Let $P$ be the product algebra $\prod_{(s,t) \in N} A_{(s,t)}$, and let $v \colon X \to |P|$ be the set map with $(s, t)$-component $v_{(s,t)}$ for each $(s, t) \in N$. It follows from its definition that this $X$-tuple $v$ satisfies none of the relations in $N$; on the other hand, since $P$ belongs to $\mathbf{V}$, it must satisfy all relations *not* in $N$. It is easily deduced that the subalgebra $F \subseteq P$ generated by this $X$-tuple is isomorphic to the free algebra $F_{\mathbf{V}}(X)$.                                                   $\square$

The last sentence of Theorem 9.6.1 is often expressed in operator language:

$$(9.6.2) \qquad\qquad \mathbf{Var}(C) = \mathbf{H}\,\mathbf{S}\,\mathbf{P}(C).$$

To make this precise, let us fix a type $\Omega$, and let $L_\Omega$ denote the large lattice of all subsets $C \subseteq \mathrm{Ob}(\Omega\text{-}\mathbf{Alg})$ which are closed under going to isomorphic algebras (i.e., satisfy $T \cong S \in C \implies T \in C$. Thus $L_\Omega$ is isomorphic to the lattice of all subsets of the set of isomorphism classes of algebras in $\Omega\text{-}\mathbf{Alg}$.) For each $C \in |L_\Omega|$, let us define

$$\mathbf{H}(C) = \{\,\text{homomorphic images of algebras in } C\,\},$$
$$\mathbf{S}(C) = \{\,\text{subalgebras of algebras in } C\,\},$$
$$\mathbf{P}(C) = \{\,\text{products of algebras in } C\,\}.$$

Then (9.6.2) indeed expresses the last sentence of Theorem 9.6.1. (Except that $\mathbf{Var}(C)$ should, more precisely, be $\mathrm{Ob}(\mathbf{Var}(C))$. But we will ignore that distinction in this discussion, to give this statement the form in which it is usually stated.)

(The restriction to classes closed under isomorphism is not assumed in all discussions of this topic, leading to somewhat capricious behavior of the above operators: For $C$ a class of algebras not necessarily closed under isomorphism, $\mathbf{H}(C)$ is nevertheless closed under going to isomorphic algebras, by the definition of "homomorphic image", though it loses this property if the definition of this operator is changed to "quotients of members of $C$ by congruences". On the other hand, $\mathbf{S}(C)$ is not generally closed under going to isomorphic algebras if $C$ is not, but it acquires that property if one changes the criterion to "algebras embeddable in members of $C$". Whether $\mathbf{P}(C)$ is closed under isomorphism depends on whether one defines "product" to mean "any object which can be given a family of 'projection' maps having the appropriate universal property", as we have done here, or as the "standard" set-theoretic product, whose underlying set consists of tuples of elements of the given algebras. Since these distinctions are irrelevant to the algebraic questions involved, it seems best to eliminate them by restricting attention to isomorphism-closed classes. These are called "abstract classes" by some authors, though I do not favor that term. Incidentally, while discussing this topic, we will, obviously, temporarily set aside our habit of using $\mathbf{P}$ for "power set".)

In view of (9.6.2), it is natural to examine the monoid of operators on $|L_\Omega|$ generated by $\mathbf{H}$, $\mathbf{S}$ and $\mathbf{P}$. We see from (9.6.2) that the product $\mathbf{HSP}$ acts as a closure operator, and hence is *idempotent*: $(\mathbf{HSP})^2 = \mathbf{HSP}$. From this we can deduce further equalities, e.g., $\mathbf{SHSP} = \mathbf{HSP}$. This deduction is clear when we think of $\mathbf{H}$, $\mathbf{S}$ and $\mathbf{P}$ as closure operators; to abstract the argument, let $Z$ denote the monoid of all operators $\mathbf{A} \colon |L_\Omega| \to |L_\Omega|$ satisfying

(a)     $(\forall\, C \in |L_\Omega|)\ \mathbf{A}(C) \supseteq C$     ($\mathbf{A}$ is increasing),

(b)     $(\forall\, C,\, D \in |L_\Omega|)\ C \supseteq D \implies \mathbf{A}(C) \supseteq \mathbf{A}(D)$     ($\mathbf{A}$ is isotone).

This monoid $Z$ can be partially ordered by writing $\mathbf{A} \geq \mathbf{B}$ if and only if for all $C$, $\mathbf{A}(C) \supseteq \mathbf{B}(C)$. By (a), all elements of $Z$ are $\geq$ the identity operator, which we shall denote $\mathbf{I}$; we see from (b) that $\mathbf{B} \geq \mathbf{C} \implies \mathbf{AB} \geq \mathbf{AC}$, and we see from the definition of $\geq$ that $\mathbf{B} \geq \mathbf{C} \implies \mathbf{BA} \geq \mathbf{CA}$. Hence knowing only that $\mathbf{H}, \mathbf{S}, \mathbf{P} \in Z$, we can say that $(\mathbf{HSP})^2 \geq \mathbf{SHSP} \geq \mathbf{HSP}$; hence, as claimed, the equality $(\mathbf{HSP})^2 = \mathbf{HSP}$ implies $\mathbf{SHSP} = \mathbf{HSP}$.

Having illustrated how to calculate with these operators, we invite the reader to combine these methods with considerations of structures of $\Omega$-algebras in

**Exercise 9.6:1.** Describe explicitly the partially ordered monoid generated by the operators $\mathbf{H}$, $\mathbf{S}$ and $\mathbf{P}$ on classes of $\Omega$-algebras for general $\Omega$; i.e., determine the distinct products of these operators, their composition, and the order-relations among them. Are there finitely or infinitely many distinct operators? Which such operators are idempotent?

(When I say "for general $\Omega$", I mean that a relation $\leq$ or $=$ should be considered to hold if and only if it holds for *all* $\Omega$. Special cases will be looked at in the next exercise.)

The above is a large task, but an interesting one. To carry it out fully, you need counterexamples showing that each equality or inclusion that you do *not* assert actually fails to hold for some set of algebras. However, a counterexample to one relation often turns out to be a counterexample to several, so the task is not unreasonably difficult.

There are numerous modifications of this problem. For example.

**Exercise 9.6:2.** Suppose we restrict the operators **H**, **S**, **P** to classes of algebras in a particular variety **V**; then some additional inclusions and equalities may occur among the composites of these restricted operators. Investigate the partially ordered monoids of operators obtained when **V** is **Set**, respectively **Group**, respectively **Ab**. You may add to this list.

One could enlarge the set of operators considered above, introducing, for instance, $\mathbf{E} = \{\text{equalizers}\}$ (i.e., $\mathbf{E}(C) = $ the set of equalizers of pairs of homomorphisms among algebras of $C$; thus, $\mathbf{E} \leq \mathbf{S}$), $\mathbf{P}_{\text{fin}} = \{\text{products of finite families}\}$, and $\mathbf{L} = \{\text{direct limits of directed systems}\}$. (In considering these last two, we should restrict attention to finitary algebras, or else replace "finite families" and "directed systems" by "families of $< \gamma_1$ objects" and "$<\gamma_1$-directed systems" for $\gamma_1$ as in the preceding section.) Results on the structure of the monoid generated by any subset of $\{\mathbf{H}, \mathbf{S}, \mathbf{P}, \mathbf{E}, \mathbf{P}_{\text{fin}}, \mathbf{L}\}$, or any other such family of natural operators, can be turned in as homework, but I will merely pose as an exercise the questions

**Exercise 9.6:3.** Can one in general strengthen (9.6.2) to

(i)     $\mathbf{Var}(C) = \mathbf{HEP}(C)$ ?

(ii)    $\mathbf{Var}(C) = \mathbf{HSP}_{\text{fin}}(C)$ ?

**Exercise 9.6:4.** Since every variety of $\Omega$-algebras is closed under $<\gamma_1$-directed direct limits, it is natural to ask whether such direct limits can be obtained using the operators **H**, **S**, **P** in a more direct way than by calling on Birkhoff's theorem. They can, as sketched in

(i)     Given a $<\gamma_1$-directed system $(A_i, (f_{i,j})_{i,j\in I})$ of $\Omega$-algebras, obtain the direct limit of the $A_i$ by roughly the following construction, filling in or fixing details where necessary: Within $\prod_i A_i$, call an element $(a_i)$ "acceptable" if the relations $f_{ij}(a_i) = a_j$ hold "eventually", i.e., for all "sufficiently large" $i$ and $j$, in an appropriate sense. Define an equivalence relation on acceptable elements, making two such elements equivalent if their components are "eventually" equal. Verify that using these ideas, one can express the direct limit of the $A_i$ as a homomorphic image of a subalgebra of a product of the $A_i$. (Check that your proof handles the case where some of the $A_i$ are empty algebras, and that it actually uses the $<\gamma_1$-directedness condition. If it fails either of these tests, it needs some tweaking.)

Though the emphasis of this course is on varieties of algebras, general algebra also studies classes of algebras more general than varieties (e.g., the class of torsion-free groups). For this, it is useful to have a version of the above result that obtains the direct limit using more restricted operators, which such classes may be closed under even if they are not closed under the full operators **H**, **S** and **P**. Hence

(ii) Show in fact that a class of $\Omega$-algebras will be closed under $<\gamma_1$-directed direct limits if it is closed under products, under taking equalizers (a subclass of the subalgebras), and under taking direct limits of $<\gamma_1$-directed systems with surjective maps, and that this last condition is strictly weaker than closure under homomorphic images.

The next part shows that the above version of our result covers some cases that one may want to look at, but not all:

(iii) Verify that the class of torsion-free groups is closed under the operations of (ii) above. On the other hand, show that the class of integral domains (commutative rings without zero-divisors) is not closed under all those operations, but is nevertheless closed under direct limits.

The proof of Birkhoff's Theorem leads us to examine the class of $\Omega$-algebras that are free in *some* variety.

**Proposition 9.6.3.** *Let $\Omega$ be a type, $F$ an $\Omega$-algebra, $X$ a set, and $u$ an $X$-tuple of elements of $|F|$. Then the following conditions are equivalent:*

(i) *$(F, u)$ is a free algebra on the set $X$ in some variety $\mathbf{V}$ of $\Omega$-algebras.*

(ii) *$(F, u)$ is a free algebra on the set $X$ in the variety generated by $F$.*

(iii) *$F$ is generated by the set $u(X)$, and there exists some full subcategory $\mathbf{C}$ of $\Omega$-$\mathbf{Alg}$ containing $F$ such that $(F, u)$ is free in $\mathbf{C}$ on the set $X$.*

(iv) *$F$ is generated by the set $u(X)$, and for every set map $v\colon X \to |F|$, there exists an endomorphism $e$ of $F$ such that $v = eu$. (If we assume $u$ is an inclusion map, this latter condition can be stated, "Every map of $X$ into $|F|$ extends to an endomorphism of $F$.")*

(v) *$F$ is isomorphic to the quotient of $F_\Omega(X)$ by a congruence $E$ which is carried into itself by every endomorphism $f$ of $F_\Omega(X)$ (i.e., which for every such $f$ satisfies $(s, t) \in |E| \implies (f(s), f(t)) \in E$); and $u$ is the composite of universal maps $X \to |F_\Omega(X)| \to |F_\Omega(X)/E| \cong |F|$.*

*Proof.* We have (i) $\implies$ (ii) because a free algebra in a given concrete category is easily seen to remain free in any full subcategory which contains it; (ii) $\implies$ (iii) is immediate. The universal property of a free object gives (iii) $\implies$ (iv). To see (iv) $\implies$ (v), identify $F$ with the quotient of $F_\Omega(X)$ by the congruence $E$ consisting of all relations satisfied by the $X$-tuple $u$. Then if $f$ is an endomorphism of $F_\Omega(X)$, (iv) implies that $f$ induces an endomorphism $e$ of $F = F_\Omega(X)/E$, which can be seen to be equivalent to the condition that $E$ is carried into itself by $f$, which is the assertion of (v).

Finally, given (v) we claim that the relations satisfied by $u$ are satisfied by every $X$-tuple $v$ of elements of $F$. Indeed, since $F$ is a homomorphic image of $F_\Omega(X)$, every $X$-tuple $v$ of elements of $F$ is the image of some $X$-tuple $w$ of elements of $F_\Omega(X)$; and by the universal property of $F_\Omega(X)$, $w$ is the image of the free generating set of $F_\Omega(X)$ under some endomorphism $f$ of $F_\Omega(X)$. By assumption, the congruence $E$ is preserved under $f$; hence every relation satisfied in $F$ by the $X$-tuple $u$ is satisfied by every $X$-tuple $v$ of elements of $F$; i.e., is an identity of $\mathbf{Var}(\{F\})$. Conversely the identities of $\mathbf{Var}(\{F\})$ are necessarily satisfied by $u$. Hence $F$, being generated by an $X$-tuple $u$ which satisfies precisely those relations which are identities of $\mathbf{Var}(\{F\})$ in an $X$-tuple of variables, is the free $\mathbf{Var}(\{F\})$-algebra on $X$, proving (i).                                                                                 □

**Exercise 9.6:5.** Suppose $\mathbf{V} = \mathbf{Monoid}$ and $\mathbf{C}$ is the class of monoids all of whose elements are invertible.

(i)   Show that $\mathbf{C}$ has free algebras (i.e., that its underlying-set functor has a left adjoint), but that these are not the free algebras of $\mathbf{Var}(\mathbf{C})$.

(ii)   Show using this example that if we remove from conditions (iii) and (iv) of Proposition 9.6.3 the requirement that $F$ be generated by the image of $X$, these conditions no longer imply condition (i).

From Proposition 9.6.3 we can deduce the corresponding result with $\Omega$-**Alg** replaced by an arbitrary variety $\mathbf{V}$. In particular, we record

**Proposition 9.6.4.** *Let $\mathbf{V}$ be a variety. Then*

(i)   *If $X$ is a set and $u$ an $X$-tuple of elements of an object $F$ of $\mathbf{V}$, then $(F, u)$ is a free algebra in a subvariety $\mathbf{W}$ of $\mathbf{V}$ if and only if $F$ is isomorphic to a quotient of the free $\mathbf{V}$-algebra $F_\mathbf{V}(X)$ by a congruence invariant under all endomorphisms of $F_\mathbf{V}(X)$, and $u$ is the composite of the universal map $X \to F_\mathbf{V}(X)$ with the factor map $F_\mathbf{V}(X) \to F$.*

(ii)   *If $\gamma_0$ is an infinite cardinal greater than or equal to the arities of all operations of $\Omega$, then the subvarieties of $\mathbf{V}$ are in bijective correspondence with congruences on $F_\mathbf{V}(\gamma_0)$ which are invariant under all endomorphisms of this algebra, each subvariety $\mathbf{W}$ corresponding to the congruence determined by the natural map $F_\mathbf{V}(\gamma_0) \to F_\mathbf{W}(\gamma_0)$, and each endomorphism-invariant congruence $E$ corresponding to the subvariety generated by $F_\mathbf{V}(\gamma_0)/E$.*   □

Point (ii) above, in the case where $\mathbf{V} = \Omega$-**Alg**, solves the problem of characterizing equational theories (Definition 9.4.6):

**Theorem 9.6.5.** *Let $\Omega$ be a type, and $\gamma_0$ an infinite cardinal greater than or equal to the arities of all operations of $\mathbf{V}$. Then a subset $J \subseteq |F_\Omega(\gamma_0)| \times |F_\Omega(\gamma_0)|$ is an equational theory if and only if it is a congruence on $F_\Omega(\gamma_0)$ which is carried into itself by all endomorphisms of $F_\Omega(\gamma_0)$; in other words, if and only if it satisfies the following five conditions for all $s, t, v$, etc. in $|F_\Omega(\gamma_0)|$ and $\sigma \in |\Omega|$. (In (d) and (e), $\sigma_{F_\Omega(\gamma_0)}$, $s_{F_\Omega(\gamma_0)}$ and $t_{F_\Omega(\gamma_0)}$ denote the derived operations on $F_\Omega(\gamma_0)$ induced by $\sigma$, $s$ and $t$.)*

(a) $(s, s) \in J$.

(b) $(s, t) \in J \implies (t, s) \in J$.

(c) $(s, t) \in J, (t, v) \in J \implies (s, v) \in J$.

(d) $(t_i, v_i)_{i \in \mathrm{ari}(\sigma)} \in J^{\mathrm{ari}(\sigma)} \implies (\sigma_{F_\Omega(\gamma_0)}(t_i), \sigma_{F_\Omega(\gamma_0)}(v_i)) \in J$.

(e) $(s, t) \in J, (v_i)_{i \in \gamma_0} \in |F_\Omega(\gamma_0)|^{\gamma_0} \implies (s_{F_\Omega(\gamma_0)}(v_i), t_{F_\Omega(\gamma_0)}(v_i)) \in J$.

*Proof.* Parts (a)–(c) express the statement that $J$ is an equivalence relation, and (d) says that the operations of $F_\Omega(\gamma_0)$ respect this relation, i.e., that it is a congruence. Condition (e) expresses the fact that this congruence is preserved by all endomorphisms of $F_\Omega(\gamma_0)$, since every such endomorphism is determined by the $\gamma_0$-tuple $(v_i)$ to which it takes the universal $\gamma_0$-tuple of elements of $F_\Omega(\gamma_0)$. □

**Definition 9.6.6.** *A pair $(F, u)$ satisfying the equivalent conditions of Proposition 9.6.3 (in particular, condition (i)) is called a* relatively free $\Omega$-algebra.

*If $\mathbf{V}$ is any subvariety of $\Omega$-$\mathbf{Alg}$ and $F$ is any relatively free algebra in $\mathbf{V}$ (equivalently, if $F$ satisfies Proposition 9.6.4 (i) with some subvariety $\mathbf{W} \subseteq \mathbf{V}$ in place of the $\mathbf{V}$ of that statement), then $(F, u)$ can also be called a* relatively free $\mathbf{V}$-algebra.

By Lemma 9.4.2 an algebra relatively free on $\gamma_0$ generators uniquely determines the corresponding variety. But a relatively free algebra $(F, u)$ on an $\alpha$-tuple of generators for $\alpha < \gamma_0$ may be free in more than one variety; for example, the free group on one generator is also a free *abelian* group on one generator. The variety $\mathbf{Var}(\{F\})$ used in the proof of Proposition 9.6.3 (v) $\implies$ (i) will clearly be the *smallest* variety in which $(F, u)$ is free. The *largest* such variety is the variety defined by the identities in $\alpha$ variables satisfied by $F$; equivalently, by the relations satisfied by the universal $\alpha$-tuple $u$ in $F$, regarded as identities. The details, and some examples, are indicated in

**Exercise 9.6:6.**

(i) Let $\mathbf{V}$ be a variety, and suppose $F \in \mathrm{Ob}(\mathbf{V})$ is relatively free on an $\alpha$-tuple $u$ of indeterminates. Show that $(F, u)$ is a free algebra in precisely those subvarieties $\mathbf{U} \subseteq \mathbf{V}$ which contain the variety $\mathbf{Var}(F)$ (defined by all identities satisfied in $F$), and are contained in the subvariety of $\mathbf{V}$ defined by the identities in $\leq \alpha$ variables holding in $F$.

(ii) For $\mathbf{V} = \mathbf{Group}$, $\alpha = 1$, and $(F, u)$ the infinite cyclic group $Z = \langle x \rangle$, with $u$ selecting $x$ as free generator, characterize group-theoretically those subvarieties of $\mathbf{V}$ in which $(F, u)$ is free.

(iii) Show that if again $\mathbf{V} = \mathbf{Group}$, but we now take $\alpha = 2$, and for $(F, u)$ either the free group on two generators or the free abelian group on two generators, then in each case, the greatest and least subvarieties of $\mathbf{V}$ in which this group is free coincide.

(iv)  Are there any relatively free groups $(F, u)$ on two generators such that the greatest and least varieties of groups in which $(F, u)$ is free are distinct?

(v)  If $\Omega$ is the type of groups, and $(F, u)$ is either the free group on two generators or the free abelian group on two generators, show that the greatest and the least varieties of $\Omega$-*algebras* in which $(F, u)$ is free do not coincide, but that if $(F, u)$ is the free group or free abelian group on three generators, they again coincide.

Here are some exercises on subvarieties of familiar varieties.

**Exercise 9.6:7.** (If you do both parts, give the proof of one in detail, and for the other only give details where the proofs differ.)

(i)  Let $G$ be a group, and $G$-**Set** the variety of all $G$-sets. Show that subvarieties of $G$-**Set** other than the least subvariety (characterized in Remark 9.4.7) are in one-to-one correspondence with the normal subgroups $N$ of $G$, in such a way that the subvariety corresponding to $N$ is equivalent to the variety $(G/N)$-**Set**, by an equivalence which respects underlying sets.

(ii)  Prove the analogous result for subvarieties of $R$-**Mod**, where $R$ is an arbitrary ring. (In that case, the least subvariety is not an exceptional case.)

**Exercise 9.6:8.**

(i)  Let **CommRing**$^1$ denote the category of commutative rings. Show that if $\mathbf{V}$ is a proper subvariety of **CommRing**$^1$ generated by an *infinite integral domain*, then $\mathbf{V}$ is the variety $\mathbf{V}_p$ determined by the 0-variable identity $p = 0$ for some prime $p$, where the symbol "$p$" in this identity is an abbreviation for $1 + 1 + \cdots + 1$ with $p$ summands.

(ii)  Show that the subvariety **Bool**$^1 \subseteq$ **CommRing**$^1$ is a *proper* subvariety of the variety $\mathbf{V}_2$ defined as in (i).

**Exercise 9.6:9.** Let $F = F_{\mathbf{Ring}^1}(\omega)$, the free associative (noncommutative) ring on indeterminates $x_0, x_1, \ldots$. For each positive integer $n$, let

$$ S_n = \sum_\pi (-1)^\pi \, x_{\pi(0)} \cdots x_{\pi(n-1)} \in |F|, $$

where $\pi$ ranges over the permutations on $n$ elements, and $(-1)^\pi$ denotes $+1$ if $\pi$ is an even permutation, $-1$ if $\pi$ is odd.

(i)  Show that any ring satisfying $S_n = 0$ also satisfies $S_{n'} = 0$ for all $n' \geq n$; i.e., that $(S_{n'}, 0) \in \{(S_n, 0)\}^{**}$ under the Galois connection introduced at the beginning of this section.

(ii)  Show that for every $d > 0$ there exists $n > 0$ such that for every commutative ring $k$, the ring $M_d(k)$ of $d \times d$ matrices over $k$ satisfies the identity $S_n = 0$.

(iii) Show that for every $n > 0$ there exists $d > 0$ such that $M_d(k)$ does not satisfy $S_n = 0$ for any nontrivial commutative ring $k$.

(iv) Deduce that there is an infinite chain of distinct varieties of rings of the form $\mathbf{V}(\{(S_n, 0)\})$, and an infinite chain of distinct varieties of rings of the form $\mathbf{Var}(\{M_d(\mathbb{Z})\})$. (In these symbols, the expressions in set-brackets denote singletons.)

Note on the above exercise: The *least* $n$ such that all $d \times d$ matrix rings $M_d(k)$ over commutative rings $k$ satisfy $S_n = 0$ is $2d$. The hard part of this result, namely that $M_d(k)$ satisfies $S_{2d} = 0$, is known as the Amitsur-Levitzki Theorem [39]. All known proofs are either messy (e.g., by graph theory [135]) or tricky (e.g., using exterior algebras [128]). The reader is invited to attempt to find a new proof! Part (ii) of the above exercise can be done relatively easily, however, using a larger-than-optimal $n$.

The study of subvarieties of $\mathbf{Ring}^1$ is called the theory of *rings with polynomial identity*, affectionately known as PI rings. See [129, Chapter 6] for an introduction to this subject.

Here is a curious variety closely related to the variety of groups.

**Exercise 9.6:10.** Let $\Omega$ be the type defined by a single ternary (i.e., "3-ary") operation-symbol, $\tau$. Let $H \colon \mathbf{Group} \to \Omega\text{-}\mathbf{Alg}$ be the functor taking a group $G$ to the $\Omega$-algebra with underlying set $|G|$ and operation

$$(9.6.7) \qquad \tau(x, y, z) = x\, y^{-1} z.$$

(i)   Show that the objects $H(G)$ are the nonempty algebras in a certain subvariety of $\Omega\text{-}\mathbf{Alg}$, and give a set $J$ of identities defining this variety.

   The algebras (empty and nonempty) in this variety are called *heaps*, so let us call the variety $\mathbf{Heap}$.

(ii)   Show that for groups $G$ and $G'$, one has

$$H(G) \cong H(G') \text{ in } \mathbf{Heap} \iff G \cong G' \text{ in } \mathbf{Group}.$$

(iii) Show, however, that not every isomorphism between $H(G)$ and $H(G')$ has the form $H(i)$ for $i$ an isomorphism between $G$ and $G'$!

(iv) Show that the following categories are equivalent: (a) $\mathbf{Group}$, (b) the variety of algebras $(|A|, \tau, \iota)$ where $(|A|, \tau)$ is a heap, and $\iota$ is a zeroary operation, subject to no further identities (intuitively, "heaps with distinguished elements $\iota$"), (c) $\mathbf{Heap}^{\mathrm{pt}}$, where the construction $\mathbf{C}^{\mathrm{pt}}$ is defined as in Exercise 7.8:4.

(v)   Show that if $X$, $Y$ are two objects of any category $\mathbf{C}$, then the set of isomorphisms $X \to Y$ forms a heap under the operation $\tau(x, y, z) = x\, y^{-1} z$. How is the structure of this heap related to those of the groups $\mathrm{Aut}(X)$ and $\mathrm{Aut}(Y)$?

The concept of heap is not very well known, and many mathematicians have from time to time rediscovered it and given it other names (myself included). Heaps were apparently first studied by Prüfer [125] and Baer [41], under the name *Schar* meaning "crowd" or "flock", a humorous way of saying "something like a group". The term was rendered into Russian by Suškevič [134] as груда, meaning "heap", which gave both a loose approximation of the meaning of *Schar* and a play on the sounds of the Russian words: "group" = *gruppa*, "heap" = *gruda*. Since the concept and its generalizations have gotten most attention in Russian-language works, it has come back into Western European languages via translations of this Russian term rather than of the original German. (Incidentally, there is an unrelated notion with the name "heap" in the theory of data structures [146, p. 72].)

Part (ii) of the above exercise shows that there is no need for a separate theory of the *structure* of heaps; this is essentially contained in that of groups. However, the variety of heaps is both a taking-off point for various generalizations ("semiheaps" etc.), and a source of examples in general algebra and category theory.

Part (iv) of the preceding exercise suggests

**Exercise 9.6:11.**

(i)   For what varieties **V** is it true that the category $\mathbf{V}^{\mathrm{pt}}$ can be identified with the variety gotten by adding to **V** one zeroary operation, and no additional identities?

(ii)   What varieties **V** satisfy the conditions (a)–(d) of Exercise 7.8:3? (Note that for varieties these conditions are all equivalent, by the last part of that exercise.)

The above two questions have fairly straightforward answers. But I don't know an elegant answer to the next one. We have just seen that **Group** is an example to which it applies.

(iii)   Which varieties **V** can be obtained from some variety **W** by adjoining one zeroary operation and no identities?

Let us remark that in stretching the concept of "variety" from its classical definition as a class of algebras defined by identities to our present category-theoretic use, we have pulled it over a lot of ground, so that care is needed in using the term. For example, when should we think of two varieties as being "essentially the same"? If they are precisely equal? If we can establish a bijection between their types such that they are defined by the corresponding identities? If they are equivalent as categories? If there is a category-theoretic equivalence which also respects the underlying-set functors of the varieties?

There is no right answer, but these four conditions are all inequivalent.

**Exercise 9.6:12.**   What implications exist among the above four conditions on a pair of varieties? Give examples showing that no two of those conditions are equivalent.

## 9.7. Lie algebras

Let me digress here to introduce a variety important in algebra, geometry, and differential equations, that of *Lie algebras*. I have referred to these in previous chapters in a few comments "for the reader familiar with the concept". The reader who prefers to remain unfamiliar with it for the time being may skip this section, and perhaps come back later on. In subsequent sections, Lie algebras will be again be referred to only in occasional exercises and remarks.

To motivate the definition, consider an associative algebra $A$ over a field (or more generally, over a commutative ring) $k$, and suppose we look at the underlying set of $A$ together with the operations that make it a *k-vector-space* (or *k-module*), and the *commutator bracket* operation,

$$(9.7.1) \qquad\qquad [x, y] = x\,y - y\,x.$$

These operations obviously satisfy the identities saying

$$(9.7.2) \quad \begin{array}{l} +, \ -, \ 0 \text{ and the scalar multiplications by members of } k \text{ make} \\ |A| \text{ a } k\text{-module, and } [-, -] \text{ is a } k\text{-bilinear operation with respect to this } k\text{-module structure.} \end{array}$$

There is a further obvious identity satisfied by $[-, -]$, and another that, though not so obvious, is straightforward to verify:

$$(9.7.3) \qquad\qquad [x, x] = 0 \qquad\qquad \text{(alternating identity)},$$

$$(9.7.4) \quad [x, [y, z]] + [y, [z, x]] + [z, [x, y]] = 0 \quad \text{(Jacobi identity)}.$$

Note that if in (9.7.3) we substitute $x + y$ for $x$, apply (9.7.2) to expand the result as a sum of four terms, and then apply (9.7.3) again to drop the summands $[x, x]$ and $[y, y]$, we get

$$(9.7.5) \qquad\qquad [x, y] + [y, x] = 0 \qquad\qquad \text{(anticommutativity)}.$$

(If 2 is invertible in $k$, then, still assuming (9.7.2), we also have the converse implication, (9.7.5) $\implies$ (9.7.3), as can be seen by setting $y = x$ in (9.7.5); so (9.7.5) and (9.7.3) become equivalent.)

The straightforward way of verifying (9.7.4) for the operation (9.7.1) is to expand the left-hand side of (9.7.4) using (9.7.1). One gets 12 terms, and applying associativity one finds that they all cancel. But the following alternative verification gives some useful insight. Recall that a *derivation* on a $k$-algebra (associative or not) means a $k$-linear map $D$ satisfying the identity

$$(9.7.6) \qquad\qquad D(y\,z) = D(y)\,z + y\,D(z).$$

Now it is easy to check that if $x$ is any element of an associative $k$-algebra $A$, then the operation $[x, -]$ is a derivation on $A$; that is, for all $y$ and $z$,

(9.7.7)                                  $[x,\, y\, z]\ =\ [x,\, y]\, z + y\, [x,\, z].$

A map that is a derivation with respect to a given multiplication is also a derivation with respect to the opposite multiplication, $y * z = z\, y$. Subtracting the opposite multiplication from the original multiplication gives the commutator map (9.7.1); so by (9.7.7) $[x,\, -]$ also acts as a derivation with respect to that map; i.e., we have

(9.7.8)                          $[x,\, [y,\, z]]\ =\ [[x,\, y],\, z] + [y,\, [x,\, z]].$

We can use anticommutativity to rearrange this identity so that the bracket arrangement of the second term, like that of the other two, becomes $[-,\, [-,\, -]]$, and so that the last term has the same cyclic order of $x$, $y$ and $z$ as the first two terms. Bringing all three terms to the same side, we find that the above formula becomes precisely (9.7.4). Thus, the Jacobi identity (9.7.4) is equivalent to the condition (9.7.8), saying that for each element $x$, the unary operation $[x,\, -]$ is a derivation with respect to the commutator bracket operation.

**Definition 9.7.9.** *Let $k$ be a commutative ring (often assumed a field). Then a Lie algebra over $k$ means a set $|A|$ given with operations $+$, $-$, $0$, and also a set of unary "scalar multiplication" operations corresponding to the elements of $k$, and a binary operation $[-,\, -]$, which together satisfy (9.7.2)–(9.7.4). (In brief, a $k$-module given with a $k$-bilinear operation $[-,\, -]$ which is alternating and satisfies the Jacobi identity.) The variety of Lie algebras over $k$ will be denoted* $\mathbf{Lie}_k$.

*For $L$ a Lie algebra and $x$ an element of $L$, the map $[x,\, -]\colon |L| \to |L|$ is often denoted* $\mathrm{ad}_x$. *(This stands for "adjoint", but for obvious reasons we will not call it by that name in these notes.)*

In view of the way (9.7.2)–(9.7.4) arose above, we see that if we write $\mathbf{Ring}_k^1$ for the category of associative $k$-algebras, we have a functor

$$B\colon\ \mathbf{Ring}_k^1\ \longrightarrow\ \mathbf{Lie}_k$$

taking each associative $k$-algebra $A$ to the Lie algebra with the same underlying $k$-module, and with bracket operation given by the commutator bracket (9.7.1).

It is not hard to verify

**Exercise 9.7:1.** Show that $B$ has a left adjoint

$$E\colon\ \mathbf{Lie}_k\ \longrightarrow\ \mathbf{Ring}_k^1.$$

This is called the *universal enveloping algebra* construction.

When $L$ is free as a $k$-module (as is automatic if $k$ is a field), the algebra $E(L)$ has an elegant normal form, given by the Poincaré-Birkhoff-Witt

Theorem ([95, § V.2], [46, § 3]) from which one easily sees that for such $L$, the map giving the unit of the above adjunction, $\eta(L)\colon L \to B(E(L))$, is one-to-one. Thus, every Lie algebra over a field can be "embedded in" an associative algebra. An important consequence is

**Exercise 9.7:2.** Suppose $k$ is a field.

(i)    Assuming, as asserted above, that for all Lie algebras $L$, the map $\eta(L)$ is one-to-one, show that the Lie algebras of the form $B(A)$, as $A$ ranges over all associative $k$-algebras, generate the variety $\mathbf{Lie}_k$.

(ii)   Deduce that every identity satisfied by the $k$-module structure and the derived operation $[-, -]$ in all associative $k$-algebras $A$ is a consequence of the identities (9.7.2)–(9.7.4).

(iii)  Assuming the above results, describe how one can use the normal form for free associative $k$-algebras found in § 4.12 to test whether two terms in the Lie operations and an $X$-tuple of generator-symbols represent the same element of the free Lie algebra $F_{\mathbf{Lie}_k}(X)$.

(This is not quite the same as having a normal form in $F_{\mathbf{Lie}_k}(X)$, but it is useful in many of the same ways. Normal forms for $F_{\mathbf{Lie}_k}(X)$ are known, but they are messy. Incidentally, the result that $\eta(L) : L \to B(E(L))$ is one-to-one is actually known to hold for a much wider class of Lie algebras than those that are free as $k$-modules; but there are also examples for which it fails; see [71].)

If $R$ is any $k$-algebra (which for the moment we do not assume associative or Lie), and $H(R)$ is the *associative* $k$-algebra of all $k$-linear maps (i.e., $k$-module homomorphisms) $R \to R$, then it is easy to verify that if $s, t \in H(R)$ are both derivations of $R$ (i.e., satisfy (9.7.6)), then $s\,t - t\,s$ is also a derivation. Thus, the $k$-derivations on $R$ form a Lie subalgebra $\mathrm{Der}_k(R) \subseteq B(H(R))$; we will write this $\mathrm{Der}(R)$ when there is no danger of ambiguity.

For $R$ a Lie algebra or an associative algebra, a derivation of the form $\mathrm{ad}_x = [x, -]$ is called an *inner derivation*. (Here in the Lie algebra case, $[\,,\,]$ denotes the Lie bracket of $R$, while in the associative algebra case, it denotes the operation (9.7.1).)

**Exercise 9.7:3.** Let $R$ be a not necessarily associative $k$-algebra, with multiplication denoted $*$, and for $x \in |R|$ let $\mathrm{Ad}_x : |R| \to |R|$ denote the map $y \mapsto x * y - y * x$. (Thus if $R$ is associative, $\mathrm{Ad}_x$ coincides with the operation $\mathrm{ad}_x$ of the Lie algebra $B(R)$, while if $R$ is a Lie algebra, so that $*$ denotes $[-, -]$, $\mathrm{Ad}_x$ will be $2\,\mathrm{ad}_x$.)

Write down the identity that $R$ must satisfy for all of the maps $\mathrm{Ad}_x$ to be derivations. Show that if $R$ is anticommutative (satisfies $x*y+y*x = 0$) and 2 is invertible in $k$, this identity is equivalent to the Jacobi identity $x * (y * z) + y * (z * x) + z * (x * y) = 0$, but that in general (in particular, if $R$ is associative) it is not.

In terms of the "ad" notation, we can get yet another interpretation of the Jacobi identity. It is not hard to check that (9.7.8) is equivalent to

$$(9.7.10) \qquad \qquad \mathrm{ad}_{[y,\,z]} \ = \ \mathrm{ad}_y\,\mathrm{ad}_z - \mathrm{ad}_z\,\mathrm{ad}_y.$$

Thus the Jacobi identity also tells us that $\mathrm{ad}: L \to \mathrm{Der}(L)$ is a homomorphism of Lie algebras.

If $R$ is a commutative algebra, we see that the Lie algebra $B(R)$ has trivial bracket operation. However, even for such $R$, the associative $k$-algebra of $k$-module endomorphisms of $R$ is in general noncommutative, hence the Lie algebra $\mathrm{Der}(R)$, a sub-Lie-algebra thereof, can have nonzero bracket operation; and indeed, such Lie algebras $\mathrm{Der}(R)$ are important in commutative ring theory and differential geometry.

For example, let us take for $k$ the field $\mathbb{R}$ of real numbers, and for $R$ the commutative $\mathbb{R}$-algebra of all $C^\infty$ (i.e., infinitely differentiable) real-valued functions on $\mathbb{R}^n$, for some positive integer $n$. The next exercise will show that the derivations on $R$ are precisely the left $R$-linear combinations of the $n$ derivations $\partial/\partial x_0, \dots, \partial/\partial x_{n-1}$. Geometers identify the derivation

$$(9.7.11) \qquad \qquad D \ = \ \sum a_i(x)\,\partial/\partial x_i \quad (a_i(x) \in R)$$

with the $C^\infty$ vector field

$$(9.7.12) \qquad \qquad a(x) \ = \ (a_0(x),\ \dots,\ a_{n-1}(x)),$$

because for $f \in R$, $D(f)$ gives, at each point $p$, the rate of change of $f$ that would be seen by a particle at $p$ moving with velocity $a(p)$. In this way they make the space of such vector fields into a Lie algebra; more generally, they do this with the vector fields on any $C^\infty$ manifold.

**Exercise 9.7:4.** Let $R$ be the ring of $C^\infty$ functions on $\mathbb{R}^n$, and $D: R \to R$ any $\mathbb{R}$-linear derivation. For $i = 0, \dots, n-1$, let $a_i = D(x_i) \in R$, where $x_i \in R$ denotes the $i$-th projection map $\mathbb{R}^n \to \mathbb{R}$. You will show below that $D$ is given by (9.7.11); in other words, that for all $f \in R$ and $p \in \mathbb{R}^n$,

$$(9.7.13) \qquad \qquad D(f)(p) \ = \ \sum a_i(p)(\partial f/\partial x_i)(p).$$

The plan of the proof is to verify that for each $f \in R$, $f(x)$ can be written near $p$ as $f(p) + \sum (\partial f/\partial x_i)(p)\,(x_i - p_i)$ plus a second order remainder term ((9.7.14) below). One also shows that $D$ takes such a remainder term to a function that is $0$ at $p$. Hence, when one applies $D$ to the resulting expression for $f$ and evaluates at $p$, one gets (9.7.11).

We begin with two easy general facts.

(i)  Show that if $R$ is a $k$-algebra with multiplicative neutral element $1$, then any $k$-linear derivation on $R$ has $k$ in its kernel (where by $k$, I here mean $k\,1_R \subseteq R$).

(ii)  Show that if $I$ is an ideal of a commutative ring $R$, and we write $I^2$ for the ideal spanned by $\{gh \mid g, h \in I\}$, then for any derivation $D\colon R \to R$, we have $D(I^2) \subseteq I$.

Now let $R$, $D$, $f$ and $p$ be as in the first two paragraphs above, and let $I_p = \{a \in R \mid a(p) = 0\}$, an ideal of $R$. The next step will show that

(9.7.14)      $f(x) = f(p) + \sum (\partial f / \partial x_i)(p)(x_i - p_i) +$ term in $I_p^2$.

(iii)  For any point $x \in \mathbb{R}^n$, evaluate $f(x) - f(p)$ by the Fundamental Theorem of Calculus, applied along the line-segment from $p$ to $x$ parametrized by $t \in [0, 1]$. From the summand of this integral involving each operator $\partial / \partial x_i$, extract a factor $x_i - p_i$. Show that the integral remaining as the coefficient of $x_i - p_i$ is, as a function of $x$, a member of $R$ whose value at $p$ is $(\partial f / \partial x_i)(p)$; i.e., that the integral equals this constant plus a member of $I_p$. (In showing that these functions are $C^\infty$, you can use the fact that if $g(x_0, \ldots, x_{n-1}, t)$ is a $C^\infty$ function of $n+1$ variables, then $\int_0^1 g(x_0, \ldots, x_{n-1}, t) \, dt$ is a $C^\infty$ function of $n$ variables.)

(iv)  Complete the proof of (9.7.11).

(v)  Conclude that the Lie algebra of $\mathbb{R}$-derivations on $R$ is free as an $R$-module on the basis $\{\partial / \partial x_0, \ldots, \partial / \partial x_{n-1}\}$.

An analogous purely algebraic construction is to start with any commutative ring $k$, think of the polynomial ring $R = k[x_1, \ldots, x_n]$ as "functions on affine $n$-space over $k$," and think of its Lie algebra of derivations as "polynomial vector fields". These are easily shown (e.g., by calculation on monomials) to have the form $\sum a_i(x) \, \partial / \partial x_i$, where the derivations $\partial / \partial x_i$ are this time the operations of formal partial differentiation, and again $a_i(x) \in R$.

It turns out that for each $n$, the Lie algebra of vector fields on $n$-dimensional space, in either the geometric or algebraic sense, satisfies some additional identities beyond those satisfied by all Lie algebras. The case $n = 1$ is examined in the next exercise.

**Exercise 9.7:5.**

(i)   Let $C^\infty(\mathbb{R}^1)$ denote the ring of $C^\infty$ functions on the real line $\mathbb{R}^1$, and consider the Lie algebra of vector fields $L(\mathbb{R}^1) = \{f \, d/dx \mid f \in C^\infty(\mathbb{R}^1)\}$. Verify that the Lie bracket operation on $L(\mathbb{R}^1)$ is given by the formula

$$[f \, d/dx, \; g \, d/dx] = (f \, g' - g \, f') \, d/dx.$$

For notational convenience, let us regard this as a Lie algebra structure on $|C^\infty(\mathbb{R}^1)|$ :

$$[f, g] = f \, g' - g \, f'.$$

We shall continue to denote this Lie algebra $L(\mathbb{R}^1)$.

(ii)  Show that $L(\mathbb{R}^1)$ does not generate $\mathbf{Lie}_\mathbb{R}$. (You will want to find an identity satisfied in $L(\mathbb{R}^1)$, but not in all Lie algebras. To see how to prove the latter property, cf. Exercise 9.7:2(iii).)

The above result requires some computational dirty-work. On the other hand, even if you do not do that part, a little ingenuity will allow you to do the remaining parts.

(iii)  Show that for every positive integer $n$, $L(\mathbb{R}^1)$ contains a subalgebra which is free on $n$ generators in $\mathbf{Var}(L(\mathbb{R}^1))$.

(iv)  Show that $\mathbf{Var}(\mathrm{Der}(\mathbb{R}[x])) = \mathbf{Var}(L(\mathbb{R}^1))$, i.e., that polynomial vector fields satisfy no identities not satisfied by $C^\infty$ vector fields. However, show that in contrast to (iii), $\mathrm{Der}(\mathbb{R}[x])$ does *not* contain a subalgebra which is free on more than one generator in this variety.

You can carry the idea of part (ii) farther if you are interested, letting $n$ be a positive integer and looking at the variety generated by the Lie algebra $L(\mathbb{R}^n)$ of $C^\infty$ vector fields on $\mathbb{R}^n$, and possibly various subalgebras, such as the subalgebra of vector fields with zero divergence (in the sense defined in multivariable calculus).

One of the most important interpretations of Lie algebras lies outside the scope of this course, and I will only sketch it: the connection with Lie groups.

A *Lie group* is a topological group $G$ whose underlying topological space is a finite-dimensional manifold. Typical examples are the rotation group of real 3-space, which is a three-dimensional compact Lie group, and the group of motions of 3-space generated by rotations and translations, which is six-dimensional and noncompact. Some degenerate but important examples are the additive group of the real line, which is one-dimensional; its compact homomorphic image, the circle group $\mathbb{R}/\mathbb{Z}$, and finally, the discrete groups, which are the zero-dimensional Lie groups. It is known that every Lie group admits a unique $C^\infty$ structure, in fact, an analytic structure, respected by the group operations, [118, §4.10]. (Once this is known, one can choose to *define* a Lie group as an analytic space with a group structure defined by analytic operations. The authors of [118] make this choice; so the result cited, as they express it, is that "Every locally Euclidean topological group is a Lie group".)

If $G$ is a Lie group, $e$ its identity element, and $T_e$ the tangent space to $G$ at $e$, then every tangent vector $t \in T_e$ extends by left translation to a left-translation-invariant vector field on $G$. Hence the space of left-invariant vector fields may be identified in a natural manner with $T_e$. The commutator bracket of two left-invariant vector fields is left-invariant, so such vector fields form a Lie algebra; hence the above identification gives us a Lie algebra structure on $T_e$.

Here is another way of arriving at the same Lie algebra. Let us think of the *additive* structure of $T_e$ as the "first order approximation to the group structure of $G$ in the neighborhood of $e$." This approximation is abelian, which corresponds to the fact that the commutator of two elements of $G$

both of which are close to $e$ deviates from $e$ only "to second order". To measure the second-order *noncommutativity* of $G$ near the identity, let us identify a neighborhood of $0 \in T_e$ with a neighborhood of $e \in G$ in a $C^\infty$ manner, and on this identified neighborhood use vector-space notation for the operations of $T_e$, and $\circ$ for the multiplication of $G$. Then for $x, y \in T_e$ and real variables $s$ and $t$, that second-order noncommutativity is measured by the limit

$$\lim_{s,\, t \to 0} \frac{(s\,x) \circ (t\,y) - (t\,y) \circ (s\,x)}{s\,t}.$$

This limit turns out to exist for all $x, y \in T_e$; and writing its value as $[x, y] \in T_e$, one finds that this operation on $T_e$ coincides with the operation constructed above using left invariant vector fields. One can discover the identities (9.7.2)–(9.7.4) (with $\mathbb{R}$ for $k$) directly by examining the properties of the above limit; this gives another standard motivation of the concept of Lie algebra. (In proving (9.7.4), the group identity of Exercise 3.4:2(ii) can be useful.) Elements of this Lie algebra are thus often viewed heuristically as "infinitesimal" elements of the Lie group $G$.

For a familiar case, let $G$ be the rotation group of Euclidean 3-space, so that elements of $G$ represent rotations of space through various angles about various axes. Then elements of its Lie algebra $L$, heuristically, rotations differing infinitesimally from the identity, correspond to *angular velocities* about various axes. As a vector space, $L$ may be identified with $\mathbb{R}^3$, each $x \in L$ being described by a vector pointing along the axis of rotation, with magnitude equal to the angular velocity. The Lie bracket on $L$ is an operation on $\mathbb{R}^3$ known to all math, physics, and engineering students: the "cross product" of vectors.

It can be shown that the structure of a Lie group $G$ is determined "near $e$" by its Lie algebra: two Lie groups with isomorphic Lie algebras have neighborhoods of the identity which are isomorphic under the restrictions of the group operations to partial operations on that set.

**Exercise 9.7:6.** The ideas of Exercise 3.3:2 and the discussion preceding it showed that in the variety generated by a *finite* algebra, a free object on finitely many generators is finite. Is it similarly true that in the variety **Var**$(A)$ generated by any finite-dimensional associative or Lie algebra $A$ over a field $k$, a free object on finitely many generators is finite-dimensional? If not, can you prove some related condition (e.g., a condition of small "growth-rate" in the sense of Exercises 5.2:2–5.2:9)?

Can you at least show that such a variety **Var**$(A)$ must be distinct from the whole variety **Ring**$^1_k$, respectively **Lie**$_k$? In the Lie case, if $k = \mathbb{R}$, can you show it distinct from the subvariety **Var**$(L(\mathbb{R}^1))$ of Exercise 9.7:5(ii)?

If we combine the observation that the derivations on a $k$-algebra form a Lie algebra over $k$ with the intuition that elements of the Lie algebra associated with a Lie group represent "infinitesimal" elements of that group, we get the heuristic principle that a derivation on an algebra $A$ may be regarded as representing the difference between the identity automorphism of $A$ and an automorphism "very close" thereto. This relation with automorphisms suggests that every derivation on $A$ should be determined by what it does on a generating set, and, in the case of a free algebra, that it should be possible to specify it in an arbitrary way on the free generators. The next exercise obtains results of these sorts.

**Exercise 9.7:7.** Let $A$ be a not necessarily associative algebra over a commutative ring $k$.

(i)   Show that the *kernel* of any derivation $d \colon A \to A$ is a subalgebra of $A$. (This is analogous to the *fixed subalgebra* of an automorphism.) Deduce that two derivations which agree on a generating set for $A$ are equal.

The other result we want, about derivations on free algebras, requires a trick to turn derivations into something to which we can apply the universal property of free algebras. For this purpose, let $A'$ denote the $k$-algebra whose $k$-module structure is that of $A \times A$, and whose multiplication is given by

$$(9.7.15) \qquad\qquad (a,\, x)(b,\, y) \;=\; (a\,b,\ a\,y + x\,b).$$

(Thus, any two elements with first component $0$ have product $0$. We think of elements with first component zero as infinitesimals.) To apply the universal property of free algebras in (certain) varieties of $k$-algebras, we need the next two observations:

(ii)   Verify that if $A$ is associative with 1, respectively associative and commutative with 1, respectively Lie, then $A'$ has the same property.

(iii) Show that a map $d \colon A \to A$ is a derivation if and only if the map $a \mapsto (a, d(a))$ is a homomorphism $A \to A'$ as $k$-algebras, and that in this case, if $A$, and hence $A'$ has 1, this map will preserve that element.

(iv) Deduce that if $A$ is the free nonassociative $k$-algebra, the free associative $k$-algebra, the free associative commutative $k$-algebra, or the free Lie algebra over $k$ on a set $X$, then every set-map $X \to |A|$ extends uniquely to a derivation $A \to A$.

Returning to the last sentence of (i), there is the following related result.

(v)   Show that if $A$ is a field or a division ring, and $X$ a subset generating $A$ *as a field or division ring*, then any derivation $A \to A$ is determined by its restriction to $X$. Can you generalize this result?

We remark that the concept of a derivation from a $k$-algebra $A$ into itself is a case of the more general concept of a derivation $A \to B$, where $A$ is a

$k$-algebra and $B$ is an $A$-module (if $A$ is commutative and associative, or Lie) or an $A$-bimodule (in the general associative or nonassociative case); but we will not go into the details of these concepts here.

Some general references for the theory of Lie algebras are [91, 95, 131].

I will end this section by briefly mentioning some other concepts related to that of Lie algebra.

Our observation that for $A$ a $k$-algebra, the set of derivations $A \to A$ is closed under $k$-module operations and commutator brackets, and thus forms a Lie algebra, makes the concept of Lie algebra a useful tool for studying derivations when $k$ is a field of characteristic 0. But when $k$ has characteristic $p$, one finds that the set of derivations is also closed under the operation of taking $p$-th powers, and this fact needs to be taken into account in studying them. This leads one to study, for such $k$, Lie algebras $L$ given with an additional unary operation $a \mapsto a^{(p)}$ related to the other operations by the identities satisfied by the $p$-th power operation on derivations. Such structures are called *restricted Lie algebras* or *p-Lie algebras;* see [95].

If an associative algebra $A$ is given with an additional "bracket operation", written $\{x, y\}$, not assumed to be constructed from the associative multiplication, but making $L$ a Lie algebra such that each unary operation $\{a, -\}$ $(a \in L)$ is a derivation of the associative multiplication, then $A$ with this operation is called a *Poisson algebra*.

Our motivation of the definition of Lie algebra starting from (9.7.1) suggests the analogous question of what identities will be satisfied by the operation

$$(x, y) = xy + yx$$

on an associative algebra. This is the starting-point of the theory of *Jordan algebras*, though the subject is not as neat as that of Lie algebras. Jordan algebras are defined using the identities of degree $\leq 4$ satisfied by the above operation, but that operation also satisfies identities of higher degrees not implied by the Jordan identities; Jordan algebras satisfying these as well are called "semispecial". No analog of the connection between Lie *groups* and Lie algebras appears to exist for Jordan algebras. A standard reference for the theory of Jordan algebras is [96].

For our last variant of the concept of Lie algebras, suppose $A$ is a $Z_2$-*graded* associative algebra, and we define a modified bracket operation on $A$, given on homogeneous elements by

$$(9.7.16) \qquad [x, y] = xy - (-1)^{\deg(y)\deg(x)}yx.$$

On homogeneous elements, this operation turns out to satisfy identities like (9.7.4) and (9.7.5), but with appropriate sign-changes depending of the degrees of the elements. An arbitrary $Z_2$-graded $k$-module with a bracket operation satisfying these identities is called a *super Lie algebra*.

Let us now return to general algebras.

## 9.8. Some instructive trivialities

**Definition 9.8.1.** *If* $g\colon S^X \to S$ *is an* $X$-*ary operation on a set* $S$, *and* $a\colon X \to Y$ *is a set map, then by the* $Y$-*ary operation on* $S$ *induced by* $g$ *via the map* $a$ *of arity-sets, we shall mean the map* $f\colon S^Y \to S$ *defined by*

$$f((c_y)_{y\in Y}) \;=\; g((c_{a(x)})_{x\in X}).$$

The covariance of this construction in the arity-set is actually the result of two contravariances: $a\colon X \to Y$ induces a map $S^Y \to S^X$, then this gives a map $S^{(S^X)} \to S^{(S^Y)}$.

If in the above definition we take for $g$ a derived $X$-ary operation of an algebra structure on $S$, say corresponding to an element $s \in |F_\Omega(X)|$, then $f$ will be a derived $Y$-ary operation, corresponding to the image of $s$ under the homomorphism $F_\Omega(a)\colon F_\Omega(X) \to F_\Omega(Y)$. (In terms of *this* description, the covariance is straightforward.)

**Definition 9.8.2.** *If* $f\colon S^Y \to S$ *is an operation on a set, and* $X$ *is a subset of the index set* $Y$, *we shall say that* $f$ *depends only on the indices in* $X$ *if* $f$ *takes on the same value at any two* $Y$-*tuples that* (*regarded as functions* $Y \to S$) *have the same restriction to* $X$.

**Lemma 9.8.3.** *If in the context of Definition 9.8.2 either* $S$ *or* $X$ *is nonempty, then* $f$ *depends only on the indices in* $X$ *if and only if* $f$ *is induced by an* $X$-*ary operation* $g$ *on* $S$, *via the inclusion of* $X$ *in* $Y$. $\square$

**Exercise 9.8:1.** Prove Lemma 9.8.3. Your proof should show why the condition "$S$ or $X$ is nonempty" is needed.

In some works on general algebra, there is a confusion between *zeroary* derived operations, and *constant* derived operations of *nonzero* arities. The next two exercises show that these two sorts of operations carry almost, but not exactly, the same information.

**Exercise 9.8:2.**

(i) (Like Exercise 9.8:1, but for derived operations.) Show that if a derived $Y$-ary operation $s$ of an algebra $A$ depends only on indices in a subset $X \subseteq Y$, and $X$ is nonempty, then $s$ is in fact induced by an $X$-ary derived operation of $A$.

(ii) On the other hand, suppose the derived $Y$-ary operation $s$ of $A$ depends only on the empty set of indices in $Y$, i.e., is constant. If $A$ has zeroary operations, show that, as in (i), but for a different reason, $s$ is induced by a zeroary derived operation of $A$. Show, however, that if $A$ has no zeroary operations, then derived operations depending on the empty set of indices can still exist, but will not be induced by derived zeroary operations.

In particular, for $m \leq n$, derived $m$-ary operations correspond to derived $n$-ary operations depending only on the first $m$ variables, *except* for the $m = 0$ case, where this is not true unless the algebra has zeroary primitive operations.

Zeroary operations and constant unary operations look still more alike if one excludes empty algebras, as is shown by

**Exercise 9.8:3.** We have seen that the $X$-ary derived operations of a variety $\mathbf{V}$ can be characterized as the morphisms $U_\mathbf{V}^X \to U_\mathbf{V}$ where $U_\mathbf{V}$ is the underlying-set functor of $\mathbf{V}$.

Suppose now that $\mathbf{V}$ is a variety *without* zeroary operations, hence having an empty algebra $I$. Let $\mathbf{V} - \{I\}$ denote the full subcategory of $\mathbf{V}$ consisting of all nonempty algebras, and let $U_{\mathbf{V}-\{I\}}$ denote the restriction of $U_\mathbf{V}$ to this subcategory.

(i)   Show that morphisms $(U_{\mathbf{V}-\{I\}})^X \to U_{\mathbf{V}-\{I\}}$ correspond to derived $X$-ary operations of $\mathbf{V}$ *except* in the case $X = \emptyset$, in which case they can be put in natural correspondence with the constant derived unary operations.

(ii)   Show that if $\mathbf{V}$ has constant derived unary operations, then $\mathbf{V} - \{I\}$ is isomorphic in a natural way to a variety of algebras (of a different type) having zeroary operations.

As an example, suppose that (as has been proposed from time to time) one sets up a variant of the concept of "group", based only on the two operations of composition and inverse, and one axiomatizes these using the *associative law* for composition, and the following identities, which hold in ordinary groups as consequences of the inverse and neutral-element laws:

$$x = x\,y\,y^{-1} = y^{-1}y\,x.$$

(iii)   Let $\mathbf{V}$ be the variety so defined. Show that the category $\mathbf{V} - \{I\}$ is isomorphic to **Group**.

Having thrown some light, I hope, on the relationship between zeroary operations and constant unary operations, let me end this section with an exercise of a different sort that comes out of Definition 9.8.2.

**Exercise 9.8:4.** Let $S$ and $Y$ be sets and $f \colon S^Y \to S$ a $Y$-ary operation on $S$.

(i)   Suppose $W, X \subseteq Y$ are sets such that $f$ depends only on the indices in $W$, and $f$ also depends only on the indices in $X$. Show that $f$ depends only on the indices in $W \cap X$.

(ii)   On the other hand, show that given an infinite family of subsets $X_i \subseteq Y$ such that for each $i$, $f$ depends only on the indices in $X_i$, it may not be true that $f$ depends only on the indices in $\bigcap X_i$. (Suggestion: Let $S = [0, 1] \subseteq \mathbb{R}$, $Y = \omega$, and $f$ be the operation $\limsup$.)

(iii) In general, given a $Y$-ary operation $f$ on $S$, what properties must the set

$$D_f = \{X \subseteq Y \mid f \text{ depends only on indices in } X\}$$

have? Can you find conditions on a family $U$ of subsets of $Y$ which are necessary and sufficient for there to exist a set $S$ and a function $f \colon S^Y \to S$ such that $U = D_f$?

## 9.9. Clones and clonal theories

Given a family of *unary* operations on a set $S$, i.e., maps $S \to S$, the composites of these (together with the "empty composite", the identity map) form a *monoid* of maps of $S$ into itself. In this section we will look at the structure of the set of derived operations of a family of *not necessarily unary* operations, under the operations analogous to composition of unary operations.

We will limit ourselves to finitary operations. (There is no problem with the infinitary case, but I thought the concepts would come across more clearly in the familiar finitary context. The reader interested in the infinitary case can easily make the appropriate generalizations, replacing "finite" by " $\leq \gamma_0$ ", for $\gamma_0$ any regular infinite cardinal.) We will also, in this presentation, make our arities natural numbers (for the infinitary case read "cardinals $\leq \gamma_0$") rather than arbitrary finite sets, since allowing all finite sets as arities would mean that every algebra would have a *large* set of formally distinct operations.

**Definition 9.9.1.** *Let $S$ be a set. Then a* clone of operations on $S$ *will mean a set $C$ of operations on $S$, of natural-number arities, which is closed under formation of derived operations. Concretely, this says that*

(i) *For every natural number $n$, $C$ contains the $n$ projection maps $p_{n,i} \colon S^n \to S$ $(i \in n)$, defined by*

$$(9.9.2) \qquad\qquad p_{n,i}(\xi_0, \ldots, \xi_{n-1}) = \xi_i,$$

*and*

(ii) *Given natural numbers $m, n \in \omega$, an $m$-ary operation $s \in C$, and $m$ $n$-ary operations $t_0, \ldots, t_{m-1} \in C$, the set $C$ also contains the $n$-ary operation*

$$(9.9.3) \quad (\xi_0, \ldots, \xi_{n-1}) \longmapsto s(t_0(\xi_0, \ldots, \xi_{n-1}), \ldots, t_{m-1}(\xi_0, \ldots, \xi_{n-1}))$$

*i.e., the composite*

$$S^n \xrightarrow{\ (t_0, \ldots, t_{m-1})\ } S^m \xrightarrow{\ s\ } S.$$

*The least clone on $S$ containing a given set of operations will be called the clone generated by that set. Thus, for any finitary type $\Omega$ and any $\Omega$-algebra A, the set of derived operations of A of natural-number arities constitutes the clone generated by the primitive operations of A.*

Let us look at an example of how this procedure of generation works. Given a binary operation $f$ and a ternary operation $g$ on a set $S$, how do we express in terms of the constructions (9.9.2) and (9.9.3) the 6-ary operation

$$(\xi_0, \ldots, \xi_5) \longmapsto f(g(\xi_0, \xi_1, \xi_2), \ g(\xi_3, \xi_4, \xi_5)) \, ?$$

It should clearly arise as an instance of (9.9.3) with $f$ for $s$, but we cannot, as we might first think, take $g$ for $t_0$ and $t_1$. That would give the *ternary* operation $(\xi_0, \xi_1, \xi_2) \mapsto f(g(\xi_0, \xi_1, \xi_2), g(\xi_0, \xi_1, \xi_2))$. We need, rather, to use as $t_0$ and $t_1$ the two 6-ary operations $(\xi_0, \ldots, \xi_5) \mapsto g(\xi_0, \xi_1, \xi_2)$ and $(\xi_0, \ldots, \xi_5) \mapsto g(\xi_3, \xi_4, \xi_5)$. We get these, in turn, as instances of (9.9.3) with $g$ in the role of $s$, and projection maps (9.9.2) in the role of the $t_i$'s. Namely, taking for $t_0, t_1, t_2$ the projection maps $p_{6,0}$, $p_{6,1}$, $p_{6,2}$, we get the first of the above 6-ary operations, and using the remaining three 6-ary projection maps, we get the other. We can then apply (9.9.3) to $f$ and these two 6-ary operations to get the desired 6-ary operation. (In this example, each of our variables happened to appear exactly once in the final expression, and the occurrences were in ascending order of subscripts, but obviously, by different choices of projection maps, we can get expressions in which variables appear more than once, and in arbitrary orders.)

Above, we got a "new" operation by inserting into the ternary operation $g$ the 6-ary projection maps $p_{6,0}$, $p_{6,1}$, $p_{6,2}$. It is clear that if we had, instead, inserted the ternary projections $p_{3,0}$, $p_{3,1}$, $p_{3,2}$ (in that order) we would have gotten back precisely the operation $g$. Note also that if we substitute any operation $f$ into the unary projection map $p_{1,0}$, we get the operation $f$ back. These phenomena are analogs of the *neutral-element laws* in a monoid.

One also has an analog of the associative law: If $m$, $n$ and $p$ are nonnegative integers, then given an $m$-ary operation $s$, any $m$ $n$-ary operations $t_i$ $(i \in m)$, and any $n$ $p$-ary operations $u_j$ $(j \in n)$, all on the set $S$, one can either substitute the $t$'s into $s$, and the $u$'s into the resulting operation, or first substitute the $u$'s into the $t$'s, and then the resulting operations into $s$. In each case one gets the $p$-ary operation which is the composite of the set maps

$$(9.9.4) \qquad S^p \xrightarrow{\ (u_0, \ldots, u_{n-1})\ } S^n \xrightarrow{\ (t_0, \ldots, t_{n-1})\ } S^m \xrightarrow{\ s\ } S.$$

It looks as though we ought to abstract these properties, and use them as the definition of a new sort of algebraic object, which we might call a "formal substitution algebra" or a "clonal algebra". We would then have a new way of looking at varieties of algebras: Given a type $\Omega$ and a family $J$ of identities, we would construct a "clonal algebra" $\langle \Omega \mid J \rangle$ presented by these operation-symbols and identities. We could then define a "representation" of this clonal algebra on a set $|A|$ to mean a homomorphism of $\langle \Omega \mid J \rangle$ into the clone of all finitary operations on that set. Such representations of $\langle \Omega \mid J \rangle$ could be identified with $\Omega$-algebras satisfying the identities of $J$; thus, each variety of algebras could be looked at as the category of representations of a clonal algebra.

Unfortunately, these "clonal algebras" would not be algebras as we have so far defined the term. Our algebras $A$ have an underlying set $|A|$; but a "clonal algebra" would have an underlying *family* of sets, one set for each arity of the operations symbolized, with composition operations associated to appropriate combinations of these.

Now there is, in fact, a concept of *many-sorted algebra* (algebra having different "sorts" of elements), and our general theory of algebras can be adapted to that context in a fairly straightforward way; I have not done so simply for conceptual simplicity. If I had developed the theory of many-sorted algebras, we could formalize the ideas sketched above using these.

But in fact, we don't need a new kind of mathematical object to do what we have been discussing. After all, we introduced the concept of a *category* to formalize the properties of composition of maps, which is what we are dealing with here. The apparent difficulty with looking at the members of a clone of operations as morphisms in a category is that an $m$-ary operation in a clone is composed on the right, not with a single $n$-ary operation, but with a family of $m$ such operations. The solution is to define our category so that the typical morphism therein is not a single $n$-ary operation $|A|^n \to |A|$, but an $m$-tuple of $n$-ary operations, corresponding to a map $|A|^n \to |A|^m$.

Now everything falls into place! The category should have objects $X_n$ in one-to-one correspondence with the natural numbers $n$, and a morphism between $X_n$ and $X_m$ should correspond to an $m$-tuple of $n$-ary operations in our clone.

In saying "a morphism between $X_n$ and $X_m$", I have skirted the question of which way the morphism should go. This is a notational choice: whether we want to encode our structure as a certain category, or as its opposite. The development we have just seen suggests that the morphisms corresponding to $m$-tuples of $n$-ary operations should go from $X_n$ to $X_m$, since an $m$-tuple of $n$-ary operations of an algebra $A$ gives a set map $|A|^n \to |A|^m$. More globally, an $m$-tuple of derived $n$-ary operations of a variety $\mathbf{V}$ is equivalent to a morphism $U_{\mathbf{V}}^n \to U_{\mathbf{V}}^m$, so the "clone of derived operations" of $\mathbf{V}$ should be describable as the full subcategory of $\mathbf{Set}^{\mathbf{V}}$ having the functors $U_{\mathbf{V}}^n$ as objects.

But there is also motivation for the opposite choice. Recall that the derived $n$-ary operations of a variety $\mathbf{V}$ correspond to the elements of the free algebra $F_{\mathbf{V}}(n)$. An $m$-tuple of such elements is picked out by a homomorphism $F_{\mathbf{V}}(m) \to F_{\mathbf{V}}(n)$; so the full subcategory of $\mathbf{V}$ consisting of the free objects $F_{\mathbf{V}}(n)$ also embodies the structure of the operations of $\mathbf{V}$, in the manner *opposite* to way it is embodied in morphisms $U_{\mathbf{V}}^n \to U_{\mathbf{V}}^m$. This is, of course, a case of the contravariance of the Yoneda equivalence between the covariant functors $U_{\mathbf{V}}^n$ and their representing objects $F_{\mathbf{V}}(n)$.

Postponing the above question for a moment, let us note that, whichever choice we make, we will want to know *which* categories with object-set of the form $\{X_n \mid n \in \omega\}$ correspond in this way to clones of operations. Clearly, such a category should be given with a distinguished family of $n$ morphisms $p_{n,i}$ ($i \in n$) between $X_1$ and $X_n$ for each $n$ (corresponding, in one description to the $n$ projection maps $|A|^n \to |A|$, and in the other to the $n$ obvious morphisms $F_{\mathbf{V}}(1) \to F_{\mathbf{V}}(n)$). It must also have the property that the morphisms between $X_n$ and $X_m$ (in the appropriate direction) correspond, via composition with the $p_{m,i}$, to the $m$-tuples of morphisms between $X_n$ and $X_1$.

These conditions together say that in the category, each object $X_n$ is the *product* (or *coproduct*) of $n$ copies of $X_1$, with the given morphisms $p_{n,i}$ as (co)projection maps. As to the choice of direction of the morphisms, Lawvere, in his doctoral thesis [17] where he introduced these ideas, made $X_n$ a *product* of $n$ copies of $X_1$, but in later published work he switched to the definition under which it would be a coproduct, in other words, under which the category would look like the category of free algebras $F_{\mathbf{V}}(n)$. An attractive feature of the latter choice for Lawvere is that the category having *only* the maps $p_{n,i}$ for morphisms $X_1 \to X_n$ (corresponding to the variety with no primitive operations) is the full subcategory $\mathbf{N} \subseteq \mathbf{Set}$ having the set $\mathbb{N} = \omega$ of natural numbers for object-set; hence, the category corresponding to a general variety can be characterized as a certain kind of extension of $\mathbf{N}$. This fits with his project of creating a category-theoretic foundation for set theory and for mathematics, with the category $\mathbf{N}$ as a basic building-block. I prefer the other choice of variance because it leads to a covariant relationship between this category of formal operations, and the actual operations in the variety. I will include both versions in the definition below, calling them the "contravariant" and "covariant" versions, but from that point on, we will generally work with the covariant formulation.

Lawvere calls the category of formal operations of a variety $\mathbf{V}$ the "theory of $\mathbf{V}$", and any category of this form an "algebraic theory". For us this would be awkward, for though these categories carry approximately the same information as equational theories (Definition 9.4.6), the two concepts are different enough that we cannot identify them. So let us introduce a different term.

**Definition 9.9.5.** *A covariant clonal category will mean a category $\mathbf{X}$ given with a bijective indexing of its object-set by the natural numbers,*

$$\mathrm{Ob}(\mathbf{X}) = \{X_n \mid n \in \omega\},$$

*and given with morphisms*

$$p_{n,i} \colon \ X_n \ \longrightarrow \ X_1 \ \ (i \in n)$$

*which make each $X_n$ the product of $n$ copies of $X_1$, and such that $p_{1,0}$ is the identity map of $X_1$. (Equivalently, letting $\mathbf{N}$ denote the full subcategory of $\mathbf{Set}$ whose objects are the natural numbers, this means a category $\mathbf{X}$ given with a functor $\mathbf{N}^{\mathrm{op}} \to \mathbf{X}$ which is bijective on object-sets, and turns finite coproducts in $\mathbf{N}$ to products in $\mathbf{X}$.)*

*A contravariant clonal category will mean a category $\mathbf{X}$ given with the dual sort of structure, equivalently, given with a covariant clonal category structure on $\mathbf{X}^{\mathrm{op}}$, equivalently, given with a functor $\mathbf{N} \to \mathbf{X}$ which is bijective on object-sets and respects finite coproducts.*

*(More generally, for any infinite regular cardinal $\gamma_1$, one may define concepts of covariant and contravariant $<\gamma_1$-clonal category, using in place of $\mathbf{N}$ the full subcategory of $\mathbf{Set}$ having for object-set the cardinal $\gamma_1$, and in place of finite (co)products, (co)products of $< \gamma_1$ factors.)*

**Exercise 9.9:1.** Establish the equivalence of structures noted parenthetically in the first paragraph of the above definition.

A clonal category is itself a mathematical object, so we make

**Definition 9.9.6.** *By* **Clone** *we shall denote the category whose objects are the covariant clonal categories, and where a morphism $\mathbf{X} \to \mathbf{Y}$ is a functor which carries $X_n$ to $Y_n$ for each $n$, and respects the morphisms $p_{n,i}$. In other words,* **Clone** *will denote the full subcategory of the comma category $(\mathbf{N}^{\mathrm{op}} \downarrow \mathbf{Cat})$ (Exercise 7.8:30(iii)) whose objects are the clonal categories.*

Incidentally, when one forms the category of *contravariant* clonal categories, this is isomorphic to our present category **Clone**, *not* opposite thereto, since the direction of morphisms within clonal categories does not affect the direction of functors among them.

We now wish to establish the relation between clonal categories and varieties of algebras. First, given a variety $\mathbf{V}$, how shall we define the associated clonal category? The most convenient choice from the formal point of view would be to use $n$-ary derived operations of $\mathbf{V}$ as the morphisms from $X_n$ to $X_1$. (This construction was sketched for $\mathbf{V} = \mathbf{Group}$ when we were noting "nonprototypical" ways categories could arise, in the paragraph containing (7.2.1).) Unfortunately, these derived operations are not small as sets (though the set of them is quasi-small). So let us use in their stead the corresponding elements of the free $\mathbf{V}$-algebra $F_{\mathbf{V}}(n)$. Of course, we will define the composition operation of the clone so as to correspond to composition of derived operations.

**Definition 9.9.7.** *If* **V** *is a variety of finitary algebras, the covariant* clonal theory *of* **V** *will mean the clonal category* **Cl(V)**, *with objects denoted* $\mathrm{Cl}_n(\mathbf{V})$, *where a morphism from* $\mathrm{Cl}_n(\mathbf{V})$ *to* $\mathrm{Cl}_m(\mathbf{V})$ *means an m-tuple of elements of* $|F_{\mathbf{V}}(n)|$, *and composition of such morphisms*

$$\mathrm{Cl}_p(\mathbf{V}) \xrightarrow{\;(u_i)\,\in\,|F_{\mathbf{V}}(p)|^n\;} \mathrm{Cl}_n(\mathbf{V}) \xrightarrow{\;(t_i)\,\in\,|F_{\mathbf{V}}(n)|^m\;} \mathrm{Cl}_m(\mathbf{V})$$

*is defined by substitution of n-tuples of expressions in p indeterminates into expressions in n indeterminates; and where each morphism* $p_{n,i}$ *is given by i-th member of the universal n-tuple of generators of* $F_{\mathbf{V}}(n)$. *We note that this is equivalent (via a natural isomorphism) to the full subcategory of the large category* **Set**$^{\mathbf{V}}$ *having for objects the functors* $U_{\mathbf{V}}^n$ *(*$n \in \omega$*), and also to the* opposite *of the small full subcategory of* **V** *having for objects the free* **V***-algebras* $F_{\mathbf{V}}(n)$.

*Given a clonal category* **X**, *an* **X***-algebra will mean a functor* **X** $\rightarrow$ **Set** *respecting the product structures defined on the objects* $X_n$ *by the projection maps* $p_{n,i}$. *For each clonal category* **X**, *the category of all* **X***-algebras will be written* **X-Alg**. *The functor* **X-Alg** $\rightarrow$ **Set** *taking each* **X***-algebra* $A$ *to the set* $A(X_1)$ *will be written* $U_{\mathbf{X\text{-}Alg}}$, *or* $U$ *when there is no danger of ambiguity, and called the "underlying-set functor" of* **X-Alg**.

*(The analogs of the categories and objects named in this and the preceding definition with arities taken from an arbitrary regular infinite cardinal* $\gamma_1$, *rather than the natural numbers, may be written* **Clone**$^{(\gamma_1)}$, **Cl**$^{(\gamma_1)}(\mathbf{V})$, $\mathrm{Cl}_\alpha^{(\gamma_1)}(\mathbf{V})$, *etc.)*

Note that by our general conventions, unless the contrary is stated, a clonal category **X** is legitimate, hence, as it has by definition a small set of objects, it is small. Thus, the corresponding category **X-Alg** is legitimate.

We have designed these concepts so that the categories **X-Alg** are essentially the same as classical varieties of algebras. Let us state this property as

**Lemma 9.9.8.** *If* **X** *is a clonal category, then* **X-Alg** *(second paragraph of Definition 9.9.7) is equivalent to a variety* **V** *of finitary algebras, by an equivalence respecting underlying-set functors. For* **V** *so constructed,* **Cl(V)** *(first paragraph of Definition 9.9.7) is naturally isomorphic in* **Clone** *to* **X**.

*Inversely, if* **V** *is any variety of finitary algebras, then* **Cl(V)-Alg** *is equivalent to* **V**.                                                                                        □

**Exercise 9.9:2.** Prove Lemma 9.9.8.

As discussed in our motivation (paragraph containing (9.9.4)), for **X** a clonal category, an **X**-algebra can be thought of as a "representation" of the clone **X** by sets and set maps. This suggests the following more general definition, which we will find useful in the next chapter:

**Definition 9.9.9.** *If* $\mathbf{X}$ *is a clonal category, and* $\mathbf{C}$ *any category with finite products, then a* representation *of* $\mathbf{X}$ *in* $\mathbf{C}$ *will mean a covariant functor* $A\colon \mathbf{X} \to \mathbf{C}$ *respecting the product structures defined on the objects* $X_n$ *by the projection maps* $p_{n,\,i}$ *(i.e., such that for each* $n \in \omega$, *the object* $A(X_n)$ *of* $\mathbf{C}$ *is the product of* $n$ *copies of* $A(X_1)$ *via the projection maps* $A(p_{n,i})$.)

We remark that the information given by a clonal category is not quite the same as that given by a variety, in that the clonal category does not distinguish between *primitive* and *derived* operations, while under our definition, a variety does.

Lawvere *defines* a variety of algebras (in his language, an "algebraic category") to mean a category of the form $\mathbf{X}\text{-}\mathbf{Alg}$, where $\mathbf{X}$ is what we call a clonal category (and he calls a theory). This is a reasonable and elegant definition, but since we began with the classical concepts of variety and theory, and it is pedagogically desirable to hold to one definition, we shall keep to our previous definition of variety, and study the categories $\mathbf{X}\text{-}\mathbf{Alg}$ as a closely related concept.

**Exercise 9.9:3.** Let $\mathbf{2N}$ be the full subcategory of $\mathbf{Set}$ having for objects the nonnegative *even* integers. For each integer $n$, the object $2n$ of $\mathbf{2N}$ is a coproduct of $n$ copies of the object $2$, hence the opposite category $(\mathbf{2N})^{\mathrm{op}}$ can be made a covariant clonal category by an appropriate choice of maps $p_{n,\,i}$. Write down such a system of maps $p_{n,\,i}$, and obtain an explicit description of $(\mathbf{2N})^{\mathrm{op}}\text{-}\mathbf{Alg}$ as a variety $\mathbf{V}$ determined by finitely many operations and finitely many identities. Your answer should show what it means to put a $(\mathbf{2N})^{\mathrm{op}}$-algebra structure on a set.

**Exercise 9.9:4.** In defining an $\mathbf{X}$-algebra as a certain kind of functor in Definition 9.9.7, we required that this functor respect the given structures of the objects $X_n$ as $n$-fold products of $X_1$.

(i)    Show that under the conditions of that definition, to respect these distinguished products is equivalent to respecting *all* finite products that exist in $\mathbf{X}$.

(ii)    Show on the other hand that an $\mathbf{X}$-algebra may fail to respect infinite products in $\mathbf{X}$. (To do this, you must start by finding a clonal category $\mathbf{X}$ having a nontrivial infinite product of objects!)

**Exercise 9.9:5.** Show that, up to isomorphism, there are just two clonal categories $\mathbf{X}$ such that the functor $\mathbf{N}^{\mathrm{op}} \to \mathbf{X}$ is not faithful. What are the corresponding varieties?

The next exercise does not involve the concept of clonal category, and could have come immediately after the definition of a clone of operations on a set, but I didn't want to break the flow of the discussion. It requires familiarity with a bit of elementary electronics.

**Exercise 9.9:6.** (Inspired by a question of F. E. J. Linton.)

If $n$ is a positive integer, let us understand an "$n$-labeled circuit graph" to mean a finite connected graph $\Gamma$ (which may have more than one edge between two given vertices), with two distinguished vertices $v_0$ and $v_1$, and given with a function from the set of edges of $\Gamma$ to $n = \{0, \ldots, n-1\}$. To each such graph let us associate the $n$-ary operation on the nonnegative real numbers that takes each $n$-tuple $(r_0, \ldots, r_{n-1})$ of such numbers to the *resistance* that would be measured between $v_0$ and $v_1$ if for each $i$, every edge of $\Gamma$ mapped to $i$ were a resistor with resistance $r_i$.

(i)  Explain (briefly) why the set of operations on nonnegative real numbers arising in this way from labeled circuit graphs forms a *clone*.

(ii)  Let $s$ denote the binary operation in this clone corresponding to putting two resistors in *series*, $p$ the binary operation corresponding to putting two resistors in *parallel*, and $w$ the 5-ary operation corresponding to a *Wheatstone Bridge*; i.e., determined by the graph ⬦, with $v_0$ and

$v_1$ the top and bottom vertices, and distinct labels on all five edges. Show that none of these three operations is in the subclone generated by the other two. (Suggestion: Look at the behavior of these three operations with respect to the order relation on the nonnegative reals.)

A much more difficult question is

(iii)  Do the three operations listed in (ii) generate the clone of (i)?

I do not know answers to the next two questions.

(iv)  Can one *characterize* the set of operations belonging to the clone of part (i), i.e., describe some test that can be applied to an $n$-ary operation on positive real numbers to determine whether it belongs to the clone?

(v)  Can one find a generating set for the identities satisfied by the two binary operations $s$ and $p$? (This was the question of Fred Linton's which inspired this exercise. Generating sets for identities of other families of operations in this clone would, of course, likewise be of interest.)

(vi)  Suppose one is interested in more general electrical circuits; e.g., circuits containing not only resistors, but also capacitors, inductances, and possibly other elements. Can one somehow extend the "clonal" viewpoint to such circuits? (If we allow circuit components such as rectifiers, which do not behave symmetrically, we must work with directed rather than undirected graphs.)

(vii) If you succeed in extending the clonal approach to circuits composed of resistors, capacitors and inductances, is the clone you get isomorphic to the clone of part (i) (the clone one obtains assuming all elements are resistors)?

Recall that the morphisms between clonal categories are the functors respecting the indexing of the object-set and the morphisms $p_{n,i}$. What do such functors mean from the viewpoint of the corresponding varieties of algebras? If $\mathbf{V}$ is a variety of $\Omega$-algebras and $\mathbf{W}$ is a variety of $\Omega'$-algebras, we see that to specify a morphism $f \in \mathbf{Clone}(\mathbf{Cl}(\mathbf{V}), \mathbf{Cl}(\mathbf{W}))$ one must associate to every primitive operation $s$ of $\mathbf{V}$ a derived operation $f(s)$ of $\mathbf{W}$ of the same arity, so that the defining identities for $\mathbf{V}$ in those primitive operations are satisfied by the derived operations $f(s)$ in $\mathbf{W}$. We find that such a morphism $f$ determines a functor in the opposite direction, $\mathbf{W} \to \mathbf{V}$; namely, given a $\mathbf{W}$-algebra $A$, we get a $\mathbf{V}$-algebra $A_f$ with the same underlying set by using for each primitive $\mathbf{V}$-operation $s_{A_f}$ the derived operation $f(s)_A$ of the $\mathbf{W}$-structure on $|A|$. In fact we have

**Lemma 9.9.10** (Lawvere). *Functors between varieties of algebras which preserve underlying sets correspond bijectively to morphisms in the opposite direction between the clonal theories of these varieties, via the construction described above.*                                                                    □

**Exercise 9.9:7.** Prove Lemma 9.9.10.

Easy examples of such functors among varieties are the *forgetful* functors $\mathbf{Group} \to \mathbf{Monoid}$, $\mathbf{Ring}^1 \to \mathbf{Monoid}$, $\mathbf{Ring}^1 \to \mathbf{Ab}$, $\mathbf{Lattice} \to \vee\text{-}\mathbf{Semilattice}$, and similar constructions, including the underlying-set functor of every variety, and the inclusion functor of any subvariety in a larger variety, e.g., $\mathbf{Ab} \to \mathbf{Group}$. In the above list of cases, each primitive operation of the codomain variety happens to be mapped to a primitive operation of the domain variety. Some examples in which primitive operations are mapped to non-primitive operations are the functor $\mathbf{Bool}^1 \to \vee\text{-}\mathbf{Semilattice}$ under which the semilattice operation $x \vee y$ is mapped to the Boolean ring operation $x + y + xy$; the functor $H \colon \mathbf{Group} \to \mathbf{Heap}$ of Exercise 9.6:10, under which the ternary heap operation $\tau$ is mapped to the group operation $xy^{-1}z$, and the functor $B \colon \mathbf{Ring}^1_k \to \mathbf{Lie}_k$ of §9.7, under which, though the primitive $k$-module operations of $\mathbf{Lie}_k$ are mapped to the corresponding primitive operations of $\mathbf{Ring}^1_k$, the Lie bracket is mapped to the commutator operation $xy - yx$.

We have seen most of the above constructions before, as examples of functors having left adjoints. In fact, one can prove that any functor between varieties induced by a morphism of their clonal theories—in other words, every functor between varieties that preserves underlying sets—has a left adjoint! We will not stop to do this here, because it is a case of a much more general result we will prove in the next chapter, which will show us precisely which functors between varieties have left adjoints. But you can, if you wish, do this case now as an exercise:

**Exercise 9.9:8** (Lawvere). Show that any functor between varieties of finitary algebras which preserves underlying sets has a left adjoint.

(You may drop "finitary" if you wish, either using generalized versions of the results of this section, or proving the result without relying on the ideas of this section.)

**Exercise 9.9:9.** Since morphisms $\mathbf{X} \to \mathbf{Y}$ in **Clone** are defined to be certain functors, we can also look at *morphisms between two such functors*, i.e., morphisms between morphisms in **Clone**. Interpret this concept in terms of varieties of algebras. That is, given two varieties of algebra $\mathbf{V}$ and $\mathbf{W}$, and two underlying-set-preserving functors $F, G \colon \mathbf{V} \to \mathbf{W}$, corresponding to functors $f, g \colon \mathrm{Cl}(\mathbf{W}) \to \mathrm{Cl}(\mathbf{V})$, what data relating $F$ and $G$ corresponds to a morphism $f \to g$?

In particular, name one or more underlying-set preserving functors $\mathbf{Bool}^1 \to \mathbf{Semilat}$ and/or $\mathbf{Group} \to \mathbf{Monoid}$, describe the corresponding functors between clonal categories, and then find examples of morphisms between two of those functors, or nonidentity endomorphisms of one of them, and interpret these in terms of the given varieties of algebras.

Here are some exercises on particular underlying-set-preserving functors and their adjoints:

**Exercise 9.9:10.** Let $U \colon \mathbf{Group} \to \mathbf{Monoid}$ denote the forgetful functor, and $F \colon \mathbf{Monoid} \to \mathbf{Group}$ its left adjoint (called in § 4.11 the "universal enveloping group" construction).

(i)   Show that there exist proper subvarieties $\mathbf{V} \subseteq \mathbf{Group}$ such that $U(\mathbf{V})$ does not lie in a proper subvariety of **Monoid**.

A much harder problem is

(ii)   If $\mathbf{V}$ is a proper subvariety of **Monoid**, must $F(\mathbf{V})$ be contained in a proper subvariety of **Group**? Must one in fact have $UF(\mathbf{V}) \subseteq \mathbf{V}$?

**Exercise 9.9:11.** Let $H \colon \mathbf{Group} \to \mathbf{Heap}$ be the functor described by (9.6.7) (in Exercise 9.6:10), and $F \colon \mathbf{Heap} \to \mathbf{Group}$ its left adjoint. Let $A$ be a nonempty heap. We recall that $A \cong H(G)$ for some group $G$.

(i)   Describe the group $F(A)$ as explicitly as possible in terms of $G$.

(ii)   It follows from Exercise 9.6:10(iii) that in general, $A = H(G)$ has automorphisms not arising from automorphisms of the group $G$. Take an example of such an automorphism $i$ (or better, obtain a complete characterization of automorphisms of any nonempty heap $A = H(G)$ and let $i$ be a general automorphism of this form), and describe the induced automorphism $F(i)$ of the group $F(H(G))$.

**Exercise 9.9:12.** Show that there exist exactly two underlying-set-preserving functors $\mathbf{Set} \to \mathbf{Semigroup}$. (Hint: What derived operations does **Set** have?) Find their left adjoints.

The next exercise looks at clonal categories as mathematical objects:

**Exercise 9.9:13.** Show that the category **Clone** has small limits and colimits.

The approach of the paragraph preceding Lemma 9.9.10 also shows that for any clonal category $\mathbf{X}$ and any type $\Omega$, to give a morphism $\mathbf{Cl}(\Omega\text{-}\mathbf{Alg}) \to \mathbf{X}$ is simply to pick for each $s \in |\Omega|$ an appropriate morphism in $\mathbf{X}$; and gives a similar characterization of the morphisms from clonal categories $\mathbf{Cl}(\mathbf{V})$ to $\mathbf{X}$. We record these observations as

**Lemma 9.9.11.** *Let* $\Omega = (|\Omega|, \text{ari})$ *be any type. Then the functor* $\mathbf{Clone} \to \mathbf{Set}$ *associating to each clonal category* $\mathbf{X}$ *the set of maps*

$$(9.9.12) \qquad \{f : |\Omega| \to \textstyle\bigsqcup_n \mathbf{X}(X_n, X_1) \mid (\forall\, s \in |\Omega|)\ f(s) \in \mathbf{X}(X_{\text{ari}(s)}, X_1)\}$$

*is representable, with representing object* $\mathbf{Cl}(\Omega\text{-}\mathbf{Alg})$*. Thus,* $\mathbf{Cl}(\Omega\text{-}\mathbf{Alg})$ *may be regarded as a "free clonal category on an* $|\Omega|$*-tuple of formal operations of arities given by the function* $\text{ari}_\Omega$*".*

*Suppose further that* $J$ *is a set of identities for* $\Omega$*-algebras, which we will here express, not as pairs of elements of* $|F_\Omega(\omega)|$*, but as pairs of elements of* $|F_\Omega(n)|$ *for various* $n \in \omega$*; and let us identify these sets* $|F_\Omega(n)|$ *with the sets* $\mathbf{Cl}(\Omega\text{-}\mathbf{Alg})(\mathrm{Cl}_n(\Omega\text{-}\mathbf{Alg}), \mathrm{Cl}_1(\Omega\text{-}\mathbf{Alg}))$*. Let*

$$A_{\Omega,\,J} : \mathbf{Clone} \to \mathbf{Set}$$

*denote the functor associating to each clonal category* $\mathbf{X}$ *the subset of* (9.9.12) *consisting of maps* $f$ *that satisfy the additional condition:*

> *For each* $(s, t) \in J$*, the induced map* $\mathbf{Cl}(\Omega\text{-}\mathbf{Alg}) \to \mathbf{X}$ *corresponding to* $f$ *carries* $s$ *and* $t$ *to the same element.*

*Then* $A_{\Omega,\,J}$ *is representable, with representing object* $\mathbf{Cl}(\mathbf{V}(J))$*. Thus,* $\mathbf{Cl}(\mathbf{V}(J))$ *may be written* $\langle \Omega \mid J \rangle_{\mathbf{Clone}}$*, i.e., may be regarded as the clonal category "presented by the family* $\Omega$ *of formal operations, and the family* $J$ *of relations in this family".* $\qquad\square$

Let me end this section by mentioning a few related concepts on which there is considerable literature, though we will not study them further here.

One is often interested in properties of a variety $\mathbf{V}$ of algebras that do not depend on which operations are considered primitive. These can be expressed as statements about the clonal category $\mathbf{Cl}(\mathbf{V})$. The formally simplest such statements are universally or existentially quantified equations, in families of operations of specified arities. Universally quantified equations of this sort are called *hyperidentities* [137]. An example, and its interpretation in terms of ordinary identities, is noted in

**Exercise 9.9:14.** Show that the following conditions on a variety $\mathbf{V}$ are equivalent:

(a) $\mathbf{V}$ satisfies the hyperidentity saying that all derived *unary* operations are equal.

(b) All *primitive* operations $s$ of $\mathbf{V}$ (of all arities) satisfy the identity of idempotence: $s(x, \ldots, x) = x$.

(c) For every $A \in \mathrm{Ob}(\mathbf{V})$, every one-element subset of $|A|$ is the underlying set of a subalgebra of $A$.

The above hyperidentity is satisfied, for instance, by the varieties of lattices, semilattices, and heaps. On the other hand, there are many varieties that satisfy no nontrivial hyperidentities; e.g., it is shown in [137] that this is true of the variety of commutative rings. A class of varieties determined by a family of hyperidentities is called a *hypervariety*. (However, the term "hyperidentity" is used by some authors, e.g., in [119], with the similar but different meaning of an identity holding for all families of *primitive* operations of given arities in a variety.)

**Exercise 9.9:15.**

(i)   Show that for every monoid identity $s = t$ there is a hyperidentity $s' = t'$ such that for $S$ a monoid, the variety $S$-**Set** satisfies $s' = t'$ if and only if the monoid $S$ satisfies $s = t$.

(ii)   Is the inverse statement true, that for every hyperidentity there exists a monoid identity such that $S$-**Set** satisfies the hyperidentity if and only if $S$ satisfies the monoid identity? If not, is there a modified version of this statement that is correct?

(iii)   Are analogs of the result of (i), and of whatever answer you got for (ii), true for ring identities, and hyperidentities of varieties $R$-**Mod**? If not, how much can be said about the relation between hyperidentities satisfied by varieties $R$-**Mod** and identities or other conditions satisfied by $R$?

Because hyperidentities involve both universal quantification over derived operations, and universal quantification over the algebra-elements to which these operations are applied, they tend to be very strong, and hence somewhat "crude" conditions, as illustrated by the fact that the variety of commutative rings satisfies no nontrivial hyperidentities. *Existentially* quantified equations in derived operations, on the other hand, which translate to certain "$\exists\forall$" conditions on operations and elements, have proved a more versatile tool in General Algebra. An example of this sort of condition on a variety $\mathbf{V}$ is the statement that there exists a derived ternary operation $M$ of $\mathbf{V}$ satisfying the identities

$$(9.9.13) \qquad M(x, x, y) = M(x, y, x) = M(y, x, x) = x.$$

This is satisfied, for instance, by the variety of lattices, where one can take $M(x, y, z) = (x \vee y) \wedge (y \vee z) \wedge (z \vee x)$. Another such condition is gotten by replacing the final "$= x$" above with "$= y$", and is satisfied in the variety of abelian groups of exponent 2, by $M(x, y, z) = x + y + z$.

Many important technical conditions on a variety $\mathbf{V}$ (for instance, the condition that for any two congruences $E$ and $E'$ on an object $A$ of $\mathbf{V}$, one has $E \circ E' = E' \circ E$ under composition of binary relations on $|A|$; or the condition that each subalgebra $B$ of a finite direct product algebra $A_1 \times \cdots \times A_n$ in $\mathbf{V}$ is determined by its images in the pairwise products $A_i \times A_j$) turn out to be equivalent to the statement that $\mathbf{V}$ belongs to the union of a chain of classes of varieties, where each class in the chain is determined by an existentially quantified equation in derived operations. The condition of belonging to such a union is called a *Mal'cev condition*; see [6, §II.12] and [13, §60] for examples and applications.

Finally, let me sketch the idea of another sort of structure, called an *operad*, similar to a clonal category but designed to apply to a wider class of situations. To motivate this, suppose that we wish to think of an algebra over a field $k$, not as a set with operations $+$, $0$, $-$, $\cdot$, etc., but as a $k$-vector-space with a single additional $k$-*bilinear* operation "$\cdot$", and that we want to look at this in the context of other systems consisting of $k$-vector-spaces $V$ given with $k$-multilinear operations satisfying various multilinear identities. To study such entities, we would like to set up an abstract model, analogous to a clonal category, but modeling, not an unstructured set and a family of set-theoretic operations, but a vector space and a family of multilinear operations. Note that as in the situation that motivated clonal categories, one can form derived multilinear operations from given multilinear operations. However, there are things one can do in a clone of set-theoretic operations but not in this context: The projection maps $V^n \to V$ are not multilinear, so they will not appear in our structure, nor, for the same reason, will derived operations based on repeating variables, such as $s(x, x, y)$. On the other hand, there is structure in this multilinear context which one does not have for ordinary clones, namely a $k$-vector-space structure on the set of multilinear operations of each arity, under which composition of multilinear operations is given by multilinear maps. The analog of a clonal category that one gets on taking these features into account is called a $k$-*linear operad*.

Now let the role that was held in our development of the concept of clonal category by the construction of pairwise direct products of sets (since $n$-fold direct products can be obtained as iterated pairwise products), and, implicitly in the development of $k$-linear operads by pairwise tensor products (since a $k$-multilinear map $V^n \to V$ is equivalent to a vector space map $V \otimes_k \cdots \otimes_k V \to V$, and, again, $n$-fold tensor products reduce to two-fold tensor products), be filled by a general bifunctor "$\square$" on a general category $\mathbf{C}$, satisfying appropriate associativity conditions. One can write down a description of the sort of composition of operations that is possible without any more specific assumptions on $\square$. The structure one obtains in this way is called an *operad*. For more details, see [81].

## 9.10. Structure and Semantics

The results of this section will not be essential to what follows, and our presentation will be sketchy. They give, however, a useful perspective on what we have been doing, and in the next chapter we will often refer to them in noting alternative descriptions of various concepts.

Let us look back at the way we associated a clonal theory to a variety $\mathbf{V}$ in Definition 9.9.7. I claim that the various equivalent forms of that construction all reduce to an observation that is applicable in much broader contexts, namely

**Lemma 9.10.1.** *Let $\mathbf{C}$ be a category and $A$ an object of $\mathbf{C}$ such that all finite products $A \times \cdots \times A$ exist in $\mathbf{C}$, and suppose such a product $A^n = \prod_{i \in n} A$ $(n \in \omega)$ is chosen for each $n$, so that the objects $A^n$ are distinct in $\mathrm{Ob}(\mathbf{C})$. Then the full subcategory of $\mathbf{C}$ whose objects are the $A^n$, given with the projection maps $p_{n,i}: A^n \to A$, is a clonal category.* $\square$

To see that this was essentially what we were using in Definition 9.9.7, note on the one hand that each free object $F_{\mathbf{V}}(n)$ is a coproduct of $n$ copies of $F_{\mathbf{V}}(1)$, hence in $\mathbf{C}^{\mathrm{op}}$, the corresponding objects are products of $n$ copies of one object, and the full subcategory of $\mathbf{C}^{\mathrm{op}}$ with these as its objects is one of our descriptions of the clonal theory of $\mathbf{V}$. The description based on looking at the products $U_{\mathbf{V}}^n$ of copies of the functor $U_{\mathbf{V}}$ applied the same idea in the large category $\mathbf{Set}^{\mathbf{V}}$.

We may generalize this latter example by considering any category $\mathbf{C}$ given with a functor $U: \mathbf{C} \to \mathbf{Set}$. The full subcategory of $\mathbf{Set}^{\mathbf{C}}$ having for objects the functors $U^n$ will in general be large; however, in many cases it will be *quasi-small*, i.e., isomorphic to a small category $\mathbf{X}$. (This is true whenever the $U^n$ are representable, or more generally, if the solution-set condition (b) of Proposition 8.10.4 holds, even though the other conditions may not.) To formalize this class of examples, let us make

**Definition 9.10.2.** *For the remainder of this section, $\mathbf{Conc}$ will denote the large category having for objects all pairs $(\mathbf{C}, U)$, where $\mathbf{C}$ is a category, and $U$ a functor $\mathbf{C} \to \mathbf{Set}$, such that for every integer $n$, $\mathbf{Set}^{\mathbf{C}}(U^n, U)$ is quasi-small, and where a morphism $(\mathbf{C}, U) \to (\mathbf{D}, V)$ means a functor $F: \mathbf{C} \to \mathbf{D}$ such that $VF = U$.*

(I've chosen the symbol $\mathbf{Conc}$ as an abbreviation for "concrete", though that term is only an approximation, since we are not assuming that the functors to $\mathbf{Set}$ are faithful, while we *are* assuming a quasi-smallness hypothesis not in the definition of "concrete category". The point of this terminology is to make us think of $U$ (at least at the beginning) as "like an underlying-set functor", so that we can picture the morphisms of $\mathbf{Conc}$ as the underlying-set-*preserving* functors.)

If we associate to each object of **Conc** the clonal category having for object-set the powers of $U$ (Definition 7.8.5), this gives a contravariant construction (because of the way morphisms are defined in **Conc**) of clonal categories from these objects. Unfortunately, this cannot be regarded as a functor to **Clone** because the values assumed, though quasi-small, are not in general small. Hence, for each $(\mathbf{C}, U) \in \mathrm{Ob}(\mathbf{Conc})$ let us choose a small category isomorphic to the category of natural-number powers of $U$, and regard this as an object of **Clone**. (We achieved this in the preceding section, for the particular case where $(\mathbf{C}, U)$ had the form $(\mathbf{V}, U_{\mathbf{V}})$, using the opposite of the category of free **V**-algebras on the natural numbers in this way.) Thus we get a functor $\mathbf{Conc}^{\mathrm{op}} \to \mathbf{Clone}$. Since the morphisms $X_n \to X_1$ in the category constructed in this way from $(\mathbf{C}, U)$ correspond to the $n$-ary *operations* that we can put on the sets $U(C)$ $(C \in \mathrm{Ob}(\mathbf{C}))$ in a functorial manner, the category can be thought of as describing the *algebraic structure* that can be put on the values the functor $U$; hence Lawvere has named this functor "Structure". (Cf. Lemma 9.5.3 and Exercise 9.5:2 for examples.)

**Exercise 9.10:1.** Describe precisely how to make Structure a functor. (Cf. Lemma 8.2.9.)

On the other hand, Lawvere calls the construction taking a clonal category **X** to the variety **X-Alg** given with its underlying set functor, i.e., the concrete category $(\mathbf{X}\text{-}\mathbf{Alg}, U_{\mathbf{X}\text{-}\mathbf{Alg}})$ (which we have seen is also a contravariant construction) "Semantics", because it takes a category of *symbolic* operations, and *interprets* these in all possible ways as genuine operations on sets.

Consider now an arbitrary $(\mathbf{C}, U) \in \mathrm{Ob}(\mathbf{Conc})$, and let **X** be the clone Structure$(\mathbf{C}, U)$. By construction of **X**, the sets $U(C)$ $(C \in \mathrm{Ob}(\mathbf{C}))$ have structures of **X**-algebra, and these are functorial, in the sense that for $f$ a morphism of **C**, the set-map $U(f)$ is a homomorphism of **X**-algebras. This is equivalent to saying that we have an underlying-set-preserving functor $(\mathbf{C}, U) \to (\mathbf{X}\text{-}\mathbf{Alg}, U_{\mathbf{X}\text{-}\mathbf{Alg}})$. Of course, there are other clonal categories **Y** for which one can put functorial **Y**-algebra structures on the values of $U$ (e.g., clonal subcategories of **X**), but it is not hard to verify that **X** is universal for this property, i.e., that every functorial **Y**-algebra structure arises from a morphism of clones, $\mathbf{Y} \to \mathbf{X}$. This universal property is expressed in Lawvere's celebrated slogan, "Structure is adjoint to Semantics".

Since in the universal property, an arbitrary clonal category **Y** such that $U_{\mathbf{X}\text{-}\mathbf{Alg}}$ has a **Y**-algebra structure is mapped *to* the universal clonal category **X** with this property, the latter is *right* universal. So the precise statement is:

**Theorem 9.10.3** (Lawvere). *The functors*

$$\text{Structure}: \mathbf{Conc}^{\mathrm{op}} \to \mathbf{Clone} \quad and \quad \text{Semantics}: \mathbf{Clone}^{\mathrm{op}} \to \mathbf{Conc}$$

*are mutually right adjoint contravariant functors.* □

**Exercise 9.10:2.** Prove the above theorem.

As with any adjunction, we have a pair of *universal morphisms* connecting the two *composites* of these functors with the identity functors of the given categories. In the more familiar case of a *covariant* adjunction, one of these morphisms, the unit, goes from the identity functor to the composite (e.g., the map from each set $X$ to the underlying set of the free group on $X$), and the other, the counit, from the composite to the identity (e.g., from the free group on the underlying set of a group $G$ to $G$ itself). But in the case of a contravariant adjunction, they both go in the same direction; in the right-adjoint case, which we have here, from the identity functor to the composite functor. In the present example, one of these universal maps, namely

$$(9.10.4) \qquad \mathrm{Id}_{\mathbf{Clone}} \longrightarrow \mathrm{Structure} \circ \mathrm{Semantics}$$

is an isomorphism; this is essentially the last assertion of Lemma 9.5.3. Looking at the other composite,

$$(9.10.5) \qquad \mathrm{Id}_{\mathbf{Conc}} \longrightarrow \mathrm{Semantics} \circ \mathrm{Structure},$$

it is not hard to see that it will give an equivalence when applied to an object of **Conc** if and only if that object is (up to equivalence) of the form $(\mathbf{V}, U_{\mathbf{V}})$ where $\mathbf{V}$ is a variety and $U_{\mathbf{V}}$ its underlying-set functor. When we apply Semantics $\circ$ Structure to a more general object $(\mathbf{C}, U)$ of **Conc**, it can be thought of as giving a best approximation of that category by a variety and its underlying-set functor. Thus, for every given pair $(\mathbf{C}, U)$, (9.10.5) gives a "comparison functor"

$$(9.10.6) \qquad (\mathbf{C}, U) \longrightarrow \mathrm{Semantics} \circ \mathrm{Structure}(\mathbf{C}, U).$$

**Exercise 9.10:3.** Describe Structure$(\mathbf{C}, U)$ in each of the following cases (e.g., by choosing a set of "primitive operations" and identities), and determine whether the comparison functor is an equivalence.
  (i)   $\mathbf{C} = \mathbf{Set}$, $\qquad\qquad U(X) = X \times X$.
  (ii)  $\mathbf{C} = \mathbf{Set} \times \mathbf{Set}$, $\quad U(X, Y) = X \times Y$.
  (iii) $\mathbf{C} = \mathbf{Ab}$, $\qquad\qquad U(X) = U_{\mathbf{Ab}}(X \times X)$.
  (iv)  $\mathbf{C} = \mathbf{Ab} \times \mathbf{Ab}$, $\quad U(X, Y) = U_{\mathbf{Ab}}(X \times Y)$.
  (v)   $\mathbf{C} = \mathbf{POSet}$, $\qquad\qquad U = $ the underlying-set functor.
  In cases (iii) and (iv), show that the clonal category Structure$(\mathbf{C}, U)$ can be naturally identified with the clonal theory of modules over a certain ring.

**Exercise 9.10:4.**
  (i)   Same task as in the above exercise, for $\mathbf{C} = \mathbf{Set}^{\mathrm{op}}$, and $U$ the power-set functor $\mathbf{Set}^{\mathrm{op}} \to \mathbf{Set}$.

(ii)   If you are comfortable generalizing the concepts of this and the preceding section to algebras with operations of possibly infinite arities, getting in particular a functor $\mathrm{Structure}^{(\gamma)} : \mathbf{Conc}^{\mathrm{op}} \to \mathbf{Clone}^{(\gamma)}$ for $\gamma$ an infinite regular cardinal, investigate $\mathrm{Structure}^{(\gamma)}(\mathbf{C}, U)$ for the case of part (i).

**Exercise 9.10:5.** Let **CpLattice** denote the category of complete lattices, and **Cp-$\vee$-Semilattice$^0$** the category of complete upper semilattices with least element (regarded as a zeroary operation). We recall that the objects of these two categories are essentially the same, but the morphisms are not (cf. Proposition 6.2.3).

(i)   Show that the underlying-set functor on one of these categories satisfies the smallness condition in the definition of **Conc**, but that of the other does not.

(ii)   In the case that does give an object $(\mathbf{C}, U)$ of **Conc**, describe the variety $\mathrm{Semantics} \circ \mathrm{Structure}(\mathbf{C}, U)$. (Note that in contrast to part (ii) of the preceding exercise, we are here talking about finitary "Structure".)

I will end this section with a few observations on the question, "Given a category, how can one tell whether it is equivalent to a variety of algebras?" (Birkhoff's Theorem tells us which full subcategories of a category $\Omega$-**Alg** *are* varieties, but the present question, about abstract categories and *equivalence*, is of a different sort.)

By our preceding observations, a necessary and sufficient condition is that there *exist* a functor $U : \mathbf{C} \to \mathbf{Set}$ such that $(\mathbf{C}, U)$ lies in **Conc**, and the comparison functor (9.10.6) is an equivalence. Note also that the underlying-set functor of any variety is *representable* (by the free object on one generator), so if the above condition holds, $U$ can be taken to have the form $h_G$ for some object $G$ of **C**. In this situation (since by our general convention, **C** is assumed legitimate), the quasi-smallness condition on the powers of $U$ automatically holds by Yoneda's Lemma. In summary:

**Lemma 9.10.7.** *A category* **C** *is equivalent to a variety of finitary algebras if and only if there exists some* $G \in \mathrm{Ob}(\mathbf{C})$ *such that the comparison map*

$$(9.10.8) \qquad (\mathbf{C}, h_G) \longrightarrow \mathrm{Semantics} \circ \mathrm{Structure}(\mathbf{C}, h_G)$$

*is an isomorphism in* **Conc**.

(*The analogous result holds with "finitary" replaced by "having all operations of arity* $\leq \gamma_0$ *" for any fixed regular infinite cardinal* $\gamma_0$, *if we use corresponding modified functors* $\mathrm{Structure}^{(\gamma_0)}$ *and* $\mathrm{Semantics}^{(\gamma_0)}$.)          $\square$

Though this does not say very much, it gives a useful heuristic pointer: If we want to determine whether a category **C** is equivalent to a variety of algebras, we should look at possible candidates for the free object on one generator. The next exercise gives several cases where you can show that no such object exists. I do not advise trying to use the above lemma in this and the next two exercises, but only the "heuristic pointer".

**Exercise 9.10:6.** Show that none of the following categories are equivalent to varieties of algebras, even if we allow the latter to have infinitary operations (though, as always, we assume the set of all operations to be small).

(i) **POSet**. (Suggestion: For each of the situations (a) **C** a variety of algebras, and $A$ a free algebra in **C** on a nonempty set, (b) **C** = **POSet**, and $A$ a discrete partially ordered set, and (c) **C** = **POSet**, and $A$ a nondiscrete partially ordered set, investigate the relationship between the set of coequalizer maps in **C**, and the set of morphisms in **C** that $h_A$ takes to surjective set maps.)

(ii) **CompactHaus**, the category of compact Hausdorff spaces and continuous maps. (Suggestion: If **V** is a variety with all operations having arities $< \gamma_1$ for some infinite regular cardinal $\gamma_1$, what does this imply about the closure operator "subalgebra generated by –" on the underlying sets of algebras in **V** ? (Cf. Definition 6.3.7 for the case $\gamma_1 = \omega$.) Translate this into a statement involving the free object on one generator in **V**, and show that no object has this property in **CompactHaus**.)

(iii) The full subcategory of **Ab** whose objects are the torsion-free abelian groups.

(iv) The full subcategory of **Ab** whose objects are the divisible abelian groups (groups such that for every group element $x$ and nonzero integer $n$, the equation $n y = x$ has a solution $y$ in $A$).

**Exercise 9.10:7.** In contrast to the last two cases above, show that the full subcategory of **Ab** whose objects are the divisible torsion-free abelian groups *is* equivalent to a variety of algebras.

**Exercise 9.10:8.** Show that **Clone** is not equivalent to any variety of finitary algebras. (Suggestion: Show that (a) an object corresponding to a free object on one generator would have to be a finitely generated clonal category, (b) if it were generated by elements of arities $\leq n$, this would be true of all clonal categories, and (c) this is not the case.)

Can you prove that it is or is not equivalent to a variety of possibly infinitary algebras?

In contrast to Exercise 9.10:6(ii), it is proved in [115] that **CompactHaus** *can* be identified with a "variety" if we generalize that concept to allow a *large* set of operations—as we would also have to do, for instance, to speak of the "variety" of complete lattices or semilattices. Under this construction of **CompactHaus**, the operations of each cardinality $\alpha$ correspond to the points of the Stone–Čech compactification of the discrete space $\alpha$. Note that this means that, in contrast to the case of complete lattices (but as for complete upper or lower semilattices, cf. Exercise 9.10:5), for each $\alpha$, the set of $\alpha$-ary operations is small; i.e., the corresponding generalized clonal category, though not small, is legitimate. A consequence is that compact Hausdorff spaces actually behave more like ordinary algebras than do complete lattices! In particular, there is a "free compact Hausdorff space" on every small set $X$, namely, the Stone–Čech compactification of $X$ as a discrete space.

The difference between the cases of complete *semilattices* and *lattices* noted in the above paragraph has the curious consequence that though complete semilattices behave "well", the category of sets with *two* complete semilattice operations $\bigvee_1$ and $\bigvee_2$ does not, since if it behaved like a variety, then complete lattices would behave like a subvariety. As a still more striking example of this sort, though compact Hausdorff spaces are well behaved, the category of sets with a compact Hausdorff topology and a single unary operation not assumed continuous in that topology will not have a free object on one generator. Indeed any non-limit ordinal $\alpha$ can be given a compact Hausdorff topology in which each nonzero limit ordinal $\beta < \alpha$ is a topological limit of the lower ordinals; and using this topology and the unary successor operation, the whole set $\alpha$ will be generated by $\{0\}$.

Lemma 9.10.7 does *not* say that an object $G$ with the indicated properties is unique up to isomorphism if it exists. Let us examine the extent to which we can vary $G$ in a couple of familiar varieties.

**Exercise 9.10:9.**

(i)   When $\mathbf{C} = \mathbf{Ab}$, determine for what objects $G$ the functor (9.10.8) is an equivalence. Show that for every such $G$, Structure($\mathbf{Ab}$, $h_G$) can be identified with the theory of modules over some ring $R$.

(ii)   Similarly, for $\mathbf{C} = \mathbf{Set}$ determine what objects $G$ make (9.10.8) an equivalence, and try to describe the theory Structure($\mathbf{Set}$, $h_G$) in these cases.

The answer to (i) shows that $\mathbf{Ab}$ is equivalent to several different varieties $R$-$\mathbf{Mod}$, and in (ii) we similarly discover that $\mathbf{Set}$ is equivalent to several varieties of algebras.

Lawvere gives in his thesis [17, §III.2] a version of Lemma 9.10.7 which is less trivial than ours, but also more complicated to formulate; I will not present it here.

Despite the technical meaning given the word "structure" in this section, we will also continue to use it as a non-specific meta-term in our mathematical discussions.

# Part III. More on Adjunctions

Chapter 10 (the only chapter of this part) represents the culmination of the course. In it we obtain Freyd's beautiful characterization of functors among varieties of algebras that have left adjoints, and study several classes of examples, and related results.

(I had hoped to gradually add several further chapters to this part; but, at least in the short run, this is not to be.)

# Chapter 10
# Algebra and Coalgebra Objects in Categories, and Functors Having Adjoints

One of our long-range goals, since we took our "Cook's tour" of universal constructions in Chapter 4, has been to obtain general results on when algebras with given universal properties exist. We have gotten several existence results holding in all varieties $\mathbf{V}$: for free objects, limits and colimits, and objects presented by generators and relations. The result on free objects can be restated as the existence of a left adjoint to the forgetful functor $\mathbf{V} \to \mathbf{Set}$, and we have also shown that the inclusion $\mathbf{V} \to \Omega\text{-}\mathbf{Alg}$ has a left adjoint, where $\Omega$ is the type of $\mathbf{V}$. In the first four sections of this chapter, we shall develop a result of a much more sweeping sort: a characterization of *all* functors between varieties of algebras $\mathbf{V}$ and $\mathbf{W}$ which have left adjoints.

To get an idea what such a characterization should be, we should look at some typical examples. Most of the functors with left adjoints among varieties of algebras that we have seen so far have been cut from a fairly uniform mold: underlying-set-preserving constructions that forget some or all of the operations, and things close to these. We shall begin by looking at an example of a different sort, which will give us some perspective on the features that make the construction of the adjoint possible. We will then formalize these features, arriving at a pair of concepts (those of algebra and coalgebra objects in a general category) of great beauty in their own right, in terms of which we shall establish the desired condition in § 10.4.

In the remaining sections of this chapter we work out in detail several general cases, and note various related results.

## 10.1. An example: $\mathrm{SL}(n)$

Let $n$ be a positive integer. Then for any commutative ring $A$, the $n \times n$ matrices over $A$ having determinant 1 form a group, called the *special linear group* $\mathrm{SL}(n, A)$. (Recall from § 4.12 that rings are assumed to be associative and to have 1 unless the contrary is stated.) We see in fact that $\mathrm{SL}(n, -)$

© Springer International Publishing Switzerland 2015
G.M. Bergman, *An Invitation to General Algebra and Universal Constructions*, Universitext, DOI 10.1007/978-3-319-11478-1_10

is a functor **CommRing**$^1$ $\rightarrow$ **Group**. Let us simplify our name for this functor to $\mathrm{SL}(n)$, but continue to write its value at $A$ as $\mathrm{SL}(n, A)$.

Does $\mathrm{SL}(n)$ have a left adjoint? In concrete terms this asks: Given a group $G$, can we find a universal example of a commutative ring $A_G$ with a homomorphism $G \rightarrow \mathrm{SL}(n, A_G)$ ?

Let us approach this question in our standard way (first noted in Remark 3.2.13), by considering an arbitrary commutative ring $A$ with a homomorphism

$$h\colon G \longrightarrow \mathrm{SL}(n, A),$$

and asking what elements of $A$, and what relations among these, are determined by this situation.

Clearly, we can get $n^2$ elements of $A$ from each element $g$ of $G$, to wit, the entries of the matrix $h(g)$:

(10.1.1)  $$h(g)_{ij} \in |A| \quad (g \in |G|,\ i, j = 1, \ldots, n).$$

By definition of $\mathrm{SL}(n, A)$, these satisfy the relation saying that the determinant of the matrix they form is $1$:

(10.1.2)  $$\det(h(g)_{ij}) = 1 \quad (g \in |G|).$$

The condition that $h$ be a group homomorphism says that for every two elements $g, g' \in |G|$, the matrix $(h(g\,g')_{ij})$ is the product of the matrices $(h(g)_{ij})$ and $(h(g')_{ij})$. Each such matrix equation is equivalent to $n^2$ equations in the ring $A$:

(10.1.3)  $$h(g\,g')_{ik} = \sum_j h(g)_{ij}\, h(g')_{jk} \quad (g, g' \in |G|,\ i, k = 1, \ldots, n).$$

We see, in fact, that a system of elements (10.1.1) satisfying (10.1.2) and (10.1.3) is equivalent to a homomorphism $G \rightarrow \mathrm{SL}(n, A)$. Hence, if we let $A_G$ be the object of **CommRing**$^1$ presented by generators (10.1.1) and relations (10.1.2) and (10.1.3), and denote by $h\colon G \rightarrow \mathrm{SL}(n, A_G)$ the resulting group homomorphism, then the pair $(A_G, h)$ will be initial among commutative rings $A$ given with such homomorphisms, and the construction $G \mapsto A_G$ will be the desired left adjoint to $\mathrm{SL}(n)$.

What properties of the functor $\mathrm{SL}(n)$ have we used here? First, the fact that for every commutative ring $A$, the elements of $\mathrm{SL}(n, A)$ could be described as all families of elements of $A$ indexed by a certain fixed set (in this case the set $\{1, \ldots, n\} \times \{1, \ldots, n\}$) which satisfied certain equations (in this case, the single equation saying that the matrix they formed had determinant 1). It was this that allowed us to write down the generators (10.1.1) and relations (10.1.2) in the definition of $A_G$. Secondly, we used the fact that the multiplication of the group $\mathrm{SL}(n, A)$ takes a pair of matrices $s$, $t$ to a matrix $s\,t$ whose entries are given by certain fixed polynomials (i.e., derived operations) in the $2n^2$ entries of the two given matrices. This allowed us to express the condition that $h$ be a homomorphism by the equations (10.1.3).

We also used, implicitly, a fact special to the variety of groups, namely that for a map of underlying sets to be a homomorphism, it suffices that it respect multiplication. If we want to put this example into a form that generalizes to arbitrary varieties, we should note that the unary "inverse" operation and the zeroary "neutral element" operation of $SL(n, A)$ also have the property that their entries are given by polynomials in the entries of their arguments: The inverse of a matrix of determinant 1 is a matrix of determinants of minors (with certain $\pm$ signs); the identity matrix consists of 0's and 1's in certain positions, and these 0's and 1's can be regarded as polynomials in the empty set of variables. Hence if we do not wish to call on the special property of group homomorphisms mentioned, we can still guarantee the universal property of $A_G$, by supplementing (10.1.3) with relations saying that for all $g \in |G|$, the entries of $h(g^{-1})$ are given by the appropriate signed minors in the entries of $h(g)$, and that the $(i, j)$ entry of $h(e)$ has the value $\delta_{ij}$ (i.e., 1 if $i = j$, 0 otherwise).

To abstract the conditions noted above, let us now consider arbitrary varieties $\mathbf{V}$ and $\mathbf{W}$ (in general, of different types), and a functor

$$V : \mathbf{W} \longrightarrow \mathbf{V}$$

for which we hope to find a left adjoint. The analog of the first property noted for $SL(n)$ above should be that for $A \in Ob(\mathbf{W})$, the underlying set $|V(A)|$ is describable as the set of $X$-tuples of elements of $|A|$, for some fixed set $X$, which satisfy a fixed set $Y$ of relations. We recall from Lemma 9.4.17 that this is equivalent to saying that the set-valued functor $A \mapsto |V(A)|$, i.e., the functor $U_{\mathbf{V}} V$ (where $U_{\mathbf{V}} : \mathbf{V} \to \mathbf{Set}$ is the underlying-set functor of $\mathbf{V}$) is representable, with representing object the $\mathbf{W}$-algebra defined using $X$ and $Y$ as generators and relations:

$$(10.1.4) \qquad\qquad R = \langle X \mid Y \rangle_{\mathbf{W}}.$$

The object (10.1.4) thus "encodes" the functor $V$ at the set level! Is there a way to extend these observations so as to encode also the $\mathbf{V}$-algebra structures on the sets $|V(A)|$?

Let us look at this question in the case $V = SL(n)$. We see that the object representing the functor $U_{\mathbf{Group}} \circ SL(n)$ is the commutative ring $R$ presented by $n^2$ generators $r_{ij}$ and one relation $\det(r_{ij}) = 1$; in other words, the commutative ring having a *universal* $n \times n$ *matrix* $r$ of determinant 1 over it. Can we now find a universal instance of *multiplication* of such matrices? Since multiplication is a binary operation, we should multiply a universal *pair* of matrices of determinant 1. The ring with such a universal pair is the coproduct of two copies of $R$. If we denote these two matrices $r_0$, $r_1 \in |SL(n, R \amalg R)|$, then the $n^2$ entries of the product matrix $r_0 r_1 \in |SL(n, R \amalg R)|$ can, like any elements of $R \amalg R$, be expressed as polynomials in our generators for that ring, the entries of $r_0$ and $r_1$. Using the universality of $r_0, r_1 \in |\mathrm{SL}(n, R \amalg R)|$, it is not hard to show that those same polynomials, when applied to the entries of two *arbitrary* elements of $\mathrm{SL}(n, A)$ for an *arbitrary* commutative ring $A$, must also give the entries of their product. So it appears that $r_0 r_1$ does in some sense encode the multiplication operation of $\mathrm{SL}(n)$.

There is a more abstract way of looking at this encoding. By the universal property of $R$, the element $r_0 r_1 \in |\mathrm{SL}(n, R \amalg R)|$ corresponds to some morphism

$$(10.1.5) \qquad\qquad \mathbf{m}\colon R \longrightarrow R \amalg R$$

(the unique morphism taking the entries of $r$ to those of $r_0 r_1$). Now given a commutative ring $A$, any two elements $x, y \in |\mathrm{SL}(n, A)|$ arise as images of the universal element $r \in |\mathrm{SL}(n, R)|$ via unique homomorphisms $f, g\colon R \to A$. Such a pair of morphisms corresponds, by the universal property of the coproduct, to a single morphism $(f, g)\colon R \amalg R \to A$ (the morphism carrying the entries of $r_0$ to those of $x$ and the entries of $r_1$ to those of $y$). Composing with (10.1.5), we get a morphism

$$(10.1.6) \qquad\qquad R \xrightarrow{\ \mathbf{m}\ } R \amalg R \xrightarrow{\ (f,\, g)\ } A,$$

which corresponds to an element of $\mathrm{SL}(n, A)$. From the facts that $\mathbf{m}$ corresponds to (i.e., sends $r$ to) the *product* of $r_0$ and $r_1$, and that $\mathrm{SL}(n)$, applied to the map $(f, g)$ gives a *group homomorphism* $\mathrm{SL}(n, R \amalg R) \to \mathrm{SL}(n, A)$, we can deduce that the matrix given by (10.1.6) (i.e., the result of applying the ring-homomorphism (10.1.6) entrywise to $r$) is the product of $x$ and $y$. So the ring homomorphism $\mathbf{m}$ of (10.1.5) indeed "encodes" our multiplication.

We note similarly that $r^{-1} \in |\mathrm{SL}(n, R)|$ will be the image of the universal element $r$ under a certain morphism

$$(10.1.7) \qquad\qquad \mathbf{i}\colon R \longrightarrow R$$

and we find that this morphism $\mathbf{i}$ encodes the *inverse* operation on $\mathrm{SL}(n)$.

If we are going to treat the zeroary neutral-element operation similarly, it should correspond to a morphism from $R$ to the coproduct of zero copies of itself. This vacuous coproduct is the *initial object* of $\mathbf{CommRing}^1$, namely the ring $\mathbb{Z}$ of integers. And indeed, if we let

$$(10.1.8) \qquad\qquad \mathbf{e}\colon R \longrightarrow \mathbb{Z}$$

be the map sending the universal element $r \in |\mathrm{SL}(n, R)|$ to the identity matrix in $\mathrm{SL}(n, \mathbb{Z})$, we find that for every commutative ring $A$, the composite of (10.1.8) with the unique homomorphism $\mathbb{Z} \to A$ is the morphism $R \to A$ that specifies the identity matrix in $\mathrm{SL}(n, A)$.

The structure $(R, \mathbf{m}, \mathbf{i}, \mathbf{e})$ sketched above is, as we shall soon see, what is called a *cogroup* in the category $\mathbf{CommRing}^1$. The maps (10.1.5), (10.1.7), (10.1.8) are called its *comultiplication*, its *coinverse*, and its *co-neutral-element*, and the cogroup $(R, \mathbf{m}, \mathbf{i}, \mathbf{e})$ is said to *represent* the functor $\mathrm{SL}(n):$ $\mathbf{CommRing}^1 \to \mathbf{Group}$, just as $R$ alone is said to represent the functor $U_{\mathbf{Group}} \circ \mathrm{SL}(n): \mathbf{CommRing}^1 \to \mathbf{Set}$.

In the next three sections we shall develop general definitions and results, of which the particular case sketched above is an example. We shall see that given a functor $V: \mathbf{W} \to \mathbf{V}$, if the first of the two properties we called on above holds, namely that the set-valued functor $U_{\mathbf{V}} V$ is representable, then the other condition, that the operations of the algebras $V(A)$ arise from a co-$\mathbf{V}$-structure on the representing object, follows automatically. (Indeed, our development of (10.1.5) above did not use our knowledge that the group operations of $\mathrm{SL}(n)$ had this form, but deduced that fact from their functoriality.)

The fact that representability of our functor at the set level is enough to insure that the operations will likewise be "represented" by co-operations does not, however, mean that we can ignore those co-operations! Rather, since they encode the $\mathbf{V}$-algebra structure of our otherwise merely set-valued functors, they will be the key to the study of such constructions.

## 10.2. Algebra objects in a category

I will approach the concept of a coalgebra object in a category $\mathbf{C}$ by starting with the dual concept, that of an algebra object, since this has a more familiar appearance. Let us make:

**Convention 10.2.1.** *Throughout this section, $\gamma$ will be a regular cardinal, $\mathbf{C}$ will be a category admitting products indexed by all families of cardinality $< \gamma$ (which we will abbreviate to "$<\gamma$-fold products"), and $\Omega$ will be a type all of whose operations have arities $< \gamma$.*

(If you are most comfortable with finitary algebras, you may assume $\gamma = \omega$ without missing any of the ideas of this chapter.)

**Definition 10.2.2.** *For $\beta < \gamma$, a $\beta$-ary operation on an object $R$ of $\mathbf{C}$ will mean a morphism $s: R^\beta \to R$.*

By Yoneda's Lemma, such operations correspond bijectively to morphisms of the induced contravariant hom-functors, $h^{R^\beta} \to h^R$; and by the universal property of the product object $R^\beta$, we can identify $h^{R^\beta}$ with $(h^R)^\beta$, so such a map corresponds to a morphism $(h^R)^\beta \to h^R$, i.e., a $\beta$-ary operation on $h^R$. In concrete terms, if $s_R$ is a $\beta$-ary operation of $R$, then given an object $A$ of $\mathbf{C}$ and a $\beta$-tuple of elements $(\xi_\alpha)_{\alpha < \beta} \in \mathbf{C}(A, R)^\beta$, we first combine these into a single element of $\mathbf{C}(A, R^\beta)$, then compose this with $s_R \colon R^\beta \to R$ to get an element of $\mathbf{C}(A, R)$, which we may denote $s_{\mathbf{C}(A, R)}((\xi_\alpha)_{\alpha \in \beta})$. This is the category-theoretic generalization of the familiar technique of taking the set of all functions from a space $A$ to an algebra $R$, and making that set an algebra under pointwise application of the operations of $R$. These observations are summarized in the next lemma (in which the equivalence of (ii) and (iii) holds by the definition of morphism of functors).

**Lemma 10.2.3.** *Let $\beta$ be a cardinal $< \gamma$, and $R$ an object of $\mathbf{C}$. Then the following data are equivalent (via the construction just described):*

(i)    *A $\beta$-ary operation $s_R \colon R^\beta \to R$.*

(ii)    *A morphism $s_{\mathbf{C}(-, R)} \colon \mathbf{C}(-, R)^\beta \to \mathbf{C}(-, R)$ as functors $\mathbf{C}^{\mathrm{op}} \to \mathbf{Set}$, i.e., as contravariant set-valued functors on $\mathbf{C}$.*

(iii)    *A way of defining on each set $\mathbf{C}(A, R)$ $(A \in \mathrm{Ob}(\mathbf{C}))$ a $\beta$-ary operation $s_{\mathbf{C}(A, R)} \colon \mathbf{C}(A, R)^\beta \to \mathbf{C}(A, R)$, so that for every morphism $f \in \mathbf{C}(A, B)$, the induced map $\mathbf{C}(B, R) \to \mathbf{C}(A, R)$ respects these operations.* □

Recalling that $\Omega$ denotes a type all of whose operation-symbols have arities $< \gamma$, we now make

**Definition 10.2.4.** *An $\Omega$-algebra object $R$ in the category $\mathbf{C}$ (or a $\mathbf{C}$-based $\Omega$-algebra) will mean a pair $(|R|, (s_R)_{s \in |\Omega|})$, where $|R| \in \mathrm{Ob}(\mathbf{C})$, and each $s_R$ is an operation*

$$s_R \colon |R|^{\mathrm{ari}(s)} \longrightarrow |R| \quad (s \in |\Omega|).$$

*A morphism between $\Omega$-algebra objects of $\mathbf{C}$ will mean a morphism between their underlying $\mathbf{C}$-objects which forms commuting squares with these operations.*

*If $R$ is an $\Omega$-algebra object of $\mathbf{C}$, and $A$ any object of $\mathbf{C}$, then $\mathbf{C}(A, R)$ will denote the ordinary (i.e., set-based) $\Omega$-algebra with underlying set $\mathbf{C}(A, |R|)$, and operations induced by the $s_R$ as in Lemma 10.2.3.*

Below, the word "algebra" will continue to mean "set-based algebra" except when the contrary is indicated by writing "algebra object", "$\mathbf{C}$-based algebra", etc. When referring to set-based algebras, I will occasionally add the words "set-based" for emphasis.

Observe that the $|\ |$-notation introduced above is relative. For example, if $\mathbf{C}$ is itself a category of algebras, and $R$ a $\mathbf{C}$-based algebra, then $|R|$ denotes the underlying $\mathbf{C}$-*object* of $R$, and if $S$ is this $\mathbf{C}$-object, then $|S| = ||R||$ denotes its underlying *set*. I shall, in fact, sometimes, as in the above definition, use the letter $R$ and its alphabetical neighbors for algebra-objects in categories $\mathbf{C}$, and other times, as in Lemma 10.2.3, for the underlying $\mathbf{C}$-objects of such objects. Of course, in any given statement I shall be consistent about which meaning I am giving a symbol.

Also note the new use of the symbol $\mathbf{C}(A, R)$ introduced in the above definition: Though $A$ denotes an object of $\mathbf{C}$, $R$ does not; rather, it is a

**C**-based $\Omega$-algebra, and the whole symbol denotes, not a set, but a (set-based) $\Omega$-algebra. Of course, a **C**-based $\Omega$-algebra is intuitively "an object of **C** with additional structure", and an $\Omega$-algebra is likewise a set with additional structure; and modulo this additional structure, we have the old meaning of $\mathbf{C}(A, R)$. So this extended notation is "reasonable". But we need to remember when discussing algebra objects of categories that if we want to know what is meant by a symbol $\mathbf{C}(A, R)$, we have to check whether $R$ is assumed to be an object of **C**, or a **C**-based $\Omega$-algebra for some $\Omega$.

The above definition also introduced the concept of a *morphism* of **C**-based $\Omega$-algebras. Combining this with Yoneda's Lemma, we easily get

**Lemma 10.2.5.** *Let $R$ and $S$ be $\Omega$-algebra objects in **C**. Then the following data are equivalent:*

(i)    *A morphism of **C**-based algebras $R \to S$.*

(ii)    *A morphism $f \in \mathbf{C}(|R|, |S|)$ such that for every object $A$ of **C**, the induced set map $\mathbf{C}(A, |R|) \to \mathbf{C}(A, |S|)$ is a homomorphism of $\Omega$-algebras $\mathbf{C}(A, R) \to \mathbf{C}(A, S)$.*

(iii)    *A morphism $\mathbf{C}(-, R) \to \mathbf{C}(-, S)$ of functors $\mathbf{C} \to \Omega\text{-}\mathbf{Alg}$.*                                                                                 □

We next want to define, for an $\Omega$-algebra object $R$ of a category **C**, the *derived operations* of $R$ corresponding to the various derived operations of set-based $\Omega$-algebras. This will allow us to say what it means for such an object to satisfy a given *identity*; namely, that the derived operations specified by the two sides of the identity are equal.

One cannot, of course, describe a derived operation of $R$ by giving a formula for its value on a tuple of "elements of $|R|$" when **C** is a general category. An approach that is often used is to express operations and identities by diagrams. For example, observe that if $m$ is a binary operation on a *set* $|R|$, the condition that $m$ be associative can be expressed as the condition that the diagram

$$(10.2.6) \qquad \begin{array}{ccc} |R| \times |R| \times |R| & \xrightarrow{\;m \times \mathrm{id}_{|R|}\;} & |R| \times |R| \\ {\scriptstyle \mathrm{id}_{|R|} \times m} \Big\downarrow & & \Big\downarrow {\scriptstyle m} \\ |R| \times |R| & \xrightarrow{\qquad m \qquad} & |R| \end{array}$$

commute, since the path that goes through the upper right-hand corner gives the ternary derived operation $(x, y, z) \mapsto m(m(x, y), z)$, and the one through the lower left-hand corner gives $(x, y, z) \mapsto m(x, m(y, z))$. Analogously, for any object $|R|$ of a general category **C** and any binary operation $m \colon |R| \times |R| \to |R|$, the same diagram can be used to define two ternary "derived operations" on $|R|$, and their equality (the commutativity of the diagram) can be made the definition of associativity of the **C**-based algebra $R = (|R|, m)$.

The above approach is nice in simple cases, but it has the disadvantage of requiring us to figure out the diagram appropriate to every identity we want to consider. Another approach, which is equivalent to the above but avoids this dependence on diagrams, is based on considering the algebra $\mathbf{C}(A, R)$ for an appropriate *universal choice* of $A$. If we want to consider derived operations in $\beta$ variables, let us look at $\mathbf{C}(|R|^\beta, R)$. Since this is a set-based algebra, we know how to construct its derived $\beta$-ary operations from its primitive operations. Applying such a derived operation $t$ to the $\beta$ projections $p_\alpha : |R|^\beta \to |R|$ $(\alpha \in \beta)$, we get an element $t((p_\alpha)_{\alpha \in \beta}) \in \mathbf{C}(|R|^\beta, |R|)$ which we *define* to be the derived operation $t_R$ of the **C**-based algebra $R$. Identities are then defined as equalities among such derived operations.

Incidentally, although in § 9.4 we found it convenient to reduce all identities for $\Omega$-algebras to identities (pairs of terms) in a fixed $\gamma_0$-tuple of variables, we shall here revert to expressing them as identities in $\beta$-tuples of variables for various ordinals $\beta < \gamma$. (So, for instance, the diagram (10.2.6) expresses associativity using three variables, rather than countably many.) The advantage will be that we only need to assume that **C** has these $\beta$-fold products, rather than making the unnecessary stronger assumption that it has $\gamma_0$-fold products.

The above can also be put in the language of "algebras as representations of clonal categories". Lemma 9.10.1 generalizes to say that the operations of arity $< \gamma$ on an object $|R|$ of **C** (equivalently, on the functor $h^{|R|} : \mathbf{C} \to \mathbf{Set}$) yield a $<\gamma$-clonal category, and that a **C**-based $\Omega$-algebra structure on $|R|$ as defined above is equivalent to a representation of the $<\gamma$-clonal category $\mathbf{Cl}^{(\gamma)}(\Omega\text{-}\mathbf{Alg})$ in **C** which takes $X_1 \in \mathrm{Ob}(\mathbf{Cl}^{(\gamma)}(\Omega\text{-}\mathbf{Alg}))$ to $|R| \in \mathrm{Ob}(\mathbf{C})$. The condition that this **C**-based algebra $R$ satisfy the identities of a given variety **V** is equivalent to saying that this representation of $\mathbf{Cl}^{(\gamma)}(\Omega\text{-}\mathbf{Alg})$ arises from (i.e., factors through) a representation of $\mathbf{Cl}^{(\gamma)}(\mathbf{V})$ :

$$(10.2.7) \qquad \mathbf{Cl}^{(\gamma)}(\Omega\text{-}\mathbf{Alg}) \longrightarrow \mathbf{Cl}^{(\gamma)}(\mathbf{V}) \longrightarrow \mathbf{C},$$

where the first arrow is induced by the given indexing of the operations of **V** by $\Omega$.

In the next lemma and definition, we set down the observations of the preceding paragraphs, and prove the one nontrivial implication.

**Lemma 10.2.8.** *Let* $R = (|R|, (s_R)_{s \in |\Omega|})$ *be an $\Omega$-algebra object of* **C**, *and let* $t$, $t'$ *be two derived $\beta$-ary operations* $(\beta < \gamma)$ *for ordinary (i.e., set-based) algebras of type* $\Omega$. *Then the following conditions are equivalent:*

(i)     *For all* $A \in \mathrm{Ob}(\mathbf{C})$, *the algebra* $\mathbf{C}(A, R)$ *satisfies the identity* $t = t'$.

(ii)  *In the algebra* $\mathbf{C}(|R|^\beta, R)$, *one has* $t((p_\alpha)_{\alpha\in\beta}) = t'((p_\alpha)_{\alpha\in\beta})$, *where the* $p_\alpha$ $(\alpha \in \beta)$ *are the projection maps.*

(iii)  *The morphisms* $t, t' \colon \mathrm{Cl}_\beta(\mathbf{V}) \rightrightarrows \mathrm{Cl}_1(\mathbf{V})$ *in the* $<\gamma$-*clonal category* $\mathbf{Cl}^{(\gamma)}(\Omega\text{-}\mathbf{Alg})$ *fall together under the functor from* $\mathbf{Cl}^{(\gamma)}(\Omega\text{-}\mathbf{Alg})$ *to the* $<\gamma$-*clonal theory of* $|R|$ *induced by the* $|\Omega|$-*tuple of operations* $(s_R)$. *(See Lemma 9.9.11 for the universal property of* $\mathbf{Cl}^{(\gamma)}(\Omega\text{-}\mathbf{Alg})$ *which allows one to define this morphism.)*

(iv)  *The algebra object* $R$ *satisfies the "diagrammatic translation" of the identity* $t = t'$.

*Proof.* Parts (ii)–(iv) are simply different ways of stating the same condition. The implication (i) $\implies$ (ii) is seen by applying (i) with $|A| = |R|^\beta$ to the $\beta$-tuple $(p_\alpha)_{\alpha\in\beta}$. The converse implication can be gotten by Yoneda's Lemma; to see it directly, consider any object $A$ of $\mathbf{C}$ and any $\beta$-tuple $(\xi_\alpha)_{\alpha\in\beta}$ of elements of $\mathbf{C}(A, |R|)$. By the universal property of the product object $|R|^\beta$, these morphisms correspond to a single morphism $\xi \colon A \to |R|^\beta$, and applying to this the functor $\mathbf{C}(-, R) \colon \mathbf{C} \to \Omega\text{-}\mathbf{Alg}$, we get an $\Omega$-algebra homomorphism $\mathbf{C}(|R|^\beta, R) \to \mathbf{C}(A, R)$ carrying each $p_\alpha$ to $\xi_\alpha$. Hence, any equation satisfied by the $\beta$-tuple $(p_\alpha)_{\alpha\in\beta}$ is also satisfied by the $\beta$-tuple $(\xi_\alpha)_{\alpha\in\beta}$. $\qquad\square$

**Definition 10.2.9.** *If the equivalent conditions of Lemma 10.2.8 hold, the* $\Omega$-*algebra object* $R$ *of* $\mathbf{C}$ *will be said to* satisfy the identity $t = t'$.

*If* $\mathbf{V}$ *is a variety of* $\Omega$-*algebras, defined by a family* $J$ *of identities, then a* $\mathbf{V}$-*object of* $\mathbf{C}$ *will mean an* $\Omega$-*algebra object* $R$ *of* $\mathbf{C}$ *satisfying the identities in* $J$ *in this sense; equivalently, such that the induced functor* $\mathbf{C}(-, R)$ *carries* $\mathbf{C}$ *into* $\mathbf{V}$; *equivalently, such that the corresponding representation of* $\mathbf{Cl}^{(\gamma)}(\Omega\text{-}\mathbf{Alg})$ *in* $\mathbf{C}$ *arises as in* (10.2.7) *from a representation of* $\mathbf{Cl}^{(\gamma)}(\mathbf{V})$ *in* $\mathbf{C}$.

Of course, since the same subvariety $\mathbf{V} \subseteq \Omega\text{-}\mathbf{Alg}$ can be determined by more than one set of identities $J$, we need to check that the above definition of being a $\mathbf{V}$-object of $\mathbf{C}$ is independent of our choice of defining identities for $\mathbf{V}$. The equivalent formulation "$\mathbf{C}(-, R)$ carries $\mathbf{C}$ into $\mathbf{V}$" shows that this is true.

We have been discussing how to define operations on representable functors $\mathbf{C}(-, |R|) \colon \mathbf{C}^{\mathrm{op}} \to \mathbf{Set}$ $(|R| \in \mathrm{Ob}(\mathbf{C}))$ (Lemma 10.2.3 showed that such operations came from operations on the representing object $|R|$), and when such operations will satisfy the identities of a variety $\mathbf{V}$. Note that this concept of "a representable set-valued functor given with operations that make it $\mathbf{V}$-valued" can also be looked at as "a $\mathbf{V}$-valued functor, whose composite with the forgetful functor $\mathbf{V} \to \mathbf{Set}$ is representable". This yields the equivalence of the two formulations of the next definition, in which we extend the term "representable functor" to include algebra-valued constructions.

**Definition 10.2.10.** *If* **V** *is a variety of* $\Omega$*-algebras, a functor* $\mathbf{C}^{\mathrm{op}} \to \mathbf{V}$ *will be called* representable *if it is isomorphic to a functor of the form* $\mathbf{C}(-, R)$*, for* $R$ *a* **V***-object of* **C***, equivalently, if its composite with the underlying-set functor* $\mathbf{V} \to \mathbf{Set}$ *is representable in the sense of Definition 8.2.3.*

## 10.3. Coalgebra objects in a category

In the next few sections we shall study *coalgebra* objects, and the functors these represent. A **V**-coalgebra object in a category **C** will be defined simply as a **V**-algebra object in $\mathbf{C}^{\mathrm{op}}$. But psychologically, the relationship between these two concepts is tricky. The definition of algebra object is easier to think about (to begin with) because it generalizes the familiar concept of a set-based algebra. But in most naturally occurring varieties of algebras, *coalgebra* objects and the covariant functors they represent turn out to be more diverse and interesting than algebra objects and their associated contravariant representable functors, and, as suggested by our example of SL$(n)$, they will be the main object of study in this chapter. Hence our flip-flop approach of using the algebra concept to introduce the definitions and basic characterization, then moving immediately to coalgebras. However, in §§ 10.12 and 10.13 we will return briefly to algebra objects, and note some examples and results on such objects in varieties of algebras.

In this section we continue to assume that $\gamma$ is a regular infinite cardinal, and $\Omega$ a type all of whose operations have arity $< \gamma$. However, we drop here the assumption of the preceding section that **C** is a category with $<\gamma$-fold products. What we will need is the dual hypothesis, and we will state that explicitly whenever it is required, as in the following definition.

**Definition 10.3.1.** *Let* **C** *be a category having* coproducts *of all families of* $< \gamma$ *objects. Then for* $\beta < \gamma$*, a* $\beta$*-ary* co-operation *on an object* $|R|$ *of* **C** *will mean a morphism of* $|R|$ *into the coproduct of* $\beta$ *copies of* $|R|$*; in other words, a* $\beta$*-ary operation on* $|R|$ *in* $\mathbf{C}^{\mathrm{op}}$*. A pair* $R = (|R|, (\mathbf{s}_R)_{s\in|\Omega|})$ *such that* $|R| \in \mathrm{Ob}(\mathbf{C})$*, and for each* $s \in |\Omega|$*,* $\mathbf{s}_R$ *is an* ari$(s)$*-ary co-operation on* $|R|$*, will be called an* $\Omega$*-coalgebra* object in **C** *(or a* **C***-based* $\Omega$*-coalgebra). A morphism of* $\Omega$*-coalgebra objects of* **C** *will mean a morphism of underlying* **C***-objects which respects co-operations.*

*For any* $\Omega$*-coalgebra object* $R$ *and object* $A$ *of* **C***, we shall write* $\mathbf{C}(R, A)$ *for the set-based algebra whose underlying set is* $\mathbf{C}(|R|, A)$*, and whose operations are induced by the co-operations of* $R$ *under the dual of the construction of the preceding section. Explicitly, for* $s \in |\Omega|$*, the operation* $s_{\mathbf{C}(R,\,A)}$ *induced by* $\mathbf{s}_R$ *on* $\mathbf{C}(|R|, A)$ *is defined to take each* ari$(s)$*-tuple* $(\xi_\alpha) \in \mathbf{C}(|R|, A)^{\mathrm{ari}(s)}$ *to the composite morphism*

$$|R| \xrightarrow{\ \mathbf{s}_R\ } \coprod_{\mathrm{ari}(s)} |R| \xrightarrow{\ (\xi_\alpha)_{\alpha\in\mathrm{ari}(s)}\ } A,$$

*where the second arrow denotes the map whose composite with the $\alpha$-th co-projection $|R| \to \coprod_{\mathrm{ari}(s)} |R|$ is $\xi_\alpha$ for each $\alpha \in \mathrm{ari}(s)$.*

I will in general, as above, use lower-case boldface letters **s** etc. to denote co-operations corresponding to operations denoted by the corresponding italic letters, $s$ etc.

Note that (as in the parallel definition in the preceding section), the $R$ in the above definition of $\mathbf{C}(R, A)$ is not an object of $\mathbf{C}$; here it is a $\mathbf{C}$-based *coalgebra* with underlying $\mathbf{C}$-object denoted $|R|$, and $\mathbf{C}(R, A)$ is likewise not a set, but an *algebra*, with underlying set $\mathbf{C}(|R|, A)$.

Let us recall from Lemma 9.4.17 what the general covariant representable set-valued functor $\mathbf{C}(|R|, -)$ "looks like" in the important case where its domain category $\mathbf{C}$ is a variety $\mathbf{W}$ of algebras. Taking a presentation $|R| = \langle X \mid Y \rangle_{\mathbf{W}}$ for the representing object, the functor $\mathbf{C}(|R|, -)$ can be described as carrying each object $A$ to the set of all $X$-tuples of elements of $A$ that satisfy the family of relations $Y$. Let us now examine the form that a $\beta$-ary operation $s$ on such a functor takes.

We know that $s$ will be induced by a co-operation $\mathbf{s}_R \colon |R| \to \coprod_\beta |R|$ of the representing object $|R| = \langle X \mid Y \rangle_{\mathbf{W}}$. The homomorphism $\mathbf{s}_R$ will, by the universal property of $\langle X \mid Y \rangle_{\mathbf{W}}$, correspond to some $X$-tuple of elements of $\coprod_\beta |R|$ which satisfies the relations $Y$. For each $x \in X$, the $x$-th entry of this $X$-tuple, being an element of $\coprod_\beta |R|$, may be expressed in terms of the $\beta$ images of $X$ generating that coproduct algebra, using some derived operation, which we may name

$$(10.3.2) \qquad\qquad s_x \in |F_{\mathbf{W}}(\beta \times X)|.$$

Now using the universality of $\coprod_\beta |R|$ as a $\mathbf{W}$-algebra $S$ with a $\beta$-tuple of elements of $\mathbf{W}(|R|, S)$, we can deduce that if $A$ is an arbitrary $\mathbf{W}$-algebra, and we regard elements of $\mathbf{W}(|R|, A)$ as $X$-tuples $\xi$ of elements of $A$ which satisfy the relations $Y$, then for each $\beta$-tuple $(\xi_\alpha)_{\alpha \in \beta}$ of such $X$-tuples, the $x$-th coordinate of the element $s_{\mathbf{W}(R, A)}(\xi_\alpha)_{\alpha \in \beta} \in \mathbf{W}(|R|, A)$ will be expressed in terms of the coordinates of the $\beta$ $X$-tuples $\xi_\alpha$ by the same derived operation (10.3.2). In summary:

**Lemma 10.3.3.** *Let $\mathbf{W}$ be a variety of algebras, $|R|$ an object of $\mathbf{W}$, and $\langle X \mid Y \rangle_{\mathbf{W}}$ a presentation of $|R|$ by generators and relations. For any $\mathbf{W}$-algebra $A$, any element $\xi \in \mathbf{W}(|R|, A)$, and any $x \in X$, let us call the image in $A$ of the generator $x$ of $|R|$ under $\xi$ "the $x$-th coordinate of $\xi$".*

*Let $\mathbf{s} \colon |R| \to \coprod_\beta |R|$ be a $\beta$-ary co-operation on $|R|$, and for any object $A$ of $\mathbf{W}$, let us write $s$ for the operation on the set $\mathbf{W}(|R|, A)$ induced by this co-operation on $|R|$. Then there exists an $X$-tuple of $\beta \times X$-ary derived operations $(s_x)_{x \in X}$ of $\mathbf{W}$, such that for every such $A$, for every $\beta$-tuple $(\xi_\alpha)_{\alpha \in \beta}$ of elements of $\mathbf{W}(|R|, A)$, and for every $x \in X$, the $x$-th coordinate* *of $s(\xi_\alpha)$ is computed from the coordinates of the given elements $\xi_\alpha$ by the derived operation $s_x$.*

*Conversely, given an $X$-tuple of $\beta \times X$-ary derived operations $s_x$ of $\mathbf{W}$ $(x \in X)$, if the identities of $\mathbf{W}$ imply that, when applied to any $\beta$ $X$-tuples all of which satisfy the relations $Y$, the $s_x$ give (as $x$ ranges over $X$) an $X$-tuple of elements which also satisfies $Y$, then $(s_x)_{x \in X}$ determines a morphism of functors $s : \mathbf{W}(|R|, -)^\beta \to \mathbf{W}(|R|, -)$, equivalently, a $\beta$-ary co-operation $\mathbf{s} : |R| \to \coprod_\beta |R|$.*                                                          □

So, for instance, if $\mathbf{W}$ is the variety of commutative rings, and $|R|$ the commutative ring with a universal $n \times n$ matrix of determinant 1, we can take for $X$ a family of $n^2$ symbols $(x_{ij})_{i,j \leq n}$, and for $Y$ the set consisting of the single relation $\det(x_{ij}) = 1$. To describe from the above point of view the comultiplication $\mathbf{m}$ on $|R|$ sketched in §10.1, take $\beta = 2$ and for each $i, j \leq n$ let $m_{ij}$ be the polynomial in $2n^2$ indeterminates by which one computes the $(i,j)$-th entry of the product of two matrices. The multiplicativity of the determinant function implies that these operations, when applied to the entries of two matrices of determinant 1, give the entries of a third matrix of determinant 1, so the hypothesis of the last paragraph of the above lemma is satisfied. Thus, these $n^2$ derived operations yield a binary co-operation on $|R|$, which induces, in a manner described abstractly in Definition 10.3.1 and concretely in Lemma 10.3.3, a binary operation on the sets $\mathbf{CommRing}^1(|R|, A) = |\mathrm{SL}(n, A)|$, namely, multiplication of matrices of determinant 1.

Back, now, to dualizing the concepts and results of the preceding section for a general category $\mathbf{C}$ (not necessarily a variety of algebras). Dualizing Definitions 10.2.9 and 10.2.10 respectively, we get

**Definition 10.3.4.** *Let $\mathbf{C}$ be a category with $<\gamma$-fold coproducts, and $\mathbf{V}$ a variety of $\Omega$-algebras defined by a set $J$ of identities. Then a co-$\mathbf{V}$ object of $\mathbf{C}$ (or $\mathbf{V}$-coalgebra in $\mathbf{C}$) will mean an $\Omega$-coalgebra $R$ in $\mathbf{C}$ satisfying the following equivalent conditions:*

*(i)   For all objects $A$ of $\mathbf{C}$, the algebra $\mathbf{C}(R, A)$ (Definition 10.3.1) lies in $\mathbf{V}$.*

*(ii)   For each identity $(t, t') \in J$, say in $\beta$ variables, if we form the $\beta$-fold coproduct $\coprod_\beta |R|$ with its canonical coprojections $(q_\alpha)_{\alpha \in \beta}$ $(\alpha \in \beta)$, then in the algebra $\mathbf{C}(R, \coprod_\beta |R|)$, one has $t(q_\alpha)_{\alpha \in \beta} = t'(q_\alpha)_{\alpha \in \beta}$. (This equality of morphisms $|R| \to \coprod_\beta |R|$ may be called the "coidentity" corresponding to the identity $t = t'$.)*

*(iii)   Writing $\mathbf{Cl}^{(\gamma)}(|R|^{\mathrm{op}})$ for the clone of all co-operations of arities $< \gamma$ on $|R|$ (i.e., operations on $|R|$ in $\mathbf{C}^{\mathrm{op}}$), the morphism of clones $\mathbf{Cl}^{(\gamma)}(\Omega\text{-}\mathbf{Alg}) \to \mathbf{Cl}^{(\gamma)}(|R|^{\mathrm{op}})$ induced by the $\Omega$-coalgebra structure of $|R|$ factors through the canonical map from $\mathbf{Cl}^{(\gamma)}(\Omega\text{-}\mathbf{Alg})$ to $\mathbf{Cl}^{(\gamma)}(\mathbf{V})$,*

$$\mathbf{Cl}^{(\gamma)}(\Omega\text{-}\mathbf{Alg}) \longrightarrow \mathbf{Cl}^{(\gamma)}(\mathbf{V}) \longrightarrow \mathbf{Cl}^{(\gamma)}(|R|^{\mathrm{op}}).$$

(iv)  *R satisfies the dual of the diagrammatic condition corresponding to each identity in  J.*

(v)  *Regarded as an $\Omega$-algebra object of $\mathbf{C}^{\mathrm{op}}$,  R is a $\mathbf{V}$-object.*

**Definition 10.3.5.** *Let $\mathbf{C}$ be a category with $<\gamma$-fold coproducts, and $\mathbf{V}$ a variety of $\Omega$-algebras. Then a covariant functor $V\colon \mathbf{C} \to \mathbf{V}$ will be called* representable *if*

(i)  *$V$ is isomorphic to a functor of the form $\mathbf{C}(R, -)$, for $R$ a co-$\mathbf{V}$ object of $\mathbf{C}$;*

*equivalently, if*

(ii)  *the composite of $V$ with the forgetful functor $\mathbf{V} \to \mathbf{Set}$ is representable in the sense Definition 8.2.3.*

> *If $\mathbf{C}$ is a variety $\mathbf{W}$ of algebras, these conditions are also equivalent to*
>
> (iii)  *there is some set $Y$ of relations in a family $X$ of variables such that the above composite is the functor associating to every object $A$ of $\mathbf{C}$ the set of all $X$-tuples of elements of $A$ satisfying $Y$.*
>
> *The full subcategory of $\mathbf{V}^{\mathbf{C}}$ consisting of the representable covariant functors $\mathbf{C} \to \mathbf{V}$ will be denoted $\mathbf{Rep}(\mathbf{C}, \mathbf{V})$.*

The equivalence of (i) and (ii) above follows from the equivalence of the corresponding conditions of Definition 10.2.10, which, we recall, followed from Lemmas 10.2.3 and 10.2.8; the equivalence of (ii) and (iii) when $\mathbf{C}$ is a variety of algebras follows, as noted, from Lemma 9.4.17.

So, for example, $\mathrm{SL}(n)$ is an object of $\mathbf{Rep}(\mathbf{CommRing}^1, \mathbf{Group})$.

Note that $\mathbf{W}$-*algebra* objects of a category $\mathbf{C}$ represent *contra*variant functors $\mathbf{C}^{\mathrm{op}} \to \mathbf{W}$, while *co*variant functors $\mathbf{C} \to \mathbf{W}$ are represented by *coalgebra* objects. This is a consequence of the behavior of the covariant and contravariant Yoneda embeddings, discussed in Remark 8.2.8. For the same reason, *morphisms* among covariant representable functors correspond *contravariantly* to morphisms among their representing coalgebras:

**Corollary 10.3.6** (to Lemma 10.2.5). *If $\mathbf{C}$ is a category with $<\gamma$-fold coproducts, and $\mathbf{V}$ a variety of $\Omega$-algebras, then the category $\mathbf{Rep}(\mathbf{C}, \mathbf{V})$ of covariant representable functors $\mathbf{C} \to \mathbf{V}$ is equivalent to the* opposite *of the category of co-$\mathbf{V}$ objects of $\mathbf{C}$.*  □

Below, we shall mainly study representable functors among varieties of algebras. But for students with some knowledge of topology, here is a pair of topological examples.

**Exercise 10.3:1.** Let $\mathbf{HtpTop}^{(\mathrm{pt})}$ be the category whose objects are Hausdorff topological spaces with basepoint, and whose morphisms are *homotopy classes* of basepoint-preserving maps.

(i)  Show that $\mathbf{HtpTop}^{(\mathrm{pt})}$ has finite products and coproducts.

(ii)   We noted at the end of §7.5 that the functor $\mathbf{HtpTop}^{(\mathrm{pt})} \to \mathbf{Set}$ taking an object $(X, x_0)$ to $|\pi_1(X, x_0)|$ (the underlying set of its fundamental group) was representable, with representing object $(S^1, 0)$. By the above results, the structure of group on these sets must be induced by a *cogroup* structure on $(S^1, 0)$. Describe the co-operations, and verify the cogroup identities.

(iii)  Describe likewise the structure of *group* object on $(S^1, 0)$ which represents the contravariant *first cohomotopy group* functor, $\pi^1$.

(Tangential remark: In Exercise 7.8:4 we defined, for any category $\mathbf{C}$ with a terminal object $T$, the category $\mathbf{C}^{\mathrm{pt}} = (T \downarrow \mathbf{C})$. The category called $\mathbf{HtpTop}^{(\mathrm{pt})}$ above is obtained by taking $\mathbf{HausTop}^{\mathrm{pt}}$, so defined, and passing to homotopy classes of maps therein. My use of parentheses in the superscript is an ad hoc way of indicating that this is *not* the category $\mathbf{HtpTop}^{\mathrm{pt}}$ gotten by taking the category $\mathbf{HtpTop}$ of Hausdorff spaces and homotopy class of maps, and applying the $^{\mathrm{pt}}$ construction to this. In the latter category, $h_{(S^1,0)}(A)$ is the set of conjugacy classes in $\pi_1(A)$, and these have no natural group structure. In fact, it appears that that category has no coproduct of two copies of $(S^1, 0)$.)

We are now ready to relate representability and the existence of adjoints!

## 10.4. Freyd's criterion for the existence of left adjoints

In Chapter 8 we obtained some curiously similar results about the class of covariant representable $\mathbf{Set}$-valued functors, and the class of right adjoint functors (functors having left adjoints) between arbitrary categories: both sorts of functors respected limits, and in both cases, all examples of functors respecting limits that were not of the desired sort arose from the failure of a "solution-set" condition. The former sort of functors were by definition $\mathbf{Set}$-valued, while the latter could have values in any category; Exercise 8.3:3 asked you to show that if the domain category had small coproducts and the codomain category was $\mathbf{Set}$, the two classes coincided. We shall now prove the corresponding result for the more general class of representable *algebra-valued* functors that we have defined.

One direction is still easy: Suppose a functor $V : \mathbf{C} \to \mathbf{V}$, where $\mathbf{C}$ is any category and $\mathbf{V}$ is a variety of algebras, has a left adjoint $G$. Then since the forgetful functor $U_{\mathbf{V}} : \mathbf{V} \to \mathbf{Set}$ also has a left adjoint $F_{\mathbf{C}}$, their composite $U_{\mathbf{V}} V$ has the left adjoint $G F_{\mathbf{V}}$. It follows that $U_{\mathbf{V}} V$ is representable, namely, by the image under its adjoint of a one-element set. Indeed,

$$(10.4.1) \qquad U_{\mathbf{V}} V(-) \cong \mathbf{Set}(1, U_{\mathbf{V}} V(-)) \cong \mathbf{C}(G F_{\mathbf{V}}(1), -).$$

And as we saw in the last section, if $\mathbf{C}$ has $<\gamma$-fold coproducts, representability of the set-valued functor $U_{\mathbf{V}} V$ is equivalent to representability of the algebra-valued functor $V$.

When we were considering only **Set**-valued functors, the other direction was also easy: If a functor $V : \mathbf{C} \to \mathbf{Set}$ had representing object $R$, then its left adjoint $G$ could be constructed as taking each set $Z$ to the coproduct of a $Z$-tuple of copies of $R$ (Exercises 8.2:5 and 8.3:3). To adapt this construction to the case where **Set** is replaced by a general variety **V**, we will (in the proof of the next theorem) take a presentation of an arbitrary algebra $A$ in **V**, by generators and relations,

$$(10.4.2) \qquad\qquad A = \langle Z \mid S \rangle_{\mathbf{V}}.$$

We will again take the coproduct of a $Z$-tuple of copies of $R$, but we will now use a second colimit construction, essentially a coequalizer applied to this coproduct, to "impose the set $S$ of relations", and produce an object $G(A)$ of **C** representing the functor $\mathbf{V}(A, V(-))$. (We use symbols $Z$ and $S$ here rather than $X$ and $Y$ so that if one considers the case where **C** is a variety **W** of algebras, there will be no confusion between this presentation for $A$ in **V**, and the presentation in **W** for the representing object $|R|$, which was written $\langle X \mid Y \rangle_{\mathbf{W}}$ in Lemma 10.3.3.)

This is in fact essentially the construction used at the beginning of § 10.1 to get a left adjoint for $SL(n)$. However, there we could give an explicit generators-and-relations description of the universal ring, while here, with **C** not assumed a variety of algebras, what we did has been abstracted as a colimit construction.

For completeness, the statement of the theorem below shows (as conditions (ii) and (iii)) both versions of the concept of representability, whose equivalence was noted in Definition 10.3.5.

**Theorem 10.4.3** (after Freyd [12]). *Let* **C** *be a category with small colimits,* **V** *a variety of $\Omega$-algebras, and*

$$V : \ \mathbf{C} \longrightarrow \mathbf{V}$$

*a (covariant) functor. Then the following conditions are equivalent:*

(i)   *$V$ has a left adjoint $G : \mathbf{V} \to \mathbf{C}$.*

(ii)   *$V$ is representable, i.e., is isomorphic to the **V**-valued functor represented by a co-**V** object $R$ of **C** (Definition 10.3.4).*

(iii)   *The composite $U_{\mathbf{V}} V$ of $V$ with the underlying set functor $U_{\mathbf{V}} : \mathbf{V} \to$ **Set** is representable, i.e., is isomorphic to the set-valued functor $h_{|R|}$ represented by an object $|R|$ of **C**.*

*Proof.* We already know that (ii) $\iff$ (iii); and (i) $\implies$ (iii) was shown above as (10.4.1). We shall complete the proof by showing (ii) $\implies$ (i).

Given $A \in \mathrm{Ob}(\mathbf{V})$, we want a $G(A) \in \mathrm{Ob}(\mathbf{C})$ such that $\mathbf{C}(G(A), -) \cong \mathbf{V}(A, V(-))$ (Theorem 8.3.8(ii)). Let us take a presentation (10.4.2) of $A$ in **V**. Thus, $\mathbf{V}(A, V(-))$ can be described as associating to each $B \in \mathrm{Ob}(\mathbf{C})$

the set of all $Z$-tuples of elements of the **V**-algebra $V(B)$ that satisfy the relations given by $S$.

Let us form a coproduct $\coprod_{z \in Z} |R|^{(z)} \in \mathrm{Ob}(\mathbf{C})$ of a $Z$-tuple of copies, $|R|^{(z)}$ ($z \in Z$), of the underlying **C**-object $|R|$ of our representing coalgebra. Then for any object $B$ of **C**, the set $\mathbf{C}(\coprod_z |R|^{(z)}, B)$ can be naturally identified with $\mathbf{C}(|R|, B)^Z \cong |V(B)|^Z$, the set of all $Z$-tuples of elements of $V(B)$. To get the subset of $Z$-tuples satisfying the relations in our presentation (10.4.2) of $A$, we want to formally "impose" these relations on $\coprod_z |R|^{(z)}$. Hence, for each relation $(s, t) \in S$ let us form the two morphisms $|R| \rightrightarrows \coprod_z |R|^{(z)}$ corresponding to $s$ and $t$, namely $s((q_z)_{z \in Z})$ and $t((q_z)_{z \in Z})$, where $(q_z)_{z \in Z}$ is the $Z$-tuple of coprojection morphisms $|R| \to \coprod_z |R|^{(z)}$, and $s$ and $t$ are evaluated on this $Z$-tuple using the co-**V** structure on $R$ assumed in (ii). Let $G(A)$ be the colimit of the diagram built out of all these pairs of arrows (one pair for each element of $S$):

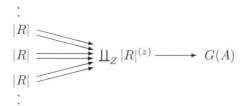

It follows from the universal property of this colimit that $G(A)$ has the desired property $\mathbf{C}(G(A), -) \cong \mathbf{V}(A, V(-))$.  $\square$

(Note that the above theorem required that **C** have arbitrary small colimits, so that we could construct $G(A)$ as above for all small **V**-algebras $A$. This requirement subsumes the condition of having $<\gamma$-fold coproducts assumed in Definition 10.3.5.)

**Exercise 10.4:1.** Verify the equivalence of the universal properties of $G(A)$ asserted in the last sentence of the above proof.

**Exercise 10.4:2.** Describe the construction used in proving (ii) $\implies$ (i) above in the particular case $\mathbf{C} = \mathbf{CommRing}^1$, $\mathbf{V} = \mathbf{Group}$, $V = SL(n)$, $A = Z_2$. (You are not asked to find a normal form for the ring obtained; simply show the generators-and-relations description that the construction gives in this case.) Show directly from your description that the result is a ring with a universal determinant-1 $n \times n$ matrix of exponent 2.

An alternative way to complete the proof of the above theorem, by showing (iii) $\implies$ (i) rather than (ii) $\implies$ (i), is indicated in

**Exercise 10.4:3.** Assuming condition (iii) of the above theorem, let **A** denote the full subcategory of **V** consisting of those objects $A$ such that the functor $\mathbf{V}(A, V(-)): \mathbf{C} \to \mathbf{Set}$ is representable, and let $G_{\mathbf{A}}: \mathbf{A} \to \mathbf{C}$

be the resulting "partial adjoint" to $V$. Show that $F_{\mathbf{V}}(1)$ belongs to $\mathbf{A}$, that $\mathbf{A}$ is closed under small colimits, and that every object of $\mathbf{V}$ can be obtained from the free object on one generator by iterated small colimits. Deduce that $\mathbf{A} = \mathbf{V}$.

We mentioned at the beginning of Chapter 9 the unfortunate ambiguity of the word "algebra", which has both a specific ring-theoretic sense, and the general sense with which this course is concerned. Note that in the ring-theoretic concept of an algebra over a commutative ring $k$, the operations other than the multiplication (and other than the multiplicative neutral element 1, if this is given), constitute a structure of $k$-module $M$, and the multiplication is then a $k$-bilinear map $M \times M \to M$, which is equivalent to a $k$-module map $M \otimes_k M \to M$. Ring-theorists also consider the dual concept of a $k$-module $M$ given with a $k$-module map $M \to M \otimes_k M$, and call this a $k$-*coalgebra*. So the ambiguity of the preceding chapter has pursued us into this material as well! A $k$-module with both a $k$-algebra and a $k$-coalgebra structure, related by certain identities, is called a $k$-*bialgebra,* and with certain additional unary, zeroary, and co-zeroary structure, a *Hopf algebra.* The study of these is an interesting field ([136], cf. [48]), but not one that we will touch on in this course. (If we did want to refer to such coalgebras here, we would call them "coalgebras in the sense of the theory of Hopf algebras".)

A much more general pair of senses of "algebra" and "coalgebra" is sometimes used. Given a category $\mathbf{C}$ and a functor $F\colon \mathbf{C} \to \mathbf{C}$, an "$F$-algebra in $\mathbf{C}$" means, in that usage, a pair $(X, a)$ consisting of an object $X$ of $\mathbf{C}$ and a morphism $a\colon F(X) \to X$; an $F$-coalgebra in $\mathbf{C}$ means, dually, a pair $(X, b)$ where $X$ is an object of $\mathbf{C}$ and $b$ a morphism $X \to F(X)$. You should not find it hard to show that *all* the senses of algebra and coalgebra that we have mentioned can be subsumed by this pair of definitions. But we will not find this degree of generality useful.

## 10.5. Some corollaries and examples

Since composites of adjunctions are adjunctions (Theorem 8.3.10), the result of the last section yields

**Corollary 10.5.1.** *A composite of representable functors among varieties of algebras is representable.*                                                                                □

Actually, this reasoning shows that a composite of representable functors $\mathbf{C} \to \mathbf{V} \to \mathbf{W}$, where $\mathbf{V}$ and $\mathbf{W}$ are varieties, and $\mathbf{C}$ is any category with small colimits, is representable, but I have given the above more limited statement because of its simplicity.

What does the representing object for a composite of representable functors among varieties look like? Suppose we have representing coalgebras: $\qquad\qquad R \qquad\quad S$

right adjoints: $\qquad\qquad\qquad \overset{V}{\longrightarrow} \qquad \overset{W}{\longrightarrow}$

$$\mathbf{X} \underset{D}{\overset{V}{\rightleftarrows}} \mathbf{V} \underset{E}{\overset{W}{\rightleftarrows}} \mathbf{W},$$

left adjoints:

so that the composite functor $WV$ has left adjoint $DE$. To determine the underlying **X**-object of the **W**-coalgebra representing $WV$, we note that this object of **X** will represent the set-valued functor $U_{\mathbf{W}}WV$. The factor $U_{\mathbf{W}}W$ is represented by $|S|$, so by Theorem 8.7.1, the object representing $U_{\mathbf{W}}WV$ can be obtained by applying to $|S|$ the left adjoint of $V$. Thus, the underlying **X**-object of our desired representing object is $D(|S|)$.

Let us combine this observation with the description of $D$ in the proof of Theorem 10.4.3. $D$ takes a **V**-algebra $A$ to an **X**-algebra obtained by "pasting together" a family of copies of $|R|$ indexed by the generators in any presentation of $A$, using "pasting instructions" obtained from the relations in that presentation. Hence the representing object $D(|S|)$ for $U_{\mathbf{W}}WV$ can be obtained by "pasting together" a family of copies of $|R|$ in a way prescribed by any presentation of $|S|$. From this one can deduce that if $|R| = \langle X \mid Y \rangle_{\mathbf{X}}$ and $|S| = \langle X' \mid Y' \rangle_{\mathbf{V}}$, then the representing object for $U_{\mathbf{W}}WV$ can be presented in **X** by a generating set indexed by $X \times X'$, and a set of relations indexed by $(Y \times X') \sqcup (X \times Y')$. (Equivalently, if we look at $V$ as taking each **X**-algebra $A$ to a **V**-algebra whose elements are $X$-tuples of elements of $A$ satisfying a certain $Y$-tuple of equations, and similarly regard $W$ as taking each **V**-algebra $B$ to a **W**-algebra whose elements are $X'$-tuples of elements of $B$ satisfying a certain $Y'$-tuple of equations, then their composite can be described as taking each **X**-algebra $A$ to a **W**-algebra whose elements are all $X \times X'$-tuples of elements of $A$ that satisfy a $(Y \times X') \sqcup (X \times Y')$-tuple of relations.)

Of course, we also want to know the co-**W** structure on this object. Not unexpectedly, this arises from the co-**W** structure on the object $|S|$. We shall see some examples of representing objects of composite functors in §10.9. I won't work out the details of the general description of such objects, but if you are interested, you can do this, as

**Exercise 10.5:1.** Describe precisely how to construct a presentation of the object representing $WV$, and a description of its co-**W** structure, in terms of presentations of $|R|$ and $|S|$ and their co-**V** and co-**W** structures.

Theorem 10.4.3 has the following special case (which was Exercise 9.9:8 in the last chapter); though it is unfortunate that this case is better known than the theorem, and is thought by many to be the "last word" on the subject!

**Corollary 10.5.2.** *Any functor* $V: \mathbf{W} \to \mathbf{V}$ *between varieties of algebras which respects underlying sets has a left adjoint.*

*Proof.* By Theorem 10.4.3(i) $\implies$ (iii), to show $V$ has a left adjoint it suffices to show that $U_{\mathbf{V}}V: \mathbf{W} \to \mathbf{Set}$ is representable. But by hypothesis,

$U_{\mathbf{V}}V = U_{\mathbf{W}}$, which is clearly representable, by any of our three criteria (representing object: $F_{\mathbf{W}}(1)$; description: sends each object $A$ to the set of 1-tuples of elements of $A$ satisfying the empty set of relations; left adjoint: $F_{\mathbf{W}}$). □

This corollary applies to such constructions as (i) the underlying-set functor $U_{\mathbf{W}} \colon \mathbf{W} \to \mathbf{Set}$ of any variety $\mathbf{W}$, the left adjoint of which is, we already know, the *free algebra* construction; (ii) the inclusion of any variety $\mathbf{W}$ in a larger variety $\mathbf{V}$ of algebras of the same type (i.e., one defined by a subset of the identities of $\mathbf{W}$), the left adjoint of which is the construction of "imposing the additional identities of $\mathbf{W}$" on algebras in $\mathbf{V}$; (iii) the functor $\mathbf{Set} \to G\text{-}\mathbf{Set}$ (for any group $G$) which takes a set $A$ and regards it as a $G$-set with trivial action; this has for left adjoint the *orbit-set* functor $G\text{-}\mathbf{Set} \to \mathbf{Set}$ (cf. Exercise 8.6:1); (iv) the functor taking an associative ring $A$ to its underlying additive group, whose left adjoint is the *tensor ring* construction, and similarly (v) the functor taking an associative ring $A$ to its underlying multiplicative monoid, whose left adjoint is the *monoid-ring* construction (both these left adjoint constructions were discussed in terms of their universal properties in §4.12), and (vi) the "commutator brackets" functor $B \colon \mathbf{Ring}_k^1 \to \mathbf{Lie}_k$ of §9.7, taking associative algebras to Lie algebras, whose left adjoint is the *universal enveloping algebra* construction $E$.

On the other hand, the functor $\mathrm{SL}(n) \colon \mathbf{CommRing}^1 \to \mathbf{Group}$ with which we began this chapter certainly does not preserve underlying sets. That was a good example for getting away from functors represented by free algebras on one generator, because the representing algebra both requires more than one generator, and requires nontrivial relations, i.e., is nonfree. There are also important examples where a representing algebra is free, but on more than one generator (equivalently, where the functor has the property that the underlying set $|V(A)|$ of the constructed algebra is a fixed power $|A|^X$ of the underlying set of the given algebra $A$), or can be generated by one element, but subject to some relations (equivalently, where $|V(A)|$ can be described as the subset of $|A|$ itself consisting of those elements which satisfy certain equations). Among constructions of the first type are the $n \times n$ matrix ring functor $M_n \colon \mathbf{Ring}^1 \to \mathbf{Ring}^1$, the representing object for which is free on $n^2$ generators, and the formal power series functor taking a ring $A$ to the ring $A[[t]]$ (either $\mathbf{Ring}^1 \to \mathbf{Ring}^1$ or $\mathbf{CommRing}^1 \to \mathbf{CommRing}^1$), whose representing object is free on countably many generators. The left adjoints of these have no standard names, but can be described as taking a ring $B$ to the ring over which one has a "universal $n \times n$ matrix representation of $B$", respectively a "universal representation of $B$ by formal power series". A functor with representing algebra presented by one generator and a nonempty set of relations is the construction $\mathbf{CommRing}^1 \to \mathbf{Bool}^1$ taking a ring $A$ to the set of its idempotent elements, made a Boolean ring as described in Exercise 4.14:3. The underlying ring of its representing coalgebra is presented by a generator $x$ and the relation $x^2 = x$, and can be described as $\mathbb{Z} \times \mathbb{Z}$, with $x = (1, 0)$. Another example with one generator and a nonempty relation-set is the functor $\mathbf{Ab} \to \mathbf{Ab}$ taking any abelian group to its subgroup of elements of exponent $n$ (for any fixed $n > 0$), which is represented by the cyclic group of order $n$. Still another is the functor $G$-$\mathbf{Set} \to \mathbf{Set}$ (for $G$ any nontrivial group) represented by the one-element $G$-set. This takes a $G$-set $A$ to the set of fixed points of the action of $G$; its left adjoint is the functor $\mathbf{Set} \to G$-$\mathbf{Set}$ mentioned in point (iii) of the preceding paragraph, which thus has both a left and a right adjoint!

We saw in Chapter 4 that every monoid has both a universal map into a group, and a universal map of a group into it. This says that the forgetful functor

$$U: \mathbf{Group} \longrightarrow \mathbf{Monoid}$$

also has both a left and a right adjoint. That it has a left adjoint is now clear from that fact that it preserves underlying sets. Our present results do not say anything about why it should have a right adjoint, but they do say that that right adjoint must be a representable functor. Let us find its representing cogroup.

We recall that that right adjoint is the functor

$$G: \mathbf{Monoid} \longrightarrow \mathbf{Group}$$

taking every monoid $A$ to its group of invertible elements. Since the invertible elements of a monoid $A$ form a subset of $|A|$, one might at first glance expect that $U_{\mathbf{Group}}\, G$, when expressed in the form described in Lemma 9.4.17(ii), should have $X$ a singleton, i.e., should be represented by a monoid presented by one generator and some relations. But at second glance, we see that this cannot be so: the condition that an element of a monoid be invertible is not an equation in that element alone. By the considerations in the paragraph containing (10.4.1), we can find the representing monoid for $G$ by applying its left adjoint $U$ to the free group on one generator. The result is this same group, regarded as a monoid, and as such, it has presentation

$$(10.5.3) \qquad R = \langle x, y \mid x\,y = e = y\,x \rangle.$$

Thus for any monoid $A$, the description of $|G(A)|$ in the form described in Lemma 9.4.17(ii) is

$$(10.5.4) \qquad \{(\xi, \eta) \in |A| \times |A| \mid \xi\eta = e = \eta\xi\}.$$

Since two-sided inverses to monoid elements are unique when they exist, every element $(\xi, \eta)$ of $|G(A)|$ is determined by its first component, subject to the condition that this have an inverse. So up to functorial isomorphism, (10.5.4) is indeed the set of invertible elements of $A$. (We noted this example briefly in the paragraph following Lemma 9.4.17.)

Let us write down the cogroup structure on the representing monoid (10.5.3). If we write the coproduct of two copies of this monoid as

$$R \amalg R \;=\; \langle x_0,\, y_0,\, x_1,\, y_1 \mid x_0\, y_0 = e = y_0\, x_0,\; x_1\, y_1 = e = y_1\, x_1 \rangle,$$

then we find that the comultiplication is given by

$$\mathbf{m}(x) \;=\; x_0\, x_1, \qquad \mathbf{m}(y) \;=\; y_1\, y_0.$$

(If you are uncertain how I got these formulas, stop here and think it out. If you are still not sure, ask in class! Note the reversed multiplication of the $y$'s, a consequence of the fact that when one multiplies two invertible monoid elements, their inverses multiply in the reverse order.) It is also easy to see that the coinverse operation $\mathbf{i}\colon R \to R$ is given by

$$\mathbf{i}(x) \;=\; y, \qquad \mathbf{i}(y) \;=\; x,$$

and, finally, that the co-neutral-element map, from $R$ to the initial object of **Monoid**, namely $\{e\}$, is the unique element of $\mathbf{Monoid}(R, \{e\})$, characterized by

$$\mathbf{e}(x) \;=\; e \;=\; \mathbf{e}(y).$$

**Exercise 10.5:2.** Describe explicitly the co-operations of the coalgebras representing two of the other examples discussed above, as we have done for the group-of-units functor **Monoid** $\to$ **Group**.

**Exercise 10.5:3.** We noted above that we might naively have expected the group-of-invertible-elements functor **Monoid** $\to$ **Group** to be represented by a one-generator monoid, but that it was not. Let us look more closely at this type of situation. Suppose $W\colon \mathbf{V} \to \mathbf{W}$ is a representable functor among varieties of algebras, with representing **W**-coalgebra $R$.

(i)   Show that $U_{\mathbf{W}} W\colon \mathbf{V} \to \mathbf{Set}$ is isomorphic to a subfunctor of $U_{\mathbf{V}}$ if and only if there exists a map $F_{\mathbf{V}}(1) \to |R|$ which is an *epimorphism* in $\mathbf{V}$ (but not necessarily surjective).

(ii)   Describe the epimorphism implicit in our discussion of the group-of-invertible-elements functor.

(iii)  Generalize the result of (i) in one way or another.

We can get other examples of representable functors by composing some of those we have described. For instance, if we start with the $n \times n$ matrix ring functor $\mathbf{Ring}^1 \to \mathbf{Ring}^1$, follow it by the underlying multiplicative monoid functor $\mathbf{Ring}^1 \to \mathbf{Monoid}$, and this by the group-of-units functor $\mathbf{Monoid} \to \mathbf{Group}$, we get a functor $\mathbf{Ring}^1 \to \mathbf{Group}$ which takes every ring $A$ to the group of all invertible $n \times n$ matrices over $A$, denoted $GL(n, A)$.

Let us record a couple of other general results on representability of functors, equivalently, on existence of adjoints. As we noted in example (ii) following Corollary 10.5.2, that corollary implies

**Corollary 10.5.5.** *The inclusion of any subvariety* **U** *in a variety* **V** *has a left adjoint.*                                                                    □

Combining this with Corollary 10.5.1 (composites of representable functors are representable), we get

**Corollary 10.5.6.** *If a functor* $W : \mathbf{V} \to \mathbf{W}$ *between varieties of algebras is representable, then so is its restriction to any subvariety of* $\mathbf{U} \subseteq \mathbf{V}$. □

For instance, having observed that $\mathrm{GL}(n)$ is a representable functor on **Ring**[1], we know automatically that it gives a representable functor on **CommRing**[1]. (What is the relation between the representing objects for these two functors?)

When a functor between varieties of algebras $W : \mathbf{V} \to \mathbf{W}$ is representable, this representability is usually easy to see and to prove—the construction of the underlying set of $W(A)$ is easily expressed in the form described in Lemma 9.4.17(ii). On the other hand, when we want to prove that a functor $V$ is *not* representable, that criterion is clearly not as helpful; the more useful criterion here is Proposition 8.10.4, which says that $W$ is representable if and only if it respects limits and satisfies a certain solution-set condition. As we noted in §§ 8.7–8.10, most cases of nonrepresentability reveal themselves through failure of the functor to respect limits of one sort or another. For example:

**Exercise 10.5:4.** Verify that *none* of the following covariant functors from abelian groups to abelian groups is representable:

(i)    $F(A) = A \otimes A$.

(ii)    $G(A) =$ the torsion subgroup of $A$ (the subgroup of all elements of finite order).

(iii)    $H(A) = A/nA$ ($n$ a fixed integer).

(iv)    $J(A) = nA$ ($n$ a fixed integer).

In Exercises 8.10:5 and 8.10:6 we saw examples of the rarer situation in which some left universal construction was impossible only because the solution-set condition was not satisfied. Those examples were of nonexistence of *initial objects* and of *free objects*, so by Theorem 9.4.14 the categories in question were, necessarily, not varieties (though in one of the examples the category, that of complete lattices, failed to be a variety only in that it had a large set of operations). The following exercise shows that in the case of the criterion for *representability*, there are counterexamples where the domain *is* a variety.

**Exercise 10.5:5.** Let us call an object $S$ of a variety **V** *simple* if the only congruences on $S$ are the trivial congruence and the total congruence (the least and the greatest equivalence relations on $|S|$).

(i)   Find a variety $\mathbf{V}$ having the properties that (a) for every cardinal $\alpha$ there exists a simple algebra $S_\alpha$ in $\mathbf{V}$ of cardinality $\geq \alpha$, and (b) every algebra in $\mathbf{V}$ contains a unique one-element subalgebra. (Suggestion: Either show that there are simple groups of arbitrarily large cardinalities, or that there are fields of arbitrarily large cardinalities. In the latter case you must also say how to regard fields as simple objects of a variety satisfying (b).)

Now assume we have chosen such a $\mathbf{V}$, and for each $\alpha$ some $S_\alpha$, as above. For every object $A$ of $\mathbf{V}$, define $V(A)$ to be the hom-set $\mathbf{V}(\coprod_{\alpha \leq \mathrm{card}(|A|)} S_\alpha, A) \in \mathrm{Ob}(\mathbf{Set})$; equivalently (up to natural isomorphism) $V(A) = \prod_{\alpha \leq \mathrm{card}(|A|)} \mathbf{V}(S_\alpha, A)$.

(ii)   Show how to make $V$ a functor, and show that this functor respects small limits, but is not representable. (You may either get these results directly, or with the help of part (iii) below.)

(iii)   Recall that the variety we are writing $\mathbf{V}$ could be more precisely written as $\mathbf{V}_{(\mathbb{U})}$, the category of $\mathbb{U}$-small objects of a certain type that satisfy a certain system of identities. Letting $\mathbb{U}'$ be any universe properly larger than $\mathbb{U}$, show that $\mathbf{V}_{(\mathbb{U}')}$ contains an object $S$ such that the restriction to $\mathbf{V}_{(\mathbb{U})}$ of the functor $h_S \colon \mathbf{V}_{(\mathbb{U}')} \to \mathbf{Set}_{(\mathbb{U}')}$ is isomorphic to the functor $V$ of (ii) above.

Thus, intuitively, this example is based on a functor which is representable, but by an object outside our universe. What was tricky was to find such a functor which nevertheless took $\mathbb{U}$-small algebras to $\mathbb{U}$-small sets.

Curiously, in the condition from Chapter 8 for the existence of *right* adjoint functors (the dual of Theorem 8.10.5), one *can* drop the solution-set condition when the domain category is a variety:

**Exercise 10.5:6.** Show that if $\mathbf{V}$ is a variety of algebras and $\mathbf{C}$ any category, then every functor $F \colon \mathbf{V} \to \mathbf{C}$ which respects small colimits has a right adjoint; i.e., is the left adjoint to a representable functor.

Knowing that representable functors from a variety $\mathbf{W}$ to a variety $\mathbf{V}$ correspond to $\mathbf{V}$-coalgebra objects of $\mathbf{W}$, it is natural to try, for various choices of $\mathbf{V}$ and $\mathbf{W}$, to find *all* $\mathbf{V}$-coalgebra objects of $\mathbf{W}$, hence all such functors. How difficult this task is depends on the varieties in question. At the easy extreme are certain large classes of cases for which we shall see in §10.10 that there can be no nontrivial representable functors. At the other end are cases such as that of representable functors from the variety of commutative rings (or commutative algebras over a fixed commutative ring $k$) to **Group**. Such functors are called "affine algebraic groups", and are an important area of research in algebraic geometry.

In the next three sections, we shall tackle some cases of an intermediate level of difficulty, for which the problem is nontrivial, but where, with a reasonable amount of work, we can get a complete classification.

## 10.6. Representable endofunctors of Monoid

Let us consider representable functors from the variety **Monoid** into itself.

A representable functor from an arbitrary category $\mathbf{C}$ with finite coproducts to **Monoid** is represented by a comonoid, which we shall for convenience write as a 3-tuple $(R, \mathbf{m}, \mathbf{e})$ (rather than as a pair $(R, (\mathbf{m}, \mathbf{e}))$), with $R$ an object of $\mathbf{C}$, and the other two components a binary *comultiplication*

$$\mathbf{m} \colon R \longrightarrow R \amalg R$$

and a zeroary *co-neutral-element*

$$\mathbf{e} \colon R \longrightarrow I.$$

Here $I$ denotes the initial object of $\mathbf{C}$, that is, the coproduct of the empty family. These co-operations must satisfy the coassociative law, and the right and left coneutral laws. The coassociative law can be shown diagrammatically as the dual to (10.2.6); thus, it says that the diagram

(10.6.1)

$$
\begin{array}{ccc}
R & \xrightarrow{\ \ \mathbf{m}\ \ } & R \amalg R \\[2mm]
\Big\downarrow{\scriptstyle \mathbf{m}} & & \Big\downarrow{\scriptstyle \mathrm{id}_R \amalg \mathbf{m}} \\[2mm]
R \amalg R & \xrightarrow{\ \mathbf{m} \amalg \mathrm{id}_R\ } & R \amalg R \amalg R
\end{array}
$$

commutes. The two coneutral laws likewise say that if we write $i_R$ for the unique map from the initial object $I$ to $R$, then the composite maps

(10.6.2)
$$
\begin{array}{l}
R \xrightarrow{\ \mathbf{m}\ } R \amalg R \xrightarrow{\ (i_R\,\mathbf{e},\, \mathrm{id}_R)\ } R \\[2mm]
R \xrightarrow{\ \mathbf{m}\ } R \amalg R \xrightarrow{\ (\mathrm{id}_R,\, i_R\,\mathbf{e})\ } R
\end{array}
$$

are each the identity morphism of $R$, where in each diagram of (10.6.2), the parenthesized pair shown above the second arrow is an abbreviation for the morphism obtained from its two entries via the universal property of the coproduct $R \amalg R$.

Let us now specialize to the case $\mathbf{C} = \mathbf{Monoid}$. Then the initial object $I$ is the trivial monoid $\{e\}$; hence the homomorphism $\mathbf{e}$ can only be the map taking every element of $R$ to $e$. (Contrast this with the case of $SL(n)$ discussed in §10.1, where $\mathbf{e}$ had for codomain the initial object $\mathbb{Z}$ of $\mathbf{CommRing}^1$, a nonzero ring, so that the description of the identity matrix gave nontrivial information.) Nonetheless, the fact that this unique zeroary co-operation satisfies the coneutral laws (10.6.2) will be a nontrivial condition.

To study (10.6.1) and (10.6.2), we need to recall the structure of a coproduct of monoids. We noted in § 4.10 that such a coproduct

$$(10.6.3) \qquad\qquad \coprod_{\alpha \in I} R^\alpha$$

could be described in essentially the same way as for groups; namely, assuming for notational convenience that the sets $|R^\alpha| - \{e\}$ are disjoint, each element of (10.6.3) can be written uniquely as a product

$$(10.6.4) \qquad \begin{array}{l} r_0 r_1 \ldots r_{h-1}, \quad \text{with } h \geq 0, \text{ each } r_i \text{ in some } |R^{\alpha_i}| - \{e\}, \\ \qquad\qquad \text{and } \alpha_i \neq \alpha_{i+1} \text{ for } 0 \leq i < h - 1. \end{array}$$

(Here the neutral element $e$ of (10.6.3) is understood to be the case $h = 0$ of (10.6.4).)

But in the case of the coproduct $R \amalg R$ we are interested in now, the two monoids being put together are *not* disjoint. Let us therefore distinguish our two canonical images of $R$ in $R \amalg R$ as $R^\lambda$ and $R^\rho$ (the superscripts corresponding to the "left" and "right" arguments of the comultiplication we want to study). We shall thus write $R \amalg R$ as $R^\lambda \amalg R^\rho$, i.e., as the coproduct of these two copies of $R$, and write the images of an element $x \in |R|$ under the two coprojections $R \rightrightarrows R^\lambda \amalg R^\rho$ as $x^\lambda$ and $x^\rho$ respectively.

The coassociative law involves three variables, hence in (10.6.1), $R$ is ultimately mapped into a three-fold coproduct of copies of itself; let us write this object $R^\lambda \amalg R^\mu \amalg R^\rho$, the $\mu$ standing for the "middle" variable in the associativity identity.

A natural first step in describing an element (10.6.4) is to specify the sequence of indices $(\alpha_0, \ldots, \alpha_{h-1})$; so let us define an *index-string* to mean a finite (possibly empty) sequence of members of $\{\lambda, \mu, \rho\}$, with no two successive terms equal. We shall call $h$ the *length* of the index-string $(\alpha_0, \ldots, \alpha_{h-1})$. For every index-string $\sigma = (\alpha_0, \ldots, \alpha_{h-1})$, we shall denote by $|R|^\sigma$ the set of all products (10.6.4) with that sequence of superscripts, i.e.,

$$|R|^\sigma = (|R^{\alpha_0}| - \{e\}) \ldots (|R^{\alpha_{h-1}}| - \{e\}).$$

The underlying set of each of the monoids $R^\lambda \amalg R^\rho$ and $R^\lambda \amalg R^\mu \amalg R^\rho$ is thus the disjoint union of its subsets $|R|^\sigma$. We define the *height* $\mathrm{ht}(s)$ of $s \in |R^\lambda \amalg R^\rho|$ as the length of the unique $\sigma$ such that $s \in |R|^\sigma$. Finally, to study our comultiplication **m**, let us define the *degree* of an element of $R$ itself by

$$\deg(x) = \mathrm{ht}(\mathbf{m}(x)).$$

We note that for each $h > 0$, there are precisely two index-strings of length $h$ consisting only of $\rho$'s and $\lambda$'s: one beginning with $\rho$ and the other beginning with $\lambda$. Thus, if $x \in |R|$ is an element of positive degree $h$, then $\mathbf{m}(x)$ either belongs to $|R|^{(\lambda, \rho, \lambda, \, \cdots)}$ ($h$ entries in the superscript) i.e., has the form $y_0^\lambda z_1^\rho y_2^\lambda \ldots$, or it belongs to $|R|^{(\rho, \lambda, \rho, \, \cdots)}$, and has the form $z_0^\rho y_1^\lambda z_2^\rho \ldots$.

It is easy to see that the two coneutral laws (10.6.2) say

$$(10.6.5) \qquad \text{If } \mathbf{m}(x) = \ldots y_i^\lambda \, z_{i+1}^\rho \, y_{i+2}^\lambda \, z_{i+3}^\rho \cdots, \qquad \text{then}$$
$$x = \ldots y_i \, y_{i+2} \cdots = \ldots z_{i+1} \, z_{i+3} \cdots.$$

(Note that the way we have written $\mathbf{m}(x)$ here covers both the cases $x \in |R|^{(\lambda, \rho, \lambda, \cdots)}$ and $x \in |R|^{(\rho, \lambda, \rho, \cdots)}$.) In particular, (10.6.5) implies

$$(10.6.6) \qquad \text{If } x \neq e, \quad \text{then} \quad \deg(x) \geq 2.$$

On the two possible sorts of elements of degree exactly 2, we see that (10.6.5) precisely determines the action of $\mathbf{m}$:

$$(10.6.7) \qquad \begin{cases} \text{If } \mathbf{m}(x) \in |R|^{(\lambda, \rho)}, \text{ then } \mathbf{m}(x) = x^\lambda \, x^\rho. \\ \text{If } \mathbf{m}(x) \in |R|^{(\rho, \lambda)}, \text{ then } \mathbf{m}(x) = x^\rho \, x^\lambda. \end{cases}$$

Let us also record what (10.6.5) tells us about the degree 3 case:

$$(10.6.8) \qquad \begin{cases} \text{If } \mathbf{m}(x) \in |R|^{(\lambda, \rho, \lambda)}, \text{ then } \mathbf{m}(x) = y_0^\lambda \, x^\rho \, y_2^\lambda \text{ where } y_0 \, y_2 = x. \\ \text{If } \mathbf{m}(x) \in |R|^{(\rho, \lambda, \rho)}, \text{ then } \mathbf{m}(x) = z_0^\rho \, x^\lambda \, z_2^\rho \text{ where } z_0 \, z_2 = x. \end{cases}$$

We now turn to the coassociative law. This says that for any $x \in |R|$,

$$(10.6.9) \qquad (\mathrm{id}_{R^\lambda}, \mathbf{m}) \, \mathbf{m}(x) = (\mathbf{m}, \mathrm{id}_{R^\rho}) \, \mathbf{m}(x) \quad \text{in } R^\lambda \amalg R^\mu \amalg R^\rho.$$

Let us note the explicit descriptions of the left-hand factor on each side of the above equation. The homomorphism $(\mathrm{id}_{R^\lambda}, \mathbf{m}) : R^\lambda \amalg R^\rho \to R^\lambda \amalg R^\mu \amalg R^\rho$ leaves each element of the form $y^\lambda \in |R^\lambda \amalg R^\rho|$ unchanged, while it takes an element $z^\rho \in |R^\lambda \amalg R^\rho|$ to the element $\mathbf{m}(z)$, but with all the superscripts "$\lambda$" changed to "$\mu$" (because of the way we label our three-fold coproduct). Likewise, $(\mathbf{m}, \mathrm{id}_{R^\rho})$ leaves each $z^\rho$ unchanged, and takes each $y^\lambda$ to the element $\mathbf{m}(y)$, with all superscripts "$\rho$" changed to "$\mu$".

Now let $x \in |R| - \{e\}$, suppose that $\mathbf{m}(x)$ belongs to the set $|R|^\sigma$ ($\sigma$ a string of $\lambda$'s and $\rho$'s), and let the common value of the two sides of (10.6.9) belong to the set $|R|^\tau$ ($\tau$ a string of $\lambda$'s, $\mu$'s and $\rho$'s). Note that each "$\lambda$" in $\sigma$ yields a single $\lambda$ in $\tau$ on evaluating the left-hand side of (10.6.9), but looking at the right-hand side of (10.6.9), it gives *at least* one $\lambda$ in $\tau$, because of (10.6.6). Since the two sides of (10.6.9) are equal, all of these "at least one"s must be exactly one. For this to happen, the elements $y_i$ in the expansion (10.6.5) must all have degree $\leq 3$. By a symmetric argument (comparing occurrences of $\rho$ in $\sigma$ and in $\tau$) we get the same conclusion for the elements $z_i$. Note also that if $\tau$ begins with $\mu$, then the right-hand side of (10.6.9) tells us $\sigma$ must begin with a $\lambda$, while the left-hand side says it must begin with a $\rho$, a contradiction. Hence $\tau$ can only begin with a $\lambda$ or a $\rho$. In the former case, $\sigma$ must begin with a $\lambda$ which expands to $\lambda \mu$ on the right-hand side of (10.6.9) (so as not to yield more than one $\lambda$); in the latter case it must begin with a $\rho$ which expands to $\rho\mu$ on the left-hand side. In either case, we conclude that the first factor in the expansion of $\mathbf{m}(x)$ must have degree 2. The same arguments apply to the last factor. In summary:

(10.6.10)  For all $x \in |R|$, all elements $y_i$ and $z_i$ in (10.6.5) have degree $\leq 3$; hence by (10.6.5), every element of $R$ is a product of elements of degree $\leq 3$. Moreover, the elements giving the *first* and *last* factors of $\mathbf{m}(x)$ have degree 2.

But the observation about first and last factors, applied to the final equation in each line of (10.6.8), gives

(10.6.11)  Every element of $R$ of degree 3 is a product of two elements of degree 2.

Equations (10.6.10) and (10.6.11) together allow one to express every element of $R$ as a product of elements of degree 2, showing that $R$ is generated by these elements. We can prove still more:

**Lemma 10.6.12.** *Let* $(R, \mathbf{m}, e)$ *be a co-**Monoid** object in **Monoid**. Then every element* $x \in |R|$ *has an expression as a product*

$$x_0 \ldots x_{h-1} \quad (h \geq 0),$$

*where all* $x_i$ *are of degree 2, and this expression is* unique *subject only to the condition that there be no two successive factors* $x_i$, $x_{i+1}$ *such that one of* $\mathbf{m}(x_i)$, $\mathbf{m}(x_{i+1})$ *belongs to* $|R|^{(\lambda, \rho)}$, *the other belongs to* $|R|^{(\rho, \lambda)}$, *and* $x_i \, x_{i+1} = e$.

*Proof.* Since $R$ is generated by elements of degree 2, and since any expression involving two successive factors whose product is $e$ can be simplified to a shorter expression, we can clearly express every element in the indicated form subject to the conditions noted. To show that this form is unique, it suffices to prove that given an element and such an expression for it,

$$x = x_0 \ldots x_{h-1} \quad (\deg(x_i) = 2, \ i = 0, \ldots, h-1),$$

we can recover the factors $x_i$ from $x$ and $\mathbf{m}$. I claim in fact that if for this $x$ we write the common value of the two sides of (10.6.9) as a reduced product of elements of $R^\lambda$, $R^\mu$ and $R^\rho$, i.e., as in (10.6.4), then the sequence of factors belonging to $R^\mu$ will be precisely $x_0^\mu, \ldots, x_{h-1}^\mu$, recovering the $x_i$, as required.

Indeed, let us note that for any $x$ such that $\mathbf{m}(x) \in |R|^{(\lambda, \rho)}$, the common value of the two sides of (10.6.9), computed using (10.6.7), is $x^\lambda \, x^\mu \, x^\rho$, while when $\mathbf{m}(x) \in |R|^{(\rho, \lambda)}$ it is $x^\rho \, x^\mu \, x^\lambda$. Hence when we evaluate the common value of the two sides of (10.6.9) for $x = x_0 \ldots x_{h-1}$, the factors with superscript $\mu$ comprise, *initially*, the sequence claimed. They will continue to do so after we reduce this product to the form (10.6.4) by combining any successive factors that may belong to the same monoid $|R|^\lambda$, $|R|^\mu$ or $|R|^\rho$, unless, in the course of this reduction, the factors with superscript $\rho$ and/or $\lambda$ separating some pair of successive $\mu$-factors cancel, allowing these $\mu$-factors to merge. Now if $\mathbf{m}(x_i)$ and $\mathbf{m}(x_{i+1})$ both belong to $|R|^{(\lambda,\,\rho)}$ or both belong to $|R|^{(\rho,\,\lambda)}$, then between $x_i^\mu$ and $x_{i+1}^\mu$ we will have exactly one $\lambda$-factor and one $\rho$-factor, and these cannot cancel. In the case where one belongs to $|R|^{(\lambda,\,\rho)}$ and the other to $|R|^{(\rho,\,\lambda)}$, we get adjacent factors $x_i^\rho\, x_{i+1}^\rho$ or $x_i^\lambda\, x_{i+1}^\lambda$ in the same set $R^\rho$ or $R^\lambda$. When we merge these, they will cancel only if $x_i\, x_{i+1} = e$ in $R$; but this is the case excluded by our hypothesis.    □

Note that in the above argument, we could have asserted that every element can be reduced to a unique product of the indicated form in which no two successive factors *whatever* have product $e$. However, we have proved uniqueness subject to a *weaker* condition than this, so we have a *stronger* uniqueness result. Indeed, this result implies (as the weaker uniqueness statement would not):

**Corollary 10.6.13.** *If $(R, \mathbf{m}, \mathbf{e})$ is a co-**Monoid** object in **Monoid**, then the monoid $R$ has a presentation $\langle X \mid Y \rangle$, where $X$ is the set of elements of $R$ having degree 2 with respect to the comultiplication $\mathbf{m}$, and $Y$ is the set of all relations of the form $x_0\, x_1 = e$ holding in $R$ such that one of $\mathbf{m}(x_0)$, $\mathbf{m}(x_1)$ lies in $|R|^{(\lambda,\,\rho)}$, and the other in $|R|^{(\rho,\,\lambda)}$.*

*Proof.* Lemma 10.6.12 shows that $X$ generates $R$, and by definition the relations comprising $Y$ are satisfied by these generators. It remains to verify that if two words $w_0$ and $w_1$ in the elements of $X$ are equal in $R$, then this equality follows from the relations in $Y$.

Now if $w_i$ ($i = 0$ or $1$) contains a substring which is the left-hand side of some relation in $Y$, then by applying that relation, we can reduce $w_i$ to a shorter word. Hence a finite number of applications of such relations will transform $w_0$ and $w_1$ to words $w_0'$ and $w_1'$ that contain no such substrings. The values of these words in $R$ are still equal; hence the uniqueness statement of Lemma 10.6.12 tells us they are the same word. Thus, by applying relations in $Y$, we have obtained the equality of $w_0$ and $w_1$ in $R$, as required.    □

Clearly the next step in studying our comonoid should be to examine the properties of the set of pairs of elements of $R$ of degree 2 satisfying $x_0\, x_1 = e$. So let us make

**Definition 10.6.14.** *If $(R, \mathbf{m}, \mathbf{e})$ is a co-**Monoid** object in **Monoid**, then $P(R, \mathbf{m}, \mathbf{e})$ will denote the 4-tuple $(X^+, X^-, E, u)$, where*

$$X^+ = \{x \in |R| \mid \mathbf{m}(x) = x^\lambda x^\rho\} = \{x \in |R| \mid \mathbf{m}(x) \in |R|^{(\lambda,\,\rho)}\} \cup \{e\},$$
$$X^- = \{x \in |R| \mid \mathbf{m}(x) = x^\rho x^\lambda\} = \{x \in |R| \mid \mathbf{m}(x) \in |R|^{(\rho,\,\lambda)}\} \cup \{e\},$$
$$E = \{(x_0, x_1) \in |R|^2 \mid \deg(x_0),\, \deg(x_1) \leq 2,\ x_0\, x_1 = e\} \subseteq$$
$$(X^+ \times X^-) \cup (X^- \times X^+),$$

$u = e$, *the neutral element of $R$.*

Thus, $X^+$ and $X^-$ are sets intersecting in the singleton $\{u\}$, and $E$ is a binary relation on the union of these sets, which relates certain elements of $X^+$ to certain elements of $X^-$, and vice versa. We note a key property of this relation: If both $(x_0,\, x_1)$ and $(x_1,\, x_2)$ belong to it, then since $x_1$ has $x_0$ as a left inverse and $x_2$ as a right inverse in $R$, $x_0$ must equal $x_2$.

Let us formalize the type of combinatorial object we have obtained.

**Definition 10.6.15.** *An $E$-system will mean a 4-tuple $(X^+,\, X^-,\, E,\, u)$, where $X^+$ and $X^-$ are sets, and $u$ an element such that*

$$X^+ \cap X^- = \{u\},$$

*and where*

$$E \subseteq (X^+ \times X^-) \cup (X^- \times X^+)$$

*is a relation such that*

$$(10.6.16) \qquad\qquad\qquad u\, E\, u,$$

$$(10.6.17) \qquad\qquad x_0\, E\, x_1 \wedge x_1\, E\, x_2 \implies x_0 = x_2.$$

*A morphism of $E$-systems $(X^+,\, X^-,\, E,\, u) \to (X'^+,\, X'^-,\, E',\, u)$ will mean a map $X^+ \cup X^- \to X'^+ \cup X'^-$ carrying $X^+$ into $X'^+$, $X^-$ into $X'^-$, the relation $E$ into the relation $E'$, and $u$ to $u'$.*

Thus, the objects $P(R,\, \mathbf{m},\, \mathbf{e})$ that we constructed in Definition 10.6.14 are $E$-systems.

When an $E$-system $(X^+,\, X^-,\, E,\, u)$ arises as in that definition from a co-**Monoid** object $(R,\, \mathbf{m},\, \mathbf{e})$ of **Monoid**, Corollary 10.6.13 tells us how to recover the monoid $R$ from $(X^+,\, X^-,\, E,\, u)$, and (10.6.7) tells us how to recover $\mathbf{m}$. (As noted earlier, there is no choice regarding $\mathbf{e}$.) If the concept of $E$-system does a good enough job of capturing the structure of the co-**Monoid** objects of **Monoid**, then every $E$-system should arise from such an object, and Corollary 10.6.13 and (10.6.7) should allow us to construct that object. Let us try to see whether this is so.

Given any $E$-system $(X^+,\, X^-,\, E,\, u)$, let us form the monoid with the presentation suggested by Corollary 10.6.13:

$$(10.6.18) \qquad R = \langle X^+ \cup X^- - \{u\} \mid x_0\, x_1 = e \text{ whenever } x_0\, E\, x_1 \rangle.$$

On this monoid we have a unique zeroary co-operation $\mathbf{e}$, namely the trivial map $R \to \{e\}$. We would now like to define a comultiplication homomorphism from this monoid into the coproduct of two copies of itself by setting

$$(10.6.19) \qquad \mathbf{m}(x) = \begin{cases} x^\lambda x^\rho & \text{if } x \in X^+ - \{u\}, \\ x^\rho x^\lambda & \text{if } x \in X^- - \{u\}. \end{cases}$$

The next two exercises will show that this is possible, and that this construction indeed inverts that of Definition 10.6.14, a result which we will then summarize as a theorem. You should therefore read these exercises through, and think about what is involved, even if you do not work out all the details.

**Exercise 10.6:1.**

(i)   Show that for any $E$-system $X = (X^+, X^-, E, u)$, if we define $R$ by (10.6.18), then (10.6.19) gives a well-defined homomorphism $\mathbf{m} \colon R \to R^\lambda \amalg R^\rho$.

(ii)   Show that this $\mathbf{m}$ and the trivial morphism $\mathbf{e}$ make $R$ a comonoid object of **Monoid**. Let us denote this object $Q(X)$.

The next observation will make some subsequent results easier to state:

(iii)   Verify that the presentation (10.6.18) is equivalent to the modified presentation with $u$ included among the generators, and $u = e$ added to the relations; and that (10.6.19) then holds with the "$-\{u\}$"s deleted.

(iv)   Show that the construction $P$ of Definition 10.6.14, and the construction $Q$ of (i), (ii) above, may be made functors in obvious ways, and that $Q$ is then left adjoint to $P$.

(v)   Deduce from Corollary 10.6.13 that the counit of this adjunction, i.e., the canonical morphism from the functor $QP$ to the identity functor of the category of co-**Monoid** objects of **Monoid**, is an isomorphism. In particular, every comonoid object of **Monoid** arises under $Q$ from an $E$-system.

We have not yet shown that every $E$-system arises from a comonoid. We shall prove this by showing that the *unit* of the above adjunction, i.e., the canonical morphism from the identity functor of the category of $E$-systems to $PQ$, is also an isomorphism.

(To banish any suspicion that our desired conclusion might follow automatically from (v) above, consider the analogous situation where $P$ is the forgetful functor **Group** $\to$ **Monoid**, and $Q$ its left adjoint, taking every monoid to its universal enveloping group. Then the counit $QP \to \mathrm{Id}_{\mathbf{Group}}$ is an isomorphism, but the unit $\mathrm{Id}_{\mathbf{Monoid}} \to PQ$ is not: monoids containing noninvertible elements do not appear as values of $P$, and each such monoid falls together under $Q$ with a monoid that *is* a value of $P$.)

We will get this result by obtaining a normal form for monoids $Q(X)$:

**Exercise 10.6:2.**

(i)   Show that given any $E$-system $X = (X^+, X^-, E, u)$, the monoid $R$ with presentation (10.6.18) has for normal form the set of words in the indicated generators (including the empty word) that contain no subsequences $x_0 \, x_1$ with $x_0 \, E \, x_1$. (Suggestion: van der Waerden's trick.)

(ii)   Deduce that the unit of the adjunction between $P$ and $Q$ is an isomorphism.

The above results are summarized in the first sentence of the next theorem. The second sentence translates the comonoid structure (10.6.18) and (10.6.19) into a description of the functor represented, and the final sentence follows by Corollary 10.3.6.

**Theorem 10.6.20.** *The isomorphism classes of representable functors* $V :$ **Monoid** $\rightarrow$ **Monoid** *are in natural bijective correspondence with the isomorphism classes of* $E$-*systems. The functor corresponding to the* $E$-*system* $(X^+, X^-, E, u)$ *can be described as a subfunctor* (*in the sense of Lemma 7.9.4 and Definition 9.4.11*) *of a direct product of copies of the* identity *functor and of the* opposite-monoid *functor; namely, as the construction taking each monoid* $A$ *to the submonoid of* $A^{(X^+ - \{u\})} \times (A^{\mathrm{op}})^{(X^- - \{u\})}$ *consisting of those elements* $b$ *such that for all* $(x, y) \in E - \{(u, u)\}$, *the coordinate* $b_x$ *is a left inverse to the coordinate* $b_y$.

*Writing* $E$-**System** *for the category of* $E$-*systems, the above construction yields a* contravariant *equivalence*

$$E\text{-}\mathbf{System}^{\mathrm{op}} \rightarrow \mathbf{Rep}(\mathbf{Monoid}, \mathbf{Monoid}). \qquad \square$$

For the purpose of describing the morphism of representable functors induced by a given morphism of $E$-systems, it is actually most convenient to treat the functor $V :$ **Monoid** $\rightarrow$ **Monoid** corresponding to the $E$-system $(X^+, X^-, E, u)$ as taking a monoid $A$ to a submonoid of

$$A^{(X^+ - \{u\})} \times \{e\} \times (A^{\mathrm{op}})^{(X^- - \{u\})};$$

i.e., to introduce into our description of the functor an extra slot, indexed by the element $u$ of the $E$-system, such that the coordinate of $V(A)$ in that slot is required to be the neutral element $e$ of $A$. (Cf. Exercise 10.6:1(iii).) We can then say that if $\mathbf{f} : E \rightarrow E'$ is a morphism of $E$-systems, and $f : V' \rightarrow V$ the corresponding morphism of representable functors, then for a monoid $A$ and an element $\xi \in |V'(A)|$, the image $f(A)(\xi)$ has for $x$-th coordinate the $\mathbf{f}(x)$-th coordinate of $\xi$, whether $\mathbf{f}(x)$ happens to be $u$, or to be a member of $X^+ \cup X^- - \{u\}$.

Let us look at some simple examples of $E$-systems and the corresponding representable functors. We shall display an $E$-system by showing the elements of $X^+ - \{u\}$ and $X^- - \{u\}$ respectively as points in two boxes,

, and indicating a condition $x_0 E\, x_1$ by an arrow from the point $x_0$ to the point $x_1$. (The element $u$ will not be shown; it may be thought of as embedded in the dividing line between the boxes.)

By (10.6.18) and (10.6.19), the comonoid $R$ corresponding to this $E$-system is the free monoid on one generator $x$, with the comultiplication under which $\mathbf{m}(x) = x^\lambda x^\rho$. We see that the functor this represents is (up to isomorphism) the *identity* functor **Monoid** $\to$ **Monoid**. This description of the functor represented can also be seen from the second sentence of the above theorem.

You should verify that this $E$-system similarly gives the functor taking each monoid $A$ to its *opposite* monoid $A^{\mathrm{op}}$, i.e., the monoid with the same underlying set, but with the order of multiplication reversed.

(the relation $E - \{(u,\, u)\}$ still being empty). This gives the direct product of the above two functors, i.e., the functor associating to every monoid $A$ the monoid

$$\{(\alpha,\, \beta) \mid \alpha,\, \beta \in |A|\},$$

with multiplication

(10.6.21)                 $(\alpha_0,\, \beta_0)(\alpha_1,\, \beta_1) \;=\; (\alpha_0\, \alpha_1,\, \beta_1\, \beta_0).$

This corresponds to the subfunctor of the preceding example determined by adding to the description of its underlying set the conditions

$$\alpha\, \beta = e = \beta\, \alpha.$$

Since under these conditions $\alpha$ uniquely determines $\beta$, the second coordinate provides no new information, and we can describe this functor, up to isomorphism, as associating to $A$ its *group of invertible elements* $\alpha$, regarded as a monoid.

As above, except that only the condition $\alpha\, \beta = e$, and not $\beta\, \alpha = e$ is imposed. Right inverses are *not* generally unique, so we must describe this functor as associating to $A$ the monoid of elements $\alpha \in |A|$ given with a *specified* right inverse $\beta$. The multiplication is again as in (10.6.21).

This associates to $A$ the monoid of elements $\alpha$ given with a specified *left inverse* $\beta$, again multiplied as in (10.6.21). Set-theoretically, this construction is isomorphic to the preceding, via $(\alpha,\, \beta) \leftrightarrow (\beta,\, \alpha)$, but the monoid structures are opposite to one another. (I have indicated this in the paraphrases by naming, after the words "monoid of", the elements which are multiplied as in $A$, while those with the opposite multiplication are referred to as specified inverses of these elements.)

 "The monoid of *pairs* of elements of $A$ with a specified *common right inverse*".

And so on. We note that for a general diagram such as

the associated functor is the direct product of the functors associated with the graph-theoretic connected components of the diagram. Each of these components, *except* those of the form ⟨·⇄·⟩ , must have, by (10.6.17), the property that arrows, if any, all go in the same direction, i.e., from left to right or from right to left. Subject to this restriction, the arrows are independent.

Let us pause to note the curious fact that, although for every *nonzero* cardinal $r$, the construction that associates to a monoid $A$ the monoid of its right invertible elements given with a specified $r$-tuple of right inverses is a representable functor, this is false for $r = 0$:

**Exercise 10.6:3.** Let $H$: **Monoid** $\to$ **Monoid** be the functor associating to a monoid $A$ its submonoid of right invertible elements (a subfunctor of the identity functor).

(i)  Show that $H$ is not representable, if you did not already do so as Exercise 8.2:2(i).

(ii)  Show, however, that the composite functor $HH$ is representable, and concisely describe this functor.

(iii)  Show that $H$ can be written as a direct limit of representable functors. (Hint: can you write the empty set as an inverse limit of nonempty sets?)

It is natural to ask how to compose two representable functors expressed in terms of $E$-systems.

**Exercise 10.6:4.** In this exercise, "functor" will mean "representable functor **Monoid** $\to$ **Monoid**".

(i)  Define precisely what is meant by the connected components of an $E$-system, and prove the assertion made above that the functor associated with an $E$-system is the direct product of the functors associated with its connected components. Using this result, reduce the problem of describing the $E$-system of the composite of two functors to the case where the $E$-systems of the given functors are connected.

(ii)  Characterize in terms of $E$-systems the results of composing an arbitrary functor on the right and on the left with the functors having the diagrams [ · [ ]] and [[ · ] . (Thus, four questions are asked, though two of them are trivial to answer.)

This leaves us with the problem of describing the composite of two functors whose associated diagrams are both connected, and each have more than one element. The answer is quite simple, but the argument requires two preliminary observations:

(iii) Show that if $s$, $t$ are two left invertible elements of a monoid $A$, or two right invertible elements, then the condition $st = e$ implies that they are both invertible.

(iv) Let $V$ be a functor whose diagram is connected. Show that if some $\xi \in |V(A)|$ has an invertible element of $A$ in at least one coordinate, then it has invertible elements in all coordinates, and these are determined by that one coordinate. Show that the set of elements $\xi$ with these properties forms a submonoid of $V(A)$, isomorphic to the group of invertible elements of $A$. (In writing "at least one coordinate" above, I am taking our description of $V$ to be that of Theorem 10.6.20, which does not include a coordinate indexed by $u$.)

(v)  Deduce from (iii) and (iv) a description for the composite of any two functors whose diagrams are both connected and each have more than one element (not counting $u$ as an element of our diagrams).

**Exercise 10.6:5.** Suppose $f\colon V \to V'$ is a morphism of representable functors **Monoid** $\to$ **Monoid**, and $W$ is another such functor. Assuming the results of the preceding exercise, show how to describe the map of $E$-systems corresponding to $f \circ W\colon VW \to V'W$, respectively $W \circ f\colon WV \to WV'$, in terms of the map of $E$-systems corresponding to $f$.

**Exercise 10.6:6.** We saw in the discussion following Corollary 10.5.1 that the object representing a composite of representable functors among varieties could be constructed from presentations $\langle X \mid Y \rangle_{\mathbf{U}}$ and $\langle X' \mid Y' \rangle_{\mathbf{V}}$ of representing objects for those functors, using a set of generators indexed by $X \times X'$ and a set of relations indexed by $(Y \times X') \sqcup (X \times Y')$. See whether you can get the results of the preceding two exercises by applying this idea to presentations of the representing objects for functors **Monoid** $\to$ **Monoid** induced by given $E$-systems. (If you did Exercise 10.5:1, you will be able to apply the results of that exercise here; if not, you can still work out the corresponding results for this particular case.)

## 10.7. Functors to and from some related categories

The above characterization of representable functors **Monoid** $\to$ **Monoid** can be used to characterize various classes of representable functors involving the category **Group** as well.

For example, if $V$ is a representable functor **Group** $\to$ **Monoid**, then writing $G$: **Monoid** $\to$ **Group** for the group-of-invertible-elements functor, we see that $V$ will be determined by the composite $VG$: **Monoid** $\to$ **Monoid**. Since $G$ is representable (cf. (10.5.3)), $VG$ will also be, and so can be described as in the preceding section. Which representable functors **Monoid** $\to$ **Monoid** occur as such composites $VG$? Clearly, those which depend on their argument only via its group of invertible elements. If we write $U$ for the forgetful functor **Group** $\to$ **Monoid**, of which $G$ is the right adjoint, this means that the functors **Monoid** $\to$ **Monoid** that come up will be those that are invariant under right composition with $UG$; the precise statement turns out to be that they are isomorphic with their composite with $UG$, the isomorphism being given by the counit $UG \to \mathrm{Id}_{\mathbf{Monoid}}$ of the above adjunction.

A similar approach, using the left adjoint $F$ of $U$ (the construction of the enveloping group of a monoid) gives a characterization of the representable functors **Monoid** $\to$ **Group**; and a combination of the two gives a characterization of representable functors **Group** $\to$ **Group**.

The next exercise gives the details of the approach sketched above, in the general context of relating representable functors **Group** $\to$ **V** and **V** $\to$ **Group** to representable functors **Monoid** $\to$ **V** and **V** $\to$ **Monoid**. The exercise after that combines this result with our characterization of representable functors **Monoid** $\to$ **Monoid** to get very simple descriptions of the functors we are looking for.

First, some notational details. As above,

(10.7.1)
$U$: **Group** $\to$ **Monoid** will denote the "forgetful" functor.

$F$: **Monoid** $\to$ **Group** will denote the left adjoint of $U$, the "universal enveloping group" functor.

$G$: **Monoid** $\to$ **Group** will denote the right adjoint of $U$, the "group of invertible elements" functor.

It is clear that the counit of the first adjunction and the unit of the second are isomorphisms

$$\varepsilon_{U,F} : FU \cong \mathrm{Id}_{\mathbf{Group}} \quad \text{and} \quad \eta_{G,U} : \mathrm{Id}_{\mathbf{Group}} \cong GU.$$

In general, when two maps compose to the identity in one order, their composite in the opposite order is a retraction of one object to a subobject isomorphic to the other. In this case, we see that the composites of our two adjoint pairs in the reverse order, $UF$ and $UG$, are retractions of **Monoid** onto $U(\textbf{Group})$, which is clearly a full subcategory of **Monoid** isomorphic to **Group**. The other unit and counit of our adjunctions relate each monoid to its image in this subcategory under the corresponding retraction; let us write these

$$\eta = \eta_{U,F}\colon \text{Id}_{\textbf{Monoid}} \longrightarrow UF \quad \text{and} \quad \varepsilon = \varepsilon_{G,U}\colon UG \longrightarrow \text{Id}_{\textbf{Monoid}}$$

(breaking the convention that $\eta$ and $\varepsilon$ generally denote the unit and counit of the same adjunction).

**Exercise 10.7:1.**

(i)   Show that the monoids $S$ of the form $U(A)$ ($A$ a group) are precisely those monoids for which the universal map $\eta(S)\colon S \to UF(S)$ is an isomorphism, and also precisely those monoids for which the universal map $\varepsilon(S)\colon UG(S) \to S$ is an isomorphism.

(ii)   Show that $UF$ is left adjoint to $UG$.

(iii)   Show that for any variety **V**, the representable functors **Group** $\to$ **V** can be identified with the representable functors $V\colon$ **Monoid** $\to$ **V** which are invariant under composition on the right with $UG$ (i.e., those $V$ such that the induced map $V\varepsilon\colon VUG \to V$ is an isomorphism).

(iv)   Show similarly that the representable functors **V** $\to$ **Group** can be identified with the representable functors **V** $\to$ **Monoid** which are invariant under composition on the left with $UG$ (i.e., such that the induced map $\varepsilon V\colon UGV \to V$ is an isomorphism).

   Though we shall not need it, you may also

(v)   Show that the functors **Group** $\to$ **V**, respectively **V** $\to$ **Group** which have right adjoints (i.e., the left adjoints of representable functors) can be identified with the functors **Monoid** $\to$ **V**, respectively **V** $\to$ **Monoid** which have right adjoints and are invariant under composition on the right, respectively on the left with $UF$.

**Exercise 10.7:2.** Using the preceding exercise,

(i)   Show that every representable functor **Group** $\to$ **Monoid** is a power (i.e., product of copies) of the forgetful functor $U$. (First proved by Kan [97].)

(ii)   Show that every representable functor **Monoid** $\to$ **Group** is a power of the group-of-invertible-elements functor $G$.

(iii)   Show that every representable functor **Group** $\to$ **Group** is a power of the identity functor.

   Thus, in each of these three cases, all representable functors arise as powers of one "basic" functor, $U$, $G$ or $\text{Id}_{\textbf{Group}}$ respectively. Calling this functor $B$ in each case, so that the general representable functor between the categories in question has the form $B^X$, let us observe that for any set map $X \to Y$ we get a map $B^Y \to B^X$. Are these the only morphisms among these functors?

Not quite. For instance, in the case of functors **Group** → **Group**, if we take $X = Y = 1$, so that we are considering endomorphisms of the identity functor of **Group**, there is not only the identity morphism, associating to every group its identity map, and arising from the unique set map $1 \to 1$, but also the trivial morphism, associating to every group the endomorphism under which all elements go to $e$. To correctly describe the morphisms among our functors, let $\mathbf{Set}^{\mathrm{pt}}$ denote the category of *pointed sets*, whose objects are sets given with a single distinguished element, and whose morphisms are set maps sending distinguished element to distinguished element. (This may be identified with the variety $\Omega$-**Alg** with $\Omega$ consisting of a single zeroary operation.) The next exercise shows that this is the right category for parametrizing these functors.

**Exercise 10.7:3.**

(i)   Let $L: E$-**System** → $\mathbf{Set}^{\mathrm{pt}}$ denote the functor taking every $E$-system $X = (X^+, X^-, E, u)$ to the pointed set $(X^+, u)$. Show that when restricted to the full subcategory of $E$-systems whose "box pictures" have

all connected components of the form $\boxed{\;\bullet \rightleftarrows \bullet\;}$ , the functor $L$ gives an

equivalence of categories.

(ii)   Deduce that in each of the cases of the preceding exercise, the indicated category of representable functors is equivalent to $(\mathbf{Set}^{\mathrm{pt}})^{\mathrm{op}}$. Precisely, letting $B$ denote the indicated "basic" functor in each case, show that morphisms $B^X \to B^Y$ correspond to the morphisms of pointed sets $(Y \cup \{u\}, u) \to (X \cup \{u\}, u)$ where $u$ denotes an element not in $X$ or $Y$.

Let us return to a point mentioned at the beginning of the preceding section. In the description of a comonoid object of **Monoid**, the co-neutral-element was uniquely determined, and hence provided no information; nevertheless, the coidentities it was required to satisfy played an important role in our arguments. The next exercise shows that without these coidentities, those results fail.

**Exercise 10.7:4.** Consider the following two representable functors from **Monoid** to the variety of semigroups with a distinguished element (zeroary operation) $e$ subject to no additional identities.

(a) The functor $V$ taking $A \in \mathrm{Ob}(\mathbf{Monoid})$ to the semigroup with underlying set $|A|$, multiplication given by $x * y = x$ for all $x$ and $y$, and distinguished element the neutral element $e$ of $A$.

(b) The functor $W$ specifying the same underlying set and distinguished element, but with multiplication given by $x * y = e$.

Verify that in both cases the operation $*$ is indeed associative (so that the functors take values in the variety claimed), and also that in both cases the distinguished element $e$ is an *idempotent* with respect to $*$ (i.e., satisfies $e * e = e$). Show that in case (a), this element also satisfies the *right* neutral law, but not the left neutral law, while in case (b), neither neutral law is satisfied.

Note that in case (b) of the above exercise, the distinguished element satisfies the identities $e * x = e = x * e$. An element with this property is called a *zero* element of a semigroup, because these identities hold for 0 in the multiplicative semigroup of a ring. An element of a semigroup satisfying only the first of these identities is called a *left zero* element. We see that in case (a) *every* element is a left zero. The unique multiplication with the latter property on any set is therefore called the *left zero multiplication.*

Little is known about general representable functors from **Monoid** to **Semigroup**. Dropping the zeroary co-operations **e**, the above exercise gives examples that are noteworthy in that construction (a) used nothing about the given monoid $A$ but its underlying set, while (b) used only its structure of set with distinguished element $e$. The next exercise displays some constructions that do use the monoid operation, but in peculiar—almost random—ways.

**Exercise 10.7:5.**

(i)   Show that one can define a representable functor **Monoid** $\rightarrow$ **Semigroup** by associating to every monoid $A$ the set of pairs $(\xi, \eta)$ such that $\xi$ is an invertible and $\eta$ an arbitrary element of $A$, with the operation $(\xi, \eta)(\xi', \eta') = (e, \xi^{-1}\xi'^{-1}\xi\xi')$.

(ii)   Show that if we impose on the ordered pairs in the description of the above functor the additional condition that $\xi^n = e$ for a fixed positive integer $n$, and/or the condition $\xi\eta = \eta$, the resulting subsets are still closed under the above operation, and hence define further representable functors.

The above observations lead to

**Exercise 10.7:6** (Open question [3, Problem 21.7, p. 94]). Find a description of (or other strong results about) all representable functors **W** $\rightarrow$ **Semigroup**, where **W** is any of the varieties **Monoid**, **Group** or **Semigroup**.

The following questions may be easy or hard to answer; I have not thought about them:

**Exercise 10.7:7.** Let $V:$ **Monoid** $\rightarrow$ **Monoid** be a representable functor whose $E$-system has a single connected component, and is not one of

$\boxed{\;\cdot\;|\;\;} \;,\; \boxed{\;\;|\;\cdot\;}\;$ , or $\boxed{\;\cdot\rightleftarrows\cdot\;}$ . What can one say about the class of monoids of the form $V(A)$ $(A \in \mathrm{Ob}(\textbf{Monoid}))$? How much does this class depend on the choice of $V$? How does it compare with the class of monoids that are embeddable in groups? With the class of monoids $H(A)$, where $H$ is the functor of Exercise 10.6:3?

One may likewise ask these questions for the classes of monoids arising as values of the *left adjoints* of such functors.

Some results on representable functors to and from the variety of *heaps* and related varieties are given in [3, §22].

## 10.8. Representable functors among categories of abelian groups and modules

Let us next consider representable functors from abelian groups to monoids. Let

$$V: \ \mathbf{Ab} \longrightarrow \mathbf{Monoid}$$

be such a functor, with representing coalgebra $(R, \mathbf{m}, \mathbf{e})$. Since coproducts of abelian groups are direct sums, we may write the coproduct of two copies of $R$ as $R^\lambda \oplus R^\rho$; thus, every element of this group has the form $y^\lambda + z^\rho$ for unique $y, z \in |R|$. In particular, for each $x \in |R|$ there exist $y, z \in |R|$ such that

$$\mathbf{m}(x) \ = \ y^\lambda + z^\rho.$$

As in the case of functors on **Monoid**, the co-neutral-element must be the trivial map. Applying the coneutral laws to the above equation, we immediately get $x = y = z$, that is,

$$\mathbf{m}(x) \ = \ x^\lambda + x^\rho \quad (x \in |R|).$$

Given any two elements $a, b \in |V(A)| = \mathbf{Ab}(R, A)$, this says that their "product" in $V(A)$ is the homomorphism taking each $x \in |R|$ to $a(x)+b(x)$. In other words, the induced "multiplication" of homomorphisms is just the familiar addition of homomorphisms of abelian groups.

It is clear that, conversely, for every abelian group $R$ this operation on homomorphisms with domain $R$ does make $h_R$ a **Monoid**-valued functor. So for each $R \in \mathrm{Ob}(\mathbf{Ab})$, there is a unique representable functor $V: \mathbf{Ab} \to$ **Monoid** whose representing coalgebra has underlying object $R$.

In view of the form $V$ takes, it is natural to call the binary co-operation on $R$ a "coaddition" rather than a "comultiplication". Of course, it is well known that on the sets $\mathbf{Ab}(R, A)$, addition of maps is actually an operation of *group*, and, indeed, of *abelian* group, with the unique inverse operation described in the obvious way. Thus, our determination of all representable functors **Ab** → **Monoid** also determines all representable functors **Ab** → **Group** and **Ab** → **Ab**. That is,

**Lemma 10.8.1.** *For every object $R$ of **Ab**, there is a unique co-**Monoid** object, a unique co-**Group** object and a unique co-**Ab** object of **Ab**, with underlying object $R$. Each of these has coaddition given by the diagonal map*

(10.8.2)  $$\mathbf{a}(x) \ = \ x^\lambda + x^\rho \quad (x \in |R|),$$

*and co-neutral-element given by* $\mathbf{e}(x) = 0$ $(x \in |R|)$. *In the co-* **Group** *and co-* **Ab** *structures, the co-inverse operation is given by*

$$\mathbf{i}(x) = -x. \qquad \qquad \square$$

Since this result was so easy to prove, let's make some more work for ourselves, and try to generalize it!

Recall that an abelian group is equivalent to a left $\mathbb{Z}$-module, and that for any ring $K$, a left $K$-module $M$ can be described as an abelian group with a family of abelian group endomorphisms, called "scalar multiplications", indexed by the elements of $K$, such that sums of these endomorphisms, composites of these endomorphisms, and the identity endomorphism are the endomorphisms indexed by the corresponding sums and products of elements of $K$ and by the multiplicative neutral element $1 \in |K|$. (As usual, unless the contrary is stated our rings are members of the variety **Ring**[1] of associative rings with multiplicative neutral element $1$.)

Let us write $K$-**Mod** for the variety of left $K$-modules. It is easy to see that the argument giving Lemma 10.8.1 goes over immediately to give the same conclusions for representable functors from $K$-**Mod** to the varieties **Monoid**, **Group** and **Ab**.

What about functors from $K$-**Mod** to $K$-**Mod** —or better, to $L$-**Mod**, for another ring $L$?

To study this question, let us write out explicitly the identities for the scalar multiplication operations of $K$-**Mod** which we stated above in words. The identities saying that each such multiplication is an abelian group endomorphism say that for all $c \in |K|$ and $x, x' \in |M|$,

$$(10.8.3) \qquad \qquad c\,(x + x') \;=\; c\,x + c\,x'.$$

(We are here taking advantage of the fact that group homomorphisms can be characterized as set-maps respecting the binary group operation alone.) The identities characterizing sums and composites of scalar multiplications, and scalar multiplication by $1 \in |K|$, say that for $c, c' \in |K|$, $x \in |M|$,

$$(10.8.4) \qquad \qquad (c + c')\,x \;=\; c\,x + c'\,x$$

$$(10.8.5) \qquad \qquad (c\,c')\,x \;=\; c\,(c'\,x)$$

$$(10.8.6) \qquad \qquad 1\,x \;=\; x.$$

Now suppose $L$ is another ring, and $(R, \mathbf{a}, \mathbf{i}, \mathbf{e}, (\mathbf{s}_d)_{d \in |L|})$ a co-$L$-module object in $K$-**Mod**, where $R$ is a $K$-module, the morphisms $\mathbf{a}$, $\mathbf{i}$ and $\mathbf{e}$ give the co-abelian-group structure of $R$, and for each $d \in |L|$, $\mathbf{s}_d$ is the co-operation corresponding to scalar multiplication by $d$. The co-abelian-group structure will, as we have noted, have the form described in Lemma 10.8.1. The $s_d$ will be unary co-operations, i.e., $K$-module homomorphisms $R \to R$, which can thus be looked at as unary *operations* on the set $|R|$. We now need some basic observations:

**Exercise 10.8:1.** Let $R$ be any $K$-module, and **a**, **i**, **e** the coaddition, coinverse and cozero morphisms defining the unique co- **Ab** structure on $R$ in $K$-**Mod**.

(i)   Show that *every* $K$-module endomorphism $\mathbf{s}\colon R \to R$ satisfies the coidentity corresponding to the identity (10.8.3); i.e., show that the unary operation induced by such an **s** on each $h_R(A)$ is an abelian group endomorphism.

(ii)   Show that such an operation **s** induces the identity operation on each $h_R(A)$ (cf. (10.8.6)) if and only if it is the identity endomorphism of $R$.

(iii)   Show that if $\mathbf{s}_d$, $\mathbf{s}_{d'}$ and $\mathbf{s}_{d''}$ are three endomorphisms of $R$, then the operations on the abelian groups $h_R(A)$ induced by $\mathbf{s}_d$ and $\mathbf{s}_{d'}$ sum to the operation induced by $\mathbf{s}_{d''}$ if and only if $\mathbf{s}_d + \mathbf{s}_{d'} = \mathbf{s}_{d''}$.

(iv)   Show likewise that the operation induced by $\mathbf{s}_{d''}$ is the composite in a given order of the operations induced by $\mathbf{s}_d$ and $\mathbf{s}_{d'}$ (cf. (10.8.5)) if and only if $\mathbf{s}_{d''}$ is the composite of $\mathbf{s}_d$ and $\mathbf{s}_{d'}$ in the *opposite* order.

From the above results we deduce that

> If $K$ and $L$ are rings, and $R$ a left $K$-module, then a co-left-$L$-module structure on $R$ is equivalent to a system of $R$-module endomorphisms $(\mathbf{s}_d)_{d \in |L|}$ which for all $d$, $d' \in |L|$ satisfy

(10.8.7)    (10.8.8)

$$\mathbf{s}_1 = \mathrm{id}_R$$

(10.8.9)

$$\mathbf{s}_{d+d'} = \mathbf{s}_d + \mathbf{s}_{d'}$$

(10.8.10)

$$\mathbf{s}_{dd'} = \mathbf{s}_{d'}\,\mathbf{s}_d.$$

This is a nice result, but we can make it more elegant with a change of notation. The reversal of the order of composition in (10.8.10) can be cured if we write the operators $\mathbf{s}_d$ on the *right* of their arguments, instead of on the left, and compose them accordingly. Moreover, once the operation of elements of $L$ (by co-scalar-multiplications) is written on a different side from the operation of elements of $K$ (by scalar multiplication), there is no real danger of confusion if we drop the symbols **s**, i.e., replace the above notation $\mathbf{s}_d(x)$ by $x\,d$ ($x \in |R|$, $d \in |L|$). We now find that the scalar multiplications by elements of $K$ and the co-scalar-multiplications by elements of $L$ satisfy a very symmetrical set of conditions, namely, that for all $c$, $c' \in |K|$, $x$, $x' \in |R|$, $d$, $d' \in |L|$,

$$(10.8.11) \qquad\qquad 1\,x \,=\, x \qquad\qquad\qquad\qquad x\,1 \,=\, x$$

$$(10.8.12) \qquad c\,(x + x') \,=\, c\,x + c\,x' \qquad (x + x')\,d \,=\, x\,d + x'd$$

$$(10.8.13) \qquad (c + c')\,x \,=\, c\,x + c'x \qquad x\,(d + d') \,=\, x\,d + x\,d'$$

$$(10.8.14) \qquad\quad (c\,c')\,x \,=\, c\,(c'x) \qquad\qquad x\,(d\,d') \,=\, (x\,d)\,d'$$

$$(10.8.15) \qquad\qquad\qquad\qquad c\,(x\,d) \,=\, (c\,x)\,d.$$

Here (10.8.15), and the right hand equation of (10.8.12), say that the co-scalar-multiplications are endomorphisms of the left $K$-module $R$. The conditions in the left-hand column, together with the identities for the abelian group structure of $R$, constitute the identities of a left $K$-module, while the remaining three conditions on the right say that the co-scalar-multiplication endomorphisms behave as required to give a co-left-$L$-module structure. (Only three such conditions are needed, as against the four on the left, because of Exercise 10.8:1(i).)

We have, in fact, rediscovered a standard concept of ring theory:

**Definition 10.8.16.** *An abelian group on which one ring $K$ operates by maps written on the left and another ring $L$ operates by maps written on the right so that* (10.8.11)–(10.8.15) *are satisfied is called a* $(K, L)$*-bimodule.*

*For given $K$ and $L$, the variety of $(K, L)$-bimodules will be denoted* $K$*-**Mod**-$L$.*

Note that given two arbitrary varieties of algebras $\mathbf{V}$ and $\mathbf{W}$, the category of $\mathbf{V}$-coalgebra objects of $\mathbf{W}$ cannot in general be regarded as a variety of algebras, because the co-operations $\mathbf{s}\colon R \to \coprod_{\mathrm{ari}(s)} R$ do not have the form of maps $|R|^\beta \to |R|$, *unless* $\mathrm{ari}(s) = 1$. In the present case, it happened that the two *non-unary* co-operations of our coalgebras, the coaddition and the cozero, were uniquely determined, so that the structure could be defined wholly by unary co-operations, and so, atypically, the category of these coalgebras could be identified with a variety of algebras, to wit, $K$-**Mod**-$L$.

Ring-theorists often write a $(K, L)$-bimodule $R$ as $_K R_L$. Here the subscripts are not part of the "name" of the object, but reminders that $K$ operates on the left and $L$ on the right. (Actually, ring-theorists more often use other letters, such as $B$, for "bimodule", or $M$, for "module", reserving $R$ for rings. But in this chapter we are using $R$ wherever possible for "representing object".) That such a bimodule structure makes $R$ a co-$L$-module in $K$-**Mod** corresponds to the result familiar to ring-theorists, that the set of left $K$-module homomorphisms from a $(K, L)$-bimodule to a left $K$-module,

$$(10.8.17) \qquad\qquad K\text{-}\mathbf{Mod}(_K R_L,\ _K A)$$

has a natural structure of left $L$-module. Let us describe without using the language of coalgebras how this $L$-module structure arises. If we regard the actions of the elements of $L$ on $R$ as $K$-module endomorphisms, then the functoriality of $K\text{-}\mathbf{Mod}(-, -)$ in its first variable turns these into endomorphisms of the abelian group $K\text{-}\mathbf{Mod}(_K R, \, _K A)$, and since this functoriality is contravariant, the order of composition of these endomorphisms is reversed; so from the right $L$-module structure on $R$, we get a left $L$-module structure on that hom-set. Explicitly, given any $f \in K\text{-}\mathbf{Mod}(R, A)$ and $d \in |L|$, the action of $d$ on $f$ in this induced left $L$-module structure is given by

$$(10.8.18) \qquad\qquad (d\,f)(x) \,=\, f(x\,d).$$

This takes a more elegant form if we adopt

$(10.8.19)$ (Frequent convention in ring theory.) If possible, write homomorphisms of *left* modules on the *right* of their arguments, and homomorphisms of *right* modules on the *left* of their arguments, and use the notation for composition of such homomorphisms appropriate to the side on which they are written.

For a discussion of (10.8.19), and its advantages and disadvantages, see [56]. We have already applied this idea once, in the change of notation introduced immediately after (10.8.7). In our present situation, it suggests that we should write elements $f \in K\text{-}\mathbf{Mod}(R, A)$ on the right of elements $x \in |A|$. When we do so, (10.8.18) takes the form

$$(10.8.20) \qquad\qquad x\,(d\,f) \,=\, (x\,d)f.$$

In summary:

**Lemma 10.8.21.** *If $K$ and $L$ are unital associative rings, then a co-$L$-$\mathbf{Mod}$ object $R$ of $K$-$\mathbf{Mod}$ is essentially the same as a $(K, L)$-bimodule $_K R_L$. When $R$ is regarded in this way, the left $L$-module structure on the functor $K$-$\mathbf{Mod}(R, -)$ consists of the usual abelian group structure on hom-sets, together with the scalar multiplications (10.8.18), or in right-operator notation, (10.8.20).* ☐

It is not hard to verify that the above correspondence yields an equivalence of categories $K\text{-}\mathbf{Mod}\text{-}L \approx \mathbf{Rep}(K\text{-}\mathbf{Mod}, L\text{-}\mathbf{Mod})^{\mathrm{op}}$.

## 10.9. More on modules: left adjoints of representable functors

Let us now find the left adjoint to the functor induced as above by a $(K, L)$-bimodule $R$. This must take a left $L$-module $B$ to a left $K$-module $A$ with a universal left $L$-module homomorphism

$$(10.9.1) \qquad\qquad h\colon B \longrightarrow K\text{-}\mathbf{Mod}(R, A).$$

To find this object $A$, we will apply our standard heuristic approach: Consider an arbitrary left $K$-module $A$ with an $L$-module homomorphism (10.9.1), and see what elements of $A$, and what relations among these elements, this map gives us.

For each $y \in |B|$, (10.9.1) gives a homomorphism $h(y) \colon R \to A$; and such a homomorphism gives us, for each $x \in |R|$, an element of $A$. With (10.8.19) in mind, let us write this as

$$x * y = x\, h(y) \quad (x \in |R|,\, y \in |B|).$$

I claim that the conditions that these elements must satisfy are that for all $x,\, x' \in |R|$, $y,\, y' \in |B|$, $c \in |K|$, $d \in |L|$,

(10.9.2)    $(x + x') * y = x * y + x' * y \qquad x * (y + y') = x * y + x * y'$

(10.9.3)    $c\,(x * y) = (c\,x) * y \qquad\qquad\qquad\quad —$

(10.9.4)    $\qquad\qquad\qquad x * (d\,y) = (x\,d) * y$

Indeed, the two equations on the left are the conditions for the maps $h(y)$ to be left $K$-module homomorphisms, while the equations on the right and at the bottom are the conditions for the map (10.9.1) to be a homomorphism of left $L$-modules with respect to the given $L$-module structure on $B$ and the operations (10.8.20) on $K$-**Mod**$(R, A)$. We note the gap on the right-hand side of (10.9.3); since nothing acts on the *right* on the $L$-module $B$, there is nothing to put there. (But do not lose heart; this gap will eventually be filled.) So the universal $A$ with a homomorphism (10.9.1) will be presented by generators $x * y$ ($x \in |R|$, $y \in |B|$) and relations (10.9.2)–(10.9.4).

Again we have discovered a standard concept. The $K$-module presented by this system of generators and relations is denoted

$$R \otimes_L B,$$

and called the *tensor product over $L$* of the $(K, L)$-bimodule $R$ and the left $L$-module $B$. The generators of this module corresponding to the elements $x * y$ of the above discussion are written $x \otimes y$ ($x \in |R|$, $y \in |B|$).

We reiterate that $R \otimes_L B$ is only a left $K$-module. Intuitively, when we form the tensor product $({}_K R_L) \otimes_L ({}_L B)$, the operation of tensoring over $L$ "eats up" the two $L$-module structures, leaving only a $K$-module structure. This is dual to the situation of (10.8.17), where the construction of taking the hom-set over $K$ "ate up" the two $K$-module structures, leaving only an $L$-module structure.

We have shown:

**Lemma 10.9.5.** *If we regard a $(K, L)$-bimodule ${}_K R_L$ as a co-$L$-**Mod** object of $K$-**Mod**, then the functor it represents,*

$$K\text{-}\mathbf{Mod}(R, -)\colon\ K\text{-}\mathbf{Mod} \longrightarrow L\text{-}\mathbf{Mod}$$

*has for left adjoint the functor*

$$R \otimes_L - : \quad L\text{-}\mathbf{Mod} \longrightarrow K\text{-}\mathbf{Mod}.$$

*Thus, given the bimodule $R$, a left $K$-module $A$, and a left $L$-module $B$, we have a functorial bijection, which is in fact an isomorphism of abelian groups,*

$$L\text{-}\mathbf{Mod}(B, K\text{-}\mathbf{Mod}(R, A)) \;\cong\; K\text{-}\mathbf{Mod}(R \otimes_L B, A). \qquad \square$$

An interesting consequence of Lemmas 10.8.21 and 10.9.5 is that every *representable* functor between module categories, and likewise the left adjoint of every such functor, respects the **Ab**-structures of these categories, i.e., sends sums of morphisms to sums of morphisms. This is not true of general functors between module categories, as the reader can see from the functor $A \mapsto A \otimes A$ of Exercise 10.5:4(i).

In defining the tensor product $R \otimes_L B$, I said that one presents it as a left $K$-module using the relations (10.9.2)–(10.9.4). But another standard definition is to first present it as an *abelian group* using only the relations corresponding to (10.9.2) and (10.9.4), and then to use (10.9.3) to define a left $K$-module structure on this group. Not every abelian group with elements $x * y$ $(x \in |R|, y \in |B|)$ satisfying (10.9.2) and (10.9.4) has a left $K$-module structure satisfying (10.9.3); but the *universal* abelian group with these properties does, because the universal construction is functorial in $R$ as a right $L$-module, and the left $K$-module structure of $R$ constitutes a system of right-$L$-module endomorphisms; these induce endomorphisms of the constructed abelian group which make it a $K$-module.

This approach shows that the underlying abelian group structure of $R \otimes_L B$ depends only on the right $L$-module structure of $R$ and the left $L$-module structure of $B$; this is again analogous to the situation for the hom-set $K\text{-}\mathbf{Mod}(_K R_L, {}_K A)$, which starts out as an abelian group constructed using only the left $K$-module structures of $R$ and $A$, and then acquires a left $L$-module structure from the right $L$-module structure of $R$, by functoriality.

We should now learn how to compose the representable functors we have described. Suppose we have three rings, $H$, $K$, $L$, and adjoint pairs determined by an $(H, K)$-bimodule $R$, and a $(K, L)$-bimodule $S$:

$$(10.9.6) \qquad H\text{-}\mathbf{Mod} \underset{{}_H R_K \otimes_K -}{\overset{H\text{-}\mathbf{Mod}(_H R_K, -)}{\rightleftarrows}} K\text{-}\mathbf{Mod} \underset{{}_K S_L \otimes_L -}{\overset{K\text{-}\mathbf{Mod}(_K S_L, -)}{\rightleftarrows}} L\text{-}\mathbf{Mod}.$$

By observations we made in § 10.5, the underlying left $K$-module of the coalgebra determining the composite adjoint pair can be gotten by applying the left adjoint functor $R \otimes_K -$ to the underlying object of the coalgebra determining the other adjoint pair; hence it is the left $H$-module $R \otimes_K S$. It remains to find the coalgebra structure, i.e., the right $L$-module structure, on this object; this arises from the right $L$-module structure on $S$, by the same "functoriality" effect noted above for the left module structure of $R \otimes_L B$. So the composite of the adjoint pairs shown above is determined by an $(H, L)$-bimodule $_H(R \otimes_K S)_L$.

At this point, we have discussed enough kinds of structure on tensor products that we are ready to put them all into a definition, after which we will state formally the above characterization of representing objects for composite functors.

**Definition 10.9.7.** *If $K$ is a ring, $R$ a right $K$-module and $S$ a left $K$-module, then*

$$R \otimes_K S$$

*will denote the* abelian group *presented by generators $x \otimes y$ ($x \in |R|$, $y \in |S|$) and the relations (for all $x$, $x' \in |R|$, $d \in |K|$, $y$, $y' \in |S|$)*

$$(10.9.8) \quad (x + x') \otimes y \ = \ x \otimes y + x' \otimes y, \quad x \otimes (y + y') \ = \ x \otimes y + x \otimes y'$$

$$(10.9.9) \qquad\qquad\qquad (x \, d) \otimes y \ = \ x \otimes (d \, y).$$

*If $R$ is in fact an $(H, K)$-bimodule, respectively if $S$ is a $(K, L)$-bimodule, respectively if both are true (by which we mean, if the right $K$-module structure of $R$ is given as part of an $(H, K)$-bimodule structure, and/or if the left $K$-module structure of $S$ is given as part of a $(K, L)$-bimodule structure, for some rings $H$, $L$), then the abelian group $R \otimes_K S$ becomes a left $H$-module, respectively a right $L$-module, respectively an $(H, L)$-bimodule, with scalar multiplications characterized by (one or both of) the following formulas for $c \in |H|$, $e \in |L|$:*

$$(10.9.10) \qquad c \, (x \otimes y) \ = \ (c \, x) \otimes y, \qquad (x \otimes y) \, e \ = \ x \otimes (y \, e)$$

We see that (10.9.10) has supplied the symmetry that was missing in (10.9.3)!

Note that the four cases of the above definition (tensoring a right module $R_K$ or a bimodule $_H R_K$ with a left module $_K S$ or a bimodule $_K S_L$) reduce to a single case, if for every $K$ we identify **Mod-**$K$ with $\mathbb{Z}$-**Mod-**$K$ and $K$-**Mod** with $K$-**Mod-**$\mathbb{Z}$, and likewise **Ab** with $\mathbb{Z}$-**Mod-**$\mathbb{Z}$. So in further considerations, it will suffice to talk about the case where both objects are bimodules.

Let us now set down the result sketched before this definition.

**Lemma 10.9.11.** *In the situation shown in (10.9.6), the composite of the functors among left module categories represented by bimodules $_H R_K$ and $_K S_L$ is represented by the $(H, L)$-bimodule $R \otimes_K S$.* □

*Terminological note*: Given bimodules $_H R_K$ and $_K S_L$, we may call a map $*$ from $|R| \times |S|$ into an $(H, L)$-bimodule $_H T_L$ satisfying the equations corresponding to (10.9.8)–(10.9.10) (with $*$ for $\otimes$) a "bilinear map $R \times S \to T$", generalizing the term we have already used in the case of abelian groups (§ 4.9), so that we may describe $R \otimes_K S$ as an $(H, L)$-bimodule with a universal bilinear map of these bimodules into it. However, most ring theorists feel that the term "bilinear" logically only means "left $H$-linear and right $L$-linear", i.e., the conditions of (10.9.8) and (10.9.10), and they use the adjective "balanced" to express the remaining condition (10.9.9). So they would call $R \otimes_K S$ an $(H, L)$-bimodule with a universal $K$-*balanced* bilinear map of $R \times S$ into it.

The results on modules and bimodules developed above are sometimes regarded as a "model case" in terms of which to think of the general theory of representable functors among varieties of algebras, and their representing coalgebras. Thus, Freyd entitled the paper [12] in which he introduced the theory of such functors and their representing coalgebras, *On algebra-valued functors in general, and tensor products in particular*, and he called the coalgebra that represents a composite of representable functors between arbitrary varieties of algebras the "tensor product" of the coalgebras representing the given functors. I recommend that paper to the interested student, though with one word of advice: Ignore the roundabout way the author treats zeroary operations. Instead, consider them, as we have done, to be morphisms from the empty product (the terminal object of the category) into the object in question.

Further remarks: In the paragraph following (10.8.10), we chose a notation which "separated" the actions of elements of $K$ and elements of $L$, writing them on opposite sides of elements of $R$. It is also worth seeing what happens if we do not separate them, but continue to write them both to the left of their arguments. The actions of elements of $L$ will then compose in the opposite way to the multiplication of those elements in the ring $L$. This can be thought of as making $R$ a left module over $L^{\mathrm{op}}$, the opposite of the ring $L$ (defined as the ring with the same underlying set and additive group structure as $L$, but with the order of multiplication reversed). Thus we have on $R$ both a left $K$-module structure and a left $L^{\mathrm{op}}$-module structure, related by the conditions that the additive group operations of the two module structures are the same, and that the scalar multiplications of the $L^{\mathrm{op}}$-module structure are endomorphisms of the $K$-module structure. The latter condition says that the images of the elements of $K$ and of elements of $L^{\mathrm{op}}$ in the endomorphism ring of the common abelian group $R$ *commute* with one another. Now we saw in § 4.13 that given two rings $P$ and $Q$, if we form the tensor product of their underlying abelian groups, this can be given a ring structure such that the maps $p \mapsto p \otimes 1$ and $q \mapsto 1 \otimes q$ are homomorphisms of $P$ and $Q$ into $P \otimes Q$, whose images centralize one another, and which is universal among rings given with such a pair of homomorphisms from $P$ and $Q$. Thus, in our present situation, the mutually commuting left $K$-module and left $L^{\mathrm{op}}$-module structures on $R$ are equivalent to a single structure of left $K \otimes L^{\mathrm{op}}$-module. That is

(10.9.12)                     $K\text{-}\mathbf{Mod}\text{-}L \cong (K \otimes L^{\mathrm{op}})\text{-}\mathbf{Mod}.$

Hence one can study bimodules with the help of the theory of tensor product rings, and vice versa.

This also shows us that if we want to study representable functors to or from categories of *bimodules*, we do not need to launch a new investigation, but can reduce this situation to that of modules by using rings $K_0 \otimes K_1^{\mathrm{op}}$, etc., in place of $K$, etc.

**Exercise 10.9:1.**

(i)  If you did Exercise 4.13:4(ii), translate the results you got in that exercise into a partial or complete description of all $(\mathbb{Q}(2^{1/3}), \mathbb{Q}(2^{1/3}))$-bimodules.

(ii)  If you did Exercise 4.13:4(i), translate the results you got there to a partial or complete description of all $\mathbb{R}$-centralizing $(\mathbb{C}, \mathbb{C})$-bimodules $B$, where "$\mathbb{R}$-centralizing" means satisfying the identity $r\,x = x\,r$ for all $r \in \mathbb{R},\ x \in B$.

The student familiar with the theory of modules over *commutative* rings may have been surprised at my saying earlier that when we form a hom-set $K\text{-}\mathbf{Mod}(_K R,\ _K A)$ or a tensor product $R_L \otimes_L {}_L B$, the $K$-module structure, respectively the $L$-module structure, was "eaten up" in the process, since in the commutative case, these objects inherit natural $K$- and $L$-module structures. You can discover the general statement, of which these apparently contradictory observations are cases, by doing the next exercise. (The answer comes out in parts (iii) and (iv).)

**Exercise 10.9:2.** Let $K$ be a ring (not assumed commutative) and $M$ a left $K$-module.

(i)  Determine the *structure*, in the sense of §9.10, of the functor $K\text{-}\mathbf{Mod}(_K M, -) : K\text{-}\mathbf{Mod} \to \mathbf{Set}$.

(ii)  Determine similarly the structure of $K\text{-}\mathbf{Mod}(-,\ _K M) : K\text{-}\mathbf{Mod}^{\mathrm{op}} \to \mathbf{Set}$.

(iii)  Determine the structure of $K\text{-}\mathbf{Mod}(-, -) : K\text{-}\mathbf{Mod}^{\mathrm{op}} \times K\text{-}\mathbf{Mod} \to \mathbf{Set}$.

(iv)  Answer the corresponding three questions with tensor products in place of hom-sets (with or without the help of Corollary 8.11.6).

Let us note another basic ring-theoretic tool that we can understand with the help of the results of this section. Suppose $f : L \to K$ is a ring homomorphism. Then we can make any left $K$-module $A$ into a left $L$-module by keeping the same abelian group structure, and defining the new scalar multiplication by $d \cdot x = f(d)\,x \ (d \in |L|)$. This functor preserves underlying sets, hence it is representable. It is called "restriction of scalars along $f$", and its left adjoint is called "extension of scalars along $f$". (When $f$ is the inclusion of a subring $L$ in a ring $K$ these are natural terms to use. The usage in the case of arbitrary homomorphisms $f$ is a generalization from that case.) You should find the first part of the next exercise straightforward, and the second not too hard.

**Exercise 10.9:3.** Let $f : L \to K$ be a ring homomorphism.

(i)   Describe the bimodule representing the restriction-of-scalars functor, and deduce a description of the extension-of-scalars construction $L$-**Mod** $\to K$-**Mod** as a tensor product operation.

(ii)   If $K$ and $L$ are commutative, we may also consider the "restriction of base-ring" functor from $K$-algebras to $L$-algebras, defined to preserve underlying ring-structures, and act as restriction of scalars on module structures. (You may here take "algebras" over $K$ and $L$ either to mean commutative algebras, or not-necessarily commutative (but, as always when the contrary is not stated, associative) algebras, depending on what you are comfortable with.) We know this functor is representable. (Why?) Describe its representing coalgebra. Show that the left adjoint of this functor acts on the underlying modules of algebras by extension of scalars. How is the ring structure on the resulting modules defined?

I will close our discussion of abelian groups and modules with an observation that goes back to the beginning of the preceding section: When we determined the form of the general comonoid object of **Ab**, our argument used only the fact that we had a binary co-operation satisfying the coneutral laws with respect to the unique zeroary co-operation—the coassociative law was never called on! Thus, if we let **Binar**$^e$ denote the variety of sets with a binary operation and a neutral element $e$ for that operation, then co-**Binar**$^e$ objects of **Ab** are automatically co-**Monoid** objects, and even co-**AbMonoid** objects, and, as we noted, these have unique coinverse operations making them co-**Group** and co-**Ab** objects.

On the other hand, if we drop the co-neutral-element, then associativity and commutativity conditions do make a difference:

**Exercise 10.9:4.** Characterize all representable functors from **Ab** to each of the following varieties:

(i)   **Binar**, the variety of sets given with a binary operation.

(ii)   **Semigroup** (a subvariety of **Binar**).

(iii)   **AbSemigroup** (a subvariety of **Semigroup**).

In the last two cases, you should discover that every such functor decomposes as a direct sum of a small number of functors whose structures are easily described.

## 10.10. Some general results on representable functors, mostly negative

As we mentioned in Exercise 10.7:6, the form of the general representable functor **Monoid** → **Semigroup** is not known. What about representable functors going the other way, from **Semigroup** to **Monoid**?

It is easy to show that in this case there are no nontrivial examples. The idea is that in working in **Semigroup**, one has no distinguished elements available, so there is no way to pin down a zeroary "neutral element" operation.

Before having you prove this, let me indicate the exception implied in the word "nontrivial". If $\mathbf{C}$ is any category with an initial object $I$, then the functor $h_I$ takes every object of $\mathbf{C}$ to a one-element set, which of course has a unique structure of $\mathbf{V}$-algebra for every variety $\mathbf{V}$; hence for every variety $\mathbf{V}$, the object $I$ admits co-operations making it a $\mathbf{V}$-coalgebra. Let us call a functor represented by such a coalgebra, which takes every object of $\mathbf{C}$ to a one-element algebra (terminal object) of $\mathbf{V}$, a *trivial* functor $\mathbf{C} \to \mathbf{V}$. (Loosely, we could say *the* trivial functor, since it is unique up to isomorphism.)

**Exercise 10.10:1.** Show that if $\mathbf{W}$ is a variety without zeroary operations, and $\mathbf{V}$ a variety with at least one zeroary operation, then there is no nontrivial representable functor $\mathbf{W} \to \mathbf{V}$.

More generally, can you give a condition on a general category $\mathbf{C}$ with finite coproducts that insures that there are no nontrivial representable functors from $\mathbf{C}$ to a variety $\mathbf{V}$ with at least one zeroary operation?

Here is another observation about specific varieties from which we can extract a general principle. We began this chapter with an example of a representable functor from rings to groups; but if one looks for a nontrivial representable functor from groups to rings, it is hard to imagine how one might be constructed, because a nontrivial ring must have distinct 0 and 1, and we have only one distinguished group element, $e$, to use in the coordinates of a distinguished element of a ring we are constructing. This argument can be made precise. To get the result from general considerations, we make

**Definition 10.10.1.** *If $\mathbf{C}$ is a category with a terminal object $T$, let us (as in Exercise 7.8:4) define a* pointed object *of $\mathbf{C}$ to mean a pair $(A, p)$ where $A$ is an object of $\mathbf{C}$ and $p$ a morphism $T \to A$. (Thus, since $T$ is the product of the empty family of copies of $A$, such a pair is an object of $\mathbf{C}$ given with a single zeroary operation.) A morphism $(A, p) \to (A', p')$ of such objects will mean a morphism $A \to A'$ making a commuting triangle with $p$ and $p'$. The category of pointed objects of $\mathbf{C}$, with these morphisms, will be denoted $\mathbf{C}^{\mathrm{pt}}$.*

*Dually, if $\mathbf{C}$ is a category with an initial object $I$, then an* augmented *object of $\mathbf{C}$ will mean a pair $(A, a)$ where $A$ is an object of $\mathbf{C}$ and $a$ is*

*a morphism $A \to I$ (an "augmentation map"), equivalently, a zeroary co-operation on $A$. Again using the obvious commuting triangles as morphisms, we denote the category of augmented objects of $\mathbf{C}$ by $\mathbf{C}^{\mathrm{aug}}$.*

*Thus, in comma category notation, $\mathbf{C}^{\mathrm{pt}} = (T \downarrow \mathbf{C})$, and $\mathbf{C}^{\mathrm{aug}} = (\mathbf{C} \downarrow I)$.*

*A category $\mathbf{C}$ will be called "pointed" if it has a zero object (an object that is both initial and terminal; Definition 7.8.1).*

Exercise 7.8:4 shows that if $\mathbf{C}$ is a category with a terminal object, then $\mathbf{C}^{\mathrm{pt}}$ is a pointed category. By duality, if $\mathbf{C}$ is a category with an initial object, then $\mathbf{C}^{\mathrm{aug}}$ is likewise pointed. The next exercise begins with a few more observations of the same sort, then gets down to business.

**Exercise 10.10:2.**

(i)   Let $\mathbf{C}$ be a category with a terminal (respectively initial) object, so that, as noted above, $\mathbf{C}^{\mathrm{pt}}$ (respectively $\mathbf{C}^{\mathrm{aug}}$) is a pointed category. Show that the forgetful functor $\mathbf{C}^{\mathrm{pt}} \to \mathbf{C}$ (respectively $\mathbf{C}^{\mathrm{aug}} \to \mathbf{C}$) is an equivalence if and only if $\mathbf{C}$ is a pointed category.

(ii)   Show that if $\mathbf{V}$ is a variety of algebras, then $\mathbf{V}^{\mathrm{pt}}$ is equivalent to a variety of algebras.

(iii) Show that a variety of algebras $\mathbf{V}$ is a pointed category if and only if $\mathbf{V}$ has a zeroary operation, and all derived zeroary operations of $\mathbf{V}$ are equal.

Now suppose $\mathbf{V}$ is a variety, and $\mathbf{C}$ a category having small coproducts.

(iv)   Show that $\mathbf{C}^{\mathrm{aug}}$ also has small coproducts, and that

$$\mathbf{Rep}(\mathbf{C}, \mathbf{V}^{\mathrm{pt}}) \approx \mathbf{Rep}(\mathbf{C}^{\mathrm{aug}}, \mathbf{V}^{\mathrm{pt}}) \approx \mathbf{Rep}(\mathbf{C}^{\mathrm{aug}}, \mathbf{V}).$$

(If you don't see how to begin, you might think first about the case $\mathbf{V} = \mathbf{Set}$.)

(v)   Show that **Group** is pointed, and that $(\mathbf{Ring}^1)^{\mathrm{pt}}$ consists only of the trivial ring with its unique pointed structure. Deduce that there are no nontrivial representable functors from **Group** or any of its subvarieties to $\mathbf{Ring}^1$ or any of its subvarieties (e.g., $\mathbf{CommRing}^1$).

(vi)   Also deduce from (iv) the result of Exercise 10.10:1.

The term "augmented" comes from ring theory, where an "augmentation" on a $k$-algebra $R$ means a $k$-algebra homomorphism $\varepsilon \colon R \to k$. This ring-theoretic concept probably originated in algebraic topology, where the cohomology of a pointed space acquires, by contravariance of the cohomology ring functor, such an augmentation.

Here is another sort of nonexistence result.

**Exercise 10.10:3.** Let $R$ be an object of a variety $\mathbf{V}$, and let $\tau \colon R^{\lambda} \amalg R^{\rho} \to R^{\lambda} \amalg R^{\rho}$ denote the automorphism that interchanges $x^{\lambda}$ and $x^{\rho}$ for all $x \in |R|$. Denote by $\mathrm{Sym}(R^{\lambda} \amalg R^{\rho})$ the fixed-point algebra of $\tau$; i.e., the algebra of "$(\lambda, \rho)$-symmetric" elements of $R^{\lambda} \amalg R^{\rho}$.

(i)   Show that a binary co-operation $\mathbf{m}\colon R \to R^\lambda \amalg R^\rho$ is cocommutative (i.e., satisfies the coidentity making the induced operations on all sets $\mathbf{V}(R, A)$ commutative) if and only if it carries $R$ into $\mathrm{Sym}(R^\lambda \amalg R^\rho)$.

(ii)  Show that in the variety **Group**, one has $\mathrm{Sym}(R^\lambda \amalg R^\rho) = \{e\}$ for all objects $R$.

(iii) Deduce that there are no nontrivial representable functors **Group** $\to$ **Ab**, hence also no nontrivial representable functors **Group** $\to$ **Ring**$^1$; and that there are no nontrivial representable functors **Group** $\to$ **Semilattice**, hence also no nontrivial representable functors **Group** $\to$ **Lattice**.

Having exhausted the subject of representable functors from groups to lattices, we may ask, what about functors in the reverse direction? The category **Lattice** has no zeroary operations, so there can be no nontrivial functors from it or any of its subvarieties to **Group** by Exercise 10.10:1; but suppose we get out of this hole by considering lattices with one or more distinguished elements. I do not know the answer to the first part of the next exercise, though I do know the answer to the second.

**Exercise 10.10:4.**

(i)   Is there a variety **L** of lattices for which there exists a nontrivial representable functor $\mathbf{L}^{\mathrm{pt}} \to \mathbf{Group}$?

(ii)  For **C** a category with a terminal object $T$, let $\mathbf{C}^{2\text{-pt}}$ denote the category of 3-tuples $(A, p_0, p_1)$ where $A$ is an object of **C**, and $p_0$, $p_1$ are morphisms $T \rightrightarrows A$. Is there any variety **L** of lattices for which there exists a nontrivial representable functor $\mathbf{L}^{2\text{-pt}} \to \mathbf{Group}$?

Of course, not every plausible heuristic argument restricting the properties of representable functors is valid. For instance, every primitive operation of lattices, and hence also every derived operation of lattices, is isotone with respect to the natural ordering of the underlying set, while Boolean rings have the operation of complementation, which is not isotone. Nevertheless, we have the construction of the following exercise.

**Exercise 10.10:5.** Let $\mathbf{DistLat}^{0,1}$ denote the variety of distributive lattices (Exercise 6.1:15) with least element $0$ and greatest element $1$. An element $x$ of such a lattice $L$ is called *complemented* if there exists $y \in |L|$ such that $x \wedge y = 0$ and $x \vee y = 1$.

Show that for $L \in \mathrm{Ob}(\mathbf{DistLat}^{0,1})$, the set of complemented elements of $L$ can be made a Boolean ring, whose natural partial ordering is the restriction of the natural partial ordering of $L$, and that this construction yields a representable functor $C\colon \mathbf{DistLat}^{0,1} \to \mathbf{Bool}^1$. Give a description of this functor in terms of "tuples of elements satisfying certain relations", and describe the Boolean operations on such tuples.

Here is a triviality question of a different sort.

**Exercise 10.10:6.** If $\mathbf{U}$, $\mathbf{V}$, $\mathbf{W}$ are varieties such that there exist nontrivial representable functors $\mathbf{W} \to \mathbf{V}$ and $\mathbf{V} \to \mathbf{U}$, must there exist a nontrivial representable functor $\mathbf{W} \to \mathbf{U}$?

Let us turn to positive results. We recall from Exercise 8.3:5 that every equivalence of categories is also an adjunction. Note also that a category $\mathbf{C}$ equivalent to a variety with small colimits will have small colimits, hence in particular, will have small coproducts. Thus we can make sense of the concept of a representable functor from $\mathbf{C}$ to varieties of algebras. We deduce

**Lemma 10.10.2.** *Suppose* $\mathbf{C} \underset{j}{\overset{i}{\rightleftarrows}} \mathbf{V}$ *is an equivalence between an arbitrary category* $\mathbf{C}$ *and a variety of algebras* $\mathbf{V}$. *Then* $i: \mathbf{C} \to \mathbf{V}$ *is representable, and has a representing coalgebra with underlying object* $j(F_{\mathbf{V}}(1))$.  □

The above fact is used in [60] to study the *self-equivalences* of the variety of rings, and more generally, of the variety of algebras over a commutative ring $k$. (The self-equivalences of any category $\mathbf{C}$, modulo isomorphism of functors, form a group, called the *automorphism class group* of $\mathbf{C}$. When $\mathbf{C} = \mathbf{Ring}^1$, this group is shown in [60] to be isomorphic to $Z_2$, the nonidentity element arising from the self-equivalence $K \mapsto K^{\mathrm{op}}$. For $k$ a commutative ring, the variety of $k$-algebras has a more complicated automorphism class group if $k$ has nontrivial idempotent elements or nontrivial automorphisms.)

**Exercise 10.10:7.** We saw in Exercise 7.9:17 that for $K$ a ring, the varieties $K$-**Mod** and $M_n(K)$-**Mod** were equivalent. By the above lemma, both functors occurring in the equivalence must be representable. Determine the bimodules that yield these functors.

(This suggests the question: Given rings $K$ and $L$ and an object $R$ of $K$-**Mod**-$L$, under what conditions is the functor $K$-**Mod** $\to L$-**Mod** represented by $R$ an equivalence? That is the subject of *Morita theory* [107, §§18 and 19].)

A challenging related problem is

**Exercise 10.10:8.** Characterize those functors between module categories, $F: K$-**Mod** $\to L$-**Mod**, which have both a left and a right adjoint.

Another useful result is given in

**Exercise 10.10:9.** Let $\mathbf{C}$ be a category with small colimits, and $\mathbf{V}$ a variety of algebras. Show that the category $\mathbf{Rep}(\mathbf{C}, \mathbf{V})$ is closed under taking small limits within the functor category $\mathbf{V}^{\mathbf{C}}$.

As an example, let $n$ be a positive integer, and consider the functors $\mathrm{GL}(n)$, $\mathrm{GL}(1) \in \mathbf{Rep}(\mathbf{CommRing}^1, \mathbf{Group})$. (Note that $\mathrm{GL}(1)$ is just the "group of units" functor.) We can define morphisms

$$e, \det: \ \mathrm{GL}(n) \ \rightrightarrows \ \mathrm{GL}(1),$$

where the first takes every invertible matrix to 1, and the second takes every invertible matrix to its determinant. By the preceding exercise, the limit (equalizer) of this diagram of functors and morphisms is representable. This equalizer is the functor $\mathrm{SL}(n)$, with which we began this chapter.

**Exercise 10.10:10.** In §10.6 we described the general object of **Rep(Monoid, Monoid)**. Find a finite family $S$ of such objects with the property that every object of this category is the limit of a system of objects in $S$ and morphisms among these.

Note that on taking the ring $K$ of Exercise 10.10:7 to be finite, one gets examples showing that an *equivalence* of varieties need not preserve *cardinalities* of algebras. The next two exercises explore this topic further.

**Exercise 10.10:11.** Give an example of a variety $\mathbf{V}$ with a self-equivalence $i: \mathbf{V} \to \mathbf{V}$ and an algebra $A$ such that $\mathrm{card}(|i(A)|) \neq \mathrm{card}(|A|)$.

**Exercise 10.10:12.** Show that if $i: \mathbf{V} \to \mathbf{W}$ is an equivalence between varieties of algebras, and $A$ an object of $\mathbf{V}$ such that $\mathrm{card}(|A|)$ is infinite, then $\mathrm{card}(|i(A)|) = \mathrm{card}(|A|)$.

## 10.11. A few ideas and techniques

In §§10.7–10.9, we considered some cases of the problem, "Given varieties $\mathbf{V}$ and $\mathbf{W}$, find all representable functors $\mathbf{W} \to \mathbf{V}$". We can turn this question around and ask, "Given an object $R$ of a variety $\mathbf{W}$, what kinds of algebras can we make out of the values of the functor $h_R$?" This question asks for the *structure* on the set-valued functor $h_R$, in the sense of §9.10, i.e., for the operations admitted by that functor and the identities that they satisfy.

I gave some examples of this question in Exercise 9.10:3; we can now see what you probably discovered if you did that exercise (though you did not then have the terminology to state it nicely): that to find the operations on such a functor and the identities they satisfy, one needs to look for the co-operations admitted by its representing object, and the coidentities satisfied by those co-operations.

Let us work out an example. Suppose we are interested in the algebraic structure one can put, in a functorial way, on the set of *elements of exponent* 2 in a general group $G$. This means we want to study the structure on the functor taking $G$ to the set of such elements, i.e., the set-valued functor represented by the group $Z_2$; so our task is equivalent to describing the clone of co-operations admitted by $Z_2$ in **Group**.

An $n$-ary co-operation on $Z_2$ means a group homomorphism $Z_2 \to Z_2 \amalg \ldots \amalg Z_2$, and hence corresponds to an element of exponent 2 in the latter group. Though in $Z_2$ one often uses additive notation, these coproduct groups are noncommutative, so let us write $Z_2$ multiplicatively, calling the identity element $e$ and the nonidentity element $t$. Then the coproduct of $n$ copies of $Z_2$ will be generated by elements $t_0, \ldots, t_{n-1}$ of exponent 2, and (by the description of coproduct groups in Proposition 4.6.5), the general element of this coproduct can be written uniquely

(10.11.1)   $t_{\alpha_0} t_{\alpha_1} \ldots t_{\alpha_{h-1}}$, where $h \geq 0$, all $\alpha_i \in n$, and $\alpha_i \neq \alpha_{i+1}$
for $0 \leq i < h - 1$.

Let us begin by seeing what structure on $h_{Z_2}$ is apparent to the naked eye, and translating it into the above terms. Since the identity element of every group is of exponent 2, $h_{Z_2}$ admits

(10.11.2)   the zeroary operation $e$, determined by the unique homomorphism $Z_2 \to \{e\}$.

Also, any conjugate of an element of exponent 2 has exponent 2, hence $h_{Z_2}$ admits

(10.11.3)   the binary operation $(x, y) \mapsto x^y = y^{-1}xy = yxy$, determined by the homomorphism $Z_2 \to Z_2 \amalg Z_2$ taking $t$ to $t_1 t_0 t_1$.

To see whether these generate all functorial operations on $h_{Z_2}$, consider the general element (10.11.1) of the $n$-fold coproduct of copies of $Z_2$. If (10.11.1) has exponent 2, all factors must cancel when we square it, which we see means that we must have $\alpha_0 = \alpha_{h-1}$, $\alpha_1 = \alpha_{h-2}$, etc. If $h$ is even and positive, this gives, in particular, $\alpha_{(h/2)-1} = \alpha_{h/2}$, which contradicts the final condition of (10.11.1). Hence the only element (10.11.1) of even length having exponent 2 is $e$. This element induces the constant $n$-ary operation $e$, and we see that for each $n$, this is a derived $n$-ary operation arising from the zeroary operation (10.11.2).

On the other hand, for $h = 2k + 1$, we see that (10.11.1) will have exponent 2 if and only if it has the form

(10.11.4)   $t_{\alpha_0} \ldots t_{\alpha_{k-1}} t_{\alpha_k} t_{\alpha_{k-1}} \ldots t_{\alpha_0}$ with $k \geq 0$, and $\alpha_i \neq \alpha_{i+1}$ for $0 \leq i < k$.

The $n$-ary operation that such an element induces on elements of exponent 2 can clearly be expressed in terms of the operation (10.11.3); it is

(10.11.5)        $(x_0, \ldots, x_n) \longmapsto (\ldots (((x_{\alpha_k})^{x_{\alpha_{k-1}}})^{x_{\alpha_{k-2}}}) \ldots)^{x_{\alpha_0}}.$

So the operations (10.11.2) and (10.11.3) do indeed generate the clone of operations on $h_{Z_2}$.

How can we find a generating set for the *identities* these operations satisfy? This is shown in

**Exercise 10.11:1.** Note that the set of all terms in the two operations (10.11.2) and (10.11.3) includes terms not of the form $e$ or (10.11.5); e.g., $e^{x_0}$, $x_0^{(x_1^{x_2})}$, $(x_0^{x_1})^{x_1}$. (The last is not of the form (10.11.5) because it fails to satisfy the condition $\alpha_i \neq \alpha_{i+1}$ of (10.11.4).)

(i)    For each of the above three terms, show how the resulting *derived operation* of $h_{Z_2}$ can be expressed either as $e$ or in the form (10.11.5), and extract from each such observation an identity satisfied by (10.11.2) and (10.11.3). Do the same with other such terms, until you can show that you have enough identities to reduce every term in the operations (10.11.2) and (10.11.3) either to $e$ or to the form (10.11.5).

(ii)    Deduce that all identities of the operations (10.11.2) and (10.11.3) of $h_{Z_2}$ are consequences of the identities in your list.

**Exercise 10.11:2.** Let **V** denote the variety defined by a zeroary operation $e$ and a binary operation $(-)^-$, subject to the identities of our two operations on $h_{Z_2}$, and let $V:$ **Group** $\to$ **V** be the functor represented by $Z_2$ with the co-operations defined above.

(i)    Is every object of **V** embeddable in an object of the form $V(G)$ for $G$ a group?

(ii)    Translate your answer to (i) into a property of the functor $V$ and its left adjoint. (Or if you haven't fully answered (i), translate the question into a *question* about these functors.)

Let me present, next, an interesting problem which, though not at first obviously related to the concepts of this chapter, turns out, like the question examined above, to be approachable by studying the structure on a functor.

If $y$ is an element of a group $G$, recall that the map $x \mapsto y^{-1}xy$ is an automorphism of $G$, and that an automorphism that has this form for some $y \in |G|$ is called an *inner* automorphism of $G$. Now suppose one is handed a group $G$ and an automorphism $\alpha \in$ **Group**$(G, G)$. Is it possible to decide whether $\alpha$ is inner, by looking only at its properties within the category **Group**, i.e., conditions statable in terms of objects and morphisms, without reference to the "internal" nature of the objects?

Well, observe that if $\alpha$ is an inner automorphism of $G$, induced as above by an element $y \in |G|$, then given any homomorphism $h$ from $G$ to another group $H$, there exists an automorphism of $H$ which yields a commuting square with $\alpha$:

$$\begin{array}{ccc} G & \xrightarrow{\phantom{xx}h\phantom{xx}} & H \\ \Big\downarrow{\alpha} & & \Big\downarrow \\ G & \xrightarrow{\phantom{xx}h\phantom{xx}} & H, \end{array}$$

namely, the inner automorphism of $H$ induced by $h(y)$. In fact, this construction associates to every such pair $(H, h)$ an automorphism $\alpha_{(H, h)}$ in a "coherent manner", in the sense that given two such pairs $(H_0, h_0)$ and $(H_1, h_1)$, and a morphism $f \colon H_0 \to H_1$ such that $h_1 = f\,h_0$, the automorphisms $\alpha_{(H_0, h_0)}$ of $H_0$ and $\alpha_{(H_1, h_1)}$ of $H_1$ form a commuting square with $f$.

I claim, conversely, that any automorphism $\alpha$ of a group $G$ which can be "extended coherently", in the above sense, to all groups $H$ with maps of $G$ into them, is inner. The next exercise formalizes this "coherence" property in a general category-theoretic setting, then asks you to prove this characterization of inner automorphisms.

**Exercise 10.11:3.** Given an object $C$ of a category $\mathbf{C}$, recall that $(C \downarrow \mathbf{C})$ denotes the category whose objects are pairs $(D, d)$, $(D \in \mathrm{Ob}(\mathbf{C})$, $d \in \mathbf{C}(C, D))$ and where a morphism $(D_0, d_0) \to (D_1, d_1)$ is a morphism $D_0 \to D_1$ making a commuting triangle with $d_0$ and $d_1$. (Cf. Definition 7.8.12.) Let $U_C$ denote the forgetful functor $(C \downarrow \mathbf{C}) \to \mathbf{C}$ sending $(D, d)$ to $D$.

Call an endomorphism $\alpha$ of an object $C$ "functorializable" if there exists an endomorphism $a$ of the above forgetful functor $U_C$ which, when applied to the initial object of $(C \downarrow \mathbf{C})$, namely $(C, \mathrm{id}_C)$, yields $\alpha$.

Show that for $\mathbf{C} = \mathbf{Group}$, an automorphism of an object $G$ is functorializable if and only if it is an inner automorphism. In fact, determine the monoid of endomorphisms of $U_G \colon (G \downarrow \mathbf{Group}) \to \mathbf{Group}$ and its image in the monoid of endomorphisms of $G$. (Suggestion: consider a universal example of a pair $(X, x)$ where $X$ is an object of $(G \downarrow \mathbf{Group})$ and $x$ an element of the underlying group of $X$.)

Some related questions you can also look at: How does the above monoid compare with the monoid of endomorphisms of the *identity* functor of $(G \downarrow \mathbf{Group})$? Can you characterize functorializable endomorphisms of objects of other interesting varieties?

On to another topic. The next exercise is unexpectedly hard (unless there is a trick I haven't found), but is interesting.

**Exercise 10.11:4.** Let $\mathbf{V}$ and $\mathbf{W}$ be varieties of algebras (finitary if you wish). Show that the category $\mathbf{Rep}(\mathbf{V}, \mathbf{W})$ has an *initial object*.

(I obtain the result of Exercise 10.11:3 and more in [54], and similarly for Exercise 10.11:4 in [53]. If you read these, then, of course, any solutions you submit to those exercises has to bring in some improvement or some alternative approach to what is done in those papers.)

The next exercise develops some results and examples regarding these initial representable functors (also found in [53]).

**Exercise 10.11:5.** Suppose we classify varieties into three sorts: (a) those with no zeroary operations, (b) those with a unique derived zeroary operation, and (c) those with more than one derived zeroary operation. Applying this classification to both of the varieties $\mathbf{V}$ and $\mathbf{W}$ in the preceding exercise, we get nine cases.

(i)   Show that in *most of* these cases, the initial object of **Rep(V, W)** must be trivial, in the weak sense that it takes every object $A$ *either* to the one-element algebra *or* to the empty algebra.

(ii)   Determine the initial object of **Rep(Set, Semigroup)**.

(iii) Determine the initial object of **Rep(Set, Binar)**, where **Binar** is the variety of sets with a single (unrestricted) binary operation.

(iv) Interpret the result of Exercise 9.3:11 as describing the initial object of **Rep(Binar, Binar)**.

(v)   The three preceding examples all belong to the same one of the nine cases referred to at the start of this exercise. Give an example belonging to a different case, in which **Rep(V, W)** also has nontrivial initial object.

When I first learned about the concept of "coidentities" in coalgebra objects of a category **C**, I was a little disappointed that the possible coidentities merely corresponded to the identities of set-based algebras of the same type— I thought it would have been more interesting if this "exotic" version of the concept of algebra led to "exotic" sorts of identities as well. But perhaps there is hope for something exotic if the question is posed differently. Recall that in § 9.6 we characterized varieties of algebras as those classes of algebras that were closed under three operators **H**, **S** and **P**. I don't know the answer to the question asked in

**Exercise 10.11:6.** Define analogs of the operators **H**, **S** and **P** for classes of objects of **Rep(C, Ω-Alg)**. Presumably, for every variety **V** of Ω-algebras, **Rep(C, V)** will be closed in **Rep(C, Ω-Alg)** under your operators; but will these be the only closed classes?

If not, try to characterize the classes closed under your operators (possibly assuming some restrictions on **C** and Ω).

## 10.12. Contravariant representable functors

In § 10.2 we defined the concept of an *algebra*-object of a category, but we immediately passed to that of a *coalgebra* object in § 10.3, and showed in § 10.4 that a covariant functor has a left adjoint if and only if it is represented by such an object. Let us now look at the version of this result for algebra objects, and the contravariant functors these represent. We recall from § 8.12 that a contravariant adjunction involves a pair of *mutually right adjoint* or *mutually left adjoint* functors. Algebra objects in a category give the former, as we see by putting "**C**$^{op}$" in place of "**C**" in the definition of an adjunction

$$\mathbf{C} \, \underset{j}{\overset{i}{\rightleftarrows}} \, \mathbf{V}.$$ Theorem 10.4.3 thus translates to give

**Theorem 10.12.1.** *Let* **C** *be a category with small limits,* **V** *a variety of algebras, and* $V \colon \mathbf{C}^{\mathrm{op}} \to \mathbf{V}$ *a functor. Then the following conditions are equivalent:*

(i)    *V has a right adjoint* $W \colon \mathbf{V}^{\mathrm{op}} \to \mathbf{C}$ *(so that V and W form a pair of mutually right adjoint contravariant functors).*

(ii)   $V \colon \mathbf{C}^{\mathrm{op}} \to \mathbf{V}$ *is representable, i.e., is isomorphic to* $\mathbf{C}(-, R)$ *for some* **V***-algebra object R of* **C** *(Definition 10.2.9).*

(iii)  *The composite of V with the underlying-set functor* $U_{\mathbf{V}} \colon \mathbf{V} \to \mathbf{Set}$ *is representable, i.e., is isomorphic to* $h^{|R|} = \mathbf{C}(-, |R|)$ *for some object* $|R|$ *of* **C**.                                                                         □

Now suppose that in the above situation we take for **C** another variety of algebras, **W**. What will a **V**-object $R$ of **W** look like? Its **V**-operations will be **W**-algebra homomorphisms $t_R \colon |R|^{\mathrm{ari}(t)} \to |R|$; that is, set maps $||R||^{\mathrm{ari}(t)} \to ||R||$ which respect the **W**-operations of $|R|$. The condition for an $n$-ary operation $t$ on a set to "respect" an $m$-ary operation $s$ is

$$s(t(x_{0,0}, \ldots, x_{0,n-1}), \ldots, t(x_{m-1,0}, \ldots, x_{m-1,n-1}))$$
$$= t(s(x_{0,0}, \ldots, x_{m-1,0}), \ldots, s(x_{0,n-1}, \ldots, x_{m-1,n-1})).$$

The above equation assumes the arities $m$ and $n$ are natural numbers. For operations of arbitrary arities, the condition may be written

$$(10.12.2) \qquad s((t(x_{ij})_{j \in \mathrm{ari}(t)})_{i \in \mathrm{ari}(s)}) = t((s(x_{ij})_{i \in \mathrm{ari}(s)})_{j \in \mathrm{ari}(t)}).$$

Note that this condition is symmetric in $s$ and $t$, and that when $s$ and $t$ are both *unary*, it says that $s(t(x)) = t(s(x))$, i.e., that as elements of the monoid of set maps $||R|| \to ||R||$, $s$ and $t$ commute. Generalizing this usage, one calls operations $s$ and $t$ of arbitrary arities which satisfy (10.12.2) *commuting* operations. This condition is equivalent to commutativity of the diagram

$$
\begin{array}{ccc}
||R||^{\mathrm{ari}(s) \times \mathrm{ari}(t)} & \xrightarrow{\ t^{\mathrm{ari}(s)}\ } & ||R||^{\mathrm{ari}(s)} \\
{\scriptstyle s^{\mathrm{ari}(t)}} \downarrow & & \downarrow {\scriptstyle s} \\
||R||^{\mathrm{ari}(t)} & \xrightarrow{\quad t \quad} & ||R||
\end{array}
$$

where $t^{\mathrm{ari}(s)}$ and $s^{\mathrm{ari}(t)}$ act in the "obvious" ways on $\mathrm{ari}(s) \times \mathrm{ari}(t)$-tuples of elements of $||R||$. In picturing these "obvious" actions, it can be helpful to visualize an $\mathrm{ari}(s) \times \mathrm{ari}(t)$-tuple of elements of $||T||$ as arranged in an $\mathrm{ari}(s)$-by-$\mathrm{ari}(t)$ rectangle, with $s$ sending a column to a value written below it, and $r$ sending a row to a value written to its right; it is then not hard to understand why $s$ respecting $t$ is equivalent to commutativity of the above diagram. Of course, if $\mathrm{ari}(s)$ and/or $\mathrm{ari}(t)$ is 0, the rectangle has no entries, but the commutativity still has a nontrivial content, as noted in part (i) of the next exercise. That and the remaining parts can give one further feel for the concept of commuting operations.

**Exercise 10.12:1.**

(i)    Show that two zeroary operations commute if and only if they are equal. More generally, when will an $n$-ary operation $s$ commute with a zeroary operation $t$?

(ii)   Verify that every unary operation on a set commutes with itself.

(iii)  Show that not every binary operation $s$ on a set $X$ commutes with itself. In fact, consider the following four conditions on a binary operation $s$: (a) $s$ commutes with itself, (b) $s$ satisfies the commutativity identity $s(x, y) = s(y, x)$, (c) $s$ satisfies the associativity identity $s(s(x, y), z) = s(x, s(y, z))$, and (d) there exists a neutral element $e \in X$ for $s$, i.e., an element satisfying the identities $s(x, e) = x = s(e, x)$. Determine which of the 16 possible combinations of truth values for these conditions can be realized.

Summarize your results as one or more implications which hold among these conditions, and such that any combination of truth-values consistent with those implications can be realized.

We see that if **V** is a variety of $\Omega$-algebras and **W** a variety of $\Omega'$-algebras, then a **V**-algebra object of **W** is equivalent to a set-based algebra $R = (|R|, (s_R)_{s\in|\Omega'|\sqcup|\Omega|})$, where the operations indexed by $|\Omega'|$, respectively, $|\Omega|$, are of the arities specified in $\Omega'$, respectively, $\Omega$, and satisfy the identities of **W**, respectively **V**, and where, moreover, for every $s \in |\Omega'|$ and $t \in |\Omega|$, the commutativity identity (10.12.2) is satisfied. Since all these conditions are identities, the category of such objects forms a variety!

Given such an object $R$, and an ordinary object $A$ of **W**, we see that the resulting **V**-algebra operations on the set $\mathbf{W}(A, R)$ are given by "pointwise" application of the **V**-operations of $R$ to **W**-homomorphisms $A \to R$. Now in general, if $A$ and $B$ are objects of a variety **W** and one combines a family of algebra homomorphisms $f_\alpha : A \to B$ $(\alpha \in \beta)$ by pointwise application of a $\beta$-ary operation $t$ on the set $|B|$, the result need not be a homomorphism of **W**-algebras. What makes this true here is the fact that $t$ is an operation on $R$ as an object of the category **W**, i.e., that it commutes with all the **W**-operations.

Since the functor $\mathbf{W}(-, R)\colon \mathbf{W}^{\mathrm{op}} \to \mathbf{V}$ belongs to a *mutually right adjoint* pair, its *adjoint* will *also* satisfy condition (i) of Theorem 10.12.1, and hence the other two equivalent conditions; that is, this adjoint will also be a representable contravariant functor, but going the other way, $\mathbf{V}^{\mathrm{op}} \to \mathbf{W}$. As the next exercise shows, the representing object for this functor is gotten by trivially modifying the representing object $R$ for the original functor.

**Exercise 10.12:2.** Let $V\colon \mathbf{W}^{\mathrm{op}} \to \mathbf{V}$ be a representable contravariant functor, whose representing **V**-algebra object $R$ is, in the above formulation $(|R|, (s_R)_{s\in|\Omega'|\sqcup|\Omega|})$. Show that the right adjoint to $V$ is the functor

$\mathbf{V}(-, R')$, where $R'$ has the same underlying set as $R$, and the same operations, but with the roles of the $\mathbf{W}$-operations and the $\mathbf{V}$-operations as "primary" and "secondary" interchanged, so that it becomes a $\mathbf{W}$-algebra object of $\mathbf{V}$.

A basic contrast between covariant and contravariant representable functors on a variety $\mathbf{W}$ is that the former, as we saw in §10.3, define their operations *using* derived operations of $\mathbf{W}$, while the objects representing the latter have operations that must *commute* with those of $\mathbf{W}$. A consequence is that, generally speaking, the "richer" the structure of $\mathbf{W}$, the richer is the class of covariant representable functors on $\mathbf{W}$, but the scarcer are the contravariant representable functors. Thus, the case in which it is easiest to get contravariant representable functors is when $\mathbf{V}$ is the variety with the smallest family of operations, namely **Set**.

A $\mathbf{V}$-algebra object of **Set** is just an ordinary $\mathbf{V}$-algebra. Let us take the smallest nontrivial object in **Set**, and find the richest algebra structure we can put on it, and the functor this represents.

**Exercise 10.12:3.**

(i)   Show that the clone of all finitary operations on the object $2 = \{0, 1\}$ of **Set** can be described as the clone of derived operations of the *ring* $\mathbb{Z}/2\mathbb{Z}$, and that this is isomorphic to the clone of operations of the variety $\mathbf{Bool}^1$.

(ii)   Describe the contravariant adjunction between **Set** and $\mathbf{Bool}^1$ determined by this $\mathbf{Bool}^1$-structure on 2.

As an interesting sideline,

(iii)   Regarding $\mathbf{Bool}^1$ as the variety generated by the 2-element Boolean ring, obtain a cardinality bound for the free Boolean ring on $n$ generators by considerations analogous to those applied to the free group on three generators in $\mathbf{Var}(S_3)$ in the discussion leading up to Exercise 3.3:2. If you did that exercise and Exercise 4.14:1, compare these two cases with respect to how close the resulting bounds are to the actual cardinalities of these free algebras.

More generally,

**Exercise 10.12:4.** For $n$ any integer $> 1$, let $\mathbf{X}^{[n]}$ denote the clone of all finitary operations on the set $n = \{0, \ldots, n-1\}$.

(i)   Show that for $p$ a prime, $\mathbf{X}^{[p]}$ can be described as the clone of derived operations of the ring $\mathbb{Z}/p\mathbb{Z}$. Show moreover that the variety $\mathbf{X}^{[p]}$-**Alg**, regarded as a subvariety of $\mathbf{CommRing}^1$, is equivalent to $\mathbf{Bool}^1$ by the "Boolean ring of idempotent elements" functor (Exercise 4.14:3). Describe the functor going the other way.

(ii)   Show that if $n$ is not a prime, then $\mathbf{X}^{[n]}$-**Alg** does not coincide with the clone of derived operations of the ring $\mathbb{Z}/n\mathbb{Z}$.

(iii) For $n$ not a prime, is it still true that $\mathbf{X}^{[n]}\text{-}\mathbf{Alg}$ is equivalent to $\mathbf{Bool}^1$?

The next exercise looks at a contravariant representable algebra-valued functor on a category $\mathbf{C}$ other than a variety of algebras, which nonetheless has properties similar to those of functors $\mathbf{V}^{\mathrm{op}} \to \mathbf{W}$ as discussed above. It leads you to an important result of lattice theory.

**Exercise 10.12:5.** In the category **POSet** of partially ordered sets and isotone maps, let 2 denote the object with underlying set $\{0, 1\}$, ordered so that $0 < 1$. (It is natural to speak of this as a "structure of partially ordered set" on 2; but beware confusion with Lawvere's technical sense of "structure", i.e., the operations which an object admits, which are the subject of (i) below.)

(i)    Show that the finitary structure on this object of **POSet**, i.e., the clone of all operations $2^n \to 2$ that are morphisms of **POSet**, is a structure of *distributive lattice* (Exercise 6.1:15) with *least* element 0 and *greatest* element 1, regarded as zeroary operations. Describe the resulting functor $\mathbf{POSet}^{\mathrm{op}} \to \mathbf{DistLat}^{0,1}$. (You will need to know the form that products take in **POSet**; for this see Definition 5.1.4.)

(ii)    Verify that **POSet** has small limits, so that Theorem 10.12.1 is applicable to this functor.

(iii)    Show that the adjoint to this functor, a functor $(\mathbf{DistLat}^{0,1})^{\mathrm{op}} \to \mathbf{POSet}$, can be characterized as taking every object of $\mathbf{DistLat}^{0,1}$ to the set of its morphisms into 2, now regarded as a distributive lattice with greatest and least element, and where the partial ordering on 2 is now used to get a partial ordering on the set of morphisms. (Cf. Exercise 7.6:5.)

(iv)    Suppose instead that we consider $2 = \{0, 1\}$ as an object of $\mathbf{POSet}^{0,1}$, the category whose objects are partially ordered sets with least and greatest elements, and whose morphisms are the isotone maps that respect those elements. Show that the structure on 2 in this category leads to a contravariant right adjunction with the variety **DistLat**.

(v)    What if you start with $\mathbf{POSet}^0$ or $\mathbf{POSet}^1$?

It is clear from Lemma 10.10.2 that any contravariant *equivalence* $i\colon \mathbf{C}^{\mathrm{op}} \to \mathbf{V}$, where $\mathbf{C}$ is a category and $\mathbf{V}$ a variety of algebras, will be representable. In such a situation, can $\mathbf{C}$ also be a variety of algebras? This is addressed in the next exercise.

**Exercise 10.12:6.** Let us call a variety "nontrivial" if it does not satisfy the identity $x = y$.

(i)    Show that there can exist no contravariant equivalences between nontrivial varieties.

(Suggestion: find a condition on categories which is invariant under equivalence of categories, and is satisfied by all nontrivial varieties, but is not satisfied by the *opposite* of any nontrivial variety. Essentially, any condition on categories that does not refer to how many isomorphic copies an object has will be invariant under equivalence. What is hard is finding one that distinguishes between varieties and their opposites. I know some ways to do this, but they are not obvious. Perhaps you can find a more natural one. If you wish, take "variety" to mean "finitary variety".)

(ii)  Does your criterion also show that $(\mathbf{POSet})^{\mathrm{op}}$ and $(\mathbf{POSet}^{0,1})^{\mathrm{op}}$ are not equivalent to varieties? If not, can you nonetheless prove this?

Nonetheless, some of the contravariant representable functors considered above come surprisingly close to being equivalences; namely, when restricted to the *finitely generated* objects of one category they yield finitely generated objects of the other, and they *do* give equivalences between these subcategories of finitely generated objects. In the case of duality of vector spaces, this is a category-theoretic translation of some well-known facts of linear algebra. In the cases of Boolean rings (Exercise 10.12:3) and of distributive lattices (Exercise 10.12:5), the results in question are translations of classical structure theorems about these two kinds of object ([4, §III.3]; cf. also Exercise 7.9:16 above). For the variety $\mathbf{Ab}$ the functor $\mathbf{Ab}(-, \mathbb{Q}/\mathbb{Z})$ is a self-duality on the category of finite (though not on the category of finitely generated) abelian groups (see [27, §4.6], noting the comment after [27, Theorem 6.2]).

It turns out, moreover, that the dualities we have described for finite objects can be extended to equivalences between *all* objects of one category and certain *topologized* objects of the other. The reader interested in learning about a large class of such results might look at [40], and at [98], which generalizes the results of the former paper and puts them in category-theoretic language. The result on $\mathbf{Ab}(-, \mathbb{Q}/\mathbb{Z})$ does not fall within the scope of those papers, but it, too, has a generalization to topological abelian groups, the theory of *Pontryagin duality* of locally compact abelian groups, via morphisms into the topological group $\mathbb{R}/\mathbb{Z}$ [130]. The topological approach to duality of not necessarily finite-dimensional vector spaces is implicit in Exercises 6.5:6 and 8.5:18. An interesting book on dualities is [7].

### Exercise 10.12:7.

(i)  Show from Exercise 4.14:5 that our functors connecting $\mathbf{Bool}^1$ and $\mathbf{Set}$ do indeed induce a contravariant equivalence between the subcategories of finite objects.

(ii)  Deduce that if $\mathbf{V}$ is any variety of finitary algebras, and $A$ a finite object of $\mathbf{V}$, then there exists a $\mathbf{V}$-*coalgebra* object $R$ of $\mathbf{Bool}^1$ such that $\mathbf{Bool}^1(R, 2) \cong A$.

If you or the class succeeded in characterizing derived operations of the "majority vote function" $M_3$ on $\{0, 1\}$ in Exercise 2.7:1, you can now try:

### Exercise 10.12:8.

(i)  Can you find some structure (in the nontechnical sense, i.e., not necessarily given by operations!) on $\{0, 1\}$, such that the clone of operations

generated by the majority vote function $M_3$ is precisely the clone of finitary operations respecting that structure?

(ii)   Does one in fact have a duality result, to the effect that the set $\{0,1\}$,
with this structure on the one hand, and with the operation $M_3$ on the
other, induces an adjunction, which, when restricted to finite objects, gives
a contravariant equivalence between finite algebras in the variety generated
by $(\{0,1\}, M_3)$, and finite objects of an appropriate category?

I have not thought hard about the following question:

**Exercise 10.12:9.** Suppose **V** and **W** are varieties, and we have a contravariant equivalence between their subcategories of finite (finitely generated? finitely presented?) objects. Will this necessarily be the restriction
of a pair of mutually right adjoint representable functors between all of **V**
and all of **W**?

What can we say about *composites* involving contravariant representable
functors? We know that for adjoint pairs of *covariant* functors

$$\mathbf{C} \underset{F}{\overset{U}{\rightleftarrows}} \mathbf{D} \underset{G}{\overset{V}{\rightleftarrows}} \mathbf{E},$$

the composites $\mathbf{C} \underset{FG}{\overset{VU}{\rightleftarrows}} \mathbf{E}$ are also adjoint; so let us look at the results we
get on replacing some subset of the three categories **C**, **D**, **E** in this result
by their opposites. This will give eight statements, saying that composites of
certain combinations of covariant adjoint pairs, contravariant right adjoint
pairs, and contravariant left adjoint pairs, are again adjoint pairs of one sort
or another.

These statements will break into pairs of statements which have the same
translations after some relabeling, because Theorem 8.3.10 itself is invariant
under replacing all three categories by their opposites and interchanging the
roles of **C** and **E**. Of the resulting four statements, one is, of course, the
original Theorem 8.3.10. Two of the others involve contravariant *left* adjunctions, of which, as I have mentioned, there are no interesting cases among
varieties of algebras [49]. I state the one remaining case as the next corollary.
In that corollary, for a functor between arbitrary categories, $A \colon \mathbf{C} \to \mathbf{D}$, the
"same" functor regarded as going from $\mathbf{C}^{\mathrm{op}}$ to $\mathbf{D}^{\mathrm{op}}$ is written $A^{\mathrm{op}}$ (though
for most purposes, it is safe to write this $A$).

**Corollary 10.12.3** (to Theorem 8.3.10). *Let* **C**, **D**, **E** *be arbitrary cate-*
*gories* (*not necessarily varieties of algebras*). *Suppose*

$$\mathbf{C}^{\mathrm{op}} \xrightarrow{\ V\ } \mathbf{D}$$
$$\mathbf{C} \xleftarrow[\ V'\ ]{} \mathbf{D}^{\mathrm{op}}$$

*is a pair of mutually right adjoint contravariant functors, and*

$$\mathbf{D} \underset{F}{\overset{U}{\rightleftarrows}} \mathbf{E}$$

*is a pair of covariant adjoint functors, with $U$ the right adjoint and $F$ the left adjoint. Then the composite functors $UV$ and $V'F^{\mathrm{op}}$ (in less discriminating notation, $V'F$) :*

$$
\begin{array}{ccc}
\mathbf{C}^{\mathrm{op}} \xrightarrow{\ V\ } \mathbf{D} \xrightarrow{\ U\ } \mathbf{E} \\
\mathbf{C} \xleftarrow[\ V'\ ]{} \mathbf{D}^{\mathrm{op}} \xleftarrow[\ F^{\mathrm{op}}\ ]{} \mathbf{E}^{\mathrm{op}}
\end{array}
$$

*are also mutually right adjoint contravariant functors.*

*In particular, the class of contravariant functors admitting right adjoints is closed under postcomposition with right adjoint covariant functors, and under precomposition with left adjoint covariant functors.* □

**Exercise 10.12:10.**

(i)   Derive the above result from Theorem 8.3.10, and also derive the two other statements referred to in the paragraph before the corollary which involve contravariant *left* adjunctions.

(ii)   Give a (nontrivial) example of Corollary 10.12.3, verifying directly the adjointness.

**Exercise 10.12:11.** Suppose in the context of the above corollary that $\mathbf{C}$ and $\mathbf{E}$ are both varieties of algebras. Thus the pair of mutually right adjoint functors $UV$ and $V'F^{\mathrm{op}}$ are induced by some object with commuting $\mathbf{C}$- and $\mathbf{E}$-algebra structures. Describe this object and its $\mathbf{C}$- and $\mathbf{E}$-algebra structures in terms of the representing objects $R$ and $S$ for the given functors $V : \mathbf{C}^{\mathrm{op}} \to \mathbf{D}$ and $U : \mathbf{D} \to \mathbf{E}$.

Corollary 10.12.3 does *not* say anything about a composite of two contravariant representable functors. This will be a covariant functor, but as the first part of the next exercise shows, it need not have an adjoint on either side.

**Exercise 10.12:12.**

(i)   Let $K$ be a field, and $V : (K\text{-}\mathbf{Mod})^{\mathrm{op}} \to K\text{-}\mathbf{Mod}$ the contravariant representable functor taking each $K$-vector space to its dual. Show that the composite of $V$ with itself, $VV$, or more accurately, $VV^{\mathrm{op}}$, a covariant functor $K\text{-}\mathbf{Mod} \to K\text{-}\mathbf{Mod}$, has no left or right adjoint.

(ii)   Show by examples that the class of representable contravariant functors between varieties is closed neither under precomposition with right adjoint covariant functors nor under postcomposition with left adjoint covariant functors.

The "double dual" functor of part (i) above *does* belong to a class of functors which have interesting properties, namely, composites of functors (covariant or contravariant) with their own adjoints [2, §§ 4–7].

I have mentioned the principle that the richer the structure of a variety of algebras, the more covariant representable functors it admits, and the fewer contravariant representable functors, and we then looked at contravariant representable functors on the variety with the least algebraic structure, namely **Set**. In the opposite direction, rings have a particularly rich structure; thus, as the next exercise shows, they are quite poor when it comes to contravariant representable functors.

**Exercise 10.12:13.** Let $R$ be a nonzero ring (commutative if you wish).

(i)   Show that if $R$ has no zero divisors, then any finitary operation on $R$ as an object of the category of rings, i.e., any ring homomorphism $R^n \to R$, can be expressed as a composite $a\, p_{i,n}$, where $i \in n$, $p_{i,n} \colon R^n \to R$ is the $i$-th projection map, and $a$ is an endomorphism of $R$, Deduce that any clone of finitary operations on $R$ as an object of **Ring**[1] or of **CommRing**[1] is generated by unary operations.

(ii)   Can you generalize these observations to a wider class of rings than those without zero divisors?

(iii) Choose a simple example of a ring $R$ with zero divisors for which the conclusion of (i) fails, and see whether you can describe the clone of operations on that ring.

## 10.13. More on commuting operations

We have seen that for varieties **V** and **W**, the **V**-algebra objects of **W** correspond to sets given with two families of operations which commute with one another in the sense of (10.12.2). Let us look further at this concept of commuting operations.

**Lemma 10.13.1.** *If $s$ is an operation on a set $A$, then the set of operations on $A$ which commute with $s$ forms a clone.*

*Idea of proof.* If the map $s \colon A^{\mathrm{ari}(s)} \to A$ is a homomorphism for all members of some set $T$ of operations on $A$, it will clearly be a homomorphism for all derived operations of that family.    □

**Exercise 10.13:1.** Give a detailed proof of the above lemma. (Remember that proving a set of operations to be a clone includes proving that it contains the projection maps.)

**Definition 10.13.2.** *If* $s$ *is an operation on a set* $A$, *then the clone of operations on* $A$ *which commute with* $s$ *will be called the* centralizer *of* $s$. *If* $S$ *is a set of operations on* $A$, *the intersection of the centralizers of these operations will be called the* centralizer *of* $S$.

*If* $C$ *is a clone of operations on* $A$ *and* $S$ *a set of operations on* $A$ *(which may or may not be contained in* $C$*), then the intersection of* $C$ *with the centralizer of* $S$ *will be called the* centralizer *of* $S$ *in* $C$. *The centralizer of* $C$ *in* $C$ *will be called the* center *of* $C$. *A clone which is its own center will be called* commutative.

Let us fix a notation for a construction we defined in the preceding section.

**Definition 10.13.3.** *If* $\Omega$ *and* $\Omega'$ *are types, then* $\Omega \sqcup \Omega'$ *will denote the type whose set of operation-symbols is* $|\Omega| \sqcup |\Omega'|$, *and where the arity function on this set is induced in the obvious way by the arity functions of* $\Omega$ *and* $\Omega'$.

*If* $\mathbf{V}$ *and* $\mathbf{W}$ *are varieties of algebras, of types* $\Omega$ *and* $\Omega'$ *respectively, then the variety of algebras of type* $\Omega' \sqcup \Omega$ *such that the operations from* $\Omega$ *satisfy the identities of* $\mathbf{V}$, *the operations from* $\Omega'$ *satisfy identities of* $\mathbf{W}$, *and all* $\mathbf{V}$*-operations commute with all* $\mathbf{W}$*-operations, will be denoted* $\mathbf{V} \bigcirc \mathbf{W}$.

Note that in the above definition, $\mathbf{V}$ and $\mathbf{W}$ are specified as *varieties*, i.e., in terms of given primitive operations. However, even if we are not interested in distinguishing "primitive" from "derived" operations, e.g., if we are interested in varieties as categories of representations of given clonal categories, the above construction " $\bigcirc$ " also induces a construction on these, since by Lemma 10.13.1, the property that two sets of primitive operations centralize one another is equivalent to the property that their clones of derived operations centralize one another. Finally, if we are interested in varieties only up to equivalence as categories, without reference to concretization (e.g., if we are not interested in distinguishing the varieties $K$-**Mod** and $M_n(K)$-**Mod**), then $\mathbf{V} \bigcirc \mathbf{W}$ is also determined up to equivalence on these, namely, as the category of contravariant right adjunctions between $\mathbf{V}$ and $\mathbf{W}$. (To describe the morphisms of this category, consider how Corollary 8.11.6 should be adapted to contravariant right adjunctions.)

Freyd introduces essentially the construction we have called $\mathbf{V} \bigcirc \mathbf{W}$ in [12, pp. 93–95], but rather than naming the resulting variety, he names its clonal theory $T_1 \otimes T_2$, where $T_1$ and $T_2$ are the clonal theories of the given varieties. But here I have chosen to minimize the dependence of this chapter on the view of a variety as the category of representations of a clonal theory.

**Exercise 10.13:2.** Given varieties $\mathbf{V}$ and $\mathbf{W}$, an object $A$ of $\mathbf{V}$, and an object $B$ of $\mathbf{W}$, it is easy to construct by generators and relations an object $C$ of $\mathbf{V} \bigcirc \mathbf{W}$ having a map $m: |A| \times |B| \to |C|$ such that for each $y \in |B|$, the map $m(-, y): |A| \to |C|$ is a homomorphism with respect to $\mathbf{V}$-structures, and for each $x \in |A|$, the map $m(x, -): |B| \to |C|$ is a homomorphism with respect to $\mathbf{W}$-structures.

Characterize the contravariant representable functors between $\mathbf{V}$ and $\mathbf{W}$ determined by this object $C$, in terms of one or more universal properties.

In the case of *covariant* representable functors, we saw in §10.10 that certain *differences* between two varieties $\mathbf{V}$ and $\mathbf{W}$ regarding the number of derived zeroary operations led to restrictions on representable functors between these varieties. For contravariant functors, on the other hand, it is when both varieties *have* such operations that one gets a restriction:

**Lemma 10.13.4** ([12, p. 94]). *Suppose $\mathbf{V}$ and $\mathbf{W}$ are varieties of algebras, each having at least one zeroary operation. Then $\mathbf{V} \bigcirc \mathbf{W}$ satisfies identities saying that every derived zeroary operation of $\mathbf{V}$ and every derived zeroary operation of $\mathbf{W}$ fall together. The resulting derived zeroary operation of $\mathbf{V} \bigcirc \mathbf{W}$ defines a one-element subalgebra of every $\mathbf{V} \bigcirc \mathbf{W}$-object.*

*Proof.* The fact that each derived zeroary operation coming from $\mathbf{V}$ commutes with each derived zeroary operation coming from $\mathbf{W}$ means that each of the former is equal to each of the latter (Exercise 10.12:1(i)). Hence, as both families are nonempty, all of these derived zeroary operations are equal. Since zeroary operations from $\mathbf{V}$ commute with arbitrary operations from $\mathbf{W}$ and vice versa, the resulting zeroary operation of $\mathbf{V} \bigcirc \mathbf{W}$ is central. It is easy to verify that this means that it defines a one-element subalgebra of every algebra, equivalently, is the *unique* derived zeroary operation of $\mathbf{V} \bigcirc \mathbf{W}$. $\square$

**Exercise 10.13:3.** Deduce from the above lemma that if $\mathbf{V}$ is a variety having at least one zeroary operation, then the variety $\mathbf{V} \bigcirc \mathbf{Ring}^1$ is trivial; equivalently, that there is no nontrivial contravariant representable functor $\mathbf{V}^{\mathrm{op}} \to \mathbf{Ring}^1$ or $(\mathbf{Ring}^1)^{\mathrm{op}} \to \mathbf{V}$. (So, for instance, there is no nontrivial contravariant representable functor $(\mathbf{Ring}^1)^{\mathrm{op}} \to \mathbf{Ring}^1$.)

**Exercise 10.13:4.** Show that if $\mathbf{V}$ is any variety of algebras, then $\mathbf{V}^{\mathrm{pt}} \cong \mathbf{V} \bigcirc \mathbf{Set}^{\mathrm{pt}}$. (Cf. Definition 10.10.1.)

The next result shows a similar phenomenon for binary operations with neutral element.

**Lemma 10.13.5** ([12, p. 94]). *Suppose $\mathbf{V}$ and $\mathbf{W}$ are varieties of algebras, each having at least one (not necessarily associative) derived binary operation with a neutral zeroary operation. Then in $\mathbf{V} \bigcirc \mathbf{W}$, the operations induced by all such binary operations of $\mathbf{V}$ and all such binary operations of $\mathbf{W}$ fall together, and give the unique binary operation with neutral element in this clone. The resulting binary operation and neutral element constitute a structure of abelian monoid, which is central in the clone of operations of $\mathbf{V} \bigcirc \mathbf{W}$.*

*Proof.* We shall show that if in any variety a binary operation $*$ with a neutral element and a binary operation $\circ$ with a neutral element commute, and their neutral elements likewise commute, then $* = \circ$, and their common value satisfies the commutative and associative identities. The remaining assertions follow as in the proof of the preceding lemma.

The neutral elements of $*$ and $\circ$, being commuting zeroary operations, are equal; let us write $e$ for their common value. We now write down several cases of the commutativity of $*$ with $\circ$. One such identity, $(x * e) \circ (e * y) = (x \circ e) * (e \circ y)$, immediately reduces to $x \circ y = x * y$, proving equality of the two operations. On the other hand, $(e * x) \circ (y * e) = (e \circ y) * (x \circ e)$ reduces to $x \circ y = y * x$, so the common value of $*$ and $\circ$ is abelian. Finally, $(x * y) \circ (e * z) = (x \circ e) * (y \circ z)$ yields associativity.                    $\square$

The above result fails without the assumption that *both* binary operations have a neutral element. For example, the variety **Set** has the binary "derived operation" $p_{2,0}$ (projection of an ordered pair on its first component); but it is easy to see that for every variety **V**, one has $\mathbf{V} \bigcirc \mathbf{Set} \cong \mathbf{V}$; so a binary operation of **V** with neutral element is not forced in $\mathbf{V} \bigcirc \mathbf{Set}$ to become associative, or become commutative, or to fall together with $p_{2,0}$.

Recall that we denote the variety of algebras with a single binary operation with neutral element by **Binar**$^e$. Below, when the prefix "**Ab**" is added to the name of a variety whose type involves one binary operation, the result will denote the subvariety determined by the additional identity of commutativity for that operation.

**Corollary 10.13.6.** *If each of* **V** *and* **W** *is one of* **Binar**$^e$, **Monoid**, **AbBinar**$^e$ *or* **AbMonoid**, *then* $\mathbf{V} \bigcirc \mathbf{W} \cong \mathbf{AbMonoid}$.

*Proof.* Applying the preceding lemma, we see that the given zeroary and binary operations of **V** and **W** fall together in $\mathbf{V} \bigcirc \mathbf{W}$ to give a single zeroary and a single binary operation that generate the clone of operations of $\mathbf{V} \bigcirc \mathbf{W}$ and satisfy the identities of **AbMonoid**. To show that $\mathbf{V} \bigcirc \mathbf{W}$ satisfies no other identities, it suffices to note that the multiplication and neutral element of **AbMonoid** satisfy all the identities of **V** and of **W** (clear in each case), and commute with themselves and one another (a quick calculation).                    $\square$

The above corollary shows that the representing object for any *contravariant representable functor* between any two of the varieties listed is essentially an abelian monoid.

The next result concerns the case where our abelian monoid structures turn out to give abelian group structures. If a binary derived operation $*$ of a variety has a neutral element $e$, then a *left inverse operation* (respectively a *right* inverse operation) for $*$ will mean a unary operation $\iota$ satisfying the identity $\iota(x) * x = e$ (respectively $x * \iota(x) = e$).

**Theorem 10.13.7** (cf. [12, p. 95]). *Suppose* **V** *and* **W** *are varieties of algebras, each having at least one binary operation with a neutral element, and such that at least one such operation of* **V** *or of* **W** *has a right or left inverse operation* $\iota$. *Then in* **V** ○ **W**, $\iota$ *becomes a two-sided inverse operation for the unique* **AbMonoid** *operation of this variety, making this an* **Ab** *structure, again central in the clone of operations.*

*Moreover, any clone of finitary operations admitting a homomorphism of the clone of operations of* **Ab** *into its center is, up to isomorphism, the clone of operations of a variety* $K$*-Mod, where* $K$ *is the set of unary operations of the clone, made a ring in a natural way.*    □

**Exercise 10.13:5.** With the help of previous results,

(i)   prove the first paragraph of the above theorem,

(ii)   prove the second paragraph of the above theorem.

Where the above result characterizes clones with a central image of **Ab**, Freyd [12, p. 95] gives the analogous characterization of clones with a central image of **AbMonoid**, with "half-ring" in place of ring. (The term "half-ring" is not standard. He presumably means an abelian monoid given with a bilinear multiplication having a neutral element 1. The more common term would be "semiring with 0 and 1". A module over such a semiring $K$ means an abelian monoid $R$ with a 0- and 1-respecting homomorphism of $K$ into its semiring of endomorphisms.)

**Exercise 10.13:6.**

(i)   Deduce from Theorem 10.13.7 that **Group** ○ **Group** ≅ **Ab**. Translate this result into a description of all representable functors **Group**$^{\mathrm{op}}$ → **Group**.

(ii)   Your proof of (i) should also show that **Ab** ○ **Ab** ≅ **Ab**. Thus, every abelian group yields a contravariant right adjunction between **Ab** and **Ab**. Describe the functors involved, and express the universal property of the adjunction as a certain bijection of hom-sets.

**Exercise 10.13:7.**

(i)   If $K$, $L$ are rings, describe $(K\text{-}\mathbf{Mod}) \bigcirc (L\text{-}\mathbf{Mod})$, and determine the general form of a representable functor $(K\text{-}\mathbf{Mod})^{\mathrm{op}} \to L\text{-}\mathbf{Mod}$.

(ii)   Bring the above result into conformity with (10.8.19) by turning it into a characterization of representable functors $(K\text{-}\mathbf{Mod})^{\mathrm{op}} \to \mathbf{Mod}\text{-}L$. Write the associated contravariant right adjunctions as functorial isomorphisms of hom-sets.

(iii)  If $K$ is any ring, the natural $(K, K)$-bimodule structure of $|K|$ induces, via the result of (ii), a functor $(K\text{-}\mathbf{Mod})^{\mathrm{op}} \to \mathbf{Mod}\text{-}K$. Describe this functor, and show that in the case where $K$ is a field, it is ordinary "duality of vector spaces".

(iv) Given any pair of contravariant mutually right adjoint functors among categories, $U : \mathbf{C}^{\mathrm{op}} \to \mathbf{D}$, $V : \mathbf{D}^{\mathrm{op}} \to \mathbf{C}$, one has universal maps $\mathrm{Id}_{\mathbf{C}} \to VU$, $\mathrm{Id}_{\mathbf{D}} \to UV$. Determine these in case (iii) above.

Here is an important way of getting sets with two mutually commuting algebra structures.

**Lemma 10.13.8.** *Let* $\mathbf{V}$ *and* $\mathbf{W}$ *be varieties of algebras in which all operations have arities less than some regular cardinal* $\gamma$, *let* $\mathbf{C}$ *be any category having* $<\gamma$-*fold products and* $<\gamma$-*fold coproducts, and let* $R$ *and* $S$ *be a* $\mathbf{V}$-*coalgebra object and a* $\mathbf{W}$-*algebra object of* $\mathbf{C}$ *respectively. Then* $\mathbf{C}(|R|, |S|)$ *has a natural structure of* $\mathbf{V} \bigcirc \mathbf{W}$-*algebra (which we may denote* $\mathbf{C}(R, S)$). $\qquad\qquad\square$

**Exercise 10.13:8.**

(i)   Prove the above lemma.

(ii)   If you are familiar with basic algebraic topology, deduce from that lemma and Theorem 10.13.7 and Exercise 10.3:1(ii) that the fundamental group of any topological group is abelian.

(You will need to verify that a topological group induces a group object of $\mathbf{HtpTop}^{(\mathrm{pt})}$. A key step is to verify that the forgetful functor $\mathbf{Top}^{\mathrm{pt}} \to \mathbf{HtpTop}^{(\mathrm{pt})}$ respects products.)

In fact, the method of part (ii) above shows that all $\mathbf{Binar}^e$-objects of $\mathbf{HtpTop}^{(\mathrm{pt})}$ (called "$H$-spaces" by topologists) have abelian fundamental group. For a brute force proof see [90, Proposition II.11.4, p. 81].

**Exercise 10.13:9.** Describe $\mathbf{Heap} \bigcirc \mathbf{Heap}$. (Hint: If $A$ is a nonempty object of $\mathbf{Heap} \bigcirc \mathbf{Heap}$, show that any choice of a zeroary operation allows one to regard $A$ as an object of $\mathbf{Group} \bigcirc \mathbf{Group}$.)

If possible, generalize your result; i.e., show that conditions weaker than the heap identities are enough to force two commuting ternary operations on a set to coincide, and to satisfy the identities you established for $\mathbf{Heap} \bigcirc \mathbf{Heap}$.

**Exercise 10.13:10.** Recall that $\mathbf{Semilattice}$ denotes the variety of sets with a single idempotent commutative associative binary operation.

(i)   Show that in $\mathbf{Semilattice} \bigcirc \mathbf{Semilattice}$, the two binary operations fall together (even though $\mathbf{Semilattice}$ has no 0-ary operations).

(ii)   Deduce that $\mathbf{Semilattice} \bigcirc \mathbf{Lattice}$ and $\mathbf{Lattice} \bigcirc \mathbf{Lattice}$ are trivial.

(iii)   Show that $\mathbf{Semilattice} \bigcirc \mathbf{AbMonoid} \cong \mathbf{Semilattice}^0$, the variety of semilattices with neutral element. (In writing that neutral element as "$0$", I am arbitrarily interpreting the semilattice operation as "join".)

(iv)   Again, can you get similar results using a smaller set of identities than the full identities of $\mathbf{Semilattice}$ and/or $\mathbf{AbMonoid}$?

In this section we have seen several parallel results; let us put in abstract form what they involve.

**Exercise 10.13:11.** Let **CommClone** denote the full subcategory of **Clone** consisting of all *commutative* clonal categories (Definition 10.13.2).

(i)    Show that for any variety **V**, the following conditions are equivalent:

(a) The two underlying-set-preserving functors $\mathbf{V} \circ \mathbf{V} \to \mathbf{V}$, which act by forgetting the one or the other of these families of **V**-operations, are equivalences.

(b) The clone of operations of **V** is commutative, and is an epimorph of the initial object in **CommClone** (i.e., the morphism from the initial object to that object is an epimorphism).

(ii)    Show that if the two functors of (a) above are equal, then the equivalent conditions (a) and (b) hold.

In connection with condition (b) above, we note

**Exercise 10.13:12.** Show that the initial objects of **CommClone** and **Clone** are the same, but that this initial object has no proper epimorphs in **Clone** other than the theory of the trivial variety, but does have nontrivial proper epimorphs in **CommClone**.

(This contrasts with the result proved for rings in Exercise 7.7:9(ii).)

Let us call a variety "$\circ$-idempotent" if it satisfies the equivalent conditions of Exercise 10.13:11(i). It would be interesting to see whether one can determine all such varieties. The epimorphs of the clone of operations of **Ab** in **CommClone** are the clones of operations of the varieties $K$-**Mod** for all epimorphs $K$ of $\mathbb{Z}$ in **CommRing**[1] (cf. Exercise 7.7:8(i)). For a nice classification of these rings $K$, of which there are uncountably many, see [67]. More generally, if $K$ is a semiring with 0 and 1 (cf. paragraph following Exercise 10.13:5) which is an epimorph of the semiring $\mathbb{N}$ of natural numbers in the category of such semirings, then the clonal theory of the variety of $K$-modules is an epimorph of the clonal theory of **AbMonoid**. (This class of clonal theories includes those arising from epimorphs of $\mathbb{Z}$, since $\mathbb{Z}$ is an epimorph of $\mathbb{N}$ in the semiring category.)

In most of the results in this section that yielded $\circ$-idempotent varieties **V**, we also found larger classes of varieties, necessarily noncommutative, whose $\circ$-products with themselves and each other gave **V**. I don't know what is going on here; the phenomenon is described in

**Exercise 10.13:13.** If **V** is a variety of algebras, let $\mathbf{V}^{\mathrm{ab}}$ denote the subvariety obtained by imposing on **V** the identities making all operations of **V** commute, and $A(\mathbf{V}) \colon \mathbf{V}^{\mathrm{ab}} \to \mathbf{V}$ the inclusion functor. For each positive integer $n$, let $\mathbf{V}^{\circ n}$ denote the variety $\mathbf{V} \circ \ldots \circ \mathbf{V}$ with $n$ "**V**"s, and $Q(\mathbf{V}, n, i)$ $(i = 0, \ldots, n - 1)$ be the natural $n$-tuple of forgetful functors $\mathbf{V}^{\circ n} \to \mathbf{V}$.

Show that for any variety $\mathbf{V}$ and integer $n > 1$ the following conditions are equivalent: (a) $Q(\mathbf{V}, n, 0) = Q(\mathbf{V}, n, 1)$. (b) $\mathbf{V}^{\mathrm{ab}}$ and $\mathbf{V}^{\circ n}$ are isomorphic, by a functor making a commuting triangle with the functors $A(\mathbf{V})$ and $Q(\mathbf{V}, n, 0)$.

Show that when these conditions hold, $\mathbf{V}^{\mathrm{ab}}$ is $\circ$-idempotent.

A question I also don't know the answer to is

**Exercise 10.13:14.** Let $\mathbf{V}$ be a variety of algebras.

(i)   If $\mathbf{V}^{\circ 3}$ has commutative clone of operations, must the clone of operations of $\mathbf{V}^{\circ 2}$ also be commutative?

(ii)   If $\mathbf{V}$ satisfies the equivalent conditions of Exercise 10.13:13 for $n = 3$, must it also satisfy those conditions for $n = 2$?

On an easier note, recall from Exercise 7.9:5 that the *monoid of endomorphisms* of the identity functor of any category is commutative. This generalizes to

**Lemma 10.13.9.** *If* $\mathbf{C}$ *is a category with finite products, then the clone of operations of the identity functor of* $\mathbf{C}$ *is commutative.*    $\square$

**Exercise 10.13:15.** Prove the above lemma, and characterize the clone in question in the case where $\mathbf{C}$ is a variety.

We have noted that examples of contravariant representable functors between varieties are limited because of the difficulty of getting interesting operations to commute with one another. But it is much easier to get operations that respect interesting *relational* structure. (A well-known example of this was sketched in Exercise 10.12:5.) Thus, the natural direction in which to pursue the study of contravariant adjunctions is that of categories defined in terms of a combination of operations and relations. But that is outside the scope of this course.

## 10.14. Some further reading on representable functors, and on General Algebra

Covariant representable functors among particular varieties of algebras are studied extensively in [3]. Indeed, §§ 10.1–10.5 above were adapted from the introductory sections of [3], and §§ 10.6 and 10.7 from a couple of later sections. Most of [3] deals with representable functors on varieties of associative and commutative rings; for the former case, the representable functors to many varieties are precisely determined. Thus, [3] may be considered a natural sequel to this chapter. Many open questions are also noted there. (The notation, language, and viewpoint of [3] are close to those of these notes. One difference is that where I here use the word "monoid", in that work my coauthor and I wrote "semigroup with neutral element", and called the variety of those objects **Semigp**$^e$. Also, the references in *that* work to *these* notes refer to an earlier version.)

Several other texts in General Algebra, most of which include some major topics not covered in these notes, were listed in the first paragraph of §1.6. A classical topic I have not touched on which is particularly striking and useful is that of ultraproducts and ultrapowers, and can be found in most General Algebra texts.

An active area of research since the late 1980's has been the study of lattices of congruences of algebras, and their relationship to the general properties of varieties of algebras [10, 16]. More recently, there has been considerable work on a class of algorithmic questions called *constraint satisfaction problems,* in connection with the structures of finite algebras [43, 92].

We have made use in these notes of partially ordered sets, and of equivalence relations (and discussed briefly their common generalization, preordered sets); but as noted at the end of the last section, we have not developed a general theory of structures consisting of a set and one or more *relations* of specified arities, alongside operations of various arities. Many results parallel to those considered here are applicable to such structures; but, like manysorted algebras, we have left them out for the sake of simplicity.

# References

Numbers in angle brackets at the end of each listing show pages of these notes on which the work is referred to. "MR" refers to the review of the work in *Mathematical Reviews*, readable online at http://www.ams.org/mathscinet/.

## Works related to major topics of this course

1. Clifford Bergman, *Universal algebra. Fundamentals and selected topics*, Pure and Applied Mathematics, 301. CRC Press, 2012. MR2839398 ⟨5, 6⟩

2. George M. Bergman, *Some category-theoretic ideas in algebra,* pp. 285–296 of v.I of the *Proceedings of the 1974 International Congress of Mathematicians (Vancouver)*, Canadian Mathematical Congress, 1975. MR **58** #22222. ⟨522⟩

3. George M. Bergman and Adam O. Hausknecht, *Cogroups and Co-rings in Categories of Associative Rings*, AMS Math. Surveys and Monographs, v.45, 1996. MR **97k** :16001. ⟨209, 494, 495, 529⟩

4. Garrett Birkhoff, *Lattice Theory*, third edition, AMS Colloq. Publications, v. XXV, 1967. MR **37** #2638. ⟨6, 43, 105, 171, 178, 519⟩

5. Andreas Blass, *The interaction between category theory and set theory*, pp. 5–29 in *Mathematical applications of category theory*, 1983. MR **85m** :03045. ⟨237⟩

6. Stanley Burris and H. P. Sankappanavar, *A Course in Universal Algebra*, Springer GTM, v.78, 1981. MR **83k** :08001. ⟨5, 6, 448⟩

7. David M. Clark and Brian A. Davey, *Natural Dualities for the Working Algebraist*, Cambridge University Press, 1998. MR **2000d** :18001. ⟨519⟩

8. P. M. Cohn, *Universal Algebra,* second edition, Reidel, 1981. MR **82j** :08001. ⟨5, 6, 7, 86, 105⟩

9. Samuel Eilenberg and Saunders Mac Lane, *General theory of natural equivalences,* Trans. AMS **58** (1945) 231–294. MR **7** 109d. ⟨6, 280⟩

10. Ralph Freese and Ralph McKenzie, *Commutator theory for congruence modular varieties*, London Math. Soc. Lecture Note Series, v.125, 1987, 227 pp. MR **89c** :08006. ⟨530⟩

11. Peter Freyd, *Abelian Categories,* Harper and Row, 1964. Out of print, but accessible online, with a new Foreword, at http://www.tac.mta.ca/tac/reprints/articles/3/tr3.pdf . MR **29** #3517. ⟨6, 292, 367⟩

12. Peter Freyd, *Algebra valued functors in general and tensor products in particular,* Colloquium Mathematicum (Wrocław) **14** (1966) 89–106. MR **33** #4116. ⟨6, 7, 471, 503, 523, 524, 526⟩

13. George Grätzer, *Universal Algebra,* 2nd ed., Springer, 1979. MR **40** #1320, **80g** : 08001. ⟨5, 6, 448⟩

14. George Grätzer, *Lattice theory: foundation,* Birkhäuser/Springer, 2011. MR **2012f** : 06001. ⟨6, 105⟩

15. Paul Halmos, *Naive Set Theory,* Van Nostrand University Series in Undergraduate Mathematics, 1960; Springer Undergraduate Texts in Mathematics, 1974. MR **22** # 5575, MR **56** #11794. ⟨143, 144⟩

16. David Hobby and Ralph McKenzie, *The structure of finite algebras*, Contemp. Math., **76**, 1988, 203 pp. MR **89m** :08001. ⟨530⟩

17. F. William Lawvere, *Functorial Semantics of Algebraic Theories,* doctoral thesis, Columbia University, 1963. (Summarized without proofs, under the same title, in Proc. Nat. Acad. Sci. U. S. A., **50** (1963) 869–872. MR **28** #2143.) ⟨439, 454⟩

© Springer International Publishing Switzerland 2015

G.M. Bergman, *An Invitation to General Algebra and Universal Constructions*, Universitext, DOI 10.1007/978-3-319-11478-1

18. F. William Lawvere, *The category of categories as a foundation for mathematics*, pp. 1–20 in *Proc. Conf. Categorical Algebra (La Jolla, Calif., 1965)* ed. S. Eilenberg et al. Springer-Verlag, 1966. MR **34** #7332. (Note corrections to this paper in the MR review.) $\langle 237 \rangle$

19. Carl E. Linderholm, *Mathematics Made Difficult,* World Publishing, N.Y., 1972. (Out of print.) MR **58** #26623. $\langle 6 \rangle$

20. Saunders Mac Lane, *Categories for the Working Mathematician,* Springer GTM, v.5, 1971. MR **50** #7275. $\langle 6, 7, 233, 258, 275, 291, 293, 336, 345, 363, 367, 373, 374, 407 \rangle$

21. Ralph McKenzie, George McNulty and Walter Taylor, *Algebras, Lattices, Varieties, volume 1,* Wadsworth and Brooks/Cole, 1987. MR **88e** :08001. $\langle 5, 6, 384 \rangle$

22. J. Donald Monk, *Introduction to Set Theory,* McGraw-Hill, 1969. MR **44** #3877. $\langle 143 \rangle$

23. Richard S. Pierce, *Introduction to the Theory of Abstract Algebras,* Holt Rinehart and Winston, Athena Series, 1968. MR **37** #2655. $\langle 5, 6 \rangle$

24. Robert L. Vaught, *Set Theory, an Introduction,* Birkhäuser, 1985; second edition 1995. MR **95k** :03001. $\langle 143, 144, 151, 153 \rangle$

## General references in algebra

25. Nicolas Bourbaki, *Éléments de Mathématique. Algèbre Commutative, Ch. 3–4,* Hermann, Paris, 1961. MR **30** #2027. $\langle 321 \rangle$

26. P. M. Cohn, *Algebra,* second edition, v. 1, Wiley & Sons, 1982. (first edition, MR **50** #12496) MR **83e** :00002. $\langle 6, 214 \rangle$

27. P. M. Cohn, *Algebra,* second edition, v. 2, Wiley & Sons, 1989. (first edition, MR **58** #26625) MR **91b** :00001. $\langle 6, 319, 321, 519 \rangle$

28. P. M. Cohn, *Algebra,* second edition, v. 3 Wiley & Sons, 1991. MR **92c** :00001. $\langle 6 \rangle$

29. David S. Dummit and Richard M. Foote, *Abstract Algebra,* Prentice-Hall, 1991. MR **92k** :00007. $\langle 6, 49, 214 \rangle$

30. Marshall Hall, Jr., *The Theory of Groups,* MacMillan, 1959; Chelsea, 1976. MR **21** #1996, MR **54** #2765. $\langle 61, 214 \rangle$

31. Israel N. Herstein, *Noncommutative Rings,* AMS Carus Math. Monographs, No. 15, 1968. MR **37** #2790. $\langle 60 \rangle$

32. Thomas W. Hungerford, *Algebra,* Springer GTM, v. 73, 1974. MR **50** #6693. $\langle 5, 6, 22, 49, 140, 158, 207, 214 \rangle$

33. D. L. Johnson, *Presentations of Groups,* London Math. Soc. Student Texts, vol.15, Cambridge University Press, 1990, second edition, 1997. MR **91h** :20001. $\langle 55 \rangle$

34. Serge Lang, *Algebra,* Addison-Wesley, third edition, 1993. Reprinted as Springer GTM v. 211, 2002. MR **2003e** :00003. $\langle 5, 6, 22, 49, 87, 93, 97, 140, 158, 164, 207, 251, 296, 319 \rangle$

35. Joseph J. Rotman, *An Introduction to the Theory of Groups,* 4th edition, Springer GTM, v. 148, 1995. MR **95m** :20001. $\langle 55, 214 \rangle$

## Other works cited

36. S. A. Adeleke, A. M. W. Glass and L. Morley, *Arithmetic permutations,* J. London Math. Soc. (2) **43** (1991) 255–268. MR **92h** :20041. $\langle 414 \rangle$

37. S. I. Adyan, *Burnside's Problem and Identities in Groups* (Russian), Nauka, 1975. MR **55** #5753. $\langle 60, 61 \rangle$

38. S. I. Adyan, *The Burnside problem and related questions* (Russian), Uspekhi Mat. Nauk **65** (2010) 5–60; translation in Russian Math. Surveys **65** (2010) 805–855. MR **2011m** :20086. $\langle 61 \rangle$

39. A. S. Amitsur and J. Levitzki, *Minimal identities for algebras,* Proc. AMS **1** (1950) 449–463. MR **12**, 155d. $\langle 423 \rangle$

40. Richard F. Arens and Irving Kaplansky, *Topological representations of algebras,* Trans. AMS **63** (1948) 457–481. MR **10**, 7c. $\langle 519 \rangle$

41. Reinhold Baer, *Zur Einführung des Scharbegriffs*, J. reine und angew. Math. **160** (1929) 199–207. ⟨424⟩

42. Joan Bagaria, Carles Casacuberta, A. R. D. Mathias, and Jiří Rosický, *Definable orthogonality classes in accessible categories are small*, preprint, Jan. 2013, 38 pp., http://arxiv.org/abs/1101.2792. ⟨414⟩

43. Libor Barto and Marcin Kozik, *Constraint satisfaction problems solvable by local consistency methods*, J. ACM **61** (2014) no.1, Art. 3, 19 pp. MR3167919. ⟨530⟩

44. George M. Bergman, *Centralizers in free associative algebras*, Trans. AMS **137** (1969) 327–344. MR **38** #4506. ⟨92⟩

45. George M. Bergman, *Boolean rings of projection maps*, J. London Math. Soc. (2) **4** (1972) 593–598. MR **47** #93. ⟨58, 168⟩

46. George M. Bergman, *The diamond lemma for ring theory*, Advances in Math. **29** (1978) 178–218. MR **81b**:16001. ⟨43, 312, 427⟩

47. George M. Bergman, *Modules over coproducts of rings*, Trans. AMS **200** (1979) 1–32. MR **50** #9970. ⟨100⟩

48. George M. Bergman, *Everybody knows what a Hopf algebra is*, pp. 25–48 in *Group Actions on Rings* (Proceedings of a conference on group actions, Bowdoin College, Bowdoin, Maine, July 18–24, 1984) ed. S. Montgomery, Contemporary Mathematics, **43** (1985). MR **87e**:16024. ⟨473⟩

49. George M. Bergman, *On the scarcity of contravariant left adjunctions*, Algebra Universalis **24** (1987) 169–185. MR **88k**:18003. ⟨376, 410, 520⟩

50. George M. Bergman, *Supports of derivations, free factorizations, and ranks of fixed subgroups in free groups*, Trans. AMS **351** (1999) 1531–1550. MR **99f**:20036. ⟨227⟩

51. George M. Bergman, *Constructing division rings as module-theoretic direct limits*, Trans. AMS. **354** (2002) 2079–2114. MR **2003b**:16017. ⟨204⟩

52. George M. Bergman, *Direct limits and fixed point sets*, J.Alg. **292** (2005) 592–614. MR **2006k**:08017. ⟨358⟩

53. George M. Bergman, *Colimits of representable algebra-valued functors*, Theory and Applications of Categories, **20** (2008) 334–404. MR **2009c**:18012. ⟨513⟩

54. George M. Bergman, *An inner automorphism is only an inner automorphism, but an inner endomorphism can be something strange*, Publicacions Matemàtiques, **56** (2012) 91–126. http://dx.doi.org/10.5565/PUBLMAT_56112_04. MR2918185. ⟨513⟩

55. George M. Bergman, *Some notes on sets, logic and mathematical language*, supplementary course notes, 12 pp., accessible in several versions from http://math.berkeley.edu/~gbergman/ug.hndts/#sets_etc. ⟨211⟩

56. George M. Bergman, *Notes on composition of maps*, supplementary course notes, 6 pp., at http://math.berkeley.edu/~gbergman/grad.hndts/left+right.ps. ⟨228, 248, 499⟩

57. George M. Bergman, *Infinite Galois theory, Stone spaces, and profinite groups*, supplementary course notes, 12 pp., at http://math.berkeley.edu/~gbergman/grad.hndts/infGal+profin.ps. ⟨332⟩

58. George M. Bergman, *Tensor algebras, exterior algebras, and symmetric algebras*, supplementary course notes, 10 pp., at http://math.berkeley.edu/~gbergman/grad.hndts/OX+ext+sym.ps. ⟨93⟩

59. George M. Bergman, *Why do we have three sorts of adjunctions?* unpublished note, 3 pp., 2008, at http://math.berkeley.edu/~gbergman/papers/unpub/3adjs.pdf. ⟨377⟩

60. George M. Bergman and W. Edwin Clark, *The automorphism class group of the category of rings*, J. Alg. **24** (1973) 80–99. MR **47** #210. ⟨509⟩

61. George M. Bergman and P. M. Cohn, *Symmetric elements in free powers of rings*, J. London Math. Soc. (2) **1** (1969) 525–534. MR **40** #4301. ⟨140⟩

62. William Blake, *The Marriage of Heaven and Hell*, 1825. ⟨317⟩

63. G. R. Brightwell, S. Felsner and W. T. Trotter, *Balancing pairs and the cross product conjecture*, Order **12** (1995), 327–349. MR **96k**:06004. ⟨127⟩

64. R. Brown, D. L. Johnson and E. F. Robertson, *Some computations on nonabelian tensor products of groups,* J. Alg. **111** (1987) 177–202. MR **88m** :20071. ⟨77⟩

65. W. Burnside, *On an unsettled question in the theory of discontinuous groups,* Quarterly J. Pure and Applied Math., **33** (1902) 230–238. ⟨60⟩

66. W. Burnside, *Theory of Groups of Finite Order,* 2nd edition, 1911. ⟨60⟩

67. T. Cheatham and E. Enochs, *The epimorphic images of a Dedekind domain,* Proc. AMS **35** (1972) 37–42. MR **46** #1784. ⟨528⟩

68. A. H. Clifford and G. B. Preston, *The algebraic theory of semigroups, v. I.,* Math. Surveys No. 7, AMS 1961, xv+224 pp.. MR **24** #A2627. ⟨386⟩

69. Stephen D. Cohen, *The group of translations and positive rational powers is free,* Quart. J. Math. Oxford (2) **46** (1995) 21–93. MR **96e** :20033. ⟨414⟩

70. Stephen D. Cohen and A. M. W. Glass, *Free groups from fields,* J. London Math. Soc. (2) **55** (1997) 309–319. MR **98c** :12003. ⟨414⟩

71. P. M. Cohn, *A remark on the Birkhoff-Witt theorem,* J. London Math. Soc., **38** (1963) 197–203. MR **26**#6233. ⟨427⟩

72. P. M. Cohn, *Free Rings and their Relations,* second ed., London Math. Soc. Monographs v.19, Academic Press, 1985. MR **87e** :16006. (First ed., MR **51** #8155. The next item constitutes the first half of a planned third edition.) ⟨92⟩

73. P. M. Cohn, *Free ideal rings and localization in general rings.* New Mathematical Monographs, 3., Cambridge University Press, 2006. MR **2007k** :16020. ⟨92⟩

74. Thomas Delzant and Misha Gromov, *Courbure mésoscopique et théorie de la toute petite simplification,* J. Topol. **1** (2008) 804–836. MR **2011b** :20093. ⟨61⟩

75. A. Dundes, *Interpreting Folklore,* Indiana University Press, 1980. ⟨203⟩

76. S. Peter Farbman, *Non-free two-generator subgroups of* $SL_2(\mathbb{Q})$, Publicacions Matemàtiques (Univ. Autònoma, Barcelona) **39** (1995) 379–391. MR **96k** :20090. ⟨44⟩

77. Solomon Feferman, *Set-theoretical foundations of category theory,* pp. 201–247 in *Reports of the Midwest Category Seminar,* Springer LNM, v.106, 1969. MR **40** #2727. ⟨237⟩

78. Pierre Gabriel, *Des catégories abéliennes,* Bull. Soc. Math. France **90** (1962) 323–448. MR **38** #1144. ⟨233⟩

79. Haim Gaifman, *Infinite Boolean polynomials. I,* Fundamenta Mathematica **54** (1964) 229–250. MR **29** #5765. ⟨367⟩

80. Leonard Gillman and Meyer Jerison, *Rings of Continuous Functions,* Springer GTM, v.43, 1976. MR **22** #6994. ⟨112⟩

81. V. Ginzburg and M. Kapranov, *Koszul duality for operads,* Duke Math. J. **76** (1994), 203–272. (Erratum regarding §2.2 at **80** (1995), p. 90.) MR **96a** : 18004. ⟨448⟩

82. A. M. W. Glass and W. Charles Holland (eds.), *Lattice-Ordered Groups. Advances and Techniques,* Kluwer, Mathematics and its Applications v.48, 1989. MR **91i** : 06017. ⟨105⟩

83. E. S. Golod and I. R. Shafarevich, *On the tower of class fields,* Izv. ANSSSR **28** (1964), 261–272. MR **28** #5056. ⟨60⟩

84. Alfred W. Hales, *On the nonexistence of free complete Boolean algebras,* Fundamenta Mathematica **54** (1964) 45–66. MR **29** #1162. ⟨367⟩

85. Philip Hall, *Some word-problems,* J. London Math. Soc. **33** (1958) 482–496. MR **21** # 1331. ⟨43⟩

86. Frank Harary, *A survey of the reconstruction conjecture,* in *Graphs and combinatorics* (Proc. Capital Conf., George Washington Univ., Washington, D.C., 1973), pp. 18–28. Lecture Notes in Math., Vol, 406, Springer, 1974. MR **50** #12818. ⟨128⟩

87. Leon Henkin, J. Donald Monk and Alfred Tarski, *Cylindric algebras. Part I.* Studies in Logic and the Foundations of Math., v.64. North-Holland, 1971; reprinted 1985. MR **47** #3171. ⟨105⟩

88. Wilfrid Hodges, *Six impossible rings,* J. Algebra **31** (1974) 218–244. MR **50** #315. ⟨170⟩

89. E. Hotzel, *On semigroups with maximal conditions*, Semigroup Forum **11** (1975/76), 337–362. MR **52** #10921. ⟨386⟩

90. Sze-Tsen Hu, *Homotopy Theory*, Academic Press Series in Pure and Applied Math., v.8, 1959. MR **21** #5186. ⟨114, 527⟩

91. James E. Humphreys, *Introduction to Lie Algebras and Representation Theory*, Springer GTM, vol. 9, 1972, 1978. MR **48** #2197, **81b**:17007. ⟨433⟩

92. Paweł Idziak, Petar Marković, Ralph McKenzie, Matthew Valeriote, Ross Willard, *Tractability and learnability arising from algebras with few subpowers*, SIAM J. Comput. **39** (2010) 3023–3037. MR2678065 ⟨530⟩

93. T. Ihringer, *Congruence Lattices of Finite Algebras: the Characterization Problem and the Role of Binary Operations*, Algebra Berichte v.53, Verlag Reinhard Fischer, München, 1986. MR **87c**:08003. ⟨193⟩

94. Sergei V. Ivanov, *The free Burnside groups of sufficiently large exponents*, Internat. J. Algebra Comput. **4** (1994) ii+308 pp. MR **95h**:20051. ⟨60⟩

95. Nathan Jacobson, *Lie Algebras,* Interscience Tracts in Pure and Applied Math., vol. 10, 1962. (Dover Edition, 1979.) MR **26** #1345. ⟨105, 427, 433⟩

96. Nathan Jacobson, *Structure and Representations of Jordan Algebras*, AMS Colloq. Pub., vol. 39, 1968. MR **40** #4330. ⟨105, 433⟩

97. Daniel M. Kan, *On monoids and their dual*, Boletín de la Sociedad Matemática Mexicana (2) **3** (1958) 52–61. MR **22** #1900. ⟨492⟩

98. Klaus Keimel and Heinrich Werner, *Stone duality for varieties generated by quasi-primal algebras*, pp. 59–85 in *Recent advances in the representation theory of rings and C\*-algebras by continuous sections* (Sem., Tulane Univ., New Orleans, La., 1973), Mem. AMS No. 148, 1974. MR **50** #12861. ⟨519⟩

99. John L. Kelley, *General Topology*, Van Nostrand, University Series in Higher Mathematics, 1955; Springer GTM, v.27, 1975. MR **16** 1136c, MR **51** #6681. ⟨111⟩

100. Gregory Maxwell Kelly, *Basic concepts of enriched category theory*, London Math. Soc. Lecture Note Series, 64, 1982, 245 pp. Readable at http://www.tac.mta.ca/tac/reprints/articles/10/tr10abs.html . ⟨291⟩

101. O. Kharlampovich, *The word problem for the Burnside varieties*, J. Alg. **173** (1995) 613–621. MR **96b**:20040. ⟨61⟩

102. E. W. Kiss, L. Márki, P. Pröhle and W. Tholen, *Categorical algebraic properties. A compendium on amalgamation, congruence extension, epimorphisms, residual smallness, and injectivity*, Studia Scientiarum Mathematicarum Hungarica, **18** (1983) 79–141. MR **85k**:18003. ⟨258⟩

103. I. B. Kozhukhov, *Semigroups with certain conditions on congruences*, J. Math. Sci. (N.Y.) **114** (2003), 1119–1126. MR **2004b**:20087. ⟨386⟩

104. G. R. Krause and T. H. Lenagan, *Growth of Algebras and Gelfand-Kirillov Dimension*, Research Notes in Mathematics Series, v. 116, Pitman, 1985. MR **86g**:16001. ⟨132⟩

105. A. H. Kruse, *Grothendieck universes and the super-complete models of Shepherdson*, Compositio Math., **17** (1965) 96–101. MR **31** #4716. ⟨237⟩

106. Hans Kurzweil, *Endliche Gruppen mit vielen Untergrupppen*, J. reine u. angewandte Math. **356** (1985) 140–160. MR **86f**:20024. ⟨193⟩

107. T. Y. Lam, *Lectures on modules and rings*, Springer GTM, v.189, 1999. MR **99i**: 16001. ⟨286, 509⟩

108. Solomon Lefschetz, *Algebraic Topology*, AMS Colloq. Pub. No. 27, 1942, reprinted 1963. MR **4**, 84f. ⟨209⟩

109. Hendrik W. Lenstra Jr., *Math 250A, Groups, Rings and Fields*, notes taken at Berkeley by Megumi Harada and Joe Fendel, http://websites.math.leidenuniv.nl/algebra/topics.pdf . ⟨98⟩

110. Lynn H. Loomis, *An Introduction to Abstract Harmonic Analysis*, University Series in Higher Mathematics, Van Nostrand, 1953. MR **14**, 883c. ⟨113⟩

111. I. G. Lysënok, *Infinite Burnside groups of even exponent,* Izv. Ross. Akad. Nauk Ser. Mat. **60** (1996) 3–224. English transl. at Izv. Math. 60:3 (1996), 453–654. MR **97j**:20037. ⟨60, 61⟩

112. Saunders Mac Lane, *One universe as a foundation for category theory,* pp. 192–200 in *Reports of the Midwest Category Seminar,* Springer LNM, v.106, 1969. MR **40** #2731. ⟨237⟩

113. Anatoliy I. Mal'cev, *Über die Einbettung von assoziativen Systemen in Gruppen* (Russian, German summary), Mat. Sb. N.S. **6** (1939) 331–336. MR **2**, 7d. ⟨86⟩

114. Anatoliy I. Mal'cev, *Über die Einbettung von assoziativen Systemen in Gruppen, II* (Russian, German summary), Mat. Sb. N.S. **8** (1940) 251–264. MR **2**, 128b. ⟨86⟩

115. Ernest Manes, *A triple theoretic construction of compact algebras,* pp. 91–118 in *Seminar on Triples and Categorical Homology Theory (ETH, Zürich, 1966/67),* Springer LNM, v.80, 1969. MR **39** #5657. ⟨453⟩

116. Edward J. Maryland, ed., *Problems in knots and 3-manifolds* (collected at a special session at the 80th Summer meeting of the AMS), Notices of the AMS **23** (1976) 410–411. ⟨54⟩

117. J. L. Mennicke, ed., *Burnside Groups, Proceedings of a workshop held at the University of Bielefeld, Germany, June-July 1977,* Springer LNM, v.806, 1980. MR **81j**:20002. ⟨61⟩

118. Deane Montgomery and Leo Zippin, *Topological Transformation Groups,* Interscience Tracts in Pure and Applied Mathematics, v.1, 1955, reprinted by R. E. Krieger Pub. Co., 1974. MR **17**, 383b, **52** #644. ⟨332, 430⟩

119. Yu. M. Movsisyan, *Introduction to the Theory of Algebras with Hyperidentities,* (Russian) Erevan. Univ., Erevan, 1986. 240 pp. MR **8f**:08001. ⟨447⟩

120. G. Nöbeling, *Verallgemeinerung eines Satzes von Herrn E. Specker,* Inventiones Math. **6** (1968) 41–55. MR **38** #233. ⟨58⟩

121. A. Ju. Ol'šanskii, *An infinite group with subgroups of prime orders,* Izv. Akad. Nauk SSSR Ser. Mat. **44** (1980) 309–321, 479. MR **82a**:20035. ⟨61⟩

122. Donald Passman, *The Algebraic Structure of Group Rings,* Wiley Series in Pure and Applied Mathematics, 1977; Robert E. Krieger Publishing, Melbourne, FL, 1985. MR **81d**:16001, MR **86j**:16001. ⟨93⟩

123. Marcin Peczarski, *The Gold Partition Conjecture,* Order **23** (2006) 89–95. MR **2007i**: 06005. ⟨127⟩

124. A. G. Pinus, *Vopenka's principle and skeletons of varieties* (Russian), Izv. Vyssh. Uchebn. Zaved. Mat. (1993) no.3, 68–71, *translation in* Russian Math (Iz. VUZ) **37** (1993) 66–69. MR **95j**:08005. ⟨414⟩

125. Heinz Prüfer, *Theorie der abelschen Gruppen. I,* Math. Z. **20** (1924) 165–187. ⟨424⟩

126. Pavel Pudlák and Jiří Tůma, *Every finite lattice can be embedded in a finite partition lattice,* Algebra Universalis **10** (1980) 74–95. MR **81e**:06013. ⟨193⟩

127. Jean-Xavier Rampon, *What is reconstruction for ordered sets?* Discrete Math. **291** (2005) 191–233. MR **2005k**:06010. ⟨128⟩

128. Shmuel Rosset, *A new proof of the Amitsur-Levitski identity,* Israel J. Math. **23** (1976), 187–188. MR **53** #5631. ⟨423⟩

129. Louis H. Rowen, *Ring Theory, v. II,* Academic Press Series in Pure and Applied Math., v.128, 1988. MR **89h**:16002. ⟨105, 423⟩

130. Walter Rudin, *Fourier Analysis on Groups,* Interscience Tracts in Pure and Applied Mathematics, No. 12, 1962. MR **27** #2808. ⟨251, 519⟩

131. Jean-Pierre Serre, *Lie Algebras and Lie Groups,* Benjamin, 1965. MR **36** #1582. ⟨433⟩

132. Robert M. Solovay, *New proof of a theorem of Gaifman and Hales,* Bull. AMS **72** (1966) 282–284. MR **32** #4057. ⟨367⟩

133. E. Specker, *Additive Gruppen von Folgen ganzer Zahlen,* Portugaliae Math. **9** (1950) 131–140. MR **12**, 587b. ⟨58⟩

134. A. K. Suškevič, *Theory of Generalized Groups,* Gos. Naučno-Tehn. Izdat. Ukrainy, Kharkov, 1937. ⟨424⟩

135. Richard G. Swan, *An application of graph theory to algebra,* Proc. AMS **14** (1963) 367–373. Correction at *ibid.* **21** (1969) 379–380. MR **26** #6956, **41** #101. ⟨423⟩
136. Moss E. Sweedler, *Hopf Algebras,* Math. Lecture Note Series, Benjamin, N.Y., 1969. MR **40** #5705. ⟨105, 473⟩
137. Walter Taylor, *Hyperidentities and hypervarieties,* Aequationes Mathematicae, **23** (1981) 30–49. MR **83e**:08021a. ⟨446, 447⟩
138. Wolfgang J. Thron, *Topological Structures,* Holt, Rinehart and Winston, 1966. MR **34** #778. ⟨107⟩
139. V. Trnková, J. Adámek and J. Rosický, *Are all limit-closed subcategories of locally presentable categories reflective?* pp. 1–18 in *Categorical algebra and its applications: proceedings of a conference, held in Louvain-La-Neuve, Belgium, July 26-August 1, 1987,* F. Borceux, ed.; Springer LNM, v.1348, 1988. MR **90b**:18002. ⟨414⟩
140. Michael R. Vaughan-Lee, *The Restricted Burnside Problem,* 2nd edition, London Math. Soc. Monographs, New Series, 8 1993. MR **98b** 20047. ⟨61⟩
141. Michael R. Vaughan-Lee, *On Zelmanov's solution of the Restricted Burnside Problem,* J. Group Theory **1** (1998) 65–94. MR **99a** 20019. ⟨61⟩
142. B. L. van der Waerden, *Free products of groups,* Amer. J. Math. **70** (1948) 527–528. MR **10**, 9d. ⟨43, 64⟩
143. Alan G. Waterman, *The free lattice with 3 generators over* $N_5$, Portugal. Math. **26** (1967) 285–288. MR **42** #147. ⟨183⟩
144. D. J. A. Welsh, *Matroid Theory,* Academic Press, 1976. MR **55** #148. ⟨204⟩
145. Samuel White, *The group generated by* $x \mapsto x + 1$ *and* $x \mapsto x^p$ *is free,* J. Alg. **118** (1988) 408–422. MR **90a**:12014. ⟨414⟩
146. Niklaus Wirth, *Algorithms + Data Structures = Programs,* Series in Automatic Computation, Prentice-Hall, 1976. MR **55** #13850. ⟨424⟩
147. Joseph A. Wolf, *Growth of finitely generated solvable groups and curvature of Riemannian manifolds,* J. Diff. Geom. **2** (1968) 421–446. MR **40** #1939. ⟨131⟩
148. W. Hugh Woodin, *The continuum hypothesis. I,* Notices AMS **48** (2001) 567–576. MR **2002j**:03056. ⟨170⟩
149. *The World of Groups,* http://www.grouptheory.info/. ⟨55⟩
150. Efim Zelmanov, *On some open problems related to the restricted Burnside problem,* pp. 237–243 in *Recent progress in algebra* (Taejon/Seoul, 1997), Contemp. Math., **224**, 1999. MR **99k**:20082. ⟨60⟩

# List of Exercises

I hope the telegraphic descriptions given below will help you recall in most cases roughly what the exercises are. When they don't, you can always look back to the page in question (shown at the left). If you find any of the descriptions incorrect, or think of a more effective wording to briefly describe some exercise, let me know.

© Springer International Publishing Switzerland 2015

G.M. Bergman, *An Invitation to General Algebra and Universal Constructions*, Universitext, DOI 10.1007/978-3-319-11478-1

# Symbol Index

Boldface numbers indicate pages where definitions are given. If a symbol is defined in one place and used again without explanation more than a page or so away, I show the page(s) where it is defined, and often some of the pages where it is used or where the entity it symbolizes is discussed; but I do not attempt to show all significant occurrences of each subject. For that, the Word Index, with its headings and subheadings, is more useful.

Symbols are ordered alphabetically. For Greek letters, I use their Latin spelling, e.g., "Omega" for $\Omega$. Other symbols are alphabetized in various ways; e.g., $\vee$ and $\wedge$ are alphabetized as "vee" and "wedge" based on their LaTeX names; the symbol $=$, and related symbols such as $\cong$, are alphabetized, in an arbitrary order, under "equals"; and $\mapsto$ and $\downarrow$ are similarly alphabetized under "arrow". Fortunately, you do not have to know all the details; some symbols will require more search than others, but this index is only a few pages long.

Font-differences, capitalization, and "punctuation" such as brackets, do not affect ordering unless everything else is equal. Operator-symbols are often shown in combination with letters with which they are commonly used, e.g., $\langle X \mid Y \rangle$ is alphabetized under XY.

Category theory, introduced in Chapter 6, brings with it a proliferation of symbols for categories. I do not record below cases where the meaning is obvious, like **Group**, nor cases discussed only briefly, like **GermAnal** (germs of analytic functions, which can be found in the Word Index), but only category names used in more than one place, for which some aspect of the definition (e.g., the associativity assumption in **Ring**[1]) or the abbreviation (as with **Ab**) is not obvious.

© Springer International Publishing Switzerland 2015

G.M. Bergman, *An Invitation to General Algebra and Universal Constructions*, Universitext, DOI 10.1007/978-3-319-11478-1

| | |
|---|---|
| $\mathrm{Ar}(\mathbf{C})$ | the class of morphisms ('arrows') of the category $\mathbf{C}$, **217**. |
| $\mathrm{ari}_\Omega$, $\mathrm{ari}$ | arity function (cf. $\Omega$ below), 20, **379**. |
| $\downarrow$ | see "$(S \downarrow T)$, $(S \downarrow \mathbf{C})$, $(\mathbf{C} \downarrow T)$" below. |
| $\mapsto$ | action of a function on elements, **11**. |
| $\mathrm{Aut}(X)$ | automorphism group of $X$, **59**, 213, 241–242. |
| $\beta \colon (A, B) \to C$ | bilinear map (temporary notation), **75**–79. |
| **Binar**, **Binar**$^e$ | variety of sets with a binary operation, resp., a binary operation and neutral element, **505**, **514**, 525. |
| $\mathbb{C}$ | the complex numbers. |
| $\mathbf{C}^{\mathbf{D}}$ | category of all functors $\mathbf{D} \to \mathbf{C}$, **278**, 287–290. |
| $\mathbf{C}^{\mathrm{pt}}$ | category of pointed objects of $\mathbf{C}$, **264**, 275, **506**, 524. |
| $\mathbf{C}(X, Y)$ | set (or in Chapter 10, algebra) of morphisms $X \to Y$ in the category $\mathbf{C}$, **218**, **462**, **466**. |
| $\mathrm{card}(X)$ | cardinality of the set $X$, **156**. |
| **Cat**, **Cat**$_\mathbb{U}$ | category of all $\mathbb{U}$-small categories, **243**. |
| cl | general symbol for a closure operator, **194**. |
| $\mathrm{Cl}(\mathbf{V})$, $\mathrm{Cl}_n(\mathbf{V})$ | clonal theory of variety $\mathbf{V}$, and its $n$th object, **441**. |
| **Clone**, **Clone**$^{(\gamma)}$ | category of all covariant ($\leq \gamma$-)clonal categories, **440**–**441**–453, 528. |
| **CommRing**$^1$ | variety of all commutative associative rings with unity, **221**, 314, 457–461, 468, 472. |
| $\deg(x)$ | in § 10.6, degree of element $x$ in a comonoid object of **Monoid**, **481**–485. |
| $\Delta$ | diagonal functor $\mathbf{C} \to \mathbf{C}^{\mathbf{D}}$, **313**, **337**, 340, 346, 349. |
| $\mathbf{E}(X)$ | lattice of equivalence relations on $X$, **191**–193, 299. |
| $\mathrm{End}(X)$ | monoid of endomorphisms of $X$, **59**, 214. |
| $\approx$ | equivalence of categories, **285**. |
| $\cong$ | isomorphism (of algebras, categories, etc.). |
| $\eta$, $\varepsilon$ | unit and counit of an adjunction, **309**. |
| $\exists$ | "there exists", **11**, 211. |
| $\exists 1$ | "there exists a unique", **47**. |
| $F_\Omega$, $F_\mathbf{V}$ | free-algebra functors (see mainly "free" in Word Index), **393**, **408**, 416. |
| $f|X$ | restriction of the function $f$ to the set $X$, 40, **135**. |

# Word and Phrase Index

I have tried to include in this index not only the locations where terms are defined, but also all significant occurrences of the concepts in question; but it has not been easy to decide which occurrences are significant. I would welcome readers' observations on the types of cases they would find it useful to have in the index, and on entries that are erroneous, unnecessary, or missing.

Pages where terms are defined or where conventions are made relating to them are shown with boldface page numbers. (Sometimes a formal definition occurs after the first page of discussion of a topic, and sometimes more than one version of a concept is defined, leading to occasional entries such as 172-**173**-**187**-234.)

Terms used by other authors for which different words are used here are, if referenced, put in single quotes; e.g., 'free product', for what we call a coproduct.

In cross-references, I often truncate multiword entries; e.g., I may say "*see* rings" though the actual heading is "rings and $k$-algebras". But where this would be confusing, I add "...". E.g., under "associative algebra" I write "*see* rings ...".

In referring to secondary headings, I use a colon, writing "*see* main-heading: subheading". Within subheadings, the main current heading is abbreviated "–". In particular, "*see* –: subheading" points to another subheading under the current main heading.

In the few cases where an entry ends with a numeral (e.g., "matrices: with determinant 1"), I have put the numeral in double quotes, so that it cannot be mistaken for a page number.

G.M. Bergman, *An Invitation to General Algebra and Universal Constructions*, Universitext, DOI 10.1007/978-3-319-11478-1